Studies in Logic
Mathematical Logic and Foundations
Volume 17

Reasoning in Simple Type Theory
Festschrift in Honour of Peter B. Andrews on His 70th Birthday

Volume 7
Fallacies – Selected Papers 1972-1982
John Woods and Douglas Walton, with a Foreword by Dale Jacquette

Volume 8
A New Approach to Quantum Logic
Kurt Engesser, Dov M. Gabbay and Daniel Lehmann

Volume 9
Handbook of Paraconsistency
Jean-Yves Béziau, Walter Carnielli and Dov Gabbay, editors

Volume 10
Automated Reasoning in Higher-Order Logic. Set Comprehension and Extensionality in Church's Type Theory
Chad E. Brown

Volume 11
Foundations of the Formal Sciences V: Infinite Games
Stefan Bold, Benedikt Löwe, Thoralf Räsch and Johan van Benthem, editors

Volume 12
Second-Order Quantifier Elimination: Foundations, Computational Aspects and Applications
Dov M. Gabbay, Renate A. Schmidt and Andrzej Szałas

Volume 13
Knowledge in Flux. Modeling the Dynamics of Epistemic States
Peter Gärdenfors. With a foreword by David Makinson

Volume 14
New Approaches to Classes and Concepts
Klaus Robering, editor

Volume 15
Logic, Navya-Nyāya and Applications. Homage to Bimal Krishna Matilal
Mihir K. Chakraborti, Benedikt Löwe, Madhabendra Nath Mitra and Sundar Sarukkai, eds.

Volume 16
Foundations of the Formal Sciences VI. Probabilistic Reasoning and Reasoning with Probabilities.
Benedikt Löwe, Eric Pacuit and Jan-Willem Romejin, eds.

Volume 17
Reasoning in Simple Type Theory. Festschrift in Honour of Peter B. Andrews on His 70th Birthday.
Christoph Benzmüller, Chad E. Brown and Jörg Siekmann, eds.

Studies in Logic Series Editor
Dov Gabbay
dov.gabbay@kcl.ac.uk

Reasoning in Simple Type Theory

Festschrift in Honour of Peter B. Andrews on His 70th Birthday

Edited by

Christoph Benzmüller,

Chad E. Brown

Jörg Siekmann

and

Richard Statman

ISBN 978-1-904987-70-3

College Publications
Scientific Director: Dov Gabbay
Managing Director: Jane Spurr
Department of Computer Science
King's College London, Strand, London WC2R 2LS, UK

http://www.collegepublications.co.uk

Original cover design by orchid creative www.orchidcreative.co.uk
Printed by Lightning Source, Milton Keynes, UK

Contents

Introduction

THE EDITORS

We are pleased to introduce this Festschrift in honor of Peter B. Andrews on his 70th birthday. He turned 70 on November 1, 2007. Peter Andrews obtained his Ph.D. in 1964 from Princeton University where his advisor was Alonzo Church. In 1963 (before his thesis was submitted) he joined the faculty of Carnegie Institute of Technology (now Carnegie Mellon University) where he became a professor in the Department of Mathematical Sciences. As an expert in higher-order logic and a pioneer in automating higher-order theorem proving, Peter B. Andrews has contributed significant theoretical results and has initiated the development of practical theorem proving systems. The higher-order theorem prover TPS (Theorem Proving System) is the result of decades of research conducted by Professor Andrews and his students. Students taking his logic course at Carnegie Mellon University have used, as a part of the course, a subsystem of TPS called ETPS (Educational Theorem Proving System) to interactively construct proofs. Professor Andrews was one of the first teachers to make use of a theorem prover in a logic course, and has consistently integrated the theorem prover with the course over many years. In the first article Peter B. Andrews provides an overview of his various research results along with some interesting historical background.

In 2003 Peter B. Andrews was the recipient of the prestigious Herbrand Award for his contributions to the field of automated reasoning.

This Festschrift includes reprints of several papers as well as several new contributed articles by a variety of authors. For the reprints we have chosen some of the most important papers by Peter Andrews, a paper by Alonzo Church (introducing his simple theory of types) and a paper by Leon Henkin (proving completeness of Church's type theory relative to Henkin's semantics). The new articles were contributed by Peter Andrews' students and collaborators as well as a number of researchers his work has influenced.

Christoph Benzmüller
Chad E. Brown
Jörg Siekmann
Richard Statman

Part I

Historical Background

Some Historical Reflections

PETER B. ANDREWS

1 Introduction

In this paper[1] I will discuss some historical matters, particularly matters related to Herbrand's Theorem, higher-order logic, my work, and the development of the automated theorem proving system TPS. In the process I will have opportunities to make it clear how contributions by others have been very helpful to me.

First of all, I'd like to express my deep appreciation for all the wonderful work that's been done by those who worked as research assistants on the TPS Theorem Proving System over the years. Here they are, in chronological order:

Eve Longini Cohen	1974 - 1980
Dale A. Miller	1978 - 1983
Frank Pfenning	1980 - 1986
Sunil Issar	1984 - 1990
Carl Klapper	1984 - 1987
Dan Nesmith	1987 - 1991
Hongwei Xi	1992 - 1995
Matthew Bishop	1992 - 1999
Chad E. Brown	1999 - 2004

Some of them never met each other, but they were essentially able to work effectively as a team in developing TPS over a period of three decades. This was greatly facilitated by the fine organization for our file system which Frank Pfenning developed when we were converting everything to Common Lisp.

Sometimes different people have different ideas about the best ways to do various things. We often resolved such situations by implementing both ideas, and having flags in TPS which allowed us to experiment with the alternatives.

One of the great advantages of having bright research assistants is that your ambition is not inhibited by concerns about the difficulties of the

[1] This is an expanded version of [18].

project you have in mind. I no longer remember who implemented proofwindows in TPS, but I do remember that when I realized that it would be very nice to have a proofwindow to display and update a proof as one worked on constructing it interactively, I had no idea how such a thing could be done. Nevertheless, the research assistant helping me at the time soon came up with a way of doing it.

The research assistants working on TPS did far more than implement ideas which I proposed. They also developed and implemented their own ideas, which have contributed enormously to TPS. Dale Miller, Frank Pfenning, Sunil Issar, Matthew Bishop, and Chad Brown all wrote theses [41, 46, 71, 76, 81] and associated papers [40, 45, 70, 77, 78, 80, 82] whose ideas have been implemented in TPS. The contributions of the research assistants to TPS are also reflected in a variety of joint papers [21, 22, 23, 24, 25, 26, 28, 29, 30, 31, 32, 33, 34, 42, 79].

2 My Introduction to Logic

When I was an undergraduate at Dartmouth, I was concerned about how various problems in the world might be solved, or at least alleviated, and I became aware of how complicated many of these problems are. Attempts to solve them can have unforeseen side-effects, and create new problems. We need very sophisticated methods of thinking about complex problems. Our technology and scientific knowledge progress steadily, but are we any better at thinking than Socrates or Pythagoras?

At about this time I took my first logic course, which was taught by John Kemeny. Kemeny was a remarkable person. While a student at Princeton he was Einstein's assistant. At Dartmouth he was Chairman of the Mathematics Department and later President of the college. He developed the programming language Basic, and was very active teaching and writing. When he was a graduate student at Princeton, he undertook to prove rigorously that Zermelo Set Theory was equivalent to Type Theory in logical strength. This was generally believed at the time. He found it difficult to do this, and in frustration he exclaimed to his advisor, Alonzo Church, "I don't even think it's true." To this Church replied, "All right, prove that", and Kemeny soon had a proof that the consistency of Type Theory could be proved in Zermelo Set Theory, from which it follows by Gödel's Second Theorem that Zermelo Set Theory must be stronger than Type Theory. Kemeny was a wonderful teacher, and I was enthralled by the discovery that one can actually study the mysterious process of reasoning in a mathematically rigorous way.

3 Type Theory

I went to Princeton for graduate work, where Alonzo Church was also my advisor. Of all the things I have to thank Church for, I think the most important one was inventing the system of simple type theory [49] which he introduced in 1940.

Of course, Bertrand Russell had developed type theory [89, 104], and I'm very grateful for this. However, Russell was concerned about avoiding semantic paradoxes such as Grelling's paradox[2] as well as more mathematical paradoxes such as Russell's paradox, so in Principia Mathematica he used ramified type theory, with Axioms of Reducibility to alleviate some of the ramification. The result of this and other features of Russell's formulation of type theory made it seem quite complicated, so in spite of the enormous influence of Principia Mathematica as a landmark in the development of logic, it was the book that everyone talked about, but practically no one read.

Church's type theory is much simpler, and is at the same time a richer, more expressive language, since it recognizes functions as first-class objects which do not have to be regarded as sets of ordered n-tuples, and it has λ-notation for functions and sets. It permits one to express mathematical ideas in ways that are very close to traditional mathematical notation. Nevertheless, the feeling that type theory is complicated persists, and many people who are otherwise quite logically sophisticated shy away from it. A simple introduction to type theory can be found in the last three chapters of my book [17].

Leon Henkin spent some time in Princeton while I was there as a graduate student. Henkin was writing a paper [63] in which he presented a formulation of Church's type theory (with types restricted to propositional types, though this is incidental to much of the paper) in which the only primitive logical constant was equality. (Church had introduced constants for negation, disjunction, and quantifiers, and defined equality in terms of these.) I was very interested in this, and Henkin graciously allowed me to read a draft of the paper. I discovered that the system of axioms for Henkin's system could be substantially simplified [2].

Henkin's paper [62] was a landmark in the history of type theory. It was clearly a consequence of Gödel's Incompleteness Theorem [58] that no system of type theory could be complete in the sense that its theorems

---

[2]An *autological* adjective applies to itself. For example, the word "polysyllabic" is polysyllabic, so it is an autological word. A *heterological* adjective is one which does not apply to itself. For example, the word "long" is not long, so it is a heterological word. Grelling's paradox concerns the question whether the word "heterological" is heterological.

are precisely the wffs which are valid in all models of the sort we now call *standard* models. Henkin considered the problem of clarifying in what sense the axioms for Church's type theory are complete, formulated the notion of a *general* model, and proved that the theorems are the wffs which are valid in all general models.[3] I developed an enduring interest in the problem of understanding general models better, and in [8] I developed a characterization of general models for type theory and showed how it could be used to prove certain independence results. While working on this, I was surprised to discover a small error in Henkin's paper [62]. A proof that the point in question was actually an error required constructing a special non-standard model which is described in [7], along with a suggestion for correcting the error.

4 Herbrand's Theorem

I was in graduate school in the early 60's, when pioneering work in automated theorem proving was being done, but the resolution method hadn't yet been invented. As I came to appreciate what a marvelously expressive language Church's type theory is, I realized that what interested me most was the development of a sufficiently deep understanding of how to prove theorems of this system that one could, in principle, automate the process. I knew that one of the fundamental theorems underlying proof procedures and decision procedures for first-order logic was Herbrand's Theorem.

As we seek to understand how to prove theorems efficiently, one of the things we can ask is "*Why* is a particular wff a theorem?" A theorem is not simply a wff which happens to have a proof. It has special structural properties which guarantee that it is true in all interpretations. The essential idea underlying Herbrand's Theorem is to focus on this structural aspect of theorems.

Herbrand's Theorem plays a fundamental role in the pioneering papers of Quine [87], Gilmore [56], Prawitz [83], Davis and Putnam [50], and Davis [51]. In his paper [88] introducing the resolution method, Robinson referred to his Resolution Theorem as a form of Herbrand's Theorem.

Herbrand was by all accounts a brilliant mathematician. He died in a mountaineering accident in the Alps at the age of 23. This was a tragic loss to logic and mathematics. Herbrand's proof of his theorem was in his thesis [65].

While I don't want to get into too many technical details, let's review

[3]Henkin observed (see [64]) that the method he used for his completeness proof could also be applied to first-order logic, and he presented a completeness proof for first-order logic [61] which is simpler than Gödel's original proof [57] and which is used, with some variations, in all modern presentations of the theorem.

the most important results in this thesis.

Herbrand introduced a system of first-order logic which we shall call $\mathcal{H}$. It can be described as follows:

(1) All quantifier-free tautologies are axioms of $\mathcal{H}$.

The rules of inference of $\mathcal{H}$ are the following:

(2) Rules of Passage [66, pp. 74, 225] for pulling out or pushing in quantifiers, as when transforming to prenex normal form or miniscope form.

(3) Universal Generalization.

(4) Existential Generalization.

(5) Simplification: From $[\mathbf{P} \lor \mathbf{P}]$ infer $\mathbf{P}$.

(6) Modus Ponens: From $\mathbf{P}$ and $[\mathbf{P} \supset \mathbf{Q}]$, infer $\mathbf{Q}$.

The rule of Alphabetic Change of Bound Variables is also an implicit rule of inference of $\mathcal{H}$.

It is easy to see that $\mathcal{H}$ is equivalent to a traditional Hilbert-style system of first-order logic.

We shall use $\mathcal{G}$ as a name for the system obtained from $\mathcal{H}$ by deleting Modus Ponens from the list of rules of inference and replacing the Simplification rule by:

(5') Generalized Simplification: Replace $[\mathbf{P} \lor \mathbf{P}]$ in a theorem by $\mathbf{P}$.

Thus, $\mathcal{G}$ can be described as follows:

(1) All quantifier-free tautologies are axioms of $\mathcal{G}$.

The rules of inference of $\mathcal{G}$ are the following:

(2) Rules of Passage for pulling out or pushing in quantifiers, as when transforming to prenex normal form or miniscope form.

(3) Universal Generalization.

(4) Existential Generalization.

(5') Generalized Simplification: Replace $[\mathbf{P} \lor \mathbf{P}]$ in a theorem by $\mathbf{P}$.

The rule of Alphabetic Change of Bound Variables is also an implicit rule of inference of $\mathcal{G}$.

Note that $\mathcal{G}$ is a cut-free system, but unlike a Gentzen-style system, it has no rule of Conjunction Introduction.

Herbrand showed how to associate with a wff $\mathbf{P}$ of first-order logic certain quantifier-free wffs which we shall call *Herbrand expansions* of $\mathbf{P}$. Let us say that a wff $\mathbf{P}$ has the *Herbrand property* iff some Herbrand expansion of $\mathbf{P}$ is tautologous. Actually, for technical reasons the Herbrand property appears in three forms in Herbrand's thesis: Property A, Property B, and Property C. As part of the proof it is established that they are equivalent to each other.

Herbrand's proof involved establishing the following claims about any wff $\mathbf{P}$ of first-order logic:

(1) If $\mathbf{P}$ has the Herbrand property, then $\vdash_{\mathcal{G}} \mathbf{P}$.

(2) If $\vdash_{\mathcal{G}} \mathbf{P}$, then $\vdash_{\mathcal{H}} \mathbf{P}$.

(3) If $\vdash_{\mathcal{H}} \mathbf{P}$, then $\mathbf{P}$ has the Herbrand property.

The first claim follows from an analysis of the relation between a tautologous Herbrand expansion of a wff and the wff itself. The second claim is a trivial consequence of the fact that the rules of inference of $\mathcal{G}$ are all primitive or derived rules of inference of $\mathcal{H}$. The third claim is the most difficult to establish; the proof involves showing that each rule of inference of $\mathcal{H}$ preserves the Herbrand property.

Thus, Herbrand asserted that the following are equivalent:

(a) $\mathbf{P}$ has the Herbrand property.

(b) $\vdash_{\mathcal{G}} \mathbf{P}$

(c) $\vdash_{\mathcal{H}} \mathbf{P}$

Note that this implies that Modus Ponens is a derived rule of inference of the system $\mathcal{G}$, a result that is closely related to Gentzen's Cut-Elimination Theorem [54].

It seemed natural to try to extend Herbrand's Theorem to Church's type theory. There are elegant proofs of Herbrand's Theorem using semantical concepts. For example, there is one in section 35 of my book [17]; this proof is inspired by Quine's 1955 paper [87]. However, the semantics of type theory involve certain matters which do not arise for the semantics of first-order logic, so I thought I should see if a purely syntactic proof of Herbrand's Theorem could be extended to higher-order logic. This meant working with Herbrand's original proof.

I started work developing a proof of a Herbrand Theorem for type theory, using Herbrand's proof of his theorem for first-order logic as a guide. However, I kept running into difficulties, and reformulating my approach. Finally, I decided that I didn't understand Herbrand's proof well enough, so I looked at it more carefully. As van Heijenoort remarks in his anthology, "Herbrand's thesis bears the marks of hasty writing. ... Herbrand's thoughts are not nebulous, but they are so hurriedly expressed that many a passage is ambiguous or obscure." [102, p. 525].

I kept trying to make sense of Herbrand's proof, but finally I told Professor Church that there seemed to be a gap in the proof.

I should remark that the problem in Herbrand's proof was with Lemma 3.3 of Chapter 5. For each positive integer p, Herbrand defined what it meant for a wff to have Property C of order p. One can describe this at least roughly by saying that a wff has Property C of order p iff the Herbrand expansion of the wff using all terms from the Herbrand universe with depth of nesting less than p is a tautology. The lemma asserted that for each positive integer p, the Rules of Passage preserve Property C of order p.

Professor Church advised me to consult Burton Dreben at Harvard, who was the greatest authority on Herbrand's work, at least in the United States, and whose work involved applications of Herbrand's Theorem to solvable cases of the decision problem. So I wrote to Dreben. For the sake of historical clarity, some of our correspondence is shown in the figures below.[4]

In my letter of April 9 (Figures 1 and 2), I described the problem with the proof, and provided an example where a certain part of Herbrand's argument did not work.

In his reply of May 18 (Figures 3 and 4), Dreben ascribed my difficulty understanding Herbrand's argument to a slight ambiguity in it.

On May 31 I pointed out (Figure 5) that there still seemed to be a problem with the proof.

We agreed that we needed to discuss this matter face-to-face, so I drove to Cambridge and we had a long discussion of Herbrand's proof on June 19, 1962. Dreben tried to show me that while Herbrand's argument was obscure at some points, it was essentially correct. I've learned over the years that many people think faster than I do, but this was one discussion for which I was well prepared. Every time Dreben proposed another interpretation of Herbrand's argument, I already knew what was wrong with it. We discussed it for hours, but finally Dreben realized that there really was a problem with Herbrand's proof. Our discussion turned to ways of patching up the proof, but we didn't find a way to do this, and I went back to Princeton.

[4]Dreben's letters to the author are published here with the kind consent of his widow, Juliet Floyd.

WAlnut 1-9534
116 Linden Lane
Princeton, New Jersey
1962 April 9

Professor Burton S. Dreben
Philosophy Department
Harvard University
Cambridge 38, Massachusetts

Dear Professor Dreben,

My thesis advisor, Professor Church, has suggested that I write to you about what I believe is a mistake in Herbrand's thesis, <u>Recherches sur la Théorie de la Démonstration</u>. I have spent a considerable amount of time and effort trying to justify Herbrand's argument or to find an alternative finitary argument proving the same result, and have not thus far succeeded. I therefore would be most grateful if you would tell me whether you believe my objections to Herbrand's argument are valid, and if so, whether you see any way the proof may be patched up. If you do not regard my objections as valid, I hope you will point out to me what I have overlooked.

Let us consider Chapitre 5, 3.3, pp. 101-104, the proof that if a proposition has the property C of order p, then a proposition obtained from that one by replacing a well formed part ((Ex)∅x)vp, where the individual variable x does not occur in p, by (Ex)(∅x v p), also has the property C of the same order. I am concerned about the third paragraph on p. 103:

"Remplaçons pour cela la réduite de P_1 par une autre, obtenue comme suit: $a_1, a_2, \ldots, a_N$ étant les éléments des champs $C^{(2)}_1, \ldots, C^{(2)}_p$ on définira dans ces champs $f_y(x_1, \ldots, x_n)$ comme étant $f_y(x_1, \ldots, x_n, a_i)$ pour un i que nous choisirons tout à l'heure et qui dépendra de $x_1, \ldots, x_n$. Ceci conduirait à prendre une partie de ces champs comme champs $C^{(1)}_i$."

When one looks at the manner in which a_i is chosen for a given choice of values for $x_1, \ldots, x_n$, as described on page 104, one sees no reason to believe that if $x_1, \ldots, x_n$ are all in $C^{(1)}_1 \cup C^{(1)}_2 \cup \ldots \cup C^{(1)}_r$, then a_i is in $C^{(2)}_1 \cup \ldots \cup C^{(2)}_r$. That is, from the fact that $f_y(x_1, \ldots, x_n)$ is in $C^{(1)}_1 \cup \ldots \cup C^{(1)}_s$ one (apparently) cannot infer that [illegible]. $f_y(x_1, \ldots, x_n, a_i)$ is in $C^{(2)}_1 \cup \ldots \cup C^{(2)}_s$. (You will recognize that I am following Herbrand's convention of letting the same symbol stand for an element of $C^{(1)}_{i1}$ and for the corresponding member of $C^{(2)}_j$. I trust it will be evident that the mild

Figure 1. First page of 1962 April 9 letter from Andrews to Dreben.

ambiguities inherent in this convention are in no way involved in my argument.) In particular, it might happen that $f_y(x_1, \ldots, x_n)$ is in $C^{(1)}_1 \cup \ldots \cup C^{(1)}_p$, but $f_y(x_1, \ldots, x_n, a_i)$ is not in $C^{(2)}_1 \cup \ldots \cup C^{(2)}_p$. This is the heart of the difficulty.

Suppose, for example, that n is 1, the number p is 4, and a_i is a_N (a member of $C^{(2)}_4$) in all cases which we shall consider below. Let a_1 be a member of $C^{(1)}_1$ (and hence of $C^{(2)}_1$). Then $f_y(a_1)$ is in $C^{(1)}_2$, but the corresponding element $f_y(a_1, a_N)$ is in $C^{(2)}_5$. Also $f_y(f_y(a_1))$ is in $C^{(1)}_3$, but $f_y(f_y(a_1, a_N), a_N)$ is in $C^{(2)}_6$. $f_y(f_y(f_y(a_1)))$ is in $C^{(1)}_4$, but $f_y(f_y(f_y(a_1, a_N), a_N), a_N)$ is in $C^{(2)}_7$. In order that Herbrand's proof work it seems to be necessary to show that this sort of thing cannot happen. For if it does happen, the proof that [illegible] implies [illegible] does not work.

For many purposes it would be adequate to know that if P_1 has the property C of order p, then P_2 has the property C of order q, where q is an explicitly given function of p. I do not see how Herbrand's proof can be used to yield even this result, however.

I hope that with Herbrand's thesis before you, my remarks above will be sufficiently clear to be understandable. I am sorry that I did not have a chance to talk with you when you were in Princeton.

Herbrand's proof is of more than historical interest to me, for I have been looking into the problem of generalizing Herbrand's theorem in some form to type theory, and it at one time seemed that Herbrand's finitary proof might generalize more readily than other proofs of the theorem. I have recently become more thoroughly aware of the severe difficulties involved in attacking the problem from this direction, but I still have not ruled it out as a reasonable approach. Therefore I hope very much that the proof can be patched up, or that I have been misreading it and that it needs no patching up.

I shall be looking forward to hearing from you.

Respectfully,

Peter Andrews

Peter B. Andrews

Figure 2. Second page of 1962 April 9 letter from Andrews to Dreben.

HARVARD UNIVERSITY
DEPARTMENT OF PHILOSOPHY

EMERSON HALL
CAMBRIDGE 38, MASSACHUSETTS

May 18, 1962

Dear Mr. Andrews,

The difficulty you are having with pages 103 and 104 of Herbrand results from a slight ambiguity in his argument.

Let us distinguish between the réduite $R(P)_p$ of an expression P over a domain $D_p = C_1 \cup C_2 \cup \ldots \cup C_p$ and the evaluation $\Pi(P)_p$ of $R(P)_p$ over D_p. In $R(P)_p$ there are no quantifiers but, in general, there are function letters. In $\Pi(P)_p$ the functional expressions no longer appear. However in the forming of $\Pi(P)_p$ from $R(P)_p$, if an argument a_{i_j} of $f_y(a_{i_1}, \ldots, a_{i_n})$ is an element of C_p, then the expression $f_y(a_{i_1}, \ldots, a_{i_n})$ must be replaced by (the name of) an element in C_{p+1} and not by an element in D_p. See page 101 lines 11-13.

Now let $R(P_1)_p$ be the réduite of P_1 over $D_p^{(1)}$, let $R(P_2)_p$ be the réduite of P_2 over $D_p^{(2)}$, and let $R'(P_1)_p$ be the reduite of P_1 over $D_p^{(2)}$. Moreover, let $\Pi(P_1)_p$ be the evaluation of $R(P_1)_p$ over $D_p^{(1)}$, and $\Pi(P_2)_p$ be the evaluation of $R(P_2)_p$ over $D_p^{(2)}$. By hypothesis, both $\Pi(P_1)_p$ and $\Pi(P_2)_p$ are given. Herbrand's problem is to specify an evaluation $\Pi'(P_1)_p$ of $R'(P_1)_p$ over $D_p^{(2)}$ in such a way that:

1) $\Pi(P_1)_p \supset \Pi'(P_1)_p$ is truth functionally valid

and 2) If $\Pi(P_2)_p$ is not truth functionally valid, then $\Pi'(P_1)_p$ is not truth functionally valid.

And this he does in a straightforward manner once we remember (1) that, since $\Pi(P_2)_p$ is given, the crucial element a_k on page 104 line 14 occurs either in $D_p^{(2)}$ or in $C_{p+1}^{(2)}$, and (2) that the

Figure 3. First page of 1962 May 18 letter from Dreben to Andrews.

HARVARD UNIVERSITY
DEPARTMENT OF PHILOSOPHY

EMERSON HALL
CAMBRIDGE 38, MASSACHUSETTS

part of $\Pi(P_2)_p$ represented by line 8 on page 164 occurs once and only once in $\Pi(P_2)_p$.

I apologize again for taking so long to answer. Do not hesitate to write me about any points in Herbrand, and, as I said in my note of May 6, if it is possible for you to come up here I should very much like to see you.

Sincerely,
Burton Dreben

Figure 4. Second page of 1962 May 18 letter from Dreben to Andrews.

A few weeks later I received the letter in Figure 6. Dreben had found an actual counterexample to Lemma 3.3. He had found that the wff

$[\forall y_1 M y_1 \vee \forall y_2 N y_2 \vee \exists x_3 \sim G x_3]$
$\vee\, [\exists x_1 \sim M x_1 \wedge \exists x_2 \sim N x_2 \wedge \forall y_3 \sim H y_3]$
$\vee\, [\exists x_4 H x_4 \wedge \forall y_4 G y_4]$

has property C of order 2, while the wff

$[\forall y_1 M y_1 \vee \forall y_2 N y_2 \vee \exists x_3 \sim G x_3]$
$\vee\, \exists x_1 \exists x_2 \,[\sim M x_1 \wedge \sim N x_2 \wedge \forall y_3 \sim H y_3]$
$\vee\, \exists x_4 \,[H x_4 \wedge \forall y_4 G y_4]\,,$

which can be obtained from it by Rules of Passage, has property C of order 4 but of no smaller order.

A simplification of Dreben's counterexample is discussed in a letter which I sent to Dreben on July 16 (Figures 7 - 8).

$\sim [\forall x\, \Phi x \vee \sim \forall y R y] \vee \forall y_1 \Phi y_1 \vee \sim \forall x_1 R x_1$

has property C of order 2, while the wff

$\sim \forall x\, [\Phi x \vee \sim \forall y R y] \vee \forall y_1 \Phi y_1 \vee \sim \forall x_1 R x_1,$

which can be obtained from it by one application of a Rule of Passage, does not have property C of order 2.

Throughout the fall Dreben kept working on weaker forms of the lemma which would still suffice to prove the theorem, and finding counterexamples

116 Linden Lane
Princeton, New Jersey
1962 May 31

Professor Burton S. Dreben
Emerson Hall
Harvard University
Cambridge 38, Massachusetts

Dear Professor Dreben,

I certainly appreciate your willingness to answer questions about Herbrand's thesis. As you probably realize, there are not many persons in a position to answer such questions authoritatively.

Your letter of May 18 seems quite clear, but there is one source of difficulty (which I had not yet noticed when I acknowledged receipt of your letter) which makes me wonder whether you wrote what you intended to write, or whether I am interpreting your letter correctly. Permit me to explain.

According to your terminology, $R(P_1)_p$ is the réduite of P_1 over $D_p^{(1)}$, and $\Pi(P_1)_p$ is the evaluation of $R(P_1)_p$ over $D_p^{(1)}$. Thus the individuals occurring in $\Pi(P_1)_p$ as arguments of the predicates are members of $D_{p+1}^{(1)}$. $R'(P_1)_p$ is the réduite of P_1 over $D_p^{(2)}$, and $\Pi'(P_1)_p$ is an evaluation of $R'(P_1)_p$ over $D_p^{(2)}$. Hence the individuals occurring in $\Pi'(P_1)_p$ are members of $D_{p+1}^{(2)}$. Thus if the members of $D_{p+1}^{(1)}$ are distinct from the members of $D_{p+1}^{(2)}$, the atomic formulas occurring in $\Pi(P_1)_p$ are distinct from the atomic formulas occurring in $\Pi'(P_1)_p$, and it is not clear why $\Pi(P_1)_p \supset \Pi'(P_1)_p$ should be a tautology. If, on the other hand, you are tacitly assuming that $D_{p+1}^{(1)}$ is embedded in $D_{p+1}^{(2)}$, the objection I raised in my letter of April 9 seems to apply.

I think it would be easier to discuss this question in conversation than by letter, and I am quite eager to hear your comments about my thesis topic. In addition, I have been hoping for some time for an opportunity to learn more about your work. Therefore I would like to accept your suggestion that I come to see you. I think I could come virtually any time this summer that is convenient for you, with the exception of the period between June 27 and July 14. Just to be concrete, let me tentatively suggest that I come about June 11, but if this is not a good time for you, please suggest another.

Figure 5. First page of 1962 May 31 letter from Andrews to Dreben.

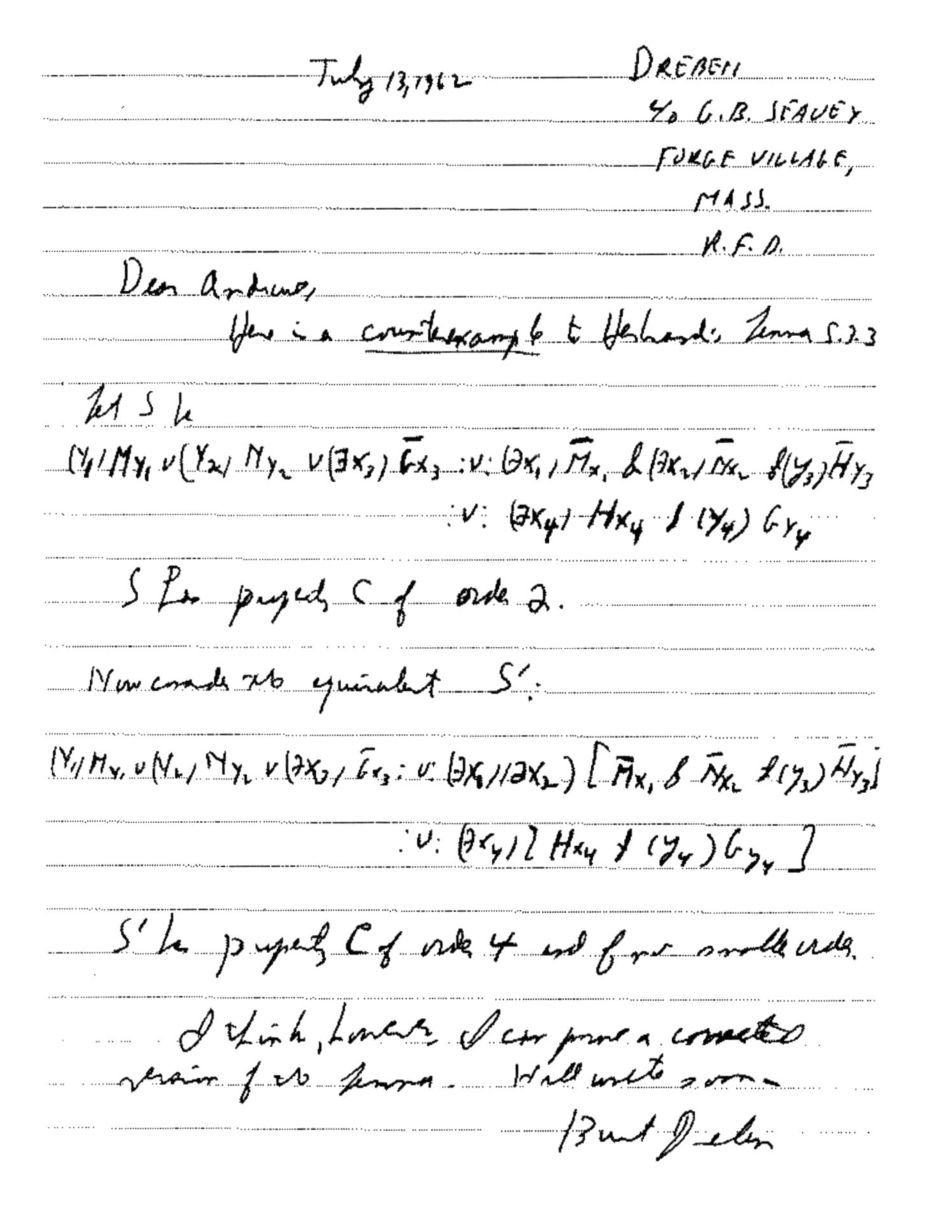

July 13, 1962

DREBEN
c/o G.B. SEAVEY
FORGE VILLAGE,
MASS.
R.F.D.

Dear Andrews,

Here is a counterexample to Herbrand's Lemma 3.3

Let S be

$(y_1)My_1 \vee (y_2)My_2 \vee (\exists x_3)\bar{G}x_3 :\vee: (\exists x_1)\bar{M}x_1 \,\&\, (\exists x_2)\bar{M}x_2 \,\&\, (y_3)\bar{H}y_3$
$:\vee: (\exists x_4)Hx_4 \,\&\, (y_4)Gy_4$

S has property C of order 2.

Now consider its equivalent S′:

$(y_1)My_1 \vee (y_2)My_2 \vee (\exists x_3)\bar{G}x_3 :\vee: (\exists x_1)(\exists x_2)[\bar{M}x_1 \,\&\, \bar{M}x_2 \,\&\, (y_3)\bar{H}y_3]$
$:\vee: (\exists x_4)[Hx_4 \,\&\, (y_4)Gy_4]$

S′ has property C of order 4 and for no smaller order.

I think, however, I can prove a corrected version of the lemma. Will write soon.

Burt Dreben

Figure 6. 1962 July 13 letter from Dreben to Andrews.

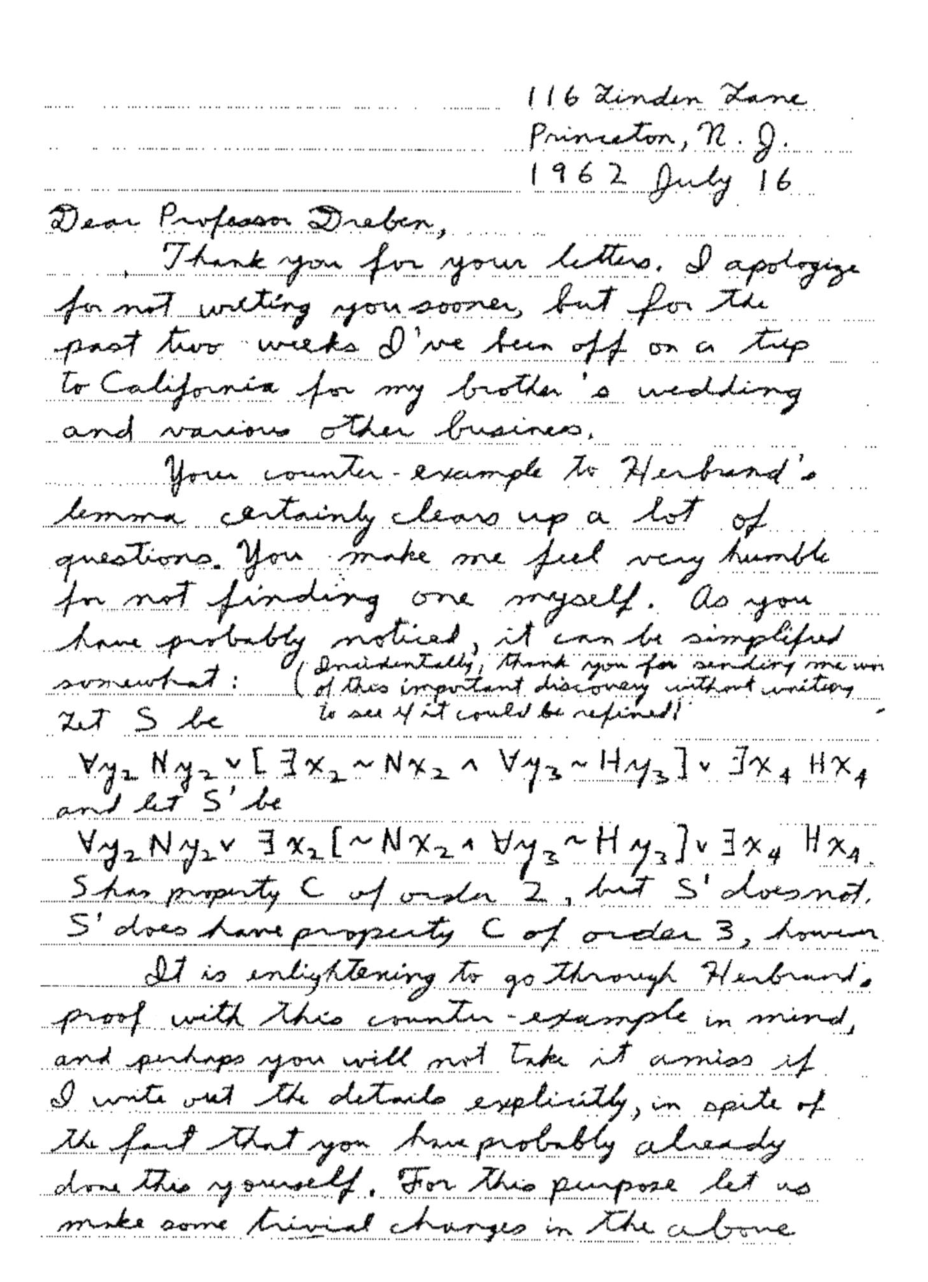

116 Linden Lane
Princeton, N. J.
1962 July 16

Dear Professor Dreben,

Thank you for your letters. I apologize for not writing you sooner, but for the past two weeks I've been off on a trip to California for my brother's wedding and various other business.

Your counter-example to Herbrand's lemma certainly clears up a lot of questions. You make me feel very humble for not finding one myself. As you have probably noticed, it can be simplified somewhat: (Incidentally, thank you for sending me word of this important discovery without waiting to see if it could be refined!)

Let S be

$$\forall y_2 \, Ny_2 \vee [\exists x_2 \sim Nx_2 \wedge \forall y_3 \sim Hy_3] \vee \exists x_4 \, Hx_4$$

and let S' be

$$\forall y_2 \, Ny_2 \vee \exists x_2 [\sim Nx_2 \wedge \forall y_3 \sim Hy_3] \vee \exists x_4 \, Hx_4.$$

S has property C of order 2, but S' does not. S' does have property C of order 3, however.

It is enlightening to go through Herbrand's proof with this counter-example in mind, and perhaps you will not take it amiss if I write out the details explicitly, in spite of the fact that you have probably already done this yourself. For this purpose let us make some trivial changes in the above

Figure 7. First page of 1962 July 16 letter from Andrews to Dreben.

2)

example:

Let P_1 be $\sim[\forall x\, \Phi x \vee \sim \forall y\, R y] \vee \forall y_1 \Phi y_1 \vee \sim \forall x_1 R x_1$

Let P_2 be $\forall x [\Phi x \vee \sim \forall y\, R y] \vee \forall y_1 \Phi y_1 \vee \sim \forall x_1 R x_1$

Let us agree to write $f_y(a_1, \ldots, a_n)$ as $y^{a_1, \ldots, a_n}$, thus letting general variables stand for their functions of index.[1]

Then

$\mathfrak{C}_1^{P_1} = \{\alpha\}$ $\mathfrak{C}_2^{P_1} = \{y_1, y\}$

The (evaluated) réduite of P_1 of order 2 is

$\sim[\Phi\alpha \wedge \Phi y_1 \wedge \Phi y \vee \sim R y] \vee \Phi y_1 \vee$
$\sim[R\alpha \wedge R y_1 \wedge R y]$,

which is a tautology.

Also

$\mathfrak{C}_1^{P_2} = \{\alpha\}$ $\mathfrak{C}_2^{P_2} = \{y_1, y^{\alpha}\}$

The (evaluated) réduite of P_2 of order 2 is

$$\Pi_2 = \sim\{[\underset{T}{\Phi\alpha} \vee \sim \underset{T}{R y^{\alpha}}] \wedge [\underset{F}{\Phi y_1} \vee \sim \underset{F}{R y^{y_1}}] \wedge [\underset{T}{\Phi y^{\alpha}} \vee \sim R y^{y^{\alpha}}]\}$$

$$\vee \underset{F}{\Phi y_1} \vee \sim[\underset{T}{R\alpha} \wedge \underset{T}{R y_1} \wedge \underset{T}{R y^{\alpha}}]$$

which can be falsified by the indicated truth value assignment.

Now according to Herbrand p. 104 (103), we must define the 'function' y in the P_1 fields as y^{a_i} in the P_2 fields for an appropriately chosen a_i. The crucial part

Figure 8. Second page of 1962 July 16 letter from Andrews to Dreben.

HARVARD UNIVERSITY
DEPARTMENT OF PHILOSOPHY

EMERSON HALL
CAMBRIDGE 38, MASSACHUSETTS

November 27, 1962

Dear Andrews,

I'm truly sorry to have taken so long on our paper, but just as I was getting back to it I became ill. However, I'll return to work in a day or two.

My new proof of Lemma 5:3.3 is quite complicated, and will entail changing our paper. Herbrand's error is much deeper than I had previously thought. About two months ago, Stal Aanderaa, a student of mine, showed me that given any number p we can construct schemata S and T such that S and S ⊃ T both have property C of order 3 but T has property C only of orders greater than p. My new proof of Lemma 5:3.3 now explains this phenomenon. No function of p alone will ever be sufficient to give in general the upper bound of a single use of the crucial rule of passage.

How is your thesis going? Is Schutte's paper in the latest J.S.L. relevant?

As ever,

Burt Dreben

Figure 9. 1962 November 27 letter from Dreben to Andrews.

to them. In November I received the letter in Figure 9, stating that Herbrand's error was much deeper than Dreben had previously thought, and mentioning some work by Aanderaa which showed this.

As a result of all this we published the paper [52] giving counterexamples to Herbrand's key lemma.

Dreben pursued this matter vigorously, and a few years later he and John Denton finally managed to prove [53] a weaker form of the lemma which was still sufficient for filling the gap in Herbrand's proof.

This finishes one chapter in the story of Herbrand's Theorem, but there are some footnotes.

While our paper on false lemmas was being written, I was considering where I should be the following year, and I heard a suggestion that I should find out if there might be any kind of suitable visiting position at the Institute for Advanced Study in Princeton. I wrote to Kurt Gödel about this, and by way of introduction mentioned that I had found a mistake in Herbrand's proof. Gödel subsequently mentioned to Church on the telephone that he had known of errors in Herbrand's work, and when I told Dreben about this, he sent a draft of our paper to Gödel. Gödel chose not to reply to this letter, but when Gödel's papers were examined after his death, it was found[5] that in the early 1940's Gödel had seen the fallacy in Herbrand's argument (though there is no evidence that he had an explicit counterexample to Herbrand's lemma), and he had devised a correction which was in all essentials the same as that in Dreben and Denton's paper [53].

Warren Goldfarb has speculated [60, p. 113] that one factor which may have contributed to Gödel's reticence to discuss his correction was that Gödel was not sure that Herbrand's lemma (which was much stronger than the corresponding lemma in Gödel's correction) was actually false.

I mentioned Herbrand's error to Georg Kreisel when he visited Princeton in the spring of 1963, and he mentioned it in a letter to Paul Bernays. Bernays replied in a letter[6] to Kreisel that Gödel had mentioned Herbrand's error to him in 1958.

It is interesting to ask where a correct proof of some form of Herbrand's Theorem was first published. A likely candidate is [67, pp. 2-33, 157ff]. Of course, Herbrand's Theorem is closely related to Gentzen's sharpened Hauptsatz [54].[7]

5 Herbrand's Theorem for Higher-Order Logic

I next turned my attention to formulating and proving a cut-elimination theorem for Church's type theory, and eventually came up with what seemed like a rather nice proof. After checking it carefully I took it to Professor Church. He read it thoroughly with me and agreed that it looked like a good proof.

However, I had heard that Gaisi Takeuti had done some work [99] which might be relevant, and I thought it was time to find out about it. Takeuti was working with a rather different formulation of type theory. He had done extensive work on the cut-elimination problem for this system, which was generally referred to as Takeuti's Conjecture. If one adds an Axiom of Infinity to type theory, one obtains a system in which one can formalize mathematical analysis and much more, so it is appropriate to use Analysis as the name for the logical system consisting of type theory with an Axiom

[5]See page 389 of [59].

[6]This letter was mentioned in a letter from Andrews to Dreben dated 1963 April 7.

[7]See page 106 of [55].

of Infinity. Takeuti had shown that a cut-free system of analysis must be consistent; thus, on the metatheoretic level Takeuti's Conjecture implies the consistency of Analysis [100].

I soon saw that the same basic ideas did indeed apply to the context in which I was working, and that my cut-elimination theorem implied the consistency of my formulation of analysis. Various technicalities involving axioms of extensionality and descriptions arise, and some years later I published a paper [10] explicating these matters. Some work by Robin Gandy plays a key role in that paper.

My proofs of the cut-elimination theorem and the derivation of the consistency of analysis from it were purely syntactic, and it was clear that they could be formalized within the formulation of analysis I was working with. Thus, it seemed clear that I had all the ingredients of a proof of the consistency of this system within itself. By Gödel's Second Theorem, this could happen only if this system of analysis were actually inconsistent. Clearly, I had a problem.

I went back and looked at my cut-elimination proof very critically, and eventually I did indeed find an error in it, well hidden in a part of the proof which seemed like the last place one would expect any difficulty. I am grateful to Takeuti and Gödel, for without the benefit of their work I probably would not have reexamined this proof, and eventually someone else would have found the error.

One sees over and over again that mistakes do occasionally occur in mathematical reasoning, and we can look forward to the day when the automated reasoning community builds tools which will make it practical for serious mathematical proofs to be checked routinely.

While thinking about the nature of formal languages in which it might be possible to prove Takeuti's Conjecture, I became interested in transfinite type theories, and developed a transfinite type theory with type variables which I presented and developed in my Ph.D. thesis (which was published in [3]). In this system one can prove a Theorem of Infinity as well as the consistency of finite type theory.

A few years later Takahashi proved Takeuti's conjecture [98] using semantical methods which could not be formalized within analysis, and by coincidence Prawitz proved the same result [84] at just about the same time. Both proofs relied on some ideas which had been developed by Schütte [90].

Ray Smullyan had developed a very elegant metatheorem which we know as Smullyan's Unifying Principle [96, 95], and I built on the ideas of Schütte, Takahashi, and Smullyan to establish [6] a version of Smullyan's Unifying Principle for the subsystem of Church's type theory which I call elementary type theory. This is Church's system minus axioms of extensionality, descriptions, choice, and infinity. Elementary type theory embodies the logic of propositional connectives, quantifiers, and λ-conversion in the context of type theory. From this Unifying Principle I derived the completeness of a rather weak form of resolution for type theory, as well as a cut-elimination theorem for elementary type theory. The Unifying Principle was also used to derive information about some very special cases of the decision problem for elementary type theory in [9]. Although elementary type theory is analogous to first-order logic in certain respects, it is actually a considerably more complex language.

Gérard Huet, who was a student at that time, read the paper [6], or a preprint of it, and came to talk to me about it. He was interested in the problem of a unification algorithm for type theory, which was still open, and I encouraged him to work on it. He soon devised a unification algorithm for type theory [69]. By coincidence, Jensen and Pietrzykowski [72] devised a similar algorithm about the same time. These were extremely important contributions to automated theorem proving in higher-order logic.

A suitable notion of a Herbrand expansion for a theorem of higher-order logic was still needed. As a start toward this, I proved the following

Theorem. A sentence is provable in elementary type theory if and only if it has a tautologous development.

Definition. A *development* of a sentence is a wff obtained from it by a sequence of the following operations, which may be applied to parts of the wff which are not in the scopes of quantifiers:

- Delete an essentially universal quantifier after assuring that its variable is not free in the current wff.
- Duplicate an essentially existential quantifier.
- Instantiate an essentially existential quantifier with an arbitrary wff.
- Apply λ-reduction.

Note that we are working here with provability rather than refutability, so we instantiate existential rather than universal quantifiers.

This theorem is a consequence of the cut-elimination theorem. It is an extension to type theory of Herbrand's Theorem inspired by Herbrand's Property A. The theorem and its proof were eventually published in [33].

This provided a significant step toward developing a general theorem proving system for type theory, but a better representation of higher-order Herbrand expansions was needed. Dale Miller developed the idea of using a tree-like structure called an *expansion tree proof*, otherwise known as an an *expansion proof*, to concisely represent the theorem, the substitution terms, the tautology, the associated mating, and the relationships between these entities. Miller proved

Miller's Expansion Proof Theorem. A sentence is provable in elementary type theory if and only if it has an expansion proof [76, 78].

This is a very elegant generalization of Herbrand's Theorem to Church's type theory. Miller's proof in [76] used Smullyan's Unifying Principle for type theory.

This completes another chapter in the story of Herbrand's Theorem.

6 Matings and Proof Transformations

In my early contributions [4, 5] to automated theorem proving, I explored a variety of ideas, but my thoughts kept returning to the desirability of focusing more directly on the fundamental structure of theorems as revealed by Herbrand's Theorem. The crucial role which Herbrand's Theorem assigns to the tautologies which underlie theorems of first- and higher-order logic makes clear the importance of developing good methods for generating, recognizing, and utilizing those tautologies (or contradictions, if one works in refutation mode).

Consideration of various approaches to theorem proving leads to the concept of a *mating* of occurrences of literals. This is a quite fundamental and general concept which can be applied to a variety of methods for recognizing or establishing theorems. In general, a mating induced by a proof of a theorem is a set of pairs of literal-occurrences in the theorem such that in the proof, literals descended from mated pairs are complementary (negations of each other), and the complementarity of these descendant literals is crucial to the structure of the proof. The terminology *mating* was used by Martin Davis [51]. There may be many proofs of a theorem which differ from each other in various ways but which induce the same mating, so it is natural to focus on searching directly for a mating which contains the essential information from which a proof can be constructed.

This leads to the abstract concept of a mating for a wff as simply a set of pairs of literal-occurrences (with some literals possibly duplicated). A mating is called *acceptable* if it actually corresponds to a proof of the theorem. A crucial problem is to devise criteria for acceptability which can be incorporated into efficient search procedures.

In [11] I discussed matings induced by resolution [88] refutations, and a criterion for acceptability of matings of sets of clauses.

In [12, 14], the concept of a mating was extended to apply not only to sets of clauses, but to arbitrary sentences of first-order logic. This approach to automated theorem proving, called *general matings*, is closely related to the *connection method* [39, 38] which was developed independently by Bibel.

The basic ideas involved here had been discussed by Prawitz [85, 86] as they applied to refuting wffs with matrices in conjunctive normal form.

General matings provide significant insight into the logical structure of theorems, but it is not always easy for people to grasp them intuitively or to relate them to other approaches to theorem proving, so a procedure for automatically transforming acceptable general matings into proofs in natural deduction style was developed [13].

When Miller developed the concept of an expansion proof, he also gave the details of an explicit algorithm for converting expansion proofs into natural deduction proofs, and proved that it works.

In his thesis [81] Frank Pfenning investigated a variety of issues in higher-order proof transformations, and developed an improved method of translating expansion proofs into natural deduction proofs based on the use of tactics. [81] and [82] also contain discussions of equality and extensionality, and methods of generating more elegant natural deduction proofs.

Some general considerations about proof transformations are discussed in [20].

7 TPS and ETPS

The ideas in [13, 14, 69] were implemented in a theorem proving system called TPS [79, 33] which we now refer to as TPS1. About 1985 work started on the design and implementation of the current version of TPS [24, 26, 27, 101], which uses expansion proofs and the proof translation methods developed by Pfenning. A MacLisp version of TPS called TPS2 was developed while we waited for Common Lisp to become available on our computer, and the current version, which runs in Common Lisp, is called TPS3. Henceforth we refer to TPS3 simply as TPS.

The ideas developed by Miller and Pfenning lie at the heart of TPS, and provide a firm foundation for automated theorem proving in higher-order logic and for the investigation of essential structural features of theorems of higher-order logic.

In automatic mode, TPS first searches for an expansion proof by setting up an expansion tree to represent the wff, expanding it (by duplicating quantifiers or instantiating higher-order variables), and searching for an acceptable mating of its literals. (An expansion tree which is appropriately expanded and has an acceptable mating is an expansion proof.) Once an expansion proof has been found, TPS transforms it into a proof in natural deduction style. Thus, the search is conducted in a context which facilitates focusing on the essential structure of the problem, but the final proof is presented in a readable format. Over the years, a number of different search procedures have been implemented in TPS. Users can choose between them, and variations of them, by setting flags. As TPS has developed, it has been tested on many examples. Some of the theorems it has proven automatically are presented in [14, 19, 22, 23, 24, 33, 40, 41, 42, 45, 46, 47, 70, 71, 79].

While most of the effort in developing TPS concentrated on procedures for proving theorems automatically, TPS can also be used to prove theorems purely interactively by applying rules of inference, or in a mixture of interactive and automatic modes, as described in [26]. The purely interactive facilities of TPS for constructing natural deduction proofs and editing formulas are available in a separate program called ETPS (Educational Theorem Proving System) [28] for use by students in logic courses.

Over the years various facilities have been added to TPS to enhance its usefulness as a research tool. TPS has a library for saving many types of entities, and there are facilities for creating and using a variety of library classification schemes. These classification schemes can themselves be saved in the library. TPS has a top level called Models which can be used to compute the semantic value of a formula in small finite standard models of type theory in which the domains of all types have cardinalities which are powers of 2. This top level also features a SOLVE command that will solve for values of "output" variables in terms of values of "input" variables which will make a given formula true.

8 Primitive Substitutions and Gensubs

To provide a method of instantiating quantifiers on set variables which is complete in principle, a procedure for applying expansion options (quantifier replications and primitive substitutions) to an expansion tree was developed [15] and implemented. Primitive substitutions introduce a small amount of new logical structure (propositional connectives and quantifiers), and contain variables for which additional substitutions can be made at a later stage. A similar idea underlies the splitting rules in [68]. Later, facilities for applying more complex substitutions for set variables, such as the *gensubs* of [24], and subformulas (possibly containing definitions) of the theorem to be proved, were implemented in TPS.

The problem of choosing appropriate instantiations for set variables is very complex. Often these instantiation terms embody key ideas of proofs. Primitive substitutions and gensubs can be applied iteratively to generate complex formulas, and higher-order unification can be used to specify substitutions for the remaining free variables, but there are many choices which must be made to specify what is done, and in what order various things are done. Various strategies for dealing with this problem have been implemented in TPS, and one can experiment with them by setting various flags.

9 Path-Focused Duplication

It is clearly desirable for expansions of an expansion tree to be motivated by the needs of the search for an acceptable mating. Sunil Issar developed a search procedure in which quantifier replications are localized to vertical paths (thus reducing the enormous growth in the number of paths which accompanies replications), and the replications for each path are generated as needed to permit the construction of a mating which spans that path. The search space grows and contracts as different vertical paths are considered. This procedure was implemented, and improved the speed of TPS very significantly. This work was reported in [70] and [71].

10 Dual Instantiation of Definitions

It is well known (see, for example, the discussion in [105]) that when searching for proofs of theorems which contain definitions, it is a significant problem to decide which instances of the definitions to instantiate (expand). Often, one needs to instantiate some, but not all, of them, and if one does instantiate all of them, one can cause the search space to expand in a very undesirable way. A partial solution [42] to this problem called dual instantiation was developed and implemented in TPS; it involves making each instance of a definition accessible to the search procedure in both its instantiated and its uninstantiated form, and letting the search procedure decide which to use, with a bias in favor of the uninstantiated form. This proved very effective in some cases.

11 Component Search

An important component of an expansion proof is an acceptable (or complete) mating [14], which is also known as a spanning set of connections [38]. An acceptable mating must contain a connection on every vertical path through the formula which represents an expanded form of the theorem to be proved. For many years, the search procedures implemented in TPS had generally built up these sets of connections by adding one connection at a time whenever a path which was not yet spanned was found. Of course, the addition of inappropriate connections would eventually lead to backtracking when it was found that the unification constraints failed.

Matt Bishop developed a quite different search procedure called *component search*, based on building up sets of connections by taking unions of smaller sets, starting from sets of connections called *components*. Components are similar to the hyper-links of [74]. Component search is largely a breadth-first search procedure, and it avoids some of the problems of previously developed depth-first search procedures.

Bishop's work also involved integrating an improved implementation of higher-order unification and rewriting rules with this search procedure. The higher-order unification algorithm [69] generates a search tree, which is usually infinite, since the decision problem for higher-order unification is unsolvable. Therefore, one must set arbitrary bounds on how far down the tree one will go to search for a unifier (unifying substitution). It has been found that a bound on the complexity of the substitution which will be applied to each variable is more useful than a bound on the depth of the whole search tree (which involves all the variables in the problem). In this context, one can compute all the possible unifying ground substitutions for the free variables in the problem before starting the search for a mating. One can represent all of these unifiers which are associated with a partial mating as a directed acyclic graph. With this representation, it is possible to compute the set of unifiers associated with the union of two partial matings very quickly from the sets of unifiers associated with the partial matings. (This representation was suggested by analogous ideas in [48]). As a result, it was possible to drastically reduce the substantial amount of time that was being devoted to higher-order unification by the search procedure.

At the same time, equality reasoning and rewriting were integrated directly into the unification procedure in TPS. (Previously, equalities were simply defined by the Leibniz definition of equality in higher-order logic, but there were no special rules for applying equality reasoning as part of the search procedure.) This allowed some or all of the equality literals to be removed from the set of literals involved in the search for a mating, further improving the performance of the search procedure.

These ideas were implemented as the search procedure MS98-1 in TPS, which was found to be generally much more effective on hard problems than search procedures previously implemented. This work was reported in Bishop's thesis [41] and the associated paper [40].

12 Set Constraints

Chad Brown developed [45] a method of finding instantiations for set variables which is more sophisticated than blindly enumerating logical possibilities, as had been done previously. This method is based on generating and solving set constraints. Just as one can use unification to delay the instantiation of first-order (and some higher-order) variables, one can use set constraints to delay the instantiation of set variables. This approach extends the former procedure in the following ways:

- Instead of instantiating set variables in a preprocessing step, the instantiations can be delayed by intertwining search with the process of instantiating set variables.

- Instead of guessing instantiations, the procedure solves for sets satisfying certain properties.

- Instead of using large terms naming sets with certain properties, lemmas asserting the existence of such sets are introduced.

This work extended and clarified Bledsoe's work [43, 44] on finding maximal and minimal solutions for predicates to reduce higher-order problems to first-order problems when this is possible.

13 Extensional Expansion Proofs

In his thesis [46, 47], Chad Brown generalized the concept of an expansion tree for a formula to the concept of an *extensional expansion dag*, and generalized the concept of an expansion proof to the concept of an *extensional expansion proof*. The structures associated with these concepts provide ways of representing more information about the structure of a proof, and incorporate ways of dealing with extensionality without simply using axioms of extensionality.

He developed two new search procedures, called MS03-7 and MS04-2, for TPS. They both extend the mating search methods previously used by TPS by including connections up to equality, by including extensional and equational reasoning, and by generating and solving set constraints. MS03-7 performs a kind of saturation search. MS04-2 performs a depth-first search (with weights to control the order of choices) with backtracking and a depth bound.

He also did extensive theoretical work motivated by the problem of trying to reduce the set of quantifiers and connectives one need consider using while constructing instantiation terms for a proof of a theorem. He developed new completeness and cut-elimination proofs, and new methods of constructing models, for various forms of Church's type theory. Using these results, he showed how to give rigorous proofs that certain theorems cannot be proven without using instantiation terms involving certain connectives and quantifiers.

14 Contacts with the Ωmega group

Our research group gained much from its extensive contacts with members of the group at Saarbrücken associated with the Ωmega system [93, 92]. We benefited greatly from visits by Jörg Siekmann, Michael Kohlhase, Christoph Benzmüller, and Mark Kaminski. The reports [73, 35] reflect activities during these visits. TPS has been integrated [36] into Ωmega. The important contribution [37] to understanding the semantics of type theory helps to clarify issues related to the completeness of the search procedures used by TPS. Mark Kaminski added a rewriting top level (which is described in [27]) to TPS.

15 Conclusion

The field of automated reasoning is a wonderful field to work in. Logical reasoning plays such a fundamental role in the spectrum of intellectual activities that advances in automating logic will inevitably have a profound impact on many intellectual disciplines. Of course, these things take time. We tend to be impatient, but we need some historical perspective. The study of logical reasoning has a very long history, going back at least as far as Aristotle. During some of this time not very much progress was made. It's gratifying to realize how much has been accomplished in the fifty years since serious efforts to mechanize logic began. Very satisfying progress has been made.

Nevertheless, much more must be done in order to realize the potential of automated reasoning. Extensive development of the infrastructure for automated reasoning is needed; we need libraries of definitions and theorems, databases, richer formal languages, formal theories, systems for translating from one formalism to another, integration of various systems, and all sorts of interfaces. We also need continuing major improvements in inference mechanisms. Procedures for proving theorems can be used as inference mechanisms, so research on automated theorem proving has relevance for applications far removed from traditional concerns about proving theorems.

If we take a historical perspective and try to imagine looking back from the future, it seems clear that we are still at a primitive stage in the development of automated deduction. In spite of the many good ideas which have been incorporated into them, current search procedures rely far too much on blind (or nearly blind) brute force. There is a great need for the development of concepts and theories which can serve as the basis for substantially more sophisticated search procedures which will be guided much more by relevant features of the context in which the search is being conducted (such as features of the theorem to be proved).

Clause-based search procedures facilitate very efficient implementations, particularly in first-order logic. However, it is natural to ask whether the development of sophisticated new concepts relevant to searches for proofs may be inhibited by a focus on searching in the context of problems represented as sets of clauses. There may be significant global aspects of theorems which are not easy to recognize or deal with in a clausal context. Expansion proofs and matings seem to provide very fundamental representations of the essential logical structure of theorems, though we need to develop much more sophisticated methods of searching for them.

The development of relevant concepts is fundamental to scientific progress, and it is not easy to find such concepts. Nevertheless, they are there, waiting to be discovered. If one thinks about the natural numbers as being defined by zero and the successor function, they seem to stretch out in a sequence of boring uniformity. However, once one defines operations such as addition and multiplication on them, the integers acquire individual characteristics (such as being prime, multiples of 3, the sum of a particular series, etc.), and a very rich theory can be developed.

Herbrand's Theorem, and various extensions, enhancements, and refinements of it, have served us very well, but what other fundamental insights important to automated deduction remain to be discovered? We may think that theorem proving is simply hard, and that all we can do is continue to find very efficient ways of searching exhaustively within the context of currently known theoretical frameworks. However, the history of science is full of surprises. In the latter part of the nineteenth century it seemed to many people that all the fundamental discoveries which could be made in physics had been made. Then along came the ideas and discoveries which radically transformed the field of physics in the twentieth century. In a similar way, the discovery of new fundamental concepts which provide foundations for rich and sophisticated theories of proof search could profoundly affect the field of automated deduction.

BIBLIOGRAPHY

[1] Peter Andrews. Church's Type Theory. In Edward N. Zalta, editor, *The Stanford Encyclopedia of Philosophy*. 2006.
http://plato.stanford.edu/archives/fall2006/entries/type-theory-church/.

[2] Peter B. Andrews. A Reduction of the Axioms for the Theory of Propositional Types. *Fundamenta Mathematicae*, 52:345–350, 1963.

[3] Peter B. Andrews. *A Transfinite Type Theory with Type Variables*. Studies in Logic and the Foundations of Mathematics. North-Holland, 1965.

[4] Peter B. Andrews. On Simplifying the Matrix of a Wff. *Journal of Symbolic Logic*, 33:180–192, 1968. Reprinted in [94].

[5] Peter B. Andrews. Resolution with Merging. *Journal of the ACM*, 15:367–381, 1968. Reprinted in [94].

[6] Peter B. Andrews. Resolution in Type Theory. *Journal of Symbolic Logic*, 36:414–432, 1971.

[7] Peter B. Andrews. General Models and Extensionality. *Journal of Symbolic Logic*, 37:395–397, 1972.

[8] Peter B. Andrews. General Models, Descriptions, and Choice in Type Theory. *Journal of Symbolic Logic*, 37:385–394, 1972.

[9] Peter B. Andrews. Provability in Elementary Type Theory. *Zeitschrift fur Mathematische Logic und Grundlagen der Mathematik*, 20:411–418, 1974.

[10] Peter B. Andrews. Resolution and the Consistency of Analysis. *Notre Dame Journal of Formal Logic*, 15(1):73–84, 1974.

[11] Peter B. Andrews. Refutations by Matings. *IEEE Transactions on Computers*, C-25:801–807, 1976.

[12] Peter B. Andrews. General Matings. In *Proceedings of the Fourth Workshop on Automated Deduction*, pages 19–25. Austin, Texas, 1979.

[13] Peter B. Andrews. Transforming Matings into Natural Deduction Proofs. In W. Bibel and R. Kowalski, editors, *Proceedings of the 5th International Conference on Automated Deduction*, volume 87 of *Lecture Notes in Computer Science*, pages 281–292, Les Arcs, France, 1980. Springer-Verlag.

[14] Peter B. Andrews. Theorem Proving via General Matings. *Journal of the ACM*, 28:193–214, 1981.

[15] Peter B. Andrews. On Connections and Higher-Order Logic. *Journal of Automated Reasoning*, 5:257–291, 1989.

[16] Peter B. Andrews. Classical Type Theory. In Alan Robinson and Andrei Voronkov, editors, *Handbook of Automated Reasoning*, volume 2, chapter 15, pages 965–1007. Elsevier Science, Amsterdam, 2001.

[17] Peter B. Andrews. *An Introduction to Mathematical Logic and Type Theory: To Truth Through Proof.* Kluwer Academic Publishers, second edition, 2002.

[18] Peter B. Andrews. Herbrand Award Acceptance Speech. *Journal of Automated Reasoning*, 31:169–187, 2003.
http://journals.kluweronline.com/article.asp?PIPS=5256402.

[19] Peter B. Andrews. Proving Theorems of Type Theory Automatically with TPS. In *Proceedings of the Twentieth National Conference on Artificial Intelligence and the Seventeenth Innovative Applications of Artificial Intelligence Conference (AAAI-05 / IAAI-05)*, pages 1676–1677. AAAI Press, 2005.

[20] Peter B. Andrews. Some Reflections on Proof Transformations. In D. Hutter and W. Stephan, editors, *Mechanizing Mathematical Reasoning: Essays in Honor of Jrg H. Siekmann on the Occasion of His 60th Birthday*, volume 2605 of *Lecture Notes in Artificial Intelligence*, pages 14–29. Springer-Verlag, 2005.
http://www.springeronline.com/sgw/cda/frontpage/0,11855,4-102-22-42171167-0,00.html?changeHeader=true.

[21] Peter B. Andrews and Matthew Bishop. On Sets, Types, Fixed Points, and Checkerboards. In Pierangelo Miglioli, Ugo Moscato, Daniele Mundici, and Mario Ornaghi, editors, *Theorem Proving with Analytic Tableaux and Related Methods. 5th International Workshop. (TABLEAUX '96)*, volume 1071 of *Lecture Notes in Artificial Intelligence*, pages 1–15, Terrasini, Italy, May 1996. Springer-Verlag.
http://dx.doi.org/10.1007/3-540-61208-4_1.

[22] Peter B. Andrews, Matthew Bishop, and Chad E. Brown. System Description: TPS: A Theorem Proving System for Type Theory. In McAllester [75], pages 164–169.
http://dx.doi.org/10.1007/10721959_11.

[23] Peter B. Andrews, Matthew Bishop, Sunil Issar, Dan Nesmith, Frank Pfenning, and Hongwei Xi. TPS: An Interactive and Automatic Tool for Proving Theorems of Type Theory. In Jeffrey J. Joyce and Carl-Johan H. Seger, editors, *Higher Order Logic Theorem Proving and Its Applications: 6th International Workshop, HUG '93*, volume 780 of *Lecture Notes in Computer Science*, pages 366–370, Vancouver, B.C., Canada, August 1994. Springer-Verlag.
http://dx.doi.org/10.1007/3-540-57826-9_148.

[24] Peter B. Andrews, Matthew Bishop, Sunil Issar, Dan Nesmith, Frank Pfenning, and Hongwei Xi. TPS: A Theorem Proving System for Classical Type Theory. *Journal of Automated Reasoning*, 16:321–353, 1996.
http://dx.doi.org/10.1007/BF00252180.

[25] Peter B. Andrews and Chad E. Brown. Tutorial: Using TPS for Higher-Order Theorem Proving and ETPS for Teaching Logic. In McAllester [75], pages 511–512.
http://dx.doi.org/10.1007/10721959_44.

[26] Peter B. Andrews and Chad E. Brown. TPS: A Hybrid Automatic-Interactive System for Developing Proofs. *Journal of Applied Logic*, 4:367–395, 2006.
http://dx.doi.org/10.1016/j.jal.2005.10.002.

[27] Peter B. Andrews, Chad E. Brown, Matthew Bishop, Sunil Issar, Dan Nesmith, Frank Pfenning, Hongwei Xi, and Mark Kaminski. TPS *User's Manual*, 2007. 133+vi pp.

[28] Peter B. Andrews, Chad E. Brown, Frank Pfenning, Matthew Bishop, Sunil Issar, and Hongwei Xi. ETPS: A System to Help Students Write Formal Proofs. *Journal of Automated Reasoning*, 32:75–92, 2004. http://journals.kluweronline.com/article.asp?PIPS=5264938.

[29] Peter B. Andrews and Eve Longini Cohen. Theorem Proving in Type Theory. In *Workshop on Automatic Deduction*, Cambridge, Mass., 1977. MIT. 5 pp.

[30] Peter B. Andrews and Eve Longini Cohen. Theorem Proving in Type Theory. In *Proceedings of the 5th International Joint Conference on Artificial Intelligence, IJCAI-77*, page 566, MIT, Cambridge, MA, 1977. IJCAI.

[31] Peter B. Andrews, Sunil Issar, Dan Nesmith, and Frank Pfenning. The TPS Theorem Proving System. In Stickel [97], pages 641–642.

[32] Peter B. Andrews, Sunil Issar, Daniel Nesmith, and Frank Pfenning. The TPS Theorem Proving System. In Ewing Lusk and Ross Overbeek, editors, *Proceedings of the 9th International Conference on Automated Deduction*, volume 310 of *Lecture Notes in Computer Science*, pages 760–761, Argonne, Illinois, 1988. Springer-Verlag.

[33] Peter B. Andrews, Dale A. Miller, Eve Longini Cohen, and Frank Pfenning. Automating Higher-Order Logic. In W. W. Bledsoe and D. W. Loveland, editors, *Automated Theorem Proving: After 25 Years*, Contemporary Mathematics series, vol. 29, pages 169–192. American Mathematical Society, 1984. Proceedings of the Special Session on Automatic Theorem Proving, 89th Annual Meeting of the American Mathematical Society, held in Denver, Colorado, January 5-9, 1983.

[34] Peter B. Andrews, Frank Pfenning, Sunil Issar, and C. P. Klapper. The TPS Theorem Proving System. In Jörg H. Siekmann, editor, *Proceedings of the 8th International Conference on Automated Deduction*, volume 230 of *Lecture Notes in Computer Science*, pages 663–664, Oxford, England, 1986. Springer-Verlag.

[35] Christoph Benzmüller. A Calculus and a System Architecture for Extensional Higher-Order Resolution. Technical Report 97-198, Department of Mathematical Sciences, Carnegie Mellon University, June 1997.

[36] Christoph Benzmüller, Matthew Bishop, and Volker Sorge. Integrating TPS and ΩMEGA. *Journal of Universal Computer Science*, 5(3):188–207, March 1999. http://www.iicm.edu/jucs_5_3/integrating_tps_and_omega.

[37] Christoph Benzmüller, Chad E. Brown, and Michael Kohlhase. Higher-Order Semantics and Extensionality. *Journal of Symbolic Logic*, 69:1027–1088, 2004.

[38] Wolfgang Bibel. *Automated Theorem Proving*. Vieweg, Braunschweig, second edition, 1987.

[39] Wolfgang Bibel and J. Schreiber. Proof Search in a Gentzen-like System of First-Order Logic. In E. Gelenbe and D. Potier, editors, *International Computing Symposium 1975*, pages 205–212. North-Holland, Amsterdam, 1975.

[40] Matthew Bishop. A Breadth-First Strategy for Mating Search. In Harald Ganzinger, editor, *Proceedings of the 16th International Conference on Automated Deduction*, volume 1632 of *Lecture Notes in Artificial Intelligence*, pages 359–373, Trento, Italy, 1999. Springer-Verlag.

[41] Matthew Bishop. *Mating Search Without Path Enumeration*. PhD thesis, Department of Mathematical Sciences, Carnegie Mellon University, April 1999. Department of Mathematical Sciences Research Report No. 99–223.

[42] Matthew Bishop and Peter B. Andrews. Selectively Instantiating Definitions. In Claude Kirchner and Hélène Kirchner, editors, *Proceedings of the 15th International Conference on Automated Deduction*, volume 1421 of *Lecture Notes in Artificial Intelligence*, pages 365–380, Lindau, Germany, 1998. Springer-Verlag. http://dx.doi.org/10.1007/BFb0054272.

[43] W. W. Bledsoe. A Maximal Method for Set Variables in Automatic Theorem Proving. In J. E. Hayes, Donald Michie, and L. I. Mikulich, editors, *Machine Intelligence 9*, pages 53–100. Ellis Harwood Ltd., Chichester, and John Wiley & Sons, 1979.

[44] W. W. Bledsoe and Guohui Feng. Set-Var. *Journal of Automated Reasoning*, 11:293–314, 1993.

[45] Chad E. Brown. Solving for Set Variables in Higher-Order Theorem Proving. In Voronkov [103], pages 408–422.

[46] Chad E. Brown. *Set Comprehension in Church's Type Theory.* PhD thesis, Department of Mathematical Sciences, Carnegie Mellon University, 2004.

[47] Chad E. Brown. *Automated Reasoning in Higher-Order Logic: Set Comprehension and Extensionality in Church's Type Theory*, volume 10 of *Studies in Logic: Logic and Cognitive Systems.* College Publications, 2007.

[48] R.E. Bryant. Graph-Based Algorithms for Boolean Function Manipulation. *IEEE Transactions on Computers*, C-35(8):677–691, 1986.

[49] Alonzo Church. A Formulation of the Simple Theory of Types. *Journal of Symbolic Logic*, 5:56–68, 1940.

[50] M. Davis and H. Putnam. A Computing Procedure for Quantification Theory. *Journal of the ACM*, 7:201–215, 1960.

[51] Martin Davis. Eliminating the Irrelevant from Mechanical Proofs. In *Experimental Arithmetic, High Speed Computing and Mathematics*, Proceedings of Symposia in Applied Mathematics XV, pages 15–30. American Mathematical Society, 1963.

[52] Burton Dreben, Peter Andrews, and Stål Aanderaa. False Lemmas in Herbrand. *Bulletin of the American Mathematical Society*, 69:699–706, 1963.

[53] Burton Dreben and John Denton. A Supplement to Herbrand. *Journal of Symbolic Logic*, 31:393–398, 1966.

[54] G. Gentzen. Untersuchungen über das Logische Schließen I und II. *Mathematische Zeitschrift*, 39:176–210,405–431, 1935. Translated in [55].

[55] G. Gentzen. Investigations into Logical Deductions. In M. E. Szabo, editor, *The Collected Papers of Gerhard Gentzen*, pages 68–131. North-Holland Publishing Co., Amsterdam, 1969.

[56] P.C. Gilmore. A Proof Method for Quantification Theory. *IBM Journal of Research and Development*, 4:28–35, 1960.

[57] Kurt Gödel. Die Vollstandigkeit der Axiome des logischen Funktionenkalküls. *Monatsh. Math. Phys.*, 37:349–360, 1930.

[58] Kurt Gödel. Über formal unentscheidbare Sätze der Principia Mathematica und verwandter Systeme I. *Monatsh. Math. Phys.*, 38:173–198, 1931.

[59] Kurt Gödel. *Collected Works, Volume IV.* Clarendon Press, 2003. Edited by Solomon Feferman, John W. Dawson, Jr., Warren Goldfarb, Charles Parsons, and Wilfried Sieg.

[60] Warren D. Goldfarb. Herbrand's error and Gödel's correction. *Modern Logic*, 3(2):103–118, February 1993.

[61] Leon Henkin. The Completeness of the First-Order Functional Calculus. *Journal of Symbolic Logic*, 14:159–166, 1949.

[62] Leon Henkin. Completeness in the Theory of Types. *Journal of Symbolic Logic*, 15:81–91, 1950.

[63] Leon Henkin. A Theory of Propositional Types. *Fundamenta Mathematicae*, 52:323–344, 1963.

[64] Leon Henkin. The discovery of my completeness proofs. *Bulletin of Symbolic Logic*, 2:127–158, 1996.

[65] Jacques Herbrand. Recherches sur la théorie de la démonstration. *Travaux de la Société des Sciences et des Lettres de Varsovie, Classe III Sciences Mathematiques et Physiques*, 33, 1930. Translated in [66].

[66] Jacques Herbrand. *Logical Writings.* Harvard University Press, 1971. Edited by Warren D. Goldfarb.

[67] David Hilbert and Paul Bernays. *Grundlagen der Mathematik*, volume 2. Springer, 1939.

[68] Gérard P. Huet. A Mechanization of Type Theory. In *Proceedings of the Third International Joint Conference on Artificial Intelligence*, pages 139–146, Stanford University, California, USA, 1973. IJCAI.

[69] Gérard P. Huet. A Unification Algorithm for Typed λ-Calculus. *Theoretical Computer Science*, 1:27–57, 1975.
[70] Sunil Issar. Path-Focused Duplication: A Search Procedure for General Matings. In *AAAI–90. Proceedings of the Eighth National Conference on Artificial Intelligence*, volume 1, pages 221–226. AAAI Press/The MIT Press, 1990.
[71] Sunil Issar. *Operational Issues in Automated Theorem Proving Using Matings*. PhD thesis, Carnegie Mellon University, 1991. 147 pp.
[72] D.C. Jensen and T. Pietrzykowski. Mechanizing ω-Order Type Theory Through Unification. *Theoretical Computer Science*, 3:123–171, 1976.
[73] Michael Kohlhase. A Unifying Principle for Extensional Higher-Order Logic. Technical Report 93–153, Department of Mathematics, Carnegie Mellon University, January 1993.
[74] Shie-Jue Lee and David A. Plaisted. Eliminating Duplication with the Hyper-Linking Strategy. *Journal of Automated Reasoning*, 9:25–42, 1992.
[75] David McAllester, editor. *Proceedings of the 17th International Conference on Automated Deduction*, volume 1831 of *Lecture Notes in Artificial Intelligence*, Pittsburgh, PA, USA, 2000. Springer-Verlag.
[76] Dale A. Miller. *Proofs in Higher-Order Logic*. PhD thesis, Carnegie Mellon University, Department of Mathematics, 1983. 81 pp.
[77] Dale A. Miller. Expansion Tree Proofs and Their Conversion to Natural Deduction Proofs. In Shostak [91], pages 375–393.
[78] Dale A. Miller. A Compact Representation of Proofs. *Studia Logica*, 46(4):347–370, 1987.
[79] Dale A. Miller, Eve Longini Cohen, and Peter B. Andrews. A Look at TPS. In Donald W. Loveland, editor, *Proceedings of the 6th International Conference on Automated Deduction*, volume 138 of *Lecture Notes in Computer Science*, pages 50–69, New York, USA, 1982. Springer-Verlag.
[80] Frank Pfenning. Analytic and Non-Analytic Proofs. In Shostak [91], pages 394–413.
[81] Frank Pfenning. *Proof Transformations in Higher-Order Logic*. PhD thesis, Carnegie Mellon University, 1987. 156 pp.
[82] Frank Pfenning and Dan Nesmith. Presenting Intuitive Deductions via Symmetric Simplification. In Stickel [97], pages 336–350.
[83] Dag Prawitz. An improved proof procedure. *Theoria*, 26:102–139, 1960.
[84] Dag Prawitz. Hauptsatz for Higher Order Logic. *Journal of Symbolic Logic*, 33:452–457, 1968.
[85] Dag Prawitz. Advances and Problems in Mechanical Proof Procedures. In Meltzer and Michie, editors, *Machine Intelligence 4*, pages 59–71. Edinburgh University Press, 1969.
[86] Dag Prawitz. A Proof Procedure with Matrix Reduction. In M. Laudet, D. Lacombe, L. Nolin, and M. Schutzenberger, editors, *Symposium on Automatic Demonstration, Versailles, France*, Lecture Notes in Mathematics 125, pages 207–214. Springer-Verlag, 1970.
[87] W. V. Quine. A Proof Procedure for Quantification Theory. *Journal of Symbolic Logic*, 20:141–149, 1955.
[88] J. A. Robinson. A Machine-Oriented Logic Based on the Resolution Principle. *Journal of the ACM*, 12:23–41, 1965.
[89] Bertrand Russell. Mathematical Logic as Based on the Theory of Types. *American Journal of Mathematics*, 30:222–262, 1908. Reprinted in [102, pp. 150–182].
[90] K. Schütte. Syntactical and Semantical Properties of Simple Type Theory. *Journal of Symbolic Logic*, 25(4):305–326, 1960.
[91] R. E. Shostak, editor. *Proceedings of the 7th International Conference on Automated Deduction*, volume 170 of *Lecture Notes in Computer Science*, Napa, California, USA, 1984. Springer-Verlag.
[92] Jörg Siekmann, Christoph Benzmüller, and Serge Autexier. Computer Supported Mathematics with OMEGA. *Journal of Applied Logic*, 4(4):533–559, 2006.

[93] Jörg Siekmann, Christoph Benzmüller, Vladimir Brezhnev, Lassaad Cheikhrouhou, Armin Fiedler, Andreas Franke, Helmut Horacek, Michael Kohlhase, Andreas Meier, Erica Melis, Markus Moschner, Immanuel Normann, Martin Pollet, Volker Sorge, Carsten Ullrich, Claus-Peter Wirth, and Jürgen Zimmer. Proof Development with Ωmega. In Voronkov [103], pages 144–149.
[94] Jörg Siekmann and Graham Wrightson, editors. *Automation of Reasoning. Vol. 2. Classical Papers on Computational Logic 1967–1970.* Springer-Verlag, 1983.
[95] R. M. Smullyan. *First-Order Logic.* Springer-Verlag, Berlin, 1968.
[96] Raymond M. Smullyan. A Unifying Principle in Quantification Theory. *Proceedings of the National Academy of Sciences, U.S.A.*, 49:828–832, 1963.
[97] M. E. Stickel, editor. *Proceedings of the 10th International Conference on Automated Deduction*, volume 449 of *Lecture Notes in Artificial Intelligence*, Kaiserslautern, Germany, 1990. Springer-Verlag.
[98] Moto-o Takahashi. A Proof of Cut-Elimination Theorem in Simple Type Theory. *Journal of the Mathematical Society of Japan*, 19:399–410, 1967.
[99] Gaisi Takeuti. On a Generalized Logic Calculus. *Japanese Journal of Mathematics*, 23:39–96, 1953. Errata: ibid, vol. 24 (1954), 149–156.
[100] Gaisi Takeuti. Remark on the fundamental conjecture of GLC. *Journal of the Mathematical Society of Japan*, 10:44–45, 1958.
[101] TPS and ETPS Homepage. http://gtps.math.cmu.edu/tps.html.
[102] Jean van Heijenoort. *From Frege to Gödel. A Source Book in Mathematical Logic 1879–1931.* Harvard University Press, Cambridge, Massachusetts, 1967.
[103] Andrei Voronkov, editor. *Proceedings of the 18th International Conference on Automated Deduction*, volume 2392 of *Lecture Notes in Artificial Intelligence*, Copenhagen, Denmark, 2002. Springer-Verlag.
[104] Alfred North Whitehead and B. Russell. *Principia Mathematica*, volume 1. Cambridge University Press, 2nd edition, 1927.
[105] Larry Wos. The Problem of Definition Expansion and Contraction. *Journal of Automated Reasoning*, 3:433–435, 1987.

The Journal of Symbolic Logic
Volume 5 (1940), pp. 56-68

A FORMULATION OF THE SIMPLE THEORY OF TYPES

ALONZO CHURCH

The purpose of the present paper is to give a formulation of the simple theory of types[1] which incorporates certain features of the calculus of λ-conversion.[2] A complete incorporation of the calculus of λ-conversion into the theory of types is impossible if we require that λx and juxtaposition shall retain their respective meanings as an abstraction operator and as denoting the application of function to argument. But the present partial incorporation has certain advantages from the point of view of type theory and is offered as being of interest on this basis (whatever may be thought of the finally satisfactory character of the theory of types as a foundation for logic and mathematics).

For features of the formulation which are not immediately connected with the incorporation of λ-conversion, we are heavily indebted to Whitehead and Russell,[3] Hilbert and Ackermann,[4] Hilbert and Bernays,[5] and to forerunners of these, as the reader familiar with the works in question will recognize.

1. The hierarchy of types. The class of *type symbols* is described by the rules that ι and o are each type symbols and that if α and β are type symbols then $(\alpha\beta)$ is a type symbol: it is the least class of symbols which contains the symbols ι and o and is closed under the operation of forming the symbol $(\alpha\beta)$ from the symbols α and β.

As exemplified in the statement just made, we shall use the Greek letters α, β, γ to represent variable or undetermined type symbols. We shall abbreviate type symbols by omission of parentheses with the convention that association is to the left—so that, for instance, $o\iota$ will be an abbreviation for $(o\iota)$, $\iota\iota\iota$ for $((\iota\iota)\iota)$, $\iota\iota(\iota\iota)$ for $((\iota\iota)(\iota\iota))$, etc. Moreover, we shall use α' as an abbreviation for $((\alpha\alpha)(\alpha\alpha))$, α'' as an abbreviation for $((\alpha'\alpha')(\alpha'\alpha'))$, etc.

The type symbols enter our formal theory only as subscripts upon variables and constants. In the interpretation of the theory it is intended that the

Received March 23, 1940.

[1] See Rudolf Carnap, ***Abriss der Logistik,*** Vienna 1929, §9. (The simple theory of types was suggested as a modification of Russell's ramified theory of types by Leon Chwistek in 1921 and 1922 and by F. P. Ramsey in 1926.)

[2] See, for example, Alonzo Church, ***Mathematical logic*** (mimeographed), Princeton, N. J., 1936, and ***The calculi of lambda-conversion,*** forthcoming monograph.

[3] Bertrand Russell, *Mathematical logic as based on the theory of types*, ***American journal of mathematics,*** vol. 30 (1908), pp. 222-262; Alfred North Whitehead and Bertrand Russell, ***Principia mathematica,*** vol. 1, Cambridge, England, 1910 (second edition 1925), vol. 2, Cambridge, England, 1912 (second edition 1927), and vol. 3, Cambridge, England, 1913 (second edition 1927).

[4] D. Hilbert and W. Ackermann, ***Grundzüge der theoretischen Logik,*** Berlin 1928 (second edition 1938).

[5] D. Hilbert and P. Bernays, ***Grundlagen der Mathematik,*** vol. 1, Berlin 1934, and vol. 2, Berlin 1939.

subscript shall indicate the type of the variable or constant, o being the type of propositions, ι the type of individuals, and $(\alpha\beta)$ the type of functions of one variable for which the range of the independent variable comprises the type β and the range of the dependent variable is contained in the type α. Functions of several variables are explained, after Schönfinkel,[6] as functions of one variable whose values are functions, and propositional functions are regarded simply as functions whose values are propositions. Thus, e.g., $o\iota\iota$ is the type of propositional functions of two individual variables.

We purposely refrain from making more definite the nature of the types o and ι, the formal theory admitting of a variety of interpretations in this regard. Of course the matter of interpretation is in any case irrelevant to the abstract construction of the theory, and indeed other and quite different interpretations are possible (formal consistency assumed).

2. Well-formed formulas. The *primitive symbols* are given in the following infinite list:

$$\lambda,\ (,\ ),\ N_{oo},\ A_{ooo},\ \Pi_{o(o\alpha)},\ \iota_{\alpha(o\alpha)},\ a_\alpha,\ b_\alpha,\ \cdots,\ z_\alpha,\ \bar{a}_\alpha,\ \bar{b}_\alpha,\ \cdots.$$

Of these, the first three are *improper symbols*, and the others are *proper symbols*. Of the proper symbols, N_{oo}, A_{ooo}, $\Pi_{o(o\alpha)}$, and $\iota_{\alpha(o\alpha)}$ are *constants*, and the remainder are *variables*.

(The inclusion of $\Pi_{o(o\alpha)}$ in this list of primitive symbols is meant in this sense, that, if α is any type symbol, $\Pi_{o(o\alpha)}$ is a primitive symbol, a proper symbol, and a constant; similarly in the case of $\iota_{\alpha(o\alpha)}$, a_α, etc.)

Any finite sequence of primitive symbols is a *formula*. Certain formulas are distinguished as being *well-formed* and as having a certain *type*, in accordance with the following rules: (1) a formula consisting of a single proper symbol is well-formed and has the type indicated by the subscript; (2) if $\mathbf{x}_\beta$ is a variable with subscript β and $\mathbf{M}_\alpha$ is a well-formed formula of type α, then $(\lambda\mathbf{x}_\beta\mathbf{M}_\alpha)$ is a well-formed formula having the type $\alpha\beta$; (3) if $\mathbf{F}_{\alpha\beta}$ and $\mathbf{A}_\beta$ are well-formed formulas of types $\alpha\beta$ and β respectively, then $(\mathbf{F}_{\alpha\beta}\mathbf{A}_\beta)$ is a well-formed formula having the type α. The well-formed formulas are the least class of formulas which these rules allow, and the type of a well-formed formula is that determined (uniquely) by these rules. An occurrence of a variable $\mathbf{x}_\beta$ in a well-formed formula is *bound* or *free* according as it is or is not an occurrence in a well-formed part of the formula having the form $(\lambda\mathbf{x}_\beta\mathbf{M}_\alpha)$. The bound variables of a well-formed formula are those which have bound occurrences in the formula, and the free variables are those which have free occurrences.

In making metamathematical (syntactical) statements, we shall use bold capital letters as variables for well-formed formulas, and bold small letters as variables for variables, employing subscripts to denote the type—as in the preceding paragraph. Moreover we shall adopt the customary, self-explanatory, usage, according to which symbols belonging to the formal language serve in

[6] M. Schönfinkel, *Über die Bausteine der mathematischen Logik*, ***Mathematische Annalen***, vol. 92 (1924), pp. 305–316.

the syntax language (English) as names for themselves, and juxtaposition serves to denote juxtaposition.

In writing well-formed formulas we shall often employ various conventions of abbreviation. In particular, we may omit parentheses ( ) when possible without ambiguity, using the convention in restoring omitted parentheses that the formula must be well-formed and that otherwise association is to the left. Thus, for instance, $a_{\iota'}b_{\iota\iota}(c_{\iota\iota}d_{\iota})$ is an abbreviation for $((a_{((\iota\iota)(\iota\iota))}b_{(\iota\iota)})(c_{\iota\iota}d_{\iota}))$, and $\lambda b_{\iota\iota}\lambda c_{\iota\iota}(a_{\iota'}b_{\iota\iota}(c_{\iota\iota}d_{\iota}))$ is an abbreviation for $(\lambda b_{\iota\iota}(\lambda c_{\iota\iota}((a_{((\iota\iota)(\iota\iota))}b_{(\iota\iota)})(c_{\iota\iota}d_{\iota}))))$.

As indicated in the examples just given, type-symbol subscripts may be abbreviated in the way described in §1. When the subscript is o it may be omitted altogether: thus a small italic letter without subscript is to be read as having the subscript o.

We introduce further the following conventions of abbreviation (reading the arrow as "stands for," or "is an abbreviation for"):

$[\sim \mathbf{A}_o] \rightarrow N_{oo}\mathbf{A}_o.$

$[\mathbf{A}_o \vee \mathbf{B}_o] \rightarrow A_{ooo}\mathbf{A}_o\mathbf{B}_o.$

$[\mathbf{A}_o\mathbf{B}_o] \rightarrow [\sim[[\sim \mathbf{A}_o] \vee [\sim \mathbf{B}_o]]].$

$[\mathbf{A}_o \supset \mathbf{B}_o] \rightarrow [[\sim \mathbf{A}_o] \vee \mathbf{B}_o].$

$[\mathbf{A}_o \equiv \mathbf{B}_o] \rightarrow [[\mathbf{A}_o \supset \mathbf{B}_o][\mathbf{B}_o \supset \mathbf{A}_o]].$

$[(\mathbf{x}_\alpha)\mathbf{A}_o] \rightarrow \Pi_{o(o\alpha)}(\lambda \mathbf{x}_\alpha \mathbf{A}_o).$

$[(\exists \mathbf{x}_\alpha)\mathbf{A}_o] \rightarrow [\sim[(\mathbf{x}_\alpha)[\sim \mathbf{A}_o]]].$

$[(\imath \mathbf{x}_\alpha)\mathbf{A}_o] \rightarrow \iota_{\alpha(o\alpha)}(\lambda \mathbf{x}_\alpha \mathbf{A}_o).$

$Q_{o\alpha\alpha} \rightarrow \lambda x_\alpha \lambda y_\alpha[(f_{o\alpha})[f_{o\alpha}x_\alpha \supset f_{o\alpha}y_\alpha]].$

$[\mathbf{A}_\alpha = \mathbf{B}_\alpha] \rightarrow Q_{o\alpha\alpha}\mathbf{A}_\alpha\mathbf{B}_\alpha.$

$[\mathbf{A}_\alpha \neq \mathbf{B}_\alpha] \rightarrow [\sim[\mathbf{A}_\alpha = \mathbf{B}_\alpha]].$

$I_{\alpha\alpha} \rightarrow \lambda x_\alpha x_\alpha.$

$K_{\alpha\beta\alpha} \rightarrow \lambda x_\alpha \lambda y_\beta x_\alpha.$

$0_{\alpha'} \rightarrow \lambda f_{\alpha\alpha}\lambda x_\alpha x_\alpha,$

$1_{\alpha'} \rightarrow \lambda f_{\alpha\alpha}\lambda x_\alpha(f_{\alpha\alpha}x_\alpha),$

$2_{\alpha'} \rightarrow \lambda f_{\alpha\alpha}\lambda x_\alpha(f_{\alpha\alpha}(f_{\alpha\alpha}x_\alpha)),$

$3_{\alpha'} \rightarrow \lambda f_{\alpha\alpha}\lambda x_\alpha(f_{\alpha\alpha}(f_{\alpha\alpha}(f_{\alpha\alpha}x_\alpha)))$, etc.

$S_{\alpha'\alpha'} \rightarrow \lambda n_{\alpha'}\lambda f_{\alpha\alpha}\lambda x_\alpha(f_{\alpha\alpha}(n_{\alpha'}f_{\alpha\alpha}x_\alpha)).$

$N_{o\alpha'} \rightarrow \lambda n_{\alpha'}[(f_{o\alpha'})[f_{o\alpha'}0_{\alpha'} \supset [[(x_{\alpha'})[f_{o\alpha'}x_{\alpha'} \supset f_{o\alpha'}(S_{\alpha'\alpha'}x_{\alpha'})]] \supset f_{o\alpha'}n_{\alpha'}]]].$

$\omega_{\alpha''\alpha'\alpha'} \rightarrow \lambda y_{\alpha'}\lambda z_{\alpha'}\lambda f_{\alpha'\alpha'}\lambda g_{\alpha'}\lambda h_{\alpha\alpha}\lambda x_\alpha(y_{\alpha'}(f_{\alpha'\alpha'}g_{\alpha'}h_{\alpha\alpha})(z_{\alpha'}(g_{\alpha'}h_{\alpha\alpha})x_\alpha)).$

$\langle \mathbf{A}_{\alpha'}, \mathbf{B}_{\alpha'}\rangle \rightarrow \omega_{\alpha''\alpha'\alpha'}\mathbf{A}_{\alpha'}\mathbf{B}_{\alpha'}.$

$P_{\alpha'\alpha'''} \rightarrow \lambda n_{\alpha'''}(n_{\alpha'''}(\lambda p_{\alpha''}\langle S_{\alpha'\alpha'}(p_{\alpha''}(K_{\alpha'\alpha'\alpha'}I_{\alpha'})0_{\alpha'}),$
$p_{\alpha''}(K_{\alpha'\alpha'\alpha'}I_{\alpha'})0_{\alpha'}\rangle)\langle 0_{\alpha'}, 0_{\alpha'}\rangle(K_{\alpha'\alpha'\alpha'}0_{\alpha'})I_{\alpha'}.$

$T_{\alpha''\alpha'} \rightarrow \lambda x_{\alpha'}[(\imath x_{\alpha''})[(N_{o\alpha''}x_{\alpha''})[x_{\alpha''}S_{\alpha'\alpha'}0_{\alpha'} = x_{\alpha'}]]].$

$P_{\alpha'\alpha'} \rightarrow \lambda x_{\alpha'}(P_{\alpha'\alpha'''}(T_{\alpha'''\alpha''}(T_{\alpha''\alpha'}x_{\alpha'}))).$

As a further abbreviation, we omit square brackets [ ] introduced by the above abbreviations, when possible without ambiguity. When, in omitting square brackets, the initial bracket is replaced by a bold dot **.**, it is to be understood that the scope of the omitted pair of brackets is from the dot forward the maximum distance which is consistent with the whole expression's being well-formed or interpretable as an abbreviation of a well-formed formula. When omitted brackets are not thus replaced by a dot, the convention in restoring omitted brackets is association to the left, except as modified by the understanding that the abbreviated formulas are well-formed and by the following relation of precedence among the different kinds of brackets. The brackets in $[\sim \mathbf{A}_o]$ and $[\mathbf{A}_o\mathbf{B}_o]$ are of lowest rank, those in $[(\mathbf{x}_\alpha)\mathbf{A}_o]$ and $[(\exists \mathbf{x}_\alpha)\mathbf{A}_o]$ and $[(\iota \mathbf{x}_\alpha)\mathbf{A}_o]$ and $[\mathbf{A}_\alpha = \mathbf{B}_\alpha]$ and $[\mathbf{A}_\alpha \neq \mathbf{B}_\alpha]$ are of next higher rank, those in $[\mathbf{A}_o \vee \mathbf{B}_o]$ are of next higher rank, and those in $[\mathbf{A}_o \supset \mathbf{B}_o]$ and $[\mathbf{A}_o \equiv \mathbf{B}_o]$ are of highest rank; in restoring omitted brackets (not represented by a dot), those of lower rank are to be put in before those of higher rank, so that the smaller scope is allotted to those of lower rank. For example,

$$\sim p \supset q \supset . \, pq \vee rs \supset \sim . \, q \vee s \supset \sim p \sim r$$

is an abbreviation for

$$[[[\sim p] \supset q] \supset [[[pq] \vee [rs]] \supset [\sim [[q \vee s] \supset [[\sim p][\sim r]]]]]],$$

which is in turn an abbreviation for

$$((A_{ooo}(N_{oo}((A_{ooo}(N_{oo}(N_{oo}p_o)))q_o)))((A_{ooo}(N_{oo}((A_{ooo}(N_{oo}((A_{ooo}(N_{oo}p_o))(N_{oo}q_o))))$$
$$(N_{oo}((A_{ooo}(N_{oo}r_o))(N_{oo}s_o))))))$$
$$(N_{oo}((A_{ooo}(N_{oo}((A_{ooo}q_o)s_o)))(N_{oo}((A_{ooo}(N_{oo}(N_{oo}p_o)))(N_{oo}(N_{oo}r_o)))))))).$$

In the intended interpretation of the formal system λ will have the rôle of an abstraction operator, N_{oo} will denote negation, A_{ooo} will denote disjunction, $\Pi_{o(o\alpha)}$ will denote the universal quantifier (as a propositional function of propositional functions), $\iota_{\alpha(o\alpha)}$ will denote a selection operator (as a function of propositional functions), and juxtaposition, between parentheses, will denote application of a function to its argument. Such a logical construction of the natural numbers in each type α' is intended that $0_{\alpha'}$ will denote the natural number 0, $1_{\alpha'}$ will denote 1, $2_{\alpha'}$ will denote 2, etc. Then $S_{\alpha'\alpha'}$ will denote the successor function of natural numbers; or, more exactly, it will denote a function which has the entire type α' as the range of its argument and which operates as a successor function in the case that the argument is a natural number. Moreover, $N_{o\alpha'}$ will denote the propositional function "to be a natural number (of type α')." If $\mathbf{N}_{\alpha''}$ denotes a natural number of type α'', then $\mathbf{N}_{\alpha''}S_{\alpha'\alpha'}0_{\alpha'}$ denotes the same (more exactly, the corresponding) natural number in the type α'. Hence if $\mathbf{N}_{\alpha'}$ denotes a natural number of type α', the same natural number in the type α'' will be denoted by $T_{\alpha''\alpha'}\mathbf{N}_{\alpha'}$. The formula $P_{\alpha'\alpha'''}$ is adapted from Kleene's formula P employed in the calculus of λ-conversion[7] and has the property that if $\mathbf{N}_{\alpha'''}$ denotes a natural number of type α''' then $P_{\alpha'\alpha'''}\mathbf{N}_{\alpha'''}$ denotes the predecessor of that natural number in the type α'. The true predecessor function, which gives the predecessor in the same type, is denoted by $P_{\alpha'\alpha'}$; it follows from the independence of the axiom of infinity (§4) that this predecessor function cannot be defined without using descriptions (i.e., the selection operator $\iota_{\beta(o\beta)}$).

[7] S. C. Kleene, *A theory of positive integers in formal logic*, **American journal of mathematics**, vol. 57 (1935), pp. 153–173, 219–244.

3. Rules of inference. The rules of inference (or rules of procedure) are the six following:

I. *To replace any part* $\boldsymbol{M}_\alpha$ *of a formula by the result of substituting* $\boldsymbol{y}_\beta$ *for* $\boldsymbol{x}_\beta$ *throughout* $\boldsymbol{M}_\alpha$, *provided that* $\boldsymbol{x}_\beta$ *is not a free variable of* $\boldsymbol{M}_\alpha$ *and* $\boldsymbol{y}_\beta$ *does not occur in* $\boldsymbol{M}_\alpha$. (I.e., to infer from a given formula the formula obtained by this replacement.)

II. *To replace any part* $((\lambda \boldsymbol{x}_\beta \boldsymbol{M}_\alpha)\boldsymbol{N}_\beta)$ *of a formula by the result of substituting* $\boldsymbol{N}_\beta$ *for* $\boldsymbol{x}_\beta$ *throughout* $\boldsymbol{M}_\alpha$, *provided that the bound variables of* $\boldsymbol{M}_\alpha$ *are distinct both from* $\boldsymbol{x}_\beta$ *and from the free variables of* $\boldsymbol{N}_\beta$.

III. *Where* $\boldsymbol{A}_\alpha$ *is the result of substituting* $\boldsymbol{N}_\beta$ *for* $\boldsymbol{x}_\beta$ *throughout* $\boldsymbol{M}_\alpha$, *to replace any part* $\boldsymbol{A}_\alpha$ *of a formula by* $((\lambda \boldsymbol{x}_\beta \boldsymbol{M}_\alpha)\boldsymbol{N}_\beta)$, *provided that the bound variables of* $\boldsymbol{M}_\alpha$ *are distinct both from* $\boldsymbol{x}_\beta$ *and from the free variables of* $\boldsymbol{N}_\beta$.

IV. *From* $\boldsymbol{F}_{o\alpha}\boldsymbol{x}_\alpha$ *to infer* $\boldsymbol{F}_{o\alpha}\boldsymbol{A}_\alpha$, *provided that* $\boldsymbol{x}_\alpha$ *is not a free variable of* $\boldsymbol{F}_{o\alpha}$.

V. *From* $\boldsymbol{A}_o \supset \boldsymbol{B}_o$ *and* $\boldsymbol{A}_o$, *to infer* $\boldsymbol{B}_o$.

VI. *From* $\boldsymbol{F}_{o\alpha}\boldsymbol{x}_\alpha$ *to infer* $\Pi_{o(o\alpha)}\boldsymbol{F}_{o\alpha}$, *provided that* $\boldsymbol{x}_\alpha$ *is not a free variable of* $\boldsymbol{F}_{o\alpha}$.

The word *part* of a formula is to be understood here as meaning *consecutive well-formed part* other than a variable immediately following an occurrence of λ. Moreover, as already explained, bold capital letters represent *well-formed* formulas and bold small letters represent variables, the subscript in each case showing the type. When (as in the rules I, II, III) we speak of replacing a part $\boldsymbol{M}_\alpha$ of a formula by something else, it is to be understood that, if there are several occurrences of $\boldsymbol{M}_\alpha$ as a part of the formula, any *one* of them may be so replaced. When we speak of the result of substituting $\boldsymbol{N}_\beta$ for $\boldsymbol{x}_\beta$ throughout $\boldsymbol{M}_\alpha$, the case is not excluded that $\boldsymbol{x}_\beta$ fails to occur in $\boldsymbol{M}_\alpha$, the result of the substitution in that case being $\boldsymbol{M}_\alpha$.

The rules I–III are called rules of λ-*conversion*, and any chain of applications of these rules is called a λ-conversion, or briefly, a conversion. Rule IV is the rule of *substitution*, Rule V is the rule of *modus ponens*, and Rule VI is the rule of *generalization*. In an application of Rule IV, we say that the variable $\boldsymbol{x}_\alpha$ is *substituted for*; and in an application of Rule VI, we say that the variable $\boldsymbol{x}_\alpha$ is *generalized upon*.

The two following rules of inference are derived rules, in the sense that the indicated inference can be accomplished in each case by a chain of applications of I–VI (the effect of IV′ can be obtained by means of λ-conversion and Rule IV, the effect of VI′ can be obtained by means of λ-conversion and Rule VI):

IV′. *From* $\boldsymbol{M}_o$ *to infer the result of substituting* $\boldsymbol{A}_\alpha$ *for the free occurrences of* $\boldsymbol{x}_\alpha$ *throughout* $\boldsymbol{M}_o$, *provided that the bound variables of* $\boldsymbol{M}_o$ *other than* $\boldsymbol{x}_\alpha$ *are distinct from the free variables of* $\boldsymbol{A}_\alpha$.

VI′. *From* $\boldsymbol{M}_o$ *to infer* $(\boldsymbol{x}_\alpha)\boldsymbol{M}_o$.

4. Formal axioms. The formal axioms are the formulas in the following infinite list:

1. $p \vee p \supset p$.
2. $p \supset p \vee q$.

3. $p \vee q \supset q \vee p.$
4. $p \supset q \supset . r \vee p \supset r \vee q.$
5$^{\alpha}$. $\Pi_{o(o\alpha)} f_{o\alpha} \supset f_{o\alpha} x_{\alpha}.$
6$^{\alpha}$. $(x_{\alpha})[p \vee f_{o\alpha} x_{\alpha}] \supset p \vee \Pi_{o(o\alpha)} f_{o\alpha}.$
7. $(\exists x_{\iota})(\exists y_{\iota}) . x_{\iota} \neq y_{\iota}.$
8. $N_{o\iota'} x_{\iota'} \supset . N_{o\iota'} y_{\iota'} \supset . S_{\iota'\iota'} x_{\iota'} = S_{\iota'\iota'} y_{\iota'} \supset x_{\iota'} = y_{\iota'}.$
9$^{\alpha}$. $f_{o\alpha} x_{\alpha} \supset . (y_{\alpha})[f_{o\alpha} y_{\alpha} \supset x_{\alpha} = y_{\alpha}] \supset f_{o\alpha}(\iota_{\alpha(o\alpha)} f_{o\alpha}).$
10$^{\alpha\beta}$. $(x_{\beta})[f_{\alpha\beta} x_{\beta} = g_{\alpha\beta} x_{\beta}] \supset f_{\alpha\beta} = g_{\alpha\beta}.$
11$^{\alpha}$. $f_{o\alpha} x_{\alpha} \supset f_{o\alpha}(\iota_{\alpha(o\alpha)} f_{o\alpha}).$

The *theorems* of the system are the formulas obtainable from the formal axioms by a succession of applications of the rules of inference. A *proof* of a theorem of the system is a finite sequence of formulas, the last of which is the theorem, and each of which is either a formal axiom or obtainable from preceding formulas in the sequence by an application of a rule of inference.

We must, of course, distinguish between *formal theorems*, or theorems of the system, and *syntactical theorems*, or theorems about the system, this and related distinctions being a necessary part of the process of using a known language (English) to set up another (more exact) language. (We deliberately use the word "theorem" ambiguously, sometimes for a proposition and sometimes for a sentence or formula meaning the proposition in some language.)

Axioms 1–4 suffice for the propositional calculus and Axioms 1–6$^{\alpha}$ for the logical functional calculus.

In order to obtain elementary number theory it is necessary to add (to 1–6$^{\alpha}$) Axioms 7, 8, and 9$^{\alpha}$. Of these, 9$^{\alpha}$ are *axioms of descriptions*, and 7 and 8 taken together have the effect of an *axiom of infinity*. The independence of Axiom 7 may be established by considering an interpretation of the primitive symbols according to which there is exactly one individual, and that of Axiom 8 by considering an interpretation according to which there are a finite number, more than one, of individuals.

In order to obtain classical real number theory (analysis) it is necessary[8] to add also Axioms 10$^{\alpha\beta}$ and 11$^{\alpha}$. Of these, 10$^{\alpha\beta}$ are *axioms of extensionality* for functions, and 11$^{\alpha}$ are *axioms of choice*.

Axioms 10$^{\alpha\beta}$, although weaker in some directions than axioms of extensionality which are sometimes employed, are nevertheless adequate. For classes may be introduced in such a way that the class associated with the propositional function denoted by $F_{o\alpha}$ is denoted by $\lambda x_{\alpha}(\imath y_{\iota'}) . (F_{o\alpha} x_{\alpha})[y_{\iota'} = 0_{\iota'}] \vee (\sim F_{o\alpha} x_{\alpha})[y_{\iota'} = 1_{\iota'}]$. We remark, however, on the possibility of introducing the additional axiom of extensionality, $p \equiv q \supset p = q$, which has the effect of imposing so broad a criterion of identity between propositions that there are in consequence only two propositions, and which, in conjunction with 10$^{\alpha\beta}$, makes possible the identification of classes with propositional functions.

Axioms 9$^{\alpha}$ obviously fail to be independent of 1–4 and 11$^{\alpha}$. We have never-

[8] Devices of contextual definition, such as Russell's methods of introducing classes and descriptions (loc. cit.), are here avoided, and assertions concerning the necessity of axioms and the like are to be understood in the sense of this avoidance.

theless included the axioms 9^α because of the desirability of considering the consequences of Axioms 1–9^α without $10^{\alpha\beta}$, 11^α.

If 1–9^α are the only formal axioms, each of the axioms 9^α is then independent, but if 10^α is added there is a sense in which those other than 9° and 9^ι, although independent, are superfluous. For, of the symbols $\iota_{\alpha(o\alpha)}$, we may introduce only $\iota_{o(oo)}$ and $\iota_{\iota(o\iota)}$ as primitive symbols and then introduce the remainder by definition (i.e., by conventions of abbreviation) in such a way that the formulas 9^α, read in accordance with these definitions (conventions of abbreviation), become theorems provable from the formal axioms 1–8, 9°, 9^ι, $10^{\alpha\beta}$. The required definitions are summarized in the following schema, which states the definition of $\iota_{\alpha\beta(o(\alpha\beta))}$ in terms of $\iota_{\alpha(o\alpha)}$:

$$\iota_{\alpha\beta(o(\alpha\beta))} \rightarrow \lambda h_{o(\alpha\beta)} \lambda x_\beta (℩ y_\alpha)(\exists f_{\alpha\beta}) \,.\, h_{o(\alpha\beta)} f_{\alpha\beta} \,.\, y_\alpha = f_{\alpha\beta} x_\beta.$$

5. The deduction theorem. Derivation of the formal theorems of the propositional calculus from Axioms 1–4 by means of Rules IV′ and V is well known and need not be repeated here.[9] In what follows we shall employ theorems of the propositional calculus as needed, assuming the proof as known.

It is also clear that, by means of Rules I and IV′, alphabetical changes of the variables (free and bound) may be made in any formal axiom, provided that the types of the variables are not altered, that variables originally the same remain the same, and that variables originally different remain different. Formal theorems obtained in this way (including the formal axioms themselves) will be called *variants* of the axioms and will be employed as needed without explicit statement of the proof.

By a *proof* of a formula $\boldsymbol{B}_o$ *on the assumption of* the formulas $\boldsymbol{A}_o^1, \boldsymbol{A}_o^2, \cdots, \boldsymbol{A}_o^n$, we shall mean a finite sequence of formulas, the last of which is $\boldsymbol{B}_o$, and each of which is either one of the formulas $\boldsymbol{A}_o^1, \boldsymbol{A}_o^2, \cdots, \boldsymbol{A}_o^n$, or a variant of a formal axiom, or obtainable from preceding formulas in the sequence by an application of a rule of inference subject to the condition that no variable shall be substituted for or generalized upon which appears as a free variable in any of the formulas $\boldsymbol{A}_o^1, \boldsymbol{A}_o^2, \cdots, \boldsymbol{A}_o^n$. In order to express that there is a proof of $\boldsymbol{B}_o$ on the assumption of $\boldsymbol{A}_o^1, \boldsymbol{A}_o^2, \cdots, \boldsymbol{A}_o^n$, we shall employ the (syntactical) notation:

$$\boldsymbol{A}_o^1, \boldsymbol{A}_o^2, \cdots, \boldsymbol{A}_o^n \vdash \boldsymbol{B}_o.$$

In the use of this notation, it is not excluded that n should be 0 and the set of formulas $\boldsymbol{A}_o^i$ vacuous; i.e., the notation $\vdash \boldsymbol{B}_o$ will be used to mean that $\boldsymbol{B}_o$ is a (formal) theorem. (This use of the sign $\vdash$ must be distinguished from the entirely different use of the assertion sign by Russell and earlier by Frege.)

The following syntactical theorem is known as the *deduction theorem*:

VII. *If* $\boldsymbol{A}_o^1, \boldsymbol{A}_o^2, \cdots, \boldsymbol{A}_o^n \vdash \boldsymbol{B}_o$, *then* $\boldsymbol{A}_o^1, \boldsymbol{A}_o^2, \cdots, \boldsymbol{A}_o^{n-1} \vdash \boldsymbol{A}_o^n \supset \boldsymbol{B}_o$ $(n = 1, 2, 3, \cdots)$.

In order to prove this, we suppose that the finite sequence of formulas $\boldsymbol{B}_o^1$, $\boldsymbol{B}_o^2, \cdots, \boldsymbol{B}_o^m$ is a proof of $\boldsymbol{B}_o$ on the assumption of $\boldsymbol{A}_o^1, \boldsymbol{A}_o^2, \cdots, \boldsymbol{A}_o^n$, the formula

[9] Cf. Hilbert and Ackermann, loc. cit.; P. Bernays, *Axiomatische Untersuchung des Aussagen-Kalkuls der "Principia Mathematica," Mathematische Zeitschrift*, vol. 25 (1926), pp. 305–320.

B_o^m being the same as B_o, and we show in succession, for each value of i from 1 to m, that

$$A_o^1, A_o^2, \cdots, A_o^{n-1} \vdash A_o^n \supset B_o^i.$$

This is done by cases, according as B_o^i is A_o^n, is one of $A_o^1, A_o^2, \cdots, A_o^{n-1}$, is a variant of an axiom, or is obtained from a preceding formula or pair of formulas by one of the rules I–VI. If B_o^i is A_o^n, we may obtain $A_o^n \supset B_o^i$ from $p \supset p$ by IV′. If B_o^i is one of $A_o^1, A_o^2, \cdots, A_o^{n-1}$ or is a variant of an axiom, we may obtain $A_o^n \supset B_o^i$ from $p \supset . q \supset p$ by a succession of applications of IV′ and V. If B_o^i is obtained from B_o^a $(a<i)$ by one of the rules I, II, III, we may obtain $A_o^n \supset B_o^i$ from $A_o^n \supset B_o^a$ by the same rule. If B_o^i is obtained from B_o^a $(a<i)$ by Rule IV, we may obtain $A_o^n \supset B_o^i$ from $A_o^n \supset B_o^a$ by IV′. If B_o^i is obtained from B_o^a and B_o^b $(a<i, b<i)$ by Rule V, we may obtain $A_o^n \supset B_o^i$ from $A_o^n \supset B_o^a$ and $A_o^n \supset B_o^b$ and $p \supset [q \supset r] \supset . p \supset q \supset . p \supset r$ by a succession of applications of IV′ and V. If B_o^i is obtained from B_o^a $(a<i)$ by Rule VI, we may obtain $A_o^n \supset B_o^i$ from $A_o^n \supset B_o^a$ and 6^α by a succession of applications of IV′, V, and VI′.

Proof of the following theorems,[10] which are consequences of the formal axioms 1–6^α, is left to the reader (it will be found convenient in most cases to abbreviate the proof by employing the deduction theorem in the rôle of a derived rule):

12^α. $(x_\alpha)f_{o\alpha}x_\alpha \supset f_{o\alpha}y_\alpha$.
13^α. $f_{o\alpha}y_\alpha \supset (\exists x_\alpha)f_{o\alpha}x_\alpha$.
14^α. $(x_\alpha)[p \supset f_{o\alpha}x_\alpha] \supset . p \supset (x_\alpha)f_{o\alpha}x_\alpha$.
15^α. $(x_\alpha)[f_{o\alpha}x_\alpha \supset p] \supset . (\exists x_\alpha)f_{o\alpha}x_\alpha \supset p$.
16^α. $x_\alpha = x_\alpha$.
17^α. $x_\alpha = y_\alpha \supset . f_{o\alpha}x_\alpha \supset f_{o\alpha}y_\alpha$.
$18^{\beta\alpha}$. $x_\alpha = y_\alpha \supset f_{\beta\alpha}x_\alpha = f_{\beta\alpha}y_\alpha$.
19^α. $x_\alpha = y_\alpha \supset y_\alpha = x_\alpha$.
20^α. $x_\alpha = y_\alpha \supset . y_\alpha = z_\alpha \supset x_\alpha = z_\alpha$.

The following theorems are consequences of the formal axioms 1–4 and $10^{\alpha\beta}$ (no use will be made of them below because we shall be concerned entirely with consequences of 1–9^α):

$21^{\alpha\beta}$. $f_{\alpha\beta} = \lambda x_\beta(f_{\alpha\beta}x_\beta)$.

6. Peano's postulates for arithmetic. Three of the five Peano postulates for arithmetic[11] are represented by the following formal theorems:

[10] The same device of typical ambiguity which was employed in stating the rules of inference and formal axioms now serves us, not only to condense the statement of an infinite number of theorems (differing only in the type subscripts of the proper symbols which appear) into a single schema of theorems, but also to condense the proof of the infinite number of theorems into a single schema of proof. Of course, in the explicit formal development of the system, a stage would never be reached at which all of the theorems 12°, 12^ι, $12^{\iota\iota}$, $\cdots$ (for example) had been proved, but by the device of a schema of proof with typical ambiguity we obtain metamathematical assurance that any required *one* of the theorems in the infinite list can be proved. Cf. the Prefatory Statement to the second volume of ***Principia mathematica***.

[11] G. Peano, *Sul concetto di numero*, ***Rivista di matematica,*** vol. 1 (1891), pp. 87–102, 256–267.

22^{α}. $N_{o\alpha'}0_{\alpha'}$.

23^{α}. $N_{o\alpha'}x_{\alpha'} \supset N_{o\alpha'}(S_{\alpha'\alpha'}x_{\alpha'})$.

24^{α}. $f_{o\alpha'}0_{\alpha'} \supset . (x_{\alpha'})[N_{o\alpha'}x_{\alpha'} \supset . f_{o\alpha'}x_{\alpha'} \supset f_{o\alpha'}(S_{\alpha'\alpha'}x_{\alpha'})] \supset . N_{o\alpha'}x_{\alpha'} \supset f_{o\alpha'}x_{\alpha'}$.

These theorems are consequences of 1–6^{α}; proofs are left to the reader.

From 24^{α} and the deduction theorem we obtain the following syntactical theorem which we shall call the *induction theorem*:

VIII. *If* $\mathbf{x}_{\alpha'}$ *is not a free variable of* $\mathbf{A}_o^1, \mathbf{A}_o^2, \cdots, \mathbf{A}_o^n, \mathbf{F}_{o\alpha'}$, *if* $\mathbf{A}_o^1, \mathbf{A}_o^2, \cdots, \mathbf{A}_o^n \vdash \mathbf{F}_{o\alpha'}0_{\alpha'}$, *and if* $\mathbf{A}_o^1, \mathbf{A}_o^2, \cdots, \mathbf{A}_o^n, N_{o\alpha'}\mathbf{x}_{\alpha'}, \mathbf{F}_{o\alpha'}\mathbf{x}_{\alpha'} \vdash \mathbf{F}_{o\alpha'}(S_{\alpha'\alpha'}\mathbf{x}_{\alpha'})$, *then* $\mathbf{A}_o^1, \mathbf{A}_o^2, \cdots, \mathbf{A}_o^n \vdash N_{o\alpha'}\mathbf{x}_{\alpha'} \supset \mathbf{F}_{o\alpha'}\mathbf{x}_{\alpha'}$.

A proof which is or can be abbreviated by employing the induction theorem in the rôle of a derived rule will be called a proof by (*mathematical*, or *complete*) *induction* on the variable $\mathbf{x}_{\alpha'}$.

Another of the Peano postulates is represented by the following formal theorems:

25^{α}. $N_{o\alpha'}x_{\alpha'} \supset S_{\alpha'\alpha'}x_{\alpha'} \neq 0_{\alpha'}$.

These theorems are consequences of 1–6^{α} and 7, as we shall show (for certain types α they are consequences of 1–6^{α} only).

The remaining Peano postulate would correspond to the following:

26^{α}. $N_{o\alpha'}x_{\alpha'} \supset . N_{o\alpha'}y_{\alpha'} \supset . S_{\alpha'\alpha'}x_{\alpha'} = S_{\alpha'\alpha'}y_{\alpha'} \supset x_{\alpha'} = y_{\alpha'}$.

These formulas are demonstrably not theorems (consistency assumed) in the case of type symbols α consisting entirely of o's with no ι's. We shall show that the formulas 26^{ι}, $26^{\iota'}$, $26^{\iota''}$, $\cdots$ are theorems—in fact they are consequences of 1–6^{α} and 8, the formula 26^{ι} being the same as 8.

A proof of the theorem,

27°. $(\exists x_o)(\exists y_o) . x_o \neq y_o$,

may be made as follows. In 17° substitute $p \vee \sim p$ for x_o, and $\sim . p \vee \sim p$ for y_o and $\lambda r \sim . p \vee \sim p \supset \sim r$ for f_{oo}, by successive applications of IV′, and then apply Rule II twice, so obtaining

$$[p \vee \sim p] = [\sim . p \vee \sim p] \supset . [\sim . p \vee \sim p \supset \sim . p \vee \sim p] \supset \sim . p \vee \sim p \supset \sim\sim . p \vee \sim p.$$

Hence using the theorems of the propositional calculus,

$$\sim . p \vee \sim p \supset \sim . p \vee \sim p,$$

$$q \supset [r \supset s] \supset . r \supset . q \supset s,$$

and the rules IV′ and V, obtain

$$[p \vee \sim p] = [\sim . p \vee \sim p] \supset \sim . p \vee \sim p \supset \sim\sim . p \vee \sim p.$$

Hence, using the theorems of the propositional calculus,

$$p \vee \sim p \supset \sim\sim . p \vee \sim p,$$

$$q \supset . r \supset \sim q \supset \sim r,$$

and IV′ and V (method of *reductio ad absurdum*), obtain

$$[p\vee\sim p]\neq[\sim.p\vee\sim p].$$

Hence by two successive uses of 13°, with I, II, III, IV′, V, obtain 27°.

In regard to proof of the theorems,

$27^\alpha.\quad (\exists x_\alpha)(\exists y_\alpha)\,.\,x_\alpha\neq y_\alpha,$

since we have a proof of 27°, and 27ᶥ is Axiom 7, it is sufficient to show how to obtain a proof of $27^{\alpha\beta}$ if a proof of 27^α is given.

By conversion $z_\alpha\neq t_\alpha \vdash K_{\alpha\beta\alpha}z_\alpha x_\beta\neq K_{\alpha\beta\alpha}t_\alpha x_\beta$.

Hence by 17^α (using II, IV′, V), $z_\alpha\neq t_\alpha,\ K_{\alpha\beta\alpha}z_\alpha=K_{\alpha\beta\alpha}t_\alpha \vdash K_{\alpha\beta\alpha}t_\alpha x_\beta\neq K_{\alpha\beta\alpha}t_\alpha x_\beta$.

Hence by the deduction theorem, $z_\alpha\neq t_\alpha \vdash K_{\alpha\beta\alpha}z_\alpha=K_{\alpha\beta\alpha}t_\alpha \supset K_{\alpha\beta\alpha}t_\alpha x_\beta\neq K_{\alpha\beta\alpha}t_\alpha x_\beta$.

By 16^α (using IV′), $K_{\alpha\beta\alpha}t_\alpha x_\beta=K_{\alpha\beta\alpha}t_\alpha x_\beta$.

Hence by *reductio ad absurdum*, as above, $z_\alpha\neq t_\alpha \vdash K_{\alpha\beta\alpha}z_\alpha\neq K_{\alpha\beta\alpha}t_\alpha$.

Hence by two successive uses of $13^{\alpha\beta}$ (with I, II, III, IV′, V), $z_\alpha\neq t_\alpha \vdash (\exists x_{\alpha\beta})(\exists y_{\alpha\beta})\,.\,x_{\alpha\beta}\neq y_{\alpha\beta}$.

Hence by the deduction theorem, $\vdash z_\alpha\neq t_\alpha\supset(\exists x_{\alpha\beta})(\exists y_{\alpha\beta})\,.\,x_{\alpha\beta}\neq y_{\alpha\beta}$.

Hence using VI′, $\vdash (t_\alpha)\,.\,z_\alpha\neq t_\alpha\supset(\exists x_{\alpha\beta})(\exists y_{\alpha\beta})\,.\,x_{\alpha\beta}\neq y_{\alpha\beta}$.

Hence by 15^α (using I, II, III, IV′, V), $\vdash (\exists t_\alpha)[z_\alpha\neq t_\alpha]\supset(\exists x_{\alpha\beta})(\exists y_{\alpha\beta})\,.\,x_{\alpha\beta}\neq y_{\alpha\beta}$.

Hence using VI′, $\vdash (z_\alpha)\,.\,(\exists t_\alpha)[z_\alpha\neq t_\alpha]\supset(\exists x_{\alpha\beta})(\exists y_{\alpha\beta})\,.\,x_{\alpha\beta}\neq y_{\alpha\beta}$.

Hence by 15^α (using I, II, III, IV′, V), $\vdash (\exists z_\alpha)(\exists t_\alpha)[z_\alpha\neq t_\alpha]\supset(\exists x_{\alpha\beta})(\exists y_{\alpha\beta})\,.\,x_{\alpha\beta}\neq y_{\alpha\beta}$.

Hence if $\vdash 27^\alpha$ then, using I and V, $\vdash 27^{\alpha\beta}$.

Thus for every type α we have a proof of 27^α. Using this, we proceed to the proof of

$28^\alpha.\quad S_{\alpha'\alpha'}x_{\alpha'}\neq 0_{\alpha'}.$

By conversion, $z_\alpha\neq t_\alpha \vdash S_{\alpha'\alpha'}x_{\alpha'}(K_{\alpha\alpha\alpha}z_\alpha)t_\alpha\neq 0_{\alpha'}(K_{\alpha\alpha\alpha}z_\alpha)t_\alpha$. Hence by the method illustrated in the preceding proof, using in order 17^α, the deduction theorem, 16^α, and *reductio ad absurdum*, $z_\alpha\neq t_\alpha\vdash S_{\alpha'\alpha'}x_{\alpha'}\neq 0_{\alpha'}$. Eliminating the assumption $z_\alpha\neq t_\alpha$ by the method of the preceding proof, using in order the deduction theorem, VI′, 15^α, VI′, 15^α, 27^α, we have $\vdash 28^\alpha$.

Having 28^α, we prove 25^α by using $p\supset.q\supset p$.

We need also the theorems:

$29^\alpha.\quad N_{o\alpha''}n_{\alpha''}\supset N_{o\alpha'}(n_{\alpha''}S_{\alpha'\alpha'}0_{\alpha'}).$

The (schema of) proof of these theorems is a simple example of proof by induction.

From 22^α by conversion, $\vdash N_{o\alpha'}(0_{\alpha''}S_{\alpha'\alpha'}0_{\alpha'})$.

By 23^α, $N_{o\alpha'}(n_{\alpha''}S_{\alpha'\alpha'}0_{\alpha'})\vdash N_{o\alpha'}(S_{\alpha'\alpha'}(n_{\alpha''}S_{\alpha'\alpha'}0_{\alpha'}))$.

Hence by conversion, $N_{o\alpha'}(n_{\alpha''}S_{\alpha'\alpha'}0_{\alpha'})\vdash N_{o\alpha'}(S_{\alpha''\alpha''}n_{\alpha''}S_{\alpha'\alpha'}0_{\alpha'})$.

Hence by the induction theorem, taking $F_{o\alpha''}$ to be $\lambda x_{\alpha''}(N_{o\alpha'}(x_{\alpha''}S_{\alpha'\alpha'}0_{\alpha'}))$ and $x_{\alpha''}$ to be $n_{\alpha''}$, and employing conversion as required, we have $\vdash 29^\alpha$.

Returning now to 26^{α}, we consider in connection with it:

30^{α}. $N_{o\alpha''}m_{\alpha''} \supset . N_{o\alpha''}n_{\alpha''} \supset . m_{\alpha''}S_{\alpha'\alpha'}0_{\alpha'} = n_{\alpha''}S_{\alpha'\alpha'}0_{\alpha'} \supset m_{\alpha''} = n_{\alpha''}$.

As in the case of 26^{α}, not all the formulas 30^{α} are theorems. We shall show that 26^{α} and 30^{α} are theorems if α is one of the types $\iota, \iota', \iota'', \cdots$. Since 26^{ι} is Axiom 8, we may do this by showing that (1) if $\vdash 26^{\alpha}$ then $\vdash 30^{\alpha}$, and (2) if $\vdash 26^{\alpha}$ and $\vdash 30^{\alpha}$ then $\vdash 26^{\alpha'}$.[12]

By $18^{\alpha'\alpha''}$, $S_{\alpha''\alpha''}x_{\alpha''} = S_{\alpha''\alpha''}y_{\alpha''} \vdash S_{\alpha''\alpha''}x_{\alpha''}S_{\alpha'\alpha'}0_{\alpha'} = S_{\alpha''\alpha''}y_{\alpha''}S_{\alpha'\alpha'}0_{\alpha'}$.

Hence by conversion, we have $S_{\alpha''\alpha''}x_{\alpha''} = S_{\alpha''\alpha''}y_{\alpha''} \vdash S_{\alpha'\alpha'}(x_{\alpha''}S_{\alpha'\alpha'}0_{\alpha'}) = S_{\alpha'\alpha'}(y_{\alpha''}S_{\alpha'\alpha'}0_{\alpha'})$.

Hence if $\vdash 26^{\alpha}$, we have by 29^{α}, $N_{o\alpha''}x_{\alpha''}$, $N_{o\alpha''}y_{\alpha''}$, $S_{\alpha''\alpha''}x_{\alpha''} = S_{\alpha''\alpha''}y_{\alpha''} \vdash x_{\alpha''}S_{\alpha'\alpha'}0_{\alpha'} = y_{\alpha''}S_{\alpha'\alpha'}0_{\alpha'}$.

Hence if $\vdash 26^{\alpha}$ and $\vdash 30^{\alpha}$, we have $N_{o\alpha''}x_{\alpha''}$, $N_{o\alpha''}y_{\alpha''}$, $S_{\alpha''\alpha''}x_{\alpha''} = S_{\alpha''\alpha''}y_{\alpha''} \vdash x_{\alpha''} = y_{\alpha''}$.

Hence by three applications of the deduction theorem, if $\vdash 26^{\alpha}$ and $\vdash 30^{\alpha}$ then $\vdash 26^{\alpha'}$. This is (2) above.

Now by conversion, $0_{\alpha''}S_{\alpha'\alpha'}0_{\alpha'} = 0_{\alpha''}S_{\alpha'\alpha'}0_{\alpha'} \vdash 0_{\alpha'} = 0_{\alpha'}$. Hence by the deduction theorem, $\vdash 0_{\alpha''}S_{\alpha'\alpha'}0_{\alpha'} = 0_{\alpha''}S_{\alpha'\alpha'}0_{\alpha'} \supset 0_{\alpha'} = 0_{\alpha'}$.

By conversion, $0_{\alpha''}S_{\alpha'\alpha'}0_{\alpha'} = S_{\alpha''\alpha''}n_{\alpha''}S_{\alpha'\alpha'}0_{\alpha'} \vdash 0_{\alpha'} = S_{\alpha'\alpha'}(n_{\alpha''}S_{\alpha'\alpha'}0_{\alpha'})$. Hence by $19^{\alpha'}$, $0_{\alpha''}S_{\alpha'\alpha'}0_{\alpha'} = S_{\alpha''\alpha''}n_{\alpha''}S_{\alpha'\alpha'}0_{\alpha'} \vdash S_{\alpha'\alpha'}(n_{\alpha''}S_{\alpha'\alpha'}0_{\alpha'}) = 0_{\alpha'}$. By 28^{α}, $\vdash S_{\alpha'\alpha'}(n_{\alpha''}S_{\alpha'\alpha'}0_{\alpha'}) \neq 0_{\alpha'}$. Hence, using $p{\sim}p \supset q$, we have $0_{\alpha''}S_{\alpha'\alpha'}0_{\alpha'} = S_{\alpha''\alpha''}n_{\alpha''}S_{\alpha'\alpha'}0_{\alpha} \vdash 0_{\alpha''} = S_{\alpha''\alpha''}n_{\alpha''}$. Hence by the deduction theorem, $\vdash 0_{\alpha''}S_{\alpha'\alpha'}0_{\alpha'} = S_{\alpha''\alpha''}n_{\alpha''}S_{\alpha'\alpha'}0_{\alpha'} \supset 0_{\alpha''} = S_{\alpha''\alpha''}n_{\alpha}$.

Hence by the induction theorem, followed by VI′, $\vdash (n_{\alpha''}) . N_{o\alpha''}n_{\alpha''} \supset . 0_{\alpha''}S_{\alpha'\alpha'}0_{\alpha'} = n_{\alpha''}S_{\alpha'\alpha'}0_{\alpha'} \supset 0_{\alpha''} = n_{\alpha''}$.

By conversion, $S_{\alpha''\alpha''}m_{\alpha''}S_{\alpha'\alpha'}0_{\alpha'} = 0_{\alpha''}S_{\alpha'\alpha'}0_{\alpha'} \vdash S_{\alpha'\alpha'}(m_{\alpha''}S_{\alpha'\alpha'}0_{\alpha'}) = 0_{\alpha'}$. By 28^{α}, $\vdash S_{\alpha'\alpha'}(m_{\alpha''}S_{\alpha'\alpha'}0_{\alpha'}) \neq 0_{\alpha'}$. Hence, using $p{\sim}p \supset q$, we have $S_{\alpha''\alpha''}m_{\alpha''}S_{\alpha'\alpha'}0_{\alpha'} = 0_{\alpha''}S_{\alpha'\alpha'}0_{\alpha'} \vdash S_{\alpha''\alpha''}m_{\alpha''} = 0_{\alpha''}$. Hence by the deduction theorem, $\vdash S_{\alpha''\alpha''}m_{\alpha''}S_{\alpha'\alpha'}0_{\alpha'} = 0_{\alpha''}S_{\alpha'\alpha'}0_{\alpha'} \supset S_{\alpha''\alpha''}m_{\alpha''} = 0_{\alpha''}$.

By conversion, $S_{\alpha''\alpha''}m_{\alpha''}S_{\alpha'\alpha'}0_{\alpha'} = S_{\alpha''\alpha''}n_{\alpha''}S_{\alpha'\alpha'}0_{\alpha'} \vdash S_{\alpha'\alpha'}(m_{\alpha''}S_{\alpha'\alpha'}0_{\alpha'}) = S_{\alpha'\alpha'}(n_{\alpha''}S_{\alpha'\alpha'}0_{\alpha'})$. Hence if $\vdash 26^{\alpha}$, we have by 29^{α}, $N_{o\alpha''}m_{\alpha''}$, $N_{o\alpha''}n_{\alpha''}$, $S_{\alpha''\alpha''}m_{\alpha''}S_{\alpha'\alpha'}0_{\alpha'} = S_{\alpha''\alpha''}n_{\alpha''}S_{\alpha'\alpha'}0_{\alpha'} \vdash m_{\alpha''}S_{\alpha'\alpha'}0_{\alpha'} = n_{\alpha''}S_{\alpha'\alpha'}0_{\alpha'}$. Hence if $\vdash 26^{\alpha}$, we have (using $12^{\alpha''}$), $N_{o\alpha''}m_{\alpha''}$, $(n_{\alpha''}) . N_{o\alpha''}n_{\alpha''} \supset . m_{\alpha''}S_{\alpha'\alpha'}0_{\alpha'} = n_{\alpha''}S_{\alpha'\alpha'}0_{\alpha'} \supset m_{\alpha''} = n_{\alpha''}$, $N_{o\alpha''}n_{\alpha''}$, $S_{\alpha''\alpha''}m_{\alpha''}S_{\alpha'\alpha'}0_{\alpha'} = S_{\alpha''\alpha''}n_{\alpha''}S_{\alpha'\alpha'}0_{\alpha'} \vdash m_{\alpha''} = n_{\alpha''}$. Hence, using $18^{\alpha''\alpha''}$ to obtain $S_{\alpha''\alpha''}m_{\alpha''} = S_{\alpha''\alpha''}n_{\alpha''}$ and then applying the deduction theorem, we have (if $\vdash 26^{\alpha}$), $N_{o\alpha''}m_{\alpha''}$, $(n_{\alpha''}) . N_{o\alpha''}n_{\alpha''} \supset . m_{\alpha''}S_{\alpha'\alpha'}0_{\alpha'} = n_{\alpha''}S_{\alpha'\alpha'}0_{\alpha'} \supset m_{\alpha''} = n_{\alpha''}$, $N_{o\alpha''}n_{\alpha''} \vdash S_{\alpha''\alpha''}m_{\alpha''}S_{\alpha'\alpha'}0_{\alpha'} = S_{\alpha''\alpha''}n_{\alpha''}S_{\alpha'\alpha'}0_{\alpha'} \supset S_{\alpha''\alpha''}m_{\alpha''} = S_{\alpha''\alpha''}n_{\alpha''}$.

Hence by the induction theorem, followed by VI′, we have (if $\vdash 26^{\alpha}$), $N_{o\alpha''}m_{\alpha''}$,

---

[12] The question suggests itself whether 30^{ι} could be used in place of Axiom 8 as the second part of the axiom of infinity. The writer has a proof (depending on the properties of $P_{\iota'\iota''}$) that 30^{ι} and $30^{\iota'}$ are together sufficient, in the presence of 1–6^{α}, to replace Axiom 8. A proof has also been carried out by A. M. Turing that, in the presence of 1–7 and 9^{α}, 30^{ι} is sufficient alone to replace Axiom 8. Whether 8 is independent of 1–7 and 30^{ι} remains an open problem (familiar methods of eliminating descriptions do not apply here).

$(n_{\alpha''}) \,.\, N_{o\alpha''}n_{\alpha''} \supset .\, m_{\alpha''}S_{\alpha'\alpha'}0_{\alpha'} = n_{\alpha''}S_{\alpha'\alpha'}0_{\alpha'} \supset m_{\alpha''} = n_{\alpha''} \vdash (n_{\alpha''}) \,.\, N_{o\alpha''}n_{\alpha''} \supset .\, S_{\alpha''\alpha''}m_{\alpha''}S_{\alpha'\alpha'}0_{\alpha'} = n_{\alpha''}S_{\alpha'\alpha'}0_{\alpha'} \supset S_{\alpha''\alpha''}m_{\alpha''} = n_{\alpha''}$.

Hence again, applying the induction theorem to preceding results, we have (if $\vdash 26^{\alpha}$), $\vdash N_{o\alpha''}m_{\alpha''} \supset .\, (n_{\alpha''}) \,.\, N_{o\alpha''}n_{\alpha''} \supset .\, m_{\alpha''}S_{\alpha'\alpha'}0_{\alpha'} = n_{\alpha''}S_{\alpha'\alpha'}0_{\alpha'} \supset m_{\alpha''} = n_{\alpha''}$.

Hence using V and $12^{\alpha''}$, we have (if $\vdash 26^{\alpha}$), $N_{o\alpha''}m_{\alpha''}$, $N_{o\alpha''}n_{\alpha''} \vdash m_{\alpha''}S_{\alpha'\alpha'}0_{\alpha'} = n_{\alpha''}S_{\alpha'\alpha'}0_{\alpha'} \supset m_{\alpha''} = n_{\alpha''}$.

Hence by two applications of the deduction theorem, if $\vdash 26^{\alpha}$ then $\vdash 30^{\alpha}$. This is (1) above.

7. Properties of $T_{\alpha''\alpha'}$. We proceed now to proofs of the following theorems:

31^{α}. $N_{o\alpha'}x_{\alpha'} \supset N_{o\alpha''}(T_{\alpha''\alpha'}x_{\alpha'})$.
32^{α}. $N_{o\alpha'}x_{\alpha'} \supset T_{\alpha''\alpha'}x_{\alpha'}S_{\alpha'\alpha'}0_{\alpha'} = x_{\alpha'}$.

The proofs require $9^{\alpha''}$ and are possible only for types α for which there is a proof of 30^{α}.

We begin by proving as a lemma:

33^{α}. $N_{o\alpha'}x_{\alpha'} \supset (\exists x_{\alpha''}) \,.\, N_{o\alpha''}x_{\alpha''} \,.\, x_{\alpha''}S_{\alpha'\alpha'}0_{\alpha'} = x_{\alpha'}$.

Proof of this requires only the axioms 1–6^{α} and is possible for an arbitrary type α.

By 16^{α}, using IV′ and conversion, we have $\vdash 0_{\alpha''}S_{\alpha'\alpha'}0_{\alpha'} = 0_{\alpha'}$. Hence using $22^{\alpha'}$ and $p \supset .q \supset pq$ and $13^{\alpha''}$, we have $\vdash (\exists x_{\alpha''}) \,.\, N_{o\alpha''}x_{\alpha''} \,.\, x_{\alpha''}S_{\alpha'\alpha'}0_{\alpha'} = 0_{\alpha'}$.

By $18^{\alpha'\alpha'}$, $x_{\alpha''}S_{\alpha'\alpha'}0_{\alpha'} = x_{\alpha'} \vdash S_{\alpha'\alpha'}(x_{\alpha''}S_{\alpha'\alpha'}0_{\alpha'}) = S_{\alpha'\alpha'}x_{\alpha'}$. Hence by conversion, $x_{\alpha''}S_{\alpha'\alpha'}0_{\alpha'} = x_{\alpha'} \vdash S_{\alpha''\alpha''}x_{\alpha''}S_{\alpha'\alpha'}0_{\alpha'} = S_{\alpha'\alpha'}x_{\alpha'}$. Also, by $23^{\alpha'}$, $N_{o\alpha''}x_{\alpha''} \vdash N_{o\alpha''}(S_{\alpha''\alpha''}x_{\alpha''})$. Hence using $pq \supset p$ and $pq \supset q$ and $p \supset .q \supset pq$, we have $N_{o\alpha''}x_{\alpha''} \,.\, x_{\alpha''}S_{\alpha'\alpha'}0_{\alpha'} = x_{\alpha'} \vdash N_{o\alpha''}(S_{\alpha''\alpha''}x_{\alpha''}) \,.\, S_{\alpha''\alpha''}x_{\alpha''}S_{\alpha'\alpha'}0_{\alpha'} = S_{\alpha'\alpha'}x_{\alpha'}$. Hence employing in order $13^{\alpha''}$, the deduction theorem, and VI′, we have $\vdash (x_{\alpha''}) \,.\, [N_{o\alpha''}x_{\alpha''} \,.\, x_{\alpha''}S_{\alpha'\alpha'}0_{\alpha'} = x_{\alpha'}] \supset (\exists x_{\alpha''}) \,.\, N_{o\alpha''}x_{\alpha''} \,.\, x_{\alpha''}S_{\alpha'\alpha'}0_{\alpha'} = S_{\alpha'\alpha'}x_{\alpha'}$. Hence by $15^{\alpha''}$, $\vdash (\exists x_{\alpha''})[N_{o\alpha''}x_{\alpha''} \,.\, x_{\alpha''}S_{\alpha'\alpha'}0_{\alpha'} = x_{\alpha'}] \supset (\exists x_{\alpha''}) \,.\, N_{o\alpha''}x_{\alpha''} \,.\, x_{\alpha''}S_{\alpha'\alpha'}0_{\alpha'} = S_{\alpha'\alpha'}x_{\alpha'}$.

Hence by the induction theorem, $\vdash 33^{\alpha}$.

Now proceeding with the proof of 31^{α} and 32^{α} (for types α for which $\vdash 30^{\alpha}$), we may—with the aid of 30^{α}—show that $N_{o\alpha''}x_{\alpha''} \,.\, x_{\alpha''}S_{\alpha'\alpha'}0_{\alpha'} = x_{\alpha'}$, $N_{o\alpha''}y_{\alpha''} \,.\, y_{\alpha''}S_{\alpha'\alpha'}0_{\alpha'} = x_{\alpha'} \vdash x_{\alpha''} = y_{\alpha''}$.

Hence by the deduction theorem and VI′, $N_{o\alpha''}x_{\alpha''} \,.\, x_{\alpha''}S_{\alpha'\alpha'}0_{\alpha'} = x_{\alpha'} \vdash (y_{\alpha''}) \,.\, [N_{o\alpha''}y_{\alpha''} \,.\, y_{\alpha''}S_{\alpha'\alpha'}0_{\alpha'} = x_{\alpha'}] \supset x_{\alpha''} = y_{\alpha''}$.

Hence by $9^{\alpha''}$ (refer to the definition of $T_{\alpha''\alpha'}$, §2), $N_{o\alpha''}x_{\alpha''} \,.\, x_{\alpha''}S_{\alpha'\alpha'}0_{\alpha'} = x_{\alpha'} \vdash N_{o\alpha''}(T_{\alpha''\alpha'}x_{\alpha'}) \,.\, T_{\alpha''\alpha'}x_{\alpha'}S_{\alpha'\alpha'}0_{\alpha'} = x_{\alpha'}$.

Hence employing in order the deduction theorem, VI′, and $15^{\alpha''}$, we have $\vdash (\exists x_{\alpha''})[N_{o\alpha''}x_{\alpha''} \,.\, x_{\alpha''}S_{\alpha'\alpha'}0_{\alpha'} = x_{\alpha'}] \supset .\, N_{o\alpha''}(T_{\alpha''\alpha'}x_{\alpha'}) \,.\, T_{\alpha''\alpha'}x_{\alpha'}S_{\alpha'\alpha'}0_{\alpha'} = x_{\alpha'}$.

Hence using 33^{α} and $p \supset q \supset .[q \supset rs] \supset .p \supset r$ and $p \supset q \supset .[q \supset rs] \supset .p \supset s$, we have $\vdash 31^{\alpha}$ and $\vdash 32^{\alpha}$.

A further property of $T_{\alpha''\alpha'}$ is contained in the following theorem (if α is a type for which there is a proof of 30^{α}):

34^{α}. $N_{o\alpha'}x_{\alpha'} \supset T_{\alpha''\alpha'}(S_{\alpha'\alpha'}x_{\alpha'}) = S_{\alpha''\alpha''}(T_{\alpha''\alpha'}x_{\alpha'})$.

Proof of this depends on using $23^{\alpha'}$ to prove $N_{o\alpha''}(S_{\alpha''\alpha''}(T_{\alpha''\alpha'}x_{\alpha'}))$ on the assumption of $N_{o\alpha'}x_{\alpha'}$, and using $16^{\alpha'}$ and conversion to prove $S_{\alpha''\alpha''}(T_{\alpha''\alpha'}x_{\alpha'})S_{\alpha'\alpha'}0_{\alpha'} = S_{\alpha'\alpha'}(T_{\alpha''\alpha'}x_{\alpha'}S_{\alpha'\alpha'}0_{\alpha'})$ and hence $S_{\alpha''\alpha''}(T_{\alpha''\alpha'}x_{\alpha'})S_{\alpha'\alpha'}0_{\alpha'} = S_{\alpha'\alpha'}x_{\alpha'}$ on the assumption of $N_{o\alpha'}x_{\alpha'}$—then using 30^{α} (with 23^{α}, 32^{α}, 31^{α}).

A similar use of 30^{α} leads to a proof of the following (where α is a type for which there is a proof of 30^{α}):

35^{α}. $T_{\alpha''\alpha'}0_{\alpha'} = 0_{\alpha''}$.

8. Definition by primitive recursion. The formalization of definition by primitive recursion requires that, given formulas $\mathbf{A}_{\alpha'}$ and $\mathbf{B}_{\alpha'\alpha'\alpha'}$, we find a formula $\mathbf{F}_{\alpha'\alpha'}$ such that the following are theorems (where $\mathbf{x}_{\alpha'}$ is not a free variable of $\mathbf{A}_{\alpha'}$, $\mathbf{B}_{\alpha'\alpha'\alpha'}$, or $\mathbf{F}_{\alpha'\alpha'}$):

$$\mathbf{F}_{\alpha'\alpha'}0_{\alpha'} = \mathbf{A}_{\alpha'}.$$

$$N_{o\alpha'}\mathbf{x}_{\alpha'} \supset \mathbf{F}_{\alpha'\alpha'}(S_{\alpha'\alpha'}\mathbf{x}_{\alpha'}) = \mathbf{B}_{\alpha'\alpha'\alpha'}\mathbf{x}_{\alpha'}(\mathbf{F}_{\alpha'\alpha'}\mathbf{x}_{\alpha'}).$$

This may be done by taking $\mathbf{F}_{\alpha'\alpha'}$ to be the following formula (where $\mathbf{x}_{\alpha'}$, $\mathbf{y}_{\alpha''}$ are not free variables of $\mathbf{A}_{\alpha'}$ or $\mathbf{B}_{\alpha'\alpha'\alpha'}$):[13]

$$\lambda \mathbf{x}_{\alpha'} \,.\, T_{\alpha'''\alpha''}(T_{\alpha''\alpha'}\mathbf{x}_{\alpha'})(\lambda \mathbf{y}_{\alpha''}\langle S_{\alpha'\alpha'}(\mathbf{y}_{\alpha''}(K_{\alpha'\alpha'\alpha'}I_{\alpha'})0_{\alpha'}),$$
$$\mathbf{B}_{\alpha'\alpha'\alpha'}(\mathbf{y}_{\alpha''}(K_{\alpha'\alpha'\alpha'}I_{\alpha'})0_{\alpha'})(\mathbf{y}_{\alpha''}(K_{\alpha'\alpha'\alpha'}0_{\alpha'})I_{\alpha'})\rangle)\langle 0_{\alpha'}, \mathbf{A}_{\alpha'}\rangle(K_{\alpha'\alpha'\alpha'}0_{\alpha'})I_{\alpha'}.$$

The definition of $P_{\alpha'\alpha'}$ already given is a particular case and may be used as an illustration. The following theorems may be proved in order:

36^{α}. $N_{o\alpha'}n_{\alpha'} \supset \lambda f_{\alpha\alpha}\lambda x_{\alpha}(n_{\alpha'}f_{\alpha\alpha}x_{\alpha}) = n_{\alpha'}$. (By induction, using $16^{\alpha'}$, $18^{\alpha'\alpha'}$, and conversion.)

37^{α}. $N_{o\alpha'}m_{\alpha'} \supset . N_{o\alpha'}n_{\alpha'} \supset \langle m_{\alpha'}, n_{\alpha'}\rangle(K_{\alpha'\alpha'\alpha'}I_{\alpha'})0_{\alpha'} = m_{\alpha'}$. (By induction on $n_{\alpha'}$, using 36^{α}.)

38^{α}. $N_{o\alpha'}m_{\alpha'} \supset . N_{o\alpha'}n_{\alpha'} \supset \langle m_{\alpha'}, n_{\alpha'}\rangle(K_{\alpha'\alpha'\alpha'}0_{\alpha'})I_{\alpha'} = n_{\alpha'}$. (By induction on $m_{\alpha'}$, using 36^{α}.)

39^{α}. $N_{o\alpha'''}n_{\alpha'''} \supset n_{\alpha'''}(\lambda p_{\alpha''}\langle S_{\alpha'\alpha'}(p_{\alpha''}(K_{\alpha'\alpha'\alpha'}I_{\alpha'})0_{\alpha'}), p_{\alpha''}(K_{\alpha'\alpha'\alpha'}I_{\alpha'})0_{\alpha'}\rangle)\langle 0_{\alpha'}, 0_{\alpha'}\rangle(K_{\alpha'\alpha'\alpha'}I_{\alpha'})0_{\alpha'} = n_{\alpha'''}S_{\alpha''\alpha''}0_{\alpha''}S_{\alpha'\alpha'}0_{\alpha'}$. (By induction, using 29^{α}, 37^{α}.)

40^{α}. $N_{o\alpha'''}n_{\alpha'''} \supset P_{\alpha'\alpha'''}(S_{\alpha'''\alpha'''}n_{\alpha'''}) = n_{\alpha'''}S_{\alpha''\alpha''}0_{\alpha''}S_{\alpha'\alpha'}0_{\alpha'}$. (By 39^{α}, 38^{α}, using 29^{α}.)

41^{α}. $P_{\alpha'\alpha'''}0_{\alpha'''} = 0_{\alpha'}$. (By $16^{\alpha'}$ and conversion.)

42^{α}. $P_{\alpha'\alpha'}0_{\alpha'} = 0_{\alpha'}$. (By 41^{α}, $35^{\alpha'}$, 35^{α}.)

43^{α}. $N_{o\alpha'}n_{\alpha'} \supset P_{\alpha'\alpha'}(S_{\alpha'\alpha'}n_{\alpha'}) = n_{\alpha'}$. (By 40^{α}, $31^{\alpha'}$, 31^{α}, $34^{\alpha'}$, 34^{α}, $32^{\alpha'}$, 32^{α}.)

PRINCETON UNIVERSITY

---

[13] This schema employs descriptions, through the appearance in it of $T_{\alpha'''\alpha''}$ and $T_{\alpha''\alpha'}$. In certain cases a formula $\mathbf{F}_{\alpha'\alpha'}$ may be obtained which does not involve descriptions. In particular, for addition and multiplication of non-negative integers we may use the definitions due to J. B. Rosser:

$s_{\alpha'\alpha'\alpha'} \rightarrow \lambda m_{\alpha'}\lambda n_{\alpha'}\lambda f_{\alpha\alpha}\lambda x_{\alpha}(m_{\alpha'}f_{\alpha\alpha}(n_{\alpha'}f_{\alpha\alpha}x_{\alpha}))$.

$B_{\alpha'\alpha'\alpha'} \rightarrow \lambda m_{\alpha'}\lambda n_{\alpha'}\lambda f_{\alpha\alpha}(m_{\alpha'}(n_{\alpha'}f_{\alpha\alpha}))$.

$[\mathbf{A}_{\alpha'}+\mathbf{B}_{\alpha'}] \rightarrow s_{\alpha'\alpha'\alpha'}\mathbf{A}_{\alpha'}\mathbf{B}_{\alpha'}$.

$[\mathbf{A}_{\alpha'}\times\mathbf{B}_{\alpha'}] \rightarrow B_{\alpha'\alpha'\alpha'}\mathbf{A}_{\alpha'}\mathbf{B}_{\alpha'}$.

THE JOURNAL OF SYMBOLIC LOGIC
Volume 15, Number 2, June 1950

COMPLETENESS IN THE THEORY OF TYPES[1]

LEON HENKIN [2]

The first order functional calculus was proved complete by Gödel[3] in 1930. Roughly speaking, this proof demonstrates that each formula of the calculus is a formal theorem which becomes a true sentence under every one of a certain intended class of interpretations of the formal system.

For the functional calculus of second order, in which predicate variables may be bound, a very different kind of result is known: no matter what (recursive) set of axioms are chosen, the system will contain a formula which is valid but not a formal theorem. This follows from results of Gödel[4] concerning systems containing a theory of natural numbers, because a finite categorical set of axioms for the positive integers can be formulated within a second order calculus to which a functional constant has been added.

By a valid formula of the second order calculus is meant one which expresses a true proposition whenever the individual variables are interpreted as ranging over an (arbitrary) domain of elements while the functional variables of degree n range over all sets of ordered n-tuples of individuals. Under this definition of validity, we must conclude from Gödel's results that the calculus is essentially incomplete.

It happens, however, that there is a wider class of models which furnish an interpretation for the symbolism of the calculus consistent with the usual axioms and formal rules of inference. Roughly, these models consist of an arbitrary domain of individuals, as before, but now an *arbitrary*[5] *class* of sets of ordered n-tuples of individuals as the range for functional variables of degree n. If we

Received March 11, 1949.

[1] The material in this paper is included in *The completeness of formal systems*, a Thesis presented to the faculty of Princeton University in candidacy for the degree of Doctor of Philosophy and accepted in October, 1947. The results were announced at the meeting of the Association for Symbolic Logic in December, 1947 (cf. this JOURNAL, vol. 13 (1948), p. 61).

[2] The author wishes to thank Professor Alonzo Church for encouragement, suggestion, and criticism in connection with the writing of his Thesis, and to acknowledge the aid of the National Research Council who supported that project with a predoctoral fellowship.

[3] Kurt Gödel, *Die Vollständigkeit der Axiome des logischen Funktionenkalküls*, ***Monatshefte für Mathematik und Physik***, vol. 37 (1930), pp. 349–360.

[4] Kurt Gödel, *Über formal unentscheidbare Sätze der Principia Mathematica und verwandter Systeme I*, ***Monatshefte für Mathematik und Physik***, vol. 38 (1931), pp. 173–198.

[5] These classes cannot really be taken in an altogether arbitrary manner if every formula is to have an interpretation. For example, if the formula $F(x)$ is interpreted as meaning that x is in the class F, then $\sim F(x)$ means that x is in the complement of F; hence the range for functional variables such as F should be closed under complementation. Similarly, if G refers to a set of ordered pairs in some model, then the set of individuals x satisfying the formula $(\exists y)G(x, y)$ is a projection of the set G; hence, we require that the various domains be closed under projection. In short, each method of compounding formulas of the calculus has associated with it some operation on the domains of a model, with respect to which the domains must be closed. The statement of completeness can be given precisely and proved for models meeting these closure conditions.

redefine the notion of valid formula to mean one which expresses a true proposition with respect to every one of *these* models, we can then prove that the usual axiom system for the second order calculus is complete: a formula is valid if and only if it is a formal theorem.[6]

A similar result holds for the calculi of higher order. In this paper, we will give the details for a system of order ω embodying a simple theory of (finite) types. We shall employ the rather elegant formulation of Church,[7] the details of which are summarized below:

Type symbols (to be used as subscripts):
1. o and ι are type symbols
2. If α, β are type symbols so is $(\alpha\beta)$.

Primitive symbols (where α may be any type symbol):
Variables: f_α , g_α , x_α , y_α , z_α , f'_α , g'_α , $\cdots$
Constants: $N_{(oo)}$, $A_{((oo)o)}$, $\Pi_{(o(o\alpha))}$, $\iota_{(\alpha(o\alpha))}$
Improper: λ, (,).

Well-formed formulas (wffs) and their *type*:
1. A variable or constant alone is a wff and has the type of its subscript.
2. If $A_{\alpha\beta}$ and B_β are wffs of type $(\alpha\beta)$ and β respectively, then $(A_{\alpha\beta}\ B_\beta)$ is a wff of type α.
3. If A_α is a wff of type α and a_β a variable of type β then $(\lambda a_\beta A_\alpha)$ is a wff of type $(\alpha\beta)$.

An occurrence of a variable a_β is *bound* if it is in a wff of the form $(\lambda a_\beta A_\alpha)$; otherwise the occurrence is *free*.

Letters A_α , B_α , C_α , will be used as syntactical variables for wffs of type α.

Abbreviations:

$(\sim A_o)$	for $(N_{(oo)}A_o)$
$(A_o \vee B_o)$	for $((A_{((oo)o)}A_o)B_o)$
(A_oB_o)	for $(\sim((\sim A_o) \vee (\sim B_o)))$
$(A_o \supset B_o)$	for $((\sim A_o) \vee B_o)$
$(a_\alpha)B_o$	for $(\Pi_{(o(o\alpha))}(\lambda a_\alpha B_o))$
$(\exists a_\alpha)B_o$	for $(\sim((a_\alpha)(\sim A_o)))$
$(\iota a_\alpha B_o)$	for $(\iota_{(\alpha(o\alpha))}(\lambda a_\alpha B_o))$
$Q_{((o\alpha)\alpha)}$	for $(\lambda x_\alpha(\lambda y_\alpha(f_{o\alpha})((f_{o\alpha}x_\alpha) \supset (f_{o\alpha}y_\alpha))))$
$(A_\alpha = B_\alpha)$	for $((Q_{((o\alpha)\alpha)}A_\alpha)(B_\alpha)$.

In writing wffs and subscripts, we shall practise the omission of parentheses and their supplantation by dots on occasion, the principal rules of restoration

[6] A demonstration of this type of completeness can be carried out along the lines of the author's recent paper, *The completeness of the first order functional calculus*, this JOURNAL, vol. 14 (1949), pp. 159–166.

[7] Alonzo Church, *A formulation of the simple theory of types*, this JOURNAL, vol. 5 (1940), pp. 56–68.

being first that the formula shall be well-formed; secondly, that association is to the left; and thirdly, that a dot is to be replaced by a left parenthesis having its mate as far to the right as possible. (For a detailed statement of usage, refer to Church.[7])

Axioms and axiom schemata:

1. $(x_o \vee x_o) \supset x_o$
2. $x_o \supset (x_o \vee y_o)$
3. $(x_o \vee y_o) \supset (y_o \vee x_o)$
4. $(x_o \supset y_o) \supset . (z_o \vee x_o) \supset (z_o \vee y_o)$

5^α. $\Pi_{o(o\alpha)} f_{o\alpha} \supset f_{o\alpha} x_\alpha$

6^α. $(x_\alpha)(y_o \vee f_{o\alpha} x_\alpha) \supset . y_o \vee \Pi_{o(o\alpha)} f_{o\alpha}$

10. $x_o \equiv y_o \supset x_o = y_o \qquad (x_\beta)(f_{\alpha\beta} x_\beta = g_{\alpha\beta} x_\beta) \supset f_{\alpha\beta} = g_{\alpha\beta}$

11^α. $f_{o\alpha} x_\alpha \supset f_{o\alpha}(\iota_{\alpha(o\alpha)} f_{o\alpha})$

Rules of Inference:

I. To replace any part A_α of a formula by the result of substituting a_β for b_β throughout A_α, provided that b_β is not a free variable of A_α and a_β does not occur in A_α.

II. To replace any part $(\lambda a_\gamma A_\beta) B_\gamma$ of a wff by the result of substituting B_γ for a_γ throughout A_β, provided that the bound variables of A_β are distinct both from a_γ and the free variables of B_γ.

III. To infer A_o from B_o if B_o may be inferred from A_o by a single application of Rule II.

IV. From $A_{o\alpha} a_\alpha$ to infer $A_{o\alpha} B_\alpha$ if the variable a_α is not free in $A_{o\alpha}$.

V. From $A_o \supset B_o$ and A_o to infer B_o.

VI. From $A_{o\alpha} a_\alpha$ to infer $\Pi_{o(o\alpha)} A_{o\alpha}$ provided that the variable a_α is not free in $A_{o\alpha}$.

A finite sequence of wffs each of which is an axiom or obtained from preceding elements of the sequence by a single application of one of the rules I–VI is called a *formal proof*. If A is an element of some formal proof, we write $\vdash A$ and say that A is a *formal theorem*.

This completes our description of the formal system. In order to discuss the question of its completeness, we must now give a precise account of the manner in which this formalism is to be *interpreted*.

By a *standard model*, we mean a family of domains, one for each type-symbol, as follows: D_ι is an arbitrary set of elements called *individuals*, D_o is the set consisting of two truth values, T and F, and $D_{\alpha\beta}$ is the set of all functions defined over D_β with values in D_α.

By an *assignment* with respect to a standard model $\{D_\alpha\}$, we mean a mapping ϕ of the variables of the formal system into the domains of the model such that for a variable a_α of type α as argument, the value $\phi(a_\alpha)$ of ϕ is an element of D_α.

We shall associate with each assignment ϕ a mapping V_ϕ of all the formulas of the formal system such that $V_\phi(A_\alpha)$ is an element of D_α for each wff A_α of type

α. We shall define the values $V_\phi(A_\alpha)$ simultaneously for all ϕ by induction on the length of the wff A_α :

(i) If A_α is a variable, set $V_\phi(A_\alpha) = \phi(A_\alpha)$. Let $V_\phi(N_{oo})$ be the function whose values are given by the table

x	$V_\phi(N_{oo})(x)$
T	F
F	T

Let $V_\phi(A_{ooo})$ be the function whose value for arguments T, F are the functions given by the tables 1, 2 respectively.

1.

x	$V_\phi(A_{ooo})(\text{T})(x)$
T	T
F	T

2.

x	$V_\phi(A_{ooo})(\text{F})(x)$
T	T
F	F

Let $V_\phi(\Pi_{o(o\alpha)})$ be the function which has the value T for just the single argument which is the function mapping D_α into the constant value T. Let $V_\phi(\iota_{\alpha(o\alpha)})$ be some fixed function whose value for any argument f of $D_{o\alpha}$ is one of the elements of D_α mapped into T by f (if there is such an element).

(ii) If A_α has the form $B_{\alpha\beta}C_\beta$ define $V_\phi(B_{\alpha\beta}C_\beta)$ to be the value of the function $V_\phi(B_{\alpha\beta})$ for the argument $V_\phi(C_\beta)$.

(iii) Suppose A_α has the form $(\lambda a_\beta B_\gamma)$. We define $V_\phi(\lambda a_\beta B_\gamma)$ to be the function whose value for the argument x of D_β is $V_\psi(B_\gamma)$, where ψ is the assignment which has the same values as ϕ for all variables except a_β , while $\psi(a_\beta)$ is x.

We can now define a wff A_o to be *valid in the standard sense* if $V_\phi(A_o)$ is T for every assignment ϕ with respect to every standard model $\{D_\alpha\}$.[8] Because the theory of recursive arithmetic can be developed within our formal system as shown by Church,[7] it follows by Gödel's methods[4] that we can construct a particular wff A_o which is valid in the standard sense, but not a formal theorem.

We can, however, interpret our formalism with respect to other than the standard models. By a *frame*, we mean a family of domains, one for each type symbol, as follows: D_ι is an arbitrary set of individuals, D_o is the set of two truth values, T and F, and $D_{\alpha\beta}$ is some class of functions defined over D_β with values in D_α .

Given such a frame, we may consider assignments ϕ mapping variables of the formal system into its domains, and attempt to define the functions V_ϕ exactly as for standard models. For an arbitrary frame, however, it may well happen that one of the functions described in items (i), (ii), or (iii) as the value of some $V_\phi(A_\alpha)$ is not an element of any of the domains.

A frame such that for every assignment ϕ and wff A_α of type α, the value $V_\phi(A_\alpha)$ given by rules (i), (ii), and (iii) is an element of D_α , is called a *general model*. Since this definition is impredicative, it is not immediately clear that any non-standard models exist. However, they do exist (indeed, there are general models for which every domain D_α is denumerable), and we shall give a method of constructing every general model without resorting to impredicative processes.

Now we define a *valid formula in the general sense* as a formula A_o such that $V_\phi(A_o)$ is T for every assignment ϕ with respect to any general model. We shall prove a completeness theorem for the formal system by showing that A_0 is valid in the general sense if and only if $\vdash A$.

By a *closed* well-formed formula (cwff), we mean one in which no occurrence of any variable is free. If Λ is a set of cwffs such that, when added to the axioms 1–6$^\alpha$, 10$^\alpha$, 11$^\alpha$, a formal proof can be obtained for some wff A_0, we write $\Lambda \vdash A_0$. If $\Lambda \vdash A_0$ for every wff A_0, we say that Λ is *inconsistent*, otherwise *consistent*.

THEOREM 1. *If* Λ *is any consistent set of* cwffs, *there is a general model (in which each domain* D_α *is denumerable) with respect to which* Λ *is satisfiable.*[8]

We shall make use of the following derived results about the formal calculus which we quote without proof:

VII. The deduction theorem holds: If $\Lambda, A_o \vdash B_o$, then $\Lambda \vdash A_o \supset B_o$ where Λ is any set of cwffs, A_o is any cwff, and B_o is any wff. (A proof is given in Church.[7])

12. $\vdash A_o \supset . \sim A_o \supset B_o$
13. $\vdash A_o \supset B_o \supset . \sim A_o \supset B_o \supset . B_o$
14. $\vdash A_\alpha = A_\alpha$
15. $\vdash A_\alpha = B_\alpha \supset B_\alpha = A_\alpha$
16. $\vdash A_\alpha = B_\alpha \supset . B_\alpha = C_\alpha \supset . A_\alpha = C_\alpha$
17. $\vdash A_o \supset . (A_o = B_o) \supset B_o$
18. $\vdash \sim A_o \supset . (A_o = B_o) \supset \sim B_o$
19. $\vdash A_o \supset . B_o \supset . A_o = B_o$
20. $\vdash \sim A_o \supset . \sim B_o \supset . A_o = B_o$
21. $\vdash A_{\alpha\beta} = A'_{\alpha\beta} \supset . B_\beta = B'_\beta \supset . A_{\alpha\beta} B_\beta = A'_{\alpha\beta} B'_\beta$
22. $\vdash A_{\alpha\beta}((\iota x_\beta) \sim (A_{\alpha\beta} x_\beta = A'_{\alpha\beta} x_\beta)) = A'_{\alpha\beta}((\imath x_\beta) \sim (A_{\alpha\beta} x_\beta = A'_{\alpha\beta} x_\beta)) \supset . A_{\alpha\beta} = A'_{\alpha\beta}$
23. $\vdash A_o \supset \sim\sim A_o$
24. $\vdash C_o \supset . C_o \vee A_o$
25. $\vdash \sim C_o \supset . A_o \supset . C_o \vee A_o$
26. $\vdash \sim C_o \supset . \sim A_o \supset . \sim(C_o \vee A_o)$
27. $\vdash \Pi_{o(o\alpha)} A_{o\alpha} \supset A_{o\alpha} C_\alpha$
28. $\vdash A_{o\alpha}((\imath x_\alpha) \sim (A_{o\alpha} x_\alpha)) \supset \Pi_{o(o\alpha)} A_{o\alpha}$
29. $\vdash A_{o\alpha} C_\alpha \supset A_{o\alpha}(\iota_{\alpha(o\alpha)} C_\alpha)$
30. $\vdash (\sim B_o \supset B_o) \supset B_o$
31. $\vdash (x_\alpha) A_o \supset A_o$

---

[8] In addition to the notion of validity, the mappings V_ϕ may be used to define the concept of the *denotation* of a wff A_α containing no free occurrence of any variable. We first show (by induction) that if ϕ and ψ are two assignments which have the same value for every variable with a free occurrence in the wff B_α, then $V_\phi(B_\alpha) = V_\psi(B_\alpha)$. Then the denotation of A_α is simply $V_\phi(A_\alpha)$ for any ϕ. We also define the notion of satisfiability. If Γ is a set of wffs and ϕ an assignment with respect some model $\{D_\alpha\}$ such that $V_\phi(A_o)$ is T for every A_o in Γ, then we say that Γ is *satisfiable with respect to the model* $\{D_\alpha\}$. If Γ is satisfiable with respect to some model, we say simply that it is *satisfiable*.

The first step in our proof of Theorem 1 is to construct a maximal consistent set Γ of cwffs such that Γ contains Λ, where by maximal is meant that if A_o is any cwff not in Γ then the enlarged set $\{\Gamma, A_o\}$ is inconsistent. Such a set Γ may be obtained in many ways. If we enumerate all of the cwffs in some standard order, we may test them one at a time, adding them to Λ and previously added formulas whenever this does not result in an inconsistent set. The union of this increasing sequence of sets is then easily seen to be maximal consistent.

Γ has certain simple properties which we shall use. If A_o is any cwff, it is clear that we cannot have both $\Gamma \vdash A_o$ and $\Gamma \vdash \sim A_o$ for then by 12 and V, we would obtain $\Gamma \vdash B_o$ for any B_o, contrary to the consistency of Γ. On the other hand, at least one of the cwffs A_o, $\sim A_o$ must be in Γ. For otherwise, using the maximal property of Γ we would have $\Gamma, A_o \vdash B_o$ and $\Gamma, \sim A_o \vdash B_o$ for any B_o. By VII, it then follows that $\Gamma \vdash A_o \supset B_o$ and $\Gamma \vdash \sim A_o \supset B_o$, whence by 13 and V $\Gamma \vdash B_o$ contrary to the consistency of Γ.

Two cwffs A_α, B_α of type α will be called *equivalent* if $\Gamma \vdash A_\alpha = B_\alpha$. Using 14, 16, and V, we easily see that this is a genuine congruence relation so that the set of all cwffs of type α is partitioned into disjoint equivalent classes $[A_\alpha]$, $[B_\alpha]$, $\cdots$ such that $[A_\alpha]$ and $[B_\alpha]$ are equal if and only if A_α is equivalent to B_α.

We now define by induction on α a frame of domains $\{D_\alpha\}$, and simultaneously a one–one mapping Φ of equivalence classes onto the domains D_α such that $\Phi([A_\alpha])$ is in D_α.

D_o is the set of two truth values, T and F, and for any cwff A_o of type o $\Phi([A_o])$ is T or F according as A_o or $\sim A_o$ is in Γ. We must show that Φ is a function of equivalence classes and does not really depend on the particular representative A_o chosen. But by 17 and V, we see that if $\Gamma \vdash A_o$ and B_o is equivalent to A_o (i.e., $\Gamma \vdash A_o = B_o$), then $\Gamma \vdash B_o$; and similarly if $\Gamma \vdash \sim A_o$ and B_o is equivalent to A_o, then $\Gamma \vdash \sim B_o$ by 18. To see that Φ is one–one, we use 19 to show that if $\Phi([A_o])$ and $\Phi([B_o])$ are both T (i.e., $\Gamma \vdash A_o$ and $\Gamma \vdash B_o$), then $\Gamma \vdash A_o = B_o$ so that $[A_o]$ is $[B_o]$. Similarly 20 shows that $[A_o]$ is $[B_o]$ in case $\Phi([A_o])$ and $\Phi([B_o])$ are both F.

D_ι is simply the set of equivalence classes $[A_\iota]$ of all cwffs of type ι. And $\Phi([A_\iota])$ is $[A_\iota]$ so that Φ is certainly one-one.

Now suppose that D_α and D_β have been defined, as well as the value of Φ for all equivalence classes of formulas of type α and of type β, and that every element of D_α, or D_β, is the value of Φ for some $[A_\alpha]$, or $[B_\beta]$ respectively. Define $\Phi([A_{\alpha\beta}])$ to be the function whose value, for the element $\Phi([B_\beta])$ of D_β, is $\Phi([A_{\alpha\beta}B_\beta])$. This definition is justified by the fact that if $A'_{\alpha\beta}$ and B'_β are equivalent to $A_{\alpha\beta}$ and B_β respectively, then $A'_{\alpha\beta}B'_\beta$ is equivalent to $A_{\alpha\beta}B_\beta$, as one sees by 21. To see that Φ is one-one, suppose that $\Phi([A_{\alpha\beta}])$ and $\Phi([A'_{\alpha\beta}])$ have the same value for every $\Phi([B_\beta])$ of D_β. Hence $\Phi([A_{\alpha\beta}B_\beta]) = \Phi([A'_{\alpha\beta}B_\beta])$ and so, by the induction hypothesis that Φ is one-one for equivalence classes of formulas of type α, $A_{\alpha\beta}B_\beta$ is equivalent to $A'_{\alpha\beta}B_\beta$ for each cwff B_β. In particular, if we take B_β to be $(\imath x_\beta)$ $\sim(A_{\alpha\beta}x_\beta = A'_{\alpha\beta}x_\beta)$, we see by 22 that $A_{\alpha\beta}$ and $A'_{\alpha\beta}$ are equivalent so that $[A_{\alpha\beta}] = [A'_{\alpha\beta}]$. The one-one function Φ having been thus completely defined, we define $D_{\alpha\beta}$ to be the set of values $\Phi([A_{\alpha\beta}])$ for all cwffs $A_{\alpha\beta}$.

Now let ϕ be any assignment mapping each variable x_α into some element $\Phi([A_\alpha])$ of D_α, where A_α is a cwff. Given any wff B_β, let B_β^ϕ be a cwff obtained from B_β by replacing all free occurrences in B_β of any variable x_α by some cwff A_α such that $\phi(x_\alpha) = \Phi([A_\alpha])$.

LEMMA. *For every* ϕ *and* B_β *we have* $V_\phi(B_\beta) = \Phi([B_\beta^\phi])$.

The proof is by induction on the length of B_β.

(i) If B_β is a variable and $\phi(B_\beta)$ is the element $\Phi([A_\beta])$ of D_β, then by definition B_β^ϕ is some cwff A_β' equivalent to A_β and $V_\phi(B_\beta) = \phi(B_\beta) = \Phi([A_\beta]) = \Phi([A_\beta']) = \Phi([B_\beta^\phi])$.

Suppose B_β is N_{oo}, whence B_β^ϕ is N_{oo}. If $\Phi([A_o])$ is T, then by definition $\Gamma \vdash A_o$ whence by 23 $\Gamma \vdash \sim N_{oo}A_o$ so that $\Phi([N_{oo}A_o])$ is F. That is, $\Phi([N_{oo}])$ maps T into F. Conversely, if $\Phi([A_o])$ is F, then by definition $\Gamma \vdash N_{oo}A_o$ so that $\Phi([N_{oo}A_o])$ is T; i.e., $\Phi([N_{oo}])$ maps F into T. Hence $V_\phi(B_\beta) = \Phi([B_\beta^\phi])$ in this case.

Suppose B_β is A_{ooo}, whence B_β^ϕ is A_{ooo}. If $\Phi([C_o])$ is T, then by definition $\Gamma \vdash C_o$ whence by 24 $\Gamma \vdash A_{ooo}C_oA_o$ for any A_o so that $\Phi([A_{ooo}C_oA_o])$ is T no matter whether $\Phi[(A_o)]$ is T or F. Similarly, using 25 and 26, we see that $\Phi([A_{ooo}C_oA_o])$ is T, or F, if $\Phi([C_o])$ is F, and $\Phi([A_o])$ is T, or F respectively. Comparing this with the definition of $V_\phi(A_{ooo})$, we see that the lemma holds in this case also.

Suppose B_β is $\Pi_{o(o\alpha)}$, whence B_β^ϕ is $\Pi_{o(o\alpha)}$. If the value of $\Phi([\Pi_{o(o\alpha)}])$ for the argument $\Phi([A_{o\alpha}])$ is T, then $\Gamma \vdash \Pi_{o(o\alpha)}A_{o\alpha}$ whence by 27 $\Gamma \vdash A_{o\alpha}C_\alpha$ for every cwff C_α so that $\Phi([A_{o\alpha}])$ maps every element of D_α into T. On the other hand, if $\Phi([A_{o\alpha}])$ maps every $\Phi([C_\alpha])$ into T, then we have, taking the particular case where C_α is $(\imath x_\alpha)\sim(A_{o\alpha}x_\alpha)$, $\Gamma \vdash A_{o\alpha}((\imath x_\alpha)\sim(A_{o\alpha}x_\alpha))$ whence by 28 $\Gamma \vdash \Pi_{o(o\alpha)}A_{o\alpha}$. That is $\Phi([\Pi_{o(o\alpha)}])$ maps $\Phi([A_{o\alpha}])$ into T. The lemma holds in this case.

Suppose B_β is $\iota_{\alpha(o\alpha)}$, whence B_β^ϕ is $\iota_{\alpha(o\alpha)}$. Let $A_{o\alpha}$ be a cwff such that $\Phi([A_{o\alpha}])$ maps some $\Phi([C_\alpha])$ into T so that $\Gamma \vdash A_{o\alpha}C_\alpha$. Then by 29 $\Gamma \vdash A_{o\alpha}(\iota_{\alpha(o\alpha)}A_{o\alpha})$ so that the value of $\Phi([\iota_{\alpha(o\alpha)}])$ for the argument $\Phi([A_{o\alpha}])$ is mapped into T by the latter. Therefore, we may take $\Phi([\iota_{\alpha(o\alpha)}])$ to be $V_\phi(\iota_{\alpha(o\alpha)})$.

(ii) Suppose that B_β has the form $B_{\beta\gamma}C_\gamma$. We assume (induction hypothesis) that we have already shown $\Phi([B_{\beta\gamma}^\phi]) = V_\phi(B_{\beta\gamma})$ and $\Phi([C_\gamma^\phi]) = V_\phi(C_\gamma)$.

Now $V_\phi(B_{\beta\gamma}C_\gamma)$ is the value of $V_\phi(B_{\beta\gamma})$ for the argument $V_\phi(C_\gamma)$, or the value of $\Phi([B_{\beta\gamma}^\phi])$ for the argument $\Phi([C_\gamma^\phi])$, which is $\Phi([B_{\beta\gamma}^\phi C_\gamma^\phi])$. But $(B_{\beta\gamma}C_\gamma)^\phi$ is simply $B_{\beta\gamma}^\phi C_\gamma^\phi$. Hence $V_\phi(B_{\beta\gamma}C_\gamma) = \Phi([(B_{\beta\gamma}C_\gamma)^\phi])$.

(iii) Suppose that B_β has the form $\lambda a_\gamma C_\alpha$ and our induction hypothesis is that $\Phi([C_\alpha^\phi]) = V_\phi(C_\alpha)$ for every assignment ϕ. Let $\Phi([A_\gamma])$ be any element of D_γ. Then the value of $\Phi([(\lambda a_\gamma C_\alpha)^\phi])$ for the argument $\Phi([A_\gamma])$ is by definition $\Phi([(\lambda a_\gamma C_\alpha)^\phi A_\gamma])$.

But by applying II to the right member of the instance $\vdash (\lambda a_\gamma C_\alpha)^\phi A_\gamma = (\lambda a_\gamma C_\alpha)^\phi A_\gamma$ of 14, we find $\vdash (\lambda a_\gamma C_\alpha)^\phi A_\gamma = C_\alpha^\psi$, where ψ is the assignment which has the same value as ϕ for every argument except the variable a_γ and $\psi(a_\gamma)$ is $\Phi([A_\gamma])$. That is, $[(\lambda a_\gamma C_\alpha)^\phi A_\gamma] = [C_\alpha^\psi]$ so that the value of $\Phi([(\lambda a_\gamma C_\alpha)^\phi])$ for the argument $\Phi([A_\gamma])$ is $\Phi([C_\alpha^\psi])$—or $V_\psi(C_\alpha)$ by induction hypothesis. Since for every argument $\Phi([(\lambda a_\gamma C_\alpha)^\phi])$ and $V_\phi(\lambda a_\gamma C_\alpha)$ have the same value, they must be equal.

This concludes the proof of our lemma.

Theorem 1 now follows directly from the lemma. In the first place, the frame of domains $\{D_\alpha\}$ is a general model since $V_\phi(B_\beta)$ is an element of D_β for every wff B_β and assignment ϕ. Because the elements of any D_α are in one-one correspondence with equivalence classes of wffs each domain is denumerable. Since for every cwff $A_o^\phi = A_o$, ϕ being an arbitrary assignment, since therefore for every cwff A_o of Γ we have $\Phi([A_o]) = T$, and since Λ is a subset of Γ, it follows that $V_\phi(A_o)$ is T for any element A_o of Λ; i.e., Λ is satisfiable with respect to the model $\{D_\alpha\}$.

THEOREM 2. *For any* wff A_o, *we have* $\vdash A_o$ *if and only if* A_o *is valid in the general sense.*

From the definition of validity, we easily see that A_o is valid if and only if the cwff $(x_{\alpha_1}) \cdots (x_{\alpha_n})A_o$ is valid, where $x_{\alpha_1}, \cdots, x_{\alpha_n}$ are the variables with free occurrences in A_o; and hence A_o is valid if and only if $V_\phi(\sim(x_{\alpha_1}) \cdots (x_{\alpha_n})A_o)$ is F for every assignment ϕ with respect to every general model $\{D_\alpha\}$. By Theorem 1, this condition implies that the set Δ whose only element is the cwff $\sim(x_{\alpha_1}) \cdots (x_{\alpha_n})A_o$ is inconsistent and hence, in particular, $\sim(x_{\alpha_1}) \cdots (x_{\alpha_n})A_o \vdash (x_{\alpha_1}) \cdots (x_{\alpha_n})A_o$. Now applying VII, 30, and 31 (several times), we see that if A_o is valid, then $\vdash A_o$. The converse can be verified directly by checking the validity of the axioms and noticing that the rules of inference operating on valid formulas lead only to valid formulas.

THEOREM 3. *A set* Γ *of* cwffs *is satisfiable with respect to some model of denumerable domains* D_α *if and only if every finite subset* Λ *of* Γ *is satisfiable.*

By Theorem 1, if Γ is not satisfiable with respect to some model of denumerable domains, then Γ is inconsistent so that, in particular, $\Gamma \vdash (x_o)x_o$. Since the formal proof of $(x_o)x_o$ contains only a finite number of formulas, there must be some finite subset $\Lambda = \{A_1, \cdots, A_n\}$ of Γ such that $A_1, \cdots, A_n \vdash (x_o)x_o$, whence by repeated applications of VII, $\vdash A_1 \supset . \cdots \supset . A_n \supset (x_o)x_o$. But then by Theorem 2, the cwff $A_1 \supset . \cdots . A_n \supset (x_o)x_o$ is valid so that we must have some $V_\phi(A_i) = F$, $i = 1, \cdots, n$, for any ϕ with respect to any model; i.e., Λ is not satisfiable. Thus, if every finite subset Λ of Γ is satisfiable, then Γ is satisfiable with respect to a model of denumerable domains. The converse is immediate.

If Γ is satisfiable, then so are its finite subsets, and hence Γ is satisfiable with respect to some model of denumerable domains. This may be taken as a generalization of the Skolem-Löwenheim theorem for the first order functional calculus.

Analogues of Theorems 1, 2, and 3 can be proved for various formal systems which differ in one respect or another from the system which we have here considered in detail. In the first place, we may add an arbitrary set of *constants* S_α as new primitive symbols. In case the set of constants is infinite, we must replace the condition of denumerability, in the statement of Theorems 1 and 3, by the condition that the domains of the model will have a cardinality not greater than that of the set of constants. The proofs for such systems are exactly like the ones given here.

In the second place, the symbols $\iota_{\alpha(o\alpha)}$ and the axioms of choice (11^α) may be dropped. In this case, we have to complicate the proof by first performing a construction which involves forming a sequence of formal systems built up from the given one by adjoining certain constants u_α^{ij}, $i, j = 1, 2, \cdots$, and providing suitable axioms for them. The details can be obtained by consulting the paper mentioned in footnote 6.

The axioms of extensionality (10^α) can be dropped if we are willing to admit models whose domains contain functions which are regarded as distinct even though they have the same value for every argument.

Finally, the functional abstraction of the present system may either be replaced by set-abstraction or dropped altogether. In the latter case, the constants $\Pi_{o(o\alpha)}$ must be replaced by a primitive notion of quantifiers.

Theorem 3 can be applied to throw light on formalized systems of number theory.

The concepts of elementary number theory may be introduced into the pure functional calculus of order ω by definition, a form particularly suited to the present formulation being given in Church.[7] Under this approach, the natural numbers are identified with certain functions. Alternatively we may choose to identify the natural numbers with the individuals making up the domain of type ι. In such a system, it is convenient to construct an applied calculus by introducing the constants 0_ι and $S_{\iota\iota}$ and adding the following formal equivalents of Peano's postulates:

P1. $(x_\iota) \,.\, \sim S_{\iota\iota}x_\iota = 0_\iota$

P2. $(x_\iota)(y_\iota) \,.\, S_{\iota\iota}x_\iota = S_{\iota\iota}y_\iota \supset x_\iota = y_\iota$

P3. $(f_{o\iota}) \,.\, f_{o\iota}0_\iota \supset \,.\, (x_\iota)[f_{o\iota}x_\iota \supset f_{o\iota}(S_{\iota\iota}x_\iota)] \supset (x_\iota)f_{o\iota}x_\iota$.

The Peano axioms are generally thought to characterize the number-sequence fully in the sense that they form a categorical axiom set any two models for which are isomorphic. As Skolem[9] points out, however, this condition obtains only if "set"—as it appears in the axiom of complete induction (our P3)—is interpreted with its standard meaning. Since, however, the scope ("all sets of individuals") of the quantifier $(f_{o\iota})$ may vary from one general model to another,[10] it follows that we may expect non-standard models for the Peano axioms.

This argument may be somewhat clearer if we consider in detail the usual proof of the categoricity of Peano's postulates. One easily shows that any model for the axioms must *contain* a sequence of the order-type of the natural numbers by considering the individuals 0_ι , $S_{\iota\iota}0_\iota$, $S_{\iota\iota}(S_{\iota\iota}0_\iota)$, $\cdots$ and using P1 and P2 to show them distinct and without other predecessors. Then the proof continues as follows.

Suppose that the domain of individuals contained elements other than those of this sequence (which we may as well identify with the natural numbers themselves). Then consider the class of individuals consisting of just the natural

[9] Thoralf Skolem, *Über einige Grundlagenfragen der Mathematik, Skrifter utgitt av Det Norske Videnskaps-Akademi*, I, no. 4 (1929), 49 pp.

[10] Here we are identifying a set X of elements of D_ι with the function (element of $D_{o\iota}$) which maps every element of X into T and every other element of D_ι into F.

numbers. Since it contains zero (0_ι) and is closed under the successor function ($S_{\iota\iota}$), we infer from the axiom of complete induction (P3) that it contains all individuals, contrary to the hypothesis that some individuals were not numbers.

By examining this proof, we see that we can conclude only that if a general model satisfies Peano's axioms and at the same time possesses a domain of individuals not isomorphic to the natural numbers, then the domain $D_{o\iota}$ of sets of individuals *cannot* contain the set consisting of just those individuals which are numbers.

Although Skolem indicates that the meaning of "natural number" is relative to the variable meaning of "set" he does not give any example of a non-standard number system satisfying all of Peano's axioms. In two later papers,[11] however, he proves that it is impossible to characterize the natural number sequence by any denumerable system of axioms formulated within the first order functional calculus (to which may be added any set of functional constants denoting numerical functions and relations), the individual variables ranging over the "numbers" themselves. Skolem makes ingenious use of a theorem on sequences of functions (which he had previously proved) to construct, for each set of axioms for the number sequence (of the type described above) a set of numerical functions which satisfy the axioms, but have a different order type than the natural numbers. This result, for axiom systems which do not involve class variables, cannot be regarded as being at all paradoxical since the claim had never been made that such systems were categorical.

By appealing to Theorem 3, however, it becomes a simple matter to construct a model containing a non-standard number system which will satisfy all of the Peano postulates as well as any preassigned set of further axioms (which may include constants for special functions as well as constants and variables of higher type). We have only to adjoin a new primitive constant u_ι and add to the given set of axioms the infinite list of formulas $u_\iota \neq 0_\iota$, $u_\iota \neq S_{\iota\iota}0_\iota$, $u_\iota \neq S_{\iota\iota}(S_{\iota\iota}0_\iota)$, $\cdots$. Since any finite subset of the enlarged system of formulas is clearly satisfiable, it follows from theorem that some denumerable model satisfies the full set of formulas, and such a model has the properties sought. By adding a non-denumerable number of primitive constants v_ι^ξ together with all formulas $v_\iota^{\xi_1} \neq v_\iota^{\xi_2}$ for $\xi_1 \neq \xi_2$, we may even build models for which the Peano axioms are valid and which contain a number system having any given cardinal.[12]

These same remarks suffice to show more generally that no mathematical axiom system can be genuinely categorical (determine its models to within isomorphism) unless it constrains its domain of elements to have some definite

[11] Thoralf Skolem, *Über die Unmöglichkeit einer vollständigen Charakterisierung der Zahlenreihe mittels eines endlichen Axiomensystems*, ***Norsk matematisk forenings skrifter***—series 2 no. 10 (1933), pp. 73–82. And *Über die Nicht-charakterisierbarkeit der Zahlenreihe mittels endlich oder abzählbar unendlich vieler Aussagen mit ausschliesslich Zahlenvariablen*, ***Fundamenta mathematicae***, vol. 23 (1934), pp. 150–161.

[12] A similar result for formulations of arithmetic within the first order functional calculus was established by A. Malcev, *Untersuchungen aus dem Gebiete der mathematischen Logik*, ***Recueil mathématique***, n.s. vol. 1 (1936), pp. 323–336. Malcev's method of proof bears a certain resemblance to the method used above. I am indebted to Professor Church for bringing this paper to my attention. (Added February 14, 1950.)

finite cardinal number—provided that the logical notions of set and function are axiomatized along with the specific mathematical notions.

The existence of non-standard models satisfying axiom-systems for number theory throws new light on the phenomenon of ω-inconsistency, first investigated by Tarski and Gödel. A formal system is ω-inconsistent if for some formula $A_{o\iota}$ the formulas $A_{o\iota}0_\iota$, $A_{o\iota}(S_{\iota\iota}0_\iota)$, $A_{o\iota}(S_{\iota\iota}(S_{\iota\iota}0_\iota))$, $\cdots$, $\sim(x_\iota)A_{o\iota}x_\iota$ are all provable. Tarski, and later Gödel, showed the existence of consistent systems which were ω-inconsistent. We can now see that such systems can and must be interpreted as referring to a non-standard number system whose elements include the natural numbers as a proper subset.

It is generally recognized that all theorems of number theory now in the literature can be formalized and proved within the functional calculus of order ω with axioms P1–P3 added. (In fact, much weaker systems suffice.) On the one hand, it follows from Theorem 1 that these theorems can be re-interpreted as true assertions about a great variety of number-systems other than the natural numbers. On the other hand, it follows from the results of Gödel[4] that there are true theorems about the natural numbers which cannot be proved by extant methods (consistency assumed).

Now Gödel's proof furnishes certain special formulas which are shown to be true but unprovable, but there is no general method indicated for establishing that a given theorem cannot be proved from given axioms. From Theorem 1, we see that such a method is supplied by the procedure of constructing non-standard models for number theory in which "set" and "function" are reinterpreted. It, therefore, becomes of practical interest to number-theorists to study the structure of such models.

A detailed investigation of these numerical structures is beyond the scope of the present paper. As an example, however, we quote one simple result: Every non-standard denumerable model for the Peano axioms has the order type $\omega + (\omega^* + \omega)\eta$, where η is the type of the rationals.

PRINCETON UNIVERSITY

Part II

Cut Elimination and Nonstandard Models

THE JOURNAL OF SYMBOLIC LOGIC
Volume 36, Number 3, Sept. 1971

RESOLUTION IN TYPE THEORY

PETER B. ANDREWS[1]

§1. Introduction. In [8] J. A. Robinson introduced a complete refutation procedure called *resolution* for first order predicate calculus. Resolution is based on ideas in Herbrand's Theorem, and provides a very convenient framework in which to search for a proof of a wff believed to be a theorem. Moreover, it has proved possible to formulate many refinements of resolution which are still complete but are more efficient, at least in many contexts. However, when efficiency is a prime consideration, the restriction to first order logic is unfortunate, since many statements of mathematics (and other disciplines) can be expressed more simply and naturally in higher order logic than in first order logic. Also, the fact that in higher order logic (as in many-sorted first order logic) there is an explicit syntactic distinction between expressions which denote different types of intuitive objects is of great value where matching is involved, since one is automatically prevented from trying to make certain inappropriate matches. (One may contrast this with the situation in which mathematical statements are expressed in the symbolism of axiomatic set theory.)

In this paper we shall introduce a refutation system $\mathscr{R}$ for type theory which may be regarded as a generalization of resolution to type theory, and prove that $\mathscr{R}$ is complete in the (weak) sense that in $\mathscr{R}$ one can refute any sentence $\sim$A such that A is provable in a more conventional system $\mathscr{T}$ of type theory. For $\mathscr{T}$ we take the elegant and expressive formulation of type theory introduced by Church in [2], but use only Axioms 1–6. It should be noted that because substitution with λ-conversion is a much more complicated operation than substitution alone, the matching problem, which was completely solved for first order logic by Robinson's Unification Theorem [8], remains a major problem in the context of the system $\mathscr{R}$. (Some appreciation of the complexity of the situation can be gained from [3].) In this sense $\mathscr{R}$ is not as useful for refuting wffs of type theory as resolution is for refuting wffs of first order logic.

In §2 we review certain facts about the system $\mathscr{T}$ and λ-conversion. In §3 we prove a theorem which is (at least in conjunction with the results of Henkin in [4]) an extension to $\mathscr{T}$ of Smullyan's Unifying Principle in Quantification Theory ([10] and [11, Chapter VI]). Our proof relies heavily on ideas of Takahashi [12] as well as Smullyan, which is not surprising since the Unifying Principle is closely related to cut-elimination. $\mathscr{T}$ is a somewhat richer formulation of type theory than Schütte's formulation in [9] which Takahashi treats in [12], since in $\mathscr{T}$ for

Received July 17, 1970.
[1] This research was partially supported by NSF Grant GJ-580.

all types α and β there is a type $(\alpha\beta)$ of functions from elements of type β to elements of type α. Therefore we verify the details of this argument rather carefully, although there is a close parallel with Takahashi's argument. We apply the theorem in §4 to prove cut-elimination for $\mathscr{T}$, and in §5 to prove the completeness of $\mathscr{R}$. (Except for the preliminary definitions, §4 can be skipped by those interested primarily in $\mathscr{R}$.) In §6 we present some examples of refutations in $\mathscr{R}$.

§2. The system $\mathscr{T}$. For the convenience of the reader we here provide a condensed description of the system $\mathscr{T}$, with a few trivial notational changes from [2]. A more complete discussion of $\mathscr{T}$ can be found in [2] or [4]. The systems $\mathscr{G}$ and $\mathscr{R}$ in §4 and §5 will have the same wffs as $\mathscr{T}$.

2.1. We use α, β, γ, etc. (but not o or ι), as syntactical variables ranging over *type symbols*, which are defined inductively as follows:

2.1.1. o is a type symbol (denoting the type of truth values).
2.1.2. ι is a type symbol (denoting the type of individuals).
2.1.3. $(\alpha\beta)$ is a type symbol (denoting the type of functions from elements of type β to elements of type α).

2.2. The *primitive symbols* of $\mathscr{T}$ are the following:

2.2.1. *Improper symbols*: [ ] λ.
2.2.2. For each α, a denumerable list of *variables* of type α:

$$f_\alpha g_\alpha h_\alpha \cdots x_\alpha y_\alpha z_\alpha f^1_\alpha g^1_\alpha \cdots z^1_\alpha f^2_\alpha \cdots.$$

We shall write *variable*$_\alpha$ as an abbreviation for *variable of type* α. We shall use $\mathbf{f}_\alpha, \mathbf{g}_\alpha, \cdots, \mathbf{x}_\alpha, \mathbf{y}_\alpha, \mathbf{z}_\alpha$, etc., as syntactical variables for variables$_\alpha$.

2.2.3. *Logical constants*: $\sim_{(oo)}$ $\vee_{((oo)o)}$ $\Pi_{(o(o\alpha))}$.
2.2.4. In addition there may be other constants of various types, which we call *nonlogical constants* or *parameters*.

2.3. We write *wff*$_\alpha$ as an abbreviation for *wff of type* α, and use $\mathbf{A}_\alpha$, $\mathbf{B}_\alpha$, $\mathbf{C}_\alpha$, etc., as syntactical variables ranging over wffs$_\alpha$, which are defined inductively as follows:

2.3.1. A primitive variable or constant of type α is a wff$_\alpha$.
2.3.2. $[\mathbf{A}_{\alpha\beta}\mathbf{B}_\beta]$ is a wff$_\alpha$.
2.3.3. $[\lambda\mathbf{x}_\beta\mathbf{A}_\alpha]$ is a wff$_{(\alpha\beta)}$.

We shall assume given a fixed enumeration of the wffs of $\mathscr{T}$. This also provides an enumeration of the variables and constants of each type.

An occurrence of $\mathbf{x}_\alpha$ is *bound* (*free*) in $\mathbf{B}_\beta$ iff it is (is not) in a wf part of $\mathbf{B}_\beta$ of the form $[\lambda\mathbf{x}_\alpha\mathbf{C}_\delta]$. A wff is *closed* iff no variable occurs free in it. A *sentence* is a closed wff$_o$.

2.4 *Definitions and abbreviations.*

2.4.1. Brackets (and parentheses in type symbols) may be omitted when no ambiguity is thereby introduced. A dot stands for a left bracket whose mate is as far to the right as is consistent with the pairing of brackets already present and with the formula being well formed. Otherwise brackets and parentheses are to be restored using the convention of association to the left.

2.4.2. Type symbols may be omitted when the context indicates what they should be. The type symbol o will usually be omitted.
2.4.3. $[\mathbf{A}_o \vee \mathbf{B}_o]$ stands for $[[\vee_{((oo)o)}\mathbf{A}_o]\mathbf{B}_o]$.
2.4.4. $[\mathbf{A}_o \supset \mathbf{B}_o]$ stands for $[[\sim_{oo}\mathbf{A}_o] \vee \mathbf{B}_o]$.
2.4.5. $[\forall \mathbf{x}_\alpha \mathbf{A}_o]$ stands for $[\Pi_{(o(o\alpha))}[\lambda \mathbf{x}_\alpha \mathbf{A}_o]]$.
2.4.6. Other propositional connectives, and the existential quantifier, are defined in familiar ways.
2.4.7. $Q_{o\alpha\alpha}$ stands for $[\lambda x_\alpha \lambda y_\alpha \forall f_{o\alpha} \centerdot f_{o\alpha} x_\alpha \supset f_{o\alpha} y_\alpha]$.
2.4.8. $[\mathbf{A}_\alpha = \mathbf{B}_\alpha]$ stands for $Q_{o\alpha\alpha}\mathbf{A}_\alpha\mathbf{B}_\alpha$.
2.4.9. $\exists_1 \mathbf{x}_\alpha \mathbf{A}_o$ stands for $[\lambda p_{o\alpha} \centerdot \exists y_\alpha \centerdot p_{o\alpha} y_\alpha \wedge \forall z_\alpha \centerdot p_{o\alpha} z_\alpha \supset z_\alpha = y_\alpha][\lambda \mathbf{x}_\alpha \mathbf{A}_o]$.
2.4.10. $S^{\mathbf{x}_\alpha}_{\mathbf{A}_\alpha}\mathbf{B}_\beta (\acute{S}^{\mathbf{x}_\alpha}_{\mathbf{A}_\alpha}\mathbf{B}_\beta)$ denotes the result of substituting $\mathbf{A}_\alpha$ for $\mathbf{x}_\alpha$ at all (all free) occurrences of $\mathbf{x}_\alpha$ in $\mathbf{B}_\beta$.
2.4.11. $\mathbf{A}_\alpha$ is *free for* $\mathbf{x}_\alpha$ *in* $\mathbf{B}_\beta$ iff no free occurrence of $\mathbf{x}_\alpha$ in $\mathbf{B}_\beta$ is in a wf part of $\mathbf{B}_\beta$ of the form $[\lambda \mathbf{y}_\gamma \mathbf{C}_\delta]$ such that $\mathbf{y}_\gamma$ is a free variable of $\mathbf{A}_\alpha$.

2.5. *Axioms of* $\mathcal{T}$.

2.5.1. $p \vee p \supset p$.
2.5.2. $p \supset p \vee q$.
2.5.3. $p \vee q \supset q \vee p$.
2.5.4. $p \supset q \supset [r \vee p \supset r \vee q]$.
2.5.5$^\alpha$. $\Pi_{o(o\alpha)} f_{o\alpha} \supset f_{o\alpha} x_\alpha$.
2.5.6$^\alpha$. $\forall x_\alpha [p \vee f_{o\alpha} x_\alpha] \supset p \vee \Pi_{o(o\alpha)} f_{o\alpha}$.

2.6. *Rules of inference of* $\mathcal{T}$.

2.6.1. *Alphabetic change of bound variables.* To replace any wf part $[\lambda \mathbf{x}_\beta \mathbf{A}_\alpha]$ of a wff by $[\lambda \mathbf{y}_\beta S^{\mathbf{x}_\beta}_{\mathbf{y}_\beta}\mathbf{A}_\alpha]$, provided that $\mathbf{y}_\beta$ does not occur in $\mathbf{A}_\alpha$ and $\mathbf{x}_\beta$ is not bound in $\mathbf{A}_\alpha$.
2.6.2. λ*-contraction.* To replace any wf part $[[\lambda \mathbf{x}_\alpha \mathbf{B}_\beta]\mathbf{A}_\alpha]$ of a wff by $S^{\mathbf{x}_\alpha}_{\mathbf{A}_\alpha}\mathbf{B}_\beta$, provided that the bound variables of $\mathbf{B}_\beta$ are distinct both from $\mathbf{x}_\alpha$ and from the free variables of $\mathbf{A}_\alpha$.
2.6.3. λ*-expansion.* To infer $\mathbf{C}$ from $\mathbf{D}$ if $\mathbf{D}$ can be inferred from $\mathbf{C}$ by a single application of 2.6.2.
2.6.4. *Substitution.* From $\mathbf{F}_{o\alpha}\mathbf{x}_\alpha$ to infer $\mathbf{F}_{o\alpha}\mathbf{A}_\alpha$, provided that $\mathbf{x}_\alpha$ is not a free variable of $\mathbf{F}_{o\alpha}$.
2.6.5. *Modus Ponens.* From $[\mathbf{A} \supset \mathbf{B}]$ and $\mathbf{A}$ to infer $\mathbf{B}$.
2.6.6. *Generalization.* From $\mathbf{F}_{o\alpha}\mathbf{x}_\alpha$ to infer $\Pi_{o(o\alpha)}\mathbf{F}_{o\alpha}$, provided that $\mathbf{x}_\alpha$ is not a free variable of $\mathbf{F}_{o\alpha}$.

Remark. It can be proved that $\Pi_{o(o\alpha)}\mathbf{F}_{o\alpha}$ is equivalent to $\forall \mathbf{x}_\alpha \mathbf{F}_{o\alpha}\mathbf{x}_\alpha$ if $\mathbf{x}_\alpha$ is not free in $\mathbf{F}_{o\alpha}$.

2.7. λ*-conversion.*

2.7.1. Rules 2.6.1–2.6.3 are λ*-conversion rules.* We write $\mathbf{A}_\alpha$ *conv* $\mathbf{B}_\alpha$ (resp. $\mathbf{A}_\alpha$ *conv-I–II* $\mathbf{B}_\alpha$) (resp. $\mathbf{A}_\alpha$ *conv-I* $\mathbf{B}_\alpha$) iff there is a sequence of applications of rules 2.6.1–2.6.3 (resp. 2.6.1–2.6.2) (resp. 2.6.1) which transforms $\mathbf{A}_\alpha$ into $\mathbf{B}_\alpha$. It is well known that *conv* and *conv-I* are equivalence relations.

2.7.2. A *contractible part* of a wff $\mathbf{C}_\gamma$ is an occurrence of a wff of the form $[[\lambda \mathbf{x}_\alpha \mathbf{B}_\beta]\mathbf{A}_\alpha]$ in $\mathbf{C}_\gamma$. We say $\mathbf{C}_\gamma$ is in λ*-normal form* iff it has no contractible parts.

2.7.3. PROPOSITION. *For each* wff $\mathbf{D}_\gamma$ *there is a* wff $\mathbf{C}_\gamma$ *in λ-normal form such that* $\mathbf{D}_\gamma$ *conv-I–II* $\mathbf{C}_\gamma$.

PROOF.[2] Define $\#[[\lambda \mathbf{x}_\alpha \mathbf{B}_\beta]\mathbf{A}_\alpha]$ to be the number of occurrences of (in ($\beta\alpha$). Let $m(\mathbf{D}_\gamma) = \max\{\#\mathbf{G}_\beta \mid \mathbf{G}_\beta$ is a contractible part of $\mathbf{D}_\gamma\}$. We say that a contractible part $\mathbf{G}_\beta$ of $\mathbf{D}_\gamma$ is *maximal* in $\mathbf{D}_\gamma$ iff $\#\mathbf{G}_\beta = m(\mathbf{D}_\gamma)$. Let $n(\mathbf{D}_\gamma)$ be the number of maximal contractible parts in $\mathbf{D}_\gamma$. The proof is by induction on $p(\mathbf{D}_\gamma) = \omega \cdot m(\mathbf{D}_\gamma) + n(\mathbf{D}_\gamma)$. Clearly $\mathbf{D}_\gamma$ is in λ-normal form iff $p(\mathbf{D}_\gamma) = 0$.

If $p(\mathbf{D}_\gamma) > 0$, let $[[\lambda \mathbf{x}_\alpha \mathbf{B}_\beta]\mathbf{A}_\alpha]$ be that maximal contractible part $\mathbf{G}_\beta$ of $\mathbf{D}_\gamma$ which occurs farthest to the right in $\mathbf{D}_\gamma$, with the position of a contractible part being determined by the leftmost occurrence of λ in it. By applying 2.6.1 if necessary we may assume that 2.6.2 may be applied to obtain from $\mathbf{D}_\gamma$ a wff $\mathbf{E}_\gamma$ in which $\mathbf{G}_\beta$ has been replaced by $\mathcal{S}^{\mathbf{x}_\alpha}_{\mathbf{A}_\alpha}\mathbf{B}_\beta$. Thus $\mathbf{D}_\gamma$ conv-I–II $\mathbf{E}_\gamma$, and it must be shown that $p(\mathbf{E}_\gamma) < p(\mathbf{D}_\gamma)$.

For the sake of brevity, we shall not explicitly distinguish wffs from occurrences of wffs at certain points in the following argument.

We first prove that

(*) for each wf part $\mathbf{C}_\delta$ of $\mathbf{B}_\beta$, $\mathcal{S}^{\mathbf{x}_\alpha}_{\mathbf{A}_\alpha}\mathbf{C}_\delta$ contains no contractible part $\mathbf{H}_\tau$ with $\#\mathbf{H}_\tau \geq m(\mathbf{D}_\gamma)$. The proof is by induction on the construction of $\mathbf{C}_\delta$.

Case (a). $\mathbf{C}_\delta$ is $\mathbf{x}_\alpha$. Then $\mathcal{S}^{\mathbf{x}_\alpha}_{\mathbf{A}_\alpha}\mathbf{C}_\delta$ is $\mathbf{A}_\alpha$, and (*) holds by virtue of the definition of $\mathbf{G}_\beta$.

Case (b). $\mathbf{C}_\delta$ is a primitive constant or variable other than $\mathbf{x}_\alpha$. $\mathcal{S}^{\mathbf{x}_\alpha}_{\mathbf{A}_\alpha}\mathbf{C}_\delta$ is $\mathbf{C}_\delta$, so (*) holds trivially.

Case (c). $\mathbf{C}_\delta$ has the form $[\lambda \mathbf{y}_\kappa \mathbf{M}_\sigma]$. Note that $\mathbf{y}_\kappa$ cannot be $\mathbf{x}_\alpha$ by the restriction on 2.6.2. $\mathcal{S}^{\mathbf{x}_\alpha}_{\mathbf{A}_\alpha}\mathbf{C}_\delta$ is $[\lambda \mathbf{y}_\kappa \mathcal{S}^{\mathbf{x}_\alpha}_{\mathbf{A}_\alpha}\mathbf{M}_\sigma]$, so (*) holds by the inductive hypothesis applied to $\mathbf{M}_\sigma$.

Case (d). $\mathbf{C}_\delta$ has the form $[\mathbf{M}_{\delta\epsilon}\mathbf{N}_\epsilon]$ Then $\mathcal{S}^{\mathbf{x}_\alpha}_{\mathbf{A}_\alpha}\mathbf{C}_\delta$ is $[(\mathcal{S}^{\mathbf{x}_\alpha}_{\mathbf{A}_\alpha}\mathbf{M}_{\delta\epsilon})\mathcal{S}^{\mathbf{x}_\alpha}_{\mathbf{A}_\alpha}\mathbf{N}_\epsilon]$, and the inductive hypothesis applies to $\mathbf{M}_{\delta\epsilon}$ and $\mathbf{N}_\epsilon$, so we need only consider the possibility that $\mathcal{S}^{\mathbf{x}_\alpha}_{\mathbf{A}_\alpha}\mathbf{C}_\delta$ is itself a contractible part $[[\lambda \mathbf{y}_\epsilon \mathbf{P}_\delta]\mathcal{S}^{\mathbf{x}_\alpha}_{\mathbf{A}_\alpha}\mathbf{N}_\epsilon]$, where $\mathcal{S}^{\mathbf{x}_\alpha}_{\mathbf{A}_\alpha}\mathbf{M}_{\delta\epsilon} = [\lambda \mathbf{y}_\epsilon \mathbf{P}_\delta]$, with $\#(\mathcal{S}^{\mathbf{x}_\alpha}_{\mathbf{A}_\alpha}\mathbf{C}_\delta) \geq m(\mathbf{D}_\gamma)$. Since $\mathbf{M}_\delta$ has one of the forms 2.3.1–2.3.3, $\mathcal{S}^{\mathbf{x}_\alpha}_{\mathbf{A}_\alpha}\mathbf{M}_{\delta\epsilon}$ can have the form $[\lambda \mathbf{y}_\epsilon \mathbf{P}_\delta]$ in only two ways:

(i) $\mathbf{M}_{\delta\epsilon}$ is $\mathbf{x}_\alpha$ and $\mathbf{A}_\alpha$ is $[\lambda \mathbf{y}_\epsilon \mathbf{P}_\delta]$. Then $\alpha = (\delta\epsilon)$ so $\#(\mathcal{S}^{\mathbf{x}_\alpha}_{\mathbf{A}_\alpha}\mathbf{C}_\delta) < m(\mathbf{D}_\gamma)$.

(ii) $\mathbf{M}_{\delta\epsilon}$ is $[\lambda \mathbf{y}_\epsilon \mathbf{Q}_\delta]$ and $\mathcal{S}^{\mathbf{x}_\alpha}_{\mathbf{A}_\alpha}\mathcal{Q}_\delta = \mathbf{P}_\delta$. In this case $\mathbf{C}_\delta = [[\lambda \mathbf{y}_\epsilon \mathcal{Q}_\delta]\mathbf{N}_\epsilon]$ and $\#\mathbf{C}_\delta \geq m(\mathbf{D}_\gamma)$. But since $\mathbf{C}_\delta$ is a part of $\mathbf{B}_\beta$ this contradicts the definition of $\mathbf{G}_\beta$. Thus neither possibility can occur, and (*) holds in case (d) also.

For each wff $\mathbf{C}_\delta$ we let $k(\mathbf{C}_\delta)$ be the number of contractible parts $\mathbf{H}_\tau$ of $\mathbf{C}_\delta$ with $\#\mathbf{H}_\tau = m(\mathbf{D}_\gamma)$. For any wf part $\mathbf{C}_\delta$ of $\mathbf{D}_\gamma$ which contains $\mathbf{G}_\beta$, we let $\mathbf{C}'_\delta$ be the result of replacing $\mathbf{G}_\beta$ in $\mathbf{C}_\delta$ by $\mathcal{S}^{\mathbf{x}_\alpha}_{\mathbf{A}_\alpha}\mathbf{B}_\beta$, and prove

(**) $k(\mathbf{C}'_\delta) + 1 = k(\mathbf{C}_\delta)$ and $\mathbf{C}'_\delta$ contains no contractible parts $\mathbf{H}_\tau$ with $\#\mathbf{H}_\tau > m(\mathbf{D}_\gamma)$. The proof is by induction on the construction of $\mathbf{C}_\delta$.

Case (a). $\mathbf{C}_\delta$ is $\mathbf{G}_\beta$, so $\mathbf{C}'_\delta$ is $\mathcal{S}^{\mathbf{x}_\alpha}_{\mathbf{A}_\alpha}\mathbf{B}_\beta$. Then (**) follows directly from (*).

Case (b). $\mathbf{C}_\delta$ is $[\mathbf{G}_\beta \mathbf{N}_\epsilon]$, and β has the form ($\delta\epsilon$). Thus $\mathbf{C}'_\delta$ is $[(\mathcal{S}^{\mathbf{x}_\alpha}_{\mathbf{A}_\alpha}\mathbf{B}_\beta)\mathbf{N}_\epsilon]$. If $\mathbf{C}'_\delta$ is

[2] This proposition is part of the folklore of type-theoretic λ-conversion. The author first heard the idea of the proof given here from Dr. James R. Guard.

contractible, $\#C'_\delta$ = the number of occurrences of (in β, which is less than $m(\mathbf{D}_\gamma)$, so

$$\begin{aligned} k(\mathbf{C}'_\delta) + 1 &= k(S^{\mathbf{x}_\alpha}_{\mathbf{A}_\alpha}\mathbf{B}_\beta) + k(\mathbf{N}_\epsilon) + 1 \\ &= (\text{by } (*))\ 0 + k(\mathbf{N}_\epsilon) + 1 \\ &= k(\mathbf{N}_\epsilon) + k(\mathbf{G}_\beta) = k(\mathbf{C}_\delta) \end{aligned}$$

and (**) is easily seen to be true.

The remaining cases involve trivial applications of the inductive hypothesis, and are left to the reader.

$\mathbf{D}'_\gamma$ is $\mathbf{E}_\gamma$, so by (**) $k(\mathbf{E}_\gamma) + 1 = k(\mathbf{D}_\gamma) = n(\mathbf{D}_\gamma)$ and $m(\mathbf{E}_\gamma) \leq m(\mathbf{D}_\gamma)$. If $m(\mathbf{E}_\gamma) = m(\mathbf{D}_\gamma)$ then $n(\mathbf{E}_\gamma) = k(\mathbf{E}_\gamma) < n(\mathbf{D}_\gamma)$; hence whether $m(\mathbf{E}_\gamma) = m(D_\gamma)$ or $m(\mathbf{E}_\gamma) < m(\mathbf{D}_\gamma)$ we have $p(\mathbf{E}_\gamma) < p(\mathbf{D}_\gamma)$. Therefore by inductive hypothesis $\mathbf{E}_\gamma$ is conv-I–II to a wff in λ-normal form, so $\mathbf{D}_\gamma$ is also.

2.7.4. CHURCH–ROSSER THEOREM. *If B_γ and $\mathbf{C}_\gamma$ are in λ-normal form and $\mathbf{B}_\gamma$ conv $\mathbf{C}_\gamma$, then $\mathbf{B}_\gamma$ conv-I $\mathbf{C}_\gamma$. That is, a λ-normal form of a* wff *is unique up to alphabetic changes of bound variables.*

This theorem was originally proved for a different system of λ-conversion without type symbols but it is known that it applies to $\mathscr{T}$ also. See [5] and the references cited therein.

2.7.5. η-wffs. A wff $\mathbf{A}_\alpha$ of $\mathscr{T}$ is an *η-wff* iff $\mathbf{A}_\alpha$ is in λ-normal form and for each wf part $[\lambda\mathbf{x}_\beta\mathbf{C}_\gamma]$ of $\mathbf{A}_\alpha$, $\mathbf{x}_\beta$ is the first variable$_\beta$ in alphabetic order which is distinct from the other free variables$_\beta$ of $\mathbf{C}_\gamma$. Using 2.7.3, 2.7.4, and 2.6.1 it is easy to see that for each wff $\mathbf{A}_\alpha$ there is a unique η-wff $\mathbf{B}_\alpha$ such that $\mathbf{A}_\alpha$ conv $\mathbf{B}_\alpha$. We write $\mathbf{B}_\alpha = \eta\mathbf{A}_\alpha$. (To convert a wff in λ-normal form into an η-wff, proceed from left to right to decide what each bound variable should be; however some additional temporary changes of bound variables may be necessary before these changes can be made.)

η-wffs have the following pleasing properties. If $\mathbf{A}_\alpha$ is an η-wff, then every wf part of $\mathbf{A}_\alpha$ is an η-wff. $\eta[\mathbf{A}_{\alpha\beta}\mathbf{B}_\beta] = [(\eta\mathbf{A}_{\alpha\beta})(\eta\mathbf{B}_\beta)]$ if $[\mathbf{A}_{\alpha\beta}\mathbf{B}_\beta]$ is not contractible.

2.8. *Wffs$_o$*.

2.8.1. A wff$_o$ [$\mathbf{A}_o$ is *atomic* (an *atom*) iff the leftmost primitive symbol of $\mathbf{A}_o$ which is not a bracket is a variable or parameter.

2.8.2. Every wff$_o$ [$\mathbf{D}_o$ of $\mathscr{T}$ in λ-normal form has one of the following forms:

(a) $\mathbf{A}$, where $\mathbf{A}$ is atomic.

(b) $\sim\mathbf{B}$.

(c) $\mathbf{B} \vee \mathbf{C}$.

(d) $\Pi_{o(o\alpha)}\mathbf{B}_{o\alpha}$.

PROOF. The leftmost primitive symbol of $\mathbf{D}$ which is not a bracket cannot be λ, so it must be a variable, parameter, $\sim_{oo}$, $\vee_{ooo}$, or $\Pi_{o(o\alpha)}$.

2.9. A set $\mathscr{S}$ of wffs$_o$ is *inconsistent* iff there is a finite subset $\{\mathbf{A}^1, \cdots, \mathbf{A}^n\}$ of $\mathscr{S}$ such that $\vdash\sim\mathbf{A}^1 \vee \cdots \vee \sim \mathbf{A}^n$; otherwise $\mathscr{S}$ is *consistent*.

§3. Abstract consistency properties, valuations, and consistency.

3.1. DEFINITION. A property Γ of finite sets of wffs$_o$ is an *abstract consistency property* iff for all finite sets $\mathscr{S}$ of wffs$_o$ the following properties hold (for all wffs $\mathbf{A}, \mathbf{B}$):

3.1.1. If $\Gamma(\mathscr{S})$, then there is no atom **A** such that $\mathbf{A} \in \mathscr{S}$ and $[\sim\mathbf{A}] \in \mathscr{S}$.
3.1.2. If $\Gamma(\mathscr{S} \cup \{\mathbf{A}\})$, then $\Gamma(\mathscr{S} \cup \{\eta\mathbf{A}\})$.
3.1.3. If $\Gamma(\mathscr{S} \cup \{\sim\sim\mathbf{A}\})$, then $\Gamma(\mathscr{S} \cup \{\mathbf{A}\})$.
3.1.4. If $\Gamma(\mathscr{S} \cup \{[\mathbf{A} \vee \mathbf{B}]\})$, then $\Gamma(\mathscr{S} \cup \{\mathbf{A}\})$ or $\Gamma(\mathscr{S} \cup \{\mathbf{B}\})$.
3.1.5. If $\Gamma(\mathscr{S} \cup \{\sim[\mathbf{A} \vee \mathbf{B}]\})$, then $\Gamma(\mathscr{S} \cup \{\sim\mathbf{A}, \sim\mathbf{B}\})$.
3.1.6. If $\Gamma(\mathscr{S} \cup \{\Pi_{o(o\alpha)}\mathbf{A}_{o\alpha}\})$, then for each wff $\mathbf{B}_\alpha$, $\Gamma(\mathscr{S} \cup \{\Pi_{o(o\alpha)}\mathbf{A}_{o\alpha}\ \mathbf{A}_{o\alpha}, \mathbf{B}_\alpha\})$.
3.1.7. If $\Gamma(\mathscr{S} \cup \{\sim\Pi_{o(o\alpha)}\mathbf{A}_{o\alpha}\})$, then $\Gamma(\mathscr{S} \cup \{\sim\mathbf{A}_{o\alpha}\mathbf{c}_\alpha\})$ for any variable or parameter $\mathbf{c}_\alpha$ which does not occur free in $\mathbf{A}_{o\alpha}$ or any wff in $\mathscr{S}$.

Remark. Satisfiability is an abstract consistency property.

The notion of an abstract consistency property is due to Smullyan. Our main theorem of this section will be that if Γ is an abstract consistency property and $\Gamma(\mathscr{S})$, then $\mathscr{S}$ is consistent. This is an analog for $\mathscr{T}$ of Smullyan's *Unifying Principle in Quantification Theory* [10].

3.2. DEFINITION. A *semivaluation* is a function V with domain some set of wffs$_o$ and range a subset of the set $\{\mathfrak{t}, \mathfrak{f}\}$ of truth values such that the following properties hold (for all wffs **A**, **B**):
3.2.1. If $V\mathbf{A}$ is defined, then $V\eta\mathbf{A} = V\mathbf{A}$.
3.2.2. If $V[\sim\mathbf{A}] = \mathfrak{t}$, then $V\mathbf{A} = \mathfrak{f}$.
3.2.3. If $V[\sim\mathbf{A}] = \mathfrak{f}$, then $V\mathbf{A} = \mathfrak{t}$.
3.2.4. If $V[\mathbf{A} \vee \mathbf{B}] = \mathfrak{t}$, then $V\mathbf{A} = \mathfrak{t}$ or $V\mathbf{B} = \mathfrak{t}$.
3.2.5. If $V[\mathbf{A} \vee \mathbf{B}] = \mathfrak{f}$, then $V\mathbf{A} = \mathfrak{f}$ and $V\mathbf{B} = \mathfrak{f}$.
3.2.6. If $V[\Pi_{o(o\alpha)}\mathbf{A}_{o\alpha}] = \mathfrak{t}$, then for each wff $\mathbf{B}_\alpha$, $V[\mathbf{A}_{o\alpha}\mathbf{B}_\alpha] = \mathfrak{t}$.
3.2.7. If $V[\Pi_{o(o\alpha)}\mathbf{A}_{o\alpha}] = \mathfrak{f}$, then there is a wff $\mathbf{B}_\alpha$ such that $V[\mathbf{A}_{o\alpha}\mathbf{B}_\alpha] = \mathfrak{f}$.
The notion of a semivaluation is due to Schütte [9].

3.3. THEOREM. *Let $\mathscr{S}$ be a finite set of* wffs$_o$ *and Γ be an abstract consistency property such that $\Gamma(\mathscr{S})$. There is a semivaluation V such that $V\mathbf{A} = \mathfrak{t}$ for all $\mathbf{A} \in \mathscr{S}$.*

PROOF (following Smullyan [10]). We may assume $\mathscr{S}$ is nonempty, since the theorem is otherwise trivial.

3.3.1. We shall inductively define finite sequences $\mathscr{S}_1, \mathscr{S}_2, \cdots$ of wffs$_o$ such that $\mathscr{S}_i$ has at least i terms and $\mathscr{S}_i$ is an initial segment of $\mathscr{S}_{i+1}$. We let $\mathbf{E}^i$ be the ith term of $\mathscr{S}_j$. For notational convenience if $\mathscr{Y}$ is a finite sequence we let $\mathscr{Y} * \mathbf{A}$ be the sequence obtained from $\mathscr{Y}$ by adding **A** as an additional term; also when we use notations which suggest that $\mathscr{Y}$ is a set we refer tacitly to the set of terms of the sequence $\mathscr{Y}$. As we define $\mathscr{S}_i$ we prove $\Gamma(\mathscr{S}_i)$.

$\mathscr{S}_1$ is to be the sequence of wffs of $\mathscr{S}$ arranged in order. $\Gamma(\mathscr{S}_1)$ since $\Gamma(\mathscr{S})$.

Given $\mathscr{S}_i$ such that $\Gamma(\mathscr{S}_i)$, we define $\mathscr{S}_{i+1}$ and prove $\Gamma(\mathscr{S}_{i+1})$ in each case below:

3.3.1.1. $\mathbf{E}^i$ is not an η-wff.
Let $\mathscr{S}_{i+1} = \mathscr{S}_i * \eta\mathbf{E}^i$. $\mathbf{E}^i \in \mathscr{S}_i$ so $\mathscr{S}_i = \mathscr{S}_i \cup \{\mathbf{E}^i\}$ so $\Gamma(\mathscr{S}_{i+1})$ by 3.1.2.

In all other cases we assume $\mathbf{E}^i$ is an η-wff.

3.3.1.2. $\mathbf{E}^i$ is an atom or the negation of an atom. Let $\mathscr{S}_{i+1} = \mathscr{S}_i * \mathbf{E}^i$.

3.3.1.3. $\mathbf{E}^i = \sim\sim\mathbf{A}$. Let $\mathscr{S}_{i+1} = \mathscr{S}_i * \mathbf{A}$. $\Gamma(\mathscr{S}_{i+1})$ by 3.1.3.

3.3.1.4. $\mathbf{E}^i = \mathbf{A} \vee \mathbf{B}$. Let $\mathscr{S}_{i+1}$ be $\mathscr{S}_i * \mathbf{A}$ if $\Gamma(\mathscr{S}_i * \mathbf{A})$; otherwise let $\mathscr{S}_{i+1} = \mathscr{S}_i * \mathbf{B}$. Then $\Gamma(\mathscr{S}_{i+1})$ by 3.1.4.

3.3.1.5. $\mathbf{E}^i = \sim[\mathbf{A} \vee \mathbf{B}]$. Let $\mathscr{S}_{i+1} = \mathscr{S}_i * \sim\mathbf{A} * \sim\mathbf{B}$. $\Gamma(\mathscr{S}_{i+1})$ by 3.1.5.

3.3.1.6. $\mathbf{E}^i = \Pi_{o(o\alpha)}\mathbf{A}_{o\alpha}$. Let $\mathbf{B}_\alpha$ be the first wff$_\alpha$ such that $[\mathbf{A}_{o\alpha}\mathbf{B}_\alpha] \notin \mathscr{S}$ and let $\mathscr{S}_{i+1} = \mathscr{S}_i * \mathbf{A}_{o\alpha}\mathbf{B}_\alpha * \mathbf{E}^i$. $\Gamma(\mathscr{S}_{i+1})$ by 3.1.6.

3.3.1.7. $\mathbf{E}^i = \sim\Pi_{o(o\alpha)}\mathbf{A}_{o\alpha}$. Let $\mathbf{x}_\alpha$ be the first variable$_\alpha$ which is not free in any wff of $\mathscr{S}_i$ and let $\mathscr{S}_{i+1} = \mathscr{S}_i * \sim\mathbf{A}_{o\alpha}\mathbf{x}_\alpha$. $\Gamma(\mathscr{S}_{i+1})$ by 3.1.7.

3.3.2. Let $\mathscr{U} = \bigcup_{i=1}^{\infty} \mathscr{S}_i$. Note that every finite subset of $\mathscr{U}$ is a subset of some set with property Γ.

3.3.3. LEMMA. *There is no* wff $\mathbf{E}$ *such that* $\mathbf{E} \in \mathscr{U}$ *and* $[\sim\mathbf{E}] \in \mathscr{U}$.

PROOF. Clearly by 3.3.1.1 if $\mathbf{E} \in \mathscr{U}$ then $\eta\mathbf{E} \in \mathscr{U}$. Also $\eta[\sim\mathbf{E}] = [\sim\eta\mathbf{E}]$, so it suffices to prove the lemma for η-wffs. We do this by induction on the number of occurrences of logical constants in $\mathbf{E}$. In each case below we suppose $\mathbf{E}$ is an η-wff and $\mathbf{E} \in \mathscr{U}$ and $\sim\mathbf{E} \in \mathscr{U}$.

3.3.3.1. $\mathbf{E}$ is atomic. By 3.3.2 there exists a set $\mathscr{Z}$ such that $\{\mathbf{E}, \sim\mathbf{E}\} \subseteq \mathscr{Z} \subseteq \mathscr{U}$ and $\Gamma(\mathscr{Z})$. This contradicts 3.1.1.

3.3.3.2. $\mathbf{E} = \sim\mathbf{A}$. Since $\sim\mathbf{E} = \sim\sim\mathbf{A} \in \mathscr{U}$, by 3.3.1.3 $\mathbf{A} \in \mathscr{U}$, which contradicts the inductive hypothesis.

3.3.3.3. $\mathbf{E} = [\mathbf{A} \vee \mathbf{B}]$. By 3.3.1.5 $\sim\mathbf{A} \in \mathscr{U}$ and $\sim\mathbf{B} \in \mathscr{U}$, and by 3.3.1.4 $\mathbf{A} \in \mathscr{U}$ or $\mathbf{B} \in \mathscr{U}$, which contradicts the inductive hypothesis.

3.3.3.4. $\mathbf{E} = \Pi_{o(o\alpha)}\mathbf{A}_{o\alpha}$. By 3.3.1.7 there is a variable $\mathbf{x}_\alpha$ such that $\sim\mathbf{A}_{o\alpha}\mathbf{x}_\alpha \in \mathscr{U}$, and from 3.3.1.6 it can be seen that $\mathbf{A}_{o\alpha}\mathbf{x}_\alpha \in \mathscr{U}$, since there are infinitely many i such that $\mathbf{E}^i = \Pi_{o(o\alpha)}\mathbf{A}_{o\alpha}$. Hence by 3.3.1.1 $\eta[\mathbf{A}_{o\alpha}\mathbf{x}_\alpha] \in \mathscr{U}$ and $[\sim\eta[\mathbf{A}_{o\alpha}\mathbf{x}_\alpha]] \in \mathscr{U}$. If $\mathbf{A}_{o\alpha}\mathbf{x}_\alpha$ is not in λ-normal form, a single contraction will make it so, and it is easy to see that $\eta[\mathbf{A}_{o\alpha}\mathbf{x}_\alpha]$ contains the same number of occurrences of logical constants as does $\mathbf{A}_{o\alpha}$. Thus the inductive hypothesis is contradicted.

3.3.4. We now define a function V which we shall show is a semivaluation.

$$V\mathbf{E} = \mathfrak{t} \quad \text{if } \mathbf{E} \in \mathscr{U}.$$
$$V\mathbf{E} = \mathfrak{f} \quad \text{if } [\sim\mathbf{E}] \in \mathscr{U}.$$

Clearly V is well defined by 3.3.3.

3.3.5. V is a semivaluation. The proof is straightforward. Each clause of 3.2 is readily verified using 3.3.4 and the appropriate case of 3.3.1.

3.3.6. If $\mathbf{A} \in \mathscr{S}$ then $\mathbf{A} \in \mathscr{U}$ so $V\mathbf{A} = \mathfrak{t}$. This proves 3.3.

3.4. THEOREM. *If V is any semivaluation, then* $\{\mathbf{A} \mid V\mathbf{A} = \mathfrak{t}\}$ *is consistent.*

PROOF (following Takahashi [12]).

3.4.1. For each type symbol γ we define the set $\mathscr{D}_\gamma$ of *V-complexes$_\gamma$* as follows by induction on γ:

3.4.1.1. $\mathscr{D}_o = \{\langle \mathbf{A}_o, \mathfrak{p}\rangle \mid \mathbf{A}_o$ is an η-wff$_o$ and $\mathfrak{p}$ is $\mathfrak{t}$ or $\mathfrak{f}$ and if $V\mathbf{A}_o$ is defined, then $\mathfrak{p} = V\mathbf{A}_o\}$.

3.4.1.2. $\mathscr{D}_\iota = \{\langle \mathbf{A}_\iota, \iota\rangle \mid \mathbf{A}_\iota$ is an η-wff$_\iota\}$.

3.4.1.3. $\mathscr{D}_{(\alpha\beta)} = \{\langle \mathbf{A}_{\alpha\beta}, \mathfrak{p}\rangle \mid \mathbf{A}_{\alpha\beta}$ is an η-wff$_{(\alpha\beta)}$ and $\mathfrak{p}$ is a function from $\mathscr{D}_\beta$ into $\mathscr{D}_\alpha$ such that if $\langle \mathbf{B}_\beta, \mathfrak{q}\rangle$ is any member of $\mathscr{D}_\beta$, then $\mathfrak{p}\langle \mathbf{B}_\beta, \mathfrak{q}\rangle = \langle \eta[\mathbf{A}_{\alpha\beta}\mathbf{B}_\beta], \mathfrak{r}\rangle$ for some $\mathfrak{r}\}$.

3.4.2. LEMMA. *For each η-wff $\mathbf{A}_\gamma$ there is an $\mathfrak{r}$ such that* $\langle \mathbf{A}_\gamma, \mathfrak{r}\rangle \in \mathscr{D}_\gamma$.

PROOF. We choose $\mathfrak{r}$ as a function of $\mathbf{A}_\gamma$ by induction on γ, and show $\langle \mathbf{A}_\gamma, \mathfrak{r}(\mathbf{A}_\gamma)\rangle \in \mathscr{D}_\gamma$. This is trivial when $\gamma = \iota$ or $\gamma = o$. (If $V\mathbf{A}_o$ is not defined, arbitrarily let $\mathfrak{r}(\mathbf{A}_o) = \mathfrak{t}$.) If $\gamma = (\alpha\beta)$, let $\mathfrak{r}(\mathbf{A}_{\alpha\beta})\langle \mathbf{B}_\beta, \mathfrak{q}\rangle = \langle \eta[\mathbf{A}_{\alpha\beta}\mathbf{B}_\beta], \mathfrak{r}(\eta[\mathbf{A}_{\alpha\beta}\mathbf{B}_\beta])\rangle$ for each $\langle \mathbf{B}_\beta, \mathfrak{q}\rangle \in \mathscr{D}_\beta$.

3.4.3. *Definitions and notations.*

3.4.3.1. If $\mathfrak{C}$ is a V-complex, let $\mathfrak{C}^1$ and $\mathfrak{C}^2$ be the first and second components of $\mathfrak{C}$, so $\mathfrak{C} = \langle \mathfrak{C}^1, \mathfrak{C}^2 \rangle$. If f is a function whose values are V-complexes, let f^1 and f^2 be functions with the same domain as f defined so that for any argument t, $f^i t = (ft)^i$ for $i = 1, 2$. Thus $ft = \langle f^1 t, f^2 t \rangle$.

3.4.3.2. An *assignment* is a function φ defined on the variables of $\mathcal{T}$ such that $\varphi \mathbf{x}_\alpha \in \mathcal{D}_\alpha$ for every variable$_\alpha$ $\mathbf{x}_\alpha$.

3.4.3.3. Given an assignment φ, a variable $\mathbf{x}_\alpha$, and $\mathfrak{C} \in \mathcal{D}_\alpha$, let $(\varphi : \mathbf{x}_\alpha / \mathfrak{C})$ be that assignment ψ such that $\psi \mathbf{y}_\beta = \varphi \mathbf{y}_\beta$ if $\mathbf{y}_\beta \neq \mathbf{x}_\alpha$ and $\psi \mathbf{x}_\alpha = \mathfrak{C}$.

3.4.3.4. If $\mathfrak{p}$ and $\mathfrak{q}$ are truth values, we denote by $\sim \mathfrak{p}$ and $\mathfrak{p} \vee \mathfrak{q}$ the (intuitive) negation of $\mathfrak{p}$ and the (intuitive) disjunction of $\mathfrak{p}$ and $\mathfrak{q}$, respectively. The context will show whether $\sim$ and $\vee$ are to be regarded as symbols of $\mathcal{T}$ or of our metalanguage.

3.4.4. $\mathcal{V}_\varphi$. For each assignment φ and wff $\mathbf{C}_\gamma$ we define $\mathcal{V}_\varphi \mathbf{C}_\gamma$ and show $\mathcal{V}_\varphi \mathbf{C}_\gamma \in \mathcal{D}_\gamma$. Thus $\mathcal{V}_\varphi \mathbf{C}_\gamma = \langle \mathcal{V}^1_\varphi \mathbf{C}_\gamma, \mathcal{V}^2_\varphi \mathbf{C}_\gamma \rangle$.

3.4.4.1. Let $\mathcal{V}^1_\varphi \mathbf{C}_\gamma = \eta[[\lambda \mathbf{x}^1 \cdots \lambda \mathbf{x}^n \mathbf{C}_\gamma](\varphi^1 \mathbf{x}^1) \cdots (\varphi^1 \mathbf{x}^n)]$, where $\mathbf{x}^1, \cdots, \mathbf{x}^n$ are the free variables of $\mathbf{C}_\gamma$. Let $\mathcal{V}^1_\varphi \mathbf{C}_\gamma = \eta \mathbf{C}_\gamma$ if $\mathbf{C}_\gamma$ has no free variables.

3.4.4.2. Note that $\mathcal{V}^1_\varphi [\mathbf{A}_{\gamma\beta} \mathbf{B}_\beta] = \eta[(\mathcal{V}^1_\varphi \mathbf{A}_{\gamma\beta}) \mathcal{V}^1_\varphi \mathbf{B}_\beta]$. This is readily established using properties of λ-conversion.

We define $\mathcal{V}^2_\varphi \mathbf{C}_\gamma$, and show $\mathcal{V}_\varphi \mathbf{C}_\gamma \in \mathcal{D}_\gamma$, simultaneously for all φ by induction on the number of occurrences of [in $\mathbf{C}_\gamma$, considering the following cases:

3.4.4.3. $\mathbf{C}_\gamma$ is a parameter. Let $\mathcal{V}^2_\varphi \mathbf{C}_\gamma = \mathfrak{r}(\mathbf{C}_\gamma)$, where $\mathfrak{r}$ is defined as in the proof of 3.4.2. Note that $\mathbf{C}_\gamma$ is an η-wff, so $\mathcal{V}_\varphi \mathbf{C}_\gamma = \langle \mathbf{C}_\gamma, \mathfrak{r}(\mathbf{C}_\gamma) \rangle \in \mathcal{D}_\gamma$ by 3.4.2.

3.4.4.4. $\mathbf{C}_\gamma$ is a variable. Let $\mathcal{V}^2_\varphi \mathbf{C}_\gamma = \varphi^2 \mathbf{C}_\gamma$. Thus $\mathcal{V}_\varphi \mathbf{C}_\gamma = \langle \varphi^1 \mathbf{C}_\gamma, \varphi^2 \mathbf{C}_\gamma \rangle = \varphi \mathbf{C}_\gamma \in \mathcal{D}_\gamma$ by 3.4.3.2, and we see that $\mathcal{V}_\varphi$ extends φ.

3.4.4.5. $\mathbf{C}_\gamma$ is $\sim_{oo}$. For any $\langle \mathbf{B}_o, \mathfrak{q} \rangle \in \mathcal{D}_o$, let $(\mathcal{V}^2_\varphi \sim_{oo}) \langle \mathbf{B}_o, \mathfrak{q} \rangle = \langle \sim_{oo} \mathbf{B}_o, \sim \mathfrak{q} \rangle$. It is clear that $\sim_{oo} \mathbf{B}_o$ is an η-wff since $\mathbf{B}_o$ is, so to check that $\mathcal{V}_\varphi \sim_{oo} \in \mathcal{D}_{oo}$ by 3.4.1.3 we must check that $\langle \sim_{oo} \mathbf{B}_o, \sim \mathfrak{q} \rangle \in \mathcal{D}_o$. By 3.4.1.1 this is trivial if $V[\sim \mathbf{B}_o]$ is not defined. If $V[\sim \mathbf{B}_o] = \mathfrak{r}$ then by 3.2.2–3.2.3 $V\mathbf{B}_o = \sim \mathfrak{r}$; but $V\mathbf{B}_o = \mathfrak{q}$ since $\langle \mathbf{B}_o, \mathfrak{q} \rangle \in \mathcal{D}_o$ so $\sim \mathfrak{q} = \mathfrak{r} = V[\sim \mathbf{B}_o]$ and $\langle \sim \mathbf{B}_o, \sim \mathfrak{q} \rangle \in \mathcal{D}_o$.

3.4.4.6. $\mathbf{C}_\gamma$ is $\vee_{(oo)o}$. For any $\langle \mathbf{B}_o, \mathfrak{q} \rangle \in \mathcal{D}_o$, let $(\mathcal{V}^2_\varphi \vee_{(oo)o}) \langle \mathbf{B}_o, \mathfrak{q} \rangle = \langle \vee_{(oo)o} \mathbf{B}_o, \mathfrak{h} \rangle$, where $\mathfrak{h}$ is that function from $\mathcal{D}_o$ into $\mathcal{D}_o$ such that for any $\langle \mathbf{E}_o, \mathfrak{r} \rangle \in \mathcal{D}_o$, $\mathfrak{h} \langle \mathbf{E}_o, \mathfrak{r} \rangle = \langle [\vee_{(oo)o} \mathbf{B}_o] \mathbf{E}_o, \mathfrak{q} \vee \mathfrak{r} \rangle$.

Since $[[\vee_{(oo)o} \mathbf{B}_o] \mathbf{E}_o]$ is an η-wff whenever $\mathbf{B}_o$ and $\mathbf{E}_o$ are η-wffs, from 3.4.1.3 it is seen that in order to verify that $\mathcal{V}_\varphi \vee_{(oo)o} \in \mathcal{D}_{(oo)o}$ one must check that $\langle [\vee_{(oo)} \mathbf{B}_o] \mathbf{E}_o, \mathfrak{q} \vee \mathfrak{r} \rangle \in \mathcal{D}_o$ whenever $\langle \mathbf{B}_o, \mathfrak{q} \rangle \in \mathcal{D}_o$ and $\langle \mathbf{E}_o, \mathfrak{r} \rangle \in \mathcal{D}_o$. If $V[\mathbf{B}_o \vee \mathbf{E}_o]$ is not defined this is trivial. If $V[\mathbf{B}_o \vee \mathbf{E}_o]$ is defined then $V\mathbf{B}_o = \mathfrak{q}$ and $V\mathbf{E}_o = \mathfrak{r}$ so by 3.2.4–3.2.5 $V[\mathbf{B}_o \vee \mathbf{E}_o] = \mathfrak{q} \vee \mathfrak{r}$.

3.4.4.7. $\mathbf{C}_\gamma$ is $\Pi_{o(o\alpha)}$. For any $\langle \mathbf{A}_{o\alpha}, \mathfrak{p} \rangle \in \mathcal{D}_{o\alpha}$, let $(\mathcal{V}^2_\varphi \Pi_{o(o\alpha)}) \langle \mathbf{A}_{o\alpha}, \mathfrak{p} \rangle = \langle \Pi_{o(o\alpha)} \mathbf{A}_{o\alpha}, \mathfrak{r} \rangle$, where $\mathfrak{r}$ is $\mathfrak{t}$ if $\mathfrak{p}^2 \mathfrak{C} = \mathfrak{t}$ for every $\mathfrak{C} \in \mathcal{D}_\alpha$, and $\mathfrak{r}$ is $\mathfrak{f}$ otherwise.

It must be shown that $V[\Pi_{o(o\alpha)} \mathbf{A}_{o\alpha}] = \mathfrak{r}$ if $V[\Pi_{o(o\alpha)} \mathbf{A}_{o\alpha}]$ is defined, so suppose it is defined.

Suppose $V[\Pi_{o(o\alpha)} \mathbf{A}_{o\alpha}] = \mathfrak{t}$, and let $\langle \mathbf{B}_\alpha, \mathfrak{q} \rangle \in \mathcal{D}_\alpha$. By 3.2.6 and 3.2.1 $\mathfrak{t} = V[\mathbf{A}_{o\alpha} \mathbf{B}_\alpha] = V\eta[\mathbf{A}_{o\alpha} \mathbf{B}_\alpha]$, so $\mathfrak{p} \langle \mathbf{B}_\alpha, \mathfrak{q} \rangle = \langle \eta[\mathbf{A}_{o\alpha} \mathbf{B}_\alpha], V\eta[\mathbf{A}_{o\alpha} \mathbf{B}_\alpha] \rangle = \langle \eta[\mathbf{A}_{o\alpha} \mathbf{B}_\alpha], \mathfrak{t} \rangle$ by 3.4.1.3 and 3.4.1.1. Thus $\mathfrak{r} = \mathfrak{t} = V[\Pi_{o(o\alpha)} \mathbf{A}_{o\alpha}]$ in this case.

Suppose $V[\Pi_{o(o\alpha)} \mathbf{A}_{o\alpha}] = \mathfrak{f}$. By 3.2.7 and 3.2.1 there is a wff $\mathbf{B}_\alpha$ such that

$V\eta[\mathbf{A}_{o\alpha}\mathbf{B}_\alpha] = \mathfrak{f}$. By 3.4.2 there is a $\mathfrak{q}$ such that $\langle\eta\mathbf{B}_\alpha, \mathfrak{q}\rangle \in \mathscr{D}_\alpha$. Thus $\mathfrak{p}\langle\eta\mathbf{B}_\alpha, \mathfrak{q}\rangle = \langle\eta[\mathbf{A}_{o\alpha}\mathbf{B}_\alpha], \mathfrak{f}\rangle$, so $\mathfrak{r} = \mathfrak{f} = V[\Pi_{o(o\alpha)}\mathbf{A}_{o\alpha}]$ in this case.

3.4.4.8. $\mathbf{C}_\gamma$ has the form $[\mathbf{A}_{\gamma\beta}\mathbf{B}_\beta]$. Let $\mathscr{V}^2_\varphi[\mathbf{A}_{\gamma\beta}\mathbf{B}_\beta] = ((\mathscr{V}^2_\varphi\mathbf{A}_{\gamma\beta})(\mathscr{V}_\varphi\mathbf{B}_\beta))^2$. Note that $\mathscr{V}_\varphi[\mathbf{A}_{\gamma\beta}\mathbf{B}_\beta] = (\mathscr{V}^2_\varphi\mathbf{A}_{\gamma\beta})(\mathscr{V}_\varphi\mathbf{B}_\beta) \in \mathscr{D}_\gamma$, since by inductive hypothesis $\mathscr{V}_\varphi\mathbf{A}_{\gamma\beta} \in \mathscr{D}_{\gamma\beta}$ and $\mathscr{V}_\varphi\mathbf{B}_\beta \in \mathscr{D}_\beta$ so

$$\begin{aligned}((\mathscr{V}^2_\varphi\mathbf{A}_{\gamma\beta})(\mathscr{V}_\varphi\mathbf{B}_\beta))^1 &= \eta[(\mathscr{V}^1_\varphi\mathbf{A}_{\gamma\beta})\mathscr{V}^1_\varphi\mathbf{B}_\beta] &&\text{(by 3.4.1.3)}\\ &= \mathscr{V}^1_\varphi[\mathbf{A}_{\gamma\beta}\mathbf{B}_\beta] &&\text{(by 3.4.4.2).}\end{aligned}$$

3.4.4.9. $\mathbf{C}_\gamma$ has the form $[\lambda\mathbf{x}_\beta\mathbf{A}_\alpha]$. Let $\mathscr{V}^2_\varphi[\lambda\mathbf{x}_\beta\mathbf{A}_\alpha]$ be that function from $\mathscr{D}_\beta$ into $\mathscr{D}_\alpha$ whose value on each $\mathfrak{C} \in \mathscr{D}_\beta$ is $\mathscr{V}_{(\varphi:\mathbf{x}_\beta/\mathfrak{C})}\mathbf{A}_\alpha$.

To satisfy 3.4.1.3 we must show that if $\mathfrak{C} = \langle\mathbf{B}_\beta, \mathfrak{q}\rangle \in \mathscr{D}_\beta$, then $(\mathscr{V}_{(\varphi:\mathbf{x}_\beta/\mathfrak{C})}\mathbf{A}_\alpha)^1 = \eta[(\mathscr{V}^1_\varphi[\lambda\mathbf{x}_\beta\mathbf{A}_\alpha])\mathbf{B}_\beta]$. Let $\mathbf{y}^1, \cdots, \mathbf{y}^n$ be the free variables of $[\lambda\mathbf{x}_\beta\mathbf{A}_\alpha]$. Then $\mathscr{V}^1_\varphi[\lambda\mathbf{x}_\beta\mathbf{A}_\alpha] = \eta[[\lambda\mathbf{y}^1\cdots\lambda\mathbf{y}^n\lambda\mathbf{x}_\beta\mathbf{A}_\alpha](\varphi^1\mathbf{y}^1)\cdots(\varphi^1\mathbf{y}^n)]$. Also whether or not $\mathbf{x}_\beta$ is free in $\mathbf{A}_\alpha$, $(\mathscr{V}_{(\varphi:\mathbf{x}_\beta/\mathfrak{C})}\mathbf{A}_\alpha)^1$ conv $[[\lambda\mathbf{y}^1\cdots\lambda\mathbf{y}^n\lambda\mathbf{x}_\beta\mathbf{A}_\alpha](\varphi^1\mathbf{y}^1)\cdots(\varphi^1\mathbf{y}^n)\mathbf{B}_\beta]$ by 3.4.4.1 and 3.4.3.3. The desired result follows by λ-conversion.

3.4.5. *Remark.* In the terminology of [4] we have now essentially shown that the set of V-complexes constitutes a general model for $\mathscr{T}$ in which the axioms of extensionality (6.1.1 below) do not necessarily hold. Of course in order to permit the axioms of extensionality to fail we have avoided making $\mathscr{D}_o = \{\mathfrak{t}, \mathfrak{f}\}$, and we have avoided making $\mathscr{D}_{\alpha\beta}$ contain genuine functions from $\mathscr{D}_\beta$ into $\mathscr{D}_\alpha$. Instead we have in essence indexed these truth values and functions $\mathfrak{p}$ by wffs $\mathbf{A}$ and called the indexed entity $\langle\mathbf{A}, \mathfrak{p}\rangle$ a V-complex.

Since the theorems of $\mathscr{T}$ are known to be valid in all general models, the unsceptical reader will readily believe Lemma 3.4.9 below, and may proceed directly to 3.4.10 after noting 3.4.8.

3.4.6. LEMMA. *If φ and ψ are assignments which agree on the free variables of $\mathbf{A}_\alpha$, then $\mathscr{V}_\varphi\mathbf{A}_\alpha = \mathscr{V}_\psi\mathbf{A}_\alpha$.*

This follows in a straightforward way from 3.4.4.

3.4.7. LEMMA. *If $\mathbf{D}_\gamma$ conv $\mathbf{E}_\gamma$ and φ is any assignment, then $\mathscr{V}_\varphi\mathbf{D}_\gamma = \mathscr{V}_\varphi\mathbf{E}_\gamma$.*

PROOF. We first establish several subsidiary lemmas.

3.4.7.1. LEMMA. *If $\mathbf{D}_\gamma$* conv *$\mathbf{E}_\gamma$ and φ is any assignment, then $\mathscr{V}^1_\varphi\mathbf{D}_\gamma = \mathscr{V}^1_\varphi\mathbf{E}_\gamma$.*

This follows easily from 3.4.4.1 and properties of λ-conversion, using the fact that if $\mathbf{y}^1, \cdots, \mathbf{y}^m$ are the variables which occur free in $\mathbf{D}_\gamma$ or $\mathbf{E}_\gamma$, then

$$\mathscr{V}^1_\varphi\mathbf{D}_\gamma \text{ conv } [[\lambda\mathbf{y}^1\cdots\lambda\mathbf{y}^m\mathbf{D}_\gamma](\varphi^1\mathbf{y}^1)\cdots(\varphi^1\mathbf{y}^m)].$$

3.4.7.2. LEMMA. *If the bound variables of $\mathbf{B}_\beta$ are distinct from $\mathbf{x}_\alpha$ and from the free variables of $\mathbf{A}_\alpha$, φ is an assignment, and $\psi = (\varphi:\mathbf{x}_\alpha/\mathscr{V}_\varphi\mathbf{A}_\alpha)$, then $\mathscr{V}_\varphi S^{\mathbf{x}_\alpha}_{\mathbf{A}_\alpha}\mathbf{B}_\beta = \mathscr{V}_\psi\mathbf{B}_\beta$.*

PROOF.

3.4.7.2.1. First treating $\mathscr{V}^1_\varphi$ we have

$$\begin{aligned}\mathscr{V}^1_\varphi S^{\mathbf{x}_\alpha}_{\mathbf{A}_\alpha}\mathbf{B}_\beta &= \mathscr{V}^1_\varphi[[\lambda\mathbf{x}_\alpha\mathbf{B}_\beta]\mathbf{A}_\alpha] &&\text{(by 3.4.7.1)}\\ &= \eta[(\mathscr{V}^1_\varphi[\lambda\mathbf{x}_\alpha\mathbf{B}_\beta])\mathscr{V}^1_\psi\mathbf{A}_\alpha] &&\text{(by 3.4.4.2)}\\ &= \eta[(\mathscr{V}^1_\psi[\lambda\mathbf{x}_\alpha\mathbf{B}_\beta])\mathscr{V}^1_\psi\mathbf{x}_\alpha] &&\text{(by 3.4.6 and 3.4.4.4)}\\ &= \mathscr{V}^1_\psi\mathbf{B}_\beta &&\text{(by 3.4.4.2 and 3.4.7.1).}\end{aligned}$$

Next we prove the lemma by induction on the number of occurrences of [in $\mathbf{B}_\beta$, and consider the following cases:

3.4.7.2.2. $\mathbf{B}_\beta$ is $\mathbf{x}_\alpha$. Then $\mathscr{V}_\varphi \mathcal{S}^{\mathbf{x}_\alpha}_{\mathbf{A}_\alpha}\mathbf{B}_\beta = \mathscr{V}_\varphi \mathbf{A}_\alpha = \psi\mathbf{x}_\alpha = \mathscr{V}_\psi \mathbf{B}_\beta$.

3.4.7.2.3. $\mathbf{x}_\alpha$ does not occur in $\mathbf{B}_\beta$. Then $\mathscr{V}_\varphi \mathcal{S}^{\mathbf{x}_\alpha}_{\mathbf{A}_\alpha}\mathbf{B}_\beta = \mathscr{V}_\varphi \mathbf{B}_\beta = \mathscr{V}_\psi \mathbf{B}_\beta$ by 3.4.6.

3.4.7.2.4. $\mathbf{B}_\beta$ has the form $[\mathbf{G}_{\beta\delta}\mathbf{H}_\delta]$. This is straightforward using 3.4.4.8 and the inductive hypothesis.

3.4.7.2.5. $\mathbf{B}_\beta$ has the form $[\lambda \mathbf{y}_\delta \mathbf{E}_\epsilon]$. Note that $\mathbf{y}_\delta$ must be distinct from $\mathbf{x}_\alpha$ and from the free variables of $\mathbf{A}_\alpha$. Let $\mathfrak{C} \in \mathscr{D}_\delta$. Let $\varphi' = (\varphi : \mathbf{y}_\delta/\mathfrak{C})$ and $\psi' = (\psi : \mathbf{y}_\delta/\mathfrak{C})$. Then $\mathscr{V}_\varphi \mathbf{A}_\alpha = \mathscr{V}_{\varphi'} \mathbf{A}_\alpha$ by 3.4.6, so $\psi' = (\varphi' : \mathbf{x}_\alpha/\mathscr{V}_{\varphi'}\mathbf{A}_\alpha)$. Thus

$$\begin{aligned}(\mathscr{V}^2_\varphi \mathcal{S}^{\mathbf{x}_\alpha}_{\mathbf{A}_\alpha}\mathbf{B}_\beta)\mathfrak{C} &= (\mathscr{V}^2_\varphi[\lambda \mathbf{y}_\delta \mathcal{S}^{\mathbf{x}_\alpha}_{\mathbf{A}_\alpha}\mathbf{E}_\epsilon])\mathfrak{C} \\ &= \mathscr{V}_{\varphi'} \mathcal{S}^{\mathbf{x}_\alpha}_{\mathbf{A}_\alpha}\mathbf{E}_\epsilon \qquad \text{(by 3.4.4.9)} \\ &= \mathscr{V}_{\psi'}\mathbf{E}_\epsilon \qquad \text{(by inductive hypothesis)} \\ &= (\mathscr{V}^2_\psi \mathbf{B}_\beta)\mathfrak{C} \qquad \text{(by 3.4.4.9).}\end{aligned}$$

Thus $\mathscr{V}^2_\varphi \mathcal{S}^{\mathbf{x}_\alpha}_{\mathbf{A}_\alpha}\mathbf{B}_\beta$ is the same function as $\mathscr{V}^2_\psi \mathbf{B}_\beta$.

3.4.7.3. LEMMA. *If the bound variables of* $\mathbf{B}_\beta$ *are distinct from* $\mathbf{x}_\alpha$ *and the free variables of* $\mathbf{A}_\alpha$, *and* φ *is an assignment, then* $\mathscr{V}_\varphi[[\lambda \mathbf{x}_\alpha \mathbf{B}_\beta]\mathbf{A}_\alpha] = \mathscr{V}_\varphi \mathcal{S}^{\mathbf{x}_\alpha}_{\mathbf{A}_\alpha}\mathbf{B}_\beta$.

PROOF. Let ψ be as in 3.4.7.2. Then

$$\begin{aligned}\mathscr{V}_\varphi[[\lambda \mathbf{x}_\alpha \mathbf{B}_\beta]\mathbf{A}_\alpha] &= (\mathscr{V}^2_\varphi[\lambda \mathbf{x}_\alpha \mathbf{B}_\beta])\mathscr{V}_\varphi \mathbf{A}_\alpha \qquad \text{(by 3.4.4.8)} \\ &= \mathscr{V}_\psi \mathbf{B}_\beta \qquad \text{(by 3.4.4.9)} \\ &= \mathscr{V}_\varphi \mathcal{S}^{\mathbf{x}_\alpha}_{\mathbf{A}_\alpha}\mathbf{B}_\beta \qquad \text{(by 3.4.7.2).}\end{aligned}$$

3.4.7.4. LEMMA. *If* $\mathbf{y}_\beta$ *does not occur in* $\mathbf{A}_\alpha$ *and* $\mathbf{x}_\beta$ *is not bound in* $\mathbf{A}_\alpha$ *and* φ *is an assignment, then* $\mathscr{V}_\varphi[\lambda \mathbf{x}_\beta \mathbf{A}_\alpha] = \mathscr{V}_\varphi[\lambda \mathbf{y}_\beta \mathcal{S}^{\mathbf{x}_\beta}_{\mathbf{y}_\beta}\mathbf{A}_\alpha]$.

PROOF. We assume $\mathbf{x}_\beta \neq \mathbf{y}_\beta$, since otherwise the result is trivial. $\mathscr{V}^1_\varphi[\lambda \mathbf{x}_\beta \mathbf{A}_\alpha] = \mathscr{V}^1_\varphi[\lambda \mathbf{y}_\beta \mathcal{S}^{\mathbf{x}_\beta}_{\mathbf{y}_\beta}\mathbf{A}_\alpha]$ by 3.4.7.1.

Considering $\mathscr{V}^2_\varphi$, for all $\mathfrak{C} \in \mathscr{D}_\beta$ we have

$$\begin{aligned}(\mathscr{V}^2_\varphi[\lambda \mathbf{x}_\beta \mathbf{A}_\alpha])\mathfrak{C} &= \mathscr{V}_{(\varphi : \mathbf{x}_\beta/\mathfrak{C})}\mathbf{A}_\alpha \qquad \text{(by 3.4.4.9)} \\ &= \mathscr{V}_{((\varphi : \mathbf{y}_\beta/\mathfrak{C}) : \mathbf{x}_\beta/\mathfrak{C})}\mathbf{A}_\alpha \qquad \text{(by 3.4.6)} \\ &= \mathscr{V}_{(\varphi : \mathbf{y}_\beta/\mathfrak{C})}\mathcal{S}^{\mathbf{x}_\beta}_{\mathbf{y}_\beta}\mathbf{A}_\alpha \qquad \text{(by 3.4.7.2)} \\ &= (\mathscr{V}^2_\varphi[\lambda \mathbf{y}_\beta \mathcal{S}^{\mathbf{x}_\beta}_{\mathbf{y}_\beta}\mathbf{A}_\alpha])\mathfrak{C} \qquad \text{(by 3.4.4.9)}\end{aligned}$$

so the indicated functions are the same.

3.4.7.5. The proof of 3.4.7 now follows easily from 3.4.7.3 and 3.4.7.4. One may assume that $\mathbf{E}_\gamma$ is obtained from $\mathbf{D}_\gamma$ by a single application of a rule of λ-conversion, and proceed by induction on the number of occurrences of [in $\mathbf{D}_\gamma$.

3.4.8. LEMMA. *Let* φ *be any assignment.*

3.4.8.1. $\mathscr{V}^2_\varphi[\sim \mathbf{A}_o] = \sim \mathscr{V}^2_\varphi \mathbf{A}_o$.

3.4.8.2. $\mathscr{V}^2_\varphi[\mathbf{A}_o \vee \mathbf{B}_o] = (\mathscr{V}^2_\varphi \mathbf{A}_o) \vee (\mathscr{V}^2_\varphi \mathbf{B}_o)$.

These follow directly from 3.4.4.8, 3.4.4.5, and 3.4.4.6.

3.4.9. LEMMA. *If* φ *is any assignment and* $\vdash_{\mathscr{T}} \mathbf{A}_o$, *then* $\mathscr{V}^2_\varphi \mathbf{A}_o = \mathsf{t}$.

PROOF. We show that $\mathscr{W} = \{\mathbf{A}_o \mid \mathscr{V}^2_\varphi \mathbf{A}_o = \mathsf{t}$ for all assignments $\varphi\}$ contains the axioms of $\mathscr{T}$ and is closed under the rules of inference. This follows immediately from 3.4.8 for Axioms 2.5.1–2.5.4 and Modus ponens, and from 3.4.7 for the rules of λ-conversion. We leave to the reader the routine calculations for Axioms 2.5.5 and 2.5.6, using 3.4.4 and 3.4.8.

For 2.6.4 (Substitution) and 2.6.6 (Generalization) we suppose $\mathbf{x}_\alpha$ is not free in $\mathbf{F}_{o\alpha}$ and that $[\mathbf{F}_{o\alpha}\mathbf{x}_\alpha] \in \mathscr{W}$; we must show $\Pi_{o(o\alpha)}\mathbf{F}_{o\alpha} \in \mathscr{W}$ and $[\mathbf{F}_{o\alpha}\mathbf{A}_\alpha] \in \mathscr{W}$. Given φ, we let $\mathfrak{C} \in \mathscr{D}_\alpha$ and $\psi = (\varphi : \mathbf{x}_\alpha/\mathfrak{C})$.

Then
$$t = \mathscr{V}^2_\psi[\mathbf{F}_{o\alpha}\mathbf{x}_\alpha]$$
$$= ((\mathscr{V}^2_\psi\mathbf{F}_{o\alpha})\mathscr{V}_\psi\mathbf{x}_\alpha)^2 \qquad \text{(by 3.4.4.8)}$$
$$= ((\mathscr{V}^2_\psi\mathbf{F}_{o\alpha})\mathfrak{C})^2$$
$$= ((\mathscr{V}^2_\varphi\mathbf{F}_{o\alpha})\mathfrak{C})^2 \qquad \text{(by 3.4.6)}$$
for all $\mathfrak{C} \in \mathscr{D}_\alpha$ so
$$t = ((\mathscr{V}^2_\varphi\Pi_{o(o\alpha)})\mathscr{V}_\varphi\mathbf{F}_{o\alpha})^2 \qquad \text{(by 3.4.4.7)}$$
$$= [\mathscr{V}^2_\varphi\Pi_{o(o\alpha)}\mathbf{F}_{o\alpha}] \qquad \text{(by 3.4.4.8)}$$
so $[\Pi_{o(o\alpha)}\mathbf{F}_{o\alpha}] \in \mathscr{W}$. Also, if we let $\mathfrak{C} = \mathscr{V}_\varphi\mathbf{A}_\alpha$ then
$$t = ((\mathscr{V}^2_\varphi\mathbf{F}_{o\alpha})\mathscr{V}_\varphi\mathbf{A}_\alpha)^2 = \mathscr{V}^2_\varphi[\mathbf{F}_{o\alpha}\mathbf{A}_\alpha] \qquad \text{(by 3.4.4.8)}$$
so $[\mathbf{F}_{o\alpha}\mathbf{A}_\alpha] \in \mathscr{W}$.

3.4.10. We now complete the proof of 3.4. If 3.4 is not true, there are η-wffs $\mathbf{A}^1, \cdots, \mathbf{A}^n$ such that $V\mathbf{A}^i = t$ for $1 \le i \le n$ but $\vdash_{\mathscr{T}} \sim \mathbf{A}^1 \vee \cdots \vee \sim \mathbf{A}^n$. By 3.4.2 we can define an assignment φ so that $\varphi^1\mathbf{x}_\alpha = \mathbf{x}_\alpha$ for all variables $\mathbf{x}_\alpha$. Then $\mathscr{V}^1_\varphi\mathbf{A} = \mathbf{A}$ for all η-wffs $\mathbf{A}$ by 3.4.4.1, so by 3.4.4 and 3.4.1.1 for $1 \le i \le n$, $\mathscr{V}_\varphi\mathbf{A}^i = \langle \mathbf{A}^i, V\mathbf{A}^i\rangle = \langle \mathbf{A}^i, t\rangle$ and $\mathscr{V}^2_\varphi\mathbf{A}^i = t$. Hence by 3.4.8 $\mathscr{V}^2_\varphi[\sim\mathbf{A}^1 \vee \cdots \vee \sim\mathbf{A}^n] = \mathfrak{f}$, contradicting 3.4.9.

3.5. THEOREM. *If* Γ *is an abstract consistency property and* $\mathscr{S}$ *is a finite set of* wffs$_o$ *such that* $\Gamma(\mathscr{S})$, *then* $\mathscr{S}$ *is consistent.*

PROOF. By 3.3 and 3.4.

Remark. Our analogy with [10] suggests that the conclusion of 3.5 should be that $\mathscr{S}$ has a denumerable general model. By the Remark 3.4.5 we have actually shown that $\mathscr{S}$ has a general model (although we have not actually defined what is meant by a general model when axioms of extensionality are not assumed). Of course we have not dealt with the question of denumerability.

§4. Cut-elimination.

4.1. Preliminary definitions.

4.1.1. The *disjunctive components* of a wff$_o$ are defined inductively as follows:

4.1.2.1. $\mathbf{A}$ and $\mathbf{B}$ are disjunctive components of $[\mathbf{A} \vee \mathbf{B}]$.

4.1.2.2. $\mathbf{A}$ is a disjunctive component of $\mathbf{A}$.

4.1.2.3. If $\mathbf{A}$ is a disjunctive component of $\mathbf{B}$, and $\mathbf{B}$ is a disjunctive component of $\mathbf{C}$, then $\mathbf{A}$ is a disjunctive component of $\mathbf{C}$. We regard disjunctive components as occurrences of wffs$_o$.

4.1.2. We now find it convenient to modify our conventions concerning syntactical variables so that $\mathbf{A} \vee \mathbf{B}$ and $\mathbf{B} \vee \mathbf{A}$ may simply stand for $\mathbf{A}$ in appropriate contexts. To this end we introduce a "pseudo-wff", the constant $\square$, which may be interpreted as the empty disjunction, and therefore denotes falsehood. We henceforth let $\mathbf{A}_o$, $\mathbf{B}_o$, $\mathbf{C}_o$, (etc.) take $\square$ as value when these syntactic variables occur as disjunctive components of an expression which stands for a wff. Then we regard $\mathbf{A} \vee \square$ and $\square \vee \mathbf{A}$ as abbreviations for $\mathbf{A}$. $\square$ standing alone may be regarded as an abbreviation for $\forall p_o p_o$.

4.2. *The system* $\mathscr{G}$.

4.2.1. AXIOMS. $\sim\mathbf{A} \vee \mathbf{A}$, *where* $\mathbf{A}$ *is atomic.*

4.2.2. *Rules of inference.*

4.2.2.1. *Conversion-I–III.* Apply 2.6.1 or 2.6.3.

4.2.2.2. *Disjunction rules.* To replace a disjunctive component **D** of a wff by **E**, where **D** is $[[\mathbf{A} \vee \mathbf{B}] \vee \mathbf{C}]$ and **E** is $[\mathbf{A} \vee [\mathbf{B} \vee \mathbf{C}]]$, or **D** is $[\mathbf{A} \vee [\mathbf{B} \vee \mathbf{C}]]$ and **E** is $[[\mathbf{A} \vee \mathbf{B}] \vee \mathbf{C}]$, or **D** is $[\mathbf{A} \vee \mathbf{B}]$ and **E** is $[\mathbf{B} \vee \mathbf{A}]$.

4.2.2.3. *Weakening.* From **M** to infer $\mathbf{M} \vee \mathbf{A}$ (where **M** is not $\Box$).

4.2.2.4. *Negation introduction.* From $\mathbf{M} \vee \mathbf{A}$ to infer $\mathbf{M} \vee \sim\sim\mathbf{A}$.

4.2.2.5. *Conjunction introduction.* From $\mathbf{M} \vee \sim\mathbf{A}$ and $\mathbf{M} \vee \sim\mathbf{B}$ to infer $\mathbf{M} \vee \sim[\mathbf{A} \vee \mathbf{B}]$.

4.2.2.6. *Existential generalization.* From $\mathbf{M} \vee \sim\Pi_{o(o\alpha)}\mathbf{A}_{o\alpha} \vee \sim\mathbf{A}_{o\alpha}\mathbf{B}_\alpha$ to infer $\mathbf{M} \vee \sim\Pi_{o(o\alpha)}\mathbf{A}_{o\alpha}$.

4.2.2.7. *Universal generalization.* From $\mathbf{M} \vee \mathbf{A}_{o\alpha}\mathbf{x}_\alpha$ to infer $\mathbf{M} \vee \Pi_{o(o\alpha)}\mathbf{A}_{o\alpha}$, provided $\mathbf{x}_\alpha$ is not free in **M** or $\mathbf{A}_{o\alpha}$.

4.3. PROPOSITION. *If* $\vdash_{\mathscr{G}}\mathbf{A}$, *then* $\vdash_{\mathscr{T}}\mathbf{A}$.

This is readily established by showing that the rules of inference of $\mathscr{G}$ are derived rules of inference of $\mathscr{T}$.

We next establish some subsidiary lemmas. We shall discuss their proofs together since they all have the same form.

4.4. LEMMA. *If* **P** conv **Q** *then* $\vdash_{\mathscr{G}}\mathbf{P}$ *iff* $\vdash_{\mathscr{G}}\mathbf{Q}$.

4.5. LEMMA. *If* $\vdash_{\mathscr{G}}\mathbf{P}$, *and* **P** *has a disjunctive component of the form* $\sim\sim\mathbf{D}$, *and* **Q** *is the result of replacing this component of* **P** *by* **D**, *then* $\vdash_{\mathscr{G}}\mathbf{Q}$.

4.6. LEMMA. *If* $\vdash_{\mathscr{G}}\mathbf{P}$, *and* **P** *has a disjunctive component of the form* $\sim[\mathbf{D} \vee \mathbf{E}]$, *and* **Q** *is the result of replacing this component of* **P** *by* $\sim\mathbf{D}$ *or by* $\sim\mathbf{E}$, *then* $\vdash_{\mathscr{G}}\mathbf{Q}$.

4.7. LEMMA. *If* $\vdash_{\mathscr{G}}\mathbf{P}$, *and* $\mathbf{y}^1, \cdots, \mathbf{y}^n$ *are distinct variables and* $\mathbf{z}^j$ *is a variable of the same type as* $\mathbf{y}^j$ *for* $1 \leq j \leq n$, *then there is a* wff **Q** *such that* **P** conv -I **Q** *and* $\mathbf{z}^j$ *is free for* $\mathbf{y}^j$ *in* **Q** *for* $1 \leq j \leq n$ *and* $\vdash_{\mathscr{G}} \mathcal{S}^{\mathbf{y}^1 \cdots \mathbf{y}^n}_{\mathbf{z}^1 \cdots \mathbf{z}^n}\mathbf{Q}$.

4.8. LEMMA. *If* $\vdash_{\mathscr{G}}\mathbf{P}$, *and* **P** *has a disjunctive component of the form* $\Pi_{o(o\beta)}\mathbf{B}_{o\beta}$, *and* **Q** *is the result of replacing this component of* **P** *by* $\mathbf{B}_{o\beta}\mathbf{z}_\beta$, *then* $\vdash_{\mathscr{G}}\mathbf{Q}$.

PROOFS OF 4.4–4.8. Note that to prove 4.4 it suffices to prove 4.4′: if $\vdash_{\mathscr{G}}\mathbf{P}$ then $\vdash_{\mathscr{G}}\eta\mathbf{P}$. For when this is established one knows that if $\vdash\mathbf{P}$, then $\vdash\eta\mathbf{P}$, so $\vdash\eta\mathbf{Q}$, so $\vdash\mathbf{Q}$ by 4.2.2.1.

To prove 4.4′ and 4.5–4.8 let $\mathbf{P}^1, \cdots, \mathbf{P}^m$ be a proof in $\mathscr{G}$. We prove by induction on i that the lemmas hold for $\mathbf{P}^i$ for $1 \leq i \leq m$. Each lemma is trivial when $\mathbf{P}^i$ is an axiom. If $\mathbf{P}^i$ is not an axiom one considers how $\mathbf{P}^i$ was inferred and applies the inductive hypothesis (if necessary) to the wff(s) from which it was inferred. The proofs of Lemmas 4.4′, 4.5, and 4.6 are routine in all cases.

The proof of 4.7 is trivial except when $\mathbf{P}^i$ is inferred by 4.2.2.1 or 4.2.2.7. Suppose $\mathbf{P}^i$ is inferred from $\mathbf{P}^k$ by 4.2.2.1. One easily defines a wff $\mathbf{Q}^i$ such that $\mathbf{P}^i$ conv-I $\mathbf{Q}^i$ and $\mathbf{z}^j$ is free for $\mathbf{y}^j$ for $1 \leq j \leq n$. Let $\mathbf{Q}^k$ be a wff whose existence is assured by the inductive hypothesis. Then $\mathbf{Q}^k$ conv $\mathbf{P}^k$ conv $\mathbf{P}^i$ conv $\mathbf{Q}^i$ so

$$\begin{aligned} \mathcal{S}^{\mathbf{y}^1 \cdots \mathbf{y}^n}_{\mathbf{z}^1 \cdots \mathbf{z}^n}\mathbf{Q}^k &\text{ conv } [[\lambda\mathbf{y}^1 \cdots \lambda\mathbf{y}^n\mathbf{Q}^k]\mathbf{z}^1 \cdots \mathbf{z}^n] \\ &\text{ conv } [[\lambda\mathbf{y}^1 \cdots \lambda\mathbf{y}^n\mathbf{Q}^i]\mathbf{z}^1 \cdots \mathbf{z}^n] \text{ conv } \mathcal{S}^{\mathbf{y}^1 \cdots \mathbf{y}^n}_{\mathbf{z}^1 \cdots \mathbf{z}^n}\mathbf{Q}^i, \end{aligned}$$

so the latter wff is a theorem of $\mathscr{G}$ by 4.4 and the inductive hypothesis.

Suppose $\mathbf{P}^i$ is $\mathbf{M} \vee \Pi_{o(o\alpha)}\mathbf{A}_{o\alpha}$ and is inferred by 4.2.2.7 from $\mathbf{M} \vee \mathbf{A}_{o\alpha}\mathbf{x}_\alpha$. Since $\mathbf{x}_\alpha$ is not free in $\mathbf{P}^i$ we may assume $\mathbf{x}_\alpha$ is distinct from $\mathbf{y}^1, \cdots, \mathbf{y}^n$, but we must allow for the possibility that some $\mathbf{z}^j$ is $\mathbf{x}_\alpha$. Let $\mathbf{g}_\alpha$ be distinct from $\mathbf{z}^1, \cdots, \mathbf{z}^n$ and all variables free in $\mathbf{P}^i$. By the inductive hypothesis 4.7 there is a wff $[\mathbf{M}' \vee \mathbf{A}'_{o\alpha}\mathbf{x}_\alpha]$

conv-I $[\mathbf{M} \vee \mathbf{A}_{o\alpha}\mathbf{x}_\alpha]$ such that $\mathbf{z}^j$ is free for $\mathbf{y}^j$ in $\mathbf{M}' \vee \mathbf{A}'_{o\alpha}\mathbf{x}_\alpha$ for $1 \leq j \leq n$ and $\vdash_{\mathscr{G}} \mathsf{S}^{\mathbf{y}^1 \ldots \mathbf{y}^n \mathbf{x}_\alpha}_{\mathbf{z}^1 \ldots \mathbf{z}^n \mathbf{z}_\alpha}[\mathbf{M}' \vee \mathbf{A}'_{o\alpha}\mathbf{x}_\alpha]$. It is readily seen that one may apply 4.2.2.7 to obtain $\vdash_{\mathscr{G}} \mathsf{S}^{\mathbf{y}^1 \ldots \mathbf{y}^n}_{\mathbf{z}^1 \ldots \mathbf{z}^n}[\mathbf{M}' \vee \Pi_{o(o\alpha)}\mathbf{A}'_{o\alpha}]$, which completes the proof of 4.7.

The proof of 4.8 is trivial except when $\mathbf{P}^i$ is inferred by 4.2.2.7, so suppose $\mathbf{P}^i$ has the form $\mathbf{M} \vee \Pi_{o(o\alpha)}\mathbf{A}_{o\alpha}$ and is inferred from $\mathbf{M} \vee \mathbf{A}_{o\alpha}\mathbf{x}_\alpha$. If the component $\Pi_{o(o\beta)}\mathbf{B}_{o\beta}$ referred to in 4.8 is the component $\Pi_{o(o\alpha)}\mathbf{A}_{o\alpha}$ introduced by this application of 4.2.2.7, one obtains $\mathbf{M} \vee \mathbf{B}_{o\beta}\mathbf{z}_\beta$ from $\mathbf{M} \vee \mathbf{A}_{o\alpha}\mathbf{x}_\alpha$ by 4.7 and 4.2.2.1. Otherwise one may assume without real loss of generality that $\mathbf{P}^i$ has the form $\mathbf{N} \vee \Pi_{o(o\beta)}\mathbf{B}_{o\beta} \vee \Pi_{o(o\alpha)}\mathbf{A}_{o\alpha}$ and is inferred from $\mathbf{N} \vee \Pi_{o(o\beta)}\mathbf{B}_{o\beta} \vee \mathbf{A}_{o\alpha}\mathbf{x}_\alpha$. Let $\mathbf{y}_\beta$ be distinct from all variables in the latter wff.

$\vdash_{\mathscr{G}} \mathbf{N} \vee \mathbf{B}_{o\beta}\mathbf{y}_\beta \vee \mathbf{A}_{o\alpha}\mathbf{x}_\alpha$	by inductive hypothesis
$\vdash_{\mathscr{G}} \mathbf{N} \vee \mathbf{B}_{o\beta}\mathbf{y}_\beta \vee \Pi_{o(o\alpha)}\mathbf{A}_{o\alpha}$	by 4.2.2.7.
$\vdash_{\mathscr{G}} \mathbf{N} \vee \mathbf{B}_{o\beta}\mathbf{z}_\beta \vee \Pi_{o(o\alpha)}\mathbf{A}_{o\alpha}$	by 4.7 and 4.2.2.1.

This completes the proof of 4.8.

4.9. LEMMA. *If* $\vdash_{\mathscr{G}} \mathbf{M} \vee \mathbf{D} \vee \mathbf{D}$ *then* $\vdash_{\mathscr{G}} \mathbf{M} \vee \mathbf{D}$.

PROOF. The proof is by induction on the number of occurrences of logical constants in $\eta\mathbf{D}$. We consider the following cases, assuming that $\mathbf{D}$ is an η-wff in cases 4.9.1–4.9.4:

4.9.1. $\mathbf{D}$ has the form $[\mathbf{B} \vee \mathbf{C}]$.

$\vdash \mathbf{M} \vee \mathbf{B} \vee \mathbf{C} \vee \mathbf{B} \vee \mathbf{C}$	given.
$\vdash \mathbf{M} \vee \mathbf{B} \vee \mathbf{B} \vee \mathbf{C}$	by 4.2.2.2 and inductive hypothesis.
$\vdash \mathbf{M} \vee \mathbf{D}$	by 4.2.2.2 and inductive hypothesis.

4.9.2. $\mathbf{D}$ has the form $\sim\sim\mathbf{E}$.

$\vdash \mathbf{M} \vee \sim\sim\mathbf{E} \vee \sim\sim\mathbf{E}$	given.
$\vdash \mathbf{M} \vee \mathbf{E} \vee \mathbf{E}$	by 4.5 (twice).
$\vdash \mathbf{M} \vee \mathbf{E}$	by inductive hypothesis.
$\vdash \mathbf{M} \vee \mathbf{D}$	by 4.2.2.4.

4.9.3. $\mathbf{D}$ has the form $\sim[\mathbf{B} \vee \mathbf{C}]$.

$\vdash \mathbf{M} \vee \sim[\mathbf{B} \vee \mathbf{C}] \vee \sim[\mathbf{B} \vee \mathbf{C}]$	given.
$\vdash \mathbf{M} \vee \sim\mathbf{B} \vee \sim\mathbf{B}$	by 4.6 (twice).
$\vdash \mathbf{M} \vee \sim\mathbf{B}$	by inductive hypothesis.
$\vdash \mathbf{M} \vee \sim\mathbf{C}$	similarly.
$\vdash \mathbf{M} \vee \mathbf{D}$	by 4.2.2.5.

4.9.4. $\mathbf{D}$ has the form $\Pi_{o(o\alpha)}\mathbf{A}_{o\alpha}$.

Let $\mathbf{x}_\alpha$ be a variable which does not occur in $\mathbf{D}$.

$\vdash \mathbf{M} \vee \Pi_{o(o\alpha)}\mathbf{A}_{o\alpha} \vee \Pi_{o(o\alpha)}\mathbf{A}_{o\alpha}$	given.
$\vdash \mathbf{M} \vee \mathbf{A}_{o\alpha}\mathbf{x}_\alpha \vee \mathbf{A}_{o\alpha}\mathbf{x}_\alpha$	by 4.8 (twice).
$\vdash \mathbf{M} \vee \eta[\mathbf{A}_{o\alpha}\mathbf{x}_\alpha] \vee \eta[\mathbf{A}_{o\alpha}\mathbf{x}_\alpha]$	by 4.4.

Since $\mathbf{A}_{o\alpha}$ is an η-wff, it is easy to see as in 3.3.3.4 that $\eta[\mathbf{A}_{o\alpha}\mathbf{x}_\alpha]$ contains the same number of occurrences of logical constants as does $\mathbf{A}_{o\alpha}$, so

$\vdash \mathbf{M} \vee \eta[\mathbf{A}_{o\alpha}\mathbf{x}_\alpha]$	by inductive hypothesis.
$\vdash \mathbf{M} \vee \mathbf{D}$	by 4.2.2.1 and 4.2.2.7.

4.9.5. $\eta\mathbf{D}$ is an atom, the negation of an atom, or is of the form $\sim\Pi_{o(o\alpha)}\mathbf{A}_{o\alpha}$.

We prove that if $\mathbf{P}^1, \cdots, \mathbf{P}^m$ is any proof in $\mathscr{G}$, and $\mathbf{P}^i$ has disjunctive components $\mathbf{H}$ and $\mathbf{K}$ such that $\eta\mathbf{H} = \eta\mathbf{K}$ and $\eta\mathbf{H}$ has one of these three forms, then the result of dropping $\mathbf{K}$ from $\mathbf{P}^i$ (i.e., replacing a component of $\mathbf{P}^i$ of the form $[\mathbf{C} \vee \mathbf{K}]$ or $[\mathbf{K} \vee \mathbf{C}]$ by $\mathbf{C}$) is a theorem of $\mathscr{G}$. The proof is straightforward by induction on i.

4.10. THEOREM. *If* $\vdash_{\mathscr{T}}\mathbf{A}$ *then* $\vdash_{\mathscr{G}}\mathbf{A}$.

PROOF.

4.10.1. Let $\Gamma\{\mathbf{C}_o^1, \cdots, \mathbf{C}_o^n\}$ mean not $\vdash_{\mathscr{G}}\sim\mathbf{C}_o^1 \vee \cdots \vee \sim\mathbf{C}_o^n$. Note that by 4.2.2.2 this definition is independent of the order in which the wffs $\mathbf{C}^i$ are listed. Also by 4.2.2.3 and 4.9, $\Gamma(\{\mathbf{C}^1, \cdots, \mathbf{C}^n\} \cup \{\mathbf{D}^1, \cdots, \mathbf{D}^m\})$ is equivalent to not $\vdash_{\mathscr{G}}\sim\mathbf{C}^1 \vee \cdots \vee \sim\mathbf{C}^n \vee \sim\mathbf{D}^1 \vee \cdots \vee \sim\mathbf{D}^m$ whether or not some $\mathbf{C}^i$ is the same as some $\mathbf{D}^j$.

4.10.2. We verify that Γ is an abstract consistency property by checking the contrapositive of 3.1.k in step 4.10.2.k below:

4.10.2.1. If $\mathbf{A}$ is an atom, $\vdash\mathbf{M} \vee \sim\sim\mathbf{A} \vee \sim\mathbf{A}$ by 4.2.1, 4.2.2.4, and 4.2.2.3.

4.10.2.2. If $\vdash\mathbf{M} \vee \sim\eta\mathbf{A}$ then $\vdash\mathbf{M} \vee \sim\mathbf{A}$ by 4.2.2.1.

4.10.2.3. If $\vdash\mathbf{M} \vee \sim\mathbf{A}$ then $\vdash\mathbf{M} \vee \sim\sim\sim\mathbf{A}$ by 4.2.2.4.

4.10.2.4. If $\vdash\mathbf{M} \vee \sim\mathbf{A}$ and $\vdash\mathbf{M} \vee \sim\mathbf{B}$ then $\vdash\mathbf{M} \vee \sim[\mathbf{A} \vee \mathbf{B}]$ by 4.2.2.5.

4.10.2.5. If $\vdash\mathbf{M} \vee \sim\sim\mathbf{A} \vee \sim\sim\mathbf{B}$ then $\vdash\mathbf{M} \vee \mathbf{A} \vee \mathbf{B}$ by 4.5, so $\vdash\mathbf{M} \vee \sim\sim[\mathbf{A} \vee \mathbf{B}]$ by 4.2.2.4.

4.10.2.6. If $\vdash\mathbf{M} \vee \sim\Pi_{o(o\alpha)}\mathbf{A}_{o\alpha} \vee \sim\mathbf{A}_{o\alpha}\mathbf{B}_\alpha$ then $\vdash\mathbf{M} \vee \sim\Pi_{o(o\alpha)}\mathbf{A}_{o\alpha}$ by 4.2.2.6.

4.10.2.7. Suppose there is a variable or parameter $\mathbf{c}_\alpha$ which does not occur free in $\mathbf{M}$ or in $\mathbf{A}_{o\alpha}$ such that $\vdash\mathbf{M} \vee \sim\sim\mathbf{A}_{o\alpha}\mathbf{c}_\alpha$. By choosing an appropriate variable $\mathbf{x}_\alpha$ and substituting it for $\mathbf{c}_\alpha$ throughout the proof we obtain $\vdash\mathbf{M} \vee \sim\sim\mathbf{A}_{o\alpha}\mathbf{x}_\alpha$, where $\mathbf{x}_\alpha$ is a variable not free in $\mathbf{M}$ or $\mathbf{A}_{o\alpha}$. Hence $\vdash\mathbf{M} \vee \mathbf{A}_{o\alpha}\mathbf{x}_\alpha$ by 4.5, so $\vdash\mathbf{M} \vee \Pi_{o(o\alpha)}\mathbf{A}_{o\alpha}$ by 4.2.2.7, so $\vdash\mathbf{M} \vee \sim\sim\Pi_{o(o\alpha)}\mathbf{A}_{o\alpha}$ by 4.2.2.4.

4.10.3. Suppose $\vdash_{\mathscr{T}}\mathbf{A}$. Then $\{\sim\mathbf{A}\}$ is inconsistent (in $\mathscr{T}$) so by 3.5, not $\Gamma\{\sim\mathbf{A}\}$, i.e., $\vdash_{\mathscr{G}} \sim\sim\mathbf{A}$, so $\vdash_{\mathscr{G}}\mathbf{A}$ by 4.5.

4.11. COROLLARY. *If* $\vdash_{\mathscr{G}}\mathbf{M} \vee \mathbf{A}$ *and* $\vdash_{\mathscr{G}} \sim\mathbf{A} \vee \mathbf{N}$ *then* $\vdash_{\mathscr{G}}\mathbf{M} \vee \mathbf{N}$.

PROOF. By 4.3 and 4.10, since this result is easy to establish for $\mathscr{T}$.

§5. The resolution system $\mathscr{R}$.

5.1. DEFINITION. Let $\mathscr{S}$ be a finite set of sentences. For each type symbol γ choose a parameter $\mathbf{c}_{\gamma(o\gamma)}$ (henceforth called an *existential parameter*) which does not occur in $\mathscr{S}$. For this choice of existential parameters, a *derivation in* $\mathscr{R}$ *of* $\mathbf{E}$ *from* $\mathscr{S}$ is a finite sequence $\mathbf{D}^1, \cdots, \mathbf{D}^n$ such that $\mathbf{D}^n$ is $\mathbf{E}$ and each $\mathbf{D}^i$ is a member of $\mathscr{S}$ or is obtained from preceding members of the sequence by one of the following *rules of inference*:

5.1.1. *Conversion-I–II.* Apply 2.6.1 or 2.6.2.

5.1.2. *Disjunction rules.* (4.2.2.2).

5.1.3. *Simplification.* From $\mathbf{M} \vee \mathbf{A} \vee \mathbf{A}$ to infer $\mathbf{M} \vee \mathbf{A}$.

5.1.4. *Negation elimination.* From $\mathbf{M} \vee \sim\sim\mathbf{A}$ to infer $\mathbf{M} \vee \mathbf{A}$.

5.1.5. *Conjunction elimination.* From $\mathbf{M} \vee \sim[\mathbf{A} \vee \mathbf{B}]$ to infer $\mathbf{M} \vee \sim\mathbf{A}$ and $\mathbf{M} \vee \sim\mathbf{B}$.

5.1.6. *Existential instantiation.* From $\mathbf{M} \vee \sim\Pi_{o(o\alpha)}\mathbf{A}_{o\alpha}$ to infer $\mathbf{M} \vee \sim\mathbf{A}_{o\alpha}[\mathbf{c}_{\alpha(o\alpha)}\mathbf{A}_{o\alpha}]$.

5.1.7. *Universal instantiation.* From $\mathbf{M} \vee \Pi_{o(o\alpha)}\mathbf{A}_{o\alpha}$ to infer $\mathbf{M} \vee \mathbf{A}_{o\alpha}\mathbf{x}_\alpha$.

5.1.8. *Substitution.* From $\mathbf{A}$ to infer $[\lambda \mathbf{x}_\alpha \mathbf{A}]\mathbf{B}_\alpha$.

5.1.9. *Cut.* From $\mathbf{M} \vee \mathbf{A}$ and $\mathbf{N} \vee \sim\mathbf{A}$ to infer $\mathbf{M} \vee \mathbf{N}$.

A derivation of $\Box$ from $\mathscr{S}$ is a *refutation of* $\mathscr{S}$. In $\mathscr{R}$ one proves a sentence $\mathbf{A}$ by refuting $\sim\mathbf{A}$ (i.e., $\{\sim\mathbf{A}\}$). More generally, one shows that $\mathbf{A}$ follows from a set $\mathscr{H}$ of sentences by refuting $\mathscr{H} \cup \{\sim\mathbf{A}\}$.

5.2. *Remarks.* For convenience, $\mathscr{R}$ has been formulated so that only sets of *sentences* may be refuted in $\mathscr{R}$, but clearly this involves no real loss of generality.

We write $\mathscr{S} \vdash_{\mathscr{R}} \mathbf{E}$ (resp. $\mathscr{S} \vdash_{\mathscr{T}} \mathbf{E}$) iff there is a derivation of $\mathbf{E}$ from $\mathscr{S}$ in $\mathscr{R}$ (resp. in $\mathscr{T}$). For $\mathscr{T}$ this notion is defined, and the deduction theorem is proved, in [2, § 5]. In a proof in $\mathscr{T}$ from assumptions $\mathscr{S}$ one may not generalize upon or substitute for a variable which is free in a wff of $\mathscr{S}$.

The reader may be bothered by the presence of the cut rule 5.1.9 among the rules of inference for $\mathscr{R}$, since we showed in 4.11 that this need not be taken as a primitive rule of inference in $\mathscr{G}$. However, since one proves wffs in $\mathscr{G}$, but refutes them in $\mathscr{R}$, the role of the cut rule is quite different in the two systems. One is tempted to establish the completeness of $\mathscr{G}$ and $\mathscr{R}$ directly with a proof by induction on i that if $\mathbf{D}^1, \cdots, \mathbf{D}^n$ is a proof in $\mathscr{T}$, then $\vdash_{\mathscr{G}} \mathbf{D}^i$ and $\sim\overline{\mathbf{D}}^i \vdash_{\mathscr{R}} \Box$, where $\overline{\mathbf{D}}^i$ is obtained from $\mathbf{D}^i$ upon replacing free variables by new parameters in one-one fashion. In each case the crucial difficulty arises when $\mathbf{D}^i$ is inferred by modus ponens. In $\mathscr{G}$ one can overcome this difficulty by proving that the cut rule is a derived rule of inference. However, in $\mathscr{R}$ the analogous metatheorem is that if $\mathscr{S} \cup \{\mathbf{A}\} \vdash \Box$ and $\mathscr{S} \cup \{\sim\mathbf{A}\} \vdash \Box$ then $\mathscr{S} \vdash \Box$.

The wffs $\mathbf{c}_{\alpha(o\alpha)}\mathbf{A}_{o\alpha}$ introduced by 5.1.6 are essentially Herbrand–Skolem functors whose arguments are the free variables of $\mathbf{A}_{o\alpha}$. Suppose one is given $\mathbf{M} \vee \exists \mathbf{x}_\alpha \mathbf{B}_{o\alpha}\mathbf{x}_\alpha$, where the free variables of $\mathbf{B}_{o\alpha}$ are $\mathbf{y}^1_{\beta_1}, \cdots, \mathbf{y}^n_{\beta_n}$ and do not include $\mathbf{x}_\alpha$. (Matters may be so arranged that one may assume $\mathbf{y}^1_{\beta_1}, \cdots, \mathbf{y}^n_{\beta_n}$ were previously introduced by 5.1.7.) The given wff is $\mathbf{M} \vee \sim\Pi_{o(o\alpha)}[\lambda \mathbf{x}_\alpha \sim \mathbf{B}_{o\alpha}\mathbf{x}_\alpha]$, so by 5.1.6, 5.1.1, and 5.1.4 one obtains $\mathbf{M} \vee \mathbf{B}_{o\alpha}[\mathbf{c}_{\alpha(o\alpha)} \centerdot \lambda \mathbf{x}_\alpha \centerdot \sim\mathbf{B}_{o\alpha}\mathbf{x}_\alpha]$. One may write $[\mathbf{c}_{\alpha(o\alpha)} \centerdot \lambda \mathbf{x}_\alpha \centerdot \sim\mathbf{B}_{o\alpha}\mathbf{x}_\alpha]$ as $\mathbf{f}_{\alpha\beta_n \cdots \beta_1}\mathbf{y}^1_{\beta_1} \cdots \mathbf{y}^n_{\beta_n}$, where $\mathbf{f}_{\alpha\beta_n \cdots \beta_1}$ is a new function symbol. Thus one replaces $\mathbf{M} \vee \exists \mathbf{x}_\alpha \mathbf{B}_{o\alpha}\mathbf{x}_\alpha$ by $\mathbf{M} \vee \mathbf{B}_{o\alpha}[\mathbf{f}_{\alpha\beta_n \cdots \beta_1}\mathbf{y}^1_{\beta_1} \cdots \mathbf{y}^n_{\beta_n}]$.

When one sets out to refute a set of sentences by resolution [8] in first order logic, one eliminates all propositional connectives except negation, conjunction, and disjunction, and pushes negations in so that they have the smallest possible scope, with double negations being dropped. Then one eliminates existential quantifiers by the method of Herbrand–Skolem functors, and drops universal quantifiers. The resulting quantifier-free wffs are put into conjunctive normal form, whose conjuncts are called *clauses*. One then derives $\Box$ from this set of clauses by an operation called resolution, which is an elegant combination of substitution and cut (with 5.1.2 and 5.1.3 used implicitly). (An important open problem concerning resolution in type theory is to find an equally elegant way of combining 5.1.8, 5.1.1, and 5.1.9.) However, in type theory one may introduce new occurrences of logical constants by the substitution rule, so one must continually have available the rules 5.1.4–5.1.7 which correspond to the preliminary processing in first order logic. However, 5.1.8 and 5.1.9 (in conjunction with the subsidiary rules 5.1.1–5.1.3) remain the crucial rules of inference.

When applying Rule 5.1.7, one might as well choose $\mathbf{x}_\alpha$ to be distinct from the free variables of $\mathbf{M} \vee \Pi_{o(o\alpha)}\mathbf{A}_{o\alpha}$, since one can identify $\mathbf{x}_\alpha$ with another variable later by a substitution, if desired. If $\mathbf{x}_\alpha$ is so chosen, one might as well apply 5.1.3–5.1.7 immediately whenever these rules are applicable, and then discard the wffs to which these rules are applied, since they need not be used again.

5.3. THEOREM. *Let $\mathscr{S}$ be a finite set of sentences. If $\mathscr{S} \vdash_{\mathscr{T}} \Box$ then $\mathscr{S} \vdash_{\mathscr{R}} \Box$.*

PROOF.

5.3.1. For any finite set $\mathscr{S}$ of wffs_o, let $\Gamma(\mathscr{S})$ mean not $\mathscr{S}' \vdash_{\mathscr{R}} \Box$, where $\mathscr{S}'$ is obtained from $\mathscr{S}$ by replacing the free variables in wffs of $\mathscr{S}$ by new parameters in a one-one fashion. We shall show that Γ is an abstract consistency property, so if $\mathscr{S}$ is a set of sentences such that $\mathscr{S} \vdash_{\mathscr{T}} \Box$, then $\mathscr{S}$ is inconsistent in $\mathscr{T}$, so by 3.5 not $\Gamma(\mathscr{S})$, i.e. $\mathscr{S} \vdash_{\mathscr{R}} \Box$.

5.3.2. We verify that Γ is an abstract consistency property by checking the contrapositive of 3.1.k in step 5.3.2.k below. For the sake of brevity we shall be rather informal about the distinction between $\mathscr{S}$ and $\mathscr{S}'$, simply assuming that wffs are closed when appropriate.

5.3.2.1. If there is an atom $\mathbf{A}$ such that $\mathbf{A} \in \mathscr{S}$ and $\sim\mathbf{A} \in \mathscr{S}$ then $\mathscr{S} \vdash_{\mathscr{R}} \Box$ by 5.1.9.

5.3.2.2. If $\mathscr{S} \cup \{\eta\mathbf{A}\} \vdash_{\mathscr{R}} \Box$ then $\mathscr{S} \cup \{\mathbf{A}\} \vdash_{\mathscr{R}} \Box$ by 5.1.1.

5.3.2.3. If $\mathscr{S} \cup \{\mathbf{A}\} \vdash_{\mathscr{R}} \Box$ then $\mathscr{S} \cup \{\sim\sim\mathbf{A}\} \vdash_{\mathscr{R}} \Box$ by 5.1.4.

5.3.2.4. Suppose $\mathscr{S} \cup \{\mathbf{A}\} \vdash_{\mathscr{R}} \Box$ and $\mathscr{S} \cup \{\mathbf{B}\} \vdash_{\mathscr{R}} \Box$. We may assume given refutations $\mathbf{C}^1, \cdots, \mathbf{C}^n$ of $\mathscr{S} \cup \{\mathbf{B}\}$ and $\mathbf{E}^1, \cdots, \mathbf{E}^m$ of $\mathscr{S} \cup \{\mathbf{A}\}$ using the same existential parameters. We define which of the wffs $\mathbf{C}^i$ are *derived from* $\mathbf{B}$ in the given refutation in the obvious inductive fashion: if $\mathbf{C}^i$ is in $\mathscr{S} \cup \{\mathbf{B}\}$, then $\mathbf{C}^i$ is derived from $\mathbf{B}$ iff $\mathbf{C}^i$ is $\mathbf{B}$; if $\mathbf{C}^i$ is inferred from $\mathbf{C}^j$ (and $\mathbf{C}^k$), then $\mathbf{C}^i$ is derived from $\mathbf{B}$ iff $\mathbf{C}^j$ (or $\mathbf{C}^k$) is derived from $\mathbf{B}$. We define $\mathbf{D}^i$ (for $1 \leq i \leq n$) to be $\mathbf{A} \vee \mathbf{C}^i$ if $\mathbf{C}^i$ is derived from $\mathbf{B}$; otherwise $\mathbf{D}^i$ is $\mathbf{C}^i$. By examining the rules of inference of $\mathscr{R}$ it is easy to see that $\mathscr{S} \cup \{\mathbf{A} \vee \mathbf{B}\} \vdash \eta\mathbf{D}^i$ for $1 \leq i \leq n$ by induction on i. If $\mathbf{D}^n$ is $\Box$ we are done. Otherwise $\mathbf{D}^n$ is $\mathbf{A}$ so $\mathscr{S} \cup \{\mathbf{A} \vee \mathbf{B}\} \vdash \eta\mathbf{A}$. Now we readily establish $\mathscr{S} \cup \{\mathbf{A} \vee \mathbf{B}\} \vdash \eta\mathbf{E}^i$ for $1 \leq i \leq m$ by induction on i, so $\mathscr{S} \cup \{\mathbf{A} \vee \mathbf{B}\} \vdash \Box$.

5.3.2.5. If $\mathscr{S} \cup \{\sim\mathbf{A}, \sim\mathbf{B}\} \vdash_{\mathscr{R}} \Box$ then $\mathscr{S} \cup \{\sim[\mathbf{A} \vee \mathbf{B}]\} \vdash_{\mathscr{R}} \Box$ by 5.1.5.

5.3.2.6. If there exists a wff $\mathbf{B}_\alpha$ such that $\mathscr{S} \cup \{\Pi_{o(o\alpha)}\mathbf{A}_{o\alpha}, \mathbf{A}_{o\alpha}\mathbf{B}_\alpha\} \vdash_{\mathscr{R}} \Box$ then $\mathscr{S} \cup \{\Pi_{o(o\alpha)}\mathbf{A}_{o\alpha}\} \vdash_{\mathscr{R}} \Box$ by 5.1.7, 5.1.8, and 5.1.1.

5.3.2.7. Suppose there is a parameter $\mathbf{d}_\alpha$ which does not occur in $\mathbf{A}_{o\alpha}$ or any wff of $\mathscr{S}$ such that $\mathscr{S} \cup \{\sim\mathbf{A}_{o\alpha}\mathbf{d}_\alpha\} \vdash_{\mathscr{R}} \Box$. Let a refutation of $\mathscr{S} \cup \{\sim\mathbf{A}_{o\alpha}\mathbf{d}_\alpha\}$ be given with existential parameter $\mathbf{c}_{\alpha(o\alpha)}$. Since $[\mathbf{c}_{\alpha(o\alpha)}\mathbf{A}_{o\alpha}]$ is a closed wff it is easy to see that one can replace $\mathbf{d}_\alpha$ by $[\mathbf{c}_{\alpha(o\alpha)}\mathbf{A}_{o\alpha}]$ everywhere in the given refutation to obtain a refutation of $\mathscr{S} \cup \{\sim\Pi_{o(o\alpha)}\mathbf{A}_{o\alpha}\}$, using 5.1.6 to infer $\sim\mathbf{A}_{o\alpha}[\mathbf{c}_{\alpha(o\alpha)}\mathbf{A}_{o\alpha}]$.

§6. Remarks and examples for $\mathscr{R}$.

6.1. When one sets out to prove in $\mathscr{R}$ a theorem of some branch of mathematics, one of course assumes as hypotheses the postulates of that branch of mathematics. In addition certain assumptions which are used in all branches of mathematics, and which in other contexts would be regarded as axioms of the underlying logic, should be taken as hypotheses. Among these we mention the *axioms of extensionality*:

6.1.1°. $$\forall p_o \forall q_o \,.\, [p_o \equiv q_o] \supset . \, p_o = q_o$$

6.1.1$^{(\alpha\beta)}$. $$\forall f_{\alpha\beta} \forall g_{\alpha\beta} \,.\, \forall x_\beta [f_{\alpha\beta} x_\beta = g_{\alpha\beta} x_\beta] \supset . \, f_{\alpha\beta} = g_{\alpha\beta}$$

and the *axiom of descriptions*:

6.1.2. $$\exists i_{\iota(o\iota)} \forall f_{o\iota} \,.\, \exists_1 x_\iota f_{o\iota} x_\iota \supset f_{o\iota}[i_{\iota(o\iota)} f_{o\iota}].$$

In addition one may wish to assume some formulation of the axiom of choice (in which case 6.1.2 is dispensable) and an axiom of infinity.

Of course there are infinitely many axioms of extensionality, and it may not be obvious which of these may be needed to prove a particular theorem. However, when implementing the system it should be possible to treat the α and β of $6.1.1^{(\alpha\beta)}$ as special variables (*type variables*, in the terminology of [1]) for which one can substitute particular type symbols as necessary.

6.2. In the examples below we shall use letters with bars over them for parameters. Thus $\bar{o}_\iota$ and $\bar{s}_{\iota\iota}$ in 6.3 are parameters. For the sake of brevity we shall introduce Herbrand–Skolem functors as abbreviations in the manner discussed in 5.2. We shall call such a functor with its arguments an *existential term*. Since applications of 5.1.1–5.1.7 are routine we shall usually leave it to the reader to determine which of these rules are being used. However, we shall indicate (at the right-hand margin) from which line(s) a given line is inferred if it is not inferred from the line immediately preceding it. The reader will quickly discover the advantage of formulating derived rules of inference to speed up these manipulations. We here discuss only two such rules, which we shall need in 6.4.

6.2.1. If $\mathscr{S} \vdash_{\mathscr{R}} \sim[\mathbf{A}_\alpha = \mathbf{A}_\alpha] \vee \mathbf{B}$ then $\mathscr{S} \vdash_{\mathscr{R}} \mathbf{B}$.

PROOF. From the given wff by 2.4.8 and 5.1.1 we obtain

.1 $\sim\forall \mathbf{f}_{o\alpha}[\sim \mathbf{f}_{o\alpha}\mathbf{A}_\alpha \vee \mathbf{f}_{o\alpha}\mathbf{A}_\alpha] \vee \mathbf{B}$ where $\mathbf{f}_{o\alpha}$ is not free in $\mathbf{A}_\alpha$.
.2 $\sim[\sim \mathbf{F}_{o\alpha}\mathbf{A}_\alpha \vee \mathbf{F}_{o\alpha}\mathbf{A}_\alpha] \vee \mathbf{B}$ where $\mathbf{F}_{o\alpha}$ is an existential term.
.3 $\mathbf{F}_{o\alpha}\mathbf{A}_\alpha \vee \mathbf{B}$ — .2.
.4 $\sim \mathbf{F}_{o\alpha}\mathbf{A}_\alpha \vee \mathbf{B}$ — .2.
.5 $\mathbf{B}$ — cut: .3, .4.

6.2.2. If $\mathbf{A}_\alpha$ and $\mathbf{B}_\alpha$ are free for $\mathbf{x}_\alpha$ in $\mathbf{C}$, and $\mathscr{S} \vdash_{\mathscr{R}} \mathbf{N} \vee \mathsf{S}^{\mathbf{x}_\alpha}_{\mathbf{A}_\alpha}\mathbf{C}$, and $\mathscr{S} \vdash_{\mathscr{R}} \mathbf{M} \vee [\mathbf{A}_\alpha = \mathbf{B}_\alpha]$ or $\mathscr{S} \vdash_{\mathscr{R}} \mathbf{M} \vee [\mathbf{B}_\alpha = \mathbf{A}_\alpha]$, then $\mathscr{S} \vdash_{\mathscr{R}} \mathbf{M} \vee \mathbf{N} \vee \mathsf{S}^{\mathbf{x}_\alpha}_{\mathbf{B}_\alpha}\mathbf{C}$.

PROOF for the case $[\mathbf{A}_\alpha = \mathbf{B}_\alpha]$. Let $\mathbf{f}_{o\alpha}$ be a variable not free in $\mathbf{A}_\alpha$, $\mathbf{B}_\alpha$, or $\mathbf{M}$.

.1 $\mathbf{M} \vee \forall \mathbf{f}_{o\alpha} \,.\, \sim \mathbf{f}_{o\alpha}\mathbf{A}_\alpha \vee \mathbf{f}_{o\alpha}\mathbf{B}_\alpha$ — given
.2 $\mathbf{M} \vee \sim \mathbf{f}_{o\alpha}\mathbf{A}_\alpha \vee \mathbf{f}_{o\alpha}\mathbf{B}_\alpha$
.3 $\mathbf{M} \vee \sim[\lambda \mathbf{x}_\alpha \mathbf{C}]\mathbf{A}_\alpha \vee [\lambda \mathbf{x}_\alpha \mathbf{C}]\mathbf{B}_\alpha$ — Sub
.4 $\mathbf{M} \vee \sim(\mathsf{S}^{\mathbf{x}_\alpha}_{\mathbf{A}_\alpha}\mathbf{C}) \vee \mathsf{S}^{\mathbf{x}_\alpha}_{\mathbf{B}_\alpha}\mathbf{C}$
.5 $\mathbf{N} \vee \mathsf{S}^{\mathbf{x}_\alpha}_{\mathbf{A}_\alpha}\mathbf{C}$ — given
.6 $\mathbf{M} \vee \mathbf{N} \vee \mathsf{S}^{\mathbf{x}_\alpha}_{\mathbf{B}_\alpha}\mathbf{C}$ — cut: .4, .5

In the case $[\mathbf{B}_\alpha = \mathbf{A}_\alpha]$ substitute $[\lambda \mathbf{x}_\alpha \sim \mathbf{C}]$ for $\mathbf{f}_{o\alpha}$ in the line corresponding to .2.

6.3. *Example.* Let $\mathrm{N}_{o\iota}$ stand for

$$[\lambda n_\iota \,.\, \forall p_{o\iota} \,.\, [p\bar{o}_\iota \wedge \forall x_\iota \,.\, px \supset p \,.\, \bar{s}_{\iota\iota}x] \supset pn].$$

$\mathrm{N}_{o\iota}$ denotes the set of natural numbers when $\bar{o}_\iota$ denotes zero and $\bar{s}_{\iota\iota}$ denotes the successor function. We prove $\forall y_\iota[\mathrm{N}y \supset \mathrm{N} \,.\, \bar{s}y]$ by refuting its negation in $\mathscr{R}$.

.1 $\sim\forall y_\iota \,.\, \mathrm{N}y \supset \mathrm{N} \,.\, \bar{s}y$ — given
.2 $\mathrm{N}\bar{y}_\iota$ — .1
.3 $\sim\mathrm{N} \,.\, \bar{s}\bar{y}_\iota$ — .1

.4	$\sim .[\bar{p}_{o\iota}\bar{o} \wedge \forall x_\iota \,.\, \bar{p}x \supset \bar{p}\bar{s}x] \supset \bar{p}\bar{s}\bar{y}_\iota$	
*.5	$\bar{p}_{o\iota}\bar{o}$	.4
*.6	$\sim\bar{p}_{o\iota}x_\iota \vee \bar{p}\bar{s}x$	.4
*.7	$\sim\bar{p}_{o\iota}\bar{s}\bar{y}_\iota$	.4
.8	$\sim p_{o\iota}\bar{o} \vee \sim\forall x_\iota[\sim px \vee p \,.\, \bar{s}x] \vee p\bar{y}_\iota$	.2
*.9	$\sim p_{o\iota}\bar{o} \vee \sim[\sim p[\bar{x}_{\iota(o\iota)}p] \vee p \,:\, \bar{s} \,.\, \bar{x}p] \vee p\bar{y}_\iota.$	

Lines .5–.7 and .9 were obtained routinely from .1, and □ must be derived from these. We could apply 5.1.5 to .9, but it is convenient to postpone this.

.10	$\sim\bar{p}_{o\iota}\bar{y}_\iota$	Sub: .6; cut: .7
.11	$\sim :\sim\bar{p}_{o\iota}[\bar{x}_{\iota(o\iota)}\bar{p}] \vee \bar{p} \,.\, \bar{s} \,.\, \bar{x}\bar{p}$	Sub: .9; cut: .5, .10
.12	$\bar{p}_{o\iota}[\bar{x}_{\iota(o\iota)}\bar{p}]$	.11
.13	$\sim\bar{p}_{o\iota} \,.\, \bar{s} \,.\, \bar{x}_{\iota(o\iota)}\bar{p}$	.11
.14	□	Sub: .6; cut: .12, .13

6.4. *Example.* For a somewhat less trivial example, we prove that if some iterate of a function f has a unique fixed point, then f has a fixed point. (This example is suggested by [6].)

Let $J_{o(\iota\iota)(\iota\iota)}$ stand for

$$[\lambda f_{\iota\iota}\lambda g_{\iota\iota}\forall p_{o(\iota\iota)} \,.\, [pf \wedge \forall h_{\iota\iota} \,.\, ph \supset p \,.\, \lambda t_\iota \,.\, f \,.\, ht] \supset pg].$$

Then $Jf_{\iota\iota}g_{\iota\iota}$ means g is an iterate of f, i.e., g is in the intersection of all sets p which contain f such that p contains $f \circ h$ whenever p contains h.

We wish to prove

(*)	$\forall f_{\iota\iota} \,.\, \exists g_{\iota\iota}[Jfg \wedge \exists_1 x_\iota \,.\, gx = x] \supset \exists y_\iota \,.\, fy = y$	
.1	$\sim(*)$	given
.2	$J\bar{f}_{\iota\iota}\bar{g}_{\iota\iota}$	.1
.3	$\exists_1 x_\iota \,.\, \bar{g}_{\iota\iota}x = x$	.1
*.4	$\sim\bar{f}_{\iota\iota}y_\iota = y$	.1
*.5	$\bar{g}_{\iota\iota}\bar{x}_\iota = \bar{x}_\iota$	.3
*.6	$\sim\bar{g}_{\iota\iota}z_\iota = z_\iota \vee z_\iota = \bar{x}_\iota$	.3
*.7	$\sim p_{o(\iota\iota)}\bar{f}_{\iota\iota} \vee \sim[\sim p[\bar{h}_{\iota\iota(o(\iota\iota))}p] \vee p \,.\, \lambda t_\iota \,.\, \bar{f} \,.\, [\bar{h}p]t] \vee p\bar{g}_{\iota\iota}$	.2

We must derive □ from .4, .5, .6, and .7. We could break down .4, .5, and .6 further using the definition of equality, but we prefer to rely on 6.2.1 and 6.2.2.

Next we substitute $[\lambda k_{\iota\iota} \,.\, k[\bar{f}_{\iota\iota}\bar{x}_\iota] = \bar{f} \,.\, k\bar{x}]$ for $p_{o(\iota\iota)}$ in .7, and write the existential term corresponding to $\bar{h}_{\iota\iota(o(\iota\iota))}p_{o(\iota\iota)}$ simply as $\bar{h}_{\iota\iota}$ to obtain

.8	$\sim\bar{f}_{\iota\iota}\bar{f}\bar{x}_\iota = \bar{f}\bar{f}\bar{x} \vee \sim[\sim\bar{h}_{\iota\iota}\bar{f}\bar{x} = \bar{f}\bar{h}\bar{x} \vee \bar{f}\bar{h}\bar{f}\bar{x} = \bar{f}\bar{f}\bar{h}\bar{x}] \vee \bar{g}_{\iota\iota}\bar{f}\bar{x} = \bar{f}\bar{g}\bar{x}$	.7

Applying 6.2.1 to .8 we obtain .9 and .10 below:

.9	$\bar{h}_{\iota\iota}\bar{f}_{\iota\iota}\bar{x}_\iota = \bar{f}\bar{h}\bar{x} \vee \bar{g}_{\iota\iota}\bar{f}\bar{x} = \bar{f}\bar{g}\bar{x}$	.8
.10	$\sim\bar{f}_{\iota\iota}\bar{h}_{\iota\iota}\bar{f}_{\iota\iota}\bar{x}_\iota = \bar{f}\bar{f}\bar{h}\bar{x} \vee \bar{g}_{\iota\iota}\bar{f}\bar{x} = \bar{f}\bar{g}\bar{x}$	.8
.11	$\sim\bar{f}_{\iota\iota}\bar{f}\bar{h}_{\iota\iota}\bar{x}_\iota = \bar{f}\bar{f}\bar{h}\bar{x} \vee \bar{g}_{\iota\iota}\bar{f}\bar{x} = \bar{f}\bar{g}\bar{x}$	6.2.2: .9, .10
.12	$\bar{g}_{\iota\iota}\bar{f}_{\iota\iota}\bar{x}_\iota = \bar{f}\bar{g}\bar{x}$	6.2.1: .11
.13	$\bar{g}_{\iota\iota}\bar{f}_{\iota\iota}\bar{x}_\iota = \bar{f}\bar{x}$	6.2.2: .5, .12
.14	$\bar{f}_{\iota\iota}\bar{x}_\iota = \bar{x}_\iota$	Sub: .6; cut: .13
.15	□	Sub: .4; cut: .14

BIBLIOGRAPHY

[1] PETER B. ANDREWS, ***A transfinite type theory with type variables***, North-Holland Publishing Company, Amsterdam, 1965.

[2] ALONZO CHURCH, *A formulation of the simple theory of types*, this JOURNAL, vol. 5 (1940), pp. 56–68.

[3] WILLIAM EBEN GOULD, *A matching procedure for ω-order logic*, Ph.D. thesis, Princeton University, 1966; reprinted as Sci. Rep. No. 4 AFCRL 66-781, Oct. 15, 1966 (Contract No. AF 19(628)-3250), AD 646 560.

[4] LEON HENKIN, *Completeness in the theory of types*, this JOURNAL, vol. 15 (1950), pp. 81–91.

[5] ROGER HINDLEY, *An abstract form of the Church-Rosser theorem. I*, this JOURNAL, vol. 34 (1969), pp. 545–560.

[6] IGNACE I. KOLODNER, *Fixed points*, ***American Mathematical Monthly***, vol. 71 (1964), p. 906.

[7] DAG PRAWITZ, *Hauptsatz for higher order logic*, this JOURNAL, vol. 33 (1968), pp. 452–457.

[8] J. A. ROBINSON, *A machine-oriented logic based on the resolution principle*, ***Journal of the Association for Computing Machinery***, vol. 12 (1965), pp. 23–41.

[9] KURT SCHÜTTE, *Syntactical and semantical properties of simple type theory*, this JOURNAL, vol. 25 (1960), pp. 305–326.

[10] RAYMOND M. SMULLYAN, *A unifying principle in quantification theory*, ***Proceedings of the National Academy of Sciences***, vol. 49 (1963), pp. 828–832.

[11] ———, ***First-order logic***, Springer-Verlag, New York Inc., 1968.

[12] MOTO-O-TAKAHASHI, *A proof of cut-elimination in simple type theory*, ***Journal of the Mathematical Society of Japan***, vol. 19 (1967), pp. 399–410.

CARNEGIE-MELLON UNIVERSITY,
PITTSBURGH, PENNSYLVANIA 15213

THE JOURNAL OF SYMBOLIC LOGIC
Volume 37, Number 2, June 1972

GENERAL MODELS, DESCRIPTIONS, AND CHOICE IN TYPE THEORY

PETER B. ANDREWS[1]

§1. Introduction. In [4] Alonzo Church introduced an elegant and expressive formulation of type theory with λ-conversion. In [8] Henkin introduced the concept of a general model for this system, such that a sentence **A** is a theorem if and only if it is true in all general models. The crucial clause in Henkin's definition of a general model $\mathcal{M}$ is that for each assignment φ of values in $\mathcal{M}$ to variables and for each wff **A**, there must be an appropriate value $\mathcal{V}_\varphi\mathbf{A}$ of **A** in $\mathcal{M}$. Hintikka points out in [10, p. 3] that this constitutes a rather strong requirement concerning the structure of a general model. Henkin draws attention to the problem of constructing non-standard models for the theory of types in [9, p. 324].

We shall use a simple idea of combinatory logic to find a characterization of general models which does not directly refer to wffs, and which is easier to work with in certain contexts. This characterization can be applied, with appropriate minor and obvious modifications, to a variety of formulations of type theory with λ-conversion. We shall be concerned with a language $\mathcal{L}$ with extensionality in which there is no description or selection operator, and in which (for convenience) the sole primitive logical constants are the equality symbols $Q_{o\alpha\alpha}$ for each type α.

We shall give two applications of this characterization. First, we show that the Axiom of Descriptions (D) is independent of $\mathcal{L}$. This axiom is very natural since a general model for $\mathcal{L}$ with a finite domain of individuals is standard if and only if D is true in it. Secondly, we show how the Fraenkel-Mostowski method [7], [11], [12] can be adapted to $\mathcal{L}$. We state our fundamental lemma concerning this method in fairly general form to facilitate possible future applications (analogous to those for axiomatic set theory mentioned in [11]), but confine ourselves here to simply showing that the Axiom of Choice is not derivable in $\mathcal{L}$, even if the Axiom of Descriptions is assumed.

When a description operator $\iota_{\iota(o\iota)}$ is included among the primitive symbols,[2] the Axiom of Descriptions may be taken in the form

$$\forall p_{o\iota}\centerdot\exists_1 x_\iota p_{o\iota}x_\iota \supset p_{o\iota}[\iota_{\iota(o\iota)}p_{o\iota}],$$

so that $\iota_{\iota(o\iota)}[\lambda\mathbf{x}_\iota\mathbf{A}_o]$ (which is abbreviated $(\imath\mathbf{x}_\iota\mathbf{A}_o)$) denotes the unique $\mathbf{x}_\iota$ such that $\mathbf{A}_o$, when there is such an $\mathbf{x}_\iota$. Church showed in [4] that description operators for higher types can be introduced by definition, using the operators for lower types. Specifically, $\iota_{\alpha\beta(o(\alpha\beta))}$ may be defined as

$$[\lambda h_{o(\alpha\beta)}\lambda x_\beta\iota_{\alpha(o\alpha)}\lambda y_\alpha\exists f_{\alpha\beta}\centerdot h_{o(\alpha\beta)}f_{\alpha\beta} \wedge y_\alpha = f_{\alpha\beta}x_\beta].$$

Received June 23, 1971.

[1] This research was partially supported by NSF Grant GJ-580.

[2] We follow the well established tradition of Church [4] in which the Greek letter ι is used as a type symbol for the type of individuals, and also, when given an appropriate type symbol subscript, as a primitive constant denoting a description operator. In practice this causes no confusion, since ι usually appears as a subscript in the former usage, but rarely in the latter.

R. O. Gandy has pointed out (in a private communication) that $\iota_{o\beta(o(o\beta))}$ can be defined as

$$[\lambda h_{o(o\beta)}\lambda x_\beta\exists f_{o\beta}.h_{o(o\beta)}f_{o\beta} \wedge f_{o\beta}x_\beta],$$

so description operators for certain higher types can be defined without using those for any other type. Also, Henkin noted in [9] that $\iota_{o(oo)}$ can be defined as

$$[\lambda h_{oo}.h_{oo} = [\lambda x_o x_o]].$$

(A number of other definitions of $\iota_{o(oo)}$ are also possible, of which the shortest is perhaps the closely related $Q_{o(oo)(oo)}[\lambda x_o x_o]$.) Thus it is seen that description operators for all types can be introduced once one has $\iota_{\iota(o\iota)}$. The argument in [2, pp. 22–24] shows that the description operator $\iota_{\iota(o\iota)}$ cannot be introduced by definition for the simple reason that there are no closed wffs of this type, and that the Axiom of Descriptions mentioned above is independent, since it is the sole axiom which describes the special characteristics of $\iota_{\iota(o\iota)}$.

If no description operator is included in the list of primitive symbols, the Axiom of Descriptions may be taken in the form

$$\exists i_{\iota(o\iota)}\forall p_{o\iota}.\exists_1 x_\iota p_{o\iota}x_\iota \supset p_{o\iota}[i_{\iota(o\iota)}p_{o\iota}],$$

or equivalently

(D) $$\exists i_{\iota(o\iota)}\forall x_\iota.i_{\iota(o\iota)}[Q_{o\iota\iota}x_\iota] = x_\iota.$$

(The equivalence results from the theorem $\exists_1 x_\iota p_{o\iota}x_\iota = \exists x_\iota.p_{o\iota} = .Q_{o\iota\iota}x_\iota$.) Since in many logical systems descriptions can be eliminated, it is very natural to ask whether the wff D, which asserts the existence of a description operator, is in fact derivable. It will be seen that our independence proof below is conceptually very simple, and is compatible with any axioms concerning the cardinality of the domain of individuals which permit it to have at least two members.

Church mentions in [5] an unpublished proof by Lagerström of a closely related independence result using a complete nonatomic Boolean algebra for the domain of truth values. It seems unlikely that Lagerström's proof applies to $\mathscr{L}$, since in $\mathscr{L}$, unlike the system treated by Lagerström, there is a strong axiom of extensionality for type o (Axiom 1 below) which permits one to derive $[p_o \equiv q_o] \supset .p_o = q_o$.

§2. The language $\mathscr{L}$. The language $\mathscr{L}$ is essentially the result of dropping the description operator from the language Q_o of [2], and is closely related to the system discussed in [9]. For the convenience of the reader we here provide a description of $\mathscr{L}$.

We use α, β, γ, etc., as syntactical variables ranging over *type symbols*, which are defined inductively as follows:

(a) o is a type symbol (denoting the type of truth values).

(b) ι is a type symbol (denoting the type of individuals).

(c) $(\alpha\beta)$ is a type symbol (denoting the type of functions from elements of type β to elements of type α).

The *primitive symbols* of $\mathscr{L}$ are the following:

(a) Improper symbols: [ ] λ.

(b) For each α, a denumerable list of *variables* of type α:

$$f_\alpha g_\alpha h_\alpha \cdots x_\alpha y_\alpha z_\alpha f^1_\alpha g^1_\alpha \cdots z^1_\alpha f^2_\alpha \cdots.$$

We shall use $\mathbf{f}_\alpha$, $\mathbf{g}_\alpha, \cdots$, $\mathbf{x}_\alpha$, $\mathbf{y}_\alpha$, $\mathbf{z}_\alpha$, etc., as syntactical variables for variables of type α.

(c) For each α, $Q_{((o\alpha)\alpha)}$ is a *constant* of type $((o\alpha)\alpha)$.

We write *wff*$_\alpha$ as an abbreviation for *wff of type* α, and use $\mathbf{A}_\alpha$, $\mathbf{B}_\alpha$, $\mathbf{C}_\alpha$, etc., as syntactical variables ranging over wffs$_\alpha$, which are defined inductively as follows:

(a) A primitive variable or constant of type α is a wff$_\alpha$.

(b) $[\mathbf{A}_{\alpha\beta}\mathbf{B}_\beta]$ is a wff$_\alpha$.

(c) $[\lambda\mathbf{x}_\beta\mathbf{A}_\alpha]$ is a wff$_{(\alpha\beta)}$.

An occurrence of $\mathbf{x}_\alpha$ is *bound* (*free*) in $\mathbf{B}_\beta$ iff it is (is not) in a wf part of $\mathbf{B}_\beta$ of the form $[\lambda\mathbf{x}_\alpha\mathbf{C}_\delta]$. $\mathbf{A}_\alpha$ is *free for* $\mathbf{x}_\alpha$ in $\mathbf{B}_\beta$ iff no free occurrence of $\mathbf{x}_\alpha$ in $\mathbf{B}_\beta$ is in a wf part of $\mathbf{B}_\beta$ of the form $[\lambda\mathbf{y}_\gamma\mathbf{C}_\delta]$ such that $\mathbf{y}_\gamma$ is a free variable of $\mathbf{A}_\alpha$.

Brackets, parentheses in type symbols, and type symbols may be omitted when no ambiguity is thereby introduced. A dot stands for a left bracket whose mate is as far to the right as is consistent with the pairing of brackets already present and with the formula being well formed. Otherwise brackets and parentheses are to be restored using the convention of association to the left.

We introduce the following definitions and abbreviations:

$[\mathbf{A}_\alpha = \mathbf{B}_\alpha]$	stands for	$[Q_{o\alpha\alpha}\mathbf{A}_\alpha\mathbf{B}_\alpha]$.
T_o	stands for	$[Q_{ooo} = Q_{ooo}]$.
F_o	stands for	$[\lambda p_o p_o] = [\lambda p_o T_o]$.
$[\forall\mathbf{x}_\alpha\mathbf{A}_o]$	stands for	$[\lambda\mathbf{x}_\alpha\mathbf{A}_o] = [\lambda\mathbf{x}_\alpha T_o]$.
$\wedge_{ooo}$	stands for	$[\lambda p_o\lambda q_o{\bullet}[\lambda g_{ooo}{\bullet}g_{ooo}p_o q_o] = [\lambda g_{ooo}{\bullet}g_{ooo}T_o T_o]]$.
$[\mathbf{A}_o \wedge \mathbf{B}_o]$	stands for	$[\wedge_{ooo}\mathbf{A}_o\mathbf{B}_o]$.
$\supset_{ooo}$	stands for	$[\lambda p_o\lambda q_o{\bullet}p_o \wedge q_o = p_o]$.
$[\mathbf{A}_o \supset \mathbf{B}_o]$	stands for	$[\supset_{ooo}\mathbf{A}_o\mathbf{B}_o]$.

Other connectives and quantifiers are introduced in familiar ways.

$K^{\alpha\beta}$ and $K_{\alpha\beta\alpha}$	stand for	$[\lambda x_\alpha\lambda y_\beta x_\alpha]$.
$S^{\alpha\beta\gamma}$ and $S_{\alpha\gamma(\beta\gamma)(\alpha\beta\gamma)}$	stand for	$[\lambda x_{\alpha\beta\gamma}\lambda y_{\beta\gamma}\lambda z_\gamma{\bullet}x_{\alpha\beta\gamma}z_\gamma{\bullet}y_{\beta\gamma}z_\gamma]$.
$B^{\alpha\beta\gamma}$ and $B_{\alpha\gamma(\beta\gamma)(\alpha\beta)}$	stand for	$[\lambda f_{\alpha\beta}\lambda g_{\beta\gamma}\lambda x_\gamma{\bullet}f_{\alpha\beta}{\bullet}g_{\beta\gamma}x_\gamma]$.
$C^{\alpha\beta\gamma}$ and $C_{\alpha\gamma\beta(\alpha\beta\gamma)}$	stand for	$[\lambda f_{\alpha\beta\gamma}\lambda x_\beta\lambda y_\gamma{\bullet}f_{\alpha\beta\gamma}y_\gamma x_\beta]$.
$W^{\alpha\beta}$ and $W_{\alpha\beta(\alpha\beta\beta)}$	stand for	$[\lambda f_{\alpha\beta\beta}\lambda x_\beta{\bullet}f_{\alpha\beta\beta}x_\beta x_\beta]$.
$\underset{\bullet}{S}{}^{\mathbf{x}_\alpha}_{\mathbf{A}_\alpha}\mathbf{B}_\beta$	stands for	the result of substituting $\mathbf{A}_\alpha$ for $\mathbf{x}_\alpha$ at all free occurrences of $\mathbf{x}_\alpha$ in $\mathbf{B}_\beta$.

$\mathscr{L}$ has a single rule of inference, which is the following:

Rule R. From $\mathbf{C}_o$ and $[\mathbf{A}_\alpha = \mathbf{B}_\alpha]$ to infer the result of replacing one occurrence of $\mathbf{A}_\alpha$ (which is not an occurrence of a variable immediately preceded by λ) in $\mathbf{C}_o$ by an occurrence of $\mathbf{B}_\alpha$.

The axioms and axiom schemata for $\mathscr{L}$ are the following:

(1) $[g_{oo}T_o \wedge g_{oo}F_o] = \forall x_o{\bullet}g_{oo}x_o$.

(2) $x_\alpha = y_\alpha \supset {\bullet}h_{o\alpha}x_\alpha = h_{o\alpha}y_\alpha$.

(3) $f_{\alpha\beta} = g_{\alpha\beta} = \forall x_\beta{\bullet}f_{\alpha\beta}x_\beta = g_{\alpha\beta}x_\beta$.

(4) $[\lambda\mathbf{x}_\alpha\mathbf{B}_\beta]\mathbf{A}_\alpha = \underset{\bullet}{S}{}^{\mathbf{x}_\alpha}_{\mathbf{A}_\alpha}\mathbf{B}_\beta$, where $\mathbf{A}_\alpha$ is free for $\mathbf{x}_\alpha$ in $\mathbf{B}_\beta$.

Let us denote by $\mathscr{T}\mathscr{E}$ the system obtained when the axioms of extensionality (6.1.1 of [3]) are added to the list of axioms of the system $\mathscr{T}$ of [3]. This is essentially the system of [8] or [4] using axioms 1–6, 10°, $10^{\alpha\beta}$, and with the selection operators deleted. $\mathcal{TE}$ differs from $\mathcal{L}$ in having primitive constants $\sim_{oo}$, $\vee_{ooo}$, and $\Pi_{o(o\alpha)}$ instead of $Q_{o\alpha\alpha}$. There are natural translations Δ from $\mathcal{L}$ into $\mathcal{TE}$ and ∇ from $\mathcal{TE}$ into $\mathcal{L}$ which involve replacing the primitive constants of one language by appropriate closed wffs of the other language. For example, if $\mathbf{A}_\alpha$ is a wff of $\mathcal{L}$, $\Delta\mathbf{A}_\alpha$ is the result of replacing each occurrence of $Q_{o\beta\beta}$ in $\mathbf{A}_\alpha$ by the wff

$$[\lambda x_\beta \lambda y_\beta \forall f_{o\beta} . f_{o\beta} x_\beta \supset f_{o\beta} y_\beta]$$

of $\mathcal{TE}$. It is easy to establish that $\mathcal{L}$ and $\mathcal{TE}$ are equivalent in the sense that for each wff $\mathbf{A}_o$ of $\mathcal{L}$ and $\mathbf{B}_o$ of $\mathcal{TE}$, $\vdash_{\mathcal{L}} \mathbf{A}_o$ iff $\vdash_{\mathcal{TE}} \Delta\mathbf{A}_o$, and $\vdash_{\mathcal{TE}} \mathbf{B}_o$ iff $\vdash_{\mathcal{L}} \nabla\mathbf{B}_o$; moreover, $\vdash_{\mathcal{L}} \mathbf{A}_o = \nabla\Delta\mathbf{A}_o$ and $\vdash_{\mathcal{TE}} \mathbf{B}_o = \Delta\nabla\mathbf{B}_o$. Hence our independence proofs below apply also to $\mathcal{TE}$.

DEFINITION. $\mathbf{C}_\gamma$ is *contractible* to $\mathbf{D}_\gamma$ ($\mathbf{C}_\gamma$ *contr* $\mathbf{D}_\gamma$) iff $\mathbf{D}_\gamma$ can be obtained from $\mathbf{C}_\gamma$ by a sequence of zero or more applications of the following two rules of λ-conversion:

I. (Alphabetic change of bound variables.) To replace any wf part $[\lambda \mathbf{x}_\alpha \mathbf{B}_\beta]$ of a wff by $[\lambda \mathbf{y}_\alpha \mathsf{S}^{\mathbf{x}_\alpha}_{\mathbf{y}_\alpha} \mathbf{B}_\beta]$, provided that $\mathbf{y}_\alpha$ is not free in $\mathbf{B}_\beta$ and $\mathbf{y}_\alpha$ is free for $\mathbf{x}_\alpha$ in $\mathbf{B}_\beta$.

II. (λ-contraction.) To replace any wf part $[[\lambda \mathbf{x}_\alpha \mathbf{B}_\beta]\mathbf{A}_\alpha]$ of a wff by $\mathsf{S}^{\mathbf{x}_\alpha}_{\mathbf{A}_\alpha} \mathbf{B}_\beta$, provided that $\mathbf{A}_\alpha$ is free for $\mathbf{x}_\alpha$ in $\mathbf{B}_\beta$.

DEFINITION. $\mathbf{E}_\delta$ is a *KS-combinatorial wff* iff every occurrence of λ in $\mathbf{E}_\delta$ is in a wf part of $\mathbf{E}_\delta$ of the form $K^{\alpha\beta}$ or $S^{\alpha\beta\gamma}$.

$\mathbf{E}_\delta$ is a *KBCW-combinatorial wff* iff every occurrence of λ in $\mathbf{E}_\delta$ is in a wf part of $\mathbf{E}_\delta$ of the form $K^{\alpha\beta}$, $B^{\alpha\beta\gamma}$, $C^{\alpha\beta\gamma}$, or $W^{\alpha\beta}$.

Clearly $K^{\alpha\beta}$, $S^{\alpha\beta\gamma}$, and all primitive constants and variables are *KS*-combinatorial wffs. Also, $[\mathbf{A}_{\alpha\beta}\mathbf{B}_\beta]$ is such a wff iff $\mathbf{A}_{\alpha\beta}$ and $\mathbf{B}_\beta$ are.

We next show that every wff of $\mathcal{L}$ is convertible to a *KS*-combinatorial wff, and to a *KBCW*-combinatorial wff. This requires only a simple translation into the present context of familiar facts about combinatory logic (see [6], [13], for example).

LEMMA 1. *For any KS-combinatorial wff* $\mathbf{B}_\beta$ *and variable* $\mathbf{x}_\gamma$ *there is a KS-combinatorial wff* $\mathbf{P}_{\beta\gamma}$ *such that* $\mathbf{P}_{\beta\gamma}$ contr $[\lambda \mathbf{x}_\gamma \mathbf{B}_\beta]$.

PROOF. By induction on the number of occurrences of [in $\mathbf{B}_\beta$.

Case a. $\mathbf{B}_\beta$ is $\mathbf{x}_\gamma$. Let $\mathbf{P}_{\gamma\gamma}$ be $S^{\gamma(\gamma\gamma)\gamma} K^{\gamma(\gamma\gamma)} K^{\gamma\gamma}$. Thus $\mathbf{P}_{\gamma\gamma}$ contr $[\lambda z_\gamma . K^{\gamma(\gamma\gamma)} z_\gamma . K^{\gamma\gamma} z_\gamma]$ contr $[\lambda z_\gamma z_\gamma]$ contr $[\lambda \mathbf{x}_\gamma \mathbf{B}_\gamma]$.

Case b. $\mathbf{B}_\beta$ does not contain $\mathbf{x}_\gamma$ free. Let $\mathbf{P}_{\beta\gamma}$ be $K^{\beta\gamma}\mathbf{B}_\beta$. Then $\mathbf{P}_{\beta\gamma}$ contr $[\lambda \mathbf{x}_\gamma \mathbf{B}_\beta]$.

Case c. $\mathbf{B}_\beta$ has the form $[\mathbf{D}_{\beta\delta}\mathbf{E}_\delta]$. By inductive hypothesis there are *KS*-combinatorial wffs $\mathbf{G}_{\beta\delta\gamma}$ and $\mathbf{H}_{\delta\gamma}$ such that $\mathbf{G}_{\beta\delta\gamma}$ contr $[\lambda \mathbf{x}_\gamma \mathbf{D}_{\beta\delta}]$ and $\mathbf{H}_{\delta\gamma}$ contr $[\lambda \mathbf{x}_\gamma \mathbf{E}_\delta]$. Let $\mathbf{P}_{\beta\gamma}$ be $[S^{\beta\delta\gamma}\mathbf{G}_{\beta\delta\gamma}\mathbf{H}_{\delta\gamma}]$. Thus

$$\mathbf{P}_{\beta\gamma} \text{ contr } [S^{\beta\delta\gamma}[\lambda \mathbf{x}_\gamma \mathbf{D}_{\beta\delta}][\lambda \mathbf{x}_\gamma \mathbf{E}_\delta]] \text{ contr } [\lambda \mathbf{x}_\gamma . [\lambda \mathbf{x}_\gamma \mathbf{D}_{\beta\delta}]\mathbf{x}_\gamma . [\lambda \mathbf{x}_\gamma \mathbf{E}_\delta]\mathbf{x}_\delta] \text{ contr } [\lambda \mathbf{x}_\gamma \mathbf{B}_\beta].$$

Since every *KS*-combinatorial wff $\mathbf{B}_\beta$ falls under at least one of these three cases, this completes the proof of the lemma. □

PROPOSITION 1. *For every wff* $\mathbf{A}_\delta$ *of* $\mathcal{L}$ *there is a KS-combinatorial wff* $\mathbf{P}_\delta$ *such that* $\mathbf{P}_\delta$ contr $\mathbf{A}_\delta$.

PROOF. By induction on the number of occurrences of [in $\mathbf{A}_\delta$.

Case a. $\mathbf{A}_\delta$ is a primitive constant or variable. Let $\mathbf{P}_\delta$ be $\mathbf{A}_\delta$.

Case b. $\mathbf{A}_\delta$ has the form $[\mathbf{D}_{\delta\beta}\mathbf{E}_\beta]$. By inductive hypothesis there are *KS*-combinatorial wffs $\mathbf{D}'_{\delta\beta}$ and $\mathbf{E}'_\beta$ such that $\mathbf{D}'_{\delta\beta}$ contr $\mathbf{D}_{\delta\beta}$ and $\mathbf{E}'_\beta$ contr $\mathbf{E}_\beta$. Let $\mathbf{P}_\delta$ be $[\mathbf{D}'_{\delta\beta}\mathbf{E}'_\beta]$.

Case c. $\mathbf{A}_\delta$ has the form $[\lambda \mathbf{x}_\gamma \mathbf{B}_\beta]$.
By inductive hypothesis there is a *KS*-combinatorial wff $\mathbf{B}'_\beta$ such that $\mathbf{B}'_\beta$ contr $\mathbf{B}_\beta$. Then by Lemma 1 there is a *KS*-combinatorial wff $\mathbf{P}_{\beta\gamma}$ such that $\mathbf{P}_{\beta\gamma}$ contr $[\lambda \mathbf{x}_\gamma \mathbf{B}'_\beta]$. Thus $\mathbf{P}_{\beta\gamma}$ contr $\mathbf{A}_\delta$. □

PROPOSITION 2. *For every wff* $\mathbf{A}_\delta$ *of* $\mathscr{L}$ *there is a KBCW-combinatorial wff* $\mathbf{D}_\delta$ *such that* $\mathbf{D}_\delta$ contr $\mathbf{A}_\delta$.

PROOF. It can be verified that

$$B^{(\alpha\gamma(\beta\gamma))(\alpha\gamma(\beta\gamma)\gamma)(\alpha\beta\gamma)}[B^{(\alpha\gamma(\beta\gamma))(\alpha\gamma\gamma(\beta\gamma))(\alpha\gamma(\beta\gamma)\gamma)}[B^{(\alpha\gamma)(\alpha\gamma\gamma)(\beta\gamma)}W^{\alpha\gamma}]C^{(\alpha\gamma)(\beta\gamma)\gamma}][B^{(\alpha\gamma(\beta\gamma))(\alpha\beta)\gamma}B^{\alpha\beta\gamma}]$$

contr $S^{\alpha\beta\gamma}$. If one replaces $S^{\alpha\beta\gamma}$ by this wff everywhere in the wff $\mathbf{P}_\delta$ of Proposition 1, one obtains the desired wff $\mathbf{D}_\delta$. □

§3. General models for $\mathscr{L}$.

We next define the general models for $\mathscr{L}$ by modifying appropriately the definition in [8].

DEFINITION. A *frame* is a collection $\{\mathscr{D}_\alpha\}_\alpha$ of nonempty domains (sets), one for each type symbol α, such that $\mathscr{D}_o = \{\mathfrak{t}, \mathfrak{f}\}$ and $\mathscr{D}_{\alpha\beta}$ is a collection of functions mapping $\mathscr{D}_\beta$ into $\mathscr{D}_\alpha$. The members of $\mathscr{D}_o$ are called *truth values* and the members of $\mathscr{D}_\iota$ are called *individuals*.

DEFINITION. Given a frame $\{\mathscr{D}_\alpha\}_\alpha$, an *assignment* (of values in the frame to variables) is a function φ defined on the set of variables of $\mathscr{L}$ such that for each variable $\mathbf{x}_\alpha$, $\varphi\mathbf{x}_\alpha \in \mathscr{D}_\alpha$. Given an assignment φ, a variable $\mathbf{x}_\alpha$, and an element $\mathfrak{z} \in \mathscr{D}_\alpha$, let $(\varphi : \mathbf{x}_\alpha/\mathfrak{z})$ be that assignment ψ such that $\psi\mathbf{x}_\alpha = \mathfrak{z}$ and $\psi\mathbf{y}_\beta = \varphi\mathbf{y}_\beta$ if $\mathbf{y}_\beta \neq \mathbf{x}_\alpha$.

If $\mathfrak{h}$ is a function of which $\mathfrak{x}$ is an argument, we write the value of $\mathfrak{h}$ at $\mathfrak{x}$ as $\mathfrak{h}\mathfrak{x}$ or $(\mathfrak{h}\mathfrak{x})$. If $\mathfrak{h}\mathfrak{x}$ is itself a function of which $\mathfrak{y}$ is an argument, we may write $(\mathfrak{h}\mathfrak{x})\mathfrak{y}$ simply as $\mathfrak{h}\mathfrak{x}\mathfrak{y}$, using the convention of association to the left in our meta-language. We shall use dots to denote parentheses in our meta-language in the manner of our convention for brackets in $\mathscr{L}$. We shall also use λ-notation informally in our meta-language. Thus when $\mathfrak{A}$ is an expression of our meta-language involving a variable $\mathfrak{x}$ of our meta-language, then $(\lambda\mathfrak{x}\mathfrak{A})$ shall serve as a name for the function whose domain is the range of the variable $\mathfrak{x}$ and whose value at each argument $\mathfrak{x}$ is $\mathfrak{A}$. In contexts where a frame has been specified, if α is a type symbol it will be understood that $\mathfrak{x}_\alpha$, $\mathfrak{y}_\alpha$, $\mathfrak{z}_\alpha$, etc., range over the domain $\mathscr{D}_\alpha$ of the frame. However, we reserve $\mathfrak{q}_{o\alpha\alpha}$ as a name for the identity relation over $\mathscr{D}_\alpha$; i.e., $\mathfrak{q}_{o\alpha\alpha}\mathfrak{x}_\alpha\mathfrak{y}_\alpha = \mathfrak{t}$ if $\mathfrak{x}_\alpha = \mathfrak{y}_\alpha$, and $\mathfrak{q}_{o\alpha\alpha}\mathfrak{x}_\alpha\mathfrak{y}_\alpha = \mathfrak{f}$ if $\mathfrak{x}_\alpha \neq \mathfrak{y}_\alpha$. We note for future reference that if $\mathfrak{x}_\alpha \in \mathscr{D}_\alpha$, then $\mathfrak{q}_{o\alpha\alpha}\mathfrak{x}_\alpha$ is $\{\mathfrak{x}_\alpha\}$, the unit set whose only member is $\mathfrak{x}_\alpha$.

DEFINITION. A frame $\{\mathscr{D}_\alpha\}_\alpha$ is a *general model* for $\mathscr{L}$ iff there is a binary function $\mathscr{V}$ such that for each assignment φ and wff $\mathbf{A}_\alpha$, $\mathscr{V}_\varphi\mathbf{A}_\alpha \in \mathscr{D}_\alpha$ and the following conditions are satisfied for all assignments φ and all wffs:

(a) $\mathscr{V}_\varphi\mathbf{x}_\alpha = \varphi\mathbf{x}_\alpha$;
(b) $\mathscr{V}_\varphi Q_{o\alpha\alpha} = \mathfrak{q}_{o\alpha\alpha}$;
(c) $\mathscr{V}_\varphi[\mathbf{A}_{\alpha\beta}\mathbf{B}_\beta] = (\mathscr{V}_\varphi\mathbf{A}_{\alpha\beta})(\mathscr{V}_\varphi\mathbf{B}_\beta)$;
(d) $\mathscr{V}_\varphi[\lambda\mathbf{x}_\alpha\mathbf{B}_\beta] = (\lambda\mathfrak{y}_\alpha\mathscr{V}_{(\varphi:\mathbf{x}_\alpha/\mathfrak{y}_\alpha)}\mathbf{B}_\beta)$.

Remark. Clearly the crucial requirement above is that $\mathscr{V}_\varphi[\lambda\mathbf{x}_\alpha\mathbf{B}_\beta] \in \mathscr{D}_{\beta\alpha}$. Note that in a general model the function $\mathscr{V}$ is uniquely determined.

DEFINITION. A frame $\{\mathscr{D}_\alpha\}_\alpha$ is a *standard model* for $\mathscr{L}$ iff for all α and β, $\mathscr{D}_{\alpha\beta}$ is the set of all functions from $\mathscr{D}_\beta$ into $\mathscr{D}_\alpha$.

Clearly a standard model is a general model, and is uniquely determined by $\mathscr{D}_\iota$.

A wff $\mathbf{A}_o$ is *valid* in a general model iff $\mathscr{V}_\varphi\mathbf{A}_o = \mathsf{t}$ for all assignments φ. It can be shown by an easy modification of the argument in [8] that a wff $\mathbf{A}_o$ is a theorem of $\mathscr{L}$ iff it is valid in every general model. Also, the rule of inference of $\mathscr{L}$ preserves validity in a general model.

DEFINITION. A wff $\mathbf{A}_\alpha$ is *significant in* a frame $\{\mathscr{D}_\alpha\}_\alpha$ iff there is a function $\mathscr{V}$ such that for every assignment φ and for every wf part $\mathbf{B}_\beta$ of $\mathbf{A}_\alpha$ (including $\mathbf{A}_\alpha$ itself), $\mathscr{V}_\varphi\mathbf{B}_\beta \in \mathscr{D}_\beta$, and $\mathscr{V}$ satisfies conditions (a)–(d) (in the definition of general model).

Thus a frame is a general model iff every wff is significant in it.

Before proving the next proposition we state the following lemmas, which can be proved by straightforward induction on the construction of $\mathbf{B}_\beta$.

LEMMA 2. *If* $\mathbf{B}_\beta$ *is significant in a frame and* φ *and* ψ *are assignments which agree on the free variables of* $\mathbf{B}_\beta$, *then* $\mathscr{V}_\varphi\mathbf{B}_\beta = \mathscr{V}_\psi\mathbf{B}_\beta$.

LEMMA 3. *If* $\mathbf{A}_\alpha$ *and* $\mathbf{B}_\beta$ *are significant in a frame and* $\mathbf{A}_\alpha$ *is free for* $\mathbf{x}_\alpha$ *in* $\mathbf{B}_\beta$, *then* $\mathsf{S}^{\mathbf{x}_\alpha}_{\mathbf{A}_\alpha}\mathbf{B}_\beta$ *is significant and for any assignment* φ, $\mathscr{V}_\varphi\mathsf{S}^{\mathbf{x}_\alpha}_{\mathbf{A}_\alpha}\mathbf{B}_\beta = \mathscr{V}_{(\varphi:\mathbf{x}_\alpha/\mathscr{V}_\varphi\mathbf{A}_\alpha)}\mathbf{B}_\beta$.

PROPOSITION 3. *If* $\mathbf{C}_\gamma$ *is significant in a frame and* $\mathbf{C}_\gamma$ contr $\mathbf{D}_\gamma$, *then* $\mathbf{D}_\gamma$ *is significant, and for any assignment* φ, $\mathscr{V}_\varphi\mathbf{C}_\gamma = \mathscr{V}_\varphi\mathbf{D}_\gamma$.

PROOF. Clearly it suffices to prove this proposition for the case where $\mathbf{D}_\gamma$ is obtained from $\mathbf{C}_\gamma$ by a single application of Rule I or II of λ-conversion. In either case the proposition follows easily by induction on the construction of $\mathbf{C}_\gamma$ once one establishes it for the wf part of $\mathbf{C}_\gamma$ to which the rule is actually applied.

Thus in the case of Rule I one may suppose $\mathbf{C}_\gamma$ is $[\lambda\mathbf{x}_\alpha\mathbf{B}_\beta]$ and $\mathbf{D}_\gamma$ is $[\lambda\mathbf{y}_\alpha\mathsf{S}^{\mathbf{x}_\alpha}_{\mathbf{y}_\alpha}\mathbf{B}_\beta]$, where $\mathbf{y}_\alpha$ is not free in $\mathbf{B}_\beta$ and $\mathbf{y}_\alpha$ is free for $\mathbf{x}_\alpha$ in $\mathbf{B}_\beta$. We may assume that $\mathbf{y}_\alpha \neq \mathbf{x}_\alpha$. $\mathbf{B}_\beta$ is significant since $\mathbf{C}_\gamma$ is, so by Lemma 3, $\mathsf{S}^{\mathbf{x}_\alpha}_{\mathbf{y}_\alpha}\mathbf{B}_\beta$ is significant.

Note that for any $\mathfrak{z}_\alpha \in \mathscr{D}_\alpha$ we have $\mathscr{V}_{(\varphi:\mathbf{y}_\alpha/\mathfrak{z}_\alpha)}\mathbf{y}_\alpha = \mathfrak{z}_\alpha$ so

$$\begin{aligned}\mathscr{V}_{(\varphi:\mathbf{y}_\alpha/\mathfrak{z}_\alpha)}\mathsf{S}^{\mathbf{x}_\alpha}_{\mathbf{y}_\alpha}\mathbf{B}_\beta &= \mathscr{V}_{((\varphi:\mathbf{y}_\alpha/\mathfrak{z}_\alpha):\mathbf{x}_\alpha/\mathfrak{z}_\alpha)}\mathbf{B}_\beta \quad \text{(by Lemma 3)}\\ &= \mathscr{V}_{(\varphi:\mathbf{x}_\alpha/\mathfrak{z}_\alpha)}\mathbf{B}_\beta \quad \text{(by Lemma 2).}\end{aligned}$$

Hence

$$\mathscr{V}_\varphi\mathbf{C}_\gamma = (\lambda\mathfrak{z}_\alpha\mathscr{V}_{(\varphi:\mathbf{x}_\alpha/\mathfrak{z}_\alpha)}\mathbf{B}_\beta) = (\lambda\mathfrak{z}_\alpha\mathscr{V}_{(\varphi:\mathbf{y}_\alpha/\mathfrak{z}_\alpha)}\mathsf{S}^{\mathbf{x}_\alpha}_{\mathbf{y}_\alpha}\mathbf{B}_\beta),$$

which is the desired value for $\mathscr{V}_\varphi\mathbf{D}_\gamma$, so $\mathbf{D}_\gamma$ is significant and $\mathscr{V}_\varphi\mathbf{C}_\gamma = \mathscr{V}_\varphi\mathbf{D}_\gamma$.

In the case of Rule II one may suppose that $\mathbf{C}_\gamma$ is $[[\lambda\mathbf{x}_\alpha\mathbf{B}_\beta]\mathbf{A}_\alpha]$ and $\mathbf{D}_\gamma$ is $\mathsf{S}^{\mathbf{x}_\alpha}_{\mathbf{A}_\alpha}\mathbf{B}_\beta$, where $\mathbf{A}_\alpha$ is free for $\mathbf{x}_\alpha$ in $\mathbf{B}_\beta$. Since $\mathbf{C}_\gamma$ is significant, $\mathbf{A}_\alpha$ and $\mathbf{B}_\beta$ are, so by Lemma 3, $\mathbf{D}_\gamma$ is significant. Also $\mathscr{V}_\varphi\mathbf{C}_\gamma = (\mathscr{V}_\varphi[\lambda\mathbf{x}_\alpha\mathbf{B}_\beta])\mathscr{V}_\varphi\mathbf{A}_\alpha = \mathscr{V}_{(\varphi:\mathbf{x}_\alpha/\mathscr{V}_\varphi\mathbf{A}_\alpha)}\mathbf{B}_\beta = \mathscr{V}_\varphi\mathbf{D}_\gamma$ by Lemma 3. □

Remark. It is not true that if $\mathbf{C}_\gamma$ is significant in a frame and $\mathbf{D}_\gamma$ contr $\mathbf{C}_\gamma$, then $\mathbf{D}_\gamma$ must be significant. For x_ι is always significant, but $[[\lambda x_\iota x_\iota]x_\iota]$ might not be.

PROPOSITION 4. *For any frame* $\mathscr{M}$, *the following conditions are equivalent*:

(a) $\mathscr{M}$ *is a general model for* $\mathscr{L}$.

(b) *every KS-combinatorial wff of* $\mathscr{L}$ *is significant in* $\mathscr{M}$.

(c) *Every KBCW-combinatorial wff of* $\mathscr{L}$ *is significant in* $\mathscr{M}$.

PROOF. By Propositions 1, 2, and 3. □

We now rephrase condition (b) to obtain a simple criterion for a frame to be a general model.

THEOREM 1. *A frame $\{\mathcal{D}_\alpha\}_\alpha$ is a general model for $\mathcal{L}$ iff it satisfies all of the following conditions (for all type symbols α, β, γ):*

(a) $\mathfrak{q}_{o\alpha\alpha} \in \mathcal{D}_{o\alpha\alpha}$.

(b) *For all* $\mathfrak{x}_\alpha \in \mathcal{D}_\alpha$, $(\lambda\mathfrak{y}_\beta\mathfrak{x}_\alpha) \in \mathcal{D}_{\alpha\beta}$.

(c) $(\lambda\mathfrak{x}_\alpha\lambda\mathfrak{y}_\beta\mathfrak{x}_\alpha) \in \mathcal{D}_{\alpha\beta\alpha}$.

(d) *For all* $\mathfrak{x}_{\alpha\beta\gamma} \in \mathcal{D}_{\alpha\beta\gamma}$ *and* $\mathfrak{y}_{\beta\gamma} \in \mathcal{D}_{\beta\gamma}$, $(\lambda\mathfrak{z}_\gamma . \mathfrak{x}_{\alpha\beta\gamma}\mathfrak{z}_\gamma . \mathfrak{y}_{\beta\gamma}\mathfrak{z}_\gamma) \in \mathcal{D}_{\alpha\gamma}$.

(e) *For all* $\mathfrak{x}_{\alpha\beta\gamma} \in \mathcal{D}_{\alpha\beta\gamma}$, $(\lambda\mathfrak{y}_{\beta\gamma}\lambda\mathfrak{z}_\gamma . \mathfrak{x}_{\alpha\beta\gamma}\mathfrak{z}_\gamma . \mathfrak{y}_{\beta\gamma}\mathfrak{z}_\gamma) \in \mathcal{D}_{\alpha\gamma(\beta\gamma)}$.

(f) $(\lambda\mathfrak{x}_{\alpha\beta\gamma}\lambda\mathfrak{y}_{\beta\gamma}\lambda\mathfrak{z}_\gamma . \mathfrak{x}_{\alpha\beta\gamma}\mathfrak{z}_\gamma . \mathfrak{y}_{\beta\gamma}\mathfrak{z}_\gamma) \in \mathcal{D}_{\alpha\gamma(\beta\gamma)(\alpha\beta\gamma)}$.

PROOF. Clearly if the frame is a general model, the conditions (a)–(f) must be satisfied. To show they are sufficient, we show they imply condition (b) of Proposition 4. Since every variable is significant in every frame, and a wff $[\mathbf{A}_{\alpha\beta}\mathbf{B}_\beta]$ is significant in a frame iff $\mathbf{A}_{\alpha\beta}$ and $\mathbf{B}_\beta$ are, it suffices to show that the wffs $Q_{o\alpha\alpha}$, $K^{\alpha\beta}$, and $S^{\alpha\beta\gamma}$ are significant in the frame. This is assured by conditions (a)–(f). (We note that condition (a) implies that for all $\mathfrak{x}_\alpha \in \mathcal{D}_\alpha$, $(\mathfrak{q}_{o\alpha\alpha}\mathfrak{x}_\alpha) \in \mathcal{D}_{o\alpha}$.) □

Remark. We leave it to the reader to state the analogous theorem using $B^{\alpha\beta\gamma}$, $C^{\alpha\beta\gamma}$, and $W^{\alpha\beta}$ in place of $S^{\alpha\beta\gamma}$. Such a theorem may be useful since $B^{\alpha\beta\gamma}$, $C^{\alpha\beta\gamma}$, and $W^{\alpha\beta}$ are each conceptually simpler than $S^{\alpha\beta\gamma}$.

§4. The Axiom of Descriptions. We remind the reader that the Axiom of Descriptions is

(D) $$\exists i_{\iota(o\iota)}\forall x_\iota . i_{\iota(o\iota)}[Q_{o\iota\iota}x_\iota] = x_\iota.$$

THEOREM 2. D *is not a theorem of* $\mathcal{L}$.

PROOF. We partition the type symbols into two sets, $\mathcal{T}_o$ and $\mathcal{T}_\iota$ as follows: $o \in \mathcal{T}_o$ but $o \notin \mathcal{T}_\iota$; $\iota \in \mathcal{T}_\iota$ but $\iota \notin \mathcal{T}_o$; $(\alpha\beta)$ is in whichever set contains α. We then let $\mathcal{C} = \{(\alpha\beta) \mid \alpha \in \mathcal{T}_\iota \text{ and } \beta \in \mathcal{T}_o\}$.

We next define a frame $\mathcal{M} = \{\mathcal{D}_\alpha\}_\alpha$ by induction on α. $\mathcal{D}_o = \{\mathfrak{t}, \mathfrak{f}\}$. $\mathcal{D}_\iota = \{\mathfrak{m}, \mathfrak{n}\}$, where $\mathfrak{m}$ and $\mathfrak{n}$ are distinct individuals. (Actually $\mathcal{D}_\iota$ may be taken to have any cardinality greater than one.) If $(\alpha\beta) \in \mathcal{C}$, $\mathcal{D}_{\alpha\beta}$ is the set of all constant functions (i.e., functions with the same value for all arguments) from $\mathcal{D}_\beta$ into $\mathcal{D}_\alpha$. If $(\alpha\beta) \notin \mathcal{C}$, $\mathcal{D}_{\alpha\beta}$ is the set of all functions from $\mathcal{D}_\beta$ into $\mathcal{D}_\alpha$.

We next use Theorem 1 to verify that $\mathcal{M}$ is a general model for $\mathcal{L}$.

(a) Since $(o\alpha) \notin \mathcal{C}$ and $(o\alpha\alpha) \notin \mathcal{C}$, $\mathfrak{q}_{o\alpha\alpha} \in \mathcal{D}_{o\alpha\alpha}$.

(b) $(\lambda\mathfrak{y}_\beta\mathfrak{x}_\alpha)$ is a constant function, and so is in $\mathcal{D}_{\alpha\beta}$.

(c) $(\alpha\beta\alpha) \notin \mathcal{C}$ whether $\alpha \in \mathcal{T}_\iota$ or $\alpha \in \mathcal{T}_o$. Hence $(\lambda\mathfrak{x}_\alpha\lambda\mathfrak{y}_\beta\mathfrak{x}_\alpha) \in \mathcal{D}_{\alpha\beta\alpha}$.

(d) We need consider only the case where $(\alpha\gamma) \in \mathcal{C}$.

We must show that if $\mathfrak{x} \in \mathcal{D}_{\alpha\beta\gamma}$ and $\mathfrak{y} \in \mathcal{D}_{\beta\gamma}$, then $(\lambda\mathfrak{z}_\gamma . \mathfrak{x}\mathfrak{z}_\gamma . \mathfrak{y}\mathfrak{z}_\gamma)$ is a constant function. So we let $\mathfrak{z}^1, \mathfrak{z}^2 \in \mathcal{D}_\gamma$ and show that $(\mathfrak{x}\mathfrak{z}^1 . \mathfrak{y}\mathfrak{z}^1) = (\mathfrak{x}\mathfrak{z}^2 . \mathfrak{y}\mathfrak{z}^2)$. Since $\alpha \in \mathcal{T}_\iota$ and $\gamma \in \mathcal{T}_o$, $(\alpha\beta\gamma) \in \mathcal{C}$ so $\mathfrak{x}\mathfrak{z}^1 = \mathfrak{x}\mathfrak{z}^2$.

Case 1. $\beta \in \mathcal{T}_\iota$. Then $(\beta\gamma) \in \mathcal{C}$ so $\mathfrak{y}\mathfrak{z}^1 = \mathfrak{y}\mathfrak{z}^2$ so $(\mathfrak{x}\mathfrak{z}^1 . \mathfrak{y}\mathfrak{z}^1) = (\mathfrak{x}\mathfrak{z}^2 . \mathfrak{y}\mathfrak{z}^2)$.

Case 2. $\beta \in \mathcal{T}_o$. Then $(\alpha\beta) \in \mathcal{C}$. Since $\mathfrak{x}\mathfrak{z}^1 = \mathfrak{x}\mathfrak{z}^2 \in \mathcal{D}_{\alpha\beta}$, $\mathfrak{x}\mathfrak{z}^1(\mathfrak{y}\mathfrak{z}^1) = \mathfrak{x}\mathfrak{z}^2(\mathfrak{y}\mathfrak{z}^2)$.

(e) Suppose $(\alpha\gamma(\beta\gamma)) \in \mathcal{C}$ and $\mathfrak{x} \in \mathcal{D}_{\alpha\beta\gamma}$.

We must show that $(\lambda\mathfrak{y}_{\beta\gamma}\lambda\mathfrak{z}_\gamma . \mathfrak{x}\mathfrak{z}_\gamma . \mathfrak{y}_{\beta\gamma}\mathfrak{z}_\gamma) \in \mathcal{D}_{\alpha\gamma(\beta\gamma)}$. So suppose $\mathfrak{y}^1, \mathfrak{y}^2 \in \mathcal{D}_{\beta\gamma}$. We must show that $(\lambda\mathfrak{z}_\gamma . \mathfrak{x}\mathfrak{z}_\gamma . \mathfrak{y}^1\mathfrak{z}_\gamma) = (\lambda\mathfrak{z}_\gamma . \mathfrak{x}\mathfrak{z}_\gamma . \mathfrak{y}^2\mathfrak{z}_\gamma)$. To do this we show that for an arbitrary $\mathfrak{z} \in \mathcal{D}_\gamma$, $(\mathfrak{x}\mathfrak{z} . \mathfrak{y}^1\mathfrak{z}) = (\mathfrak{x}\mathfrak{z} . \mathfrak{y}^2\mathfrak{z})$. But $\alpha \in \mathcal{T}_\iota$ and $\beta \in \mathcal{T}_o$ so $(\alpha\beta) \in \mathcal{C}$ and $\mathfrak{x}\mathfrak{z} \in \mathcal{D}_{\alpha\beta}$, which contains only constant functions. Hence $\mathfrak{x}\mathfrak{z}(\mathfrak{y}^1\mathfrak{z}) = \mathfrak{x}\mathfrak{z}(\mathfrak{y}^2\mathfrak{z})$.

(f) $(\alpha\gamma(\beta\gamma)(\alpha\beta\gamma)) \notin \mathscr{C}$ whether $\alpha \in \mathscr{T}_\iota$ or $\alpha \in \mathscr{T}_o$, so

$$(\lambda \mathfrak{x}_{\alpha\beta\gamma}\lambda \mathfrak{y}_{\beta\gamma}\lambda \mathfrak{z}_\gamma \centerdot \mathfrak{x}_{\alpha\beta\gamma}\mathfrak{z}_\gamma \centerdot \mathfrak{y}_{\beta\gamma}\mathfrak{z}_\gamma) \in \mathscr{D}_{\alpha\gamma(\beta\gamma)(\alpha\beta\gamma)}.$$

Now $\mathfrak{q}_{o\iota\iota}\mathfrak{m}$ and $\mathfrak{q}_{o\iota\iota}\mathfrak{n}$ are elements of $\mathscr{D}_{o\iota}$, so in order that D be valid in $\mathscr{M}$ there must be a function $\mathfrak{h} \in \mathscr{D}_{\iota(o\iota)}$ such that $\mathfrak{h}(\mathfrak{q}_{o\iota\iota}\mathfrak{m}) = \mathfrak{m}$ and $\mathfrak{h}(\mathfrak{q}_{o\iota\iota}\mathfrak{n}) = \mathfrak{n}$. However, $(\iota(o\iota)) \in \mathscr{C}$, so there is no such function in $\mathscr{D}_{\iota(o\iota)}$. Thus D is not valid in the general model $\mathscr{M}$, and so is not a theorem of $\mathscr{L}$. □

The idea behind the following theorem is contained in [9], but the proof is short, so we give it here.

THEOREM 3. *Let $\mathscr{M} = \{\mathscr{D}_\alpha\}_\alpha$ be a general model for $\mathscr{L}$ in which $\mathscr{D}_\iota$ is finite. Then $\mathscr{M}$ is a standard model iff* D *is valid in $\mathscr{M}$.*

PROOF: The domains $\mathscr{D}_\alpha$ must, of course, all be finite. If $\mathscr{M}$ is standard one can enumerate the elements in $\mathscr{D}_{\iota(o\iota)}$ to see that D is valid in $\mathscr{M}$.

Suppose D is valid in $\mathscr{M}$. We show that $\mathscr{D}_{\alpha\beta}$ must contain all functions from $\mathscr{D}_\beta$ to $\mathscr{D}_\alpha$. So let $\mathfrak{g}$ be any such function. Let $\mathscr{D}_\beta = \{\mathfrak{m}_\beta^1, \cdots, \mathfrak{m}_\beta^k\}$. By the methods mentioned in §1 one sees that there must be a description operator $\mathfrak{h}_{\alpha(o\alpha)} \in \mathscr{D}_{\alpha(o\alpha)}$ such that for each $\mathfrak{n}_\alpha \in \mathscr{D}_\alpha$, $\mathfrak{h}_{\alpha(o\alpha)}[\mathfrak{q}_{o\alpha\alpha}\mathfrak{n}_\alpha] = \mathfrak{n}_\alpha$. Let φ be an assignment with values on the variables $i_{\alpha(o\alpha)}, w_\beta^1, \cdots, w_\beta^k, \cdots, z_\alpha^1, \cdots, z_\alpha^k$ as follows: $\varphi i_{\alpha(o\alpha)} = \mathfrak{h}_{\alpha(o\alpha)}$, $\varphi w_\beta^1 = \mathfrak{m}_\beta^1, \cdots, \varphi w_\beta^k = \mathfrak{m}_\beta^k$, $\varphi z_\alpha^1 = \mathfrak{g}\mathfrak{m}_\beta^1, \cdots$, and $\varphi z_\alpha^k = \mathfrak{g}\mathfrak{m}_\beta^k$. Then

$$\mathfrak{g} = \mathscr{V}_\varphi[\lambda x_\beta \centerdot i_{\alpha(o\alpha)} \centerdot \lambda y_\alpha \centerdot [x_\beta = w_\beta^1 \wedge y_\alpha = z_\alpha^1] \vee \cdots \vee [x_\beta = w_\beta^k \wedge y_\alpha = z_\alpha^k]],$$

so $\mathfrak{g}$ must be in $\mathscr{D}_{\alpha\beta}$ since $\mathscr{M}$ is a general model. □

Remark. Theorem 3 provides a strong argument for always assuming the Axiom of Descriptions. If one does this by introducing a description operator $\iota_{\iota(o\iota)}$ and modifies the definition of general model in the natural way by introducing an appropriate requirement for $\mathscr{V}_\varphi\iota_{\iota(o\iota)}$ (thus getting closer to the definition in [8]), one can again prove that the theorems are precisely the wffs valid in all general models. Thus it appears that the language Q_o of [2] is more natural than $\mathscr{L}$.

§5. The Axiom of Choice. The Axiom of Choice (for individuals) is

(E) $$\exists i_{\iota(o\iota)}\forall p_{o\iota}\centerdot\exists x_\iota p_{o\iota}x_\iota \supset p_{o\iota}\centerdot i_{\iota(o\iota)}p_{o\iota}.$$

Clearly [E ⊃ D] is a theorem of $\mathscr{L}$. We use the Fraenkel-Mostowski method to show that its converse is not. Thus E is not provable in $\mathscr{L}$, even if D is added to the list of axioms.

We first establish the following lemma, which is fundamental for applications of the Fraenkel-Mostowski method to $\mathscr{L}$. The lemma is true but trivial if $\mathscr{D}_\iota$ is finite, since in this case the conditions on $\mathscr{F}$ assure that $\mathscr{M}$ will be the standard model over $\mathscr{D}_\iota$. We use $\circ$ to denote the composition of functions.

LEMMA 4. *Let $\mathscr{D}_\iota$ be an infinite set of individuals and P a set of permutations σ of $\mathscr{D}_\iota$ such that $\sigma \circ \sigma = (\lambda \mathfrak{x}_\iota \mathfrak{x}_\iota)$. Let $\mathscr{F}$ be a family of subsets of P such that*

(a) *for each $\mathfrak{m} \in \mathscr{D}_\iota$ there is a set $K \in \mathscr{F}$ such that $\sigma\mathfrak{m} = \mathfrak{m}$ for all $\sigma \in K$, and*

(b) *for all $H, K \in \mathscr{F}$ there is a set $J \in \mathscr{F}$ such that $J \subseteq H \cap K$.*

Let the frame $\mathscr{M} = \{\mathscr{D}_\alpha\}_\alpha$ be defined, and each permutation $\sigma \in P$ be extended to a permutation of $\mathscr{D}_\alpha$ (which we may denote by σ^α) such that $\sigma^\alpha \circ \sigma^\alpha = (\lambda \mathfrak{x}_\alpha \mathfrak{x}_\alpha)$ for each α, as follows by induction on α:

$$\mathscr{D}_o = \{\mathfrak{t}, \mathfrak{f}\}; \qquad \sigma^o = (\lambda \mathfrak{x}_o \mathfrak{x}_o) \quad \textit{for all } \sigma \in P.$$

Given $\mathcal{D}_\alpha$ and $\mathcal{D}_\beta$ and any function $\mathfrak{h}$ from $\mathcal{D}_\beta$ into $\mathcal{D}_\alpha$, let $\sigma\mathfrak{h} = \sigma^\alpha \circ \mathfrak{h} \circ \sigma^\beta$, and let $\mathcal{D}_{\alpha\beta}$ be the set of all functions $\mathfrak{h}$ from $\mathcal{D}_\beta$ into $\mathcal{D}_\alpha$ such that there is some $K \in \mathcal{F}$ such that $\sigma\mathfrak{h} = \mathfrak{h}$ for all $\sigma \in K$.

Then $\mathcal{M}$ is a general model for $\mathcal{L}$ in which D *is valid.*

PROOF. For notational convenience, if $\mathfrak{h} \in \mathcal{D}_\gamma$ we let $K_\mathfrak{h}$ denote some $K \in \mathcal{F}$ such that $\sigma\mathfrak{h} = \mathfrak{h}$ for all $\sigma \in K$. Clearly such a set $K_\mathfrak{h}$ always exists. Note that if $\mathfrak{h} \in \mathcal{D}_{\alpha\beta}$ and $\mathfrak{x} \in \mathcal{D}_\beta$, then $(\sigma^{\alpha\beta}\mathfrak{h})(\sigma^\beta\mathfrak{x}) = \sigma^\alpha(\mathfrak{h}\mathfrak{x})$.

We use Theorem 1 to verify that $\mathcal{M}$ is a general model.

(a) If $\mathfrak{x}, \mathfrak{y} \in \mathcal{D}_\alpha$ and $\sigma \in K_\mathfrak{x}$ then $(\sigma^{o\alpha}.\mathfrak{q}_{o\alpha\alpha}\mathfrak{x})\mathfrak{y} = \sigma^o(\mathfrak{q}_{o\alpha\alpha}\mathfrak{x}.\sigma\mathfrak{y}) = \mathfrak{q}_{o\alpha\alpha}\mathfrak{x}(\sigma\mathfrak{y})$, which is t iff $\sigma\mathfrak{y} = \mathfrak{x} = \sigma\mathfrak{x}$ iff $\mathfrak{x} = \mathfrak{y}$, so $(\sigma^{o\alpha}.\mathfrak{q}_{o\alpha\alpha}\mathfrak{x})\mathfrak{y} = (\mathfrak{q}_{o\alpha\alpha}\mathfrak{x})\mathfrak{y}$ for all $\mathfrak{y} \in \mathcal{D}_\alpha$ so $\sigma(\mathfrak{q}_{o\alpha\alpha}\mathfrak{x}) = \mathfrak{q}_{o\alpha\alpha}\mathfrak{x}$. Thus $(\mathfrak{q}_{o\alpha\alpha}\mathfrak{x}) \in \mathcal{D}_{o\alpha}$, and $\mathfrak{q}_{o\alpha\alpha}$ at least maps $\mathcal{D}_\alpha$ into $\mathcal{D}_{o\alpha}$. Also, for any $\sigma \in P$, $(\sigma\mathfrak{q}_{o\alpha\alpha})\mathfrak{x}\mathfrak{y} = (\sigma^{o\alpha}.\mathfrak{q}_{o\alpha\alpha}.\sigma^\alpha\mathfrak{x})\mathfrak{y} = \mathfrak{q}_{o\alpha\alpha}(\sigma^\alpha\mathfrak{x})(\sigma^\alpha\mathfrak{y})$, which is t iff $\sigma\mathfrak{x} = \sigma\mathfrak{y}$ iff $\mathfrak{x} = \mathfrak{y}$, so $\sigma\mathfrak{q}_{o\alpha\alpha} = \mathfrak{q}_{o\alpha\alpha}$ and $\mathfrak{q}_{o\alpha\alpha} \in \mathcal{D}_{o\alpha\alpha}$.

(b) For any $\mathfrak{x}_\alpha \in \mathcal{D}_\alpha$ and $\sigma \in K_\mathfrak{x}$, $\sigma(\lambda\mathfrak{y}_\beta\mathfrak{x}_\alpha) = (\lambda\mathfrak{y}_\beta\sigma\mathfrak{x}_\alpha) = (\lambda\mathfrak{y}_\beta\mathfrak{x}_\alpha)$, so $(\lambda\mathfrak{y}_\beta\mathfrak{x}_\alpha) \in \mathcal{D}_{\alpha\beta}$.

(c) For any $\sigma \in P$, $\sigma(\lambda\mathfrak{x}_\alpha\lambda\mathfrak{y}_\beta\mathfrak{x}_\alpha) = (\lambda\mathfrak{x}_\alpha\sigma.\lambda\mathfrak{y}_\beta\sigma\mathfrak{x}_\alpha) = (\lambda\mathfrak{x}_\alpha\lambda\mathfrak{y}_\beta\sigma\sigma\mathfrak{x}_\alpha) = (\lambda\mathfrak{x}_\alpha\lambda\mathfrak{y}_\beta\mathfrak{x}_\alpha)$, so $(\lambda\mathfrak{x}_\alpha\lambda\mathfrak{y}_\beta\mathfrak{x}_\alpha) \in \mathcal{D}_{\alpha\beta\alpha}$.

Before checking (d)–(f) we observe that if $\mathfrak{x} \in \mathcal{D}_{\alpha\beta\gamma}$, $\mathfrak{y} \in \mathcal{D}_{\beta\gamma}$, and $\mathfrak{z} \in \mathcal{D}_\gamma$, then $\sigma^\alpha.(\sigma\mathfrak{x})(\sigma\mathfrak{z}).(\sigma\mathfrak{y}).\sigma\mathfrak{z} = \sigma^\alpha.(\sigma^{\alpha\beta}.\mathfrak{x}\mathfrak{z}).\sigma^\beta.\mathfrak{y}\mathfrak{z} = \sigma^\alpha.\sigma^\alpha.\mathfrak{x}\mathfrak{z}.\mathfrak{y}\mathfrak{z} = \mathfrak{x}\mathfrak{z}.\mathfrak{y}\mathfrak{z}$.

(d) Suppose $\mathfrak{x} \in \mathcal{D}_{\alpha\beta\gamma}$ and $\mathfrak{y} \in \mathcal{D}_{\beta\gamma}$. Let J be a member of $\mathcal{F}$ such that $J \subseteq K_\mathfrak{x} \cap K_\mathfrak{y}$. For any $\sigma \in J$, $\sigma\mathfrak{x} = \mathfrak{x}$ and $\sigma\mathfrak{y} = \mathfrak{y}$ so $\sigma(\lambda\mathfrak{z}_\gamma.\mathfrak{x}\mathfrak{z}.\mathfrak{y}\mathfrak{z}) = \sigma(\lambda\mathfrak{z}_\gamma.(\sigma\mathfrak{x})\mathfrak{z}.(\sigma\mathfrak{y})\mathfrak{z}) = (\lambda\mathfrak{z}_\gamma.\sigma^\alpha.(\sigma\mathfrak{x})(\sigma\mathfrak{z}).(\sigma\mathfrak{y}).\sigma\mathfrak{z}) = (\lambda\mathfrak{z}_\gamma.\mathfrak{x}\mathfrak{z}.\mathfrak{y}\mathfrak{z})$, which must therefore be in $\mathcal{D}_{\alpha\gamma}$.

(e) If $\mathfrak{x} \in \mathcal{D}_{\alpha\beta\gamma}$ and $\sigma \in K_\mathfrak{x}$, then

$$\begin{aligned}\sigma(\lambda\mathfrak{y}_{\beta\gamma}\lambda\mathfrak{z}_\gamma.\mathfrak{x}\mathfrak{z}.\mathfrak{y}\mathfrak{z}) &= \sigma(\lambda\mathfrak{y}_{\beta\gamma}\lambda\mathfrak{z}_\gamma.(\sigma\mathfrak{x})\mathfrak{z}.\mathfrak{y}\mathfrak{z})\\ &= (\lambda\mathfrak{y}_{\beta\gamma}.\sigma.\lambda\mathfrak{z}_\gamma.(\sigma\mathfrak{x})\mathfrak{z}.(\sigma\mathfrak{y})\mathfrak{z})\\ &= (\lambda\mathfrak{y}_{\beta\gamma}\lambda\mathfrak{z}_\gamma.\sigma^\alpha.(\sigma\mathfrak{x})(\sigma\mathfrak{z}).(\sigma\mathfrak{y}).\sigma\mathfrak{z})\\ &= (\lambda\mathfrak{y}_{\beta\gamma}\lambda\mathfrak{z}_\gamma.\mathfrak{x}\mathfrak{z}.\mathfrak{y}\mathfrak{z}),\end{aligned}$$

which must therefore be in $\mathcal{D}_{\alpha\gamma(\beta\gamma)}$.

(f) For any $\sigma \in P$,

$$\begin{aligned}\sigma(\lambda\mathfrak{x}_{\alpha\beta\gamma}\lambda\mathfrak{y}_{\beta\gamma}\lambda\mathfrak{z}_\gamma.\mathfrak{x}\mathfrak{z}.\mathfrak{y}\mathfrak{z}) &= (\lambda\mathfrak{x}_{\alpha\beta\gamma}\lambda\mathfrak{y}_{\beta\gamma}\lambda\mathfrak{z}_\gamma.\sigma^\alpha.(\sigma\mathfrak{x})(\sigma\mathfrak{z}).(\sigma\mathfrak{y}).\sigma\mathfrak{z})\\ &= (\lambda\mathfrak{x}_{\alpha\beta\gamma}\lambda\mathfrak{y}_{\beta\gamma}\lambda\mathfrak{z}_\gamma.\mathfrak{x}\mathfrak{z}.\mathfrak{y}\mathfrak{z}),\end{aligned}$$

which must therefore be in $\mathcal{D}_{\alpha\gamma(\beta\gamma)(\alpha\beta\gamma)}$. Thus $\mathcal{M}$ is a general model for $\mathcal{L}$.

We next verify that D is valid in $\mathcal{M}$. Let $\mathfrak{n} \in \mathcal{D}_\iota$. We shall construct a description operator $\mathfrak{h}$ mapping $\mathcal{D}_{o\iota}$ to $\mathcal{D}_\iota$ as follows. For each unit set $\mathfrak{q}_{o\iota\iota}\mathfrak{x}_\iota$, we let $\mathfrak{h}(\mathfrak{q}_{o\iota\iota}\mathfrak{x}_\iota) = \mathfrak{x}_\iota$. If $\mathfrak{g} \in \mathcal{D}_{o\iota}$ is not a unit set, let $\mathfrak{h}\mathfrak{g} = \mathfrak{n}$. Now we verify that $\mathfrak{h} \in \mathcal{D}_{\iota(o\iota)}$. Let $\sigma \in K_\mathfrak{n}$. For each unit set $\mathfrak{q}_{o\iota\iota}\mathfrak{x}_\iota$, $\sigma(\mathfrak{q}_{o\iota\iota}\mathfrak{x}_\iota) = (\sigma\mathfrak{q}_{o\iota\iota})(\sigma\mathfrak{x}_\iota) = \mathfrak{q}_{o\iota\iota}(\sigma\mathfrak{x}_\iota)$, so $(\sigma\mathfrak{h})(\mathfrak{q}_{o\iota\iota}\mathfrak{x}_\iota) = \sigma(\mathfrak{h}.\sigma.\mathfrak{q}_{o\iota\iota}\mathfrak{x}_\iota) = \sigma(\mathfrak{h}.\mathfrak{q}_{o\iota\iota}.\sigma\mathfrak{x}_\iota) = \sigma\sigma\mathfrak{x}_\iota = \mathfrak{x}_\iota = \mathfrak{h}(\mathfrak{q}_{o\iota\iota}\mathfrak{x}_\iota)$. If $\mathfrak{g}_{o\iota}$ is not a unit set, then $\sigma\mathfrak{g}_{o\iota}$ (i.e. $\mathfrak{g}_{o\iota} \circ \sigma^\iota$) is not either, so $(\sigma\mathfrak{h})\mathfrak{g}_{o\iota} = \sigma(\mathfrak{h}.\sigma\mathfrak{g}_{o\iota}) = \sigma\mathfrak{n} = \mathfrak{n} = \mathfrak{h}\mathfrak{g}_{o\iota}$. Thus $\sigma\mathfrak{h} = \mathfrak{h}$, and $\mathfrak{h} \in \mathcal{D}_{\iota(o\iota)}$. It is now easy to see that D is valid in $\mathcal{M}$. □

THEOREM 4. $[\mathrm{D} \supset \mathrm{E}]$ *is not a theorem of $\mathcal{L}$.*

PROOF. Let $\mathcal{J}$ be an infinite index set and for all $j \in \mathcal{J}$ let $\mathfrak{m}^j$ and $\mathfrak{n}^j$ be distinct individuals, so chosen that $\mathfrak{m}^j \neq \mathfrak{m}^i$ and $\mathfrak{n}^j \neq \mathfrak{n}^i$ if $j \neq i$. Let $\mathcal{D}_\iota = \{\mathfrak{m}^j \mid j \in \mathcal{J}\} \cup \{\mathfrak{n}^j \mid j \in \mathcal{J}\}$. Let P be the set of all mappings σ from $\mathcal{D}_\iota$ to $\mathcal{D}_\iota$ such that for all $j \in \mathcal{J}$, $\sigma\mathfrak{m}^j = \mathfrak{m}^j$ and $\sigma\mathfrak{n}^j = \mathfrak{n}^j$, or $\sigma\mathfrak{m}^j = \mathfrak{n}^j$ and $\sigma\mathfrak{n}^j = \mathfrak{m}^j$. Thus for each $\sigma \in P$ we have $\sigma \circ \sigma = (\lambda\mathfrak{x}_\iota\mathfrak{x}_\iota)$. Let $\mathcal{F}$ be the family of all subsets K of P such that there is a finite subset $\mathscr{I}$ of $\mathscr{J}$ such that $K = \{\sigma \in P \mid \text{for all } j \in \mathscr{I}, \sigma\mathfrak{m}^j = \mathfrak{m}^j \text{ and } \sigma\mathfrak{n}^j = \mathfrak{n}^j\}$. It is easily checked that $\mathscr{F}$ satisfies the conditions of Lemma 4, so let $\mathscr{M}$ be the general model constructed as in Lemma 4.

We must see that E is false in $\mathscr{M}$. Suppose it were true. Then there would be a choice function $\mathfrak{h} \in \mathscr{D}_{\iota(o\iota)}$ such that for every nonempty set $\mathfrak{g} \in \mathscr{D}_{o\iota}$, $\mathfrak{h}\mathfrak{g}$ is in $\mathfrak{g}$, i.e., $\mathfrak{g}(\mathfrak{h}\mathfrak{g}) = \mathsf{t}$. For each $j \in \mathscr{J}$, let $\mathfrak{g}^j = (\lambda x_\iota . x_\iota = \mathfrak{m}^j \text{ or } x_\iota = \mathfrak{n}^j)$, i.e., $\mathfrak{g}^j = \{\mathfrak{m}^j, \mathfrak{n}^j\}$. It is easy to see that $\sigma\mathfrak{g}^j = \mathfrak{g}^j$ for all $\sigma \in P$, so each $\mathfrak{g}^j \in \mathscr{D}_{o\iota}$. Now for any $K \in \mathscr{F}$ there is some $j \in \mathscr{J}$ which is not in the finite subset of $\mathscr{J}$ which determines K, and hence some $\sigma \in K$ such that $\sigma\mathfrak{m}^j = \mathfrak{n}^j$ and $\sigma\mathfrak{n}^j = \mathfrak{m}^j$. Then $(\sigma\mathfrak{h})\mathfrak{g}^j = \sigma(\mathfrak{h}.\sigma\mathfrak{g}^j) = \sigma(\mathfrak{h}\mathfrak{g}^j) \neq \mathfrak{h}\mathfrak{g}^j$, so $\sigma\mathfrak{h} \neq \mathfrak{h}$. Thus there can be no choice function $\mathfrak{h} \in \mathscr{D}_{\iota(o\iota)}$, so E is false in $\mathscr{M}$.

Thus [D ⊃ E] is not valid in the general model $\mathscr{M}$ and so is not a theorem of $\mathscr{L}$. □

BIBLIOGRAPHY

[1] PETER B. ANDREWS, *A reduction of the axioms for the theory of propositional types*, ***Fundamenta Mathematicae***, vol. 52 (1963), pp. 345–350.

[2] ———, ***A transfinite type theory with type variables***, North-Holland, Amsterdam, 1965, 143 pp.

[3] ———, *Resolution in type theory*, this JOURNAL, vol. 36 (1971), pp. 414–432.

[4] ALONZO CHURCH, *A formulation of the simple theory of types*, this JOURNAL, vol. 5 (1940), pp. 56–68.

[5] ———, *Non-normal truth-tables for the propositional calculus*, ***Boletin de la Sociedad Matematica Mexicana***, vol. X (1953), pp. 41–52.

[6] HASKELL B. CURRY and ROBERT FEYS, ***Combinatory logic***, vol. 1, North-Holland, Amsterdam, 1958, 1968, 433 pp.

[7] ABRAHAM A. FRAENKEL, *Der Begriff 'definit' und die Unabhängigkeit des Auswahlaxioms*, ***Sitzungsberichte der Preussischen Akademie der Wissenschaften, Physikalisch-mathematische Klasse***, vol. 21 (1922), pp. 253–257; translated in Jean van Heijenoort, ***From Frege to Gödel***, Harvard University Press, Cambridge, 1967, pp. 284–289.

[8] LEON HENKIN, *Completeness in the theory of types*, this JOURNAL, vol. 15 (1950), pp. 81–91; reprinted in [10, pp. 51–63].

[9] ———, *A theory of propositional types*, ***Fundamenta Mathematicae***, vol. 52 (1963), pp. 323–344; errata, ibid., vol. 53 (1963), p. 119.

[10] JAAKKO HINTIKKA, editor, ***The philosophy of mathematics***, Oxford University Press, Oxford, 1969, 186 pp.

[11] AZRIEL LÉVY, *The Fraenkel-Mostowski method for independence proofs in set theory*, ***The theory of models, Proceedings of the 1963 International Symposium at Berkeley***, edited by J. W. Addison, Leon Henkin, and Alfred Tarski, North-Holland, Amsterdam, 1965, pp. 221–228.

[12] ANDRZEJ MOSTOWSKI, *Über die Unabhängigkeit des Wohlordnungssatzes vom Ordnungsprinzip*, ***Fundamenta Mathematicae***, vol. 32 (1939), pp. 201–252.

[13] LUIS E. SANCHIS, *Types in combinatory logic*, ***Notre Dame Journal of Formal Logic***, vol. 5 (1964), pp. 161–180.

CARNEGIE-MELLON UNIVERSITY
PITTSBURGH, PENNSYLVANIA 15213

THE JOURNAL OF SYMBOLIC LOGIC
Volume 37, Number 2, June 1972

GENERAL MODELS AND EXTENSIONALITY

PETER B. ANDREWS[1]

§1. Introduction. It is well known that equality is definable in type theory. Thus, in the language of [2], the equality relation between elements of type α is definable as $[\lambda x_\alpha \lambda y_\alpha \forall p_{o\alpha} . p_{o\alpha} x_\alpha \supset p_{o\alpha} y_\alpha]$, i.e., $x_\alpha = y_\alpha$ iff every set which contains x_α also contains y_α. However, in a nonstandard model of type theory, the sets may be so sparse that the wff above does not denote the true equality relation. We shall use this observation to construct a general model in the sense of [2] in which the Axiom of Extensionality is not valid. Thus Theorem 2 of [2] is technically incorrect. However, it is easy to remedy the situation by slightly modifying the definition of general model.

Our construction will show that the Axiom Schema of Extensionality is independent even if one takes $f_{\alpha\beta} = [\lambda x_\beta . f_{\alpha\beta} x_\beta]$ as an axiom schema.

We shall assume familiarity with, and use the notation of, [2] and §§2–3 of [1].

§2. A nonextensional general model. The language of [2] has primitive logical constants $N_{(oo)}$, $A_{((oo)o)}$, $\Pi_{o(o\alpha)}$, $\iota_{\alpha(o\alpha)}$, whereas the language $\mathscr{L}$ of [1] has primitive logical constants $Q_{((o\alpha)\alpha)}$. By modifying the proof of Theorem 1 of [1] in the obvious way, one obtains the following:

PROPOSITION. *A frame* $\{\mathscr{D}_\alpha\}_\alpha$ *is a general model in the sense of* [2] *iff it satisfies all of the following conditions (for all type symbols* α, β, γ*):*

(a$_1$) $\mathscr{D}_{oo}$ *contains the negation function* $\mathfrak{n}$ *such that* $\mathfrak{n}\mathfrak{t} = \mathfrak{f}$ *and* $\mathfrak{n}\mathfrak{f} = \mathfrak{t}$.

(a$_2$) $\mathscr{D}_{oo}$ *contains* $(\lambda \mathfrak{x}_o \mathfrak{t})$ *and* $(\lambda \mathfrak{x}_o \mathfrak{x}_o)$. *Also,* $\mathscr{D}_{(oo)o}$ *contains the alternation (disjunction) function* $\mathfrak{a}$ *such that* $\mathfrak{a}\mathfrak{t} = (\lambda \mathfrak{x}_o \mathfrak{t})$ *and* $\mathfrak{a}\mathfrak{f} = (\lambda \mathfrak{x}_o \mathfrak{x}_o)$.

(a$_3$) $\mathscr{D}_{o(o\alpha)}$ *contains a function* $\pi_{o(o\alpha)}$ *such that for all* $\mathfrak{g} \in \mathscr{D}_{o\alpha}$, $\pi_{o(o\alpha)}\mathfrak{g} = \mathfrak{t}$ *iff* $\mathfrak{g} = (\lambda \mathfrak{x}_\alpha \mathfrak{t})$.

(a$_4$) $\mathscr{D}_{\alpha(o\alpha)}$ *contains a function* $\mathfrak{i}_{\alpha(o\alpha)}$ *such that if* $\mathfrak{g}$ *is any nonempty set in* $\mathscr{D}_{o\alpha}$, $\mathfrak{i}_{\alpha(o\alpha)}\mathfrak{g}$ *is in* $\mathfrak{g}$.

(b) *For all* $\mathfrak{x} \in \mathscr{D}_\alpha$, $(\lambda \mathfrak{y}_\beta \mathfrak{x}) \in \mathscr{D}_{\alpha\beta}$.

(c) $(\lambda \mathfrak{x}_\alpha \lambda \mathfrak{y}_\beta \mathfrak{x}_\alpha) \in \mathscr{D}_{\alpha\beta\alpha}$.

(d) *For all* $\mathfrak{x} \in \mathscr{D}_{\alpha\beta\gamma}$ *and* $\mathfrak{y} \in \mathscr{D}_{\beta\gamma}$, $(\lambda \mathfrak{z}_\gamma . \mathfrak{x}\mathfrak{z}_\gamma . \mathfrak{y}\mathfrak{z}_\gamma) \in \mathscr{D}_{\alpha\gamma}$.

(e) *For all* $\mathfrak{x} \in \mathscr{D}_{\alpha\beta\gamma}$, $(\lambda \mathfrak{y}_{\beta\gamma} \lambda \mathfrak{z}_\gamma . \mathfrak{x}\mathfrak{z}_\gamma . \mathfrak{y}_{\beta\gamma}\mathfrak{z}_\gamma) \in \mathscr{D}_{\alpha\gamma(\beta\gamma)}$.

(f) $(\lambda \mathfrak{x}_{\alpha\beta\gamma} \lambda \mathfrak{y}_{\beta\gamma} \lambda \mathfrak{z}_\gamma . \mathfrak{x}_{\alpha\beta\gamma}\mathfrak{z}_\gamma . \mathfrak{y}_{\beta\gamma}\mathfrak{z}_\gamma) \in \mathscr{D}_{\alpha\gamma(\beta\gamma)(\alpha\beta\gamma)}$.

THEOREM. *There is a general model in the sense of* [2] *in which the Axiom of Extensionality*

$$\forall x_\iota [f_{\iota\iota} x_\iota = g_{\iota\iota} x_\iota] \supset . f_{\iota\iota} = g_{\iota\iota}$$

is not valid.

Received June 23, 1971.

[1] This research was partially supported by NSF Grant GJ-580.

Proof. We construct a frame $\mathscr{M} = \{\mathscr{D}_\alpha\}_\alpha$ by induction on α. Simultaneously we define three equivalence relations $\equiv^1$, $\equiv^2$, and $\equiv^3$ on each of the $\mathscr{D}_\alpha$. When it is more convenient to do so, we shall define $\equiv^i$ in terms of the partition (set of equivalence classes) $\mathscr{P}_i^\alpha$ of $\mathscr{D}_\alpha$ induced by $\equiv^i$. A statement about $\equiv^i$ is meant to apply to each of $\equiv^1$, $\equiv^2$, and $\equiv^3$.

$\mathscr{D}_o = \{\mathfrak{t}, \mathfrak{f}\}$. $\mathfrak{x}_o \equiv^i \mathfrak{y}_o$ iff $\mathfrak{x}_o = \mathfrak{y}_o$.

$\mathscr{D}_\iota = \{\mathfrak{l}, \mathfrak{m}, \mathfrak{n}\}$, where $\mathfrak{l}$, $\mathfrak{m}$, $\mathfrak{n}$ are distinct individuals. $\mathscr{P}_1^\iota = \{\{\mathfrak{m}, \mathfrak{n}\}, \{\mathfrak{l}\}\}$. $\mathscr{P}_2^\iota = \{\{\mathfrak{l}, \mathfrak{n}\}, \{\mathfrak{m}\}\}$. $\mathscr{P}_3^\iota = \{\{\mathfrak{l}, \mathfrak{m}\}, \{\mathfrak{n}\}\}$.

Given $\mathscr{D}_\alpha$ and $\mathscr{D}_\beta$, let $\mathscr{D}_{\alpha\beta}$ be the set of all functions $\mathfrak{g}$ from $\mathscr{D}_\beta'$ into $\mathscr{D}_\alpha$ such that for all $\mathfrak{u}$ and $\mathfrak{v}$ in $\mathscr{D}_\beta$, if $\mathfrak{u} \equiv^1 \mathfrak{v}$ then $\mathfrak{g}\mathfrak{u} \equiv^1 \mathfrak{g}\mathfrak{v}$, and if $\mathfrak{u} \equiv^2 \mathfrak{v}$ then $\mathfrak{g}\mathfrak{u} \equiv^2 \mathfrak{g}\mathfrak{v}$, and if $\mathfrak{u} \equiv^3 \mathfrak{v}$ then $\mathfrak{g}\mathfrak{u} \equiv^3 \mathfrak{g}\mathfrak{v}$. If $\mathfrak{g}$ and $\mathfrak{h}$ are in $\mathscr{D}_{\alpha\beta}$, let $\mathfrak{g} \equiv^i \mathfrak{h}$ iff for all $\mathfrak{x} \in \mathscr{D}_\beta$, $\mathfrak{g}\mathfrak{x} \equiv^i \mathfrak{h}\mathfrak{x}$.

Having defined the frame $\mathscr{M}$, we use the Proposition above to show that it is a general model.

(a) Since $\equiv^i$ is trivial on $\mathscr{D}_o$, $\mathscr{D}_{\alpha o}$ contains all functions from $\mathscr{D}_o$ into $\mathscr{D}_\alpha$. Hence (a_1) and (a_2) are satisfied. Also, if $\mathfrak{u}$ and $\mathfrak{v}$ are in $\mathscr{D}_{o\gamma}$, then $\mathfrak{u} \equiv^i \mathfrak{v}$ iff $\mathfrak{u} = \mathfrak{v}$. Hence $\mathscr{D}_{\alpha(o\gamma)}$ contains all functions from $\mathscr{D}_{o\gamma}$ into $\mathscr{D}_\alpha$. Thus (a_3) and (a_4) are satisfied.

(b) Clearly $(\lambda\mathfrak{y}_\beta\mathfrak{x}_\alpha) \in \mathscr{D}_{\alpha\beta}$ since this is a constant function.

(c) If $\mathfrak{u}, \mathfrak{v} \in \mathscr{D}_\alpha$ and $\mathfrak{u} \equiv^i \mathfrak{v}$, then $(\lambda\mathfrak{y}_\beta\mathfrak{u}) \equiv^i (\lambda\mathfrak{y}_\beta\mathfrak{v})$, so $(\lambda\mathfrak{x}_\alpha\lambda\mathfrak{y}_\beta\mathfrak{x}_\alpha) \in \mathscr{D}_{\alpha\beta\alpha}$.

(d) Suppose $\mathfrak{x} \in \mathscr{D}_{\alpha\beta\gamma}$, $\mathfrak{y} \in \mathscr{D}_{\beta\gamma}$, $\mathfrak{z}^1, \mathfrak{z}^2 \in \mathscr{D}_\gamma$, and $\mathfrak{z}^1 \equiv^i \mathfrak{z}^2$. Then $\mathfrak{x}\mathfrak{z}^1 \equiv^i \mathfrak{x}\mathfrak{z}^2$ and $\mathfrak{y}\mathfrak{z}^1 \equiv^i \mathfrak{y}\mathfrak{z}^2$ so $\mathfrak{x}\mathfrak{z}^1(\mathfrak{y}\mathfrak{z}^1) \equiv^i \mathfrak{x}\mathfrak{z}^1(\mathfrak{y}\mathfrak{z}^2) \equiv^i \mathfrak{x}\mathfrak{z}^2(\mathfrak{y}\mathfrak{z}^2)$, so $(\lambda\mathfrak{z}_\gamma\centerdot\mathfrak{x}\mathfrak{z}_\gamma\centerdot\mathfrak{y}\mathfrak{z}_\gamma) \in \mathscr{D}_{\alpha\gamma}$.

(e) Suppose $\mathfrak{x} \in \mathscr{D}_{\alpha\beta\gamma}$, $\mathfrak{y}^1, \mathfrak{y}^2 \in \mathscr{D}_{\beta\gamma}$, and $\mathfrak{y}^1 \equiv^i \mathfrak{y}^2$. Then for each $\mathfrak{z} \in \mathscr{D}_\gamma$, $\mathfrak{y}^1\mathfrak{z} \equiv^i \mathfrak{y}^2\mathfrak{z}$ so $\mathfrak{x}\mathfrak{z}(\mathfrak{y}^1\mathfrak{z}) \equiv^i \mathfrak{x}\mathfrak{z}(\mathfrak{y}^2\mathfrak{z})$, so $(\lambda\mathfrak{z}_\gamma\centerdot\mathfrak{x}\mathfrak{z}_\gamma\centerdot\mathfrak{y}^1\mathfrak{z}_\gamma) \equiv^i (\lambda\mathfrak{z}_\gamma\centerdot\mathfrak{x}\mathfrak{z}_\gamma\centerdot\mathfrak{y}^2\mathfrak{z}_\gamma)$. Hence

$$(\lambda\mathfrak{y}_{\beta\gamma}\lambda\mathfrak{z}_\gamma\centerdot\mathfrak{x}\mathfrak{z}_\gamma\centerdot\mathfrak{y}_{\beta\gamma}\mathfrak{z}_\gamma) \in \mathscr{D}_{\alpha\gamma(\beta\gamma)}.$$

(f) Suppose $\mathfrak{x}^1, \mathfrak{x}^2 \in \mathscr{D}_{\alpha\beta\gamma}$ and $\mathfrak{x}^1 \equiv^i \mathfrak{x}^2$. Then for each $\mathfrak{z} \in \mathscr{D}_\gamma$ and $\mathfrak{y} \in \mathscr{D}_{\beta\gamma}$, $\mathfrak{x}^1\mathfrak{z} \equiv^i \mathfrak{x}^2\mathfrak{z}$ so $\mathfrak{x}^1\mathfrak{z}(\mathfrak{y}\mathfrak{z}) \equiv^i \mathfrak{x}^2\mathfrak{z}(\mathfrak{y}\mathfrak{z})$ so $(\lambda\mathfrak{z}_\gamma\centerdot\mathfrak{x}^1\mathfrak{z}_\gamma\centerdot\mathfrak{y}\mathfrak{z}_\gamma) \equiv^i (\lambda\mathfrak{z}_\gamma\centerdot\mathfrak{x}^2\mathfrak{z}_\gamma\centerdot\mathfrak{y}\mathfrak{z}_\gamma)$ so

$$(\lambda\mathfrak{y}_{\beta\gamma}\lambda\mathfrak{z}_\gamma\centerdot\mathfrak{x}^1\mathfrak{z}_\gamma\centerdot\mathfrak{y}_{\beta\gamma}\mathfrak{z}_\gamma) \equiv^i (\lambda\mathfrak{y}_{\beta\gamma}\lambda\mathfrak{z}_\gamma\centerdot\mathfrak{x}^2\mathfrak{z}_\gamma\centerdot\mathfrak{y}_{\beta\gamma}\mathfrak{z}_\gamma)$$

so $(\lambda\mathfrak{x}_{\alpha\beta\gamma}\lambda\mathfrak{y}_{\beta\gamma}\lambda\mathfrak{z}_\gamma\centerdot\mathfrak{x}_{\alpha\beta\gamma}\mathfrak{z}_\gamma\centerdot\mathfrak{y}_{\beta\gamma}\mathfrak{z}_\gamma) \in \mathscr{D}_{\alpha\gamma(\beta\gamma)(\alpha\beta\gamma)}$. Thus $\mathscr{M}$ is a general model in the sense of [2].

We next examine some of the domains $\mathscr{D}_\alpha$. $\mathscr{D}_{o\iota}$ contains only the constant functions $(\lambda\mathfrak{x}_\iota\mathfrak{t})$ and $(\lambda\mathfrak{x}_\iota\mathfrak{f})$. Hence for any wffs $\mathbf{A}_\iota$ and $\mathbf{B}_\iota$ and any assignment φ, $\mathscr{V}_\varphi[\mathbf{A}_\iota = \mathbf{B}_\iota] = \mathfrak{t}$, since $[\mathbf{A}_\iota = \mathbf{B}_\iota]$ is equivalent to $\forall \mathbf{p}_{o\iota}[\mathbf{p}_{o\iota}\mathbf{A}_\iota \supset \mathbf{p}_{o\iota}\mathbf{B}_\iota]$, where $\mathbf{p}_{o\iota}$ does not occur free in $\mathbf{A}_\iota$ or $\mathbf{B}_\iota$. Consequently $\mathscr{V}_\varphi \forall x_\iota[f_{\iota\iota}x_\iota = g_{\iota\iota}x_\iota] = \mathfrak{t}$ for any assignment φ.

It can be seen that $\mathscr{D}_{\iota\iota} = \{(\lambda\mathfrak{x}_\iota\mathfrak{x}_\iota), (\lambda\mathfrak{x}_\iota\mathfrak{l}), (\lambda\mathfrak{x}_\iota\mathfrak{m}), (\lambda\mathfrak{x}_\iota\mathfrak{n})\}$. To verify this, note that $\mathfrak{g} \in \mathscr{D}_{\iota\iota}$ iff $\mathfrak{g}\mathfrak{m} \equiv^1 \mathfrak{g}\mathfrak{n}$, $\mathfrak{g}\mathfrak{l} \equiv^2 \mathfrak{g}\mathfrak{n}$, and $\mathfrak{g}\mathfrak{l} \equiv^3 \mathfrak{g}\mathfrak{m}$. One can examine the twenty-seven functions from $\mathscr{D}_\iota$ into $\mathscr{D}_\iota$ to see that only the identity and constant functions satisfy all three of these properties. Alternatively, one can reason as follows: Suppose $\mathfrak{g}\mathfrak{l} = \mathfrak{m}$. Then $\mathfrak{g}\mathfrak{l} \equiv^2 \mathfrak{g}\mathfrak{n}$ so $\mathfrak{g}\mathfrak{n} = \mathfrak{m}$. Also $\mathfrak{g}\mathfrak{m} \equiv^1 \mathfrak{g}\mathfrak{n}$ so $\mathfrak{g}\mathfrak{m} \in \{\mathfrak{m}, \mathfrak{n}\}$, and $\mathfrak{g}\mathfrak{l} \equiv^3 \mathfrak{g}\mathfrak{m}$ so $\mathfrak{g}\mathfrak{m} \in \{\mathfrak{l}, \mathfrak{m}\}$; hence $\mathfrak{g}\mathfrak{m} = \mathfrak{m}$. Thus if $\mathfrak{g}\mathfrak{l} = \mathfrak{m}$, then $\mathfrak{g} = (\lambda\mathfrak{x}_\iota\mathfrak{m})$. Similarly, if $\mathfrak{g}\mathfrak{l} = \mathfrak{n}$, then $\mathfrak{g} = (\lambda\mathfrak{x}_\iota\mathfrak{n})$. Thus if $\mathfrak{g}\mathfrak{l} \neq \mathfrak{l}$, then $\mathfrak{g}$ is a constant function. Similarly if $\mathfrak{g}\mathfrak{m} \neq \mathfrak{m}$ or $\mathfrak{g}\mathfrak{n} \neq \mathfrak{n}$, then $\mathfrak{g}$ must be a constant function. Thus the only members of $\mathscr{D}_{\iota\iota}$ are the constant and identity functions.

Note that $\mathscr{P}_1^{\iota\iota} = \{\{(\lambda\mathfrak{x}_\iota\mathfrak{m}), (\lambda\mathfrak{x}_\iota\mathfrak{n})\}, \{(\lambda\mathfrak{x}_\iota\mathfrak{l})\}, \{(\lambda\mathfrak{x}_\iota\mathfrak{x}_\iota)\}\}$, $\mathscr{P}_2^{\iota\iota} = \{\{(\lambda\mathfrak{x}_\iota\mathfrak{l}), (\lambda\mathfrak{x}_\iota\mathfrak{n})\}, \{(\lambda\mathfrak{x}_\iota\mathfrak{m})\}$ $\{(\lambda\mathfrak{x}_\iota\mathfrak{x}_\iota)\}\}$, and $\mathscr{P}_3^{\iota\iota} = \{\{(\lambda\mathfrak{x}_\iota\mathfrak{l}), (\lambda\mathfrak{x}_\iota\mathfrak{m})\}, \{(\lambda\mathfrak{x}_\iota\mathfrak{n})\}, \{(\lambda\mathfrak{x}_\iota\mathfrak{x}_\iota)\}\}$.

$\mathscr{D}_{o(\iota\iota)}$ contains a function $\mathfrak{h}$ such that $\mathfrak{h}(\lambda\mathfrak{x}_\iota\mathfrak{x}_\iota) = \mathfrak{t}$ but $\mathfrak{h}(\lambda\mathfrak{x}_\iota\mathfrak{l}) = \mathfrak{h}(\lambda\mathfrak{x}_\iota\mathfrak{m}) = \mathfrak{h}(\lambda\mathfrak{x}_\iota\mathfrak{n}) = \mathfrak{f}$. Hence if φ is an assignment such that $\varphi f_{\iota\iota} = (\lambda\mathfrak{x}_\iota\mathfrak{x}_\iota)$ and $\varphi g_{\iota\iota} = (\lambda\mathfrak{x}_\iota\mathfrak{l})$, then

$$\mathscr{V}_\varphi[f_{\iota\iota} = g_{\iota\iota}] = \mathscr{V}_\varphi\forall p_{o(\iota\iota)}[p_{o(\iota\iota)}f_{\iota\iota} \supset p_{o(\iota\iota)}g_{\iota\iota}] = \mathfrak{f}.$$

Hence $\mathscr{V}_\varphi[\forall x_\iota[f_{\iota\iota}x_\iota = g_{\iota\iota}x_\iota] \supset .f_{\iota\iota} = g_{\iota\iota}] = \mathfrak{f}$, and the Axiom of Extensionality is not valid in the general model $\mathscr{M}$. □

§3. General models. We suggest that the definition of general model in [2] should be modified by adding the following requirement:

(a_0) For each α, $\mathscr{D}_{o\alpha\alpha}$ contains the identity relation $\mathfrak{q}_{o\alpha\alpha}$ on $\mathscr{D}_\alpha$ (and hence $\mathscr{D}_{o\alpha}$ contains the unit set $\mathfrak{q}_{o\alpha\alpha}\mathfrak{x}_\alpha$ for each $\mathfrak{x}_\alpha \in \mathscr{D}_\alpha$).

Of course, if this is done, clauses (a_1), (a_2), and (a_3) of the Proposition above become redundant. Indeed, $\mathfrak{n} = \mathfrak{q}_{ooo}\mathfrak{f}$,

$$\mathfrak{a} = (\lambda\mathfrak{x}_o\lambda\mathfrak{y}_o.\mathfrak{n}.\mathfrak{q}_{o(o(ooo))(o(ooo))}(\lambda\mathfrak{g}_{ooo}.\mathfrak{g}_{ooo}\mathfrak{x}_o\mathfrak{y}_o)(\lambda\mathfrak{g}_{ooo}.\mathfrak{g}_{ooo}\mathfrak{f}\mathfrak{f})),$$

and $\pi_{o(o\alpha)} = \mathfrak{q}_{o(o\alpha)(o\alpha)}(\lambda\mathfrak{x}_\alpha\mathfrak{t})$. Thus the modified definition of general model is equivalent to the result of adding a requirement concerning $\mathscr{V}_\varphi\iota_{\alpha(o\alpha)}$ to the definition of general model in [1].

With this definition, the general models constitute sound interpretations of the system of [2]. Moreover, the model constructed in the proof of Theorem 1 of [2] actually satisfies (a_0), since it can be seen that $\Phi([Q_{o\alpha\alpha}]) = \mathfrak{q}_{o\alpha\alpha}$ (in the notation of that proof). Thus Theorem 2 of [2] becomes correct under the new definition of general model.

One of the appealing properties of the definition of general model in [2] is that it is generated in a very natural way by the formation rules for the language. Our modified definition no longer has this property for the language of [2], although it has it for a language in which $Q_{o\alpha\alpha}$ is taken as a primitive constant. Thus it appears that in contexts where one wishes to assume extensionality and discuss general models, a language such as $\mathscr{L}$ of [1], augmented by a description or selection operator, is more natural than the language of [2].

BIBLIOGRAPHY

[1] Peter B. Andrews, *General models, descriptions, and choice in type theory*, this Journal, vol. 37 (1972), pp. 385–394.

[2] Leon Henkin, *Completeness in the theory of types*, this Journal, vol. 15 (1950), pp. 81–91.

CARNEGIE-MELLON UNIVERSITY
PITTSBURGH, PENNSYLVANIA 15213

Cut Elimination with ξ-Functionality

CHRISTOPH BENZMÜLLER, CHAD E. BROWN,
MICHAEL KOHLHASE

1 Introduction

In this paper we will give a complete, cut-free sequent calculus for a fragment of higher-order logic with ξ-extensionality, a weak form of functional extensionality. When one wants a calculus appropriate for automation, a common first step is to find such a complete, cut-free sequent calculus. For example, Andrews proved cut-elimination for a sequent calculus for elementary type theory [1]. Elementary type theory is a fragment of higher-order logic with no extensionality principles. In [3], we showed there is a cube of eight model classes which vary with respect to extensionality principles. For purposes of automation, we would like to have a complete, cut-free sequent calculus for each of these eight model classes. The essential ingredients for obtaining complete, cut-free sequent calculi for three of the eight model classes are already known. The sequent calculus for elementary type theory in [1] essentially provides a complete, cut-free calculus for one of the eight model classes. (Completeness of a similar calculus will follow from the results in this paper.) There are also results for two of the other model classes (the case with η and the fully extensional case) in [6, 7], though the framework is slightly different from the one here. In this paper, we prove completeness of a cut-free sequent calculus for a fourth point on the extensionality cube, elementary type theory with ξ-extensionality. To handle the case with ξ-extensionality, we will combine the techniques used in [1] with recent results on cut-simulation [4, 5].

As mentioned above, in [1] Andrews proved cut-elimination for a sequent calculus $\mathscr{G}$ for elementary type theory (Corollary 4.11 there). In particular, Andrews proved that provability in $\mathscr{G}$ is equivalent to provability in a Hilbert-style calculus $\mathscr{T}$ (where admissibility of cut is immediate). The difficult direction of this equivalence is proving that one has $\vdash_{\mathscr{T}} \mathbf{A}$ whenever one has $\vdash_{\mathscr{G}} \mathbf{A}$ (Theorem 4.10 in [1]). A key fact in Andrews' proof is his Theorem 3.5: "If Γ is an abstract consistency property and $\mathscr{S}$ is a finite set of wffs$_o$ such that $\Gamma(\mathscr{S})$, then $\mathscr{S}$ is consistent." Here, consistency is defined with respect to the calculus $\mathscr{T}$. In order to prove Theorem 3.5 from [1], Andrews proves that $\mathscr{S}$ can be used to obtain a semivaluation V such that $V\mathbf{A} = \mathtt{T}$ for all $\mathbf{A} \in \mathscr{S}$ (see Theorem 3.3 in [1]) and then proves the set $\{\mathbf{A}|V\mathbf{A} = \mathtt{T}\}$ is consistent (see Theorem 3.4 in [1]). To prove consistency of $\{\mathbf{A}|V\mathbf{A} = \mathtt{T}\}$, Andrews constructs a structure of V-complexes (following ideas in Takahashi [11] and Prawitz [10]) and proves this structure satisfies properties similar to that of a Henkin model [9].

The Andrews structure of V-complexes is not a Henkin model. In fact, it was not until much later that a notion of a nonextensional model of higher-order logic was proposed which includes V-complexes (see [3]). Given this more recent notion of a model, the Andrews structure can be viewed in a new light, and further properties of the structure can be proven.

The V-complex construction (up to the treatment of free variables) was generalized to the notion of a possible values structure in [6, 7]. We will use several results from [6, 7] to quickly conclude that the construction (starting from a Hintikka set $\mathcal{H}$ instead of a semivaluation V) yields a model. Furthermore, the model will satisfy a weak form of functional extensionality known as property ξ (cf. Definition 3.46 in [3]). If the Hintikka set is not saturated, then the model will satisfy property $\mathfrak{q}$ (cf. Definition 3.46 in [3]). Using these facts, we will prove completeness of a sequent calculus for higher-order logic with ξ-functionality relative to the model class $\mathfrak{M}_{\beta\xi}$ (cf. Definition 3.49 in [3]) and conclude cut-elimination for the sequent calculus. We will also prove completeness of a sequent calculus for elementary type theory (similar to the one given in [1]) relative to the model class $\mathfrak{M}_{\beta}$.

2 Preliminaries

We review the fundamental framework from [3] (which can be consulted for details).

As in [8], we formulate higher-order logic ($\mathcal{HOL}$) based on the simply typed λ-calculus. The set of simple types $\mathcal{T}$ is freely generated from basic types o and ι using the function type constructor $\rightarrow$.

We start with a set $\mathcal{V}$ of (typed) variables (denoted by $X_\alpha, Y, Z, X^1_\beta, X^2_\gamma \ldots$) and a signature Σ of (typed) constants (denoted by $c_\alpha, f_{\alpha\rightarrow\beta}, \ldots$). We let $\mathcal{V}_\alpha$ (Σ_α) denote the set of variables (constants) of type α. The signature Σ of constants includes the logical constants $\neg_{o\rightarrow o}$, $\vee_{o\rightarrow o\rightarrow o}$ and $\Pi^\alpha_{(\alpha\rightarrow o)\rightarrow o}$ for each type α. All other constants in Σ are called parameters. As in [3], we assume there is an infinite cardinal $\aleph_s$ such that the cardinality of Σ_α is $\aleph_s$ for each type α (cf. Remark 3.16 in [3]). The set of $\mathcal{HOL}$-formulae (or terms) are constructed from typed variables and constants using application and λ-abstraction. We let $\mathit{wff}_\alpha(\Sigma)$ be the set of all terms of type α and $\mathit{wff}(\Sigma)$ be the set of all terms.

We use vector notation to abbreviate k-fold applications and abstractions as $\mathbf{A}\overline{\mathbf{U}^k}$ and $\lambda\overline{X^k}.\mathbf{A}$, respectively. We also use Church's dot notation so that . stands for a (missing) left bracket whose mate is as far to the right as possible (consistent with given brackets). We use infix notation $\mathbf{A} \vee \mathbf{B}$ for $((\vee\mathbf{A})\mathbf{B})$ and binder notation $\forall X_\alpha.\mathbf{A}$ for $(\Pi^\alpha \lambda X_\alpha.\mathbf{A}_o)$. We further use $\mathbf{A} \wedge \mathbf{B}$, $\mathbf{A} \Rightarrow \mathbf{B}$, $\mathbf{A} \Leftrightarrow \mathbf{B}$ and $\exists X_\alpha.\mathbf{A}$ as shorthand for formulae defined in terms of $\neg$, $\vee$ and Π^α (cf. [3]). Finally, we let $(\mathbf{A}_\alpha \doteq^\alpha \mathbf{B}_\alpha)$ denote the Leibniz equation $\forall P_{\alpha\to o}.(P\mathbf{A}) \Rightarrow (P\mathbf{B})$.

Each occurrence of a variable in a term is either bound by a λ or free. We use $free(\mathbf{A})$ to denote the set of free variables of $\mathbf{A}$ (i.e., variables with a free occurrence in $\mathbf{A}$). We consider two terms to be equal if the terms are the same up to the names of bound variables (i.e., we consider α-conversion implicitly). A term $\mathbf{A}$ is closed if $free(\mathbf{A})$ is empty. We let $cwff_\alpha(\Sigma)$ denote the set of closed terms of type α and $cwff(\Sigma)$ denote the set of all closed terms. Each term $\mathbf{A} \in wff_o(\Sigma)$ is called a proposition and each term $\mathbf{A} \in cwff_o(\Sigma)$ is called a sentence.

We denote substitution of a term $\mathbf{A}_\alpha$ for a variable X_α in a term $\mathbf{B}_\beta$ by $[\mathbf{A}/X]\mathbf{B}$. Since we consider α-conversion implicitly, we assume the bound variables of $\mathbf{B}$ avoid variable capture. Similarly, we consider simultaneous substitutions σ for the finitely many free variables in the domain $\mathbf{Dom}(\sigma)$ of σ. A substitution $\sigma, [\mathbf{A}/X]$ is the substitution such that $(\sigma, [\mathbf{A}/X])(X) \equiv \mathbf{A}$ and $(\sigma, [\mathbf{A}/X])(Y) \equiv \sigma(Y)$ for variables Y other than X. Note that here and everywhere we use $\cdot \equiv \cdot$ for syntactical identity (modulo α equivalence).

A common relation on terms is given by β-reduction. A β-redex $(\lambda X.\mathbf{A})\mathbf{B}$ β-reduces to $[\mathbf{B}/X]\mathbf{A}$. For $\mathbf{A}, \mathbf{B} \in wff_\alpha(\Sigma)$, we write $\mathbf{A} \equiv_\beta \mathbf{B}$ to mean $\mathbf{A}$ can be converted to $\mathbf{B}$ by a series of β-reductions and expansions. For each $\mathbf{A} \in wff(\Sigma)$ there is a unique β-normal form (denoted $\mathbf{A}{\downarrow_\beta}$; the set of all β-normal formulae is denoted by $wff(\Sigma){\big\downarrow_\beta}$). From this fact we know $\mathbf{A} \equiv_\beta \mathbf{B}$ iff $\mathbf{A}{\downarrow_\beta}$ and $\mathbf{B}{\downarrow_\beta}$ are syntactically equal ($\mathbf{A}{\downarrow_\beta} \equiv \mathbf{B}{\downarrow_\beta}$).

A model of $\mathcal{HOL}$ is given by four objects: a typed collection of nonempty sets $(\mathcal{D}_\alpha)_{\alpha\in\mathcal{T}}$, an application operator $@\colon \mathcal{D}_{\alpha\to\beta} \times \mathcal{D}_\alpha \longrightarrow \mathcal{D}_\beta$, an evaluation function $\mathcal{E}$ for terms and a valuation function $\upsilon\colon \mathcal{D}_o \longrightarrow \{\mathtt{T}, \mathtt{F}\}$. A pair $(\mathcal{D}, @)$ is called a Σ-applicative structure (cf. Definition 16). If $\mathcal{E}$ is an evaluation function for $(\mathcal{D}, @)$ (cf. Definition 3.18 in [3]), then we call the triple $(\mathcal{D}, @, \mathcal{E})$ a Σ-evaluation. If υ satisfies appropriate properties, then we call the tuple $(\mathcal{D}, @, \mathcal{E}, \upsilon)$ a Σ-model (cf. Definitions 3.40 and 3.41 in [3]).

Given an applicative structure $(\mathcal{D}, @)$, an assignment φ is a (typed) function from $\mathcal{V}$ to $\mathcal{D}$. An evaluation function $\mathcal{E}$ maps an assignment φ and a term $\mathbf{A}_\alpha \in wff_\alpha(\Sigma)$ to an element $\mathcal{E}_\varphi(\mathbf{A}) \in \mathcal{D}_\alpha$. Evaluations $\mathcal{E}$ are required to satisfy four properties (cf. Definition 3.18 in [3]):

1. $\mathcal{E}_\varphi|_\mathcal{V} \equiv \varphi$.

2. $\mathcal{E}_\varphi(\mathbf{FA}) \equiv \mathcal{E}_\varphi(\mathbf{F})@\mathcal{E}_\varphi(\mathbf{A})$ for any $\mathbf{F} \in \mathit{wff}_{\alpha\to\beta}(\Sigma)$ and $\mathbf{A} \in \mathit{wff}_\alpha(\Sigma)$ and types α and β.

3. $\mathcal{E}_\varphi(\mathbf{A}) \equiv \mathcal{E}_\psi(\mathbf{A})$ for any type α and $\mathbf{A} \in \mathit{wff}_\alpha(\Sigma)$, whenever φ and ψ coincide on $free(\mathbf{A})$.

4. $\mathcal{E}_\varphi(\mathbf{A}) \equiv \mathcal{E}_\varphi(\mathbf{A}{\downarrow_\beta})$ for all $\mathbf{A} \in \mathit{wff}_\alpha(\Sigma)$.

If $\mathbf{A}$ is closed, then we can simply write $\mathcal{E}(\mathbf{A})$ since the value $\mathcal{E}_\varphi(\mathbf{A})$ cannot depend on φ.

Given an evaluation $(\mathcal{D}, @, \mathcal{E})$, Figure 1 shows the definition of several properties a function $v: \mathcal{D}_o \longrightarrow \{\mathtt{T}, \mathtt{F}\}$ may satisfy (cf. Definition 3.40 in [3]). A valuation $v: \mathcal{D}_o \longrightarrow \{\mathtt{T}, \mathtt{F}\}$ is required to satisfy $\mathfrak{L}_\neg(\mathcal{E}(\neg))$, $\mathfrak{L}_\vee(\mathcal{E}(\vee))$ and $\mathfrak{L}_\forall^\alpha(\mathcal{E}(\Pi^\alpha))$ for every type α.

prop.	where	holds when			for all
$\mathfrak{L}_\neg(\mathsf{n})$	$\mathsf{n} \in \mathcal{D}_{o\to o}$	$v(\mathsf{n}@\mathsf{a}) \equiv \mathtt{T}$	iff	$v(\mathsf{a}) \equiv \mathtt{F}$	$\mathsf{a} \in \mathcal{D}_o$
$\mathfrak{L}_\vee(\mathsf{d})$	$\mathsf{d} \in \mathcal{D}_{o\to o\to o}$	$v(\mathsf{d}@\mathsf{a}@\mathsf{b}) \equiv \mathtt{T}$	iff	$v(\mathsf{a}) \equiv \mathtt{T}$ or $v(\mathsf{b}) \equiv \mathtt{T}$	$\mathsf{a}, \mathsf{b} \in \mathcal{D}_o$
$\mathfrak{L}_\forall^\alpha(\pi)$	$\pi \in \mathcal{D}_{(\alpha\to o)\to o}$	$v(\pi@\mathsf{f}) \equiv \mathtt{T}$	iff	$\forall \mathsf{a} \in \mathcal{D}_\alpha\ v(\mathsf{f}@\mathsf{a}) \equiv \mathtt{T}$	$\mathsf{f} \in \mathcal{D}_{\alpha\to o}$
$\mathfrak{L}_=^\alpha(\mathsf{q})$	$\mathsf{q} \in \mathcal{D}_{\alpha\to\alpha\to o}$	$v(\mathsf{q}@\mathsf{a}@\mathsf{b}) \equiv \mathtt{T}$	iff	$\mathsf{a} \equiv \mathsf{b}$	$\mathsf{a}, \mathsf{b} \in \mathcal{D}_\alpha$

Figure 1. Logical Properties in Σ-Models

Given a model $\mathcal{M} := (\mathcal{D}, @, \mathcal{E}, v)$, an assignment φ and a proposition $\mathbf{A}$ (or set of propositions Φ), we say $\mathcal{M}$ satisfies $\mathbf{A}$ (or Φ) and write $\mathcal{M} \models_\varphi \mathbf{A}$ (or $\mathcal{M} \models_\varphi \Phi$) if $v(\mathcal{E}_\varphi(\mathbf{A})) \equiv \mathtt{T}$ (or $v(\mathcal{E}_\varphi(\mathbf{A})) \equiv \mathtt{T}$ for each $\mathbf{A} \in \Phi$). If $\mathbf{A}$ is closed (or every member of Φ is closed), then we simply write $\mathcal{M} \models \mathbf{A}$ (or $\mathcal{M} \models \Phi$) and say $\mathcal{M}$ is a model of $\mathbf{A}$ (or Φ).

In order to define model classes which correspond to different notions of extensionality, five properties of models are defined ($\mathfrak{q}$, η, ξ, $\mathfrak{f}$, and $\mathfrak{b}$; cf. Definitions 3.46, 3.21 and 3.5 in [3]). In this paper, we will only refer to properties $\mathfrak{q}$ and ξ. Let $\mathcal{M} := (\mathcal{D}, @, \mathcal{E}, v)$ be a model. We say $\mathcal{M}$ has property

$\mathfrak{q}$ iff for all $\alpha \in \mathcal{T}$ there is some $\mathsf{q}^\alpha \in \mathcal{D}_{\alpha\to\alpha\to o}$ such that $\mathfrak{L}_=^\alpha(\mathsf{q}^\alpha)$ holds.

ξ iff $(\mathcal{D}, @, \mathcal{E})$ is ξ-functional (i.e., for each $\mathbf{M}, \mathbf{N} \in \mathit{wff}_\beta(\Sigma)$, $X \in \mathcal{V}_\alpha$ and assignment φ, we have $\mathcal{E}_\varphi(\lambda X_\alpha.\mathbf{M}_\beta) \equiv \mathcal{E}_\varphi(\lambda X_\alpha.\mathbf{N}_\beta)$ whenever $\mathcal{E}_{\varphi,[\mathsf{a}/X]}(\mathbf{M}) \equiv \mathcal{E}_{\varphi,[\mathsf{a}/X]}(\mathbf{N})$ for every $\mathsf{a} \in \mathcal{D}_\alpha$).

For each $* \in \{\beta, \beta\eta, \beta\xi, \beta\mathfrak{f}, \beta\mathfrak{b}, \beta\eta\mathfrak{b}, \beta\xi\mathfrak{b}, \beta\mathfrak{f}\mathfrak{b}\}$ there is a model class $\mathfrak{M}_*$ (cf. Definition 3.49 in [3]). Here we only consider $* \in \{\beta, \beta\xi\}$: $\mathfrak{M}_\beta$ is the class of all Σ-models $\mathcal{M}$ satisfying property $\mathfrak{q}$. $\mathfrak{M}_{\beta\xi}$ is the class of all Σ-models $\mathcal{M}$ satisfying properties $\mathfrak{q}$ and ξ.

Finally, we review the model existence theorems proved in [3]. There are three stages to obtaining a model in our framework. First, we obtain an abstract consistency class Γ_Σ (usually defined as the class of irrefutable sets of sentences with respect to some calculus). Second, given a (sufficiently pure) set of sentences Φ in the abstract consistency class Γ_Σ we construct a Hintikka set $\mathcal{H}$ extending Φ. Third, we construct a model of this Hintikka set (hence a model of Φ).

We say Γ_Σ is an abstract consistency class if it is closed under subsets and satisfies properties $\nabla_c, \nabla_\neg, \nabla_\beta, \nabla_\vee, \nabla_\wedge, \nabla_\forall$ and $\nabla_\exists$ (cf. Definitions 6.1 and 6.5 in [3]). We let $\mathfrak{Acc}_\beta$ denote the collection of all abstract consistency classes. For each $* \in \boxdot$ we refine $\mathfrak{Acc}_\beta$ to a collection $\mathfrak{Acc}_*$ where the additional properties $\{\nabla_\eta, \nabla_\xi, \nabla_\mathfrak{f}, \nabla_\mathfrak{b}\}$ indicated by $*$ are required (cf. Definition 6.7 in [3]). We say an abstract consistency class Γ_Σ is saturated if ∇_{sat} holds. The only condition we will explicitly use in this paper is ∇_ξ which is defined as follows:

∇_ξ If $\neg(\lambda X_\alpha.\mathbf{M} \doteq^{\alpha\to\beta} \lambda X_\alpha.\mathbf{N}) \in \Phi$, then $\Phi * \neg([w/X]\mathbf{M} \doteq^\beta [w/X]\mathbf{N}) \in \Gamma_\Sigma$ for any parameter $w_\alpha \in \Sigma_\alpha$ which does not occur in any sentence of Φ.

In order to obtain a Hintikka set extending a set Φ, we must have parameters which will act as witnesses. For this we require sufficient purity of Φ. A set Φ of Σ-sentences is called sufficiently Σ-pure (cf. Definition 6.3 in [3]) if for each type α there is a set $\mathcal{P}_\alpha$ of parameters of type α with cardinality $\aleph_s$ (the cardinality of $\mathit{wff}_\alpha(\Sigma)$), such that no parameter in $\mathcal{P}$ occurs in a sentence in Φ. Note that since Σ is assumed to have infinite cardinality $\aleph_s$ for each type, every finite set of Σ-sentences is sufficiently Σ-pure.

A Hintikka set is a set of sentences satisfying certain properties. The following is a list of some of the properties a set $\mathcal{H}$ of sentences may satisfy (cf. Definition 6.19 in [3]):

$\vec{\nabla}_c$ $\mathbf{A} \notin \mathcal{H}$ or $\neg\mathbf{A} \notin \mathcal{H}$.

$\vec{\nabla}_\neg$ If $\neg\neg\mathbf{A} \in \mathcal{H}$, then $\mathbf{A} \in \mathcal{H}$.

$\vec{\nabla}_\beta$ If $\mathbf{A} \in \mathcal{H}$ and $\mathbf{A} \equiv_\beta \mathbf{B}$, then $\mathbf{B} \in \mathcal{H}$.

$\vec{\nabla}_\vee$ If $\mathbf{A} \vee \mathbf{B} \in \mathcal{H}$, then $\mathbf{A} \in \mathcal{H}$ or $\mathbf{B} \in \mathcal{H}$.

$\vec{\nabla}_\wedge$ If $\neg(\mathbf{A} \vee \mathbf{B}) \in \mathcal{H}$, then $\neg\mathbf{A} \in \mathcal{H}$ and $\neg\mathbf{B} \in \mathcal{H}$.

$\vec{\nabla}_\forall$ If $\Pi^\alpha\mathbf{F} \in \mathcal{H}$, then $\mathbf{F}\mathbf{W} \in \mathcal{H}$ for each $\mathbf{W} \in \mathit{cwff}_\alpha(\Sigma)$.

$\vec{\nabla}_\exists$ If $\neg\Pi^\alpha \mathbf{F} \in \mathcal{H}$, then there is a parameter $w_\alpha \in \Sigma_\alpha$ such that $\neg(\mathbf{F}w) \in \mathcal{H}$.

$\vec{\nabla}_\xi$ If $\neg(\lambda X_\alpha.\mathbf{M} \doteq^{\alpha\to\beta} \lambda X.\mathbf{N}) \in \mathcal{H}$, then there is a parameter $w_\alpha \in \Sigma_\alpha$ such that $\neg([w/X]\mathbf{M} \doteq^\beta [w/X]\mathbf{N}) \in \mathcal{H}$.

$\vec{\nabla}_{sat}$ Either $\mathbf{A} \in \mathcal{H}$ or $\neg\mathbf{A} \in \mathcal{H}$.

[3] also defines properties $\vec{\nabla}_\eta$, $\vec{\nabla}_\mathfrak{b}$, and $\vec{\nabla}_\mathfrak{f}$, but these will not be used here. A set $\mathcal{H}$ of sentences is called a Σ-*Hintikka set* if $\vec{\nabla}_c$, $\vec{\nabla}_\neg$, $\vec{\nabla}_\beta$, $\vec{\nabla}_\vee$, $\vec{\nabla}_\wedge$, $\vec{\nabla}_\forall$ and $\vec{\nabla}_\exists$ hold. We define the following collections of Hintikka sets: $\mathfrak{Hint}_\beta$, $\mathfrak{Hint}_{\beta\eta}$, $\mathfrak{Hint}_{\beta\xi}$, $\mathfrak{Hint}_{\beta\mathfrak{f}}$, $\mathfrak{Hint}_{\beta\mathfrak{b}}$, $\mathfrak{Hint}_{\beta\eta\mathfrak{b}}$, $\mathfrak{Hint}_{\beta\xi\mathfrak{b}}$, and $\mathfrak{Hint}_{\beta\mathfrak{f}\mathfrak{b}}$, where we indicate by indices which additional properties from $\{\vec{\nabla}_\eta, \vec{\nabla}_\xi, \vec{\nabla}_\mathfrak{f}, \vec{\nabla}_\mathfrak{b}\}$ are required (cf. Definition 6.20 in [3]). We call a Hintikka set $\mathcal{H}$ *saturated* if $\vec{\nabla}_{sat}$ holds (cf. Definition 6.24 in [3]).

One of the main theorems of [3] is the Model Existence Theorem for Saturated Sets which states the following:

THEOREM 1 (Model Existence Theorem for Saturated Sets (Theorem 6.33 in [3])).
For all $* \in \boxplus$ *we have: If* $\mathcal{H}$ *is a saturated Hintikka set in* $\mathfrak{Hint}_*$, *then there exists a model* $\mathcal{M} \in \mathfrak{M}_*$ *that satisfies* $\mathcal{H}$. *Furthermore, each domain* $\mathcal{D}_\alpha$ *of* $\mathcal{M}$ *has cardinality at most* $\aleph_s$.

Since saturated abstract consistency classes give rise to saturated Hintikka sets, we conclude a corresponding model existence theorem for saturated abstract consistency classes.

THEOREM 2 (Theorem 6.34 in [3]). *For all* $* \in \boxplus$, *if* Γ_Σ *is a saturated abstract consistency class in* $\mathfrak{Acc}_*$ *and* $\Phi \in \Gamma_\Sigma$ *is a sufficiently* Σ*-pure set of sentences, then there exists a model* $\mathcal{M} \in \mathfrak{M}_*$ *that satisfies* Φ. *Furthermore, each domain of* $\mathcal{M}$ *has cardinality at most* $\aleph_s$.

3 Possible Values

We now review a framework developed in [6, 7] which is essentially a general version of Andrews construction using V-complexes given in [1]. There are slight differences between the construction here and that in [1]. One difference is that our domains are constructed using pairs $\langle \mathbf{A}, \mathsf{a}\rangle$ where $\mathbf{A}$ is *closed*, whereas in [1] $\mathbf{A}$ may contain free variables. This difference stems from the fact that we use parameters as existential witnesses and Andrews uses variables for this purpose in [1]. Another difference is that we start from a Hintikka set $\mathcal{H}$ instead of a semivaluation V.

Except for the different treatment of variables, the V-complex construction provides an instance of a *possible values structure for* β (cf. Definition 3) and a *possible values evaluation for* β (cf. Definition 8). The definitions in [6, 7] are for both the β and $\beta\eta$ cases. We repeat these definitions (specialized for the β case) and a few results here. We then prove that any possible values evaluation is ξ-functional (a new result).

The results in [7] are stated with respect to a signature of logical constants $\mathcal{S}$ which is distinct from the set of parameters $\mathcal{P}$. In order to apply the results from [7] we take $\mathcal{S}$ to be the set

$$\{\neg, \vee\} \cup \{\Pi^\alpha | \alpha \in \mathcal{T}\}$$

and $\mathcal{P}$ to be the typed family of sets of parameters (non-logical constants) in Σ. Note that $\Sigma \equiv (\mathcal{S} \cup \mathcal{P})$.

DEFINITION 3 (Definition 4.1.1 from [6, 7]). A *possible values structure for* β is an applicative structure $\mathcal{F} \equiv (\mathcal{D}, @)$ satisfying the following:

1. For each type $\alpha \in \mathcal{T}$, $\mathsf{a} \in \mathcal{D}_\alpha$ implies $\mathsf{a} \equiv \langle \mathbf{A}, a\rangle$ for some a and term $\mathbf{A} \in cwff_\alpha(\Sigma)$ such that $\mathbf{A}{\downarrow_\beta} \equiv \mathbf{A}$.

2. At each base type $\alpha \in \{o, \iota\}$, for every $\mathbf{A} \in cwff_\alpha(\Sigma)$, there exists some p with $\langle \mathbf{A}{\downarrow_\beta}, p\rangle \in \mathcal{D}_\alpha$.

3. For each function type $\alpha \to \beta$, $\langle \mathbf{G}, g\rangle \in \mathcal{D}_{\alpha\to\beta}$ iff $\mathbf{G} \in cwff_{\alpha\to\beta}(\Sigma)$, $\mathbf{G}{\downarrow_\beta} \equiv \mathbf{G}$, $g : \mathcal{D}_\alpha \longrightarrow \mathcal{D}_\beta$ and for every $\langle \mathbf{A}, a\rangle \in \mathcal{D}_\alpha$ the first component of $g(\langle \mathbf{A}, a\rangle)$ is $[\mathbf{G}\,\mathbf{A}]{\downarrow_\beta}$.

4. For each $\langle \mathbf{G}, g\rangle \in \mathcal{D}_{\alpha\to\beta}$ and $\langle \mathbf{A}, a\rangle \in \mathcal{D}_\alpha$,

$$\langle \mathbf{G}, g\rangle @ \langle \mathbf{A}, a\rangle \equiv g(\langle \mathbf{A}, a\rangle).$$

DEFINITION 4 (Definition 4.1.2 from [6, 7]). Let $\mathcal{A} \equiv (\mathcal{D}, @)$ be a possible values structure for β. We call p a *possible value* for $\mathbf{A} \in cwff_\alpha(\Sigma)$ if $\langle \mathbf{A}{\downarrow_\beta}, p\rangle \in \mathcal{D}_\alpha$.

The next lemma is similar to Lemma 3.4.2 in [1] which provided the idea for the proof by induction on types.

LEMMA 5 (Lemma 4.1.3 from [6, 7]). *Let* $\mathcal{F}$ *be a possible values structure for* β. *For each closed term* $\mathbf{A} \in cwff_\alpha(\Sigma)$, *there is a possible value* p *for* $\mathbf{A}$ *in* $\mathcal{F}$.

DEFINITION 6 (Definition 4.1.4 from [6, 7]). Let $\mathcal{A} \equiv (\mathcal{D}, @)$ be a possible values structure for β. We define

$$\mathcal{D}_\alpha^{\mathbf{A}} := \{\langle \mathbf{A}{\downarrow_\beta}, a\rangle \in \mathcal{D}_\alpha \;\big|\; a \text{ is a possible value for } \mathbf{A}\}.$$

for each $\mathbf{A} \in cwff_\alpha$.

DEFINITION 7 (Definition 4.1.5 from [6, 7]). Let $\mathcal{A} \equiv (\mathcal{D}, @)$ be a possible values structure for β and φ be an assignment into $\mathcal{A}$. For any $\mathbf{A} \in wff_\alpha(\Sigma)$, we define $\varphi_1(\mathbf{A})$ to be $\theta(\mathbf{A}) \in cwff_\alpha(\Sigma)$ where θ is the substitution with $\mathbf{Dom}(\theta) \equiv free(\mathbf{A})$ and $\varphi(x_\beta) \equiv \langle \theta(x_\beta), b \rangle \in \mathcal{D}_\beta$ for each variable $x_\beta \in free(\mathbf{A})$. We define $\varphi_1^\beta(\mathbf{A})$ to be $\varphi_1(\mathbf{A})\big\downarrow_\beta$.

DEFINITION 8 (Definition 4.1.7 from [6, 7]). We call an evaluation $\mathcal{J} \equiv (\mathcal{D}, @, \mathcal{E})$ a *possible values evaluation for* β if $(\mathcal{D}, @)$ is a possible values structure for β and $\mathcal{E}_\varphi(\mathbf{A}) \in \mathcal{D}_\alpha^{\varphi_1(\mathbf{A})}$ for every $\mathbf{A} \in wff_\alpha(\Sigma)$ and assignment φ.

We can always extend an appropriate interpretation of parameters and constants in a possible values structure to obtain a possible values evaluation.

THEOREM 9 (Theorem 4.1.8 from [6, 7]). *Let $\mathcal{A} \equiv (\mathcal{D}, @)$ be a possible values structure for β and $\mathcal{I} : \Sigma \longrightarrow \mathcal{D}$ be an interpretation of parameters and constants such that $\mathcal{I}(c_\alpha) \in \mathcal{D}_\alpha^c$ for every $c \in \Sigma$. There is an evaluation function $\mathcal{E}$ such that $\mathcal{J} :\equiv (\mathcal{D}, @, \mathcal{E})$ is a possible values evaluation for β, $\mathcal{E}(c_\alpha) \equiv \mathcal{I}(c_\alpha)$ for every $c_\alpha \in \Sigma$.*

We now verify the only new result of this section: possible values evaluations are ξ-functional.

PROPOSITION 10. *Every possible values evaluation for β is ξ-functional.*

Proof. Let $(\mathcal{D}, @, \mathcal{E})$ be a possible values evaluation. Let $\mathbf{M}, \mathbf{N} \in wff_\beta(\Sigma)$ and X_α be a variable such that $\mathcal{E}_{\varphi,[\mathsf{a}/X]}(\mathbf{M}) \equiv \mathcal{E}_{\varphi,[\mathsf{a}/X]}(\mathbf{N})$ for all $\mathsf{a} \in \mathcal{D}_\alpha$. We must verify $\mathcal{E}_\varphi(\lambda X_\alpha.\mathbf{M}) \equiv \mathcal{E}_\varphi(\lambda X_\alpha.\mathbf{N})$. We know $\mathcal{E}_\varphi(\lambda X_\alpha.\mathbf{M}) \equiv \langle \varphi_1^\beta(\lambda X_\alpha.\mathbf{M}), f \rangle$ and $\mathcal{E}_\varphi(\lambda X_\alpha.\mathbf{N}) \equiv \langle \varphi_1^\beta(\lambda X_\alpha.\mathbf{N}), g \rangle$ for some $f, g : \mathcal{D}_\alpha \to \mathcal{D}_\beta$. We first check that the first components are equal. Let w_α be a parameter which occurs neither in $\mathbf{M}$ nor in $\mathbf{N}$. By Lemma 5 there is some p such that $\langle w, p \rangle \in \mathcal{D}_\alpha$. By assumption, $\mathcal{E}_{\varphi,[\langle w,p \rangle/X]}(\mathbf{M}) \equiv \mathcal{E}_{\varphi,[\langle w,p \rangle/X]}(\mathbf{N})$. Since $\mathcal{E}$ is a possible values evaluation, the first component of $\mathcal{E}_{\varphi,[\langle w,p \rangle/X]}(\mathbf{M})$ is $(\varphi, [\langle w, p \rangle/X])_1^\beta(\mathbf{M})$. It is easy to see that this is the same as $[w/X]\varphi_1^\beta(\mathbf{M})$. Similarly, the first component of $\mathcal{E}_{\varphi,[\langle w,p \rangle/X]}(\mathbf{N})$ is $[w/X]\varphi_1^\beta(\mathbf{N})$. Hence $[w/X]\varphi_1^\beta(\mathbf{M}) \equiv [w/X]\varphi_1^\beta(\mathbf{N})$. Since w was chosen to be fresh, $\varphi_1^\beta(\mathbf{M}) \equiv \varphi_1^\beta(\mathbf{N})$ and so

$$\varphi_1^\beta(\lambda X.\mathbf{M}) \equiv \lambda X.\varphi_1^\beta(\mathbf{M}) \equiv \lambda X.\varphi_1^\beta(\mathbf{N}) \equiv \varphi_1^\beta(\lambda X.\mathbf{N}).$$

Next, we show the second components are equal. Using the properties of evaluation functions and the definition of @, we easily compute

$$f(\mathsf{a}) \equiv \mathcal{E}_{\varphi}(\lambda X.\mathbf{M})@\mathsf{a} \equiv \mathcal{E}_{\varphi,[\mathsf{a}/X]}((\lambda X.\mathbf{M})X) \equiv \mathcal{E}_{\varphi,[\mathsf{a}/X]}(\mathbf{M})$$

and

$$g(\mathsf{a}) \equiv \mathcal{E}_{\varphi}(\lambda X.\mathbf{N})@\mathsf{a} \equiv \mathcal{E}_{\varphi,[\mathsf{a}/X]}((\lambda X.\mathbf{N})X) \equiv \mathcal{E}_{\varphi,[\mathsf{a}/X]}(\mathbf{N})$$

for any $\mathsf{a} \in \mathcal{D}_\alpha$. Hence $f = g$ as desired. ■

4 Model Existence Theorems Without Saturation

Model existence theorems generally say that in order to show that a set Φ of formulae has a model $\mathcal{M}$ in a given class $\mathfrak{M}$, it is sufficient to prove that Φ is a member of suitably defined abstract consistency classes Γ. Model existence theorems are usually proven in two steps: first we show that any $\Phi \in \Gamma$ can be extended to a Hintikka set $\mathcal{H} \in \Gamma$ with $\Phi \subseteq \mathcal{H}$, and then for a given Hintikka set $\mathcal{H}$ we construct a model $\mathcal{M} \in \mathfrak{M}$ that satisfies $\mathcal{H}$. The first step is already addressed by the Abstract Extension Lemma (Lemma 6.32) in [3] and it will be reused below. The second step — for the model classes $\mathfrak{M}_\beta$ and $\mathfrak{M}_{\beta\xi}$ and without assuming saturation — is a novel contribution of this paper.

When constructing models in $\mathfrak{M}_*$ of a Hintikka set $\mathcal{H}$, we must verify property $\mathfrak{q}$. For this purpose, the assumption that $\mathcal{H}$ contains no Leibniz equations is very helpful.

DEFINITION 11. Let $\mathcal{H}$ be a set of formulae. We say $\mathcal{H}$ is *Leibniz-free* if there are no terms $\mathbf{A}_\alpha, \mathbf{B}_\alpha$ such that $(\mathbf{A} \doteq^\alpha \mathbf{B}) \in \mathcal{H}$.

We can now show every Hintikka set is either saturated (in which case we have already constructed models in [3]) or Leibniz-free. Hence we will only need to construct models for Leibniz-free Hintikka sets. This result is closely related to the fact the Leibniz equations are cut-strong (see Example 14 in [4]).

THEOREM 12. *Let $\mathcal{H}$ be a Hintikka set. Either $\mathcal{H}$ is saturated or $\mathcal{H}$ is Leibniz-free.*

Proof. Suppose $\mathcal{H}$ is not Leibniz-free. Then $(\mathbf{A} \doteq^\alpha \mathbf{B}) \in \mathcal{H}$ for some $\mathbf{A}_\alpha, \mathbf{B}_\alpha$. We show $\mathcal{H}$ satisfies $\vec{\nabla}_{sat}$. Let $\mathbf{C}_o$ be a closed formula. Since $(\forall Q_{\alpha\to o}.Q\mathbf{A} \Rightarrow Q\mathbf{B}) \in \mathcal{H}$, we know $(\neg\mathbf{C} \vee \mathbf{C}) \in \mathcal{H}$ by $\vec{\nabla}_\forall$ (with the term $\lambda X_\alpha.\mathbf{C}$) and $\vec{\nabla}_\beta$. By $\vec{\nabla}_\vee$, either $\neg\mathbf{C} \in \mathcal{H}$ or $\mathbf{C} \in \mathcal{H}$. ■

The proof of the following theorem is the main contribution in the paper. It the construction is based on Andrew's V-complexes but extends the argument by checking properties ξ (from Prop 10) and $\mathfrak{q}$ (by choosing Leibniz and using that H is Leibniz free since unsaturated).

THEOREM 13. *Let $\mathcal{H}$ be a Σ-Hintikka set which is not saturated. There is a Σ-model $\mathcal{M} \in \mathfrak{M}_{\beta\xi}$ such that $\mathcal{M} \models \mathcal{H}$.*

Proof. We first define a set $\mathcal{B}^{\mathbf{A}}_{\mathcal{H}}$ of *possible booleans* for each $\mathbf{A} \in \mathit{wff}_o(\Sigma)$:

$$\mathcal{B}^{\mathbf{A}}_{\mathcal{H}} := \begin{cases} \{\mathtt{T}\} & \text{if } \mathbf{A} \in \mathcal{H} \\ \{\mathtt{F}\} & \text{if } \neg\mathbf{A} \in \mathcal{H} \\ \{\mathtt{T}, \mathtt{F}\} & \text{otherwise.} \end{cases}$$

We define $\mathcal{D}_\alpha$ for each type $\alpha \in \mathcal{T}$ by induction:

- $\mathcal{D}_o := \{\langle \mathbf{A}_o, p\rangle | \mathbf{A} \in \mathit{wff}_o(\Sigma){\downarrow_\beta}, p \in \mathcal{B}^{\mathbf{A}}_{\mathcal{H}}\}$.
- $\mathcal{D}_\iota := \{\langle \mathbf{A}_\iota, \iota\rangle | \mathbf{A} \in \mathit{wff}_\iota(\Sigma){\downarrow_\beta}\}$.
- $\mathcal{D}_{\alpha\to\beta} := \{\langle \mathbf{F}_{\alpha\to\beta}, f\rangle | \mathbf{F} \in \mathit{wff}_{\alpha\to\beta}(\Sigma){\downarrow_\beta}, f : \mathcal{D}_\alpha \to \mathcal{D}_\beta,$

  $\forall \langle \mathbf{A}, a\rangle \in \mathcal{D}_\alpha, \langle \mathbf{B}, b\rangle \in \mathcal{D}_\beta\; f(\langle \mathbf{A}, a\rangle) \equiv \langle \mathbf{B}, b\rangle \Rightarrow \mathbf{B} \equiv (\mathbf{F}\,\mathbf{A}){\downarrow_\beta}\}$.

We define an application operator @ by setting $\langle \mathbf{F}, f\rangle @ \mathsf{a}$ to be $f(\mathsf{a})$ for each $\langle \mathbf{F}, f\rangle \in \mathcal{D}_{\alpha\to\beta}$ and $\mathsf{a} \in \mathcal{D}_\alpha$. It is easy to check that $(\mathcal{D}, @)$ is a possible values structure for β. Note that for all $\mathbf{A} \in \mathit{wff}_o(\Sigma)$ either $\mathbf{A} \notin \mathcal{H}$ or $\neg\mathbf{A} \notin \mathcal{H}$ (by $\vec{\nabla}_c$) and so either $\langle \mathbf{A}, \mathtt{F}\rangle \in \mathcal{D}_o$ or $\langle \mathbf{A}, \mathtt{T}\rangle \in \mathcal{D}_o$. (It is possible that both $\langle \mathbf{A}, \mathtt{F}\rangle \in \mathcal{D}_o$ and $\langle \mathbf{A}, \mathtt{T}\rangle \in \mathcal{D}_o$.)

For each parameter w_α, we can choose some p^w such that $\langle w, p^w\rangle \in \mathcal{D}_\alpha$ using Lemma 5. These values can be used to interpret parameters. To interpret logical constants, we must make appropriate choices so that the corresponding logical properties will hold.

$\neg$ Let $p^\neg : \mathcal{D}_o \to \mathcal{D}_o$ be defined by $p^\neg(\langle \mathbf{A}, a\rangle) := \langle \neg\mathbf{A}, b\rangle$ where b is $\mathtt{T}$ if a is $\mathtt{F}$ and b is $\mathtt{F}$ if a is $\mathtt{T}$. The $\vec{\nabla}_\neg$ and $\vec{\nabla}_c$ properties of $\mathcal{H}$ guarantees this is well-defined. So, $p^\neg$ is a possible value for $\neg$.

$\vee$ For each $\langle \mathbf{A}, \mathtt{F}\rangle \in \mathcal{D}_o$, let $p^\vee_{\langle \mathbf{A},\mathtt{F}\rangle} : \mathcal{D}_o \to \mathcal{D}_o$ be the function defined by $p^\vee_{\langle \mathbf{A},\mathtt{F}\rangle}(\langle \mathbf{B}, b\rangle) := \langle \mathbf{A} \vee \mathbf{B}, b\rangle$. For each $\langle \mathbf{A}, \mathtt{T}\rangle \in \mathcal{D}_o$, let $p^\vee_{\langle \mathbf{A},\mathtt{T}\rangle} : \mathcal{D}_o \to \mathcal{D}_o$ be the function defined by $p^\vee_{\langle \mathbf{A},\mathtt{T}\rangle}(\langle \mathbf{B}, b\rangle) := \langle \mathbf{A} \vee \mathbf{B}, \mathtt{T}\rangle$.

The properties $\vec{\nabla}_\vee$, $\vec{\nabla}_\wedge$ and $\vec{\nabla}_c$ of $\mathcal{H}$ guarantees these are well-defined and $\langle \vee\mathbf{A}, p^\vee_{\langle \mathbf{A},a\rangle}\rangle \in \mathcal{D}_{o\to o}$. Now, let $p^\vee : \mathcal{D}_o \to \mathcal{D}_{o\to o}$ be defined by $p^\vee(\langle \mathbf{A}, a\rangle) := \langle \vee\mathbf{A}, p^\vee_{\langle \mathbf{A},a\rangle}\rangle$. Clearly, $p^\vee$ is a possible value for $\vee$.

Π^α Let $p^{\Pi^\alpha} : \mathcal{D}_{\alpha\to o} \to \mathcal{D}_o$ be the function defined by $p^{\Pi^\alpha}(\mathbf{F}, f) := \langle \Pi^\alpha\mathbf{F}, p\rangle$ where $p \equiv \mathtt{T}$ if for every $\langle \mathbf{A}, a\rangle \in \mathcal{D}_\alpha$, the second component of $f(\langle \mathbf{A}, a\rangle)$ is $\mathtt{T}$, and $p \equiv \mathtt{F}$ otherwise. This is well-defined by $\vec{\nabla}_\forall$, $\vec{\nabla}_\exists$ and $\vec{\nabla}_c$, and p^{Π^α} is a possible value for Π^α.

Let $\mathcal{I}(c) := \langle c, p^c \rangle$ for each $c \in \Sigma$ and $\mathcal{E}$ be the evaluation function extending $\mathcal{I}$ guaranteed to exist by Theorem 9 so that $(\mathcal{D}, @, \mathcal{E})$ is a ξ-functional possible values evaluation.

To make this a Σ-model, we must define a valuation $\upsilon: \mathcal{D}_o \longrightarrow \{\mathtt{T}, \mathtt{F}\}$. We take the obvious choice $\upsilon(\langle \mathbf{A}, p \rangle) := p$. So let $\mathcal{M} := (\mathcal{D}, @, \mathcal{E}, \upsilon)$. To check $\mathcal{M}$ is a Σ-model, we must check the requirements for υ. Each condition is trivial:

$\neg$: $\upsilon(\mathcal{E}(\neg)@\mathsf{a}) \equiv \mathtt{T}$ iff $\upsilon(\mathsf{a}) \equiv \mathtt{F}$ by the definition of $p^\neg$.

$\vee$: $\upsilon(\mathcal{E}(\vee)@\mathsf{a}@\mathsf{b}) \equiv \mathtt{T}$ iff $\upsilon(\mathsf{a}) \equiv \mathtt{T}$ or $\upsilon(\mathsf{b}) \equiv \mathtt{T}$ by the definition of $p^\vee$.

Π: $\upsilon(\mathcal{E}(\Pi^\alpha)@\mathsf{f}) \equiv \mathtt{T}$ iff $\upsilon(\mathsf{f}@\mathsf{a}) \equiv \mathtt{T}$ for each $\mathsf{a} \in \mathcal{D}_\alpha$ by the definition of p^{Π^α}.

We verify $\mathcal{M} \models \mathcal{H}$. Suppose $\mathbf{A} \in \mathcal{H}$ and let $\mathbf{B}$ be $\mathbf{A}{\downarrow_\beta}$. Note that $\mathcal{E}(\mathbf{A}) \equiv \langle \mathbf{B}, p \rangle \in \mathcal{D}_o$ for some $p \in \mathcal{B}^{\mathbf{B}}_{\mathcal{H}}$. Since $\mathbf{A} \in \mathcal{H}$, we have $\mathbf{B} \in \mathcal{H}$ by $\vec{\nabla}_\beta$. Thus $\mathcal{B}^{\mathbf{B}}_{\mathcal{H}} \equiv \{\mathtt{T}\}$, $p = \mathtt{T}$ and so $\mathcal{M} \models \mathbf{A}$.

In general, we can use Theorem 3.62 in [3] to obtain a model of $\mathcal{H}$ satisfying property $\mathfrak{q}$, though this would not preserve property ξ (cf. Remark 3.57 in [3]). Instead, we use the assumption that $\mathcal{H}$ is not saturated and hence Leibniz-free to show the possible values model $\mathcal{M}$ *already* satisfies property $\mathfrak{q}$. To see this, for each $\langle \mathbf{A}, a \rangle \in \mathcal{D}_\alpha$, let $s_{\langle \mathbf{A}, a \rangle}: \mathcal{D}_\alpha \to \mathcal{D}_o$ be defined by

$$s_{\langle \mathbf{A}, a \rangle}(\langle \mathbf{B}, b \rangle) := \begin{cases} \langle (\mathbf{A} \doteq \mathbf{A}){\downarrow_\beta}, \mathtt{T} \rangle & \text{if } \mathbf{A} = \mathbf{B} \text{ and } a = b \\ \langle (\mathbf{A} \doteq \mathbf{B}){\downarrow_\beta}, \mathtt{F} \rangle & \text{else} \end{cases}$$

This is well-defined since we never have $\neg\, (\mathbf{A} \doteq \mathbf{A}){\downarrow_\beta} \in \mathcal{H}$, and at the same time $(\mathbf{A} \doteq \mathbf{B}){\downarrow_\beta} \notin \mathcal{H}$ since $\mathcal{H}$ is Leibniz-free. Then, $\mathfrak{q}^\alpha := \langle \doteq^\alpha, \mathsf{l} \rangle$ with $l(\langle \mathbf{A}, a \rangle) := \langle (\lambda X.\mathbf{A} \doteq x){\downarrow_\beta}, s_{\langle \mathbf{A}, a \rangle} \rangle$ witnesses that $\mathcal{M}$ satisfies property $\mathfrak{q}$. Thus, $\mathcal{M} \in \mathfrak{M}_{\beta\xi}$ as desired. ∎

THEOREM 14 (Model Existence for $\mathfrak{Hint}_\beta$ and $\mathfrak{Hint}_{\beta\xi}$). *For each* $* \in \{\beta, \beta\xi\}$ *and* Σ*-Hintikka set* $\mathcal{H} \in \mathfrak{Hint}_*$*, there is a* Σ*-model* $\mathcal{M} \in \mathfrak{M}_*$ *such that* $\mathcal{M} \models \mathcal{H}$.

Proof. If $\mathcal{H}$ is not saturated, then we can obtain such an $\mathcal{M}$ by applying Theorem 13 above. If $\mathcal{H}$ is saturated, then we can obtain such an $\mathcal{M}$ by applying the Model Existence Theorem for Saturated Sets (Theorem 1). ∎

THEOREM 15 (Model Existence for $\mathfrak{Acc}_\beta$ and $\mathfrak{Acc}_{\beta\xi}$). *For each* $* \in \{\beta, \beta\xi\}$*, abstract consistency class* $\Gamma_\Sigma \in \mathfrak{Acc}_*$ *and sufficiently* Σ*-pure* $\Phi \in \Gamma_\Sigma$*, there is a* Σ*-model* $\mathcal{M} \in \mathfrak{M}_*$ *such that* $\mathcal{M} \models \Phi$.

Proof. By the Abstract Extension Lemma (Lemma 6.32 in [3]), there is a Hintikka set $\mathcal{H} \in \mathfrak{Hint}_*$ such that $\Phi \subseteq \mathcal{H}$. By Theorem 14 above there is a Σ-model $\mathcal{M} \in \mathfrak{M}_*$ such that $\mathcal{M} \models \mathcal{H}$. ■

5 A Sequent Calculus

As in [4, 5], we consider a sequent to be a finite set Δ of β-normal sentences from $cwff_o(\Sigma)$. A sequent calculus $\mathcal{G}$ provides an inductive definition for when $\Vdash_\mathcal{G} \Delta$ holds. We say a sequent calculus rule

$$\frac{\Delta_1 \quad \cdots \quad \Delta_n}{\Delta}$$

is *admissible* if $\Vdash_\mathcal{G} \Delta$ holds whenever $\Vdash_\mathcal{G} \Delta_i$ for all $1 \leq i \leq n$. Given a sequent Δ and a model $\mathcal{M}$, we say Δ is *valid for* $\mathcal{M}$ if $\mathcal{M} \models \mathbf{D}$ for some $\mathbf{D} \in \Delta$. For a class $\mathfrak{M}$ of models, we say Δ is *valid for* $\mathfrak{M}$ if Δ is valid for every $\mathcal{M} \in \mathfrak{M}$. As for sets in abstract consistency classes, we use the notation $\Delta * \mathbf{A}$ to denote the set $\Delta \cup \{\mathbf{A}\}$ (which is simply Δ if $\mathbf{A} \in \Delta$). We adopt the notation $\neg\Phi$ for the set $\{\neg\mathbf{A} | \mathbf{A} \in \Phi\}$ where $\Phi \subseteq cwff_o(\Sigma)$. Furthermore, we assume this use of $\neg$ binds more strongly than $\cup$ or $*$, so that $\neg\Phi \cup \Delta$ means $(\neg\Phi) \cup \Delta$ and $\neg\Phi * \mathbf{A}$ means $(\neg\Phi) * \mathbf{A}$. For any sequent calculus $\mathcal{G}$, we can define a class of sets of sentences $\Gamma_\Sigma^\mathcal{G}$ as in [4, 5].

DEFINITION 16 (Definition 1 from [4]/Definition 3.1 from [5]). Let $\mathcal{G}$ be a sequent calculus. We define $\Gamma_\Sigma^\mathcal{G}$ to be the class of all finite $\Phi \subset cwff_o(\Sigma)$ such that $\Vdash_\mathcal{G} \neg \Phi{\downarrow_\beta}$ does not hold.

Under certain conditions, $\Gamma_\Sigma^\mathcal{G}$ will be an abstract consistency class. The conditions are the admissibility of certain rules given in Figures 2 and 3.

LEMMA 17 (Lemma 2 from [4]/Lemma 3.2 from [5]). *Let $\mathcal{G}$ be a sequent calculus such that $\mathcal{G}(Inv^\neg)$ is admissible. For any finite sets Φ and Δ of sentences, if $\Phi \cup \neg\Delta \notin \Gamma_\Sigma^\mathcal{G}$, then $\Vdash_\mathcal{G} \neg \Phi{\downarrow_\beta} \cup \Delta{\downarrow_\beta}$ holds.*

THEOREM 18 (Theorem 3 from [4]/Theorem 3.3 from [5]). *Let $\mathcal{G}$ be a sequent calculus. If the rules $\mathcal{G}(Inv^\neg)$, $\mathcal{G}(\neg)$, $\mathcal{G}(weak)$, $\mathcal{G}(init)$, $\mathcal{G}(\vee_-)$, $\mathcal{G}(\vee_+)$, $\mathcal{G}(\Pi_-^\mathbf{C})$ and $\mathcal{G}(\Pi_+^c)$ are admissible in $\mathcal{G}$, then $\Gamma_\Sigma^\mathcal{G} \in \mathfrak{Acc}_\beta$.*

We also have the following result relating saturation with admissibility of cut.

THEOREM 19 (Theorem 4 from [4]/Theorem 3.4 from [5]). *Let $\mathcal{G}$ be a sequent calculus.*

1. *If $\mathcal{G}(cut)$ is admissible in $\mathcal{G}$, then $\Gamma_\Sigma^\mathcal{G}$ is saturated.*

2. *If $\mathcal{G}(\neg)$ and $\mathcal{G}(Inv^\neg)$ are admissible in $\mathcal{G}$ and $\Gamma_\Sigma^\mathcal{G}$ is saturated, then $\mathcal{G}(cut)$ is admissible in $\mathcal{G}$.*

$$\frac{\mathbf{A}\text{ atomic}}{\Delta * \mathbf{A} * \neg\mathbf{A}}\,\mathcal{G}(init) \qquad \frac{\Delta * \mathbf{A}}{\Delta * \neg\neg\mathbf{A}}\,\mathcal{G}(\neg)$$

$$\frac{\Delta * \neg\mathbf{A} \quad \Delta * \neg\mathbf{B}}{\Delta * \neg(\mathbf{A} \vee \mathbf{B})}\,\mathcal{G}(\vee_-) \qquad \frac{\Delta * \mathbf{A} * \mathbf{B}}{\Delta * (\mathbf{A} \vee \mathbf{B})}\,\mathcal{G}(\vee_+)$$

$$\frac{\Delta * \neg\,(\mathbf{AC}){\downarrow}_\beta \quad \mathbf{C} \in cwff_\alpha(\Sigma)}{\Delta * \neg\Pi^\alpha\mathbf{A}}\,\mathcal{G}(\Pi_-^{\mathbf{C}})$$

$$\frac{\Delta * (\mathbf{A}c){\downarrow}_\beta \quad c_\alpha \in \Sigma\text{ fresh parameter}}{\Delta * \Pi^\alpha\mathbf{A}}\,\mathcal{G}(\Pi_+^c)$$

Figure 2. Basic Sequent Calculus Rules

$$\frac{\Delta * \neg\neg\mathbf{A}}{\Delta * \mathbf{A}}\,\mathcal{G}(Inv^\neg)$$

$$\frac{\Delta}{\Delta \cup \Delta'}\,\mathcal{G}(weak) \qquad \frac{\Delta * \mathbf{C} \quad \Delta * \neg\mathbf{C}}{\Delta}\,\mathcal{G}(cut)$$

Figure 3. Inversion Rule, Weakening Rule and Cut Rule

The proofs of the previous three results are given in the appendix of [5].

We now turn our attention to the two particular sequent calculi of interest in this paper.

DEFINITION 20 (Sequent Calculi $\mathcal{G}_\beta$ and $\mathcal{G}_{\beta\xi}$). Let $\mathcal{G}_\beta$ be the sequent calculus defined by the rules in Figure 2. Let $\mathcal{G}_{\beta\xi}$ be the sequent calculus defined by the rules in Figure 2 and the $\mathcal{G}(\xi)$ rule in Figure 4

A straightforward induction on derivations proves that $\mathcal{G}_\beta$ and $\mathcal{G}_{\beta\xi}$ are sound with respect to to the model classes $\mathfrak{M}_\beta$ and $\mathfrak{M}_{\beta\xi}$, respectively. The only case which presents any difficulty is that for $\mathcal{G}(\Pi_+^c)$ which uses a fresh

$$\frac{\Delta * (\forall X_\alpha.\mathbf{M} \doteq^\beta \mathbf{N})}{\Delta * (\lambda X_\alpha.\mathbf{M} \doteq^{\alpha\to\beta} \lambda X_\alpha.\mathbf{N})}\ \mathcal{G}(\xi)$$

Figure 4. ξ Extensionality Rule

parameter c. We will show only this case. In this case one can modify a given model by changing the value of the parameter c in the model. This is worked out in detail in [7] and we will refer to some of the results there.

THEOREM 21. *Let* $* \in \{\beta, \beta\xi\}$ *and* Δ *be a sequent. If* $\Vdash_{\mathcal{G}_*} \Delta$, *then for all* $\mathcal{M} \in \mathfrak{M}_*$ *there is some* $\mathbf{A} \in \Delta$ *such that* $\mathcal{M} \models \mathbf{A}$.

Proof. This can be proven by induction on the derivation of $\Vdash_{\mathcal{G}_*} \Delta$. Suppose $\mathcal{G}(\Pi^c_+)$ is the last rule of the derivation. Then Δ is $\Delta' * \Pi^\alpha \mathbf{A}$ and $\Vdash_{\mathcal{G}_*} \Delta' * (\mathbf{A}c){\downarrow_\beta}$ for some parameter c which occurs neither in $\mathbf{A}$ nor in any sentence in Δ'. Let $\mathcal{M} \equiv (\mathcal{D}, @, \mathcal{E}, \upsilon) \in \mathfrak{M}_*$ be given. If $\mathcal{M} \models \mathbf{B}$ for some $\mathbf{B} \in \Delta'$, then we are done. Assume there is no such $\mathbf{B} \in \Delta'$, then we must prove $\mathcal{M} \models \Pi^\alpha \mathbf{A}$, i.e. that $\upsilon(\mathcal{E}(\mathbf{A})@\mathsf{a}) \equiv \mathtt{T}$ for all $\mathsf{a} \in \mathcal{D}_\alpha$. Let $\mathsf{a} \in \mathcal{D}_\alpha$ be given. We let $\mathcal{E}^{c\mapsto\mathsf{a}}$ denote the function from Definition 3.2.16 in [7] and $\mathcal{M}^{c\mapsto\mathsf{a}}$ denote $(\mathcal{D}, @, \mathcal{E}^{c\mapsto\mathsf{a}}, \upsilon)$. We have the following:

- $\mathcal{E}^{c\mapsto\mathsf{a}}(c) \equiv \mathsf{a}$ (see Theorem 3.2.18 in [7]).
- $\mathcal{E}^{c\mapsto\mathsf{a}}(\mathbf{D}) \equiv \mathcal{E}(\mathbf{D})$ if c does not occur in $\mathbf{D}$ (see Theorem 3.2.18 in [7]).
- $\mathcal{M}^{c\mapsto\mathsf{a}} \in \mathfrak{M}_*$ (see Theorem 3.3.14 in [7]).

Applying the inductive hypothesis using $\mathcal{M}^{c\mapsto\mathsf{a}}$, we have $\mathcal{M}^{c\mapsto\mathsf{a}} \models (\mathbf{A}c){\downarrow_\beta}$. Hence $\upsilon(\mathcal{E}^{c\mapsto\mathsf{a}}(\mathbf{A}c)) \equiv \mathtt{T}$. Using the properties above, we have $\upsilon(\mathcal{E}(\mathbf{A})@\mathsf{a}) \equiv \mathtt{T}$ as desired. ■

We can also prove that $\mathcal{G}_\beta$ and $\mathcal{G}_{\beta\xi}$ are complete with respect to the model classes $\mathfrak{M}_\beta$ and $\mathfrak{M}_{\beta\xi}$, respectively. In order to apply the results from [4, 5], we begin by noting that certain rules are admissible.

LEMMA 22. $\mathcal{G}(weak)$ *and* $\mathcal{G}(Inv^\neg)$ *(see Figure 3) are admissible in* $\mathcal{G}_\beta$ *and* $\mathcal{G}_{\beta\xi}$.

Proof. Both of these follow by an induction on derivations. In the case of weakening we must also carry a parameter renaming to ensure freshness of the parameter in each application of $\mathcal{G}(\Pi^c_+)$. ■

Using this result, we can conclude that $\Gamma_\Sigma^{\mathcal{G}_\beta}$ and $\Gamma_\Sigma^{\mathcal{G}_{\beta\xi}}$ are abstract consistency classes.

PROPOSITION 23. $\Gamma_\Sigma^{\mathcal{G}_\beta} \in \mathfrak{Acc}_\beta$ *and* $\Gamma_\Sigma^{\mathcal{G}_{\beta\xi}} \in \mathfrak{Acc}_{\beta\xi}$.

Proof. By Lemma 22 and Theorem 18. we know $\Gamma_\Sigma^{\mathcal{G}_\beta} \in \mathfrak{Acc}_\beta$ and $\Gamma_\Sigma^{\mathcal{G}_{\beta\xi}} \in \mathfrak{Acc}_\beta$. To complete the proof, we must verify ∇_ξ holds in $\Gamma_\Sigma^{\mathcal{G}_{\beta\xi}}$. Suppose $\neg(\lambda X_\alpha.\mathbf{M} \doteq^{\alpha\to\beta} \lambda X_\alpha.\mathbf{N}) \in \Phi \in \Gamma_\Sigma^{\mathcal{G}_{\beta\xi}}$ but $\Phi * \neg([w/X]\mathbf{M} \doteq^\beta [w/X]\mathbf{N}) \notin \Gamma_\Sigma^{\mathcal{G}_{\beta\xi}}$ where w_α is a parameter which does not occur in any sentence in Φ. By Lemma 17, we have $\Vdash_{\mathcal{G}_{\beta\xi}} \neg\,\Phi{\downarrow_\beta} * ([w/X]\mathbf{M} \doteq^\beta [w/X]\mathbf{N})\Big\downarrow_\beta$. Using the rule $\mathcal{G}(\Pi_+^w)$, we have $\Vdash_{\mathcal{G}_{\beta\xi}} \neg\,\Phi{\downarrow_\beta} * (\forall X.(\mathbf{M}{\downarrow_\beta} \doteq^\beta \mathbf{N}{\downarrow_\beta})$. Using the rule $\mathcal{G}(\xi)$, we have $\Vdash_{\mathcal{G}_{\beta\xi}} \neg\,\Phi{\downarrow_\beta} * (\lambda X.\mathbf{M} \doteq^{\alpha\to\beta} \lambda X.\mathbf{N})\Big\downarrow_\beta$. Using the rule $\mathcal{G}(\neg)$, we have $\Vdash_{\mathcal{G}_{\beta\xi}} \neg\,\Phi{\downarrow_\beta} * \neg\neg\,(\lambda X.\mathbf{M} \doteq^{\alpha\to\beta} \lambda X.\mathbf{N})\Big\downarrow_\beta$. and so $\Vdash_{\mathcal{G}_{\beta\xi}} \neg\,\Phi{\downarrow_\beta}$ since $\neg(\lambda X_\alpha.\mathbf{M} \doteq^{\alpha\to\beta} \lambda X_\alpha.\mathbf{N}) \in \Phi$. This contradicts $\Phi \in \Gamma_\Sigma^{\mathcal{G}_{\beta\xi}}$. ■

We can now prove completeness.

THEOREM 24. *Let* $* \in \{\beta, \beta\xi\}$ *and* Δ *be a sequent. If for all* $\mathcal{M} \in \mathfrak{M}_*$ *there is some* $\mathbf{A} \in \Delta$ *such that* $\mathcal{M} \models \mathbf{A}$, *then* $\Vdash_{\mathcal{G}_*} \Delta$.

Proof. Assume Δ is a sequent such that $\not\Vdash_{\mathcal{G}_*} \Delta$. Our goal is to find a model $\mathcal{M} \in \mathfrak{M}_*$ such that $\mathcal{M} \not\models \mathbf{A}$ for all $\mathbf{A} \in \Delta$ (i.e., $\mathcal{M} \models \neg\Delta$). Since $\mathcal{G}(Inv^\neg)$ is admissible, we can apply Lemma 17 to conclude that $\neg\Delta \in \Gamma_\Sigma^{\mathcal{G}_*}$. Since $\neg\Delta$ is finite, it is sufficiently Σ-pure. Hence we obtain an $\mathcal{M} \in \mathfrak{M}_*$ such that $\mathcal{M} \models \neg\Delta$ by applying Theorem 15. ■

Consequently, cut is admissible in both calculi.

COROLLARY 25. *For each* $* \in \{\beta, \beta\xi\}$, *the cut rule* $\mathcal{G}(cut)$ *is admissible in the calculus* $\mathcal{G}_*$.

Proof. Let Δ be a sequent and $\mathbf{C}$ be a sentence such that $\Vdash_{\mathcal{G}_*} \Delta * \mathbf{C}$ and $\Vdash_{\mathcal{G}_*} \Delta * \neg\mathbf{C}$. Using Theorem 24 we can prove $\Vdash_{\mathcal{G}_*} \Delta$ by proving for every $\mathcal{M} \in \mathfrak{M}_*$ there is some $\mathbf{A} \in \Delta$ such that $\mathcal{M} \models \mathbf{A}$. Let $\mathcal{M} \in \mathfrak{M}_*$ be given. Assume $\mathcal{M} \not\models \mathbf{A}$ for all $\mathbf{A} \in \Delta$. By soundness (Theorem 21), $\mathcal{M} \models \mathbf{C}$ since $\Vdash_{\mathcal{G}_*} \Delta * \mathbf{C}$. Also, $\mathcal{M} \models \neg\mathbf{C}$ since $\Vdash_{\mathcal{G}_*} \Delta * \neg\mathbf{C}$. This is a contradiction. ■

Note that since cut is admissible, we can conclude that $\Gamma_\Sigma^{\mathcal{G}_*}$ is saturated (by Theorem 19). If we had known $\Gamma_\Sigma^{\mathcal{G}_*}$ were saturated in advance, then we could have used the model existence theorems from [3] instead of the new model existence theorems proven in this paper. However, there seems to be no easier way to prove $\Gamma_\Sigma^{\mathcal{G}_*}$ is saturated than to prove cut-elimination, and there seems to be no easier way to prove cut-elimination than construction of a V-complex/possible values style of model.

6 Conclusion and Further Work

In this paper, we have employed a construction based Peter Andrews' V-complexes to prove a model existence theorem for a form of higher-order logic with a weak form of functional extensionality.

In [3] we have introduced and studied eight different model classes (including Henkin models) for classical type theory which generalize the notion of standard models and which allow for complete calculi. These model classes were motivated by different roles of extensionality and they adequately characterize the deductive power of existing theorem-proving calculi. Unfortunately, the model existence theorems in [3] assume saturation, which makes them useless for proving completeness of machine-oriented calculi since saturation is equivalent to cut-elimination.

This paper addresses the saturation problem for two of the model classes. This gives a framework that supports the development and proof-theoretical investigation of human-oriented as well as machine-oriented (ground) calculi for the corresponding type theories. For non-ground machine-oriented calculi the lifting issue has to be additionally addressed and extending our framework by tools that may also support lifting arguments remains future work.

For a complete picture, the results reported here need to be extended for the remaining six model classes, and for logics that include primitive equality (see, e.g. [2]). As mentioned in the introduction, the essential ingredients for handling two of the other six model classes are in [6, 7]. The remaining four cases include the case with full functional extensionality but not Boolean extensionality and the three cases with Boolean extensionality but not full functional extensionality.

BIBLIOGRAPHY

[1] Peter B. Andrews. Resolution in Type Theory. *Journal of Symbolic Logic*, 36(3):414–432, 1971.

[2] Christoph Benzmüller. Extensional Higher-Order Paramodulation and RUE-Resolution. In *Proceedings of the 16th International Conference on Automated Deduction*, volume 1632 of *LNAI*, pages 399–413. Springer, 1999.

[3] Christoph Benzmüller, Chad Brown, and Michael Kohlhase. Higher-Order Semantics and Extensionality. *Journal of Symbolic Logic*, 69(4):1027–1088, 2004.

[4] Christoph Benzmüller, Chad E. Brown, and Michael Kohlhase. Cut-Simulation in Impredicate Logics. In *Third International Joint Conference on Automated Reasoning (IJCAR'06)*, volume 4130 of *LNAI*, pages 220–234. Springer, 2006.

[5] Christoph E. Benzmüller, Chad E. Brown, and Michael Kohlhase. Cut-Simulation in Impredicative Logics (Extended Version). Seki Report SR-2006-01, Fachbereich Informatik, Universität des Saarlandes, 2006.
[6] Chad E. Brown. *Set Comprehension in Church's Type Theory.* PhD thesis, Department of Mathematical Sciences, Carnegie Mellon University, 2004.
[7] Chad E. Brown. *Automated Reasoning in Higher-Order Logic: Set Comprehension and Extensionality in Church's Type Theory*, volume 10 of *Studies in Logic: Logic and Cognitive Systems.* College Publications, 2007.
[8] Alonzo Church. A Formulation of the Simple Theory of Types. *Journal of Symbolic Logic*, 5:56–68, 1940.
[9] Leon Henkin. Completeness in the Theory of Types. *Journal of Symbolic Logic*, 15(2):81–91, 1950.
[10] Dag Prawitz. Hauptsatz for Higher Order Logic. *Journal of Symbolic Logic*, 33:452–457, 1968.
[11] Moto-o Takahashi. A Proof of Cut-Elimination Theorem in Simple Type Theory. *Journal of the Mathematical Society of Japan*, 19:399–410, 1967.

Cut Elimination in the Intuitionistic Theory of Types with Axioms and Rewriting Cuts, Constructively

OLIVIER HERMANT, JAMES LIPTON

1 Introduction

We give a constructive semantic proof of cut elimination for an intuitionistic formulation of Church's Theory of Types (ICTT) extended with certain classes of axioms. The argument extends techniques of Prawitz, Takahashi and Andrews, as well as those used in [5]. To the authors' knowledge it is the first constructive semantic proof of cut elimination for ICTT, and the extensions considered.

We recall that the central problem in proving cut-elimination for certain *impredicative* higher-order logics is that Gentzen's approach, based on an induction on a measure that combines proof-depth and formula complexity, does not work because the natural subformula ordering that places instances $M[t/x]$ below quantified formulae such as $\exists x.M$ is not a well-ordering. Such instances can be more complex, as can be seen by taking $M = x$ with x of logical type o and taking $t = \exists x.M \wedge A$ for any A, for example. This problem, originally known as the Takeuti conjecture (the claim that second-order logic admits cut-elimination, 1953), was solved positively, and non-constructively, independently by Tait[24] (1966), Takahashi[25] (1967), Prawitz[21] (1968) and others using semantic means, and constructively via a strong normalization proof, in 1971 by Girard [10]. In 1970 Andrews [1] gave a non constructive proof along the lines of Takahashi's V-complex construction for Church's classical theory of types. Dragalin [8] showed how to give a constructive semantic proof for higher-order classical logic. The second author gave a semantic proof for an intuitionistic formulation of Church's type theory in [5], also non-constructive.

The proof makes use of the following components. We define a class of models, a type-theoretic version of Scott-Fourman Ω sets, and show completeness constructively for the cut-free fragment of a number of type theories discussed in this paper. This gives cut-admissibility of those fragments as an immediate corollary. The impredicativity of the formal system involved makes it impossible to define a semantics along conventional lines, in the absence, a priori, of cut, or to prove completeness. The problem is that one cannot use induction on the subformula order to define truth, or use transitivity of entailment. As a result, as in the semantic proofs cited, in particular Takahashi's and Andrews', one must start from a tableau style construction of a partial model, called a *semivaluation*, and extend to a full model in a non-deterministic fashion, by assigning candidate truth values to formulae, then using induction on types to make the construction work. In [5] a series of algebraic conditions were given for partial truth assignments which guarantee that they can be extended to models. These conditions are applied here to a new, syntactic definition –inspired by results of Okada and the first author– of semivaluation based on mapping formulae A to sets generated by contexts Γ for which cut-free proofs of $\Gamma \vdash A$ exist.

This yields a constructive proof for the theory of types. The argument is then extended to include various types of sequent axioms which encapsulate rewriting rules for formulae.

The idea of building rewriting into logic is inspired in a formal system that combines sequent proofs, higher-order equational constraints and term rewriting, called Deduction Modulo, invented by Dowek, Hardin, Kirchner and Werner [6, 7]. The aim of such a formal system is to integrate computation directly into logic in a new way. Cut elimination for various fragments of this system, which does not, in general, satisfy strong normalization, has been studied by Hermant and Dowek. This, and the fact that the strong normalization approach does not lend itself readily to the addition of axioms is why we have taken the Takahashi-Andrews' approach rather than Girard's.

1.1 Outline

In a first part, we define the semantic space we are working in, and the tools we need to prove this theorem, along the lines of [5]. The main novelty is the definition of the semi-valuations used, that makes the proof constructive.

In a second part, we show that the argument works for an extension of ICTT with non-logical axioms (as for instance $\vdash \forall P.P \vee \neg P$, which gives us back the classical version of Church's Theory of Types), under the proviso that we allow specific new sequent rules which we will call *axiomatic cuts*.

We investigate this line further, and show that if we constrain the non-logical axioms to have a specific form (they have to satisfy a so-called *sign-preserving condition*) one can restrict the *axiomatic cuts* to take the form of *rewrite cuts*, where rewriting rules are expressed as cut-style rules of inference.

We conclude with a discussion of the constructive arguments used in this

$$\frac{}{\Gamma \vdash \top} \qquad \frac{}{\Gamma, U \vdash U} \qquad \frac{}{\Gamma, \bot \vdash \bot}$$

$$\frac{\Gamma, B, C \vdash A}{\Gamma, B \wedge C \vdash A} \wedge_L \qquad \frac{\Gamma \vdash B \quad \Gamma \vdash C}{\Gamma \vdash B \wedge C} \wedge_R$$

$$\frac{\Gamma, B \vdash A \quad \Gamma, C \vdash A}{\Gamma, B \vee C \vdash A} \vee_L \qquad \frac{\Gamma \vdash B_i}{\Gamma \vdash B_1 \vee B_2} \vee_R$$

$$\frac{\Gamma \vdash B \quad \Gamma, C \vdash A}{\Gamma, B \supset C \vdash A} \supset_L \qquad \frac{\Gamma, B \vdash C}{\Gamma \vdash B \supset C} \supset_R$$

$$\frac{\Gamma, P[t/x] \vdash A}{\Gamma, \forall x.P \vdash A} \forall_L \qquad \frac{\Gamma \vdash P}{\Gamma \vdash \forall x.P} \forall_R *$$

$$\frac{\Gamma, P \vdash A}{\Gamma, \exists x.P \vdash A} \exists_L * \qquad \frac{\Gamma \vdash P[t/x]}{\Gamma \vdash \exists x.P} \exists_R$$

$$\frac{\Gamma' \vdash A'}{\Gamma \vdash A} \lambda \qquad \frac{\Gamma \vdash \bot}{\Gamma \vdash B} \bot_R$$

Figure 1. Higher-order Sequent Rules

paper, and a comparison with other non-constructive semantic proofs of cut-elimination for the theory of types.

2 The Formal System: a sketch

For definitions of types, terms and reduction in the intuitionistic formulation of Church's Theory of types, due originally to Miller et al. [16], we refer the reader to [5], and will limit ourselves to recapitulating the rules of inference, in Fig. 1, λ being $\beta\eta$ and structural rules, as contraction and weakening, being implicitly assumed. Rules do not include the cut rule:

$$\frac{\Gamma \vdash B \quad \Gamma, B \vdash A}{\Gamma \vdash A} \text{Cut}$$

When we mean a proof within the rules of Fig. 1, we use the turnstyle $\vdash^*$, and use $\vdash$ when we allow the cut rule. In the rest of the paper, we will consider a fixed language S for ICTT, i.e. for each type a set of constants.

3 From Semi-valuations to Valuations: The Takahashi-Schütte lemma

We borrow the name semi-valuation from Takahashi and Schütte [22, 23, 25] also used by Andrews [1] to describe a partial interpretation of formulae in type theory that satisfies certain consistency properties, although our adaptation to the case of intuitionistic type-theory and Heyting algebras requires a considerable reworking of the definitions. Our formulation starts from constraints giving both positive and negative partial information: semivaluations consist of a *pair* of approximations to a model, which specify lower and upper bounds to the desired full interpretation. This is an abstraction of the way both positive and negative information from a Hintikka set is used to build a model for type theory in [5].

In *op. cit.* partial valuations are defined on the carrier of type o of an arbitrary typed applicative structure, and are shown, in this general setting, to extend to a full valuation without appeal to induction on subformulae which is not possible in an impredicative theory. The admissibility of cut then follows as an easy corollary. The cited result includes partial valuations on term models as a special case. Since this is all we need here, we will restate the main definitions and results for open terms only. We will also restrict attention to global models, defined below, since they are sufficient for the partial valuations chosen later in the paper to give a constructive proof of cut-admissibility.

3.1 Applicative Structures and Global Models

We will make use of the notion of applicative structures, a well-known semantic framework for the simply-typed lambda calculus, first introduced systematically by H. Friedman in [9], although obviously implicit in one form or another in [11, 15, 20]. (See also [17] for a detailed discussion.)

DEFINITION 1. A typed applicative structure $\langle \mathsf{D}, \mathsf{App}, \mathsf{Const}\rangle$ consists of an indexed family $\mathsf{D} = \{\mathsf{D}_\alpha\}$ of sets D_α for each type α, an indexed family App of functions $\mathsf{App}_{\alpha,\beta} : \mathsf{D}_{\beta\alpha} \times \mathsf{D}_\alpha \to \mathsf{D}_\beta$ for each pair (α, β) of types, and an (indexed) interpretation function $\mathsf{Const} = \{\mathsf{Const}_\alpha\}$ taking constants of each type α to elements of D_α.

An assignment φ is a function from the free variables of the language into D which respects types, and which allows us to give meaning to open terms. Given a typed applicative structure D, an *environmental model* consists of a total function $\{\!\{\ \}\!\}_\varphi$ from the open terms of the language into D for each assignment φ respecting types, for which the following equalities hold:

$$\begin{aligned}
\{\!\{c\}\!\}_\varphi &= \mathsf{Const}(c) && \text{for constants } c\\
\{\!\{x\}\!\}_\varphi &= \varphi(x) && \text{for variables } x\\
\{\!\{(MN)\}\!\}_\varphi &= \mathsf{App}(\{\!\{M\}\!\}_\varphi, \{\!\{N\}\!\}_\varphi)\\
\{\!\{\lambda x_\alpha . M_\beta\}\!\}_\varphi &\quad \text{is the unique member of } \mathsf{D}_{\beta\alpha}\ s.t. \text{ for every } d \in \mathsf{D}_\alpha\\
&\quad \mathsf{App}(\{\!\{\lambda x_\alpha . M\}\!\}_\varphi, d) = \{\!\{M\}\!\}_{\varphi[d/x]}
\end{aligned}$$

If an environmental model exists for a given assignment, it is unique, as the reader can show by proving the relevant substitution theorem.

So far we have only supplied semantics for the underlying typed lambda-calculus. Now we must interpret the logic as well, by adjoining a Heyting algebra and some additional structure to handle the logical constants and predicates.

DEFINITION 2. A Heyting applicative structure $\langle \mathsf{D}, \mathsf{App}, \mathsf{Const}, \omega, \Omega\rangle$ for ICTT is a typed applicative structure with an associated Heyting algebra Ω and function ω from D_o to Ω such that for each f in $\mathsf{D}_{o\alpha}$, Ω contains the parametrized meets and joins

$$\bigwedge\{\omega(\mathsf{App}(f,d)) : d \in \mathsf{D}_\alpha\} \text{ and } \bigvee\{\omega(\mathsf{App}(f,d)) : d \in \mathsf{D}_\alpha\},$$

and the following conditions are satisfied:

$$\begin{aligned}
\omega(\mathsf{Const}(\top_o)) &= \top_\Omega\\
\omega(\mathsf{Const}(\bot_o)) &= \bot_\Omega\\
\omega(\mathsf{App}(\mathsf{App}(\mathsf{Const}(\wedge_{ooo}), d_1), d_2)) &= \omega(d_1) \wedge \omega(d_2)\\
\omega(\mathsf{App}(\mathsf{App}(\mathsf{Const}(\vee_{ooo}), d_1), d_2)) &= \omega(d_1) \vee \omega(d_2)\\
\omega(\mathsf{App}(\mathsf{App}(\mathsf{Const}(\supset_{ooo}), d_1), d_2)) &= \omega(d_1) \to \omega(d_2)\\
\omega(\mathsf{App}(\mathsf{Const}(\Sigma_{o(o\alpha)}), f)) &= \bigvee\{\omega(\mathsf{App}(f,d)) : d \in \mathsf{D}_\alpha\}\\
\omega(\mathsf{App}(\mathsf{Const}(\Pi_{o(o\alpha)}), f)) &= \bigwedge\{\omega(\mathsf{App}(f,d)) : d \in \mathsf{D}_\alpha\}
\end{aligned}$$

By supplying an object Ω of truth values we are able to distinguish between denotations of formulae (elements d of D_o) and their truth-values $\omega(d) \in \Omega$. A definition, with suitable further restrictions on D_o, that identified D_o with Ω (*i.e.*, restricting ω to the identity function) might seem more natural but would make, for example, $A \wedge B$ indiscernible from $B \wedge A$ in the structure and thereby identify the truth values of $P_{oo}(A_o \wedge B_o)$ and $P_{oo}(B_o \wedge A_o)$. This identity holds neither in ICTT as presented here nor in the HOHH sub-system used in the λProlog programming language.

DEFINITION 3. A global model for ICTT is a total assignment-indexed function $\mathfrak{D} = \{\mathfrak{D}(\ )_\varphi : \varphi \text{ an assignment}\}$ into a Heyting applicative structure $\langle \mathsf{D}, \mathsf{App}, \mathsf{Const}, \omega, \Omega \rangle$ which takes (possibly open) terms of type α into D_α and satisfies the environmental model conditions cited above, following Def. 1, as well as η-conversion, that is to say:

$$\begin{array}{rll} \mathfrak{D}(c)_\varphi = \mathsf{Const}(c) & & \text{for constants } c \\ \mathfrak{D}(x)_\varphi = \varphi(x) & & \text{for variables } x \\ \mathfrak{D}((MN))_\varphi = \mathsf{App}(\mathfrak{D}(M)_\varphi, \mathfrak{D}(N)_\varphi) & & \\ \mathfrak{D}(\lambda x_\alpha . M_\beta)_\varphi \quad & \text{is the unique member of } \mathsf{D}_{\beta\alpha} \ s.t. \text{ for every } d \in \mathsf{D}_\alpha & \\ & \mathsf{App}(\mathfrak{D}(\lambda x_\alpha . M_\beta)_\varphi, d) = \mathfrak{D}(M)_{\varphi[d/x]} & \\ \mathfrak{D}(M)_\varphi = \mathfrak{D}(N)_\varphi & & M \ \eta\text{-equivalent to } N \end{array}$$

Given a model $\mathfrak{D}$ and an assignment φ, we say that φ satisfies B in $\mathfrak{D}$ if $\omega(\mathfrak{D}(B_o)_\varphi) = \top_\Omega$; we abbreviate this assertion to $\mathfrak{D} \models_\varphi B_o$. We say B_o is valid in $\mathfrak{D}$ (equivalently, $\mathfrak{D} \models B_o$) if $\mathfrak{D} \models_\varphi B_o$ for every assignment φ. We abbreviate the truth-value $\omega(\mathfrak{D}(B_o)_\varphi)$ to $(B_o)^*_\varphi$. We also omit the subscript φ when our intentions are clear. We often use the word *model* just to refer to the mapping $(\_)^*$ from logical formulae to truth values in Ω.

3.2 Soundness of ICTT for Global Models

In the following we extend interpretations to sequents in a natural way.

DEFINITION 4. We define the meaning of a sequent in a model to be the truth-value in Ω given by:

$$(\Gamma \vdash \Delta)^* := (\bigwedge \Gamma \supset \Delta)^*$$

where $\bigwedge \Gamma$ signifies the conjunction of the elements of Γ and where we recall that, in an intuitionistic calculus, the consequent Δ is restricted to a single formula.

Note that $(\bigwedge \Gamma \supset \Delta)^* = \top$ if and only if $\top \leq (\bigwedge \Gamma \supset \Delta)^*$, which is to say $\top \leq (\bigwedge \Gamma)^* \to (\Delta)^*$, which by the condition on $\to$ is equivalent to $\top \wedge (\bigwedge \Gamma)^* \leq (\Delta)^*$, which in turn is equivalent to $(\bigwedge \Gamma)^* \leq (\Delta)^*$. We will abbreviate $(\bigwedge \Gamma)^*$ to $(\Gamma)^*$ and express the validity of the indicated sequent by $(\Gamma)^* \leq (\Delta)^*$ or, when referring to the environment, by $(\Gamma)^*_\varphi \leq (\Delta)^*_\varphi$ henceforth.

THEOREM 5 (Soundness). *If* $\Gamma \vdash A$ *is provable in* ICTT *then* $(\Gamma)^* \leq (A)^*$ *in every global model* $\mathfrak{E}$ *of* ICTT.

A proof can be found in [5].

A straightforward proof of completeness of ICTT for global models can be given *under the assumption that cut is admissible for ICTT* along the lines of [26, 5], i.e. by choosing Ω to be the Lindenbaum algebra of equivalence classes of formulae and then interpreting each formula as its own equivalence class. Just to show Ω is partially ordered, we need cut.

Since we are not assuming cut holds in ICTT we must proceed differently. We will choose the complete Heyting algebra Ω_{cfk} generated by "cut-free contexts", that is to say, contexts from which formulae can be proved without using cut. A partial valuation will be defined for this cHa, yielding an interpretation that establishes completeness and the admissibility of cut.

3.3 Semantic preliminaries

DEFINITION 6. Let Ω be a Heyting algebra. A global Ω **semivaluation** $\mathcal{V} = \langle \mathsf{D}, \mathsf{App}, \mathsf{Const}, \pi, \nu, \Omega \rangle$ consists of a typed applicative structure $\langle \mathsf{D}, \mathsf{App}, \mathsf{Const} \rangle$ together with a pair of maps $\pi : \mathsf{D}_o \longrightarrow \Omega$ and $\nu : \mathsf{D}_o \longrightarrow \Omega$, called the lower and upper constraints of $\mathcal{V}$, or the positive and negative constraints, satisfying the following:

1. For any $d \in \mathsf{D}_o$

$$\pi(d) \leq \nu(d)$$

2.

$$\begin{array}{rcl}
\pi(\top_o) & = & \top_\Omega \\
\pi(\bot_o) & = & \bot_\Omega \\
\pi(\mathsf{Const}(\wedge) \cdot A \cdot B) & \leq & \pi(A) \wedge_\Omega \pi(B) \\
\pi(\mathsf{Const}(\vee) \cdot A \cdot B) & \leq & \pi(A) \vee_\Omega \pi(B) \\
\pi(\mathsf{Const}(\supset) \cdot A \cdot B) & \leq & \pi(A) \rightarrow_\Omega \pi(B) \\
\pi(\mathsf{Const}(\Sigma_{o(o\alpha)}) \cdot f) & \leq & \bigvee \{\pi(f \cdot d) : d \in \mathsf{D}_\alpha\} \\
\pi(\mathsf{Const}(\Pi_{o(o\alpha)}) \cdot f_{(o\alpha)}) & \leq & \bigwedge \{\pi(f \cdot d) : d \in \mathsf{D}_\alpha\}
\end{array}$$

and

$$
\begin{aligned}
\nu(\top_o) &= \top_\Omega \\
\nu(\bot_o) &= \bot_\Omega \\
\nu(\mathsf{Const}(\wedge)\cdot A\cdot B) &\geq \nu(A)\wedge_\Omega \nu(B) \\
\nu(\mathsf{Const}(\vee)\cdot A\cdot B) &\geq \nu(A)\vee_\Omega \nu(B) \\
\nu(\mathsf{Const}(\supset)\cdot A\cdot B) &\geq \nu(A)\rightarrow_\Omega \nu(B) \\
\nu(\mathsf{Const}(\Sigma_{o(o\alpha)})\cdot f) &\geq \bigvee\{\nu(f\cdot d) : d\in \mathsf{D}_\alpha\} \\
\nu(\mathsf{Const}(\Pi_{o(o\alpha)})\cdot f_{(o\alpha)}) &\geq \bigwedge\{\nu(f\cdot d) : d\in \mathsf{D}_\alpha\}
\end{aligned}
$$

and

3. the *consistency* or *separation* conditions

$$
\begin{aligned}
\pi(\mathsf{Const}(\supset)\cdot B\cdot C)\wedge\nu(B) &\leq \pi(C) && (1)\\
\pi(B)\rightarrow_\Omega \nu(C) &\leq \nu(\mathsf{Const}(\supset)\cdot B\cdot C). && (2)
\end{aligned}
$$

REMARK 7. In this definition, the application operator App is denoted by the infix operator $\cdot$ for readability. The reader should note that some of these requirements are superfluous, i.e. follow from the others. The separation conditions and the first condition imply the $\supset$ requirements for both π and ν, as well as $\top$ requirement for π (resp. $\bot$ for ν) implies their counterpart for ν (resp. π).

The separation conditions abstract the properties of the weak and strong support sets[1] $\mathcal{H}_A^{\mathsf{T}}$ and $\mathcal{H}_A^{\neg\mathsf{F}}$ associated with a Hintikka set $\mathcal{H}$ in [5].

The definition of environment, and global structure remain the same for semivaluations. As with Heyting applicative structures, in the presence of an environment φ, a semivaluation $\mathcal{V}$ induces an interpretation $\mathfrak{V}_\varphi$ from open terms A to the carriers D as follows:

$\mathfrak{V}(c)_\varphi$	$= \mathsf{Const}(c)$	for constants c
$\mathfrak{V}(x)_\varphi$	$= \varphi(x)$	for variables x
$\mathfrak{V}(M)_\varphi$	$= \mathfrak{V}(N)_\varphi$	M eta-equivalent to N
$\mathfrak{V}((MN))_\varphi$	$= \mathsf{App}(\mathfrak{V}(M)_\varphi, \mathfrak{V}(N)_\varphi)$	
$\mathsf{App}(\mathfrak{V}(\lambda x_\alpha.M_\beta)_\varphi, d)$	$= \mathfrak{V}(M)_{\varphi[x:=d]}$	

[1]whose formulation is due to Chad Brown.

This assignment induces a *pair* of partial, or semi-truth-value assignments $[\![\_]\!]^\pi_\varphi$ and $[\![\_]\!]^\nu_\varphi$ to terms A_o of type o given by

$$\begin{aligned} \mathcal{V}[\![A]\!]^\pi_\varphi &= \pi(\mathfrak{V}(A)_\varphi) \\ \mathcal{V}[\![A]\!]^\nu_\varphi &= \nu(\mathfrak{V}(A)_\varphi) \end{aligned}$$

THEOREM 8. *Given an Ω-semivaluation $\mathcal{V} = \langle \mathsf{D}, \cdot, \mathsf{Const}, \pi, \nu, \Omega \rangle$, there is a model $\mathfrak{D} = \langle \hat{\mathsf{D}}, \odot, \hat{\mathsf{C}}, \omega, \Omega \rangle$ extending $\mathcal{V}$ in the following sense: for all closed terms A_o*

$$\mathcal{V}[\![A]\!]^\pi \leq \omega(\mathfrak{D}(A)) \leq \mathcal{V}[\![A]\!]^\nu.$$

Furthermore, there is a surjective indexed map $\delta : \hat{\mathsf{D}} \longrightarrow \mathsf{D}$ such that for any $\hat{d} \in \hat{\mathsf{D}}_o$

$$\pi(\delta(\hat{d})) \leq \omega(\hat{d}) \leq \nu(\delta(\hat{d}))$$

Proof. We recall from the constructions in [25, 1, 5] that a $\mathcal{V}$-complex of a given type γ is an ordered pair $\langle A_\gamma, u \rangle$ where A_γ is a term (of type γ) in normal form, and u is a truth value, in our case a member of a cHa, which we can think of as a candidate truth value for the desired valuation (i.e. model) $\mathfrak{D}$. As in the cited works, the carriers $\hat{\mathsf{D}}$ of this model $\langle \hat{\mathsf{D}}, \hat{\mathsf{C}}, \odot, \omega, \Omega \rangle$ will be sets of such $\mathcal{V}$ -complexes, defined by induction on the *type* structure, as follows:

- $\hat{\mathsf{D}}_o = \{\langle d, u \rangle : d \in \mathsf{D}_o \text{ and } \pi(d) \leq u \leq \nu(d)\}$
- $\hat{\mathsf{D}}_\iota = \{\langle m, \iota \rangle : m \in \mathsf{D}_\iota\}$
- $\hat{\mathsf{D}}_{\beta\alpha} = \{\langle m, \mu \rangle : m \in \mathsf{D}_{\beta\alpha}, \mu : \mathsf{D}_\alpha \longrightarrow \mathsf{D}_\beta$, and for each $\langle A, a \rangle \in \mathsf{D}_\alpha, \mu\langle A, a \rangle = \langle m \cdot A, r \rangle$ for some $r\}$.
- Application is given by $\langle M, m \rangle \odot \langle A, a \rangle = m\langle A, a \rangle$.
- Define $\omega : \hat{\mathsf{D}}_o \longrightarrow \Omega$ by projection on the second coordinate.

Projections (of an element $d \in \hat{\mathsf{D}}$) on its first and second coordinates are denoted d^1 and d^2, respectively, and $\hat{\mathsf{D}}^2$ is $\{d^2 : d \in \hat{\mathsf{D}}\}$. As in [5], we can define a *selector* function $\rho : \mathsf{D} \longrightarrow \hat{\mathsf{D}}^2$ by induction on types, to show that for every type α and every $M \in \mathsf{D}_\alpha$ there is a $\rho(M)$ such that $\langle M, \rho(M) \rangle \in \hat{\mathsf{D}}_\alpha$.

Notice that in the D_o base case, we have a degree of freedom: we can choose $\rho(M) = \pi(M)$ as well $\rho(M) = \nu(M)$. This choice can be uniform (and arbitrary) or depend on M, as we shall see later.

Now we show how to define the assignment of denotations to logical and non-logical constants.

$$\begin{array}{rl}
\hat{\mathsf{C}}(\top_o) = & \langle \mathsf{Const}(\top_o), \top_\Omega\rangle \\
\hat{\mathsf{C}}(\bot_o) = & \langle \mathsf{Const}(\bot_o), \bot_\Omega\rangle \\
\hat{\mathsf{C}}(c_\alpha) = & \langle \mathsf{Const}(c_\alpha), \rho(\mathsf{Const}(c_\alpha))\rangle \text{ for non-logical constants } c_\alpha. \\
\hat{\mathsf{C}}(\wedge) = & \langle \mathsf{Const}(\wedge), \boldsymbol{\lambda}\langle B,b\rangle.\langle \mathsf{Const}(\wedge)\cdot B, \boldsymbol{\lambda}\langle D,d\rangle.\langle \mathsf{Const}(\wedge)\cdot B\cdot D, b\wedge_\Omega d\rangle\rangle\rangle \\
\hat{\mathsf{C}}(\supset) = & \langle \mathsf{Const}(\supset), \boldsymbol{\lambda}\langle B,b\rangle.\langle \mathsf{Const}(\supset)\cdot B, \boldsymbol{\lambda}\langle D,d\rangle.\langle \mathsf{Const}(\supset)\cdot B\cdot D, b\rightarrow_\Omega d\rangle\rangle\rangle \\
\hat{\mathsf{C}}(\Sigma) = & \langle \mathsf{Const}(\Sigma), \boldsymbol{\lambda}\langle M,m\rangle.\langle \mathsf{Const}(\Sigma)\cdot M, \bigvee_{\hat{d}\in\hat{\mathsf{D}}_\alpha}(m\hat{d})^2\rangle\rangle.
\end{array}$$

where Σ abbreviates $\Sigma_{o(o\alpha)}$. The $\vee$ and Π cases are similar, and left to the reader.

We now need to show that $\hat{\mathsf{C}}$ is well-defined. This is where the separation conditions play a key role. We will work a few cases.

$\hat{\mathsf{C}}(\wedge)$ What we must show here is that if $\langle B,b\rangle$ and $\langle D,d\rangle$ are in $\hat{\mathsf{D}}_o$ then so is $\langle \mathsf{Const}(\wedge)\cdot B\cdot D, b\wedge_\Omega d\rangle$. That is to say, if we are given that $\pi(B)\le b\le\nu(B)$ and $\pi(D)\le d\le\nu(D)$ then $b\wedge_\Omega d$ lies between $\pi(\mathsf{Const}(\wedge)\cdot B\cdot D)$ and $\nu(\mathsf{Const}(\wedge)\cdot B\cdot D)$. Since $\wedge$ is monotone in both arguments this follows immediately from the defining properties of upper and lower constraints.

The argument for $\vee$ is similar.

$\hat{\mathsf{C}}(\supset)$ We must show that the second component $\hat{\mathsf{C}}(\supset)^2$, namely the term $\boldsymbol{\lambda}\langle B,b\rangle.\langle \mathsf{Const}(\supset)\cdot B, \boldsymbol{\lambda}\langle D,d\rangle.\langle \mathsf{Const}(\supset)\cdot B\cdot D, b\rightarrow_\Omega d\rangle\rangle$ maps a pair of members of $\hat{\mathsf{D}}_o$ to $\hat{\mathsf{D}}_o$. If we are given two members $\langle B,b\rangle$ and $\langle D,d\rangle$ of $\hat{\mathsf{D}}_o$, then we know $\pi(B)\le b\le\nu(B)$ and similarly $\pi(D)\le d\le\nu(D)$. But then, abbreviating $\mathsf{Const}(\supset)\cdot B\cdot D$ to $B\supset D$, we have $\pi(B\supset D)\wedge b\le\pi(B\supset D)\wedge\nu(B)$. By the first separation axiom, $\pi(B\supset D)\wedge b\le\pi(D)\le d$. But then $\pi(B\supset D)\le b\rightarrow d$.

Furthermore $b\rightarrow d\le\pi(B)\rightarrow\nu(D)$ since Heyting implication is antitone (contravariant) in its first argument and monotone in its second. By the second separation axiom (2) $b\rightarrow d\le\nu(B\supset D)$,as we wanted to show.

The Π and Σ cases are both monotone in the relevant arguments, and are easy. The surjective map δ in the conclusion of the theorem is just projection of $\mathcal{V}$-complexes onto their first component.

The rest of the proof that $\mathfrak{D}$ is a model follows just like the proof for the model constructed in [5]. ■

4 Completeness and cut elimination - the ICTT case

From Thm. 8, deriving a (cut-free) completeness theorem for ICTT requires a complete Heyting algebra Ω and an Ω-semivaluation. We first give the definition of Ω_{cfk}, the Heyting algebra of cut-free contexts, which is very different from the one given in [5].

4.1 The cut-free contexts Heyting algebra

We first define what is a cut-free context, in the same way as Okada [19, 18].

DEFINITION 9 (outer value). Let A be a closed formula. We let the *outer value* of A be:

$$[\![A]\!] = \{\Gamma \mid \Gamma \vdash^* A\}$$

So, an outer value $[\![A]\!]$ is the set of contexts proving A without cut (cut-free contexts). With this, we build Ω_{cfk}.

DEFINITION 10 (Ω_{cfk}). We let $|\Omega|$ to be the least set of sets of (finite) contexts generated by $[\![A]\!]$ for any formula A, and closed under arbitrary intersection. It is ordered by inclusion. Then define meets and joins on $|\Omega|$ as follows

- $\bigwedge$ = arbitrary set-theoretic intersection.
- $\bigvee$ = arbitrary pseudo-union, that is to say

  $$\bigvee S = \bigcap \{c \in |\Omega| : c \geq S\}$$

  where $c \geq S$ means $\forall s \in S\, c \geq s$

REMARK 11. If we expand this definition a little bit, we have:

- $\top_\Omega$ is the set of all finite contexts. It is as well $[\![\top_o]\!]$ since any context proves $\top_o$ without using cut.
- $\bot_\Omega$ is the intersection of all outer values. Equivalently, it is $[\![\bot_o]\!]$ since if $\Gamma \vdash^* \bot_o$ we can prove $\Gamma \vdash^* A$ for any A. In particular, $\bot_\Omega$ is not empty.

Notice that an element $c \in \Omega$ (by which we mean $c \in |\Omega|$) can always be written as an element of the form $\bigcap [\![A_i \mid i \in \Lambda]\!]$. So we can simplify a little bit the definition of union:

LEMMA 12 (Simplification of the definition). *We may express suprema directly in terms of generating sets:*

- $\bigvee\{a_i, i \in I\} = \bigcap\{[\![A]\!] \mid \bigcup\{a_i, i \in I\} \subseteq [\![A]\!]\}$
- $a \vee_\Omega b = \bigcap\{[\![A]\!] \mid a \cup b \subseteq [\![A]\!]\}$

Proof. Each one of the c mentioned above has the form $\bigcap\{[\![C_i]\!], i \in J\}$. ■

Taking $a \rightarrow b = \bigvee\{x : x \wedge a \leq b\}$, the resulting structure $\Omega = \langle|\Omega|, \bigvee, \bigwedge, \rightarrow\rangle$ (also written Ω_{cfk}, when ambiguity may arise) is a complete Heyting algebra. One must show that the $\wedge \bigvee$ distributivity law holds [26].

First we show that for each member $a = \bigcap_i [\![A_i]\!]$ of Ω

$$a \cap \bigvee S \leq \bigvee a \cap S$$

where $a \cap S$ means $\{a \cap s : s \in S\}$. Unfolding the definitions and using Lem. 12 above, the desired conclusion is equivalent to

$$a \cap \bigcap\{[\![B]\!] : [\![B]\!] \geq S\} \subseteq \bigcap\{[\![D]\!] : [\![D]\!] \geq a \cap S\} \tag{3}$$

where $x \geq S$ abbreviates $\forall s \in S(s \subseteq x)$. Suppose the context Γ is a member of the left hand side, i.e. for each i we have $\Gamma \vdash^* A_i$ and $\Gamma \vdash^* B$ for every B such that $[\![B]\!] \geq S$.

Let D be a formula such that $[\![D]\!] \geq a \cap S$. We must show $\Gamma \vdash^* D$ to conclude.

Let Δ be a context such that $\Delta \in s$ for some $s \in S$. By weakening $\Delta, \Gamma \vdash^* A_i$ for each i, i.e. $\Delta, \Gamma \in a$ and by the same reasoning $\Delta, \Gamma \in s$. By definition of D, we have $\Delta, \Gamma \vdash^* D$. Hence $\Delta \vdash^* \bigwedge \Gamma \supset D$. Since this is valid for any s, we have shown $[\![\Gamma \supset D]\!] \geq S$.

But then, $\Gamma \vdash^* \Gamma \supset D$ by assumption on Γ. By Kleene's Lem. 33 below and contraction on the formulae in Γ we have $\Gamma \vdash^* D$, which shows Γ is a member of the right-hand-side of 3, which proves the claim.

The other direction follows, by elementary lattice theory: for any $s \in S$ it is the case that $a \cap \bigvee S \geq a \cap s$. Now take the supremum of $a \cap s$ over all $s \in S$.

4.2 A semivaluation π and ν

Now, we need to give a definition of an Ω semivaluation to have the right to apply Thm. 8. For this, we need the following definition:

DEFINITION 13 (closure). Let S be a set of contexts, we define its closure by:

$$cl(S) = \bigcap\{[\![A]\!] \mid S \subseteq [\![A]\!]\}$$

It is the least element of Ω containing S. We also write, for a single context Γ, $cl(\Gamma)$ to mean $cl(\{\Gamma\})$.

REMARK 14. Notice that $cl(A) \subseteq d$ is equivalent to $A \in d$. Indeed, $A \in cl(A)$ and $cl(A)$ is the l.u.b. of A. The closure operator can also be understood as the *set of contexts admitting cut with all the elements of S* as shown in the following lemma.

LEMMA 15. *Let A be a formula. Then the four following formulations are equivalent:*

(i) $cl(A) = \bigcap\{[\![B]\!] \mid A \in [\![B]\!]\}$

(ii) $cl(A) = \{\Gamma \mid \Gamma \vdash^* B$ *whenever* $A \vdash^* B\}$. *Equivalently,* $\Gamma \in cl(A)$ *iff* $\Gamma \vdash^* A$ *and given any proof* $A \vdash^* B$, *a proof of* $\Gamma \vdash^* B$ *is derivable.*

(iii) $cl(A) = \{\Gamma \mid \Gamma \vdash^* B$ *whenever* $\Gamma, A \vdash^* B\}$. *Equivalently,* $\Gamma \in cl(A)$ *iff* $\Gamma \vdash^* A$ *and given any proof* $\Gamma, A \vdash^* B$ *a proof of* $\Gamma \vdash^* B$ *is derivable.*

(iv) $cl(A) = \{\Gamma \mid \Delta, \Gamma \vdash^* B$ *whenever* $\Delta, A \vdash^* B\}$. *Equivalently,* $\Gamma \in cl(A)$ *iff* $\Gamma \vdash^* A$ *and given any proof* $\Delta, A \vdash^* B$ *a proof of* $\Delta, \Gamma \vdash^* B$ *is derivable.*

Cases (ii) – (iv) *can be summarized as follows:* Γ *admits cuts with* A, *hence the terminology "*Γ *is* A*-cuttable".*

Proof. Denoting $cl(A)$ as defined at point (x) as (x) itself, we have:

- $(i) = (ii)$ is just an unfolding of $[\![B]\!]$. Moreover $A \in [\![A]\!]$, thus $\Gamma \vdash^* A$ has to hold.

- $(ii) \subseteq (iii)$. Let $\Gamma \in (ii)$, and B such that $\Gamma, A \vdash^* B$. Show that $\Gamma \vdash^* B$. By $\wedge_L$ rules and a $\supset_R$ rule, we have a proof of the sequent $A \vdash^* (\bigwedge \Gamma) \supset B$. Therefore, by hypothesis, $\Gamma \vdash^* (\bigwedge \Gamma) \supset B$. Using Kleene's inversion lemma Lem. 33 below, we get a proof of the sequent $\Gamma, \Gamma \vdash^* B$ that we contract.

- $(iii) \subseteq (iv)$. Let $\Gamma \in (iii)$ and B such that $\Delta, A \vdash^* B$. By weakening, $\wedge_L$ and $\supset_R$ rules, this yields a proof of the sequent $\Gamma, A \vdash^* \bigwedge \Delta \supset B$. $\Gamma \vdash^* \bigwedge \Delta \supset B$ then holds by hypothesis, and we conclude by applying Kleene's inversion lemma Lem. 33.

- $(iv) \subseteq (ii)$. Let $\Gamma \in (iv)$ and B such that $A \vdash^* B$. Taking $\emptyset$ for Δ in (iv) shows $\Gamma \vdash^* B$.

■

Note that in all four cases above, on the right hand side of the stated equivalence, the condition $\Gamma \vdash^* A$ is unnecessary, and can be omitted. It is an immediate consequence of the A-cuttability statement, since $A \vdash^* A$ trivially. We shall use any of the formulations given above, depending on our need. Now we are ready to give the semivaluation we work with:

DEFINITION 16 (the semivaluation). Let the typed applicative structure $\langle \mathsf{D}, \mathsf{App}, \mathsf{Const}\rangle$ be the open term model: carriers D_α are open terms in normal form of type α, application $A \cdot B$ is $[AB]$, the normal form of AB, and we interpret constants as themselves. For any formula A, we define:

$$\begin{array}{rcl} \pi(A) & = & cl(A) \\ \nu(A) & = & [\![A]\!] \end{array}$$

LEMMA 17. $\langle \mathsf{D}, \mathsf{App}, \mathsf{Const}, \pi, \nu, \Omega_{\mathsf{cfk}}\rangle$ *is a semivaluation in the sense of Def. 6.*

Proof. We have to check every statement of Def. 6, with respect to the open term model.

- $cl(A) \subseteq [\![A]\!]$. By Rem. 14, this amounts to showing $A \in [\![A]\!]$. This holds since $A \vdash^* A$.
- $cl(\top_o) = \top_\Omega$. The direct inclusion is immediate since $\top_\Omega$ is the greatest element. For the converse, we have to show that any context is $\top_o$-cuttable. So consider a proof of $\top_o \vdash^* A$ for some A. The only possible rules we can use on $\top_o$ besides contraction, weakening and conversion is the axiom. We can always replace it by:

  $$\frac{}{\vdash \top_o}\ \top\text{-right}$$

  Hence, $\vdash^* A$ and, by weakening, $\Gamma \vdash^* A$ for any context Γ.
- $cl(\bot_o) = \bot_\Omega$, since $\bot_\Omega \subseteq cl(\bot) \subseteq [\![\bot]\!] = \bot_\Omega$ holds: $\bot_\Omega$ is the least element, from a previous point and from Rem. 11.
- $cl(A \wedge B) \leq cl(A) \cap cl(B)$. This amounts to showing $A \wedge B \in cl(A) \cap cl(B)$. We prove that $A \wedge B$ is A-cuttable. Consider a proof of $A \vdash^* C$. We construct the following proof:

  $$\dfrac{\dfrac{A \vdash^* C}{A, B \vdash^* C}\ \text{weak}}{A \wedge B \vdash^* C}\ \wedge_L$$

  Hence, $A \wedge B \in cl(A)$. On the same way, $A \wedge B \in cl(B)$ and the claim is proved.

- $cl(A \vee B) \subseteq cl(A) \vee_\Omega cl(B)$. It suffices to show $A \vee B \in cl(A) \vee_\Omega cl(B)$. Let C be such that $cl(A) \cup cl(B) \subseteq [\![C]\!]$. Then $A \in [\![C]\!]$ and $B \in [\![C]\!]$. Therefore, the proof

$$\frac{A \vdash^* C \qquad B \vdash^* C}{A \vee B \vdash^* C} \vee_L$$

shows that $A \vee B \in [\![C]\!]$. This holds for any such C, hence $A \vee B \in cl(A) \vee_\Omega cl(B)$.

- $cl(A \supset B) \subseteq cl(A) \to cl(B)$ is a consequence of $cl(A \supset B) \wedge [\![A]\!] \subseteq cl(B)$ (proved later) as said in Rem. 7.

- $cl(\Sigma.f) \subseteq \bigvee\{cl(ft) \mid t \in \mathcal{T}_\alpha\}$ (where α is the suitable type). Equivalently, $\Sigma.f \in \bigvee\{cl((ft)) \mid t \in \mathcal{T}_\alpha\}$. Let t be a variable y of type α that is fresh for f. We prove that $\Sigma.f$ is fy-cuttable. Assume to have a proof $fy \vdash^* C$. The proof:

$$\frac{fy \vdash^* C}{\Sigma.f \vdash^* C} \exists_L$$

justifies the fy-cuttability. Hence $\Sigma.f \in cl(fy)$, and it is in the supremum.

- $\Pi.f \in \bigwedge\{cl(ft), t \in \mathcal{T}_\alpha\}$. Let t be a term of type α. The proof:

$$\frac{ft \vdash^* C}{\Pi.f \vdash^* C} \forall_L$$

shows that $\Pi.f$ is ft-cuttable for any t.

- $[\![\top_o]\!] = \top_\Omega$ and $[\![\bot_o]\!] = \bot_\Omega$ hold both by definition, from Rem. 11.

- $[\![A \wedge B]\!] \supseteq [\![A]\!] \wedge_\Omega [\![B]\!]$. Let Γ such that $\Gamma \vdash^* A$ and $\Gamma \vdash^* B$. The proof:

$$\frac{\Gamma \vdash^* A \qquad \Gamma \vdash^* B}{\Gamma \vdash^* A \wedge B} \wedge_R$$

shows the claim.

- $[\![A \vee B]\!] \supseteq [\![A]\!] \vee_\Omega [\![B]\!]$. We show $[\![A \vee B]\!] \supseteq [\![A]\!]$. Let $\Gamma \in [\![A]\!]$. The proof:

$$\frac{\Gamma \vdash^* A}{\Gamma \vdash^* A \vee B} \vee_R$$

shows that $\Gamma \in [\![A \vee B]\!]$. Hence $[\![A \vee B]\!]$ is an upper bound for $[\![A]\!]$ and $[\![B]\!]$, and the claim is proved.

- $[\![A \supset B]\!] \supseteq [\![A]\!] \to_\Omega [\![B]\!]$ is a consequence of $cl(A) \to [\![B]\!] \subseteq [\![A \supset B]\!]$ (proved later) as said in Rem. 7.

- $[\![\Sigma.f]\!] \supseteq \bigvee\{[\![ft]\!], t \in \mathcal{T}_\alpha\}$. Let t be any term of type α. Let $\Gamma \in [\![ft]\!]$. The proof:

$$\frac{\Gamma \vdash^* ft}{\Gamma \vdash^* \Sigma.f} \exists_R$$

shows that $[\![\Sigma.f]\!]$ is an upper bound for any $[\![ft]\!]$, hence for their supremum as well.

- $[\![\Pi.f]\!] \supseteq \bigwedge\{[\![ft]\!], t \in \mathcal{T}_\alpha\}$. Let $\Gamma \in \bigwedge\{[\![ft]\!], t \in \mathcal{T}_\alpha\}$. Let y be a fresh variable with respect to Γ and f. In particular, $\Gamma \in [\![fy]\!]$. The proof

$$\frac{\Gamma \vdash^* fy}{\Gamma \vdash^* \Pi.f} \forall_R$$

shows that $\Gamma \in [\![\Pi.f]\!]$.

- $cl(B \supset C) \wedge_\Omega [\![B]\!] \subseteq cl(C)$. Let $\Gamma \in cl(B \supset C) \cap [\![B]\!]$. We must show the C-cuttability of Γ. Consider a proof of $C \vdash^* D$. Since $\Gamma \vdash^* B$:

$$\frac{\Gamma \vdash^* B \qquad \Gamma, C \vdash^* D}{\Gamma, B \supset C \vdash^* D} \supset_L$$

By $B \supset C$-cuttability of Γ we get $\Gamma \vdash^* D$.

- $cl(B) \to_\Omega [\![C]\!] \subseteq [\![B \supset C]\!]$. Let $\Gamma \in cl(B) \to [\![C]\!]$ and show $\Gamma \vdash^* B \supset C$. Indeed, since $\Gamma \in cl(B) \to [\![C]\!]$, we have: $cl(\Gamma) \cap cl(B) \subseteq [\![C]\!]$. Furthermore from Rem. 14, $\Gamma \in cl(\Gamma)$ and $B \in cl(B)$. It follows by weakenings that Γ, B belongs to both. Hence $\Gamma, B \in [\![C]\!]$, and we can derive the desired proof:

$$\frac{\Gamma, B \vdash^* C}{\Gamma \vdash^* B \supset C} \supset_R$$

■

4.3 Completeness and cut elimination of ICTT

We now have all the results needed to establish completeness.

THEOREM 18 (cut-free completeness of ICTT). *Let Γ be a context and A be a formula. Assume that for any global model we have $\Gamma^* \leq A^*$. Then we have a cut-free proof of $\Gamma \vdash A$.*

Proof. We apply Thm. 8 with Heyting algebra Ω_{cfk} given in Def. 10 and the semivaluation π, ν of Def. 16. We get, from Rem. 14, by Thm. 8 and by hypothesis that:

$$\Gamma \in cl(\Gamma) \subseteq \Gamma^* \subseteq A^* \subseteq [\![A]\!]$$

Hence, the sequent $\Gamma \vdash A$ has a cut-free proof. ■

As an immediate corollary, we have:

COROLLARY 19 (constructive cut elimination for ICTT). *Let Γ be a context and A be a formula. If $\Gamma \vdash A$ has a proof in* ICTT, *then it has a proof without cut.*

Proof. By soundness and cut-free completeness, both of which were proved constructively. ■

See further comments on the constructive character of the proof in Sec. 9.

5 Adding non-logical axioms

Now, we allow a more liberal notion of proof, with *non-logical axioms.*

DEFINITION 20. A non-logical axiom is a sequent $A \vdash B$. A proof with non-logical axioms is a proof whose leaves are either a proper axiom rule, or a non-logical axiom and allowing the use of axiomatic cuts.

Assuming that $A \vdash B$ is a non-logical axiom, an axiomatic cut is the following implicit cut rule

$$\frac{\Gamma \vdash A \qquad \Gamma, B \vdash C}{\Gamma \vdash C}$$

In the sequel, we will work with a given set (potentially infinite) of non-logical axioms, and the proof system will be a proof ICTT with non-logical axioms. This syntactical proof system will be called L_{nla}.

The non-logical axiom and the axiomatic cut rules overlap a little bit:

LEMMA 21. *In* ICTT *with non-logical axioms, when the cut rule is allowed, one can simulate axiomatic cuts. Conversely, with axiomatic cuts one can simulate the non-logical axioms, with or without the cut rule.*

Proof. For the first statement, replace any axiomatic cut by:

$$\dfrac{\dfrac{\Gamma \vdash A \qquad \overline{A \vdash B}\ \text{non-logical axiom}}{\Gamma \vdash B}\ \text{cut} \qquad \Gamma, B \vdash C}{\Gamma \vdash C}\ \text{cut}$$

For the second statement, replace a non-logical axiom $\Gamma, A \vdash B$ by:

$$\dfrac{\overline{\Gamma, A \vdash A} \qquad \overline{\Gamma, B \vdash B}}{\Gamma, A \vdash B}\ \text{axiomatic cut}$$

■

We show in this section that we still have, by the same means, cut elimination in ICTT with non-logical axioms, but that we can not, in the general setting, eliminate axiomatic cuts. First, we need another, unsurprising, notion of model:

DEFINITION 22 (models for non-logical axioms). A global model for ICTT (Def. 3) is a model of the non-logical axioms if and only if $(A)^* \leq (B)^*$ for any non-logical axiom $A \vdash B$.

In the sequel, we will only be interested in such models.

THEOREM 23 (Soundness of ICTT with non-logical axioms). *If $\Gamma \vdash A$ in ICTT with non-logical axioms, then $\Gamma^* \leq A^*$ in any global model of the non-logical axioms.*

Proof. We assume not to have axiomatic cuts by Lem. 21. The proof is by the very same induction as the one of Thm. 5. The only additional case is the case of a non-logical axiom $A \vdash B$, trivial, since we assumed the model to be a model of the non-logical axioms. ■

Now we work towards a proof of a cut-free completeness theorem for ICTT with non-logical axioms. Cut-free means free of cuts, but not of axiomatic cuts, that we will not be able to remove.

5.1 Completeness and cut elimination in presence of axioms

As well as we defined, in ICTT, the complete Heyting algebra of cut-free contexts Ω_{cfk} (Def. 10), we can define Ω_{cfk} with respect to provability in L_{nla}, i.e. ICTT with non-logical axioms. Def. 10 does *not* depend on the syntactic system we work with. Of course the contexts in $[\![C]\!]$ depend on the logic we are in: ICTT or ICTT with non-logical axioms (L_{nla}). So both algebras are different, but generated exactly the same way. To distinguish between both algebras, we will speak about $\Omega(\mathrm{L})$ and $\Omega(\mathrm{L}_{nla})$.

We can build a semivaluation, with respect to provability in L_{nla}, exactly in the same way as in Def. 16: we can check that the proof of Lem. 17 does not depend on the presence or absence of non-logical axioms and axiomatic cuts. As well, the A-cuttability notion (Lem. 15) used there does not depend on ICTT. It appeals to Kleene's Lem. 33 below.

Since the Takahashi-Schütte lemma (Thm. 8) does not at all depend on the syntactic system (it requires only a semivaluation), we can build an interpretation, generating V-complexes [1, 25, 21] as well.

So we get a model $\mathfrak{D}_{nla}$ over the cHa $\Omega(L_{nla})$ exactly in the same way as in Sec. 4. The only thing to check is:

LEMMA 24. *The global model $\mathfrak{D}_{nla}$ is a model of the non-logical axioms.*

Proof. Let $A \vdash B$ be a non-logical axiom. Let's show that $A^* \subseteq B^*$. We know that $A^* \subseteq [\![A]\!]$ and that $cl(B) \subseteq B^*$ from the Takahashi-Schütte Thm. 8. So it is sufficient to show $[\![A]\!] \subseteq cl(B)$. Let Γ such that $\Gamma \vdash^* A$. We show that Γ is B-cuttable. So, assume given a formula C and a proof of $\Gamma, B \vdash^* C$. We can build the following proof of $\Gamma \vdash^* C$:

$$\frac{\Gamma \vdash^* A \qquad \Gamma, B \vdash^* C}{\Gamma \vdash^* C} \text{ axiomatic cut}$$

which yields the desired conclusion. ■

Therefore, we have the following proof of completeness:

THEOREM 25 (cut-free completeness of ICTT with nla). *Consider any set of non-logical axioms. Let Γ be a context and A be a formula. Assume that for any global model of the non-logical axioms, we have $\Gamma^* \leq A^*$. Then we have a cut-free proof of $\Gamma \vdash A$.*

Proof. As Thm. 18. $\Omega(L_{nla})$ is a global model of the non-logical axioms. ■

As well, we have the cut elimination theorem as a corollary:

COROLLARY 26 (constructive cut elimination for ICTT with non-logical axioms). *Consider any set of non-logical axioms. Let Γ be a context and A be a formula. If $\Gamma \vdash A$ has a proof in ICTT, then it has a proof without cut.*

Proof. By soundness and cut-free completeness, both of which were proved constructively. ■

6 ICTT and rewriting cuts

Here, we assume we have non-logical axioms as well (Def. 20), as in the previous section but with the additional assumption that the non-logical axioms satisfy a sign-preserving condition (given below). We obtain a constructive semantic cut elimination theorem that guarantees the existence of a cut-free proof with the following *rewriting cuts* only, which are a restriction of axiomatic cuts.

Before proving the strengthened cut elimination theorem, we state some useful results, and we begin by recalling briefly terminology of rewriting.

6.1 Rewrite rules

DEFINITION 27. Let $\mathcal{R} = \{l_i \rightarrow r_i : i \in I\}$ be a rewrite system where all the left and right members have type o.

A proposition C is said to $\mathcal{R}$-rewrite to C' if for some i and some $D \equiv_\lambda C$, there is a redex in D matching an instance of l_i via a unifier θ and, defining D' as D where the redex is replaced by θr_i, $D' \equiv_\lambda C'$.

We use the notation $\rightarrow^*$ to denote the transitive, reflexive closure of the $\mathcal{R}$ relation $\rightarrow$ and $\equiv_\mathcal{R}$ the congruence it generates.

A rewrite system $\mathcal{R}$ is confluent if and only if for any two formulae $A \equiv_\mathcal{R} B$, there is a formula C such that $A \rightarrow^* C$ and $B \rightarrow^* C$. A rewrite system $\mathcal{R}$ is an atomic system if every antecedent A with $A \rightarrow B \in \mathcal{R}$ is an atomic formula of type o.

The reader should consult e.g. [2] for more details on term rewriting. Notice that $\equiv_\mathcal{R}$ contains $\equiv_\lambda$ and that here we are interested only in propositional rewriting, which is where interaction between rewriting and sequent rules can be delicate (see Deduction Modulo [12]).

LEMMA 28 (Main connective). *Let $\{A_i \rightarrow B_i\}$ be an atomic and confluent rewrite system. Let C, D be non atomic formulae. If $C \equiv_\mathcal{R} D$, they have the same main connective, and their immediate subformulae are congruent.*

Proof. By confluence, we can find a E such that $C \rightarrow^* E$ and $D \rightarrow^* E$, and rewriting occur only on atomic formulae by the atomicity condition. Formally, this is done by induction on the length of rewriting paths. For more details, see [12]. ■

6.2 From rewrite rules to ICTT

There are many ways to add rewrite rules to ICTT, for instance, one can try to define a Deduction Modulo [7, 6] within the ICTT frame. Instead of that, we constrain the axiomatic cuts of L_{nla} to have a rewriting form.

6.3 Rewriting cuts

DEFINITION 29 (From rewrite rules to an axiomatic system). Let $\mathcal{R}$ be a propositional rewrite system consisting of rules of the form $A \rightarrow B$ with A, B terms of type o. We consider the associated set of non logical axioms (i.e. new initial rules) to be all the sequents $\sigma A \vdash \sigma B$ and $\sigma B \vdash \sigma A$ where $A \rightarrow B \in \mathcal{R}$ and σ is some substitution: we add all the instantiations of rewrite rules, in both ways.

DEFINITION 30 (Rewriting cut). Let $A \vdash B$ a non-logical axiom. A rewriting cut is an axiomatic cut of one of the following forms:

$$\dfrac{\Gamma \vdash A \qquad \overline{\Gamma, B \vdash B}\ \text{Axiom}}{\Gamma \vdash B} \qquad \dfrac{\text{Axiom}\ \overline{\Gamma, A \vdash A} \qquad \Gamma, B \vdash C}{\Gamma, A \vdash C}$$

We define the logic L^{nla} to be L_{nla} with axiomatic cuts restricted in this way.

This restriction makes rewriting cuts almost inoffensive, since the formula we cut on has to be immediately proved in one premise. That is very close to a rewriting of A to B (on the right hand side) and of B to A (on the left hand side).

As in Lem. 21, one can simulate the non-logical axiom rule, even in a cut-free setting, with a rewrite cut and one can simulate a rewrite cut with a cut and a non-logical axiom.

DEFINITION 31 (From rewrite rules to L^{nla}). Let $\mathcal{R}$ be a rewrite system. Consider the associated set of non-logical axioms, as in Def. 29 and consider the associated logical system L^{nla} generated by Def. 30.

If $\mathcal{R}$ is confluent, we call L^{nla} a confluent axiomatic system, and call it an atomic system if $\mathcal{R}$ is atomic.

Notice that in the logic L^{nla} we do not "rewrite" within the logic directly (as with Deduction Modulo [7, 6]). We apply axiomatic cut rules. We assume we have an atomic and confluent rewrite (or axiomatic) system.

LEMMA 32. *Let $\mathcal{R}$ an atomic and confluent rewrite system. Let $\Gamma \equiv_{\mathcal{R}} \Gamma'$ (pointwise equivalence) be contexts and $A \equiv_{\mathcal{R}} A'$ be formulae. If we have a proof θ of the sequent $\Gamma \vdash^* A$ then we can build a proof of the sequent $\Gamma' \vdash^* A'$.*

Proof. By induction on the structure of θ. We copy every rule, applying the induction hypothesis before that. The only non trivial case are:

- a λ conversion rule. Apply induction hypothesis.

- A' is atomic and we have a logical rule on A. The principle is that we rewrite A' with rewriting cuts (and λ rules) until it becomes non atomic.

  By confluence, we must have a chain of instances of rewrite rules $A_i \to B_i$, $i \leq n$, such that $A_0 \equiv_\lambda A'$, $A_i \equiv_\lambda B_{i-1}$ and B_n non atomic, having the same main connective (by Lem. 28) as A. Since $B_n \equiv_\mathcal{R} A$, we construct a proof of $\Gamma' \vdash^* B_n$ by applying the induction hypothesis. We then add successive rewriting cuts and λ-conversion rules.

- we have the same on the left hand side.

- an axiom rule. We only know the existence $A'' \in \Gamma'$ such that $A'' \equiv_\mathcal{R} A \equiv_\mathcal{R} A'$, so we can't apply the axiom rule as such since A', A'' are not atomic and the rewriting path between A' and A'' is not straight. But we know by confluence that there is a formula D such that $A'' \to^n D$ and $A' \to^m D$.

  We prove by induction over $n + m + \#D$, where $\#D$ is the number of connectives and quantifiers of D that we can build a proof of $A'' \vdash^* A'$. If D is non atomic, while A' or A'' is (assume it is A'), let $A' \to B$ the first rule used in the rewrite sequence $A' \to^n D$. We have by induction hypothesis a proof of the sequent $\Gamma' \vdash^* B$ we then make a rewriting cut with the axiom $B \vdash^* A'$, adding λ rules if necessary. Otherwise, if, say $D \equiv_\lambda B_1 \vee B_2$ and A', A'' are compound, they are equal to $B_1' \vee B_2'$ and $B_1'' \vee B_2''$. By induction hypothesis we have proofs of $B_1'' \vdash^* B_1'$ and $B_2'' \vdash^* B_2'$ and we add an $\vee_L$ and an $\vee_R$ rule (plus λ if needed) to conclude. At last, if D itself is atomic, this is just a matter of using cut and λ rules with the non-logical axioms required by the rewriting paths $A' \to^n D$ and $A'' \to^m D$.

■

We see that the ability to rewrite atoms with the help of rewriting cuts is essential, as well as confluence.

7 Kleene's lemma and rewriting

7.1 Kleene's lemma in ICTT

Kleene's lemma is a standard rule inversion lemma, saying that – for certain rules – a proof of a sequent that is the conclusion of the rule may be replaced by a proof of the premise of that rule. For instance, if we have a proof of the sequent $\Gamma \vdash^* \forall x A$ then we can construct a proof of the sequent $\Gamma \vdash^* (t/x)A$ for any t. Some rules cannot be inverted, as the $\vee_R$ rule. Indeed, we can find no proof of the sequent $A \vee B \vdash^* A \vee B$ beginning with a $\vee_R$ rule.

Other non invertible rules are $\supset_L, \forall_L, \exists_R$. We here prove it in ICTT, in L_{nla} as well as in L^{nla} when the rewrite rules are confluent and atomic. This lemma is used at some places in former sections.

LEMMA 33 (Kleene). *Let $D_1 \equiv_\lambda \ldots \equiv_\lambda D_n \equiv_\lambda B \vee C$. If we have a proof π of the sequent $\Gamma, D_1, \ldots, D_n \vdash^* A$ then we have a proof of $\Gamma, B \vdash^* A$ and $\Gamma, C \vdash^* A$. If the proof was cut-free, then the obtained proofs remain cut-free.*

Proof. Standard, by induction on the height of π (the depth of the associated tree). Notice that if $n = 0$ a use of the weakening rule is sufficient. If the last rule is a rule r on Γ or on A, then apply induction hypothesis and the same rule r. If the rule is an axiom and no D_i is an active formula, so we can replace them by B or C freely. Otherwise, assume D_1 is active. The rule can be:

- a contraction, a weakening or a λ-conversion. Apply induction hypothesis on the premise.
- an $\vee$-l rule, apply induction hypothesis on the premises, we get four proofs, and keep only the two of interest, that we contract.
- an axiom, then the proof has the shape: $\overline{\Gamma, D_1, \ldots, D_n \vdash D_1}$. We expand it into the proof:

$$\cfrac{\cfrac{\cfrac{}{\Gamma, B \vdash B}\text{Axiom}}{\Gamma, B \vdash B \vee C}\vee_R}{\Gamma, B \vdash D_1}\lambda$$

  we do the same for C.

Notice that the proof remains valid if we add axiomatic cuts and non-logical axioms. There is one more cases to consider when D_1 is an active formula: a non-logical axiom, say $D_1 \vdash E$. Then we can construct the following proof:

$$\cfrac{\cfrac{\cfrac{\cfrac{}{\Gamma, B \vdash B}\text{axiom}}{\Gamma, B \vdash B \vee C}\vee_R}{\Gamma, B \vdash D_1}\lambda \qquad \cfrac{}{\Gamma, B, E \vdash E}\text{axiom}}{\Gamma, B \vdash E}\text{rewriting cut}$$

introducing a rewriting cut, allowed even in the cut-free case. If no D_i is active, for instance in an axiomatic cut, then we simply apply the induction hypothesis. Notice that this last case can simulate, by Lem. 21, any non-logical axiom so, strictly speaking, one has not to consider that case. ■

The lemma and the proof are the same for all the other connectives, save the four mentioned above.

7.2 Kleene's lemma in $\mathbf{L}^{nla}$

Now, we prove Kleene's lemma in the confluent atomic case, in L^{nla}. The statement of the lemma must be somewhat modified to obtain the results we need.

LEMMA 34 (Kleene). *If we have a proof θ of the sequent $\Gamma, D_1, \ldots, D_n \vdash^* A$ then we have a proof of $\Gamma, B \vdash^* A$ and $\Gamma, C \vdash^* A$, where $D_i \equiv_{\mathcal{R}} B \vee C$. If the proofs is cut-free, then the obtained proofs remain cut-free (with only rewriting cuts).*

Proof. The only cases differing from those of the proof of Lem. 33 are:

- the case of a λ rule is treated by induction hypothesis, even in the case of D_i being an active formula.

- in the inductive cases we have to consider the case of a rewriting cut on Γ, on D_i or on A. Let's consider the third case, and assume the rewriting cut is with the non-logical axiom $E \vdash^* A$. We apply induction hypothesis on the proof of the premise $\Gamma, D_1, \ldots, D_n \vdash^* E$. Then we add a rewriting cut to $\Gamma, B \vdash^* E$. Similarly if the rewriting cut is done with respect to some formula in Γ (first case). If the rewriting cut is done on D_i, then it is sufficient to apply induction hypothesis thanks to the generalized hypothesis.

- the connective case on D_i: thanks to confluence and Lem. 28 it can only be a $\vee_L$ rule. After an application of induction hypothesis, we have proofs of $\Gamma, B, B' \vdash^* A$ and $\Gamma, C, C' \vdash^* A$ with $B' \equiv_{\mathcal{R}} B$ and $C' \equiv_{\mathcal{R}} C$. We then apply Lem. 32 and contract.

- an axiom involving some D_i, say D_1. We have a proof of the sequent $\Gamma, D_1, \ldots, D_n \vdash^* D_1$. By Lem. 32, we have as well a proof of $\Gamma, D_1, D_2, \ldots, D_n \vdash B \vee C$. This is easily transformed into a proof of $\Gamma, B \vdash^* B \vee C$ and $\Gamma, C \vdash^* B \vee C$. We apply Lem. 32 once again to get proofs of $\Gamma, B \vdash^* D_1$ and $\Gamma, C \vdash^* D_1$.

■

8 Atomic confluent rewriting, the sign-preserving case

In this section we show how, by carefully choosing the interpretation of atomic predicates, we can restrict axiomatic cuts to be rewriting cuts in the case of sign-preserving rewrite rules (see Def. 36).

8.1 Sign-preserving condition

DEFINITION 35. Let $\{P_i, i \in \Lambda\}$ the collection of atomic predicate symbols of a language S for ICTT. A decoration on them is a total function $p : \Lambda \to \{+, -\}$. We note P_i^+ (resp. P_i^-) whenever $p(i) = +$ (resp. $p(i) = -$). A formula A in λ normal form is said positive (resp. negative) if and only if it does not contains any flex variable and:

- A is an instantiation of a predicate P_i and $p(i) = +$ (resp. $p(i) = -$).
- $A = \top$ or $A = \bot$ (resp. idem).
- $A = B \wedge C$ and B and C are positive (resp. negative).
- $A = B \vee C$ and B and C are positive (resp. negative).
- $A = B \supset C$ and B is negative and C is positive (resp. B is positive and C is negative).
- $A = \Pi.f$ and for any term t, ft is positive (resp. negative).
- $A = \Sigma.f$ and for any term t, ft is positive (resp. negative).

A formula that is not in λ normal form is said positive (resp. negative) if its corresponding normal form is.

There are many formulae that are neither positive, nor negative. First of all, any formula containing flex variables, such as $\forall X.X$ or $P \vee X$, since every instance of X can not have the same polarity. $A \vee \neg A$ is another counterexample. Notice that we even explicitly forbid $\forall X.X$ to be analyzed in Def. 35, although it can be proved that it does not fit the pattern.

Def. 35 may seem a bit circular. It can apply to quantification over propositional types (of type $\ldots \to o$), but only if the bound variable is not at a propositional position. i.e. not "flex". For instance, $\forall X.P(X)$ is positive provided P is, but not $\forall X.(P(X) \wedge X)$. Since the flex variables are the only impredicative case, Def. 35 is well founded. So we do allow quantification over propositional types, when the bound variable X appears as an argument of a predicate symbol, as for instance $P.X$. With this definition, we will forbid rewrite rules of the form $P(X) \to X$ for any predicate P. Otherwise $\forall X P(X)$ would have a sign, whereas $\forall X.X$ has not. Although this rule seems harmless, it does not fit the pattern for now.

Def. 36 below can be applied as well to rewrite systems.

DEFINITION 36. An axiomatic system is sign-preserving if there exists a decoration on the predicates symbol such that for any axiom $A \vdash B$, A has the same sign than B.

In this section we assume we have a rewrite system that is atomic, confluent and sign-preserving, and consider the associated axiomatic system and L^{nla} logic. The atomicity and confluence conditions imply that Kleene's Lem. 33 holds, which will be of paramount importance in the proof of the main lemma 40.

8.2 Model construction

As in Sec. 4 and 5.1 we can define the complete Heyting algebra Ω_{cfk}, with respect to provability in L^{nla}, here noted $\Omega(L^{nla})$. We build the same intuitionistic semivaluation π, ν as in Def. 16, and use the Takahashi-Schütte Thm. 8, but with the following small modification in choosing the interpretation of atoms, which is of crucial importance:

DEFINITION 37. Let A be an atomic formula of type o. We let

- $\rho(A) = \nu(A)$ if A is positive.
- $\rho(A) = \pi(A)$ if A is negative.

Since any atomic formula is either positive or negative, this definition is complete. Remember that $\pi(A) = cl(A)$ and $\nu(A) = [\![A]\!]$ are both expressed in L^{nla}. The Takahashi-Schütte Thm. 8 implies then the following definition of the interpretation of formulae:

DEFINITION 38. Let A be a formula of type o. Let ϕ be an environment, associating V-complexes of the right type to variables. Let ϕ^1 the substitution associating the term $\phi(x)^1$ to any variable x. Define A^*_ϕ, the interpretation of A as:

- $A^*_\phi = \nu(\phi^1 A)$ if A is atomic and positive.
- $A^*_\phi = \pi(\phi^1 A)$ if A is atomic and negative.
- otherwise, define it inductively as in Def. 2.

REMARK 39. We could certainly switch the π and ν in the previous two definitions since the choice of polarities is symmetric in Def. 36.

As in Sec. 5.1 it remains to prove only one claim: the model we constructed is a model of the non-logical axioms. Since every time we have $A \vdash B$ as a non-logical axiom we as well have $B \vdash A$, we must show that $A^* = B^*$. The following lemma is the key to show this. It speaks more generally about any positive and negative formulae, but this includes, thanks to the conditions of Def. 36 any non-logical axiom.

LEMMA 40. *Let A a positive (resp. negative) formula. Let ϕ be an environment. Then $(A^*_\phi)^2 = \nu(\phi^1 A)$ (resp. $(A^*_\phi)^2 = \pi(\phi^1 A)$).*

Proof. By induction over the structure of A. Since A does not contain any flex variable (Def. 35), this induction is well-founded. We should consider every single case for A, and we omit to mention the environment ϕ where it is not essential. We also note that this proof does not work for every π, ν, but only for the one we give (cl and $[\![]\!]$), since they are in a sense maximal.

- if A is atomic, this comes from the very definition of A^*. If A is $\top$, this is because in this particular case $cl(\top) = [\![\top]\!]$, similarly for $\bot$.
- if $A = B \wedge C$ and A is positive, then by induction hypothesis we have $(B^*)^2 = [\![B]\!]$ and $(C^*)^2 = [\![C]\!]$, and then, by definition, $(A^*)^2 = [\![B]\!] \cap [\![C]\!]$. Since $[\![B]\!] \cap [\![C]\!] \leq [\![B \wedge C]\!]$ from Lem. 17 (ν is an upper constraint of an intuitionistic semivaluation), we must show only the converse. Let Γ such that $\Gamma \vdash^* B \wedge C$. Applying Kleene's Lem. 33 gives us proofs of the sequents $\Gamma \vdash^* B$ and $\Gamma \vdash^* C$. Hence $\Gamma \in [\![B]\!] \cap [\![C]\!]$.

  If A is negative, reasoning in the same way leads us to try to show the inclusion $cl(B) \cap cl(C) \subseteq cl(B \wedge C)$. Let $\Gamma \in cl(B) \cap cl(C)$: Γ is B and C-cuttable. Let D be such that $B \wedge C \vdash^* D$. Then, $B, C \vdash^* D$ as well, by Kleene's Lem. 33. By B, and then C cuttability, we obtain a proof of the sequent $\Gamma \vdash^* D$.
- $A = B \vee C$ and A is positive. By the same reasoning as in the previous case, showing that $[\![B \vee C]\!] \subseteq [\![B]\!] \cup [\![C]\!]$ is sufficient. Let Γ be such that $\Gamma \vdash^* B \vee C$. Also let D be an upper bound for $[\![B]\!]$ and $[\![C]\!]$: if $\Delta \vdash^* B$ or $\Delta \vdash^* C$ then $\Delta \vdash^* D$. We need to show that $[\![D]\!]$ is an upper bound for $[\![B \vee C]\!]$, i.e. $\Gamma \vdash^* D$.

  We construct a proof of $\Gamma \vdash^* D$ by induction over the proof of $\Gamma \vdash^* E$ with $E \equiv_{\mathcal{R}} B \vee C$. The base case occurs when the active formula is E. If it's a $\bot_R$ rule, then we can as well have this rule generating D instead of E. If it is a rewriting cut or a λ, we simply use the induction hypothesis. If it is a logical rule, it can be only $\vee$-right rule, by confluence. Then we use the hypothesis on D, since we get a proof of either $\Gamma \vdash^* B'$ or $\Gamma \vdash^* C'$, which is the same, by Lem. 32 as a proof of $\Gamma \vdash^* B$ (resp. $\Gamma \vdash^* C$). If it is an axiom, we can, as in the last case of the proof of Lem. 32. add $\vee$-left, $\vee$-right and two axiom rules, at the potential cost of adding rewriting cuts. So, this case boils down to the previous ones. For the induction case, the last rule is on a formula of Γ. Apply the induction hypothesis on the premises (if needed, e.g. not on the left premise of $\supset_R$), and apply the same rule. Therefore $\Gamma \in [\![D]\!]$, and it belongs to the least upper bound $[\![B]\!] \cup [\![C]\!]$.

If A is negative, we show that $cl(B) \cup cl(C) \subseteq cl(B \vee C)$. Let D such that $B \vee C \vdash^* D$. Then by Kleene's Lem. 33, $B \vdash^* D$ and $C \vdash^* D$, and $[\![D]\!]$ is an upper bound for $cl(B)$ and $cl(C)$. Since this holds for any D, the result follows.

- $A = B \supset C$ and A is positive. We prove $[\![B \supset C]\!] \subseteq cl(B) \supset_\Omega [\![C]\!]$, or equivalently $[\![B \supset C]\!] \cap cl(B) \subseteq [\![C]\!]$. Let Γ a B-cuttable context, and such that we have a proof of the sequent $\Gamma \vdash^* B \supset C$. By an application of Kleene's Lem. 33 we have a proof of the sequent $\Gamma, B \vdash^* C$. Since Γ is B-cuttable, it gives a proof of $\Gamma \vdash^* C$. Hence $\Gamma \in [\![C]\!]$.

  If A is negative, then we have to show $[\![B]\!] \supset_\Omega cl(C) \subseteq cl(B \supset C)$. Let $\Gamma \in [\![B]\!] \supset_\Omega cl(C)$, we must show that Γ is $B \supset C$-cuttable. We cannot apply Kleene's Lem. 33, as in the $\vee$ positive case. So, assume we have a proof of $\Delta, E_1, \dots, E_n \vdash^* D$ for some D, with all $E_i \equiv_{\mathcal{R}} B \supset C$ (in contrast with the $\vee$ positive case we here have to consider an arbitrary multiplicity n). We show by induction on this proof that we can build a proof of $\Delta, \Gamma \vdash^* D$. As always, if it is a rule on D or a proposition of Δ (including axiom), we apply the induction hypothesis to the premises and then the same rule. Otherwise, if it is a contraction, a weakening, a λ or a rewriting cut on E_i, we apply the induction hypothesis. We omit the axiom case on E_i, since it boils down to the other cases in the same way as in the $\vee$ positive case. So suppose we have an $\supset_R$ rule on, say, E_1. Then we have the proof:

$$\dfrac{\dfrac{\pi_1}{\Delta, E_2, \dots, E_n \vdash^* B'} \qquad \dfrac{\pi_2}{\Delta, E_2, \dots, E_n, C' \vdash^* D}}{\Delta, E_1, \dots, E_n \vdash^* D}$$

  Applying the induction hypothesis on the premises gives us proofs of the sequents: $\Delta, \Gamma \vdash^* B'$ and $\Delta, \Gamma, C' \vdash^* D$, that we convert into proofs of $\Gamma \vdash^* B$ and $\Delta, \Gamma, C \vdash^* D$ (by Lem. 32). Hence $\Delta, \Gamma \in [\![B]\!]$, and $\Delta, \Gamma \in cl(\Gamma)$. So by definition of $\supset_\Omega$, $\Delta, \Gamma \in cl(C)$: it is C-cuttable and we have directly a proof of the sequent $\Gamma \vdash^* D$.

- $A = \Pi.f$ and A is positive. Calling $\phi' = \phi + (d/x)$, we know that:

$$A^*_\phi = \bigcap_{d \in D} f^*_\phi d = \bigcap_{d \in D} (fx)^*_{\phi'} = \bigcap_{d \in D} [\![\phi'^1(fx)]\!] = \bigcap_{t \in \mathcal{T}} [\![(\phi'^1 f)t]\!] = \bigcap_{t \in \mathcal{T}} [\![(\phi^1 f)t]\!]$$

  the third equality holding by the induction hypothesis, choosing x such that it does not appear in f.

We then show $[\![\phi^1\Pi.f]\!] = [\![\Pi.(\phi^1 f)]\!] \leq \bigcap\{[\![(\phi^1 f).t]\!] \mid t \in \mathcal{T}\}$. Let Γ such that $\Gamma \vdash^* \Pi.\phi^1 f$. Then by Kleene's Lem. 33, we have a proof θ of $\Gamma \vdash^* (\phi^1 f).t$, for any term t. Hence $\Gamma \in \bigcap\{[\![(\phi^1 f).t]\!] \mid t \in \mathcal{T}\}$.

If A is negative, let B be a formula such that $\Pi.(\phi^1 f) \vdash^* B$. We show $\bigcap\{cl((\phi^1 f)t), t \in \mathcal{T}_\alpha\} \subseteq [\![B]\!]$. Let $\Gamma \in \bigcap\{cl((\phi^1 f)t), t \in \mathcal{T}_\alpha\}$. We show that $\Gamma \vdash^* B$, knowing that for any t, and for any C, if $(\phi^1 f)t \vdash^* C$, then $\Gamma \vdash^* C$.

We cannot apply Kleene's Lem. 34. So we construct by induction a proof of $\Gamma, \Delta \vdash^* B$, over the proof structure of a proof of $\Delta, (\Pi.f)^n \vdash^* B$, where n is any number of contractions of $\Pi.f$. Formally we should assume that for $E_i \equiv_{\mathcal{R}} \Pi.f$ as in the $\vee$ positive and $\supset$ negative cases. $n = 0$ is a trivial case. If the last rule is a rule r on Δ or on B, then apply the induction hypothesis and then r on the proofs we obtain. Otherwise, we then have a proof of $\Delta, ft, (\Pi.f)^{n-1} \vdash^* B$. After applying the induction hypothesis, we get a proof of $\Gamma, \Delta, ft \vdash^* B$. We can then safely replace ft by Γ, since Γ is ft-cuttable by hypothesis. We then contract on the formulae of Γ.

- $A = \Sigma.f$ and A is positive. We must show $[\![\Sigma.(\phi^1 f)]\!] \leq \bigcup\{[\![(\phi^1 f)t]\!], t \in \mathcal{T}\}$, by the same reasoning as in the previous case. We omit ϕ from now on. Let Γ such that $\Gamma \vdash^* \Sigma.f$. We cannot apply Kleene's lemma. Let C such that $[\![ft]\!] \leq [\![C]\!]$ for any term t. We show by induction on the proof of $\Gamma \vdash^* E$, with $E \equiv_{\mathcal{R}} \Sigma.f$ that we can build a proof of $\Gamma \vdash^* C$. As always, copy every left rule, applying the induction hypothesis. And when we get to a right rule, it can be either $\bot_R$, then replace it with $\bot_R$ introducing D, a rewriting cut or a λ, then apply the induction hypothesis, or an axiom, that boils down to the last case, an $\exists_R$ rule. In this case, we get a proof of $\Gamma \vdash^* ft$ for some t. But then, by definition of Γ and C we get directly a proof of $\Gamma \vdash^* C$.

  If A is negative, we show $\bigcup\{cl((\phi^1 f)t)\} \leq cl(\Sigma.(\phi^1 f))$. We omit ϕ. Let Γ such that Γ is ft-cuttable for any t. Assume that we have a proof of $\Sigma.f \vdash^* B$. By Kleene's Lem. 33 we have a proof of $fx \vdash^* B$ for a fresh x. Since Γ is as well fx-cuttable, we have directly $\Gamma \vdash^* B$, and $\Gamma \in cl(\Sigma.f)$.

■

REMARK 41. This lemma just says that the π, ν semivaluation is indeed much more than a semivaluation in the sense of Def. 6. Since we have

$\pi(A \wedge B) = \pi(A) \wedge_\Omega \pi(B)$, and so on for the other connectives (special case for $\supset$, since it has a positive and a negative part). We have the same result on ν. This is due to the choice of π, ν we have made. This is not valid any π, ν but for our very specific one.

REMARK 42. Instead of invoking Kleene's lemma, we could use everywhere an induction on the proof structure. At any rate, we would need confluence and atomicity, in order to ensure that the rewrite rules are treated properly.

8.3 Completeness and cut elimination for L^{nla} with sign-preserving axioms

Lem. 40 is stated and proved in the L^{nla} logic. We have carefully chosen the ρ function (Def. 38 and Thm. 8) using the only degree of freedom we had, the atomic formulae. With this interpretation, we have $A^* = [\![A]\!]$ (resp. $A^* = cl(A)$) in $\Omega(\mathrm{L}^{nla})$, whenever A is positive. This result holds as well for $\Omega(L)$ but we won't need this fact here.

LEMMA 43. *The global model constructed above is a model of the non-logical axioms.*

Proof. If we have, say, a positive non-logical axiom $A \vdash B$ in L^{nla}, we have $[\![A]\!] \subseteq [\![B]\!]$ by an elementary reasoning:

$$\frac{\Gamma \vdash A \qquad \overline{\Gamma, B \vdash B}}{\Gamma \vdash B}$$

The symmetry comes from the symmetric axiom $B \vdash A$ we assumed to have. Hence $A^* = [\![A]\!] = [\![B]\!] = B^*$. We as well have $A^* = cl(A) = cl(B) = B^*$, i.e. Γ is A-cuttable if and only if it is B-cuttable by a similar reasoning. Therefore, from Lem. 40 for any axiom $A \vdash B$, $A^* = B^*$. ■

THEOREM 44 (cut-free completeness of L^{nla}). *Consider a set of non-logical axioms atomic, confluent and sign-preserving. Let Γ be a context and A be a formula of S. Assume that for any global model of the non-logical axioms, we have $\Gamma^* \leq A^*$. Then we have a cut-free proof of $\Gamma \vdash A$.*

Proof. As Thm. 18 and 25. $\Omega(\mathrm{L}^{nla})$ is a model of the non-logical axioms. ■

As well, we have the cut elimination theorem as a corollary:

COROLLARY 45 (constructive cut elimination for L^{nla}). *Consider a set of non-logical axioms atomic, confluent and sign-preserving. Let Γ be a context and A be a formula. If $\Gamma \vdash A$ has a proof in L^{nla}, then it has a proof in L^{nla} without cut.*

Proof. By the soundness Thm. 23, that remains exactly the same as in Sec. 5, and the above cut-free completeness Thm. 44, both of which were proved constructively. ■

9 On the constructivity of the proof of cut admissibility

Our proof extends existing semantic proofs for cut admissibility in a number of ways, as remarked above, in particular by adding axioms, and considering the intuitionistic (rather than classical) Theory of Types.

In addition, our proof, unlike [25, 1] for the classical case or [5] for the intuitionistic case, makes no appeal to the excluded middle. The works cited (and our work as well) start from Schütte's observation [22] that cut admissibility can be proved semantically by showing completeness of the cut-free fragment with respect to semivaluations, and then showing every semivaluation gives rise to a total valuation extending it.

There are a number of pitfalls to avoid here if one wants a constructively valid proof based on this kind of argument, both in the way a semivaluation is produced and how one passes to a valuation.

Andrews shows [1] that any abstract consistency property gives rise to a semivaluation, but then builds one in a way that requires deciding whether or not a refutation exists of a given finite set of sentences. One can also exhibit a semivaluation by developing a tableau refutation of a formula (a Hintikka set) as is done in [5] but must take some care in the way the steps are formalized not to appeal to the fan theorem in order to produce an open path. No discussion of how this might be done appears in [5].

In the proof given above we appeal to the strengthened version in [5] of Schütte's lemma which uses the more liberal definition of semivaluation *pairs*, (rather than semivaluations) which provide an upper and lower bound for the truth values of the valuation eventually produced by Takahashi's V-complex construction.

As we have shown in this paper, it is possible to give an instance of such a pair (namely $cl(\_)$ and $[\![\_]\!]$) and prove they satisfy the semivaluation axioms without appeal to the excluded middle.

In particular with the Hermant-Okada cut-free context-based semantics used in this paper, we are able to avoid the construction of tableaux, and also bypass abstract consistency properties.

Constructive Completeness To begin with, a constructive proof of completeness is itself problematic, as pointed out by Gödel and discussed in [14, 27] if a sufficiently restrictive definition of validity is assumed, e.g. conventional Kripke models. However, there are a number of ways to liberalize the definition of validity to "save" constructive completeness [28, 4, 26, 13], in particular by allowing truth-values in a sufficiently broad class of structures. In our case these structures include complete Heyting Algebras (complete Boolean algebras in the classical case) *in which we cannot decide whether or not any given element is distinct from* $\top$ or even, for that matter, if the structure itself collapses to a one-element set. This appears to be a natural Heyting-valued counterpart to Veldman's exploding nodes.

In [26] completeness for an intuitionistic system *with* cut is shown constructively by mapping formulae to their own equivalence class in the Lindenbaum cHa (if the object logic is classical, as in [1], one would use the corresponding Boolean algebra). The semantics used in this paper is also over a Lindenbaum-like algebra, in which formulas are mapped to the sets of contexts that prove them without cut. In both cases, one is not required to decide the provability of formulae in order to show model existence, in contrast with the $\top, \bot$-valued semantics of [1, 25].

10 Conclusion

We have given a constructive semantic proof of cut-elimination for ICTT, the intuitionistic formulation of Church's Theory of Types introduced by Miller et. al. [16] and various extensions with non-logical axioms, using new techniques extending Takahashi, Schütte and Andrews' original ideas, based on a new formulation of semivaluations from [5] and notions of cut-free context closure and context-based Heyting algebras based on earlier work by Hermant and Okada [19, 18, 3].

The techniques are not especially dependent on the formal systems studied here, and it would be interesting to apply them to other impredicative logics.

Much of the work in the paper on cut-elimination for axiomatic extensions of ICTT is motivated by work in combining rewriting and sequent calculus by Dowek, Werner, Hardin, Kirchner, Hermant and others (deduction modulo) cited earlier in the paper. It is hoped that some of our results could be extended to full higher-order logic modulo, with a characterization of those rewrite rules that preserve cut-elimination.

Acknowledgements

The authors wish to thank Philip Scowcroft for his helpful discussions on intuitionistic semantics, and Chad Brown for comments and suggestions. This work has been partially supported by the Spanish projects Merit-Forms-UCM (TIN2005-09207-C03-03), Spanish MEC TIN-2005-09207 *MERIT*, and Promesas-CAM (S-0505/TIC/0407).

Olivier Hermant
Univ. Complutense de Madrid, Spain
ohermant@fdi.ucm.es

James Lipton
Wesleyan University, USA
and visiting Researcher,
Univ. Politécnica de Madrid, Spain
jlipton@wesleyan.edu

BIBLIOGRAPHY

[1] Peter Andrews. Resolution in Type Theory. *Journal of Symbolic Logic*, 36(3), 1971.
[2] Franz Baader and Tobias Nipkow. *Term rewriting and all that.* Cambridge University Press, New York, NY, USA, 1998.
[3] Richard Bonichon and Olivier Hermant. On Constructive Cut Admissibility in Deduction Modulo. In *Proceedings of the TYPES conference.* Springer-Verlag, 2007.
[4] H. C.M̃. de Swart. Another Intuitionistic Completeness Proof. *Journal of Symbolic Logic*, 41:644–662, 1976.
[5] M. DeMarco and J. Lipton. Completeness and Cut Elimination in the Intuitionistic Theory of Types. *Journal of Logic and Computation*, pages 821–854, November 2005.
[6] Gilles Dowek, Thérèse Hardin, and Claude Kirchner. Theorem proving modulo. *Journal of Automated Reasoning*, 31:33–72, 2003.
[7] Gilles Dowek and Benjamin Werner. Proof Normalization Modulo. *The Journal of Symbolic Logic*, 68(4):1289–1316, December 2003.
[8] Albert G. Dragalin. *Cut-Elimination Theorem for Higher-order Classical Logic*, pages 243–252. Plenum Press, New York and London, 1987.
[9] Harvey Friedman. Equality between Functionals. In R. Parikh, editor, *Logic Colloquium*, volume 453 of *Lecture Notes in Mathematics*, pages 22–37. Springer, 1975.
[10] J. Y. Girard. Une extension de l'interprétation de Gödel à l'analyse et son application à l'élimination de coupures dans l'analyse et la théorie des types. In J. E. Fenstad, editor, *Proceedings of the second Scandinavian proof theory symposium.* North–Holland, 1971.
[11] Leon Henkin. Completeness in the Theory of Types. *Journal of Symbolic Logic*, 15:81–91, June 1950.
[12] Olivier Hermant. *Méthodes Sémantiques en Déduction Modulo.* PhD thesis, Université Paris 7 – Denis Diderot, 2005.
[13] Georg Kreisel. A Remark on Free Choice Sequences and the Topological Completeness Proofs. *Journal of Symbolic Logic*, 23:369–388, 1958.
[14] Georg Kreisel. On Weak Completeness of Intuitionistic Predicate Logic. *Journal of Symbolic Logic*, 27:139–158, 1962.
[15] H. Läuchli. An Abstract Notion of Realizability for which Intuitionistic Predicate Calculus is Complete. In A. Kino, J. Myhill, and R. E. Vesley, editors, *Intuitionism and Proof Theory*, studies in logic, pages 277–234. North-holland, Amsterdam, London,

1970. Proceedings of the Conference on Intuitionism and Proof Theory, Buffalo, New York, August 1968.

[16] Dale Miller, Gopalan Nadathur, Frank Pfenning, and Andre Scedrov. Uniform Proofs as a Foundation for Logic Programming. *Annals of Pure and Applied Logic*, 51(1-2):125–157, 1991.

[17] John Mitchell. *Foundations for Programming Languages.* MIT Press, Cambridge, Massachusetts, 1996.

[18] Mitsuhiro Okada. Phase semantic cut-elimination and normalization proofs of first- and higher-order linear logic. *Theoretical Computer Science*, 227:333–396, 1999.

[19] Mitsuhiro Okada. A uniform semantic proof for cut-elimination and completeness of various first and higher order logics. *Theoretical Computer Science*, 281:471–498, 2002.

[20] Gordon Plotkin. Lambda definability in the full type hierarchy. In J.P. Seldin and J. R. Hindley, editors, *To H.B. Curry: Essays in Combinatory Logic, Lambda Calculus and Formalism.* Academic Press, New York, 1980.

[21] Dag Prawitz. Hauptsatz for Higher Order Logic. *The Journal of Symbolic Logic*, 33(3):452–457, September 1968.

[22] K. Schütte. Syntactical and semantical properties of simple Type Theory. *Journal of Symbolic Logic*, 25:305–326, 1960.

[23] K. Schütte. *Proof Theory.* Springer, 1977.

[24] W. Tait. A non-constructive proof of Gentzen's Hauptsatz for Second-Order Predicate Logic. *Bulletin of the American Mathematical Society*, 72:980–983, 1966.

[25] Moto-o Takahashi. A Proof of Cut-elimination in Simple Type Theory. *J. Math. Soc. Japan*, 19(4), 1967.

[26] A. S. Troelstra and D. van Dalen. *Constructivism in Mathematics: An Introduction*, volume 2. Elsevier Science Publishers, 1988.

[27] Dirk van Dalen. *Lectures on Intuitionism*, pages 1 – 94. Number 337 in Lecture Notes in Mathematics. Springer Verlag, 1973.

[28] W. Veldman. An Intuitionistic Completeness Theorem for Intuitionistic Predicate Logic. *Journal of Symbolic Logic*, 41:159–166, 1976.

M-Set Models

CHAD E. BROWN[1]

Dedicated to Peter Andrews

1 Introduction

In [1] Andrews studies elementary type theory, a form of Church's type theory [12] without extensionality, descriptions, choice, and infinity. Since most of the automated search procedures implemented in TPS [4] do not build in principles of extensionality, descriptions, choice or infinity, they are essentially searching for proofs in elementary type theory. In particular, search procedures based on Miller's expansion proofs correspond to proofs in elementary type theory extended with η-conversion. In [9] a model class $\mathfrak{M}_{\beta\eta}$ is defined and proven sound and complete with respect a natural deduction calculus corresponding to elementary type theory with η-conversion. One can add extensionality principles to automated search procedures [8, 10, 11] in order to target smaller (more restricted) model classes (as presented in [9]) which better approximate the class of standard models. Alternatively, one can construct interesting models in $\mathfrak{M}_{\beta\eta}$ which do not satisfy the full extensionality principles. One can then prove results about such models by proving theorems in the weaker logic. Suppose $\mathcal{M} \in \mathfrak{M}_{\beta\eta}$ and we want to know if some property P holds for $\mathcal{M}$. Suppose we can find a proposition $\mathbf{A}$ such that the property P holds if $\mathcal{M} \models \mathbf{A}$. We can conclude P holds if we prove the proposition $\mathbf{A}$ in elementary type theory with η-conversion.

Category theory can provide a Kripke-style semantics of intuitionistic higher-order logic [14, 15]. In particular, categories of presheaves are Cartesian closed (thus providing a semantics for simply-typed λ-calculus) and contain a subobject classifier (thus providing a semantics for intuitionistic logic). Since a one-object category is simply a monoid, a presheaf over a one-object category is simply a set with a monoid action (an M-set) [13, 15]. From these abstract considerations, we know that M-sets (for a fixed monoid M) provide a semantics for simply typed λ-calculus and intuitionistic higher-order logic.

[1]This research has been funded by SFB 378 Project DIALOG.

In this article we consider M-sets as a semantics for simply typed λ-calculus and fragments of *classical* higher-order logic. We can start with any M-set of interest and use this to interpret a base type of individuals. Function types are interpreted using the presheaf exponent. This will provide a means for interpreting simply typed λ-terms in a way that respects $\beta\eta$-equality. However, the ξ extensionality principle may not hold in general. The type of truth values need not be interpreted as the topos subobject classifier. Instead, the type of truth values can be any M-set $\mathcal{D}_o$ with a function ν from $\mathcal{D}_o$ into a two-element set $\{\mathtt{T}, \mathtt{F}\}$. We will consider two choices for $\mathcal{D}_o$ and ν. Once we have such an M-set model of classical higher-order logic, we could use a classical theorem prover which does not build in extensionality principles (such as TPS) to prove properties of the M-set model. To demonstrate this idea, we use TPS to prove a simple fixed point theorem and then construct an M-set model in which the fixed point theorem is meaningful. In order to appeal to a wide audience, we will exclusively use set-theoretic rather than category-theoretic language.

2 Motivation: A Proof in TPS

The higher-order theorem prover TPS has been under development under the leadership of Peter B. Andrews for several decades [7, 6, 4, 5]. TPS supports both automated proof search and interactive proof construction. The automated search procedures combine mating search with higher-order unification. The search procedures in TPS written before 2003 did not build in extensionality reasoning (except η-conversion). When TPS proves a proposition using one of these search procedures, then the proposition is a theorem of elementary type theory with η.

The logic of TPS is based on simple type theory, as described briefly below. More details are given in other sources [3, 9, 11]. We take the set $\mathcal{T}$ of *simple types* to be the same as in [12]. There are two base types o (of truth values), ι (of individuals), and a type $(\alpha\beta)$ of functions from β to α for all types α and β. The set of well-formed formulas of a type α depend on given sets of variables, parameters and logical constants. Let us fix a set $\mathcal{V}$ of typed variables and a set $\mathcal{P}$ of typed parameters. For each type α, $\mathcal{V}_\alpha$ and $\mathcal{P}_\alpha$ denote the subset of $\mathcal{V}$ and $\mathcal{P}$ of type α (respectively). We assume each $\mathcal{V}_\alpha$ is countably infinite. The logical constants we consider are those in the set $\mathcal{S}_{all}$ defined by

$$\{\top_o, \bot_o, \neg_{oo}, \wedge_{ooo}, \vee_{ooo}, \Rightarrow_{ooo}, \equiv_{ooo}\}$$
$$\cup \{\Pi^{\alpha}_{o(o\alpha)} | \alpha \in \mathcal{T}\} \cup \{\Sigma^{\alpha}_{o(o\alpha)} | \alpha \in \mathcal{T}\} \cup \{=^{\alpha}_{o\alpha\alpha} | \alpha \in \mathcal{T}\}.$$

We will consider a signature $\mathcal{S}$ of typed (logical) constants which may vary throughout the paper. We will always assume $\mathcal{S}$ is a subset of $\mathcal{S}_{all}$. The set of *well-formed formulas* (or *terms*) of type α over a signature $\mathcal{S}$ is denoted by $\mathit{wff}_\alpha(\mathcal{S})$ (or wff_α when the signature $\mathcal{S}$ is clear in context). We define each such family of sets inductively as follows:

- $x_\alpha \in \mathit{wff}_\alpha(\mathcal{S})$ for each variable $x_\alpha \in \mathcal{V}_\alpha$.
- $W_\alpha \in \mathit{wff}_\alpha(\mathcal{S})$ for each parameter $W_\alpha \in \mathcal{P}_\alpha$.
- $c_\alpha \in \mathit{wff}_\alpha(\mathcal{S})$ for each constant $c_\alpha \in \mathcal{S}_\alpha$.
- $[\mathbf{F}_{\alpha\beta}\,\mathbf{B}_\beta] \in \mathit{wff}_\alpha(\mathcal{S})$ for each $\mathbf{F} \in \mathit{wff}_{\alpha\beta}(\mathcal{S})$ and $\mathbf{B} \in \mathit{wff}_\beta(\mathcal{S})$.
- $[\lambda x_\beta \mathbf{A}_\alpha] \in \mathit{wff}_{\alpha\beta}(\mathcal{S})$ for each variable $x_\beta \in \mathcal{V}_\beta$ and $\mathbf{A} \in \mathit{wff}_\alpha(\mathcal{S})$.

The set $\mathbf{Free}(\mathbf{A}_\alpha) \subset \mathcal{V}$ of free variables in $\mathbf{A}$ is defined in the usual way. A term $\mathbf{A}_\alpha$ is *closed* if $\mathbf{Free}(\mathbf{A}_\alpha) = \emptyset$. Let $\mathit{cwff}_\alpha(\mathcal{S})$ (or cwff_α) be the set of all closed terms of type α. We use $\mathbf{A}^{\downarrow\beta}$ to refer to the β-normal form of $\mathbf{A}$ and $\mathbf{A}^{\downarrow}$ to refer to the $\beta\eta$-normal form of $\mathbf{A}$.

We now consider a simple example of a proof in TPS. Note that every individual is a fixed point of the identity function $[\lambda x_\iota\, x]$. For every individual i_ι, i is the unique fixed point of the identity function $[\lambda x_\iota\, i]$. If the only functions of type $\iota\iota$ are the constant functions and the identity function, then all such functions will have a fixed point. In fact, we can find a fixed point operator $Y_{\iota(\iota\iota)}$ taking each function $f_{\iota\iota}$ to a fixed point of f. The corresponding theorem can be proven formulated as

$$\forall P_{o(\iota\iota)}[P[\lambda x_\iota\, x] \wedge \forall i_\iota\, P[\lambda x\, i] \supset \forall f_{\iota\iota}\, P\, f] \supset \exists Y_{\iota(\iota\iota)} \forall f[f[Y\, f] = Y\, f] \quad (1)$$

This theorem can be proven automatically in TPS in less than a second. TPS also translates the proof into the natural deduction shown in Figure 1. The two nontrivial instantiations are shown in the justifications of lines (2) and (10) in Figure 1. In line (10) the fixed point operator Y is chosen to be the function taking f to $f\,u$ (where u is arbitrary). In order to prove $Y\,f$ is a fixed point, we use the hypothesis to prove $[f\,[f\,u]] = [f\,u]$. The corresponding instantiation for the predicate $P_{o\iota}$ is shown in line (2). Since this instantiation contains a logical symbol (equality at type ι), TPS must use a PRIMSUB (primitive substitution, see [2]) to prove the theorem automatically.

The conclusion of (1) may seem suspicious to many readers. Unless there is only one individual of type ι, there will of course be functions from individuals to individuals which do not have fixed points. On the other hand,

(1)	$1 \vdash$	$\forall P_{o(\iota\iota)}[P[\lambda x_\iota\, x] \wedge \forall i_\iota\, P[\lambda x\, i] \supset \forall f_{\iota\iota}\, P\, f]$	Hyp
(2)	$1 \vdash$	$[\lambda f_{\iota\iota}[f[f\, u_\iota] = f\, u]][\lambda x_\iota\, x] \wedge \forall i_\iota[\lambda f[f[f\, u] = f\, u]][\lambda x\, i]$ $\supset \forall f[\lambda f[f[f\, u] = f\, u]]\, f$	UI: $[\lambda f_{\iota\iota}[f[f\, u_\iota] = f\, u]]$ 1
(3)	$1 \vdash$	$u_\iota = u \wedge \forall i_\iota[i = i] \supset \forall f_{\iota\iota}[f[f\, u] = f\, u]$	Lambda: 2
(4)	$\vdash$	$u_\iota = u$	Assert REFL=
(5)	$\vdash$	$i_\iota = i$	Assert REFL=
(6)	$\vdash$	$\forall i_\iota[i = i]$	UGen: i_ι 5
(7)	$\vdash$	$u_\iota = u \wedge \forall i_\iota[i = i]$	RuleP: 4 6
(8)	$1 \vdash$	$\forall f_{\iota\iota}[f[f\, u_\iota] = f\, u]$	MP: 7 3
(9)	$1 \vdash$	$\forall f_{\iota\iota}[f[[\lambda f\, f\, u_\iota]\, f] = [\lambda f\, f\, u]\, f]$	Lambda: 8
(10)	$1 \vdash$	$\exists Y_{\iota(\iota\iota)} \forall f_{\iota\iota}[f[Y\, f] = Y\, f]$	EGen: $[\lambda f_{\iota\iota}\, f\, u_\iota]$ 9
(11)	$\vdash$	$\forall P_{o(\iota\iota)}[P[\lambda x_\iota\, x] \wedge \forall i_\iota\, P[\lambda x\, i] \supset \forall f_{\iota\iota}\, P\, f]$ $\supset \exists Y_{\iota(\iota\iota)} \forall f[f[Y\, f] = Y\, f]$	Deduct: 10

Figure 1. TPS Natural Deduction Proof of a Fixed Point Theorem

the hypothesis of (1) is also very strong. In standard set-theoretic semantics of type theory, both the hypothesis and conclusion of (1) can only be true if there is only one individual. However, TPS has proven (1) as a formal theorem of elementary type theory. Consequently, the theorem will be true in any model of elementary type theory. There are nontrivial models in which (1) is meaningful (in the sense that the hypothesis is valid in the model). In the next sections we will prove the existence of a class of M-set models. We will construct a particular M-set model in which (1) is meaningful in Section 6.

3 Semantics

We now summarize the semantic notions used in the paper. These notions are described in more detail in other sources [9, 11]. There are no new concepts introduced in this section.

An $\mathcal{S}$-model will be a tuple $\langle \mathcal{D}, @, \mathcal{E}, \nu \rangle$ where $\langle \mathcal{D}, @ \rangle$ is an applicative structure, $\mathcal{E}$ is an evaluation function interpreting terms in $\langle \mathcal{D}, @ \rangle$, and ν determines which members of $\mathcal{D}_o$ will be considered "true."

A (*typed*) *applicative structure* is a pair $\langle \mathcal{D}, @ \rangle$ where $\mathcal{D}$ is a typed family of nonempty sets and $@^{\alpha\beta} : \mathcal{D}_{\alpha\beta} \times \mathcal{D}_\beta \to \mathcal{D}_\alpha$ for each function type $(\alpha\beta)$. We write simply $\mathsf{f}@\mathsf{b}$ for $@^{\alpha\beta}(\mathsf{f}, \mathsf{b})$, leaving the types implicit. We call an applicative structure *functional* if for all types $\alpha, \beta \in \mathcal{T}$ and $\mathsf{f}, \mathsf{g} \in \mathcal{D}_{\alpha\beta}$, if $\mathsf{f}@\mathsf{b} = \mathsf{g}@\mathsf{b}$ for all $\mathsf{b} \in \mathcal{D}_\beta$, then $\mathsf{f} = \mathsf{g}$.

Let $\mathcal{D}$ be a typed family of nonempty sets. An *assignment* φ *into* $\mathcal{D}$ is a typed function $\varphi : \mathcal{V} \to \mathcal{D}$. $\varphi, [\mathsf{a}/x]$ denotes the assignment such that

$(\varphi, [\mathsf{a}/x])(x) = \mathsf{a}$ and $(\varphi, [\mathsf{a}/x])(y) = \varphi(y)$ for variables y other than x.

Let $\langle \mathcal{D}, @\rangle$ be an applicative structure. An $\mathcal{S}$-*evaluation function* for $\mathcal{A}$ is a function $\mathcal{E}$ taking any assignment φ into $\mathcal{D}$ and term $\mathbf{A} \in wff_\alpha(\mathcal{S})$ to $\mathcal{E}_\varphi(\mathbf{A}) \in \mathcal{D}_\alpha$ satisfying the following properties:

1. $\mathcal{E}_\varphi|_\mathcal{V} = \varphi$
2. $\mathcal{E}_\varphi([\mathbf{F}\,\mathbf{B}]) = \mathcal{E}_\varphi(\mathbf{F})@\mathcal{E}_\varphi(\mathbf{B})$ for any $\mathbf{F}$ in $wff_{\alpha\beta}(\mathcal{S})$ and $\mathbf{B}$ in $wff_\beta(\mathcal{S})$ and types α and β.
3. $\mathcal{E}_\varphi(\mathbf{A}) = \mathcal{E}_\psi(\mathbf{A})$ for any type α and $\mathbf{A} \in wff_\alpha(\mathcal{S})$, whenever φ and ψ coincide on $\mathbf{Free}(\mathbf{A})$.
4. $\mathcal{E}_\varphi(\mathbf{A}) = \mathcal{E}_\varphi(\mathbf{A}^{\downarrow\beta})$ for all $\mathbf{A} \in wff_\alpha(\mathcal{S})$.

The triple $\mathcal{J} = \langle \mathcal{D}, @, \mathcal{E}\rangle$ is an $\mathcal{S}$-*evaluation* if $\langle \mathcal{D}, @\rangle$ is an applicative structure and $\mathcal{E}$ is an $\mathcal{S}$-evaluation function for $\langle \mathcal{D}, @\rangle$. We say $\mathcal{J}$ is η-*functional* if

$$\mathcal{E}_\varphi(\mathbf{A}) = \mathcal{E}_\varphi(\mathbf{A}^{\downarrow})$$

for any type α, formula $\mathbf{A} \in wff_\alpha(\mathcal{S})$, and assignment φ. We say $\mathcal{J}$ is ξ-*functional* if for all types $\alpha, \beta \in \mathcal{T}$, $\mathbf{M}, \mathbf{N} \in wff_\beta(\mathcal{S})$, assignments φ, and variables x_α,

$$\mathcal{E}_\varphi(\lambda x_\alpha \mathbf{M}) = \mathcal{E}_\varphi(\lambda x_\alpha \mathbf{N})$$

whenever $\mathcal{E}_{\varphi,[\mathsf{a}/x]}(\mathbf{M}) = \mathcal{E}_{\varphi,[\mathsf{a}/x]}(\mathbf{N})$ for every $\mathsf{a} \in \mathcal{D}_\alpha$. We say $\mathcal{J}$ is *functional* if the underlying applicative structure is functional. As proven in [9] (for a particular $\mathcal{S}$) and [11] (for a general $\mathcal{S}$), an evaluation is functional iff it is both η-functional and ξ-functional.

For the rest of the paper, we fix two distinct values $\mathtt{T} \neq \mathtt{F}$. Let $\mathcal{A} := \langle \mathcal{D}, @\rangle$ be an applicative structure and $\nu\colon \mathcal{D}_o \to \{\mathtt{T}, \mathtt{F}\}$ be a function. For each logical constant c_α and element $\mathsf{a} \in \mathcal{D}_\alpha$, we define properties $\mathfrak{L}_c(\mathsf{a})$ with respect to ν in Table 1. Roughly speaking, the property $\mathfrak{L}_c(\mathsf{a})$ means a behaves like the logical constant c modulo ν.

Let $\mathcal{J} := \langle \mathcal{D}, @, \mathcal{E}\rangle$ be an evaluation. We call a function $\nu\colon \mathcal{D}_o \to \{\mathtt{T}, \mathtt{F}\}$ an $\mathcal{S}$-*valuation* for $\mathcal{J}$ if for every logical constant $c \in \mathcal{S}$, $\mathfrak{L}_c(\mathcal{E}(c))$ holds with respect to ν. In such a case, we call $\mathcal{M} := \langle \mathcal{D}, @, \mathcal{E}, \nu\rangle$ an $\mathcal{S}$-*model* (or simply a *model* when $\mathcal{S}$ is clear in context). A model $\mathcal{M} := \langle \mathcal{D}, @, \mathcal{E}, \nu\rangle$ is called *functional* if the applicative structure $\langle \mathcal{D}, @\rangle$ is functional. We say $\mathcal{M}$ is η-*functional* [ξ-*functional*] if the evaluation $\langle \mathcal{D}, @, \mathcal{E}\rangle$ is η-functional [ξ-*functional*]. We define five properties a model $\mathcal{M}$ can satisfy in Table 2. Properties η, ξ, and $\mathfrak{f}$ are forms of functional extensionality. Property $\mathfrak{b}$ is a

prop.	where	holds when			for all
$\mathfrak{L}_{\top}(\mathsf{a})$	$\mathsf{a} \in \mathcal{D}_o$	$\nu(\mathsf{a}) = \mathrm{T}$			
$\mathfrak{L}_{\bot}(\mathsf{b})$	$\mathsf{b} \in \mathcal{D}_o$	$\nu(\mathsf{b}) = \mathrm{F}$			
$\mathfrak{L}_{\neg}(\mathsf{n})$	$\mathsf{n} \in \mathcal{D}_{oo}$	$\nu(\mathsf{n}@\mathsf{a}) = \mathrm{T}$	iff	$\nu(\mathsf{a}) = \mathrm{F}$	$\mathsf{a} \in \mathcal{D}_o$
$\mathfrak{L}_{\vee}(\mathsf{d})$	$\mathsf{d} \in \mathcal{D}_{ooo}$	$\nu(\mathsf{d}@\mathsf{a}@\mathsf{b}) = \mathrm{T}$	iff	$\nu(\mathsf{a}) = \mathrm{T}$ or $\nu(\mathsf{b}) = \mathrm{T}$	$\mathsf{a}, \mathsf{b} \in \mathcal{D}_o$
$\mathfrak{L}_{\wedge}(\mathsf{c})$	$\mathsf{c} \in \mathcal{D}_{ooo}$	$\nu(\mathsf{c}@\mathsf{a}@\mathsf{b}) = \mathrm{T}$	iff	$\nu(\mathsf{a}) = \mathrm{T}$ and $\nu(\mathsf{b}) = \mathrm{T}$	$\mathsf{a}, \mathsf{b} \in \mathcal{D}_o$
$\mathfrak{L}_{\Rightarrow}(\mathsf{i})$	$\mathsf{i} \in \mathcal{D}_{ooo}$	$\nu(\mathsf{i}@\mathsf{a}@\mathsf{b}) = \mathrm{T}$	iff	$\nu(\mathsf{a}) = \mathrm{F}$ or $\nu(\mathsf{b}) = \mathrm{T}$	$\mathsf{a}, \mathsf{b} \in \mathcal{D}_o$
$\mathfrak{L}_{\equiv}(\mathsf{e})$	$\mathsf{e} \in \mathcal{D}_{ooo}$	$\nu(\mathsf{e}@\mathsf{a}@\mathsf{b}) = \mathrm{T}$	iff	$\nu(\mathsf{a}) = \nu(\mathsf{b})$	$\mathsf{a}, \mathsf{b} \in \mathcal{D}_o$
$\mathfrak{L}_{\Pi^\alpha}(\pi)$	$\pi \in \mathcal{D}_{o(o\alpha)}$	$\nu(\pi@\mathsf{f}) = \mathrm{T}$	iff	$\forall \mathsf{a} \in \mathcal{D}_\alpha\ \nu(\mathsf{f}@\mathsf{a}) = \mathrm{T}$	$\mathsf{f} \in \mathcal{D}_{o\alpha}$
$\mathfrak{L}_{\Sigma^\alpha}(\sigma)$	$\sigma \in \mathcal{D}_{o(o\alpha)}$	$\nu(\sigma@\mathsf{f}) = \mathrm{T}$	iff	$\exists \mathsf{a} \in \mathcal{D}_\alpha\ \nu(\mathsf{f}@\mathsf{a}) = \mathrm{T}$	$\mathsf{f} \in \mathcal{D}_{o\alpha}$
$\mathfrak{L}_{=^\alpha}(\mathsf{q})$	$\mathsf{q} \in \mathcal{D}_{o\alpha\alpha}$	$\nu(\mathsf{q}@\mathsf{a}@\mathsf{b}) = \mathrm{T}$	iff	$\mathsf{a} = \mathsf{b}$	$\mathsf{a}, \mathsf{b} \in \mathcal{D}_\alpha$

Table 1. Logical Properties of $\nu : \mathcal{D}_o \to \{\mathrm{T}, \mathrm{F}\}$

$\mathcal{M}$ satisfies property	when
η	$\mathcal{M}$ is η-functional.
ξ	$\mathcal{M}$ is ξ-functional.
$\mathfrak{f}$	$\mathcal{M}$ is functional.
$\mathfrak{b}$	ν is injective.
$\mathfrak{q}$	for all types α there is a $\mathsf{q}^\alpha \in \mathcal{D}_{o\alpha\alpha}$ such that $\mathfrak{L}_{=^\alpha}(\mathsf{q}^\alpha)$ holds.

Table 2. Properties of Models

form of Boolean extensionality. Property $\mathfrak{q}$ is a requirement that the model realize equality at all types.

4 M-set Models

Let M be a monoid with identity e. An M-set is a set A with an *action* giving an element $\mathsf{a}m \in A$ for each $\mathsf{a} \in A$ and $m \in M$ such that

- $(\mathsf{a}m)n = \mathsf{a}(mn)$
- $\mathsf{a}e = \mathsf{a}$

for all $\mathsf{a} \in A$ and $m, n \in M$. Given two M-sets A and B, we can define the M-set exponent $A^{B,M}$ as the set

$$A^{B,M} := \{\mathsf{f} : M \times B \to A | \forall k, m \in M \forall \mathsf{b} \in B. \mathsf{f}(k, \mathsf{b})m = \mathsf{f}(km, \mathsf{b}m)\} \quad (2)$$

with action taking $\mathsf{f} \in A^{B,M}$ and $m \in M$ to the function $\mathsf{f}m : M \times B \to A$ defined by

$$\mathsf{f}m(k, \mathsf{b}) := \mathsf{f}(mk, \mathsf{b}). \quad (3)$$

Note that $\mathsf{f}m$ is in $A^{B,M}$ since for $k, n \in M$ and $\mathsf{b} \in B$

$$\mathsf{f}m(k, \mathsf{b})n = \mathsf{f}(mk, \mathsf{b})n = \mathsf{f}(mkn, \mathsf{b}n) = \mathsf{f}m(kn, \mathsf{b}n).$$

To check that $A^{B,M}$ is an M-set, we must ensure $(\mathsf{f}m)n = \mathsf{f}(mn)$ and $\mathsf{f}e = \mathsf{f}$. Both facts are easily verified:

$$((\mathsf{f}m)n)(k, \mathsf{b}) = (\mathsf{f}m)(nk, \mathsf{b}) = \mathsf{f}(m(nk), \mathsf{b}) = \mathsf{f}((mn)k, \mathsf{b}) = (\mathsf{f}(mn))(k, \mathsf{b})$$

$$(\mathsf{f}e)(k, \mathsf{b}) = \mathsf{f}(ek, \mathsf{b}) = \mathsf{f}(k, \mathsf{b})$$

DEFINITION 1. Let M be a monoid with identity e. An *M-set applicative structure* is a pair $\langle \mathcal{D}, @ \rangle$ where

- $\langle \mathcal{D}, @ \rangle$ is an applicative structure,
- $\mathcal{D}_\alpha$ is an M-set for each type $\alpha \in \mathcal{T}$,
- $\mathcal{D}_{\alpha\beta}$ is the M-set ${\mathcal{D}_\alpha}^{\mathcal{D}_\beta, M}$ for all types $\alpha, \beta \in \mathcal{T}$, and
- $(\mathsf{f}@\mathsf{b}) = \mathsf{f}(e, \mathsf{b})$ for $\mathsf{f} \in \mathcal{D}_{\alpha\beta}$ and $\mathsf{b} \in \mathcal{D}_\beta$ where $\alpha, \beta \in \mathcal{T}$.

We can specify an M-set applicative structure by giving two nonempty M-sets for the two base types.

THEOREM 2. *Let M be a monoid with identity $e \in M$. If A and B are nonempty M-sets, then there is a unique M-set applicative structure $\mathcal{A} = \langle \mathcal{D}, @ \rangle$ such that $\mathcal{D}_\iota = A$ and $\mathcal{D}_o = B$.*

Proof. We define $\mathcal{D}_\alpha$ by induction on α as follows:

- $\mathcal{D}_\iota := A$
- $\mathcal{D}_o := B$
- $\mathcal{D}_{\alpha\beta} := {\mathcal{D}_\alpha}^{\mathcal{D}_\beta, M}$

Let @ be defined by $(\mathsf{f}@\mathsf{b}) := \mathsf{f}(e, \mathsf{b})$ for $\mathsf{f} \in \mathcal{D}_{\alpha\beta}$ and $\mathsf{b} \in \mathcal{D}_\beta$ where $\alpha, \beta \in \mathcal{T}$. We must verify that $\mathcal{A} := \langle \mathcal{D}, @ \rangle$ is an M-set applicative structure with $\mathcal{D}_\iota = A$ and $\mathcal{D}_o = B$. The only nontrivial property to check is that each $\mathcal{D}_\alpha$ is nonempty. We prove this by induction on types. We have assumed $\mathcal{D}_\beta$ is nonempty for base types $\beta \in \{\iota, o\}$. Assume $\mathcal{D}_\alpha$ is nonempty. Choose some $\mathsf{a} \in \mathcal{D}_\alpha$. Let $\mathsf{f} : M \times \mathcal{D}_\beta \to \mathcal{D}_\alpha$ be the function defined by $\mathsf{f}(k, \mathsf{b}) = \mathsf{a}k$.

In order to conclude $\mathcal{D}_{\alpha\beta}$ is nonempty, we check $\mathsf{f} \in \mathcal{D}_{\alpha\beta}$. We must check $\mathsf{f}(k, \mathsf{b})m = \mathsf{f}(km, \mathsf{b}m)$ for $k, m \in M$ and $\mathsf{b} \in \mathcal{D}_\beta$. This is easy:

$$\mathsf{f}(k, \mathsf{b})m = (\mathsf{a}k)m = \mathsf{a}(km) = \mathsf{f}(km, \mathsf{b}m).$$

In order to show $\mathcal{A}$ is unique, suppose $\mathcal{A}' = \langle \mathcal{D}', @' \rangle$ is an M-set applicative structure such that $\mathcal{D}'_\iota = A$ and $\mathcal{D}'_o = B$. An easy induction on α proves each M-set $\mathcal{D}_\alpha$ is equal to $\mathcal{D}'_\alpha$. Given this fact, we know @ and @' must coincide as well. ■

We now define an action on the set of assignments in an obvious way.

DEFINITION 3. Let M be a monoid and $\mathcal{A} = \langle \mathcal{D}, @ \rangle$ be an M-set applicative structure. For any assignment $\varphi : \mathcal{V} \to \mathcal{D}$ into $\mathcal{D}$ and $k \in M$, we let $\varphi k : \mathcal{V} \to \mathcal{D}$ denote the assignment given by $\varphi k(x) := \varphi(x)k$ for each variable x.

An evaluation function maps terms to values in an applicative structure. In order to obtain evaluation functions which respect the actions of an M-set applicative structure, we consider an M-indexed family of evaluation functions.

DEFINITION 4. Let M be a monoid and $\mathcal{A} = \langle \mathcal{D}, @ \rangle$ be an M-set applicative structure. An *M-set family of $\mathcal{S}$-evaluation functions for $\mathcal{A}$* is a family $(\mathcal{E}^m)_{m \in M}$ of functions satisfying the following properties:

1. $\mathcal{E}^m_\varphi(x) = \varphi(x)$ for $x \in \mathcal{V}$.
2. $\mathcal{E}^m_\varphi([\mathbf{F}_{\alpha\beta}\, \mathbf{B}_\beta]) = \mathcal{E}^m_\varphi(\mathbf{F}) @ \mathcal{E}^m_\varphi(\mathbf{B})$.
3. $\mathcal{E}^m_\varphi(w) = \mathcal{E}^e_\psi(w)m$ for $w \in \mathcal{P} \cup \mathcal{S}$ and any assignments φ and ψ.
4. $\mathcal{E}^m_\varphi(\lambda x_\beta \mathbf{A}_\alpha) = \mathsf{f} \in \mathcal{D}_{\alpha\beta}$ where f is the function such that

$$\mathsf{f}(k, \mathsf{b}) = \mathcal{E}^{mk}_{\varphi k, [\mathsf{b}/x]}(\mathbf{A}).$$

Note that in Definition 4 we have not actually required each $\mathcal{E}^m$ to be an $\mathcal{S}$-evaluation function for $\mathcal{A}$. The fact that each $\mathcal{E}^m$ is such an $\mathcal{S}$-evaluation function follows from the conditions in Definition 4. The first two conditions in Definition 4 correspond directly to the first two conditions in the definition of evaluation functions. All four conditions in Definition 4 are used (together) to verify the remaining two conditions in the definition of evaluation functions.

THEOREM 5. *Let M be a monoid with identity e, $\mathcal{A} = \langle \mathcal{D}, @ \rangle$ be an M-set applicative structure and $(\mathcal{E}^m)_{m \in M}$ be an M-set family of $\mathcal{S}$-evaluation* *functions for $\mathcal{A}$. For each $m \in M$, $\mathcal{E}^m$ is an η-functional $\mathcal{S}$-evaluation function for $\mathcal{A}$. Furthermore, for any $m, n \in M$, assignment φ, and term $\mathbf{A} \in wff_\alpha(\mathcal{S})$, we have*

$$\mathcal{E}^m_\varphi(\mathbf{A})n = \mathcal{E}^{mn}_{\varphi n}(\mathbf{A}).$$

Proof. See Appendix A. ∎

Since such evaluation functions are η-functional, they will be ξ-functional iff the underlying applicative structure is functional. It is not difficult to show that the underlying applicative structure will be functional if M is a group. The more interesting case is when M is not a group. If M is not a group then using the theorems above we can construct evaluations in which η-functionality holds but ξ-functionality fails (see Example 11).

Just as we can specify an M-set applicative structure by giving nonempty M-sets for the two base types, we can specify an M-set family of $\mathcal{S}$-evaluation functions by interpreting the parameters and constants.

THEOREM 6. *Let M be a monoid with identity e, $\mathcal{A} = \langle \mathcal{D}, @ \rangle$ be an M-set applicative structure and $\mathcal{I} : (\mathcal{P} \cup \mathcal{S}) \rightarrow \mathcal{D}$ be a typed function. There is a unique M-set family of evaluation functions $(\mathcal{E}^{\mathcal{I},m})_{m \in M}$ such that $\mathcal{E}^{\mathcal{I},e}_\varphi(w) = \mathcal{I}(w)$ for all $w \in \mathcal{P} \cup \mathcal{S}$ and assignments φ.*

Proof. See Appendix B. ∎

An M-set model will be an M-set applicative structure with an evaluation function which is part of an M-set family of evaluation functions along with a function $\nu : \mathcal{D}_o \rightarrow \{\mathtt{T}, \mathtt{F}\}$ which respects the interpretations of the logical constants in $\mathcal{S}$.

DEFINITION 7. Let M be a monoid with identity e. An *M-set $\mathcal{S}$-model* (or, *M-set model*) is an $\mathcal{S}$-model $\langle \mathcal{D}, @, \mathcal{E}, \nu \rangle$ where $\langle \mathcal{D}, @ \rangle$ is an M-set applicative structure and there is an M-set family of evaluation functions $(\mathcal{E}^m)_{m \in M}$ such that $\mathcal{E}_\varphi(\mathbf{A}) = \mathcal{E}^e_\varphi(\mathbf{A})$ for all terms $\mathbf{A} \in wff_\alpha(\mathcal{S})$ and assignments φ.

We can specify an M-set $\mathcal{S}$-model by giving the M-set applicative structure, valuation ν, and an interpretation of parameters and logical constants which respects the properties of the logical constants.

THEOREM 8. *Let M be a monoid with identity e, $\mathcal{A} = \langle \mathcal{D}, @ \rangle$ be an M-set applicative structure, $\mathcal{I} : (\mathcal{P} \cup \mathcal{S}) \rightarrow \mathcal{D}$ be a typed function and $\nu : \mathcal{D}_o \rightarrow \{\mathtt{T}, \mathtt{F}\}$ be a function such that $\mathfrak{L}_c(\mathcal{I}(c))$ holds for all $c \in \mathcal{S}$. Let $(\mathcal{E}^{\mathcal{I},m})_{m \in M}$ be the M-set family of evaluation functions given by Theorem 6. Then $\mathcal{M} := \langle \mathcal{D}, @, \mathcal{E}^{\mathcal{I},e}, \nu \rangle$ is an M-set $\mathcal{S}$-model satisfying property η.*

Proof. This is an obvious consequence of Theorems 5 and 6. ∎

5 Interpreting the Type of Truth Values

In order to apply Theorem 8, we must give an interpretation $\mathcal{I}$ for all parameters and logical constants. The interpretation $\mathcal{I}(c)$ of each logical constant $c \in \mathcal{S}$ must satisfy the corresponding logical property $\mathfrak{L}_c(\mathcal{I}(c))$ with respect to ν. Given an applicative structure $\mathcal{A}$ and function ν, it may be the case that no value a satisfies $\mathfrak{L}_c(\mathsf{a})$ with respect to ν. In such a case there are no M-set $\mathcal{S}$-models (where $c \in \mathcal{S}$) over this applicative structure using this valuation ν. If there *is* such an a, we will say $\mathcal{A}$ realizes c with respect to ν.

DEFINITION 9. Let $\mathcal{A} = \langle \mathcal{D}, @ \rangle$ be an applicative structure, $\nu : \mathcal{D}_o \rightarrow \{\mathtt{T}, \mathtt{F}\}$ be a function and c_α be a logical constant. We say $\mathcal{A}$ realizes c with respect to ν if there is some $\mathsf{a} \in \mathcal{D}_\alpha$ such that $\mathfrak{L}_c(\mathsf{a})$ holds with respect to ν.

There are several options one can choose for $\mathcal{D}_o$ and $\nu : \mathcal{D}_o \rightarrow \{\mathtt{T}, \mathtt{F}\}$. This choice affects which logical constants will be realized. If we further want the model to satisfy property $\mathfrak{b}$ (Boolean extensionality), then the following choice is all but forced upon us:

- Set: $\mathcal{D}_o := \{\mathtt{T}, \mathtt{F}\}$
- Action: $\mathtt{T}m = \mathtt{T}$ and $\mathtt{F}m = \mathtt{F}$ (the *trivial action*).
- Valuation: ν is the identity – i.e., $\nu(\mathtt{T}) = \mathtt{T}$ and $\nu(\mathtt{F}) = \mathtt{F}$

In fact, using this simple choice we can realize all logical constants except equality. We can only realize equality in such a model if a certain cancellation law holds.

THEOREM 10. *Let M be a monoid with identity e and $\mathcal{A} = \langle \mathcal{D}, @ \rangle$ be an M-set applicative structure. Suppose*

- $\mathcal{D}_o = \{\mathtt{T}, \mathtt{F}\}$,
- $\mathtt{T}m = \mathtt{T}$ *and* $\mathtt{F}m = \mathtt{F}$.

Let $\nu : \mathcal{D}_o \rightarrow \{\mathtt{T}, \mathtt{F}\}$ be the identity function. Each logical constant in the set

$$\{\top_o, \bot_o, \neg_{oo}, \wedge_{ooo}, \vee_{ooo}, \Rightarrow_{ooo}, \equiv_{ooo}\}$$
$$\cup \{\Pi^\alpha_{o(o\alpha)} | \alpha \in \mathcal{T}\} \cup \{\Sigma^\alpha_{o(o\alpha)} | \alpha \in \mathcal{T}\}$$

is realized in $\mathcal{A}$ with respect to ν. Furthermore, for each $\alpha \in \mathcal{T}$, $=^\alpha$ is realized by $\mathcal{A}$ with respect to ν iff the following cancellation law holds:

$$\forall \mathsf{a}, \mathsf{b} \in \mathcal{D}_\alpha \forall m \in M \;\; \textit{if } \mathsf{a}m = \mathsf{b}m, \textit{ then } \mathsf{a} = \mathsf{b}.$$

Proof. Obviously $\top$ and $\bot$ are realized using T and F, respectively. Let $\mathsf{n} : M \times \mathcal{D}_o \to \mathcal{D}_o$ be the function defined by

$$\mathsf{n}(m, \mathsf{b}) := \begin{cases} \mathsf{T} & \text{if } \mathsf{b} = \mathsf{F} \\ \mathsf{F} & \text{otherwise.} \end{cases}$$

We easily verify $\mathsf{n} \in \mathcal{D}_{oo}$:

$$\mathsf{n}(m, \mathsf{b})k = \mathsf{n}(m, \mathsf{b}) = \mathsf{n}(mk, \mathsf{b}) = \mathsf{n}(mk, \mathsf{b}k).$$

It is also clear that $\mathfrak{L}_\neg(\mathsf{n})$ holds with respect to ν:

$$\mathsf{n}@\mathsf{b} = \mathsf{T} \Leftrightarrow \mathsf{n}(e, \mathsf{b}) = \mathsf{T} \Leftrightarrow \mathsf{b} = \mathsf{F}.$$

Let $\mathsf{d} : M \times \mathcal{D}_o \to M \times \mathcal{D}_o \to \mathcal{D}_o$ be

$$\mathsf{d}(m, \mathsf{b})(n, \mathsf{c}) := \begin{cases} \mathsf{F} & \text{if } \mathsf{b} = \mathsf{F} = \mathsf{c} \\ \mathsf{T} & \text{otherwise.} \end{cases}$$

For any $k \in M$, $\mathsf{d}(m, \mathsf{b})(n, \mathsf{c})k = \mathsf{F}$ iff $\mathsf{b} = \mathsf{F} = \mathsf{c}$ iff $\mathsf{b} = \mathsf{F} = \mathsf{c}k$ iff $\mathsf{d}(m, \mathsf{b})(nk, \mathsf{c}k) = \mathsf{F}$. Hence $\mathsf{d}(m, \mathsf{b})(n, \mathsf{c})k = \mathsf{d}(m, \mathsf{b})(nk, \mathsf{c}k)$ and $\mathsf{d}(m, \mathsf{b}) \in \mathcal{D}_{oo}$. Similarly, for any $k \in M$, $\mathsf{d}(m, \mathsf{b})k(n, \mathsf{c}) = \mathsf{F}$ iff $\mathsf{d}(m, \mathsf{b})(kn, \mathsf{c}) = \mathsf{F}$ iff $\mathsf{b} = \mathsf{F} = \mathsf{c}$ iff $\mathsf{b}k = \mathsf{F} = \mathsf{c}$ iff $\mathsf{d}(mk, \mathsf{b}k)(n, \mathsf{c}) = \mathsf{F}$. Hence $\mathsf{d}(m, \mathsf{b})k = \mathsf{d}(mk, \mathsf{b}k)$ and $\mathsf{d} \in \mathcal{D}_{ooo}$. Clearly, $\mathfrak{L}_\vee(\mathsf{d})$ holds.

Since $\neg$ and $\vee$ are realized, we can conclude that $\wedge$, $\Rightarrow$ and $\equiv$ must also be realized. Similarly, to show that each Σ^α is realized, we can simply show each Π^α is realized.

Let $\pi : M \times \mathcal{D}_{o\alpha} \to \mathcal{D}_o$ be defined by

$$\pi(m, \mathsf{f}) := \begin{cases} \mathsf{T} & \text{if } \forall \mathsf{a} \in \mathcal{D}_\alpha\, \mathsf{f}(e, \mathsf{a}m) = \mathsf{T} \\ \mathsf{F} & \text{otherwise.} \end{cases}$$

To check $\pi \in \mathcal{D}_{o(o\alpha)}$, we must prove

$$\pi(m, \mathsf{f})k = \pi(mk, \mathsf{f}k)$$

for $m, k \in M$ and $\mathsf{f} \in \mathcal{D}_{o\alpha}$. Note that

$$\begin{aligned} \pi(m, \mathsf{f})k = \mathsf{T} \quad & \text{iff} \quad \pi(m, \mathsf{f}) = \mathsf{T} \\ & \text{iff} \quad \forall \mathsf{a} \in \mathcal{D}_\alpha\, \mathsf{f}(e, \mathsf{a}m) = \mathsf{T} \\ & \text{iff} \quad \forall \mathsf{a} \in \mathcal{D}_\alpha\, \mathsf{f}(e, \mathsf{a}m)k = \mathsf{T} \\ & \text{iff} \quad \forall \mathsf{a} \in \mathcal{D}_\alpha\, \mathsf{f}(k, \mathsf{a}mk) = \mathsf{T} \\ & \text{iff} \quad \forall \mathsf{a} \in \mathcal{D}_\alpha\, (\mathsf{f}k)(e, \mathsf{a}mk) = \mathsf{T} \\ & \text{iff} \quad \pi(mk, \mathsf{f}k) = \mathsf{T}. \end{aligned}$$

Hence $\pi(m, \mathsf{f})k = \pi(mk, \mathsf{f}k)$ and so $\pi \in \mathcal{D}_{o(o\alpha)}$. Note that $\pi@\mathsf{f} = \mathtt{T}$ iff $\forall \mathsf{a} \in \mathcal{D}_\alpha\, \mathsf{f}@\mathsf{a} = \mathsf{f}(e, \mathsf{a}) = \mathtt{T}$. Thus $\mathfrak{L}_{\Pi^\alpha}(\pi)$ holds.

Finally, we turn our attention to equality. First, suppose there is some $\mathsf{q} \in \mathcal{D}_{o\alpha\alpha}$ realizing $=^\alpha$, i.e., such that

$$\mathsf{q}@\mathsf{a}@\mathsf{b} = \mathtt{T} \Leftrightarrow \mathsf{a} = \mathsf{b}$$

for $\mathsf{a}, \mathsf{b} \in \mathcal{D}_\alpha$. Assume there is some $\mathsf{a}, \mathsf{b} \in \mathcal{D}_\alpha$ and $m \in M$ such that $\mathsf{a}m = \mathsf{b}m$ but $\mathsf{a} \neq \mathsf{b}$. We can compute $\mathtt{F} = \mathtt{T}$, a contradiction, as follows:

$$\begin{aligned} \mathtt{F} = \mathsf{q}@\mathsf{a}@\mathsf{b} = \mathsf{q}(e, \mathsf{a})(e, \mathsf{b}) &= \mathsf{q}(e, \mathsf{a})(e, \mathsf{b})m = \mathsf{q}(e, \mathsf{a})(m, \mathsf{b}m) \\ &= (\mathsf{q}(e, \mathsf{a})m)(e, \mathsf{b}m) = \mathsf{q}(m, \mathsf{a}m)(e, \mathsf{b}m) \\ &= \mathsf{q}(m, \mathsf{b}m)(e, \mathsf{b}m) = (\mathsf{q}(e, \mathsf{b})m)(e, \mathsf{b}m) \\ = \mathsf{q}(e, \mathsf{b})(m, \mathsf{b}m) &= \mathsf{q}(e, \mathsf{b})(e, \mathsf{b})m = \mathsf{q}(e, \mathsf{b})(e, \mathsf{b}) = \mathtt{T} \end{aligned}$$

Conversely, suppose that the cancellation law holds in $\mathcal{D}_\alpha$. We define $\mathsf{q} : M \times \mathcal{D}_\alpha \to M \times \mathcal{D}_\alpha \to \mathcal{D}_o$ (realizing $=^\alpha$) by

$$\mathsf{q}(m, \mathsf{a})(n, \mathsf{b}) = \begin{cases} \mathtt{T} & \text{if } \exists \mathsf{c} \in \mathcal{D}_\alpha \text{ such that } \mathsf{a} = \mathsf{c}m \text{ and } \mathsf{b} = \mathsf{c}mn \\ \mathtt{F} & \text{otherwise.} \end{cases}$$

Let $k \in M$ be given and suppose $\mathsf{q}(m, \mathsf{a})(n, \mathsf{b})k = \mathtt{T}$. Then $\mathsf{q}(m, \mathsf{a})(n, \mathsf{b}) = \mathtt{T}$ and there is some $\mathsf{c} \in \mathcal{D}_\alpha$ such that $\mathsf{a} = \mathsf{c}m$ and $\mathsf{b} = \mathsf{c}mn$. Since $\mathsf{b}k = \mathsf{c}mnk$, c also witnesses $\mathsf{q}(m, \mathsf{a})(nk, \mathsf{b}k) = \mathtt{T}$. For the converse, suppose $\mathsf{q}(m, \mathsf{a})(nk, \mathsf{b}k) = \mathtt{T}$. Then there is some $\mathsf{c} \in \mathcal{D}_\alpha$ such that $\mathsf{a} = \mathsf{c}m$ and $\mathsf{b}k = \mathsf{c}mnk$. By the cancellation law, $\mathsf{b} = \mathsf{c}mn$ and so c witnesses $\mathsf{q}(m, \mathsf{a})(n, \mathsf{b}) = \mathtt{T}$. Hence $\mathsf{q}(m, \mathsf{a})(n, \mathsf{b})k = \mathsf{q}(m, \mathsf{a})(nk, \mathsf{b}k)$ and so $\mathsf{q}(m, \mathsf{a}) \in \mathcal{D}_{o\alpha}$. Next we prove $\mathsf{q} \in \mathcal{D}_{o\alpha\alpha}$ by proving $\mathsf{q}(m, \mathsf{a})k = \mathsf{q}(mk, \mathsf{a}k)$. Suppose $\mathsf{q}(m, \mathsf{a})k(n, \mathsf{b}) = \mathtt{T}$. That is, there is some $\mathsf{c} \in \mathcal{D}_\alpha$ such that $\mathsf{a} = \mathsf{c}m$ and $\mathsf{b} = \mathsf{c}mkn$. Since $\mathsf{a}k = \mathsf{c}mk$, this proves $\mathsf{q}(mk, \mathsf{a}k)(n, \mathsf{b}) = \mathtt{T}$. Finally, suppose $\mathsf{q}(mk, \mathsf{a}k)(n, \mathsf{b}) = \mathtt{T}$. Then for some $\mathsf{c} \in \mathcal{D}_\alpha$, $\mathsf{a}k = \mathsf{c}mk$ and $\mathsf{b} = \mathsf{c}mkn$. By the cancellation law, $\mathsf{a} = \mathsf{c}m$ and so $\mathsf{q}(m, \mathsf{a})(kn, \mathsf{b}) = \mathtt{T}$. Hence $\mathsf{q}(m, \mathsf{a})k = \mathsf{q}(mk, \mathsf{a}k)$ and so $\mathsf{q} \in \mathcal{D}_{o\alpha\alpha}$. Note that $\mathfrak{L}_{=^\alpha}(\mathsf{q})$ holds since $\mathsf{q}@\mathsf{a}@\mathsf{b} = \mathtt{T}$ iff $\exists \mathsf{c} \in \mathcal{D}_\alpha$ such that $\mathsf{a} = \mathsf{c} = \mathsf{b}$ iff $\mathsf{a} = \mathsf{b}$. ■

The fact that equality may not be realized by such a model is a problem, but can be overcome. Using Theorem 3.62 from [9] we can obtain a model which realizes equality by taking a quotient by the congruence relation induced by Leibniz equality. A different problem with this choice for $\mathcal{D}_o$ is that the sets $\mathcal{D}_{o\alpha}$ may be too small. In particular, as we show in the next example, the subsets of $\mathcal{D}_\iota$ represented in $\mathcal{D}_{o\iota}$ may be quite sparse.

EXAMPLE 11. Let M_2 be the monoid $\{0, 1\}$ under multiplication where 1 is the identity. Let $\mathcal{A} = \langle \mathcal{D}, @ \rangle$ be the M_2-set applicative structure given by

Theorem 2 such that $\mathcal{D}_\iota$ is the M_2-set $\{0,1\}$ with action by multiplication and $\mathcal{D}_o$ is the M_2-set $\{\mathtt{T},\mathtt{F}\}$ with the trivial action. Let ν be the identity function. Note that the cancellation law does not hold in $\mathcal{D}_\iota$ since $1\cdot 0 = 0\cdot 0$ but $1 \neq 0$. By Theorem 10, $=^\iota$ is not realized in $\mathcal{A}$ with respect to ν. Let $\mathcal{S}$ be the set

$$\{\top_o, \bot_o, \neg_{oo}, \wedge_{ooo}, \vee_{ooo}, \Rightarrow_{ooo}, \equiv_{ooo}\}$$
$$\cup\{\Pi^\alpha_{o(o\alpha)} | \alpha \in \mathcal{T}\} \cup \{\Sigma^\alpha_{o(o\alpha)} | \alpha \in \mathcal{T}\}$$

By Theorem 10 each logical constant in $\mathcal{S}$ is realized in $\mathcal{A}$ with respect to ν. Let $\mathcal{I} : (\mathcal{P} \cup \mathcal{S}) \to \mathcal{D}$ be a function such that $\mathfrak{L}_c(\mathcal{I}(c))$ holds with respect to ν for every $c \in \mathcal{S}$. Let $(\mathcal{E}^{\mathcal{I},m})_{m\in M_2}$ be the M_2-set family of evaluation functions given by Theorem 6. By Theorem 8 $\mathcal{M} := \langle \mathcal{D}, @, \mathcal{E}^{\mathcal{I},e}, \nu\rangle$ is an M_2-set $\mathcal{S}$-model satisfying property η. Since ν is the identity function, $\mathcal{M}$ also satisfies property $\mathfrak{b}$. On the other hand, $\mathcal{M}$ does not satisfy property $\mathfrak{f}$ and hence does not satisfy property ξ. Consider the two functions $\mathsf{f}, \mathsf{g} \in \mathcal{D}_{\iota\iota}$ given by $\mathsf{f}(m, \mathsf{a}) := \mathsf{a}$ and $\mathsf{g}(m, \mathsf{a}) := m \cdot \mathsf{a}$. Clearly, $\mathsf{f}(1, \mathsf{a}) = \mathsf{g}(1, \mathsf{a})$ for all $\mathsf{a} \in \{0, 1\}$. However, $\mathsf{f} \neq \mathsf{g}$. Finally, we consider $\mathcal{D}_{o\iota}$. For all $\mathsf{p} \in \mathcal{D}_{o\iota}$, we must have $\mathsf{p}(1, 0) = \mathsf{p}(1, 0)0 = \mathsf{p}(0, 0) = \mathsf{p}(1, 1)0 = \mathsf{p}(1, 1)$. Thus, the only subsets of $\mathcal{D}_\iota$ represented by functions in $\mathcal{D}_{o\iota}$ are $\emptyset$ and $\{0, 1\}$. If we take the quotient of this model as in Theorem 3.62 from [9], then $\mathcal{D}_\iota$ will collapse to be a singleton.

If we are willing to accept models for which property $\mathfrak{b}$ fails, then we have much more flexibility in our choice of $\mathcal{D}_o$. We may want to interpret $\mathcal{D}_o$ to be an M-set so that all logical constants in $\mathcal{S}_{all}$ are realized and so that all subsets of $\mathcal{D}_\alpha$ are represented by a function in $\mathcal{D}_{o\alpha}$. We can obtain such an M-set by taking $\mathcal{D}_o$ to be the power set $\mathcal{P}(M)$ of M and defining an action taking Xm to $\{y \in M | my \in X\}$ for each $X \in \mathcal{P}(M)$ and $m \in M$. One can easily verify $(Xm)n = X(mn)$ and $Xe = X$ so that this is an M-set. We must also choose a function $\nu : \mathcal{P}(M) \to \{\mathtt{T},\mathtt{F}\}$. A natural option to consider is taking $\mathcal{G}$ to be an ultrafilter on $\mathcal{P}(M)$ and then defining $\nu(X) := \mathtt{T}$ iff $X \in \mathcal{G}$. It turns out that we obtain a model with all the properties we want by taking $\mathcal{G}$ to be the principal ultrafilter with principal element $e \in M$. That is, we define $\nu(X) := \mathtt{T}$ iff $e \in X$.

THEOREM 12. *Let M be a monoid with identity e and $\mathcal{A} = \langle \mathcal{D}, @\rangle$ be an M-set applicative structure. Suppose*

- *$\mathcal{D}_o = \mathcal{P}(M)$ and*
- *$Xm = \{y \in M | my \in X\}$.*

Let $\nu : \mathcal{D}_o \to \{\mathtt{T}, \mathtt{F}\}$ *be defined by*

$$\nu(X) := \begin{cases} \mathtt{T} & \textit{if } e \in X \\ \mathtt{F} & \textit{otherwise.} \end{cases}$$

Each logical constant in the set

$$\{\top_o, \bot_o, \neg_{oo}, \wedge_{ooo}, \vee_{ooo}, \Rightarrow_{ooo}, \equiv_{ooo}\}$$
$$\cup \{\Pi^\alpha_{o(o\alpha)} | \alpha \in \mathcal{T}\} \cup \{\Sigma^\alpha_{o(o\alpha)} | \alpha \in \mathcal{T}\} \cup \{=^\alpha_{o\alpha\alpha} | \alpha \in \mathcal{T}\}.$$

is realized in $\mathcal{A}$ *with respect to* ν. *Furthermore, for all* $S \subseteq \mathcal{D}_\alpha$ *there is some* $\mathsf{p}_S \in \mathcal{D}_{\alpha\to o}$ *such that for all* $\mathsf{a} \in \mathcal{D}_\alpha$ *we have* $\nu(\mathsf{p}_S @ \mathsf{a}) = \mathtt{T}$ *iff* $\mathsf{a} \in S$.

Proof. The constants $\top$ and $\bot$ can be realized using M and $\emptyset$, respectively.

To realize negation, we define $\mathsf{n} : M \times \mathcal{P}(M) \to \mathcal{P}(M)$ by $\mathsf{n}(m, X) := M \setminus X$. To check $\mathsf{n} \in \mathcal{D}_{oo}$, we must prove $\mathsf{n}(mk, Xk) = \mathsf{n}(m, X)k$ for all $X \in \mathcal{P}(M)$ and $m, k \in M$. This is true since $x \in \mathsf{n}(mk, Xk)$ iff $x \notin Xk$ iff $kx \notin X$ iff $x \in \mathsf{n}(m, X)k$. To prove $\mathfrak{L}_\neg(\mathsf{n})$, note that $\nu(\mathsf{n}@X) = \mathtt{T}$ iff $e \in \mathsf{n}(e, X)$ iff $e \notin X$ iff $\nu(X) = \mathtt{F}$.

Next we turn to disjunction. We define $\mathsf{d} : M \times \mathcal{P}(M) \to M \times \mathcal{P}(M) \to \mathcal{P}(M)$ by $\mathsf{d}(m, X)(n, Y) := (Xn) \cup Y$. We first prove $\mathsf{d}(m, X) \in \mathcal{D}_{oo}$ by proving $\mathsf{d}(m, X)(nk, Yk) = (\mathsf{d}(m, X)(n, Y))k$. This equation holds since $x \in \mathsf{d}(m, X)(nk, Yk)$ iff $x \in Xnk$ or $x \in Yk$ iff $kx \in Xn$ or $kx \in Y$ iff $kx \in \mathsf{d}(m, X)(n, Y)$ iff $x \in (\mathsf{d}(m, X)(n, Y))k$. We next prove $\mathsf{d} \in \mathcal{D}_{ooo}$ by proving $\mathsf{d}(mk, Xk) = \mathsf{d}(m, X)k$, which holds since $\mathsf{d}(mk, Xk)(n, Y) = (Xkn) \cup Y = \mathsf{d}(m, X)(kn, Y)$. To prove $\mathfrak{L}_\vee(\mathsf{d})$, note that $\nu(\mathsf{d}@X@Y) = \mathtt{T}$ iff $e \in X \cup Y$ iff either $\nu(X) = \mathtt{T}$ or $\nu(Y) = \mathtt{T}$. Since $\neg$ and $\vee$ are realized, so are $\wedge$, $\Rightarrow$ and $\equiv$.

Let α be a type. To prove Π^α is realized, let

$$\pi^\alpha(m, \mathsf{f}) := \{x \in M | \forall \mathsf{a} \in \mathcal{D}_\alpha \, e \in \mathsf{f}(x, \mathsf{a})\}$$

define $\pi^\alpha : M \times \mathcal{D}_{o\alpha} \to \mathcal{D}_o$. Now, $x \in \pi^\alpha(mk, \mathsf{f}k)$ iff $\forall \mathsf{a} \in \mathcal{D}_\alpha e \in (\mathsf{f}k)(x, \mathsf{a})$ iff $\forall \mathsf{a} \in \mathcal{D}_\alpha e \in \mathsf{f}(kx, \mathsf{a})$ iff $kx \in \pi^\alpha(m, \mathsf{f})$ iff $x \in \pi^\alpha(m, \mathsf{f})k$. Hence $\pi^\alpha(mk, \mathsf{f}k) = \pi^\alpha(m, \mathsf{f})k$ and so $\pi^\alpha \in \mathcal{D}_{o(o\alpha)}$. Also, $\mathfrak{L}_{\Pi^\alpha}(\pi^\alpha)$ holds since $\nu(\pi^\alpha @ \mathsf{f}) = \mathtt{T}$ iff $\forall \mathsf{a} \in \mathcal{D}_\alpha e \in \mathsf{f}(e, \mathsf{a})$ iff $\forall \mathsf{a} \in \mathcal{D}_\alpha \nu(\mathsf{f}@\mathsf{a}) = \mathtt{T}$. Since $\neg$ and Π^α are realized, so is Σ^α.

To prove $=^\alpha$ is realized, let $\mathsf{q}^\alpha(m, \mathsf{a})(n, \mathsf{b}) := \{x \in M | \mathsf{a}nx = \mathsf{b}x\}$ define $\mathsf{q}^\alpha : M \times \mathcal{D}_\alpha \to M \times \mathcal{D}_\alpha \to \mathcal{P}(M)$. We have $x \in \mathsf{q}^\alpha(m, \mathsf{a})(nk, \mathsf{b}k)$ iff $\mathsf{a}nkx = \mathsf{b}kx$ iff $kx \in (\mathsf{q}^\alpha(m, \mathsf{a})(n, \mathsf{b})$ iff $x \in (\mathsf{q}^\alpha(m, \mathsf{a})(n, \mathsf{b}))k$. Hence $\mathsf{q}^\alpha(m, \mathsf{a})(nk, \mathsf{b}k) = (\mathsf{q}^\alpha(m, \mathsf{a})(n, \mathsf{b}))k$ and so $\mathsf{q}^\alpha(m, \mathsf{a}) \in \mathcal{D}_{o\alpha}$. Next, we compute $x \in \mathsf{q}^\alpha(mk, \mathsf{a}k)(n, \mathsf{b})$ iff $\mathsf{a}knx = \mathsf{b}x$ iff $x \in \mathsf{q}^\alpha(m, \mathsf{a})(kn, \mathsf{b})$. Consequently, $\mathsf{q}^\alpha(mk, \mathsf{a}k)(n, \mathsf{b}) = \mathsf{q}^\alpha(m, \mathsf{a})(kn, \mathsf{b})$ and so $\mathsf{q}^\alpha(mk, \mathsf{a}k) = \mathsf{q}^\alpha(m, \mathsf{a})k$. Therefore, $\mathsf{q}^\alpha \in \mathcal{D}_{o\alpha\alpha}$ as desired. The property $\mathfrak{L}_{=^\alpha}(\mathsf{q}^\alpha)$ holds since

$$\nu(\mathsf{q}^\alpha @\mathsf{a}@\mathsf{b}) = \mathtt{T} \text{ iff } e \in \{x \in M | \mathsf{a}ex = \mathsf{b}\} \text{ iff } \mathsf{a} = \mathsf{b}.$$

Finally, we verify that all subsets of $\mathcal{D}_\alpha$ are represented in $\mathcal{D}_{o\alpha}$. Let $S \subseteq \mathcal{D}_\alpha$ be given. We define $\mathsf{p}_S(m, \mathsf{a}) := \{x | \mathsf{a}x \in S\}$. We compute $x \in \mathsf{p}_S(mk, \mathsf{a}k)$ iff $\mathsf{a}kx \in S$ iff $kx \in \mathsf{p}_S(m, \mathsf{a})$ iff $x \in \mathsf{p}_S(m, \mathsf{a})k$. Hence $\mathsf{p}_S \in \mathcal{D}_{o\alpha}$. Note that $\nu(\mathsf{p}_S@\mathsf{a}) = \mathtt{T}$ iff $e \in \mathsf{p}_S(e, \mathsf{a})$ iff $\mathsf{a} \in S$, as desired. ■

We now use Theorem 12 to modify Example 11 and obtain an M-set model in $\mathfrak{M}_{\beta\eta}$. As in Example 11 we take M_2 to be the monoid $\{0, 1\}$ under multiplication where 1 is the identity. For the reasons discussed in Example 11 we will not choose $\mathcal{D}_o$ to be the two element set $\{\mathtt{T}, \mathtt{F}\}$. Instead we let $\mathcal{D}_o$ be $\mathcal{P}(\{0, 1\})$ with action and ν given as in Theorem 12. Using Theorems 2, 12 and 8 we know we can obtain an $\mathcal{S}_{all}$-model by specifying an M_2-set $\mathcal{D}_\iota$ and a value $\mathcal{I}(w_\alpha) \in \mathcal{D}_\alpha$ for each parameter w.

EXAMPLE 13. Let A be a set with an equivalence relation $\sim$. Suppose for each $\sim$-equivalence class we choose a canonical element. Let $C : A \to A$ be the function taking each a to the canonical element $C(a)$. Note that for all $a \in A$, $a \sim C(a)$ and for all $a, b \in A$, $a \sim b$ iff $C(a) = C(b)$. In particular, $C(C(a)) = C(a)$ for all $a \in A$. We can consider A an M_2-set by defining the action $a1 = a$ and $a0 := C(a)$. This is an M_2-set since C is idempotent. Let us take $\mathcal{D}_\iota$ to be this M_2-set and apply Theorem 2 to obtain an M_2-set applicative structure $\langle \mathcal{D}, @ \rangle$.

The equivalence relation $\sim$ on A and canonical form function C extends to all types. For each type α, we can define an idempotent function $C^\alpha : \mathcal{D}_\alpha \to \mathcal{D}_\alpha$ by $C^\alpha(\mathsf{a}) := \mathsf{a}0$. This clearly induces an equivalence relation $\sim^\alpha$ on each $\mathcal{D}_\alpha$ given by $\mathsf{a} \sim^\alpha \mathsf{b}$ iff $C^\alpha(\mathsf{a}) = C^\alpha(\mathsf{b})$. Only the functions $f : \mathcal{D}_\beta \to \mathcal{D}_\alpha$ which respect the equivalence relations are represented in $\mathcal{D}_{\alpha\beta}$. To see this, suppose $g \in \mathcal{D}_{\alpha\beta}$ and $(g@\mathsf{b}) = f(\mathsf{b})$ for all $\mathsf{b} \in \mathcal{D}_\beta$. If $\mathsf{a} \sim^\beta \mathsf{b}$, then

$$f(\mathsf{a})0 = g(1, \mathsf{a})0 = g(0, \mathsf{a}0) = g(0, \mathsf{b}0) = g(1, \mathsf{b})0 = f(\mathsf{b})0$$

and so $f(\mathsf{a}) \sim^\alpha f(\mathsf{b})$. Next, suppose $f : \mathcal{D}_\beta \to \mathcal{D}_\alpha$ is such that $f(\mathsf{a}) \sim^\alpha f(\mathsf{b})$ whenever $\mathsf{a} \sim^\beta \mathsf{b}$. Define $g : M \times \mathcal{D}_\beta \to \mathcal{D}_\alpha$ by $g(m, \mathsf{a}) := f(\mathsf{a})m$. Clearly, $g@\mathsf{a} = g(1, \mathsf{a}) = f(\mathsf{a})$. Since $g(m0, \mathsf{a}0) = f(\mathsf{a}0)0 = f(\mathsf{a})0 = g(m, \mathsf{a})0$ (using the fact that $\mathsf{a}0 \sim^\alpha \mathsf{a}$), we have $g \in \mathcal{D}_{\alpha\beta}$.

In order to apply Theorem 8 we need a typed function $\mathcal{I} : (\mathcal{P} \cup \mathcal{S}_{all}) \to \mathcal{D}$ interpreting parameters and logical constants. We know by Theorem 12 that for all $c \in \mathcal{S}_{all}$ there is some $\mathcal{I}(c)$ such that $\mathfrak{L}_c(\mathcal{I}(c))$ holds. Let $\mathcal{I}(w)$ be chosen arbitrarily for parameters. Using Theorem 8 we can conclude that $\mathcal{M} := \langle \mathcal{D}, @, \mathcal{E}^{\mathcal{I},e}, \nu \rangle$ is an M_2-set model satisfying property η. As in Example 11, property ξ fails in $\mathcal{M}$ since the applicative structure is not functional. To see this, the reader may consider $\mathsf{f}, \mathsf{g} \in \mathcal{D}_{oo}$ given by $\mathsf{f}(m, X) := X$ and $\mathsf{g}(m, X) := Xm$. Since $\mathcal{D}_o$ has four elements, property $\mathfrak{b}$ fails. Since the model realizes equality at all types, $\mathfrak{q}$ holds and so $\mathcal{M} \in \mathfrak{M}_{\beta\eta}$.

We know by Theorem 12 for all types α and sets $S \subseteq \mathcal{D}_\alpha$, there is a function in $\mathcal{D}_{o\alpha}$ representing S. In particular, $\mathcal{D}_{o\iota}$ is rich enough to represent all subsets of A. On the other hand, only the functions from A to A respecting $\sim$ are represented in $\mathcal{D}_{\iota\iota}$.

6 An Example Model

We now construct a concrete M-set model in which the formal theorem (1) from Section 2 is meaningful. Let $\mathbb{N} = \{0, 1, 2, \ldots\}$ be the set of nonnegative integers. Let $M := \mathbb{N}^{\mathbb{N}}$ be the set of all functions from $\mathbb{N}$ into itself and let $e \in M$ be the identity function. For each $m, n \in M$ we define $mn \in M$ to be $(mn)(i) := n(m(i))$ (reverse composition). Clearly M is a monoid under this operation with identity e. We consider $\mathbb{N}$ as an M-set with action taking $a \in \mathbb{N}$ and $m \in M$ to $am := m(a)$.

By Theorem 2 there is an M-set applicative structure $\langle \mathcal{D}, @ \rangle$ such that $\mathcal{D}_\iota = \mathbb{N}$ (with action given above) and $\mathcal{D}_o = \mathcal{P}(M)$ (with action as in Theorem 12). Let $\nu : \mathcal{D}_o \to \{\mathtt{T}, \mathtt{F}\}$ be defined by

$$\nu(X) := \begin{cases} \mathtt{T} & \text{if } e \in X \\ \mathtt{F} & \text{otherwise.} \end{cases}$$

By Theorem 12 every logical constant in $\mathcal{S}_{all}$ is realized in the applicative structure with respect to ν. Let $\mathcal{I} : \mathcal{S}_{all} \cup \mathcal{P} \to \mathcal{D}$ be defined so that $\mathfrak{L}_c(\mathcal{I}(c))$ holds with respect to ν for each $c \in \mathcal{S}_{all}$. By Theorem 6 there is a unique M-set family of evaluation functions $(\mathcal{E}^{\mathcal{I},m})_{m \in M}$ such that $\mathcal{E}^{\mathcal{I},e}_\varphi(w) = \mathcal{I}(w)$ for $w \in \mathcal{S}_{all} \cup \mathcal{P}$ and assignments φ. By Theorem 8 $\mathcal{M} := \langle \mathcal{D}, @, \mathcal{E}^{\mathcal{I},e}, \nu \rangle$ is an M-set $\mathcal{S}$-model satisfying property η. Furthermore, since $\mathcal{S}_{all}$ includes equality at each type, property $\mathfrak{q}$ must hold and so $\mathcal{M} \in \mathfrak{M}_{\beta\eta}$.

Since $\mathcal{M}$ is a model of elementary type theory with η, the theorem (1) from Section 2

$$\forall P_{o(\iota\iota)} [P[\lambda x_\iota\, x] \wedge \forall i_\iota\, P[\lambda x\, i] \supset \forall f_{\iota\iota}\, P\, f] \supset \exists Y_{\iota(\iota\iota)} \forall f [f [Y\, f] = Y\, f]$$

must be true in the model. Consequently, if we can prove

$$\nu(\mathcal{E}^{\mathcal{I},e}(\forall P_{o(\iota\iota)} [P[\lambda x_\iota\, x] \wedge \forall i_\iota\, P[\lambda x\, i] \supset \forall f_{\iota\iota}\, P\, f])) = \mathtt{T} \tag{4}$$

then we can conclude

$$\nu(\exists Y_{\iota(\iota\iota)} \forall f[f[Y f] = Y f]) = \mathtt{T} \tag{5}$$

In order to interpret (4), fix an arbitrary assignment φ, let

$$\mathtt{I} := \mathcal{E}_{\varphi}^{\mathcal{I},e}(\lambda x_{\iota}\, x) \in \mathcal{D}_{\iota\iota}$$

and for each $i \in \mathbb{N}$ let

$$\mathtt{K}^{i} := \mathcal{E}_{\varphi,[i/y]}^{\mathcal{I},e}(\lambda x_{\iota}\, y) \in \mathcal{D}_{\iota\iota}.$$

We easily compute

$$\mathtt{I}(m, a) = \mathcal{E}_{\varphi m,[a/x]}^{\mathcal{I},m}(x) = a$$

and

$$\mathtt{K}^{i}(m, a) = \mathcal{E}_{\varphi m,[im/y],[a/x]}^{\mathcal{I},m}(y) = im = m(i).$$

Using the properties of the interpretations of logical constants in $\mathcal{M}$, we have (4) is valid in $\mathcal{M}$ if for all $\Phi \in \mathcal{D}_{o(\iota\iota)}$, $e \in \Phi(e, f)$ whenever $e \in \Phi(e, \mathtt{I})$ and $e \in \Phi(e, \mathtt{K}^{i})$ for all $i \in \mathbb{N}$. This is the case if $\mathcal{D}_{\iota\iota} = \{\mathtt{I}\} \cup \{\mathtt{K}^{i} | i \in \mathbb{N}\}$. To verify this equation, it is enough to show every $f \in \mathcal{D}_{\iota\iota}$ is either $\mathtt{I}$ or $\mathtt{K}^{i}$ for some i.

Let $S \in M$ be the successor function. For each $m \in M$ and $a \in \mathbb{N}$, let $[a \cdot m] \in M$ be the function such that $[a \cdot m](0) := a$ and $[a \cdot m](i+1) := m(i)$ for $i \in \mathbb{N}$. Clearly $0[a \cdot m] = a$ and $S[a \cdot m] = m$.

Let $f \in \mathcal{D}_{\iota\iota}$ be given. We have

$$f(m, a) = f(S[a \cdot m], 0[a \cdot m]) = f(S, 0)[a \cdot m] = [a \cdot m](f(S, 0)).$$

If $f(S, 0) = 0$, then

$$f(m, a) = [a \cdot m](0) = a$$

for all $a \in \mathbb{N}$ and so $f = \mathtt{I}$. If $f(S, 0) = i + 1$ for $i \in \mathbb{N}$, then

$$f(m, a) = [a \cdot m](i + 1) = m(i)$$

and so $f = \mathtt{K}^{i}$.

Consequently, (4) is valid in $\mathcal{M}$. Since (1) is also valid in $\mathcal{M}$ (as a theorem of elementary type theory), the conclusion (5) must be valid in $\mathcal{M}$. We conclude the existence of a fixed point operator $Y \in \mathcal{D}_{\iota(\iota\iota)}$ such that $f(e, Y(e, f)) = Y(e, f)$ for all $f \in \mathcal{D}_{\iota\iota}$.

7 Conclusion and Future Work

We have used M-sets to construct models (in the sense of [9]) of fragments of classical higher-order logic. These models always satisfy property η, but may not satisfy property ξ. Property $\mathfrak{b}$ depends on the choice of $\mathcal{D}_o$ and ν. As we have demonstrated, there is a tradeoff between satisfying property $\mathfrak{b}$ and realizing equality (property $\mathfrak{q}$) in M-set models. We have used these abstract results to obtain a model in which $\mathcal{D}_{\iota\iota}$ is sparsely populated and there is a fixed point operator in $\mathcal{D}_{\iota(\iota\iota)}$. In future work we hope to consider more interesting choices for monoids M and M-sets. Such models may provide novel applications of nonextensional higher-order theorem proving.

BIBLIOGRAPHY

[1] Peter B. Andrews. Resolution in Type Theory. *Journal of Symbolic Logic*, 36:414–432, 1971.

[2] Peter B. Andrews. On Connections and Higher-Order Logic. *Journal of Automated Reasoning*, 5:257–291, 1989.

[3] Peter B. Andrews. Classical Type Theory. In Alan Robinson and Andrei Voronkov, editors, *Handbook of Automated Reasoning*, volume 2, chapter 15, pages 965–1007. Elsevier Science, 2001.

[4] Peter B. Andrews, Matthew Bishop, Sunil Issar, Dan Nesmith, Frank Pfenning, and Hongwei Xi. TPS: A Theorem Proving System for Classical Type Theory. *Journal of Automated Reasoning*, 16:321–353, 1996.

[5] Peter B. Andrews, Chad E. Brown, Frank Pfenning, Matthew Bishop, Sunil Issar, and Hongwei Xi. ETPS: A System to Help Students Write Formal Proofs. *Journal of Automated Reasoning*, 32:75–92, 2004. Available at http://journals.kluweronline.com/article.asp?PIPS=5264938.

[6] Peter B. Andrews, Sunil Issar, Dan Nesmith, and Frank Pfenning. The TPS Theorem Proving System. *Journal of Symbolic Logic*, 57:353–354, 1992. (abstract).

[7] Peter B. Andrews, Dale A. Miller, Eve Longini Cohen, and Frank Pfenning. Automating Higher-Order Logic. In W. W. Bledsoe and D. W. Loveland, editors, *Automated Theorem Proving: After 25 Years*, Contemporary Mathematics series, vol. 29, pages 169–192. American Mathematical Society, 1984. Proceedings of the Special Session on Automatic Theorem Proving, 89th Annual Meeting of the American Mathematical Society, held in Denver, Colorado, January 5-9, 1983.

[8] Christoph Benzmüller. *Equality and Extensionality in Automated Higher-Order Theorem Proving*. PhD thesis, Universität des Saarlandes, 1999.

[9] Christoph Benzmüller, Chad E. Brown, and Michael Kohlhase. Higher Order Semantics and Extensionality. *Journal of Symbolic Logic*, 69:1027–1088, 2004.

[10] Chad E. Brown. *Set Comprehension in Church's Type Theory*. PhD thesis, Department of Mathematical Sciences, Carnegie Mellon University, 2004.

[11] Chad E. Brown. *Automated Reasoning in Higher-Order Logic: Set Comprehension and Extensionality in Church's Type Theory*, volume 10 of *Studies in Logic: Logic and Cognitive Systems*. College Publications, 2007.

[12] Alonzo Church. A Formulation of the Simple Theory of Types. *Journal of Symbolic Logic*, 5:56–68, 1940.

[13] Robert Goldblatt. *Topoi: The Categorial Analysis of Logic*. Number 98 in Studies in Logic and the Foundations of Mathematics. North-Holland, 1979. Third edition, 1983.

[14] J. Lambek and P. Scott. *Introduction to Higher Order Categorial Logic*. Cambridge University Press, Cambridge, UK, 1986.

[15] Saunders MacLane and Ieke Moerdijk. *Sheaves in Geometry and Logic: A First Introduction to Topos Theory.* Springer-Verlag, 1992.

A Proof of Theorem 5

Let M be a monoid with identity e, $\mathcal{A} = \langle \mathcal{D}, @ \rangle$ be an M-set applicative structure, and $(\mathcal{E}^m)_{m\in M}$ be an M-set family of $\mathcal{S}$-evaluation functions for $\mathcal{A}$. We will prove a series of results allowing us to conclude Theorem 5.

LEMMA 14. *For any term* $\mathbf{A} \in \mathit{wff}_\alpha(\mathcal{S})$, $m, n \in M$, *and assignment* φ, *we have*

$$\mathcal{E}^m_\varphi(\mathbf{A})n = \mathcal{E}^{mn}_{\varphi n}(\mathbf{A}).$$

Proof. This follows by induction on $\mathbf{A}$. The variable, parameter and constant cases are easy. The application case is verified by computing

$$\begin{array}{rcll} \mathcal{E}^m_\varphi(\mathbf{FB})n & = & (\mathcal{E}^m_\varphi(\mathbf{F})@\mathcal{E}^m_\varphi(\mathbf{B}))n & \text{by Definition 4(2)} \\ & = & \mathcal{E}^m_\varphi(\mathbf{F})(e, \mathcal{E}^m_\varphi(\mathbf{B}))n & \text{by Definition 1(1)} \\ & = & \mathcal{E}^m_\varphi(\mathbf{F})(n, \mathcal{E}^m_\varphi(\mathbf{B})n) & \text{by (2) and Definition 1(1)} \\ & = & (\mathcal{E}^m_\varphi(\mathbf{F})n)(e, \mathcal{E}^m_\varphi(\mathbf{B})n) & \text{by (3)} \\ & = & (\mathcal{E}^{mn}_{\varphi n}(\mathbf{F}))(e, \mathcal{E}^{mn}_{\varphi n}(\mathbf{B})) & \text{by the inductive hypothesis} \\ & = & \mathcal{E}^{mn}_{\varphi n}(\mathbf{F})@\mathcal{E}^{mn}_{\varphi n}(\mathbf{B}) & \text{by Definition 1(1).} \end{array}$$

For the abstraction case, we verify $\mathcal{E}^m_\varphi(\lambda x_\beta \mathbf{C}_\gamma)n = \mathcal{E}^{mn}_{\varphi n}(\lambda x_\beta \mathbf{C}_\gamma)$ by computing

$$\begin{array}{rcll} (\mathcal{E}^m_\varphi(\lambda x\mathbf{C})n)(k, \mathsf{b}) & = & \mathcal{E}^m_\varphi(\lambda x\mathbf{C})(nk, \mathsf{b}) & \text{by (3)} \\ & = & \mathcal{E}^{mnk}_{\varphi nk, [\mathsf{b}/x]}(\mathbf{C}) & \text{by the inductive hypothesis} \\ & = & \mathcal{E}^{mn}_{\varphi n}(\lambda x\mathbf{C})(k, \mathsf{b}) & \text{by Definition 4(4).} \end{array}$$

■

In the next lemma, we combine property (3) of Definition 4 with Lemma 14 to verify the third property of evaluation functions.

LEMMA 15. *For any term* $\mathbf{A} \in \mathit{wff}_\alpha(\mathcal{S})$, $m \in M$, *and assignments* φ *and* ψ, *if* $\varphi(x) = \psi(x)$ *for all* $x \in \mathbf{Free}(\mathbf{A})$, *then* $\mathcal{E}^m_\varphi(\mathbf{A}) = \mathcal{E}^m_\psi(\mathbf{A})$.

Proof. The proof is by an easy induction on $\mathbf{A}$. Note that if $\mathbf{A}$ is a parameter or constant w, then we have $\mathcal{E}^m_\varphi(\mathbf{A}) = \mathcal{E}^e_\psi(\mathbf{A})m = \mathcal{E}^m_\psi(\mathbf{A})$ by using property (3) of Definition 4 twice. ■

We now prove a Substitution-Value Lemma similar to Lemma 3.20 in [9] for evaluations. In this case we must prove the result before we know $\mathcal{E}^m$ is an evaluation function.

LEMMA 16 (Substitution-Value Lemma). *For any term* $\mathbf{A} \in \mathit{wff}_\alpha(\mathcal{S})$, $\mathbf{B} \in \mathit{wff}_\beta(\mathcal{S})$, $x \in \mathcal{V}_\beta$, $m \in M$, *and assignment* φ,

$$\mathcal{E}^m_{\varphi,[\mathcal{E}^m_\varphi(\mathbf{B})/x]}(\mathbf{A}) = \mathcal{E}^m_\varphi([\mathbf{B}/x]\mathbf{A}).$$

Proof. The proof is another straightforward induction on $\mathbf{A}$. If $\mathbf{A}$ is a parameter or constant w, we use Lemma 15 to conclude

$$\mathcal{E}^m_{\varphi,[\mathcal{E}^m_\varphi(\mathbf{B})/x]}(w) = \mathcal{E}^m_\varphi(w) = \mathcal{E}^m_\varphi([\mathbf{B}/x]w).$$

Suppose $\mathbf{A}$ is $[\lambda y_\delta \mathbf{C}_\gamma]$ where y and x are distinct and $y \notin \mathbf{Free}(\mathbf{B})$. In this case, we have

$$\begin{array}{lll} & \mathcal{E}^m_{\varphi,[\mathcal{E}^m_\varphi(\mathbf{B})/x]}([\lambda y_\delta \mathbf{C}])(k,\mathsf{d}) & \\ = & \mathcal{E}^{mk}_{\varphi k,[\mathcal{E}^m_\varphi(\mathbf{B})k/x],[\mathsf{d}/y]}(\mathbf{C}) & \text{by Definition 4(4)} \\ = & \mathcal{E}^{mk}_{\varphi k,[\mathcal{E}^{mk}_{\varphi k}(\mathbf{B})/x],[\mathsf{d}/y]}(\mathbf{C}) & \text{by Lemma 14} \\ = & \mathcal{E}^{mk}_{\varphi k,[\mathsf{d}/y],[\mathcal{E}^{mk}_{\varphi k,[\mathsf{d}/y]}(\mathbf{B})/x]}(\mathbf{C}) & \text{by Lemma 15 since } y \notin \mathbf{Free}(\mathbf{B}) \cup \{x\} \\ = & \mathcal{E}^{mk}_{\varphi k,[\mathsf{d}/y]}([\mathbf{B}/x]\mathbf{C}) & \text{by the inductive hypothesis} \\ = & \mathcal{E}^m_\varphi([\lambda y_\delta [\mathbf{B}/x]\mathbf{C}])(k,\mathsf{d}) & \text{by Definition 4(4)} \\ = & \mathcal{E}^m_\varphi([\mathbf{B}/x][\lambda y_\delta \mathbf{C}])(k,\mathsf{d}). & \end{array}$$

The remaining cases are left to the reader. ■

We next check that $\mathcal{E}^m$ respects a single, top-level β-reduction.

LEMMA 17. *For any term* $\mathbf{A} \in \mathit{wff}_\alpha(\mathcal{S})$, $\mathbf{B} \in \mathit{wff}_\beta(\mathcal{S})$, $x \in \mathcal{V}_\beta$, $m \in M$, *and assignment* φ,

$$\mathcal{E}^m_\varphi([[\lambda x\, \mathbf{A}]\, \mathbf{B}]) = \mathcal{E}^m_\varphi([\mathbf{B}/x]\mathbf{A}).$$

Proof. We compute

$$\mathcal{E}^m_\varphi([[\lambda x\, \mathbf{A}]\, \mathbf{B}]) = \mathcal{E}^m_\varphi([\lambda x\, \mathbf{A}])(e, \mathcal{E}^m_\varphi(\mathbf{B})) = \mathcal{E}^m_{\varphi,[\mathcal{E}^m_\varphi(\mathbf{B})/x]}(\mathbf{A}) = \mathcal{E}^m_\varphi([\mathbf{B}/x]\mathbf{A})$$

using Lemma 16. ■

We also check $\mathcal{E}^m$ respects a single, top-level η-reduction.

LEMMA 18. *For any term* $\mathbf{F} \in wff_{\alpha\beta}(\mathcal{S})$, $x \in \mathcal{V}_\beta$, $m \in M$, *and assignment* φ, *if* $x \notin \mathbf{Free}(\mathbf{F})$, *then*

$$\mathcal{E}^m_\varphi([\lambda x\, [\mathbf{F}\, x]]) = \mathcal{E}^m_\varphi(\mathbf{F}).$$

Proof. We verify this fact by computing

$$\begin{array}{lll} & \mathcal{E}^m_\varphi([\lambda x\, [\mathbf{F}\, x]])(k, \mathsf{b}) & \\ = & \mathcal{E}^{mk}_{\varphi k,[\mathsf{b}/x]}([\mathbf{F}\, x]) & \text{by Definition 4(4)} \\ = & \mathcal{E}^{mk}_{\varphi k,[\mathsf{b}/x]}(\mathbf{F})@\mathsf{b} & \text{by Definition 4(1 and 2)} \\ = & \mathcal{E}^{mk}_{\varphi k}(\mathbf{F})@\mathsf{b} & \text{by Lemma 15 since } x \notin \mathbf{Free}(\mathbf{F}) \\ = & \mathcal{E}^{mk}_{\varphi k}(\mathbf{F})(e, \mathsf{b}) & \text{by Definition 1(1)} \\ = & (\mathcal{E}^m_\varphi(\mathbf{F})k)(e, \mathsf{b}) & \text{by Lemma 14} \\ = & \mathcal{E}^m_\varphi(\mathbf{F})(k, \mathsf{b}) & \text{by (3).} \end{array}$$

■

Using the previous two lemmas, we can prove $\mathcal{E}^m$ respects one step $\beta\eta$-reductions inside a term.

LEMMA 19. *For any terms* $\mathbf{A}, \mathbf{B} \in wff_\alpha(\mathcal{S})$, $m \in M$, *and assignment* φ, *if* $\mathbf{A}$ $\beta\eta$*-reduces to* $\mathbf{B}$ *in one step, then*

$$\mathcal{E}^m_\varphi(\mathbf{A}) = \mathcal{E}^m_\varphi(\mathbf{B}).$$

Proof. This follows by an induction on the position of the redex in $\mathbf{A}$ using Lemmas 17 and 18 for the base cases. ■

Finally, $\mathcal{E}^m$ respects $\beta\eta$-normalization. This establishes both the final property of evaluation functions and η-functionality.

LEMMA 20. *For any term* $\mathbf{A} \in wff_\alpha(\mathcal{S})$, $m \in M$, *and assignment* φ, *we have*

$$\mathcal{E}^m_\varphi(\mathbf{A}) = \mathcal{E}^m_\varphi(\mathbf{A}^\downarrow).$$

Proof. The proof is by induction on the number of reductions using Lemma 19 at each step. ■

We now have the necessary results to conclude Theorem 5. By Definition 4, we know $\mathcal{E}^m_\varphi(x) = \varphi(x)$ for each variable x and we know $\mathcal{E}^m_\varphi([\mathbf{FB}]) = \mathcal{E}^m_\varphi(\mathbf{F})@\mathcal{E}^m_\varphi(\mathbf{B})$. By Lemma 15, we know $\mathcal{E}^m_\varphi(\mathbf{A}) = \mathcal{E}^m_\psi(\mathbf{A})$ whenever φ and ψ agree on $\mathbf{Free}(\mathbf{A})$. By Lemma 20 twice, we know

$$\mathcal{E}^m_\varphi(\mathbf{A}) = \mathcal{E}^m_\varphi(\mathbf{A}^{\downarrow}) = \mathcal{E}^m_\varphi(\mathbf{A}^{\downarrow\beta})$$

(since the $\beta\eta$-normal form of $\mathbf{A}^{\downarrow\beta}$ is $\mathbf{A}^{\downarrow}$). Hence $\mathcal{E}^m$ is an $\mathcal{S}$-evaluation function. Furthermore, $\mathcal{E}^m$ is η-functional by Lemma 20. Finally, we know

$$\mathcal{E}^m_\varphi(\mathbf{A})n = \mathcal{E}^{mn}_{\varphi n}(\mathbf{A})$$

holds by Lemma 14.

B Proof of Theorem 6

Let M be a monoid with identity e, $\mathcal{A} = \langle \mathcal{D}, @ \rangle$ be an M-set applicative structure and $\mathcal{I} : (\mathcal{P} \cup \mathcal{S}) \to \mathcal{D}$ be a typed function. If $(\mathcal{E}^{\mathcal{I},m})_{m\in M}$ and $(\mathcal{F}^{\mathcal{I},m})_{m\in M}$ are both M-set families of evaluation functions such that $\mathcal{E}^{\mathcal{I},e}(w) = \mathcal{I}(w) = \mathcal{F}^{\mathcal{I},e}(w)$ for all $w \in \mathcal{P} \cup \mathcal{S}$, then an easy induction on $\mathbf{A}$ proves

$$\forall \mathbf{A}, \varphi, m \text{ we have } \mathcal{E}^{\mathcal{I},m}_\varphi(\mathbf{A}) = \mathcal{F}^{\mathcal{I},m}_\varphi(\mathbf{A}).$$

Hence we have uniqueness and must only prove existence of such a family $(\mathcal{E}^{\mathcal{I},m})_{m\in M}$.

The construction of $(\mathcal{E}^{\mathcal{I},m})_{m\in M}$ for the proof of Theorem 6 requires some work. We could try to define $\mathcal{E}^{\mathcal{I},m}_\varphi(\mathbf{A})$ by induction on $\mathbf{A}$. Unfortunately, for the λ-abstraction case we must somehow know the function f where

$$\mathsf{f}(k, \mathsf{b}) = \mathcal{E}^{\mathcal{I},mk}_{\varphi k,[\mathsf{b}/x]}(\mathbf{A})$$

is in $\mathcal{D}_{\alpha\beta}$. Instead of defining $\mathcal{E}^{\mathcal{I},m}_\varphi(\mathbf{A}_\alpha) \in \mathcal{D}_\alpha$ by induction, we define set valued function $\mathbf{Eval}^{\mathcal{I}}(\mathbf{A}_\alpha, m, \varphi) \subseteq \mathcal{D}_\alpha$ by induction, prove each value of $\mathbf{Eval}^{\mathcal{I}}(\mathbf{A}, m, \varphi)$ is a singleton, and define $\mathcal{E}^{\mathcal{I},m}_\varphi(\mathbf{A}_\alpha)$ to be the unique value in this singleton.

We define $\mathbf{Eval}^{\mathcal{I}}(\mathbf{A}_\alpha, m, \varphi) \subseteq \mathcal{D}_\alpha$ by induction as follows:

- $\mathbf{Eval}^{\mathcal{I}}(x_\alpha, m, \varphi) := \{\varphi(x)\}$ for $x \in \mathcal{V}$.
- $\mathbf{Eval}^{\mathcal{I}}(w_\alpha, m, \varphi) := \{\mathcal{I}(w)m\}$ for $w \in \mathcal{P} \cup \mathcal{S}$.
- $\mathbf{Eval}^{\mathcal{I}}([\mathbf{F}_{\alpha\beta}\,\mathbf{B}_\beta], m, \varphi) :=$

  $$\{\mathsf{f}@\mathsf{b} | \mathsf{f} \in \mathbf{Eval}^{\mathcal{I}}(\mathbf{F}, m, \varphi), \mathsf{b} \in \mathbf{Eval}^{\mathcal{I}}(\mathbf{B}, m, \varphi)\}$$

  for $\mathbf{F} \in \mathit{wff}_{\alpha\beta}(\mathcal{S})$ and $\mathbf{B} \in \mathit{wff}_\beta(\mathcal{S})$

- $\mathbf{Eval}^{\mathcal{I}}([\lambda x_\beta \, \mathbf{A}_\alpha], m, \varphi) :=$

$$\{\mathsf{f} \in \mathcal{D}_{\alpha\beta} | \forall k \in M, \mathsf{b} \in \mathcal{D}_\beta \, \mathsf{f}(k, \mathsf{b}) \in \mathbf{Eval}^{\mathcal{I}}(\mathbf{A}, mk, (\varphi k, [\mathsf{b}/x]))\}$$

  for $x \in \mathcal{V}_\beta$ and $\mathbf{A} \in \mathit{wff}_\alpha(\mathcal{S})$.

LEMMA 21. *For every* $\mathbf{A} \in \mathit{wff}_\alpha(\mathcal{S})$, *assignment* φ, $m, n \in M$ *and* $\mathsf{a} \in \mathbf{Eval}^{\mathcal{I}}(\mathbf{A}, m, \varphi)$, *we have* $\mathsf{a}n \in \mathbf{Eval}^{\mathcal{I}}(\mathbf{A}, mn, \varphi n)$.

Proof. This follows by an easy induction on $\mathbf{A}$ using the definition of

$$\mathbf{Eval}^{\mathcal{I}}(\mathbf{A}, m, \varphi).$$

If $\mathbf{A}$ is a variable x, then $\mathsf{a} = \varphi(x)$ and so

$$\mathsf{a}n = \varphi(x)n = (\varphi n)(x) \in \mathbf{Eval}^{\mathcal{I}}(x, mn, \varphi n).$$

If $\mathbf{A}$ is a parameter or constant w, then $\mathsf{a} = \mathcal{I}(w)m$ and so

$$\mathsf{a}n = \mathcal{I}(w)mn \in \mathbf{Eval}^{\mathcal{I}}(w, mn, \varphi n).$$

Suppose $\mathbf{A}$ is $[\mathbf{F}_{\alpha\beta}\mathbf{B}_\beta]$. Then a is $\mathsf{f}@\mathsf{b}$ for some $\mathsf{f} \in \mathbf{Eval}^{\mathcal{I}}(\mathbf{F}, m, \varphi)$ and $\mathsf{b} \in \mathbf{Eval}^{\mathcal{I}}(\mathbf{B}, m, \varphi)$. By the inductive hypothesis, $\mathsf{f}n \in \mathbf{Eval}^{\mathcal{I}}(\mathbf{F}, mn, \varphi n)$ and $\mathsf{b}n \in \mathbf{Eval}^{\mathcal{I}}(\mathbf{B}, mn, \varphi n)$. We compute

$$\mathsf{a}n = (\mathsf{f}@\mathsf{b})n = \mathsf{f}(e, \mathsf{b})n = \mathsf{f}(n, \mathsf{b}n) = (\mathsf{f}n)(e, \mathsf{b}n) = ((\mathsf{f}n)@(\mathsf{b}n))$$

and conclude $\mathsf{a}n \in \mathbf{Eval}^{\mathcal{I}}([\mathbf{F}\,\mathbf{B}], mn, \varphi n)$.

Suppose $\mathbf{A}$ is $[\lambda x_\beta \mathbf{C}_\gamma]$. Then $\mathsf{a} \in \mathcal{D}_{\gamma\beta}$ is a function such that $\mathsf{a}(k, \mathsf{b}) \in \mathbf{Eval}^{\mathcal{I}}(\mathbf{C}, mk, (\varphi k, [\mathsf{b}/x]))$ for all $k \in M$ and $\mathsf{b} \in \mathcal{D}_\beta$. Note that $\mathsf{a}n(k, \mathsf{b}) = \mathsf{a}(nk, \mathsf{b})$. Hence $\mathsf{a}n(k, \mathsf{b}) \in \mathbf{Eval}^{\mathcal{I}}(\mathbf{C}, mnk, (\varphi nk, [\mathsf{b}/x]))$. Thus

$$\mathsf{a}n \in \mathbf{Eval}^{\mathcal{I}}([\lambda x\, \mathbf{C}], mn, \varphi n)$$

as desired. ■

LEMMA 22. *For every* $\mathbf{A} \in \mathit{wff}_\alpha(\mathcal{S})$, *assignment* φ, *and* $m \in M$, *there is a unique* $\mathsf{a} \in \mathcal{D}_\alpha$ *such that* $\mathsf{a} \in \mathbf{Eval}^{\mathcal{I}}(\mathbf{A}, m, \varphi)$.

Proof. This also follows by induction on $\mathbf{A}$. The cases for variables, parameters, logical constants and applications are easy. We only show the λ-abstraction case. Suppose $\mathbf{A}$ is $[\lambda x_\beta \mathbf{C}_\gamma]$. By the inductive hypothesis, for all $k \in M$ and $\mathsf{b} \in \mathcal{D}_\beta$, there is a unique value in $\mathbf{Eval}^{\mathcal{I}}(\mathbf{C}, mk, (\varphi k, [\mathsf{b}/x]))$. Let $\mathsf{f} : M \times \mathcal{D}_\beta \to \mathcal{D}_\gamma$ be the function taking (k, b) to this unique value. In order to conclude $\mathsf{f} \in \mathbf{Eval}^{\mathcal{I}}(\mathbf{A}, m, \varphi)$, we must verify that $\mathsf{f} \in \mathcal{D}_{\gamma\beta}$. On the one hand, $\mathsf{f}(kn, \mathsf{b}n) \in \mathbf{Eval}^{\mathcal{I}}(\mathbf{C}, mkn, (\varphi kn, [\mathsf{b}n/x]))$ by the choice of f. On the other hand, $\mathsf{f}(k, \mathsf{b})n \in \mathbf{Eval}^{\mathcal{I}}(\mathbf{C}, mkn, (\varphi kn, [\mathsf{b}n/x]))$ by Lemma 21. Using the inductive hypothesis,

$$\mathbf{Eval}^{\mathcal{I}}(\mathbf{C}, mkn, (\varphi kn, [\mathsf{b}n/x]))$$

is a singleton and we must have

$$\mathsf{f}(k, \mathsf{b})n = \mathsf{f}(kn, \mathsf{b}n)$$

which verifies $\mathsf{f} \in \mathcal{D}_{\gamma\alpha}$. Finally, suppose $\mathsf{g} \in \mathbf{Eval}^{\mathcal{I}}(\mathbf{A}, m, \varphi)$. For any $k \in M$ and $\mathsf{b} \in \mathcal{D}_{\beta}$, we must have $\mathsf{g}(k, \mathsf{b}) \in \mathbf{Eval}^{\mathcal{I}}(\mathbf{C}, mk, (\varphi k, [\mathsf{b}/x]))$ and so $\mathsf{g}(k, \mathsf{b}) = \mathsf{f}(k, \mathsf{b})$. Thus $\mathsf{g} = \mathsf{f}$ and $\mathbf{Eval}^{\mathcal{I}}(\mathbf{A}, m, \varphi)$ is a singleton. ■

We can now prove Theorem 6. Let $\mathcal{E}^{\mathcal{I},m}_{\varphi}(\mathbf{A})$ be the unique member of $\mathbf{Eval}^{\mathcal{I}}(\mathbf{A}, m, \varphi)$ for all $\mathbf{A} \in \mathit{wff}_{\alpha}(\mathcal{S})$, $m \in M$ and assignments φ into $\mathcal{D}$. Note that $\mathcal{E}^{\mathcal{I},m}_{\varphi}(w) = \mathcal{I}(w)m$ since $\mathbf{Eval}^{\mathcal{I}}(w, m, \varphi) = \{\mathcal{I}(w)m\}$ and so $\mathcal{E}^{\mathcal{I},e}_{\varphi}(w) = \mathcal{I}(w)$ for each parameter or constant w. It remains to check the conditions in Definition 4. For each variable x, $\mathcal{E}^{m}_{\varphi}(x) = \varphi(x)$ since $\mathbf{Eval}^{\mathcal{I}}(x, m, \varphi) = \{\varphi(x)\}$. For each parameter or constant w, we have $\mathcal{E}^{m}_{\varphi}(w) = \mathcal{I}(w)m = \mathcal{E}^{e}_{\psi}(w)m$. Since $\mathcal{E}^{m}_{\varphi}(\mathbf{F}) \in \mathbf{Eval}^{\mathcal{I}}(\mathbf{F}, m, \varphi)$ and $\mathcal{E}^{m}_{\varphi}(\mathbf{B}) \in \mathbf{Eval}^{\mathcal{I}}(\mathbf{B}, m, \varphi)$, we know $\mathcal{E}^{m}_{\varphi}(\mathbf{F})@\mathcal{E}^{m}_{\varphi}(\mathbf{B}) \in \mathbf{Eval}^{\mathcal{I}}([\mathbf{FB}], m, \varphi)$. Hence

$$\mathcal{E}^{m}_{\varphi}(\mathbf{FB}) = \mathcal{E}^{m}_{\varphi}(\mathbf{F})@\mathcal{E}^{m}_{\varphi}(\mathbf{B}).$$

Finally,

$$\mathcal{E}^{m}_{\varphi}(\lambda x_{\beta}\mathbf{A}_{\alpha}) = \mathsf{f} \in \mathbf{Eval}^{\mathcal{I}}([\lambda x_{\beta}\mathbf{A}_{\alpha}], m, \varphi) \subseteq \mathcal{D}_{\alpha\beta}$$

where $\mathsf{f}(k, \mathsf{b}) \in \mathbf{Eval}^{\mathcal{I}}(\mathbf{A}, mk, (\varphi k, [\mathsf{b}/x]))$, i.e., $\mathsf{f}(k, \mathsf{b}) = \mathcal{E}^{\mathcal{I},mk}_{\varphi k,[\mathsf{b}/x]}(\mathbf{A})$. Thus all the conditions in Definition 4 hold and the proof is complete.

Part III

Higher-Order Automated Reasoning in Theory

Notre Dame Journal of Formal Logic
Volume XV, Number 1, January 1974
NDJFAM

RESOLUTION AND THE CONSISTENCY OF ANALYSIS

PETER B. ANDREWS

§1. *Introduction.** In [2] we formulated a system $\mathcal{R}$, called a Resolution system, for refuting finite sets of sentences of type theory, and proved that $\mathcal{R}$ is complete in the (weak) sense that every set of sentences which can be refuted in the system $\mathcal{T}$ of type theory due to Church [5] can also be refuted in $\mathcal{R}$. The statement that $\mathcal{R}$ is in this sense complete is a purely syntactic one concerning finite sequences of wffs. However, it is clear that there can be no purely syntactic proof of the completeness of $\mathcal{R}$, since the completeness of $\mathcal{R}$ is closely related to Takeuti's conjecture [9] (since proved by Takahashi [8] and Pravitz [7]) concerning cut-elimination in type theory. As Takeuti pointed out in [9] and [10], cut-elimination in type theory implies the consistency of analysis. Indeed, Takeuti's conjecture implies the consistency of a formulation of type theory with an axiom of infinity; in such a system classical analysis and much more can be formalized. Hence, to avoid a conflict with Gödel's theorem, any proof of the completeness of resolution in type theory must involve arguments which cannot be formalized in type theory with an axiom of infinity. Indeed, the proof in [2] does involve a semantic argument. Nevertheless, it must be admitted that anyone who does not find the line of reasoning sketched above completely clear will have difficulty finding a unified and coherent exposition of the entire argument in the published literature. We propose to remedy this situation here.

We presuppose familiarity with §2 (The System $\mathcal{T}$) and Definitions 4.1 and 5.1 (The Resolution System $\mathcal{R}$) of [2], and follow the notation used there. In particular, $\Box$ stands for the contradictory sentence $\forall p_o p_o$. To distinguish between formulations of $\mathcal{T}$ with different sets of parameters, we henceforth assume $\mathcal{T}$ has no parameters, and denote by $\mathcal{T}(\mathbf{A}^1, \ldots, \mathbf{A}^n)$ a formulation of the system with parameters $\mathbf{A}^1, \ldots, \mathbf{A}^n$. If $\mathcal{H}$ is a set of sentences, $\mathcal{H} \vdash_{\mathcal{S}} \mathbf{B}$ shall mean that $\mathbf{B}$ is derivable from some finite subset of $\mathcal{H}$ in system $\mathcal{S}$. The deduction theorem is proved in §5 of [5]. We shall

*This research was partially supported by NSF Grant GJ-28457X.

Received October 2, 1971

incorporate into our argument Gandy's results in §3 of [6] with some minor modifications. We also wish to thank Professor Gandy for the basic idea (attributed by him to Turing) used below in showing the relative consistency of the axiom of descriptions. (This idea is mentioned briefly at the top of page 48 of [6].) We shall have occasion to refer to the following wffs:

The set $\mathcal{E}$ of *axioms of extensionality*:

E^{o}: $\forall p_o \forall q_o \cdot p_o \equiv q_o \supset . p_o = q_o$.
$E^{(\alpha\beta)}$: $\forall f_{\alpha\beta} \forall g_{\alpha\beta} \cdot \forall x_\beta [f_{\alpha\beta} x_\beta = g_{\alpha\beta} x_\beta] \supset . f_{\alpha\beta} = g_{\alpha\beta}$.

The *axiom of descriptions* for type α:

D^{α}: $\forall f_{o\alpha} \cdot \exists_1 x_\alpha f_{o\alpha} x_\alpha \supset f_{o\alpha} [\iota_{\alpha(o\alpha)} f_{o\alpha}]$.

An *axiom of infinity* for type α:

J^{α}: $\exists r_{o\alpha\alpha} \forall x_\alpha \forall y_\alpha \forall z_\alpha \cdot \exists w_\alpha r_{o\alpha\alpha} x_\alpha w_\alpha \wedge$
$\sim r_{o\alpha\alpha} x_\alpha x_\alpha \wedge . \sim r_{o\alpha\alpha} x_\alpha y_\alpha \vee \sim r_{o\alpha\alpha} y_\alpha z_\alpha \vee r_{o\alpha\alpha} x_\alpha z_\alpha$.

We let $\mathcal{A}$ denote the system obtained when one adds to $\mathcal{T}(\iota_{\iota(o\iota)})$ the axioms $\mathcal{E}$, D^{ι}, and J^{ι}. (Description operators and axioms for higher types are not needed, since Church showed [5] that they can be introduced by definition. This matter is also discussed in [3]).

In §4 we shall show how the natural numbers can be defined, and Peano's Postulates can be proved, in $\mathcal{A}$. The basic ideas here go back to Russell and Whitehead [11], of course, but our simple axiom of infinity is not that of Principia Mathematica, but is due to Bernays and Schönfinkel [4]. The natural numbers can be treated in a variety of ways in type theory (e.g., as in [5]), but we believe that the treatment given here has certain advantages of simplicity and naturalness. The simplicity of the axiom of infinity J^{ι} is essential to our program in §3.

Once one has represented the natural numbers in $\mathcal{A}$, one can easily represent the primitive recursive functions. (With minor changes in type symbols, the details can be found in Chapter 3 of [1].) Syntactic statements about wffs can be represented in the usual way by wffs of $\mathcal{A}$ via the device of Gödel numbering. Thus there is a wff *Consis* of $\mathcal{A}$ whose interpretation is that $\mathcal{A}$ is consistent, and by Gödel's theorem it is not the case that $\vdash_{\mathcal{A}}$ *Consis*. Nevertheless, much of mathematics can be formalized in $\mathcal{A}$.

The completeness theorem for $\mathcal{R}$ (Theorem 5.3 of [2]) is also a purely syntactic statement, and hence can be represented by a wff R of $\mathcal{A}$. After preparing the ground in §2 with some preliminary results, in §3 we shall show that by using the completeness of $\mathcal{R}$ we can prove the consistency of $\mathcal{A}$. This argument will be purely syntactic, and could be formalized in $\mathcal{A}$, so $\vdash_{\mathcal{A}} [R \supset Consis]$. Thus it is not the case that $\vdash_{\mathcal{A}} R$, so any proof of the completeness of resolution in type theory must transcend the rather considerable means of proof available in $\mathcal{A}$. Of course such a proof can be formalized in transfinite type theory or in Zermelo set theory.

§2. *Preliminary Definitions and Lemmas.* We first establish some preliminary results which will be useful in §3. The reader may wish to postpone the proofs of this section and proceed rapidly to §3. In presenting proofs of theorems of $\mathcal{T}$ (and extensions of $\mathcal{T}$), we shall make extensive use of proofs from hypotheses and the deduction theorem. Each line of a proof will have a number, which will appear at the left hand margin in parentheses. For the sake of brevity, this number will be used as an abbreviation for the wff which is asserted in that line. At the right hand margin we shall list the number(s) of the line(s) from which the given line is inferred (unless it is simply inferred from the preceding line). We use "hyp" to indicate that the wff is inferred with the aid of one or more of the hypotheses of the given line. Thus in

(.1)	$\vdash \mathbf{A}$	
(.2)	$\mathbf{B} \vdash \mathbf{B}$	hyp
(.3)	$\mathbf{B} \vdash \mathbf{C}$	.1, .2
(.4)	$\mathbf{D} \vdash \mathbf{C}$	.1, hyp

the hypothesis $\mathbf{B}$ is introduced in line .2, and $\mathbf{C}$ is inferred from $\mathbf{B}$ and the theorem $\mathbf{A}$ in line .3; $\mathbf{C}$ is also inferred from $\mathbf{A}$ and a different hypothesis $\mathbf{D}$ in line .4. However, if the wffs $\mathbf{B}$ and $\mathbf{C}$ are long, we may write this proof instead as follows:

(.1)	$\vdash \mathbf{A}$	
(.2)	$.2 \vdash \mathbf{B}$	hyp
(.3)	$.2 \vdash \mathbf{C}$	.1, .2
(.4)	$\mathbf{D} \vdash .3$	.1, hyp

A generally useful derived rule of inference is that if $\mathcal{H}$ is a set of hypotheses such that $\mathcal{H} \vdash \exists \mathbf{x} \mathbf{A}$ and $\mathcal{H}, \mathbf{A} \vdash \mathbf{B}$, where $\mathbf{x}$ does not occur free in $\mathbf{B}$ or any wff of $\mathcal{H}$, then $\mathcal{H} \vdash \mathbf{B}$. We shall indicate applications of this rule in the following fashion:

(.17)	$\mathcal{H} \vdash \exists \mathbf{x} \mathbf{A}$	...
(.20)	$\mathcal{H}, .20 \vdash \mathbf{A}$	choose $\mathbf{x}$ (.17)
(.23)	$\mathcal{H}, .20 \vdash \mathbf{B}$	...
(.24)	$\mathcal{H} \vdash \mathbf{B}$	.17, .23

If the wff $\mathbf{A}$ is long, we might write step (.17) as follows:

(.17) $\mathcal{H} \vdash \exists \mathbf{x}.20$

We shall present only abstracts of proofs, omitting many steps and using familiar laws of quantification theory, equality, and λ-conversion quite freely. We shall usually omit type symbols on occurrences of variables after the first.

Definition. For each wff $\mathbf{A}$ of $\mathcal{T}(\iota_{o\iota(o(o\iota))})$, let $\#\mathbf{A}$ be the wff of $\mathcal{T}$ which is the result of replacing the primitive constant $\iota_{o\iota(o(o\iota))}$ everywhere by the wff

$$[\lambda f_{o(o\iota)} \lambda z_\iota \,.\, \exists x_{o\iota} \,.\, f_{o(o\iota)} x_{o\iota} \wedge x_{o\iota} z_\iota].$$

Lemma 1. $\mathrm{E}^{o}, \mathrm{E}^{o\iota} \vdash_{\mathcal{T}} \#\mathrm{D}^{o\iota}$.

Proof: First note that $\#\mathrm{D}^{o\iota}$ conv $\forall f_{o(o\iota)} \,.\, \exists_1 x_{o\iota} f x \supset f[\lambda z_\iota \,.\, \exists x_{o\iota} \,.\, f x \wedge x z]$

(.1)	$.1 \vdash \exists_1 x_{o\iota} f_{o(o\iota)} x_{o\iota}$	hyp
(.2)	$.1, .2 \vdash f_{o(o\iota)} x_{o\iota} \wedge \forall u_{o\iota} . fu \supset u = x$	choose x(.1)
(.3)	$.1, .2 \vdash x_{o\iota} z_\iota \equiv \exists x_{o\iota} . f_{o(o\iota)} x \wedge xz$	.2
(.4)	$E^o, .1, .2 \vdash \forall z_\iota . x_{o\iota} z_\iota = \exists x_{o\iota} . f_{o(o\iota)} x \wedge xz$	.3, E^o
(.5)	$E^o, E^{o\iota}, .1, .2 \vdash x_{o\iota} = [\lambda z_\iota . \exists x_{o\iota} . f_{o(o\iota)} x \wedge xz]$	.4, $E^{o\iota}$
(.6)	$E^o, E^{o\iota}, .1, .2 \vdash f_{o(o\iota)} [\lambda z_\iota . \exists x_{o\iota} . fx \wedge xz]$	.2, .5
(.7)	$E^o, E^{o\iota}, .1 \vdash .6$	.1, .6
(.8)	$E^o, E^{o\iota} \vdash \# D^{o\iota}$	.7

Lemma 2. $J^\iota \vdash J^{o\iota}$

Proof: We assume J^ι.

(.1)	$.1 \vdash \forall x_\iota \forall y_\iota \forall z_\iota . \exists w_\iota r_{o\iota\iota} xw \wedge \sim rxx \wedge . \sim rxy \vee \sim ryz \vee rxz$	choose $r_{o\iota\iota}$

Let $K_{o(o\iota)(o\iota)}$ be

$$[\lambda u_{o\iota} \lambda v_{o\iota} . \exists t_\iota v_{o\iota} t_\iota \wedge . \sim \exists s_\iota u_{o\iota} s_\iota \vee \exists s_\iota . u_{o\iota} s_\iota \wedge \forall t_\iota . v_{o\iota} t_\iota \supset r_{o\iota\iota} s_\iota t_\iota].$$

We shall establish in lines (.11), (.16) and (.31) that K has the properties necessary to establish $J^{o\iota}$. To attack (.11) we consider two cases, (.2) and (.5).

(.2)	$.2 \vdash \sim \exists s_\iota x_{o\iota} s$	hyp (case 1)
(.3)	$.2 \vdash K x_{o\iota} [\lambda t_\iota . t_\iota = t_\iota]$	.2, def. of K
(.4)	$.2 \vdash \exists w_{o\iota} K x_{o\iota} w$	.3
(.5)	$.5 \vdash \exists s_\iota x_{o\iota} s_\iota$	hyp (case 2)
(.6)	$.5, .6 \vdash x_{o\iota} s_\iota$	choose s (.5)
(.7)	$.1, .5, .6, .7 \vdash r_{o\iota\iota} s_\iota w_\iota$	choose w_ι (.1)
(.8)	$.1, .5, .6, .7 \vdash K x_{o\iota} [\lambda t_\iota . w_\iota = t_\iota]$	.6, .7, def. of K
(.9)	$.1, .5, .6, .7 \vdash \exists w_{o\iota} K x_{o\iota} w$	.8
(.10)	$.1, .5 \vdash .9$	.9, .1, .5
(.11)	$.1 \vdash \exists w_{o\iota} K x_{o\iota} w$	.4, .10

Next we attack (.16). The proof is by contradiction.

(.12)	$.12 \vdash K x_{o\iota} x_{o\iota}$	hyp
(.13)	$.12 \vdash \exists s_\iota . x_{o\iota} s \wedge \forall t_\iota . xt \supset r_{o\iota\iota} st$	.12, def. of K
(.14)	$.12 \vdash \exists s_\iota r_{o\iota\iota} ss$	.13 (instantiate t with s)
(.15)	$.1 \vdash \forall s_\iota \sim r_{o\iota\iota} ss$	.1
(.16)	$.1 \vdash \sim K x_{o\iota} x_{o\iota}$	.14, .15

Finally we attack (.31).

(.17)	$.17 \vdash K x_{o\iota} y_{o\iota} \wedge K y_{o\iota} z_{o\iota}$	hyp
(.18)	$.17 \vdash \exists t_\iota y_{o\iota} t \wedge \exists t_\iota z_{o\iota} t$	.17, def. of K
(.19)	$.17 \vdash \sim \exists s_\iota x_{o\iota} s \vee \exists s_\iota . xs \wedge \forall q_\iota . y_{o\iota} q \supset r_{o\iota\iota} sq$	.17, def. of K

In (.20) and (.21) we consider the two possibilities set forth in (.19).

(.20)	$.17, \sim \exists s_\iota x_{o\iota} s \vdash K x_{o\iota} z_{o\iota}$	.18, hyp, def. of K
(.21)	$.17, .21 \vdash \exists s_\iota . x_{o\iota} s \wedge \forall q_\iota . y_{o\iota} q \supset r_{o\iota\iota} sq$	hyp
(.22)	$.17, .21, .22 \vdash x_{o\iota} s_\iota \wedge \forall q_\iota . y_{o\iota} q \supset r_{o\iota\iota} sq$	choose s (.21)

(.23)	$.17 \vdash \exists q_\iota .24$	.17, .18, def. of K
(.24)	$.17, .24 \vdash y_{o\iota} q_\iota \wedge \forall t_\iota .\ z_{o\iota} t \supset r_{o\iota\iota} qt$	choose q (.23)
(.25)	$.17, .21, .22, .24, z_{o\iota} t_\iota \vdash r_{o\iota\iota} s_\iota q_\iota \wedge rqt_\iota$	hyp, .22, .24
(.26)	$.1, .17, .21, .22, .24, z_{o\iota} t_\iota \vdash r_{o\iota\iota} s_\iota t_\iota$	.1, .25
(.27)	$.1, .17, .21, .22, .24 \vdash \forall t_\iota .\ z_{o\iota} t \supset r_{o\iota\iota} s_\iota t$	.26
(.28)	$.1, .17, .21, .22, .24 \vdash \mathsf{K} x_{o\iota} z_{o\iota}$	.18, .22, .27, def. of K
(.29)	$.1, .17, .21 \vdash .28$	.23, .21, .28
(.30)	$.1, .17 \vdash .28$	.19, .20, .29
(.31)	$.1 \vdash \sim \mathsf{K} x_{o\iota} y_{o\iota} \vee \sim \mathsf{K} y z_{o\iota} \vee \mathsf{K} xz$	.30
(.32)	$.1 \vdash \mathrm{J}^{o\iota}$	.11, .16, .31
(.33)	$\mathrm{J}^{\iota} \vdash \mathrm{J}^{o\iota}$	.32

We next repeat Gandy's definitions in [6] with some minor modifications.

Definition. By induction on γ, we define wffs $\mathsf{Mod}_{o\gamma}$ and $\mathsf{M}_{o\gamma\gamma}$ for each type symbol γ.

$\mathbf{A}_\gamma \overset{\mathsf{M}}{=} \mathbf{B}_\gamma$ stands for $\mathsf{M}_{o\gamma\gamma} \mathbf{A}_\gamma \mathbf{B}_\gamma$.
$\mathsf{Mod}_{o\kappa}$ stands for $[\lambda x_\kappa \exists p_o p_o]$ for $\kappa = o, \iota$.
M_{ooo} stands for $[\lambda p_o \lambda q_o .\ p_o \equiv q_o]$.
$\mathsf{M}_{o\iota\iota}$ stands for $[\lambda x_\iota \lambda y_\iota .\ x_\iota = y_\iota]$.
$\mathsf{Mod}_{o(\alpha\beta)}$ stands for $[\lambda f_{\alpha\beta} .\ \forall x_\beta \forall y_\beta .\ \mathsf{Mod}_{o\beta} x_\beta \wedge \mathsf{Mod}_{o\beta} y_\beta \wedge x_\beta \overset{\mathsf{M}}{=} y_\beta \supset .\ \mathsf{Mod}_{o\alpha}[f_{\alpha\beta} x_\beta] \wedge .\ f_{\alpha\beta} x_\beta \overset{\mathsf{M}}{=} f_{\alpha\beta} y_\beta]$.
$\mathsf{M}_{o(\alpha\beta)(\alpha\beta)}$ stands for $[\lambda f_{\alpha\beta} \lambda g_{\alpha\beta} .\ \forall x_\beta .\ \mathsf{Mod}_{o\beta} x_\beta \supset .\ f_{\alpha\beta} x_\beta \overset{\mathsf{M}}{=} g_{\alpha\beta} x_\beta]$.

Lemma 3. $\vdash_{\mathcal{T}} x_\alpha \overset{\mathsf{M}}{=} x_\alpha \wedge .\ x_\alpha \overset{\mathsf{M}}{=} y_\alpha \supset .\ z_\alpha \overset{\mathsf{M}}{=} x_\alpha \equiv .\ z_\alpha \overset{\mathsf{M}}{=} y_\alpha$.

Proof: By induction on α.

Definition. For each wff $\mathbf{A}$ of $\mathcal{T}$, $\mathbf{A}^{\mathsf{T}}$ is the result of replacing $\Pi_{o(o\alpha)}$ by $[\lambda f_{o\alpha} .\ \forall x_\alpha .\ \mathsf{Mod}_{o\alpha} x_\alpha \supset f_{o\alpha} x_\alpha]$ everywhere in $\mathbf{A}$.

Lemma 4. *If* $\mathbf{A}^1, \ldots, \mathbf{A}^n$, *and* $\mathbf{B}$ *are sentences of* $\mathcal{T}$ *such that* $\mathbf{A}_1, \ldots, \mathbf{A}^n \vdash_{\mathcal{T}} \mathbf{B}$, *then* $(\mathbf{A}^1)^{\mathsf{T}}, \ldots, (\mathbf{A}^n)^{\mathsf{T}} \vdash_{\mathcal{T}} \mathbf{B}^{\mathsf{T}}$.

Proof: This is an immediate consequence of Theorem 3.26 of [6], since Gandy's full translation $\mathbf{C}^{\mathsf{F}}$ of $\mathbf{C}$ is $\mathbf{C}^{\mathsf{T}}$ when $\mathbf{C}$ is a sentence. Our modifications of Gandy's definitions do not injure the proof.

Lemma 5. $\vdash_{\mathcal{T}} \mathsf{Mod}[\mathsf{M}_{o\alpha\alpha} z_\alpha]$.

Proof: $\mathsf{Mod}[\mathsf{M}_{o\alpha\alpha} z_\alpha]$ is equivalent to

$\forall x_\alpha \forall y_\alpha [\mathsf{Mod}\ x_\alpha \wedge \mathsf{Mod}\ y_\alpha \wedge x \overset{\mathsf{M}}{=} y \supset .\ \mathsf{Mod}[\mathsf{M}_{o\alpha\alpha} z_\alpha x_\alpha] \wedge .\ \mathsf{M}_{o\alpha\alpha} z_\alpha x_\alpha \equiv \mathsf{M}_{o\alpha\alpha} z_\alpha y_\alpha]$.

This is readily proved using the definition of Mod_{oo} and Lemma 3.

Lemma 6. $\vdash_{\mathcal{T}} (\mathrm{E}^\gamma)^{\mathsf{T}}$ *for each* E^γ *in* $\mathcal{E}$.

Proof: $(\mathrm{E}^o)^{\mathsf{T}}$ is equivalent to

$\forall p_o [\mathsf{Mod}\ p_o \supset \forall q_o .\ \mathsf{Mod}\ q_o \supset .\ [p_o \equiv q_o] \supset \forall f_{oo} .\ \mathsf{Mod}\ f_{oo} \supset .\ f_{oo} p_o \supset f_{oo} q_o]$,

which is easily proved using the definition of $\mathsf{Mod}\ f_{oo}$. $(\mathrm{E}^{\alpha\beta})^{\mathsf{T}}$ is equivalent to

$\forall f_{\alpha\beta}[\text{Mod } f \supset \forall g_{\alpha\beta}.\ \text{Mod } g \supset.\ \forall x_\beta[\text{Mod } x \supset \forall h_{o\alpha}.\ \text{Mod } h \supset.\ h[fx] \supset h.gx]$
$\supset \forall k_{o(\alpha\beta)}.\ \text{Mod } k \supset.\ kf \supset kg]$,

which we prove as follows:

(.1) $.1 \vdash \text{Mod } f_{\alpha\beta} \wedge \text{Mod } g_{\alpha\beta}$ hyp

(.2) $.2 \vdash \forall x_\beta[\text{Mod } x \supset \forall h_{o\alpha}.\ \text{Mod } h \supset .h[fx] \supset h.gx]$ hyp

(.3) $.3 \vdash \text{Mod } k_{o(\alpha\beta)}$ hyp

(.4) $\vdash \text{Mod}_{(o\alpha)}\ .\ \mathsf{M}_{o\alpha\alpha} \cdot f_{\alpha\beta}x_\beta$ Lemma 5

(.5) $.2,\ \text{Mod } x_\beta \vdash [\mathsf{M}_{o\alpha\alpha}\cdot f_{\alpha\beta}x_\beta]\ [f_{\alpha\beta}x_\beta] \supset.\ [\mathsf{M}_{o\alpha\alpha}\cdot f_{\alpha\beta}x_\beta]\ .g_{\alpha\beta}x_\beta$
.2, .4 (instantiate $h_{o\alpha}$ with $\mathsf{M}[fx]$)

(.6) $\vdash \mathsf{M}_{o\alpha\alpha}[f_{\alpha\beta}x_\beta]\ [f_{\alpha\beta}x_\beta]$ Lemma 3

(.7) $.2,\ \text{Mod } x_\beta \vdash f_{\alpha\beta}x_\beta \overset{\mathsf{M}}{=} g_{\alpha\beta}x_\beta$ .5, .6

(.8) $.2 \vdash f_{\alpha\beta} \overset{\mathsf{M}}{=} g_{\alpha\beta}$ .7, def. of $\mathsf{M}_{o(\alpha\beta)(\alpha\beta)}$

(.9) $.1,\ .2,\ .3 \vdash k_{o(\alpha\beta)}f_{\alpha\beta} \equiv k_{o(\alpha\beta)}g_{\alpha\beta}$ .3, def. of $\text{Mod } k_{o(\alpha\beta)}$, .1, .8

(.10) $\vdash (\mathrm{E}^{\alpha\beta})^{\mathsf{T}}$ .9

Lemma 7. $\vdash_{\mathcal{T}} \text{Mod } r_{o\iota\iota}$.

Proof: $\text{Mod } z_{o\iota}$ is equivalent to

$\forall x_\iota\ \forall y_\iota[\text{Mod } x_\iota \wedge \text{Mod } y_\iota \wedge x_\iota = y_\iota \supset.\ \text{Mod}[z_{o\iota}x_\iota] \wedge.\ z_{o\iota}x_\iota \equiv z_{o\iota}y_\iota]$

so $\vdash \forall z_{o\iota}\ \text{Mod } z_{o\iota}$. $\text{Mod } r_{o\iota\iota}$ is equivalent to

$\forall x_\iota\ \forall y_\iota[\text{Mod } x_\iota \wedge \text{Mod } y_\iota \wedge x_\iota = y_\iota \supset.\ \text{Mod}[r_{o\iota\iota}x_\iota] \wedge \forall w_\iota.\ \text{Mod } w_\iota \supset.$
$r_{o\iota\iota}x_\iota w_\iota \equiv r_{o\iota\iota}y_\iota w_\iota]$,

which is easily proved.

Lemma 8. $\mathrm{J}^\iota \vdash_{\mathcal{T}} (\mathrm{J}^\iota)^{\mathsf{T}}$.

Proof: $(\mathrm{J}^\iota)^{\mathsf{T}}$ is equivalent to

$\exists r_{o\iota\iota}[\text{Mod } r_{o\iota\iota} \wedge \forall x_\iota.\ \text{Mod } x_\iota \supset \forall y_\iota.\ \text{Mod } y_\iota \supset \forall z_\iota.\ \text{Mod } z_\iota \supset.$
$\exists w_\iota[\text{Mod } w_\iota \wedge r_{o\iota\iota}x_\iota w_\iota] \wedge \sim r_{o\iota\iota}x_\iota x_\iota \wedge.\ \sim r_{o\iota\iota}x_\iota y_\iota \vee \sim r_{o\iota\iota}y_\iota z_\iota \vee r_{o\iota\iota}x_\iota z_\iota]$.

This is easily derived from J^ι with the aid of Lemma 7.

Definition: Let θ be the substitution $\mathsf{S}^{\mathbf{x}^1 \ldots \mathbf{x}^n}_{\mathbf{A}^1 \ldots \mathbf{A}^n}$, i.e., the simultaneous substitution of $\mathbf{A}^i$ for all free occurrences of $\mathbf{x}^i$ for $1 \leqslant i \leqslant n$, where $\mathbf{x}^1, \ldots, \mathbf{x}^n$ are distinct variables and $\mathbf{A}^i$ has the same type as $\mathbf{x}^i$ for $1 \leqslant i \leqslant n$. If $\mathbf{B}$ is any wff, we let $\theta * \mathbf{B}$ denote $\eta[[\lambda\mathbf{x}^1 \ldots \lambda\mathbf{x}^n\mathbf{B}]\,\mathbf{A}^1 \ldots \mathbf{A}^n]$. If θ is the null substitution (i.e., $n = 0$), then $\theta * \mathbf{B}$ denotes $\eta\mathbf{B}$.

Note that if $\mathbf{x}_\alpha$ and $\mathbf{y}_\beta$ are distinct variables, $[[\lambda\mathbf{x}_\alpha\lambda\mathbf{y}_\beta\mathbf{B}]\,\mathbf{A}_\alpha\mathbf{C}_\beta]$ conv $[[\lambda\mathbf{y}_\beta\lambda\mathbf{x}_\alpha\mathbf{B}]\mathbf{C}_\beta\mathbf{A}_\alpha]$, so the definition above is unambiguous. Clearly, if there are no conflicts of bound variables, $\theta * \mathbf{B}$ is simply $\eta\,\theta\mathbf{B}$, the η-normal form of the result of applying the substitution θ to $\mathbf{B}$. From the definition it is evident that if $\mathbf{B}$ conv $\mathbf{C}$, then $\theta * \mathbf{B} = \theta * \mathbf{C}$.

§3. *The Consistency of $\mathcal{A}$.*

Theorem. *$\mathcal{A}$ is consistent.*

Proof: The proof is by contradiction, so we suppose $\mathcal{A}$ is inconsistent. Thus

(1) $J^{\iota}, \mathcal{E}, D^{\iota} \vdash_{\mathcal{T}(\iota_{\iota(o\iota)})} \Box$.
(2) $J^{o\iota}, \mathcal{E}, D^{o\iota} \vdash_{\mathcal{T}(\iota_{o\iota(o(o\iota))})} \Box$.

Proof: Replace the type symbol ι by the type symbol $(o\iota)$ everywhere in the sequence of wffs which constitutes a proof of $\Box$ whose existence is asserted in step 1. By checking the axioms and rules of inference of $\mathcal{T}$ one easily sees that a proof of $\Box$ satisfying the requirements of step 2 is obtained.

(3) $J^{o\iota}, \mathcal{E}, \#D^{o\iota} \vdash_{\mathcal{T}} \Box$.

Proof: The replacement of **A** by #**A** everywhere in the proof whose existence is asserted in step 2 yields a proof satisfying step 3, possibly after the insertion of a few applications of the rule of alphabetic change of bound variables.

(4) $J^{o\iota}, \mathcal{E} \vdash_{\mathcal{T}} \Box$ by Lemma 1.
(5) $J^{\iota}, \mathcal{E} \vdash_{\mathcal{T}} \Box$ by Lemma 2.
(6) $(J^{\iota})^{\mathsf{T}}, \{(E^{\gamma})^{\mathsf{T}} \mid E^{\gamma} \in \mathcal{E}\} \vdash_{\mathcal{T}} \Box$

Proof: By Lemma 4, since $\vdash_{\mathcal{T}} \Box^{\mathsf{T}} \supset \Box$.

(7) $(J^{\iota})^{\mathsf{T}} \vdash_{\mathcal{T}} \Box$ by Lemma 6.
(8) $J^{\iota} \vdash_{\mathcal{T}} \Box$ by Lemma 8.

We next introduce parameters $\bar{r}_{o\iota\iota}$ and $\bar{g}_{\iota\iota}$. Let:

$\mathcal{J} = \{\forall x_\iota \bar{r}_{o\iota\iota} x_\iota [\bar{g}_{\iota\iota} x_\iota],\ \forall x_\iota \sim \bar{r}_{o\iota\iota} x_\iota x_\iota,\ \forall x_\iota \forall y_\iota \forall z_\iota . \sim \bar{r}_{o\iota\iota} x_\iota y_\iota \vee \sim \bar{r}_{o\iota\iota} y_\iota z_\iota \vee r_{o\iota\iota} x_\iota z_\iota\}$.

(9) $\mathcal{J} \vdash_{\mathcal{T}(\bar{r}_{o\iota\iota}, \bar{g}_{\iota\iota})} \Box$.

Proof: $J^{\iota} \vdash_{\mathcal{T}(\bar{r}, \bar{g})} \Box$ by (8), and $\mathcal{J} \vdash_{\mathcal{T}(\bar{r}, \bar{g})} J^{\iota}$.

(10) $\mathcal{J} \vdash_{\mathcal{R}} \Box$

Proof: This follows from (9) by the completeness of resolution in type theory, i.e., Theorem 5.3 of [2]. The proof of this theorem is the one non-syntactic step in our present proof of the consistency of $\mathcal{A}$.

(11) It is not the case that $\mathcal{J} \vdash_{\mathcal{R}} \Box$.

Proof: An η-wff of the form $\bar{r}_{o\iota\iota} \mathbf{A}_\iota \mathbf{B}_\iota$ will be called *positive* if the number of occurrences of $\bar{g}_{\iota\iota}$ in $\mathbf{A}_\iota$ is strictly less than the number of occurrences of $\bar{g}_{\iota\iota}$ in $\mathbf{B}_\iota$, and otherwise *negative*. An η-wff of the form $\sim r_{o\iota\iota} \mathbf{A}_\iota \mathbf{B}_\iota$ will be called positive iff $\bar{r}_{o\iota\iota} \mathbf{A}_\iota \mathbf{B}_\iota$ is negative, and negative iff $\bar{r}_{o\iota\iota} \mathbf{A}_\iota \mathbf{B}_\iota$ is positive. Let $\mathcal{G}$ be the set of wffs **G** having one of the following six forms:

(a) $\forall \mathbf{x}_\iota\ \bar{r}\ \mathbf{x}[\bar{g}\mathbf{x}]$
(b) $\forall \mathbf{x}_\iota \sim \bar{r}\mathbf{x}\mathbf{x}$
(c) $\forall \mathbf{x}_\iota \forall \mathbf{y}_\iota \forall \mathbf{z}_\iota\ [\sim \bar{r}\mathbf{x}\mathbf{y} \vee \sim \bar{r}\mathbf{y}\mathbf{z} \vee \bar{r}\mathbf{x}\mathbf{z}]$ where $\mathbf{x}_\iota$, $\mathbf{y}_\iota$, and $\mathbf{z}_\iota$ are distinct variables.

(d) $\forall y_\iota \forall z_\iota\, [\sim \bar{r}\, A_\iota y \vee \sim \bar{r} yz \vee \bar{r}\, A_\iota z]$ where y_ι and z_ι are distinct from one another and from the free variables of A_ι.
(e) $\forall z_\iota\, [\sim \bar{r}\, A_\iota B_\iota \vee \sim \bar{r}\, B_\iota z \vee \bar{r}\, A_\iota z]$ where z_ι is distinct from the free variables of A_ι and of B_ι.
(f) G is a disjunction of wffs, each of the form $\bar{r}\, A_\iota B_\iota$ or $\sim \bar{r}\, AB$, at least one of which is positive.

Let $\mathcal{C}$ be the set of wffs C such that for each substitution θ, $\theta * C$ is in $\mathcal{I}$. We assert that if $\mathcal{J} \vdash_{\mathcal{R}} C$, then $C \in \mathcal{C}$. Clearly $\mathcal{J} \subseteq \mathcal{C}$, so it suffices to show that $\mathcal{C}$ is closed under the rules of inference of $\mathcal{R}$. For each rule of inference of $\mathcal{R}$ and any substitution θ, we show that $\theta * E \in \mathcal{I}$ for any wff E derived from wff(s) of $\mathcal{C}$ by that rule.

Suppose $M \vee A$ and $N \vee \sim A$ are in $\mathcal{C}$, and $M \vee N$ is obtained from them by cut. Then $\theta * [M \vee A]$ and $\theta * [N \vee \sim A]$ must each have form (f). (For $\theta * [N \vee \sim A] = [(\theta * N) \vee \sim (\theta * A)]$; even if N is null, this cannot have any of the forms (a)-(e), so $\theta * A$ must have the form $\bar{r} B_\iota C_\iota$.) $\theta * [M \vee A] = [(\theta * M) \vee \theta * A]$; if $\theta * A$ is negative, $\theta * M$ must contain a positive wff (so M cannot be null), so $\theta * [M \vee N]$ does also. If $\theta * A$ is positive, then $\theta * [\sim A]$ is negative, so $\theta * N$ must contain a positive wff, so $\theta * [M \vee N]$ does also, and hence has form (f).

Suppose D is in $\mathcal{C}$, and $[\lambda x_\alpha D] B_\alpha$ is obtained from D by substitution. Let ρ be the substitution $\mathop{S}^{x_\alpha}_{B_\alpha}$, and let $\theta \circ \rho$ be the substitution which is the composition of θ with ρ (i.e., $(\theta \circ \rho) * C = \theta * (\rho * C)$ for each wff C). Then $\theta * [[\lambda x_\alpha D] B_\alpha] = \theta * \eta [[\lambda x_\alpha D] B_\alpha] = \theta * (\rho * D) = (\theta \circ \rho) * D \in \mathcal{I}$ since $D \in \mathcal{C}$, so $[[\lambda x_\alpha D] B_\alpha] \in \mathcal{C}$.

Suppose $D \in \mathcal{C}$ and E is derived from D by universal instantiation. Thus D has the form $M \vee \Pi_{o(o\alpha)} A_{o\alpha}$, where M may be null. By considering the null substitution we see that $\eta D \in \mathcal{I}$, so D has the form $\Pi_{o(o\iota)} A_{o\iota}$ and E has the form $A_{o\iota} x_\iota$. It is easily checked by examining forms (a)-(e) that if H is any wff obtained from a wff of $\mathcal{I}$ by universal instantiation, then $(\theta * H) \in \mathcal{I}$. But $(\eta\, A_{o\iota}) x_\iota$ is obtained from ηD by universal instantiation, so $\theta * E = \theta * [(\eta\, A_{o\iota}) x_\iota]$ is in $\mathcal{I}$.

The verification that $\mathcal{C}$ is closed under the remaining rules of inference of $\mathcal{R}$ is trivial, so our assertion is proved. Now $\Box$ is not in $\mathcal{C}$, so it is not the case that $\mathcal{J} \vdash_{\mathcal{R}} \Box$.

(12) The contradiction between (10) and (11) proves our theorem.

§4. *The Natural Numbers in $\mathcal{A}$.* We shall define the natural numbers to be equivalence classes of sets of individuals having the same finite cardinality. We let σ denote the type symbol $(o(o\iota))$. σ is the type of natural numbers.

Definitions:

0_σ stands for $[\lambda p_{o\iota} \forall x_\iota \sim p_{o\iota} x_\iota]$.
$S_{\sigma\sigma}$ stands for $[\lambda n_{o(o\iota)} \lambda p_{o\iota} . \exists x_\iota . p_{o\iota} x_\iota \wedge n_{o(o\iota)} [\lambda t_\iota . t_\iota \neq x_\iota \wedge p_{o\iota} t_\iota]]$.
$N_{o\sigma}$ stands for $[\lambda n_\sigma \forall p_{o\sigma} . [p_{o\sigma} 0_\sigma \wedge \forall x_\sigma . p_{o\sigma} x_\sigma \supset p_{o\sigma} S_{\sigma\sigma} x_\sigma] \supset p_{o\sigma} n_\sigma]$.

$\dot{\forall} x_\sigma A$ stands for $\forall x_\sigma[\mathsf{N}_{o\sigma} x_\sigma \supset A]$.
$\dot{\exists} x_\sigma A$ stands for $\exists x_\sigma[\mathsf{N}_{o\sigma} x_\sigma \wedge A]$.

Thus zero is the collection of all sets with zero members, i.e., the collection containing just the empty set $[\lambda x_\iota \Box]$. S represents the successor function. If $n_{(o(o\iota))}$ is a finite cardinal (say 2), then a set $p_{o\iota}$ (say $\{a, b, c\}$) is in $\mathsf{S}n$ iff there is an individual (say c) which is in $p_{o\iota}$ and whose deletion from $p_{o\iota}$ leaves a set ($\{a, b\}$) which is in n. $\mathsf{N}_{o\sigma}$ represents the set of natural numbers, i.e., the intersection of all sets which contain O and are closed under S.

We now prove Peano's Postulates (Theorems 1, 2, 3, 4, and 7 below.) In this section $\vdash B$ means B is a theorem of $\mathcal{A}$.

1 $\vdash \mathsf{N}_{o\sigma}\mathsf{O}_\sigma$ — by the def. of N

2 $\vdash \forall x_\sigma.\ \mathsf{N}_{o\sigma} x_\sigma \supset \mathsf{N}_{o\sigma}.\ \mathsf{S}_{\sigma\sigma} x_\sigma$

Proof:

(.1)	$\mathsf{N}x_\sigma,\ .1 \vdash p_{o\sigma}\mathsf{O} \wedge \forall x_\sigma.\ px \supset p.\ \mathsf{S}x$	hyp
(.2)	$\mathsf{N}x_\sigma,\ .1 \vdash p_{o\sigma}x_\sigma$	.1, hyp, def. of N
(.3)	$\mathsf{N}x_\sigma,\ .1 \vdash p_{o\sigma}.\ \mathsf{S}x_\sigma$	.1, .2
(.4)	$\mathsf{N}x_\sigma \vdash \mathsf{N}.\ \mathsf{S}x_\sigma$	.3, def. of N

3 The Induction Theorem:

$$\vdash \forall p_{o\sigma}.\ [p_{o\sigma}\mathsf{O}_\sigma \wedge \dot{\forall} x_\sigma.\ p_{o\sigma}x_\sigma \supset p_{o\sigma}.\ \mathsf{S}_{\sigma\sigma}x_\sigma] \supset \dot{\forall} x_\sigma p_{o\sigma}x_\sigma$$

Proof: Let $\mathrm{P}_{o\sigma}$ be $[\lambda t_\sigma.\ \mathsf{N}t \wedge p_{o\sigma}t]$.

(.1)	$.1 \vdash p_{o\sigma}\mathsf{O} \wedge \forall x_\sigma.\ \mathsf{N}x \supset.\ px \supset p.\ \mathsf{S}x$	hyp
(.2)	$\mathsf{N}y_\sigma \vdash [\mathrm{P}\mathsf{O} \wedge \forall x_\sigma.\ \mathrm{P}x \supset \mathrm{P}.\ \mathsf{S}x] \supset \mathrm{P}y_\sigma$	hyp, def. of N
(.3)	$.1 \vdash \mathrm{P}\mathsf{O}$	def. of P, .1, Theorem 1
(.4)	$.1 \vdash \forall x_\sigma.\ \mathrm{P}x \supset \mathrm{P}.\ \mathsf{S}x$	def. of P, .1, Theorem 2
(.5)	$.1, \mathsf{N}y_\sigma \vdash \mathrm{P}y_\sigma$	.2, .3, .4
(.6)	$.1 \vdash \dot{\forall} y_\sigma p y_\sigma$	.5, def. of $\dot{\forall}$, P

4 $\vdash \dot{\forall} n_\sigma.\ \mathsf{S}_{\sigma\sigma}n_\sigma \neq \mathsf{O}_\sigma$

Proof by contradiction:

(.1)	$.1 \vdash \mathsf{S}n_\sigma = \mathsf{O}$	hyp
(.2)	$\vdash \mathsf{O}_\sigma[\lambda x_\iota \Box]$	def. of O
(.3)	$.1 \vdash \mathsf{S}n_\sigma[\lambda x_\iota \Box]$	.1, .2
(.4)	$.1 \vdash \exists x_\iota \Box$	.3, def. of S
(.5)	$\vdash \mathsf{S}n_\sigma \neq \mathsf{O}$	.4
(.6)	$\vdash \dot{\forall} n_\sigma.\ \mathsf{S}n \neq \mathsf{O}$	.5, def. of $\dot{\forall}$

Our first step in proving Theorem 7 is to show that if we remove any element from a set of cardinality $\mathsf{S}n$ we obtain a set of cardinality n.

5 $\vdash \dot{\forall} n_\sigma \forall p_{o\iota}.\ \sim p_{o\iota}w_\iota \wedge \mathsf{S}_{\sigma\sigma}n_\sigma[\lambda t_\iota.\ t_\iota = w_\iota \vee p_{o\iota}t_\iota] \supset n_\sigma p_{o\iota}$

The proof is by induction on n. First we treat the case $n = \mathsf{O}$.

(.1)	$.1 \vdash \sim p_{o\iota} w_\iota \wedge \mathsf{SO}[\lambda t_\iota . t = w \vee pt]$	hyp
(.2)	$.1 \vdash \exists x_\iota . 3$	.1, def. of S
(.3)	$.1, .3 \vdash [x_\iota = w_\iota \vee p_{o\iota} x] \wedge \mathsf{O}[\lambda t_\iota . t \neq x \wedge . t = w \vee pt]$	choose x (.2)
(.4)	$.1, .3 \vdash \sim . w_\iota \neq x_\iota \wedge . w = w \vee p_{o\iota} w_\iota$	.3, def. of O
(.5)	$.1, .3 \vdash w_\iota = x_\iota$	.4
(.6)	$.1, .3 \vdash \forall t_\iota . p_{o\iota} t \equiv . t \neq x_\iota \wedge . t = w_\iota \vee pt$	.1, .5
(.7)	$.1, .3 \vdash p_{o\iota} = [\lambda t_\iota . t \neq x_\iota \wedge . t = w_\iota \vee pt]$	.6, E^o, $\mathrm{E}^{o\iota}$
(.8)	$.1, .3 \vdash \mathsf{O}\, p_{o\iota}$	.3, .7
(.9)	$\vdash \forall p_{o\iota} . \sim p_{o\iota} w_\iota \wedge \mathsf{SO}[\lambda t_\iota . t = w \vee pt] \supset \mathsf{O}p$	.2, .8

Next we treat the induction step

(.10)	$.10 \vdash \mathsf{N} n_\sigma \wedge \forall p_{o\iota} . \sim p w_\iota \wedge \mathsf{S}n[\lambda t_\iota . t = w \vee pt] \supset np$	(inductive) hyp
(.11)	$.11 \vdash \sim p_{o\iota} w_\iota \wedge [\mathsf{SS}n_\sigma][\lambda t_\iota . t = w \vee pt]$	hyp
(.12)	$.11 \vdash \exists x_\iota . 13$	.11, def. of S
(.13)	$.11, .13 \vdash [x_\iota = w_\iota \vee p_{o\iota} x] \wedge . \mathsf{S}n_\sigma[\lambda t_\iota . t \neq x \wedge . t = w \vee pt]$	choose x (.12)

From (.11) we must prove $[\mathsf{S}n]p$. We consider two cases in (.14) and (.17).

(.14)	$.14 \vdash x_\iota = w_\iota$	hyp (case 1)
(.15)	$.11, .13, .14 \vdash p_{o\iota} = [\lambda t_\iota . t \neq x_\iota \wedge . t = w_\iota \vee pt]$	.11, .14
(.16)	$.11, .13, .14 \vdash [\mathsf{S}n_\sigma] p_{o\iota}$	.13, .15

In case 2 we shall use the inductive hypothesis.

(.17)	$.17 \vdash x_\iota \neq w_\iota$	hyp (case 2)
(.18)	$.17 \vdash [\lambda t_\iota . t \neq x_\iota \wedge . t = w_\iota \vee p_{o\iota} t] = [\lambda t_\iota . t = w_\iota \vee . t \neq x_\iota \wedge p_{o\iota} t]$	.17
(.19)	$.11, .13, .17 \vdash \mathsf{S}n_\sigma[\lambda t_\iota . t = w_\iota \vee . t \neq x_\iota \wedge p_{o\iota} t]$	.13, .18
(.20)	$.10, .11, .13, .17 \vdash n_\sigma[\lambda t_\iota . t \neq x_\iota \wedge p_{o\iota} t]$	.10, .11, .19
(.21)	$.11, .13, .17 \vdash p_{o\iota} x_\iota$	.13, .17
(.22)	$.10, .11, .13, .17 \vdash [\mathsf{S}n_\sigma] p_{o\iota}$	def. of S, .20, .21
(.23)	$.10, .11 \vdash [\mathsf{S}n_\sigma] p_{o\iota}$	.16, .22, .12
(.24)	$.10 \vdash \forall p_{o\iota} . \sim p w_\iota \wedge [\mathsf{SS}n_\sigma][\lambda t_\iota . t = w \vee pt] \supset [\mathsf{S}n_\sigma] p$	.23

This completes the induction step. The theorem now follows from .9 and .24 by the Induction Theorem.

It will be observed that so far in this section we have not used the axiom of infinity J^ι. We shall use it in proving the next theorem, which will also be used to prove Theorem 7.

6 $\vdash \dot{\forall} n_\sigma . n_\sigma p_{o\iota} \supset \exists w_\iota \sim p_{o\iota} w_\iota$

(.1)	$.1 \vdash \forall x_\iota \forall y_\iota \forall z_\iota . \exists w_\iota r_{o\iota\iota} x w \wedge \sim r x x \wedge . \sim r x y \vee \sim r y z \vee r x z$	choose r (J^ι)

Let $\mathsf{P}_{o\sigma}$ be $[\lambda n_\sigma \forall p_{o\iota} . np \supset \exists z_\iota \forall w_\iota . r_{o\iota\iota} z w \supset \sim p w]$.

We may informally interpret rzw as meaning that z is below w. Thus $\mathsf{P}n$ means that if p is in n, then there is an element z which is below no member of p. We shall prove $\dot{\forall} n_\sigma \mathsf{P} n$ by induction on n.

(.2)	$\mathsf{O} p_{o\iota} \vdash \sim p_{o\iota} w_\iota$	def. of O
(.3)	$\vdash \mathsf{PO}$	.2, def. of P

Next we treat the induction step.

(.4)	$.4 \vdash \mathsf{N} n_\sigma \wedge \mathsf{P} n$	(inductive) hyp
(.5)	$.5 \vdash \mathsf{S} n_\sigma p_{o\iota}$	hyp
(.6)	$.5 \vdash \exists x_\iota\ .7$	.5, def. of S
(.7)	$.5,\ .7 \vdash p_{o\iota} x_\iota \wedge n_\sigma [\lambda t_\iota .\ t \neq x \wedge pt]$	choose x (.6)
(.8)	$.4,\ .5,\ .7 \vdash \exists z_\iota\ .9$	.4, def. of P, .7
(.9)	$.4,\ .5,\ .7,\ .9 \vdash \forall w_\iota .\ r_{o\iota\iota} z_\iota w \supset .\ w = x_\iota \vee \sim p_{o\iota} w$	choose z (.8)

Thus from the inductive hypothesis we see that there is an element z which is under nothing in $p - \{x\}$. We must show that there is an element which is under nothing in p. We consider two cases, (.10) and (.14).

(.10)	$.10 \vdash \sim r_{o\iota\iota} z_\iota x_\iota$	hyp (case 1)
(.11)	$.4,\ .5,\ .7,\ .9,\ .10 \vdash r_{o\iota\iota} z_\iota w_\iota \supset w \neq x_\iota$	.10
(.12)	$.4,\ .5,\ .7,\ .9,\ .10 \vdash \forall w_\iota .\ r_{o\iota\iota} z_\iota w \supset \sim p_{o\iota} w$	.9, .11
(.13)	$.4,\ .5,\ .7,\ .9,\ .10 \vdash \exists z_\iota\ .12$	.12

Next we consider case 2, and show that x is under nothing in p.

(.14)	$.14 \vdash r_{o\iota\iota} z_\iota x_\iota$	hyp (case 2)
(.15)	$.1,\ .14,\ r_{o\iota\iota} x_\iota w_\iota \vdash r_{o\iota\iota} z_\iota w_\iota$	.14, hyp, .1
(.16)	$.1,\ .4,\ .5,\ .7,\ .9,\ .14,\ r_{o\iota\iota} x_\iota w_\iota \vdash w_\iota = x_\iota \vee \sim p_{o\iota} w$	.9, .15
(.17)	$\vdash w_\iota = x_\iota \supset .\ r_{o\iota\iota} x w \supset r x x$	
(.18)	$.1 \vdash \sim r_{o\iota\iota} x_\iota x$	.1
(.19)	$.1,\ .4,\ .5,\ .7,\ .9,\ .14 \vdash \forall w_\iota .\ r_{o\iota\iota} x_\iota w \supset \sim p_{o\iota} w$	.16, .17, .18
(.20)	$.1,\ .4,\ .5,\ .7,\ .9,\ .14 \vdash \exists z_\iota \forall w_\iota .\ r_{o\iota\iota} z w \supset \sim p_{o\iota} w$	.19
(.21)	$.1,\ .4,\ .5 \vdash .20$	.13, .20, .8, .6
(.22)	$.1 \vdash \mathsf{N} n_\sigma \wedge \mathsf{P} n \supset \mathsf{PS} n$	.21, def. of P
(.23)	$.1 \vdash \dot{\forall} n_\sigma\, \mathsf{P} n_\sigma$	.3, .22, Theorem 3

Having finished the inductive proof, we proceed to prove the main theorem.

(.24)	$.24 \vdash \mathsf{N} n_\sigma \wedge n p_{o\iota}$	hyp
(.25)	$.1,\ .24 \vdash \exists z_\iota \forall w_\iota .\ r_{o\iota\iota} z w \supset \sim p_{o\iota} w$	.23, .24, def. of P
(.26)	$.1 \vdash \forall z_\iota\ \exists w_\iota r_{o\iota\iota} z w$	.1
(.27)	$.1,\ .24 \vdash \exists w_\iota \sim p_{o\iota} w_\iota$	.25, .26
(.28)	$.1 \vdash \dot{\forall} n_\sigma .\ n p_{o\iota} \supset \exists w_\iota \sim p_{o\iota} w_\iota$	.27
(.29)	$\vdash .28$	J^{ι}

7 $\vdash \dot{\forall} n_\sigma \dot{\forall} m_\sigma .\ \mathsf{S}_{\sigma\sigma} n_\sigma = \mathsf{S}_{\sigma\sigma} m_\sigma \supset n_\sigma = m_\sigma$

Proof:

(.1)	$.1 \vdash \mathsf{N} n_\sigma \wedge \mathsf{N} m_\sigma \wedge \mathsf{S} n = \mathsf{S} m$	hyp
(.2)	$.2 \vdash n_\sigma p_{o\iota}$	hyp
(.3)	$.1,\ .2, \vdash \exists w_\iota \sim p_{o\iota} w_\iota$	.1, .2, Theorem 6
(.4)	$.1,\ .2,\ .4 \vdash \sim p_{o\iota} w_\iota$	choose w (.3)
(.5)	$.1,\ .2,\ .4 \vdash p_{o\iota} = [\lambda t_\iota .\ t \neq w_\iota \wedge .\ t = w \vee pt]$	.4, E^{o}, $\mathrm{E}^{o\iota}$
(.6)	$.1,\ .2,\ .4 \vdash n_\sigma [\lambda t_\iota .\ t \neq w_\iota \wedge .\ t = w \vee p_{o\iota} t]$	.2, .5
(.7)	$.1,\ .2,\ .4 \vdash \mathsf{S} n_\sigma [\lambda t_\iota .\ t = w_\iota \vee p_{o\iota} t]$	.6, def. of S

(.8)	.1, .2, .4 $\vdash \mathrm{S}\, m_\sigma[\lambda t_\iota .\ t = w_\iota \vee p_{o\iota}t]$	.1, .7
(.9)	.1, .2, .4 $\vdash m_\sigma p_{o\iota}$	.1, .4, .8, Theorem 5
(.10)	.1 $\vdash n_\sigma p_{o\iota} \supset m_\sigma p$	.3, .9
(.11)	.1 $\vdash m_\sigma p_{o\iota} \supset n_\sigma p$	proof as for .10
(.12)	.1 $\vdash \forall p_{o\iota} .\ n_\sigma p \equiv m_\sigma p$	.10, .11
(.13)	.1 $\vdash n_\sigma = m_\sigma$	.12, E^o, $E^{o\iota}$

REFERENCES

[1] Andrews, Peter B., *A Transfinite Type Theory with Type Variables*, North-Holland Publishing Company, Amsterdam (1965).

[2] Andrews, Peter B., "Resolution in type theory," *The Journal of Symbolic Logic*, vol. 36 (1971), pp. 414-432.

[3] Andrews, Peter B., "General models, descriptions, and choice in type theory," *The Journal of Symbolic Logic*, vol. 37 (1972), pp. 385-394.

[4] Bernays, Paul, and Moses Schönfinkel, "Zum Entscheidungsproblem der mathematischen Logik," *Mathematische Annalen*, vol. 99 (1928), pp. 342-372.

[5] Church, Alonzo, "A formulation of the simple theory of types," *The Journal of Symbolic Logic*, vol. 5 (1940), pp. 56-68.

[6] Gandy, R. O., "On the axiom of extensionality, Part I," *The Journal of Symbolic Logic*, vol. 21 (1956), pp. 36-48.

[7] Pravitz, Dag, "Hauptsatz for higher order logic," *The Journal of Symbolic Logic*, vol. 33 (1968), pp. 452-457.

[8] Takahashi, Moto-o, "A proof of cut-elimination theorem in simple type-theory," *Journal of the Mathematical Society of Japan*, vol. 19 (1967), pp. 399-410.

[9] Takeuti, Gaisi, "On a generalized logic calculus," *Japanese Journal of Mathematics*, vol. 23 (1953), pp. 39-96; errata: *ibid.*, vol. 24 (1954), pp. 149-156.

[10] Takeuti, Gaisi, "Remark on the fundamental conjecture of GLC," *Journal of the Mathematical Society of Japan*, vol. 10 (1958), pp. 44-45.

[11] Whitehead, Alfred North, and Bertrand Russell, *Principia Mathematica*, Cambridge University Press (1913).

Carnegie-Mellon University
Pittsburgh, Pennsylvania

Journal of Automated Reasoning **5**: 257–291, 1989.

On Connections and Higher-Order Logic

PETER B. ANDREWS
Mathematics Department, Carnegie-Mellon University, Pittsburgh, PA 15213, U.S.A.

(Received: 13 December 1988)

Abstract. This is an expository introduction to an approach to theorem proving in higher-order logic based on establishing appropriate *connections* between subformulas of an expanded form of the theorem to be proved. *Expansion trees* and *expansion proofs* play key roles.

Key words. Higher-order logic, type theory, mating, connection, expansion proof.

1. Introduction

Theorem proving is difficult and deals with complex phenomena. The difficulties seem to be compounded when one works with higher-order logic, but the rich expressive power of Church's formulation [18] of this language makes research on theorem proving in this realm very worthwhile. In order to make significant progress on this problem, we need to try many approaches and ideas, and explore many questions. The purpose of this paper is to provide an informal and expository introduction to an approach based on establishing appropriate *connections* between subformulas of an expanded form of the theorem to be proved. We hope to create an awareness of the many problems and questions which await investigation in this research area.

A question which is highly relevant to theorem proving is 'What *makes* a logical formula valid?'. Of course, there can be many ways of answering this question, and each can be the basis for a method of proof.

One approach to the above question is semantic. Theorems express essential truths and thus are true in all models of the language in which they are expressed. Truth can be perceived from many perspectives, so there may be many essentially different proofs of theorems. This point of view is very appealing, but it does not shed much light on the basic question of *what makes* certain sentences true in all models, while others are not. It's certainly a good idea to take advantage of semantic insights wherever these can be used to guide the theorem-proving process, but thus far significant progress in this direction has been made only in a few special contexts, such as geometry theorem proving.

Of course, theorems are formulas which have proofs, and every proof in any logical system may provide some insight. This suggests seeing what one can learn by studying the forms proofs can take. While this may be helpful, many of the most prominent

This is an extended version of a lecture presented to the 8th International Conference on Automated Deduction in Oxford, England on 27 July 1986. This material is based upon work supported by the National Science Foundation under grants DCR-8402532 and CCR-8702699.

features of proofs seem to be influenced as much by the logical system in which the proof is given as by the theorem that is being proved.

Let's focus on trying to understand what there is about the syntactic structures of theorems that makes them valid.

2. The Structure of Tautologies of Propositional Calculus

In the case of formulas of propositional calculus, we can certainly make good use of semantic insights, since in classical two-valued propositional calculus (to which we restrict our attention here) the theorems are precisely the tautologies, and we can test a formula for being a tautology in an explicit syntactic way. However, simply checking each line of a truth table is not really very enlightening, and we may still find ourselves asking "What is there about the structure of this formula which makes it a tautology?".

Clearly the pattern of occurrences of positive and negative literals in a tautology is very important. One can gain much insight into the structure of a wff by putting it into negation normal form (nnf). A wff is in negation normal form, and is called a negation normal formula (nnf), iff it contains no propositional connectives other than $\vee$, $\wedge$, and $\sim$, and the scope of each occurrence of $\sim$ is atomic. In order to examine a nnf, it's helpful to arrange it in a two-dimensional diagram which we call a vpform (vertical path form). We display disjunctions horizontally, and conjunctions vertically.

For example, consider the wff below, which we call THM52:

$$[[P \equiv Q] \equiv R] \supset [P \equiv .Q \equiv R]$$

(A dot in wff stands for a left bracket whose mate is as far to the right as possible without changing the pairing of brackets already present.)

THM52 with $\equiv$ eliminated is:

$$\begin{array}{ll} & [[[[P \supset Q] \wedge .Q \supset P] \supset R] \\ & \wedge .R \supset .[P \supset Q] \wedge .Q \supset P] \\ \supset & .[P \supset .[Q \supset R] \wedge .R \supset Q] \\ & \wedge .[[Q \supset R] \wedge .R \supset Q] \supset P \end{array}$$

A negation normal form of THM52 is:

$$\begin{array}{l} [[\sim P \vee Q] \wedge [\sim Q \vee P] \wedge \sim R] \\ \vee [R \wedge .[P \wedge \sim Q] \vee .Q \wedge \sim P] \\ \vee .[\sim P \vee .[\sim Q \vee R] \wedge .\sim R \vee Q] \\ \quad \wedge .[Q \wedge \sim R] \vee [R \wedge \sim Q] \vee P \end{array}$$

The vpform of this is:

$$\begin{bmatrix} \sim P \vee Q \\ \sim Q \vee P \\ \sim R \end{bmatrix} \vee \begin{bmatrix} R \\ \begin{bmatrix} P \\ \sim Q \end{bmatrix} \vee \begin{bmatrix} Q \\ \sim P \end{bmatrix} \end{bmatrix} \vee \begin{bmatrix} \sim P \vee \begin{bmatrix} \sim Q \vee R \\ \sim R \vee Q \end{bmatrix} \\ \begin{bmatrix} Q \\ \sim R \end{bmatrix} \vee \begin{bmatrix} R \\ \sim Q \end{bmatrix} \vee P \end{bmatrix}$$

In this form the structure of the wff is much easier to comprehend. Let us also look at the negation of this wff. A negation normal form of $\sim$THM52 is:

$$\begin{aligned}
&[[P \wedge \sim Q] \vee [Q \wedge \sim P] \vee R] \\
&\wedge [\sim R \vee .[\sim P \vee Q] \wedge .\sim Q \vee P] \\
&\wedge .[P \wedge .[Q \wedge \sim R] \vee .R \wedge \sim Q] \\
&\quad \vee .[\sim Q \vee R] \wedge [\sim R \vee Q] \wedge \sim P
\end{aligned}$$

The vpform of this is:

$$\begin{bmatrix}
\begin{bmatrix} P \\ \sim Q \end{bmatrix} \vee \begin{bmatrix} Q \\ \sim P \end{bmatrix} \vee R \\
\sim R \vee \begin{bmatrix} \sim P \vee Q \\ \sim Q \vee P \end{bmatrix} \\
\begin{bmatrix} P \\ \begin{bmatrix} Q \\ \sim R \end{bmatrix} \vee \begin{bmatrix} R \\ \sim Q \end{bmatrix} \end{bmatrix} \vee \begin{bmatrix} \sim Q \vee R \\ \sim R \vee Q \\ \sim P \end{bmatrix}
\end{bmatrix}$$

There is an obvious relationship between the two vpforms above.

A *disjunctive* (*horizontal*) *path* through a nnf W is a set of literal-occurrences in W corresponding to a disjunction of literals obtained from W by repeatedly replacing subformulas of W of the form $[A^1 \wedge \ldots \wedge A^n]$ by one of the A^i until no conjunctions remain. Similarly, a *conjunctive* (*vertical*) *path* is a set of literal-occurrences corresponding to a conjunction of literals obtained by repeatedly replacing subformulas of the form $[A^1 \vee \ldots \vee A^n]$ by one of the A^i until no disjunctions remain. (For example, one vertical path consists of the leftmost literals in the vpform above.) A conjunctive normal form of a wff is the conjunction of the disjunctions of literals in its horizontal paths, and a disjunctive normal form of a wff is the disjunction of the conjunctions of literals in its vertical paths. (See [7].)

A *connection* is a pair of literal-occurrences, and a *mating* is a set of connections, i.e., a relation between occurrences of literals. The terminology *mating* was used by Martin Davis [19]. In propositional calculus we require that connected (or mated) literals be complementary, and when these ideas are lifted to first- or higher-order logic we require that there be some substitution which simultaneously makes the pairs in each connection complementary. When speaking about wffs of propositional calculus we usually tacitly assume that literals are connected iff they are complementary.

We say that a mating *spans* a path iff there is a connection in the mating between two literals on the path. This means that two literals on the path are complementary. A wff is a tautology if and only if all the disjunctions in its conjunctive normal form are tautologies, which is equivalent to saying that all its horizontal paths are spanned. Similarly, a wff is a contradiction if and only if all the conjunctions in its disjunctive normal form are contradictions, which is equivalent to saying that all its vertical paths are spanned. Depending on whether we are looking for a tautology or a contradiction, we say that a mating *spans* a nnf iff it spans all its horizontal paths or all its vertical paths. Of course, when checking whether a mating spans a nnf, it's not always necessary to examine each path completely and individually. (By examining the paths in the vpforms above it is easy to see that THM52 is a tautology and that $\sim$THM52 is a contradiction.)

For many purposes it doesn't matter whether one proves that a wff is a tautology, or proves that its negation is a contradiction. In general, one should be able to display the work in whichever format one prefers, and translate easily between them.

When lifted to first-order logic, these ideas are the foundation of the method of theorem proving which has been called the *connection* method [12] and the *mating* method [5]. Bibel's book [12] contains a wealth of information and ideas, many of which appeared earlier in his many papers on this subject. Of course, one can see the antecendents of this approach to theorem proving in certain early papers in this field, most notably [31].

A mating which spans a nnf is also called *p-acceptable* (*path-acceptable*). We may also use the word *acceptable* in a more general sense and say that a mating is acceptable if it satisfies some criterion which guarantees that the formula is a tautology (or a contradiction). For wffs in conjunctive normal form, another example of a criterion which guarantees that a wff is contradictory is that every cycle must contain a merge (see [3] and [32]). It would be valuable to develop additional criteria for acceptability. Of course, such criteria are most useful if they lift to higher levels of logic, as with resolution.

Note that if we have a number of conditions which must be satisfied by an acceptable mating, we may be able to apply them simultaeously to limit our search, or we may be able to use different criteria in different situations to motivate mating certain literal-occurrences. Perhaps much more can be said on this subject.

3. The Structure of Valid Formulas of Higher-Order Logic

We turn our attention to higher-order logic, which is synonymous with type theory, and was developed by Bertrand Russell. We shall use a very elegant and useful formulation of type theory due to Alonzo Church [18] which is known as the typed λ-calculus. A description of this can be found in [7], but we give a brief introduction to some basic features of the notation before proceeding.

All entities have types, and variables and constants are letters with subscripts which indicate their types. However, we often omit the subscript from a symbol after its first occurrence in a wff. If α and β are types, then $(\alpha\beta)$ is the type of functions from elements of type β to elements of type α. The type of truth values is o, and we may regard entities of type $(o\alpha)$ as sets whose elements are of type α or as properties of objects of type α. In higher-order logic, all types of variables can be quantified. Of course, propositional calculus and first-order logic are parts of type theory.

One view is that in higher-order logic as well as in first-order logic, formulas are valid because they can be expanded into tautologies. (Of course, this is the fundamental idea underlying various forms of Herbrand's Theorem.) The simple tautological structure of a valid wff can be hidden in various ways. Valid formulas may be regarded as tautologies which have been abbreviated by existential generalization, disjunctive simplification, λ-expansion, and the introduction of definitions. Their basically tautological structure has been disguised, and to prove them we must unmask the disguises.

Let us consider an example of how one can start with a tautology and disguise it by using these four operations to obtain a meaningful theorem of type theory. Start with a tautology:

$$\begin{array}{l} \quad [Q_o \wedge \sim S_o] \\ \vee\ [N_o \wedge \sim R_o] \\ \vee\ [M_o \wedge \sim Q] \\ \vee\ .[\sim M \vee S] \wedge .\sim N \vee R \end{array} \tag{1}$$

Substitute to get another tautology:

$$\begin{array}{l} \quad [P_{o\iota}[H_{\iota\iota} X_\iota] \wedge \sim P\,.G_{\iota\iota}\,.HX] \\ \vee\ [P Y_\iota \wedge \sim P\,.G Y] \\ \vee\ [PX \wedge \sim P\,.HX] \\ \vee\ .[\sim PX \vee P.G.HX] \wedge .\sim PY \vee P.GY \end{array} \tag{2}$$

Existentially generalize on $[H_{\iota\iota} X_\iota]$:

$$\begin{array}{l} \quad \exists X_\iota[P_{o\iota} X \wedge \sim P\,.G_{\iota\iota} X] \\ \vee\ [P Y_\iota \wedge \sim P\,.G Y] \\ \vee\ [PX \wedge \sim P\,.H_{\iota\iota} X] \\ \vee\ .[PX \vee P.G.HX] \wedge .\sim PY \vee P.GY \end{array} \tag{3}$$

Existentially generalize again:

$$\begin{array}{l} \quad \exists X_\iota[P_{o\iota} X \wedge \sim P\,.G_{\iota\iota} X] \\ \vee\ \exists X[PX \wedge \sim P.GX] \\ \vee\ [PX \wedge \sim P\,.H_{\iota\iota} X] \\ \vee\ .[\sim PX \vee P.G.HX] \wedge .\sim PY_\iota \vee P.GY \end{array} \tag{4}$$

Apply $\vee$ simplification:

$$\begin{array}{l} \quad \exists X_\iota[P_{o\iota} X \wedge \sim P\,.G_{\iota\iota} X] \\ \vee\ [PX \wedge \sim P\,.H_{\iota\iota} X] \\ \vee\ .[\sim PX \vee P.G.HX] \wedge .\sim PY_\iota \vee P.GY \end{array} \tag{5}$$

Existentially generalize:

$$\begin{array}{l} \quad \exists X_\iota[P_{o\iota} X \wedge \sim P\,.G_{\iota\iota} X] \\ \vee\ \exists X[PX \wedge \sim P.H_{\iota\iota} X] \\ \vee\ .[\sim PX \vee P.G.HX] \wedge .\sim PY_\iota \vee P.GY \end{array} \tag{6}$$

Next we perform two trivial changes of the wff which we don't really regard as disguising its structure in any significant way. Introduce $\forall$:

$$\begin{aligned}\forall G_{\iota\iota}\forall H_{\iota\iota}.\exists X_\iota[P_{o\iota}X \wedge \sim P.GX] \\ \vee \exists X[PX \wedge \sim P.HX] \\ \vee .\forall X[\sim PX \vee P.G.HX] \wedge \forall Y_\iota.\sim PY \vee P.GY\end{aligned} \tag{7}$$

Rewrite $\exists$ as $\sim\forall\sim$ and apply De Morgan's Laws:

$$\begin{aligned}\forall G_{\iota\iota}\forall H_{\iota\iota}.\sim\forall X_\iota[\sim P_{o\iota}X \vee P.GX] \\ \vee \sim\forall X[\sim PX \vee P.HX] \\ \vee .\forall X[\sim PX \vee P.G.HX] \wedge \forall Y_\iota.\sim PY \vee P.GY\end{aligned} \tag{8}$$

λ-expand:

$$\begin{aligned}\forall G_{\iota\iota}\forall H_{\iota\iota}.\sim[\lambda G.\forall X_\iota.\sim P_{o\iota}X \vee P.GX]G \\ \vee \sim[\lambda G.\forall X.\sim PX \vee P.GX]H \\ \vee .[\lambda G.\forall X.\sim PX \vee P.GX][\lambda Z_\iota.G.HZ] \wedge \forall Y_\iota.\sim PY \vee P.GY\end{aligned} \tag{9}$$

Existentially generalize on $[\lambda G_{\iota\iota}.\forall X_\iota.\sim PX \vee P.GX]$:

$$\begin{aligned}\exists M_{o(\iota\iota)}\forall G_{\iota\iota}\forall H_{\iota\iota}.\sim MG \vee \sim MH \\ \vee .M[\lambda Z_\iota.G.HZ] \wedge \forall Y_\iota.\sim P_{o\iota}Y \vee P.GY\end{aligned} \tag{10}$$

λ-expand:

$$\begin{aligned}\exists M_{o(\iota\iota)}\forall G_{\iota\iota}\forall H_{\iota\iota}.\sim MG \vee \sim MH \\ \vee .M[[\lambda G\lambda H\lambda Z_\iota.G.HZ]GH] \wedge \forall Y_\iota.\sim P_{o\iota}Y \vee P.GY\end{aligned} \tag{11}$$

Introduce the definition $\circ$ for $[\lambda G_{\iota\iota}\lambda H_{\iota\iota}\lambda Z_\iota.G.HZ]$, writing $[G\circ H]$ (the composition of G and H) as an abbreviation for $[\circ GH]$:

$$\begin{aligned}\exists M_{o(\iota\iota)}\forall G_{\iota\iota}\forall H_{\iota\iota}.\sim MG \vee \sim MH \\ \vee .M[G\circ H] \wedge \forall Y_\iota.\sim P_{o\iota}Y \vee P.GY\end{aligned} \tag{12}$$

Introduce $\forall$:

$$\begin{aligned}\forall P_{o\iota}\exists M_{o(\iota\iota)}\forall G_{\iota\iota}\forall H_{\iota\iota}.\sim MG \vee \sim MH \\ \vee .M[G\circ H] \wedge \forall Y_\iota.\sim PY \vee P.GY\end{aligned} \tag{13}$$

Again we make a trivial notational change, introducing $\supset$:

$$\begin{aligned}\forall P_{o\iota}\exists M_{o(\iota\iota)}\forall G_{\iota\iota}\forall H_{\iota\iota}.[MG \wedge MH] \\ \supset .M[G\circ H] \wedge \forall Y_\iota.PY \supset P.GY\end{aligned} \tag{14}$$

We have thus arrived at the

THEOREM. *For any set P, there is a set M of functions on P which is closed under composition.*

We shall call this theorem THM112. It's a trivial theorem, but it provides a nice simple example of many basic phenomena, so it will be used to illustrate a number of points.

Of course, the process leading to THM112 above provides one form of a proof of the theorem, and this suggests an approach to the problem of searching for proofs of theorems of higher-order logic. Starting with the theorem to be proved, applying the inverse of the operations used above to find the tautology hidden in the theorem.

4. Expansion Trees and Expansion Proofs

The tautology hidden in a theorem reminds us of a Herbrand expansion, and indeed, it essentially is one in the context of higher-order logic, and it sheds a lot of light on the essential syntactic structure of the theorem. However, it's a little hard to regard the tautology as a form of proof of the theorem, because we can't recover the theorem from the tautology; indeed, the tautology could be used to prove many theorems, including the ultimately weak theorem $\exists p_o p$, which is an immediate consequence of every theorem.

To understand in this sense why a formula is valid one needs more than the tautology hidden within it; one needs to know how to expand the formula into the tautology. If you haven't thought much about how to extend various methods of automated theorem proving to higher-order logic, you may not have worried much about the fact that it can be rather awkward to deal with a situation in which new quantifiers and propositional connectives are continuously introduced into the formula you're working on. To deal with such a situation efficiently, and to make rigorous arguments about the processes involved, it's important to have a nice clean representation of the relevant logical structures.

A very elegant, concise, and nonredundant way of representing the theorem, the tautology, and the relationship between them is an *expansion tree proof*, otherwise known as an *ET-proof* or an *expansion proof*. This concept, which grew out of ideas in [6], was developed by Dale Miller in his thesis [25], and the details of the definition can also be found in [27].

We need not repeat these details here, but let us give an informal discussion of this concept which should convey the main ideas involved in it. A familiar way of representing a wff of first-order logic is to regard it as a tree (i.e., a connected graph with no cycles) growing downward from a topmost node which is called its root. With each node of the tree is associated a propositional connective, quantifier, or atom, and if the node is not associated with an atom, there are subtrees below the node which represent the scopes of the connective or quantifier. Starting from this idea, let us enrich it to obtain what Miller calls an *expansion tree* for a wff of first- or higher-order logic.

Let us call the wff represented by a tree Q as described above the *shallow formula* $\mathrm{Sh}(Q)$ of the tree (and of the node which is its root). With each expansion tree Q is also associated a *deep formula* $\mathrm{Dp}(Q)$ which represents the result of instantiating the quantifier-occurrences in Q with terms which are attached as labels to the arcs descending from their nodes.

Let us call a node of a tree an *expansion node* if it corresponds to an essentially existential (see p. 123 of [7]) quantifier, and a *selection node* if it corresonds to an essentially universal quantifier. (In *dual expansion trees*, the roles of universal and existential quantifiers are interchanged.) Let finitely many arcs labeled with terms descend from each expansion node, so that if the expansion node has shallow formula $\exists \mathbf{x}\mathbf{B}$, and if $\mathbf{t}$ is the term labeling an arc descending from that node, then the type of $\mathbf{t}$ is the same as that of $\mathbf{x}$, and the node at the lower end of that arc has as shallow formula the λ-normal form of $[[\lambda \mathbf{x}\mathbf{B}]\mathbf{t}]$, i.e., the result of instantiating the quantifier with the term $\mathbf{t}$. The term $\mathbf{t}$ is called an *expansion term*. Selection nodes satisfy a similar condition, except that only one arc may descend from a selection node, and the term labeling it must be a suitably chosen parameter called a *selected parameter*.

In an expansion tree, if Q is an expansion node and $Q_1, \ldots, Q_n$ are the nodes immediately below Q, then $\mathrm{Dp}(Q)$ is $[\mathrm{Dp}(Q_1) \vee \ldots \vee \mathrm{Dp}(Q_n)]$. (In a dual expansion tree, however, $\mathrm{Dp}(Q)$ is $[\mathrm{Dp}(Q_1) \wedge \ldots \wedge \mathrm{Dp}(Q_n)]$.) The deep and shallow formulas are the same for the leaves of these trees.

In order to deal effectively with definitions, it is useful to add one more condition to the inductive definition of an expansion tree. If Q is a node whose shallow formula contains an abbreviation (such as the wff $[\circ GH]$ of the example above), then one may call Q a *definition node* and create a node Q_1 below Q whose shallow formula is the result of replacing the abbreviation by the wff for which it stands ($[[\lambda G \lambda H \lambda Z_{\iota} . G . HZ]GH]$ in the example above) and putting the result into λ-normal form. Q_1 and Q have the same deep formulas.

An example of an expansion tree which represents the proof of THM112 given above may be seen in Figure 4.1. The shallow formula for each node is written to the right of the node. The deep formulas are not displayed, but it is not hard to see that the deep formula for the whole expansion tree is formula (2) of the previous section.

It may be helpful to visualize expansion trees as being laid out in three dimensions. Go downward along the z-axis to do selections, expansions, and definition instantations, display conjunctions parallel to the y-axis, and display disjunctions parallel to the x-axis. Note how appropriate the words *deep* and *shallow* are from this point of view.

An expansion tree which satisfies certain conditions is called an *expansion proof*. In such a tree, the shallow formula is the wff being proved, and the deep formula is the underlying tautology. The conditions are:

1. The deep formula must be a tautology.
2. Selections must be done correctly. (We discuss this more thoroughly below.)

If one uses dual expansion trees instead of expansion tees, an expansion proof (which might also be called an *expansion refutation*) must have for its deep formula a contradiction instead of a tautology. In an expansion tree, which represents a proof which works its way up from the bottom of the tree to the top node, expansion corresponds to inverse exstential generalization, whereas in a dual tree, which represents

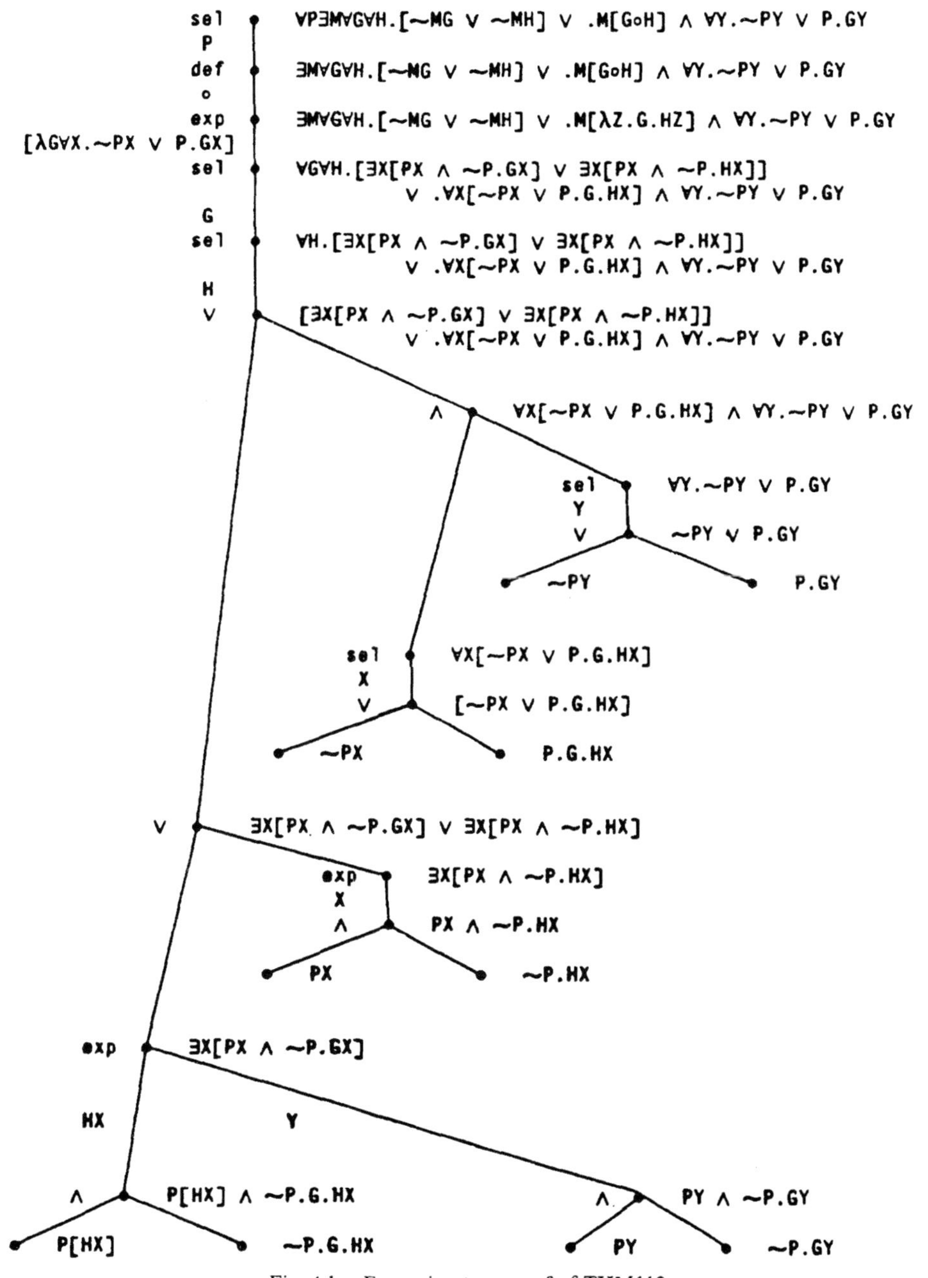

Fig. 4.1. Expansion tree proof of THM112.

a refutation which starts at the top node and works down to a contradiction, expansion corresponds to universal instantiation. Actually, from now on we'll mainly be concerned with dual trees. A dual expansion tree for a wff equivalent to $\sim$ THM112 is in Figure 4.2, which should be compared with Figure 4.1. Note that its

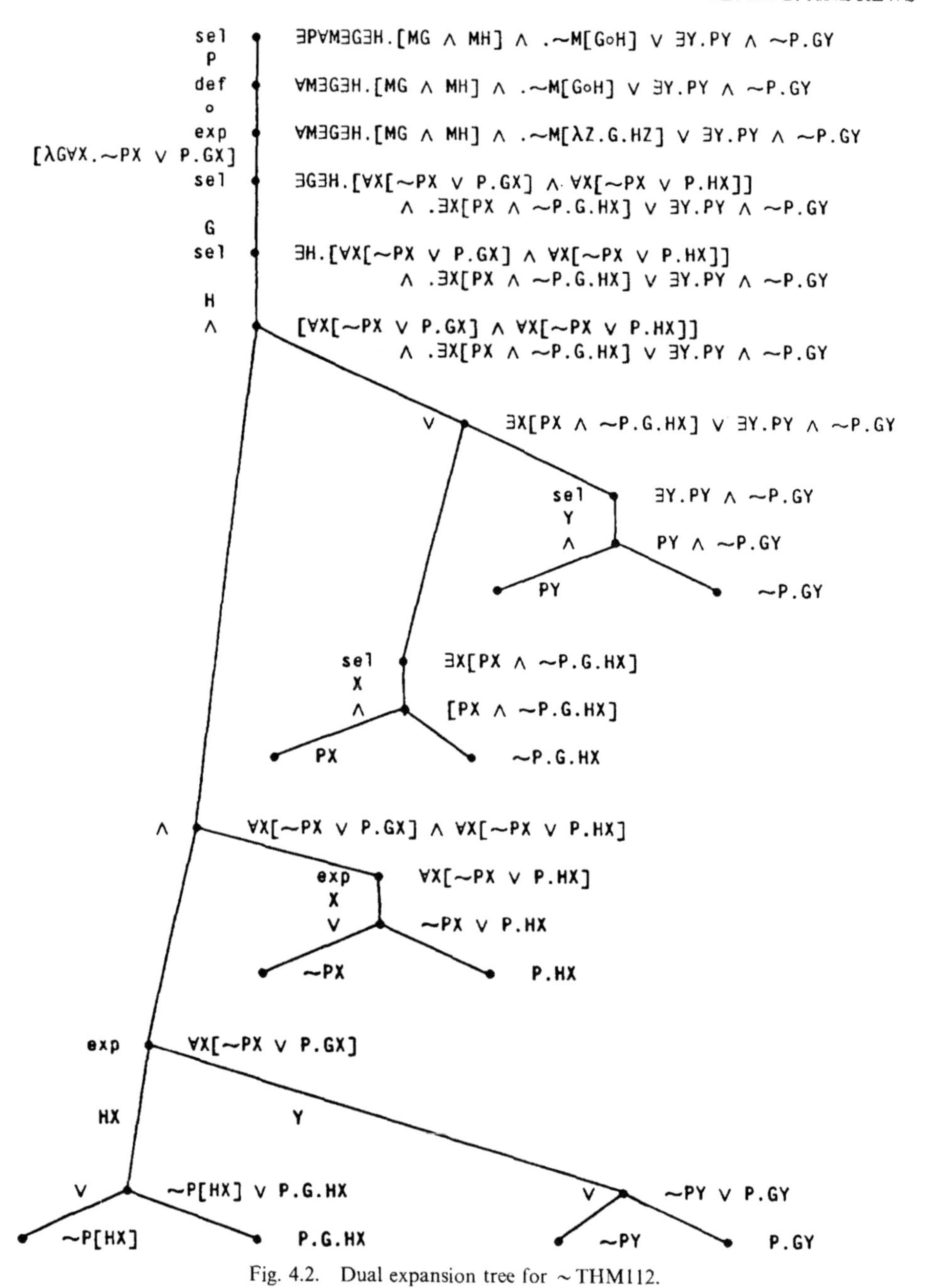

Fig. 4.2. Dual expansion tree for ~THM112.

deep formula is the contradiction

$$[\sim P_{o\iota}[H_{\iota\iota}X_\iota] \vee P[G_{\iota\iota}.HX]] \wedge [\sim PY_\iota \vee P[GY]] \wedge [\sim PX \vee P[HX]]$$
$$\wedge\ .[PX \wedge \sim P.G.HX] \vee .PY \wedge \sim P.GY.$$

One can make the condition that 'selections must be done correctly' precise in a variety of ways, and the technical details need not concern us much right now. If one generates the expansion tree in a step-by-step process starting from the formula to be proved, then it suffices to choose a new parameter which occurs nowhere in the tree each time a selection is made.

Of course, if one is just presented with an expansion tree, and asked whether it is an expansion proof, one needs a criterion to check. One such criterion, which is in Miller's thesis and is similar to one in Bibel's book, is that the imbedding relation on selected parameters be irreflexive. The imbedding relation is the transitive closure of the relation $<^0$ such that $\mathbf{a} <^0 \mathbf{b}$ iff there is an expansion term occurrence $\mathbf{t}$ such that $\mathbf{a}$ occurs in $\mathbf{t}$ and $\mathbf{b}$ is selected below $\mathbf{t}$ in Q.

When searching for an expansion proof, we generally won't know just what expansion terms we should use, so the expansion terms will contain free variables for which we will make substitutions later. Of course, when we make a substitution, we must do it in a way that does not violate the imbedding condition. Just as in first-order logic, we can do this by introducing Skolem terms in place of the selected parameters. In higher-order logic, we may introduce new quantifiers in the expansion terms, so we can't do all the Skolemization in a preprocessing stage. This is why we need the selection operation even if we use Skolemization.

Each Skolem term consists of a new function symbol with arguments which are the expansion terms above its selection node in the tree. Note that if $\mathbf{a}$ and $\mathbf{b}$ are selected parameters, $\mathbf{a} < \mathbf{b}$ means that the Skolem term for $\mathbf{a}$ is imbedded in the Skolem term for $\mathbf{b}$. If one Skolemizes naively in higher-order logic, the Skolem functions can serve as choice functions, and one can derive certain consequences of the Axiom of Choice even if one did not intend to enrich one's logic in this way. The right way to do Skolemization in higher-order logic is discussed in Miller's thesis. Skolem terms often get awkwardly long, but when we use Skolemization, we can simplify the notation by just writing a single label for each Skolem term, and keeping track of what it stands for. We will do this in examples below.

Of course, even if one wants to substitute for variables in the expansion tree, one can use selected parameters instead of Skolem terms if one takes care not to violate the imbedding condition. In particular, one can integrate the imbedding relation into the unification algorithm so that only appropriate substitutions are produced. This is discussed for first-order logic in Bibel's book, and the details for higher-order logic should be fully worked out soon.

Note that an expansion proof can be used to provide an extremely economical representation of the essential information needed to prove a theorem. This information consists of the theorem itself, the general structure of the expansion tree, the expansion terms at each expansion node, the selected parameter at each selection node, and information about which occurrences of definitions are instantiated at definition nodes. The wffs associated with the nodes of the tree other than the root

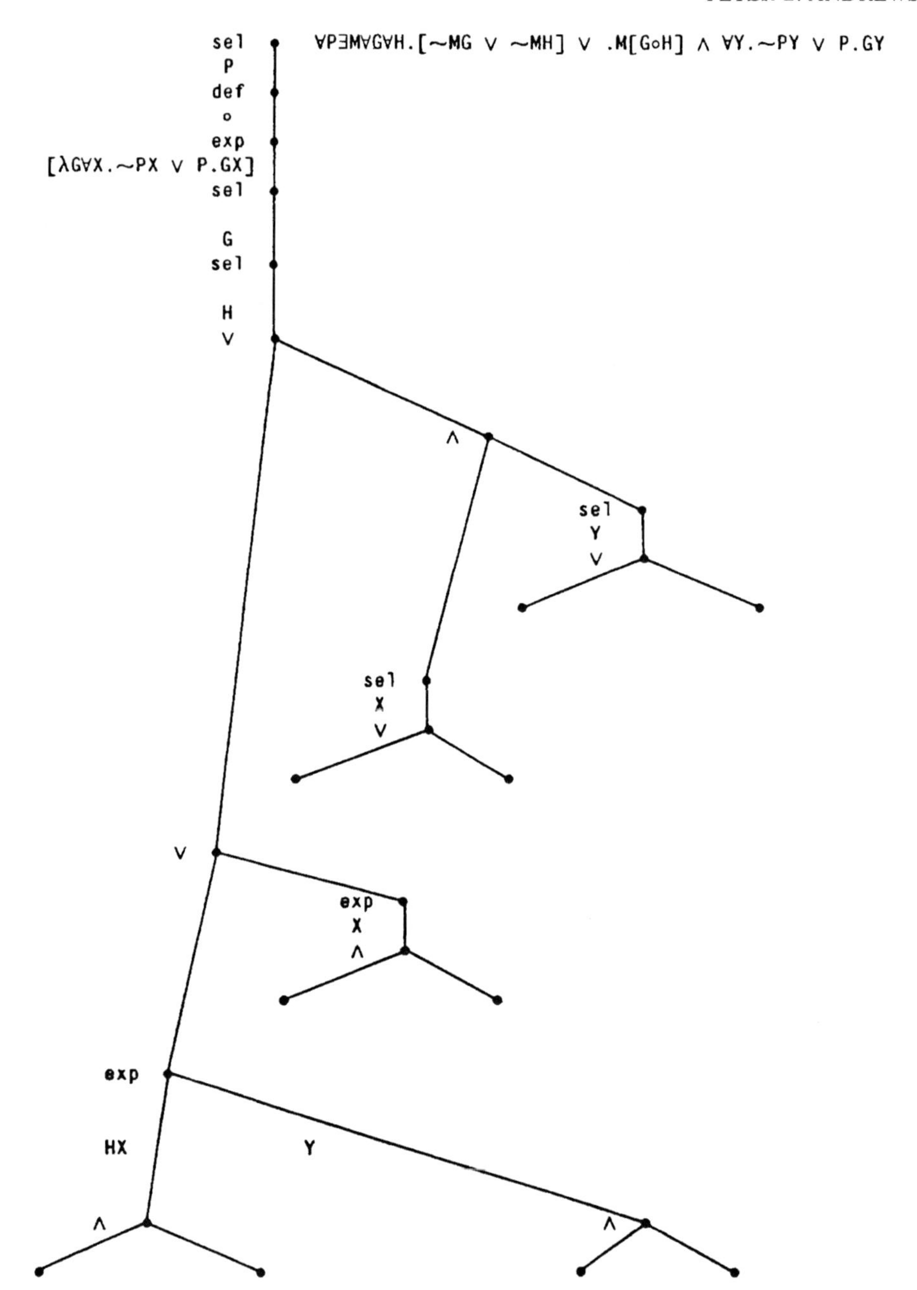

Fig. 4.3. Expansion tree proof of THM112.

can be calculated. Thus, the expansion tree displayed in Figure 4.1 can also be displayed as in Figure 4.3.

The idea of an expansion tree is really very simple and natural. However, I want to emphasize its importance as a conceptual advance. It provides a fundamental conceptual building block on which we can base the extension of the connection or mating approach to theorem proving from first- to higher-order logic.

5. Translations Between Formats

As research on automated deduction progresses, we shall probably reach a point where we are less concerned with what format is being used, and more with deeper issues. Nevertheless, certain ideas do occur more naturally in one context than in another, and it's well known in artificial intelligence, mathematics, and other disciplines that the way one represents a problem can be very important, so it's nice to be able to translate easily from one proof format to another. The translation process may in principle be trivial in the sense that no search need be involved, but it's often highly nontrivial to design a good algorithm to do this and to prove that it always works.

A familiar format for presenting proofs is natural deduction. An example of a natural deduction proof for THM112 is in Figure 5.1. (See [4] or §30 of [7] for more details about this formulation of natural deduction.) Note how a natural deduction proof differs from building up the wff from a single tautology. In natural deduction proofs we may have hypotheses introduced and eliminated, conjunctions introduced and eliminated, proofs by cases, indirect proofs, the transitive law of implication, etc. In general, a truly 'natural' deduction breaks up the proof and the theorem into parts on which we can focus.

As you can see, a natural deduction proof contains a great deal of redundancy, since certain formulas occur as subformulas of many lines of the proof. Redundancy

```
(1)1 ⊢ ∀X_ι [P_οι X ⊃ P.G_ιι X] ∧ ∀X.P X ⊃ P.H_ιι X                      Hyp
(2)1 ⊢ ∀X_ι.P_οι X ⊃ P.G_ιι X                                          Conj: 1
(3)1 ⊢ ∀X_ι.P_οι X ⊃ P.H_ιι X                                          Conj: 1
(4)1 ⊢ P_οι [H_ιι X_ι] ⊃ P.G_ιι.H X                              UI: [H X] 2
(5)1 ⊢ P_οι X_ι ⊃ P.H_ιι X                                            UI: X 3
(6)1 ⊢ P_οι X_ι ⊃ P.G_ιι.H_ιι X                                    RuleP: 4 5
(7)1 ⊢ ∀X_ι.P_οι X ⊃ P.G_ιι.H_ιι X                                  UGen: X 6
(8)1 ⊢ P_οι Y_ι ⊃ P.G_ιι Y                                            UI: Y 2
(9)1 ⊢ ∀Y_ι.P_οι Y ⊃ P.G_ιι Y                                       UGen: Y 8
(10)1 ⊢  ∀X_ι [P_οι X ⊃ P.G_ιι.H_ιι X] ∧ ∀Y_ι.P Y ⊃ P.G Y            Conj: 7 9
(11)  ⊢     ∀X_ι [P_οι X ⊃ P.G_ιι X] ∧ ∀X [P X ⊃ P.H_ιι X]
         ⊃ ∀X [P X ⊃ P.G.H X] ∧ ∀Y_ι.P Y ⊃ P.G Y                    Deduct: 10
(12)  ⊢ ∀H_ιι.    ∀X_ι [P_οι X ⊃ P.G_ιι X] ∧ ∀X [P X ⊃ P.H X]
               ⊃   ∀X [P X ⊃ P.G.H X] ∧ ∀Y_ι.P Y ⊃ P.G Y               UGen: H 11
(13)  ⊢ ∀G_ιι ∀H_ιι.    ∀X_ι [P_οι X ⊃ P.G X] ∧ ∀X [P X ⊃ P.H X]
                   ⊃   ∀X [P X ⊃ P.G.H X] ∧ ∀Y_ι.P Y ⊃ P.G Y           UGen: G 12
(14)  ⊢ ∀G_ιι ∀H_ιι.   [λG ∀X_ι.P_οι X ⊃ P.G X] G ∧ [λG ∀X.P X ⊃ P.G X] H
                   ⊃   [λG ∀X.P X ⊃ P.G X][λZ_ι G.H Z] ∧ ∀Y_ι.P Y ⊃ P.G Y
                                                                    Lambda: 13
(15)  ⊢ ∃M_ο(ιι) ∀G_ιι ∀H_ιι. M G ∧ M H  ⊃  M [λZ_ι G.H Z] ∧ ∀Y_ι.  P_οι Y ⊃ P.G Y
                                               EGen: [λG ∀X_ι.P X ⊃ P.G X] 14
(16)  ⊢ ∃M_ο(ιι) ∀G_ιι ∀H_ιι. M G ∧ M H  ⊃  M [G ∘ H] ∧ ∀Y_ι.  P_οι Y ⊃ P.G Y
                                                                       Def: 15
(17)  ⊢ ∀P_οι ∃M_ο(ιι) ∀G_ιι ∀H_ιι. M G ∧ M H  ⊃  M [G ∘ H] ∧ ∀Y_ι.  P Y ⊃ P.G Y
                                                                   UGen: P 16
```

Fig. 5.1. Natural deduction proof of THM112.

occurs naturally in human discourse, since our short-term memories have limited capacity, and there are limitations on the ways we can restructure information in our heads. It's quite appropriate that proofs in natural deduction style also contain this redundancy. However, in a computer system redundancy is often a nuisance or a problem. It wastes memory, and may create the need for additional processing. When we see redundancy, we should ask whether there is some good reason for it. What does it cost, and what does it achieve? Of course, redundancy in the final proof format should be distinguished from redundancies and inefficiencies in the search for the proof and associated data structures. The latter are much more serious.

When one is *searching* for a proof, it is desirable to work within a context where one can focus on the essential features of the problem as directly and economically as possible. Thus, it makes a lot of sense to search for ways of proving theorems of higher-order logic by searching for expansion proofs.

Miller showed [25] [26] [27] that once an expansion proof has been found, it can be converted without further search into a natural deduction proof. This extended related work for first-order logic in [4] and [12]. Of course, being able to make this translation justifies searching for a proof by looking for an expansion proof.

Some choices have to be made to determine how the natural deduction proof is arranged, since different natural deduction proofs may correspond to the same expansion proof, and heuristics can be used to make these choices in ways that will tend to produce stylistically pleasing proofs. Miller has done some work [25] under the name *focusing* on this problem too, but more needs to be done.

Of course, when we are trying to construct examples of proofs interactively for research purposes, it's usually easiest for us to construct them first in natural deduction style, so we'd like to be able to translate such proofs automatically into expansion proofs. However, there are some pitfalls when one tries to translate in this direction.

When Herbrand was proving the famous theorem which bears his name, he needed to show that every theorem of first-order logic has what we now call a tautologous Herbrand expansion. He presented an apparently straightforward inductive argument that every formula in any proof in a certain system of first-order logic has such an expansion. However, he ran into trouble [20] when he tried to prove that if A and $[A \supset B]$ both have tautologous Herbrand expansions, then B has one too. This error was not repaired [21] for thirty years.

In higher-order logic, a purely syntactic proof that every natural deduction proof can be translated into an expansion proof runs up against Gödel's Second Incompleteness Theorem, just as a purely syntactic proof of cut-elimination in type theory does. To explain this, let us introduce an axiom of infinity (a sentence whose models are all infinite) which we shall call Infin:

$$\exists R_{o\iota\iota} \forall X_\iota \forall Y_\iota \forall Z_\iota . \exists W_\iota R X W \wedge {\sim} R X X \wedge . {\sim} R X Y \vee {\sim} R Y Z \vee R X Z$$

'Analysis' is the name for the logical system of type theory with added axioms of extensionality, descriptions, and Infin. Most of mathematics can be formalized in this system. Now let us suppose for a moment that analysis is inconsistent. It can be shown by a purely syntactic argument [2] that if analysis is inconsistent, one can prove $\sim$ Infin in type theory without axioms of extensionality and descriptions. Now for the sake of simplicity, consider a stronger axiom of infinity which we shall call SkInfin:

$$\exists G_{\iota\iota} \exists R_{o\iota\iota} \forall X_\iota \forall Y_\iota \forall Z_\iota .\ RX[GX] \wedge \sim RXX \wedge .\sim RXY \vee \sim RYZ \vee RXZ$$

SkInfin implies Infin, so if analysis is inconsistent, $\sim$SkInfin can be proved in type theory. If every proof in type theory can be translated into an expansion proof, there must be a dual expansion tree for SkInfin which has a contradictory deep formula. Here is the vpform for SkInfin:

$$\exists GR \left[\begin{array}{l} \forall XYZ \\ \left[\begin{array}{c} RX[GX] \\ \sim RXX \\ \sim RXY \vee \sim RYZ \vee RXZ \end{array} \right] \end{array} \right]$$

Let's write $\lambda(A)$ for the (essentially unique) λ-normal form [1] of a wff A. It is not hard to see that the deep formula must be a conjunction of formulas which will be true if we make the atom $[RAB]$ true whenever the number of occurrences of G in $\lambda(A)$ is less than the number of occurrences in $\lambda(B)$, so it cannot be contradictory. Thus, we reach a contradiction.

If the proof that the translation process above always works could be formalized in analysis, we could formalize the argument above to obtain a proof within analysis that analysis is consistent, contradicting Gödel's Second Theorem. Thus, the proof that the translation process always works cannot be formalized in analysis, and the problem of proving that there is such a translation in intrinsically difficult.

Frank Pfenning showed in [28] how to translate a natural deduction proof in first-order logic into an expansion proof, and deals with the problem for higher-order logic in his thesis [29]. Pfenning can prove that his translation for type theory works when applied to cut-free proofs, and it's been known for a long time (by a non-constructive argument) that natural deduction proofs in type theory can be made cut-free. Pfenning has also defined a translation process for proofs which do contain cuts, but he has no proof that the translation process terminates. (Of course, any such proof would have to use methods which could not be formalized in analysis; for examples of such proofs, see references cited in [2].)

It is our thesis that proofs of formulas of first- and higher-order logic in virtually any format induce expansion proofs, and can be constructed from expansion proofs. Frank Pfenning has amassed considerable evidence for this in his thesis, and has shown how expansion trees can be extended to accommodate equality and extensionality. Expansion trees are important tools for comparing proofs and methods of searching for proofs. An expansion proof reveals how a tautology is hidden in its theorem, and provides a nice answer to the question "What is the essential reason that this wff is valid?"

An expansion proof embodies essential ideas which can be expressed in many ways in different proofs; the expansion proof is a key to translating between them. If we have a clumsy natural deduction proof, we can transform it into the expansion proof which it induces and then (if we have sufficiently good translation methods) translate that back into a well structured natural deduction proof.

Obviously, proofs which induce the same expansion proofs are equivalent in a very important and fundamental sense. It's conceivable that by looking even further in this direction of abstracting the essential ideas out of theorems we can eventually devise a significant scheme for syntactically classifying theorems. At present we have no sensible way to organize a universal library of theorems which have been proved, or considered. Mathematicians discover from time to time that they have worked hard to prove theorems which others proved earlier.

6. A Challenge to this Point of View

Of course, we mustn't let our zeal in pursuit of truth blind us to the fact that truth is complicated and elusive. Let me digress a moment to discuss a challenge to the point of view that expansion proofs provide the essential reasons why wffs are valid.

Once one has a proof in virtually any format, one can almost always construct another proof of the same theorem by adding certain redundancies and irrelevancies, and this is certainly true of expansion proofs. However, quite aside from this, one cannot ignore the fact that certain theorems have several fundamentally different expansion proofs, which express fundamentally different ideas.

Examples of two proofs of a theorem I call THM76 are in Figures 6.1 and 6.2. Do we regard such a theorem as two different theorems which happen to manifest

(1)	1	$\vdash \forall P_{o\iota} .P\ Y_\iota \supset P\ X_\iota$	Hyp
(2)	1	$\vdash [\lambda Z_\iota .\sim R_{o\iota}\ Z]\ Y_\iota \supset [\lambda Z .\sim R\ Z]\ X_\iota$	UI: $[\lambda Z.\sim R\ Z]$ 1
(3)	1	$\vdash \sim R_{o\iota}\ Y_\iota \supset \sim R\ X_\iota$	Lambda: 2
(4)	1	$\vdash R_{o\iota}\ X_\iota \supset R\ Y_\iota$	RuleP: 3
(5)	1	$\vdash \forall R_{o\iota} .R\ X_\iota \supset R\ Y_\iota$	UGen: R 4
(6)		$\vdash \forall P_{o\iota} [P\ Y_\iota \supset P\ X_\iota] \supset \forall R_{o\iota} .R\ X \supset R\ Y$	Deduct: 5

Fig. 6.1. First proof of THM76.

(1)	1	$\vdash \forall P_{o\iota} .P\ Y_\iota \supset P\ X_\iota$	Hyp
(2)	1	$\vdash [\lambda Z_\iota .R_{o\iota}\ Z \supset R\ Y_\iota]\ Y \supset [\lambda Z .R\ Z \supset R\ Y]\ X_\iota$	UI: $[\lambda Z.R\ Z \supset R\ Y]$ 1
(3)	1	$\vdash R_{o\iota}\ Y_\iota \supset R\ Y \supset .R\ X_\iota \supset R\ Y$	Lambda: 2
(4)	1	$\vdash R_{o\iota}\ X_\iota \supset R\ Y_\iota$	RuleP: 3
(5)	1	$\vdash \forall R_{o\iota} .R\ X_\iota \supset R\ Y_\iota$	UGen: R 4
(6)		$\vdash \forall P_{o\iota} [P\ Y_\iota \supset P\ X_\iota] \supset \forall R_{o\iota} .R\ X \supset R\ Y$	Deduct: 5

Fig. 6.2. Second proof of THM76.

themselves as the same wff, like multiple roots of an equation? One can start with two different tautologies, disguise them in different ways, and somehow end up with the same wff. This seems a remarkable coincidence.

Maybe someday we will have a much deeper perspective which will allow us to approach automated theorem proving on a more abstract conceptual level. At present, we really don't have this, so let's turn to the question of how to prove theorems by searching for expansion proofs.

7. Introduction to Searching for Expansion Proofs

The fundamental concepts of expansion proofs and connections are important basic tools which we can use in fashioning a general approach to automated theorem proving in higher-order logic. If one wants to construct an expansion proof for a theorem by working down from the theorem to a tautology, or from its negation to a contradiction, one has certain operations which can be applied. Ideally, the process of applying them should be guided not by blind search, but by an understanding of what makes a formula tautological or contradictory. Let's consider an example which will illustrate the general procedure of applying the operations. We'll try to provide some motivation for applying the operations the way we do, but considerable research is needed to devise a purely automatic procedure which would apply the operations this way reasonably early in its search process. Our purpose here is simply to illustrate the kind of syntactic analysis which is the basis for the kind of search procedure we have in mind.

We'll work with the dual tree, start with the negation of the theorem to be proven, and create an expansion proof with a contradictory deep formula. We shall represent the process of gradually creating the expansion proof by a sequence of vpforms, which show how the negation of the theorem can gradually be expanded into the contradictory deep formula. The reader will easily see how this sequence can be regarded as representing an expansion tree which evolves as we make expansions and selections, instantiate definitions, and substitute for variables which we introduced earlier.

Recall that THM112 is $\forall P_{o\iota}\exists M_{o(o\iota)}\forall G_{\iota\iota}\forall H_{\iota\iota}.MG \wedge MH \supset .M[G\circ H] \wedge \forall Y_{\iota}.PY \supset P.GY$.

$\sim$THM112 is:

$$
\begin{array}{l}
\exists P \\
\left[\begin{array}{l}
\forall M \\
\left[\begin{array}{l}
\exists GH \\
\left[\begin{array}{c}
MG \\
MH \\
\qquad\qquad\qquad \exists Y \\
\sim M[G\circ H] \vee \left[\begin{array}{c} PY \\ \sim P[G\,Y] \end{array}\right]
\end{array}\right]
\end{array}\right]
\end{array}\right]
\end{array}
$$

Select P:

$$
\begin{array}{l}
\forall M \\
\left[\begin{array}{l}
\exists G H \\
\left[\begin{array}{c}
M\,G \\
M\,H \\
\sim M[G \circ H] \vee \begin{array}{l} \exists Y \\ \left[\begin{array}{c} P\,Y \\ \sim P[G\,Y] \end{array}\right] \end{array}
\end{array}\right]
\end{array}\right]
\end{array}
$$

Since the head letters are P and M, and P is a constant, we must use the expansion operator which instantiates the quantifier $\forall M$. We need to span the vertical path on the right. If Y did not depend on M, we could instantiate M with $[\lambda F_{\iota\iota}. \sim P_{o\iota} Y_{\iota}]$. This would create connections on both paths. Since Y does depend on M, we introduce a quantifier which we can instantiate after selecting Y by expanding on $M_{o(\iota\iota)}$ with $[\lambda F_{\iota\iota} \forall W_{\iota}. \sim P_{o\iota} W]$. We thus obtain:

$$
\begin{array}{l}
\exists G H \\
\left[\begin{array}{c}
[\lambda F\ \forall W. \sim P\,W]G \\
[\lambda F\ \forall W. \sim P\,W]H \\
\sim [\lambda F\ \forall W. \sim P\,W][G \circ H] \vee \begin{array}{l} \exists Y \\ \left[\begin{array}{c} P\,Y \\ \sim P[G\,Y] \end{array}\right] \end{array}
\end{array}\right]
\end{array}
$$

λ-normalize and select G, H, and Y. Note that these variables depend on M, but the term we substituted for M contains no free variables.

$$
\left[\begin{array}{l}
\qquad \forall W \\
\qquad \quad \sim P\,W \\
\qquad \forall W \\
\qquad \quad \sim P\,W \\
\begin{array}{l} \exists W \\ \quad P\,W \end{array} \vee \left[\begin{array}{c} P\,Y \\ \sim P[G\,Y] \end{array}\right]
\end{array}\right]
$$

In an expansion tree a connection is a pair of nodes of the tree, and the shallow formulas of these nodes need not be literals. We shall portray matings simply by drawing lines between formulas to represent the connections between the corresponding nodes. Note that the mating indicated by the lines in the vpform below spans both vertical paths through this wff.

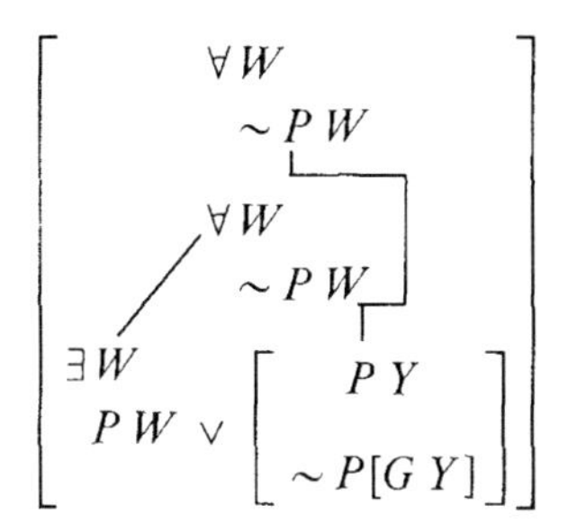

Now we instantiate the top quantifier as dictated by the mating, i.e., by the substitution which makes mated pairs complementary.

$$\left[\begin{array}{ll} & \sim P\,Y \\ & \forall W \\ & \quad \sim P\,W \\ \exists W & \\ \quad P\,W \vee & \left[\begin{array}{c} P\,Y \\ \sim P[G\,Y] \end{array}\right] \end{array}\right]$$

We have reached a contradiction. Note that this is a correct proof, since in each case $[\lambda F_{\iota\iota} \forall W_\iota . \sim P_{o\iota} W]$ is a set of functions from P into P which is closed under composition. If P is empty, this is the set of all functions of the appropriate type; if P is nonempty, it is the empty set of functions.

Of course, the proof just given is somewhat trivial, so let's look for another proof. We return to the stage where we were ready to expand on $\forall M$.

$$\begin{array}{l} \forall M \\ \left[\begin{array}{l} \exists GH \\ \left[\begin{array}{ll} & M\,G \\ & M\,H \\ & \quad \exists Y \\ \sim M[G \circ H] \vee & \left[\begin{array}{c} P\,Y \\ \sim P[G\,Y] \end{array}\right] \end{array}\right] \end{array}\right] \end{array}$$

This time let's try to mate the literal MG with the entire wff on the lower right. Consideration of the negation of this wff suggests that we instantiate $\forall M$ with the term $[\lambda F_{\iota\iota} \forall W_\iota . \sim P_{o\iota} W \vee P . GW]$. However, we cannot quite do this, since G depends on M, so we use the term $[\lambda F_{\iota\iota} \forall W_\iota . \sim P_{o\iota} W \vee P . FW]$. Of course, this wff denote the set of functions which map P into P; thus, this connection between parts of the wff has led us rather directly to an expansion term which contains the key idea in the most obvious proof of this theorem. When we make this expansion and

λ-normalize we obtain:

$$
\begin{array}{l}
\exists G H \\
\left[\begin{array}{c}
\forall W \\
\quad [\sim P\, W \vee P[G W]] \\
\forall W \\
\quad [\sim P\, W \vee P[H\, W]] \\
\begin{array}{ccc}
\exists W & & \exists Y \\
\left[\begin{array}{c} P\, W \\ \sim P[[G \circ H] W] \end{array}\right] & \vee & \left[\begin{array}{c} P\, Y \\ \sim P[G\, Y] \end{array}\right]
\end{array}
\end{array}\right]
\end{array}
$$

We rename some variables and select G, H, X, and Y:

$$
\left[\begin{array}{c}
\forall U \\
\quad [\sim P\, U \vee P[G\, U]] \\
\forall W \\
\quad [\sim P\, W \vee P[H\, W]] \\
\left[\begin{array}{c} P\, X \\ \sim P[[G \circ H] X] \end{array}\right] \vee \left[\begin{array}{c} P\, Y \\ \sim P[G\, Y] \end{array}\right]
\end{array}\right]
$$

Although our intuition tells us that we have already incorporated the expansion term representing the key idea for our proof into the expansion tree, there is still some work to be done. We can clearly mate the literals containing U with those containing Y in such a way as to span the paths passing through the wff on the lower right. We might try to mate $[PX]$ with $[\sim PW]$, but this leaves some paths unspanned. We note that the lower-left literal is not yet being used in this mating. To try to find a mate for it, we need to duplicate one of the top two lines. We do this by quantifier duplication, which is really just the introduction of a new variable as an expansion term for the node of the expansion tree which corresponds to the quantifier. At this point the vpform we display will depart a little from the expansion tree it represents, since we'll display two quantfiers where there's really only one in the expansion tree. However, it's convenient to display quantifiers for the variables to keep track of where and how the variables are quantified. After duplicating the quantifier $\forall U$ and instantiating the definition we obtain:

$$
\left[\begin{array}{c}
\forall U \\
\quad [\sim P\, U \vee P[G\, U]] \\
\forall V \\
\quad [\sim P\, V \vee P[G\, V]] \\
\forall W \\
\quad [\sim P\, W \vee P[H\, W]] \\
\left[\begin{array}{c} P\, X \\ \sim P[G . H\, X] \end{array}\right] \vee \left[\begin{array}{c} P\, Y \\ \sim P[G\, Y] \end{array}\right]
\end{array}\right]
$$

Now let's try the following mating:

$$\begin{bmatrix} \forall U \\ \quad [\sim P\,U \vee P[G\,U]] \\ \forall V \\ \quad [\sim P\,V \vee P[G\,V]] \\ \forall W \\ \quad [\sim P\,W \vee P[H\,W]] \\ \begin{bmatrix} P\,X \\ \sim P[G\,.\,H\,X] \end{bmatrix} \vee \begin{bmatrix} P\,Y \\ \sim P[G\,Y] \end{bmatrix} \end{bmatrix}$$

Instantiate the remaining universal quantifiers using the substitutions dictated by the mating:

$$\begin{bmatrix} \sim P\,Y \vee P[G\,Y] \\ \sim P[H\,X] \vee P[G\,.\,H\,X] \\ \sim P\,X \vee P[H\,X] \\ \begin{bmatrix} P\,X \\ \sim P[G\,.\,H\,X] \end{bmatrix} \vee \begin{bmatrix} P\,Y \\ \sim P[G\,Y] \end{bmatrix} \end{bmatrix}$$

This wff is contradictory.

We now have another example of a theorem with two proofs. This example illustrates a general method for constructing such theorems: assert the existence of some sort of entity, where many examples exist.

Of course, THM112 is fairly trivial, and we can't expect to guess the right expansion terms so easily in general. Nevertheless, this example illustrates the fact that the basic processes involved in constructing an expansion proof are expanding the formula and searching for an acceptable mating. The formula can be expanded by duplicating quantifiers and by instantiating quantifiers on higher-order variables. The mating-search process involves building up a spanning set of connections which is *compatible*, i.e., a set for which there is a substitution which simultaneously makes each connection a complementary pair.

Connections, or matings, arise naturally as one seeks to find the tautology buried in a theorem. One knows that once one has a tautology, certain literals will be complementary. Once one has enough copies of all the appropriate literals and one decides which literals should be complementary, one can systematically generate the substitutions which will make them so by applying a unification algorithm. In working back to the tautology, the connections generate the appropriate substitutions.

8. General Considerations about the Search Process

Let us consider certain aspects of the problem of searching for expansion proofs, and make some general suggestions about the design of a computer system for finding such proofs. Our purpose here is not to discuss the details of any existing system, but to bring relevant problems and questions into focus. We start by considering the kind of computer environment which will facilitate the process.

Since many difficult mathematical questions can be expressed as formulas of higher-order logic, it is clear that a useful theorem proving system in this realm will allow for human input, so it should be able to operate in a mixture of automatic and interactive modes. The system should be able to stop and restart gracefully, permitting human intervention and interaction in the process. It should be possible to instantiate definitions whenever this seems appropriate. To facilitate interaction, facilities should be available for displaying information and partial proofs in a variety of formats, and translating from one to another. The user should be able to control the level of detail that is displayed on the screen and recorded in various files.

A search procedure which is in principle complete will achieve early success only on relatively simple problems, and should be augmented by general heuristic search processes. We will never finish searching for, and experimenting with, heuristics to guide the decisions about what steps to take, so the system should be able to accommodate a variety of heuristics, and readily incorporate new heuristics. We will have both absolute and heuristic prescriptions and restrictions. The absolute ones are justified by metatheorems, the heuristic ones by our experience and intuition. The heurstic ones may conflict with each other, and we will need heuristic principles for resolving these conflicts.

The search for an expansion proof in higher-order logic may involve a number of simultaneous unbounded searches, so it will be particularly appropriate to use multiprocessors as suitable ones become available. (Note [15].) The search process should deal with data structures which are modified in systematic ways by processors which can work independently and interact in controlled and fruitful ways under the direction of a master processor. (For example, we should be able to expand the wff at any time and continue the matingsearch.)

We would really like to organize the search process so that we can imagine we are moving about in some well defined infinite search space. Our motion can be generated by arbitrary heuristics, and we always know what parts of the space we've already explored. We have methods for keeping track of parts of the space we need not explore because of things we have learned from earlier explorations. Of course, we would like to express all this information very concisely, and be able to access it very quickly and flexibly. Obviously, there are problems in designing such a system.

9. Matingsearch

The matingsearch process, which attempts to find an acceptable mating of the current formula, can be carried on by several processors working simultaneously.

One processor maintains the *connection graph* for the current formula, which is the set of connections which are candidates for inclusion in the mating. In order to be in the connection graph, a pair of nodes must share some vertical path, and must have the potential of becoming complementary under some substitution. Associated with each connection is the currently known information about its unification problem. Connections may be removed from the connection graph if they are eventually discovered to be incompatible.

Unification problems associated with sets of connections will be analyzed by processors which apply Huet's unification algorithm [23]. Ultimately the unification processors should accommodate any constraints on unification terms which the user wishes to impose and recognize special cases and certain unification problems where processing is guaranteed to terminate (such as those of first-order logic). There should be a special processor to recognize when certain nodes in the unification search tree are subsumed by others. (Such subsumptions actually occur with surprising frequency.)

Certain processors, which we call *matingsearch processors*, will build up sets of connections from the connection graph in an attempt to construct an acceptable mating. Unless an acceptable mating is quickly found, most of the useful information acquired in this process concerns what will *not* work, i.e., which sets of connections are incompatible. A number of matingsearch processors can profitably operate simultaneously if each contributes to, and makes use of, a collection of incompatible sets of connections which we shall call the *failure record*. When the formula is expanded, the matingsearch processors may have to re-examine what must be done to achieve an acceptable mating, but the information embodied in the failure record remains useful.

The failure record can be constantly augmented not only by incompatibilities discovered by the matingsearch processors, but also by analyses of symmetries in the formula and other methods. Symmetries may occur in the original theorem, and are always introduced when quantifiers are duplicated. They can cause significant redundancy in the search process if they are ignored.

Ideally, one of the matingsearch processors will apply a search procedure which is in principle complete, but others may apply heuristic procedures which are incomplete but likely to achieve success more quickly in favorable cases. Matingsearch heuristics can be based on the need to establish connections between nodes in such a way that an acceptable mating can eventually be achieved, and to discover incompatibilities quickly.

10. Expansion

Let's consider the problem of expanding a wff (i.e., introducing or changing expansion terms in such a way as to enlarge the deep formula of the expansion tree). In a multiprocessor environment, one or more processors could work on this while others work on matingsearch.

One way of expanding a wff is by quantifier duplication. As mentioned above, this operation involves adding an expansion term consisting of a new variable at an expansion node. (Of course, by a new variable we mean one that does not already occur in any wff associated with the expansion tree.) In first-order logic this is the only expansion operation that is necessary; once we have duplicated the quantifiers sufficiently, all we need to do (in principle) is to make an exhaustive search of the possible matings, applying the unification algorithm to generate the required substitution terms. In higher-order logic, however, quantifier duplication does not suffice. Indeed, the need to expand the formula by instantiating quantifiers with terms which cannot be generated by unification of existing subformulas is one of the vexing problems which distinguishes higher-order logic from first-order logic.

Other methods of expanding a wff in higher-order logic involve making substitutions for higher-order variables, so let's take a moment to consider what we mean by applying a substitution to a variable in an expansion tree. We shall assume that the shallow formula of the expansion tree contains no free variables. Also, since we wish to regard symbols as variables only if it is appropriate to substitute for them, we shall regard selected parameters as constants, although Miller calls them selected variables. Thus, free variables can be introduced in an expansion tree only in expansion terms, and if we compute all deep and shallow formulas of nodes (except for the shallow formula of the root) from the expansion terms and selected parameters, we can substitute for a free variable simply by substituting for its free occurrences in expansion terms.

Of course, we shall always wish to work with expansion trees in which selections are done correctly, so we assume without further discussion that new expansion terms will be introduced and substitutions will be applied in such a way as to guarantee this. If Skolemization is used this is generally automatic, but if it is not one may need to rename certain selected parameters and forbid certain substitutions so that the imbedding relation remains irreflexive.

Naturally, one can generate expansion terms incrementally by using substitutions which introduce single connectives and quantifiers in a general way, thus achieving the

$$\lambda X_{o\alpha\xi} \lambda Y_{o\beta} \lambda Z_{\alpha} . \sim R^{1}_{o\alpha(o\beta)(o\alpha\xi)} X Y Z$$

$$\lambda X_{o\alpha\xi} \lambda Y_{o\beta} \lambda Z_{\alpha} . R^{2}_{o\alpha(o\beta)(o\alpha\xi)} X Y Z \wedge R^{3}_{o\alpha(o\beta)(o\alpha\xi)} X Y Z$$

$$\lambda X_{o\alpha\xi} \lambda Y_{o\beta} \lambda Z_{\alpha} . R^{4}_{o\alpha(o\beta)(o\alpha\xi)} X Y Z \vee R^{5}_{o\alpha(o\beta)(o\alpha\xi)} X Y Z$$

$$\lambda X_{o\alpha\xi} \lambda Y_{o\beta} \lambda Z_{\alpha} \exists W_{\varphi} R^{6}_{o\varphi\alpha(o\beta)(o\alpha\xi)} X Y Z W$$

$$\lambda X_{o\alpha\xi} \lambda Y_{o\beta} \lambda Z_{\alpha} \forall W_{\varphi} R^{7}_{o\varphi\alpha(o\beta)(o\alpha\xi)} X Y Z W$$

$$\lambda X_{o\alpha\xi} \lambda Y_{o\beta} \lambda Z_{\alpha} Y . R^{8}_{\beta\alpha(o\beta)(o\alpha\xi)} X Y Z$$

$$\lambda X_{o\alpha\xi} \lambda Y_{o\beta} \lambda Z_{\alpha} X [R^{9}_{\xi\alpha(o\beta)(o\alpha\xi)} X Y Z] . R^{10}_{\alpha\alpha(o\beta)(o\alpha\xi)} X Y Z$$

$$\lambda X_{o\alpha\xi} \lambda Y_{o\beta} X . R^{11}_{\xi(o\beta)(o\alpha\xi)} X Y$$

Fig. 10.1. Primitive substitutions for $R_{o\alpha(o\beta)(o\alpha\xi)}$.

$\lambda F_{\iota\iota}. \sim M^1_{o(\iota\iota)}F$

$\lambda F_{\iota\iota}.M^2_{o(\iota\iota)}F \wedge M^3_{o(\iota\iota)}F$

$\lambda F_{\iota\iota}.M^4_{o(\iota\iota)}F \vee M^5_{o(\iota\iota)}F$

$\lambda F_{\iota\iota}\exists W_\delta M^1_{o\delta(\iota\iota)}F\,W$

$\lambda F_{\iota\iota}\forall W_\delta M^1_{o\beta(\iota\iota)}F\,W$

Special case of the above:

$\lambda F_{\iota\iota}\forall W_\iota M^1_{o\iota(\iota\iota)}F\,W$

Fig. 10.2. Primitive substitutions for $M_{o(\iota\iota)}$.

$\lambda F_{\iota\iota}\lambda W_\iota. \sim M^2_{o\iota(\iota\iota)}F\,W$

$\lambda F_{\iota\iota}\lambda W_\iota.M^2_{o\iota(\iota\iota)}F\,W \wedge M^3_{o\iota(\iota\iota)}F\,W$

$\lambda F_{\iota\iota}\lambda W_\iota.M^2_{o\iota(\iota\iota)}F\,W \vee M^3_{o\iota(\iota\iota)}F\,W$

$\lambda F_{\iota\iota}\lambda W_\iota\exists U_\varepsilon M^2_{o\varepsilon\iota(\iota\iota)}F\,W\,U$

$\lambda F_{\iota\iota}\lambda W_\iota\forall U_\varepsilon M^2_{o\varepsilon\iota(\iota\iota)}F\,W\,U$

Fig. 10.3. Primitive substitutions for $M^1_{o\iota(\iota\iota)}$.

effects of Huet's splitting rules [22]. It turns out that in the context of the search process we envision, projections such as those used in [23] are also needed. We shall refer to such substitutions for a single variable as *primitive substitutions*. Examples of primitive substitutions for certain variables may be seen in Figures 10.1, 10.2, 10.3, and 10.4, where the terms to be substituted for the variable are given. Note that they are determined by the types of the variables for which the substitution is to be made. A substitution term has the form $\lambda\mathbf{x}^1 \ldots \lambda\mathbf{x}^n\mathbf{B}_o$; we shall refer to the wff $\mathbf{B}_o$ as its *body*.

Before discussing primitive substitutions further, let's see how THM112 can be proved quite systematically by using them. We shall ignore all the incorrect choices which a real search procedure would probably make before finding the correct ones.

$\lambda S_{o\alpha}. \sim P^1_{o(o\alpha)} S$

$\lambda S_{o\alpha}. P^2_{o(o\alpha)} S \wedge P^3_{o(o\alpha)} S$

$\lambda S_{o\alpha}. P^4_{o(o\alpha)} S \vee P^5_{o(o\alpha)} S$

$\lambda S_{o\alpha} \exists X_\varepsilon P^6_{o\varepsilon(o\alpha)} S X$

$\lambda S_{o\alpha} \forall X_\varepsilon P^7_{o\varepsilon(o\alpha)} S X$

This one is a projection:

$\lambda S_{o\alpha} S . P^8_{\alpha(o\alpha)} S$

Fig. 10.4. Primitive substitutions for $P_{o(o\alpha)}$.

Recall that THM112 is:

$$\forall P_{o\iota} \exists M_{o(\iota\iota)} \forall G_{\iota\iota} \forall H_{\iota\iota} . M G \wedge M H \supset M[G \circ H] \wedge \forall Y_\iota . P Y \supset P . G Y$$

THM112 after negating and selecting P:

$$\forall M \left[\exists GH \left[\begin{array}{c} M G \\ M H \\ \sim M[G \circ H] \vee \exists Y \left[\begin{array}{c} P Y \\ \sim P[G Y] \end{array} \right] \end{array} \right] \right]$$

Expand on $M_{o(\iota\iota)}$ with the primitive substitution term $\lambda F_{\iota\iota} \forall W_\iota M^1_{o\iota(\iota\iota)} F W$:

$$\exists GH \left[\begin{array}{c} [\lambda F \ \forall W M^1 F W] G \\ [\lambda F \ \forall W M^1 F W] H \\ \sim [\lambda F \ \forall W M^1 F W][G \circ H] \vee \exists Y \left[\begin{array}{c} P Y \\ \sim P[G Y] \end{array} \right] \end{array} \right]$$

λ-normalize:

$$
\exists G H
\left[
\begin{array}{l}
\qquad\qquad \forall W \\
\qquad\qquad\quad M^1 G\, W \\[1ex]
\qquad\qquad \forall W \\
\qquad\qquad\quad M^1 H\, W \\[1ex]
\exists W \qquad\qquad\qquad\qquad\quad \exists Y \\
\quad \sim M^1[G \circ H]W \;\vee\; \left[\begin{array}{c} P\,Y \\ \sim P[G\,Y] \end{array}\right]
\end{array}
\right]
$$

Expand the formula again by substituting for $M^1_{o\iota(\iota\iota)}$ the primitive substitution term $[\lambda F_{\iota\iota}\,\lambda W_\iota M^2_{o\iota(\iota\iota)} F W \vee M^3_{o\iota(\iota\iota)} F W]$ (see Figure 10.3) and λ-normalize:

$$
\exists G H
\left[
\begin{array}{l}
\qquad\qquad \forall W \\
\qquad\qquad\quad [M^2 G\, W \vee M^3 G\, W] \\[1ex]
\qquad\qquad \forall W \\
\qquad\qquad\quad [M^2 H\, W \vee M^3 H\, W] \\[1ex]
\exists W \qquad\qquad\qquad\qquad\quad \exists Y \\
\quad \left[\begin{array}{l} \sim M^2[G \circ H]W \\ \sim M^3[G \circ H]W \end{array}\right] \vee \left[\begin{array}{c} P\,Y \\ \sim P[G\,Y] \end{array}\right]
\end{array}
\right]
$$

Next we shall duplicate a quantifier, rename bound variables, and eliminate the definition.

$$
\exists G H
\left[
\begin{array}{l}
\qquad\qquad \forall U \\
\qquad\qquad\quad [M^2 G\, U \vee M^3 G\, U] \\[1ex]
\qquad\qquad \forall V \\
\qquad\qquad\quad [M^2 G\, V \vee M^3 G\, V] \\[1ex]
\qquad\qquad \forall W \\
\qquad\qquad\quad [M^2 H\, W \vee M^3 H\, W] \\[1ex]
\exists X \qquad\qquad\qquad\qquad\qquad \exists Y \\
\quad \left[\begin{array}{l} \sim M^2[\lambda Z\; G\,.\,H\,Z]X \\ \sim M^3[\lambda Z\; G\,.\,H\,Z]X \end{array}\right] \vee \left[\begin{array}{c} P\,Y \\ \sim P[G\,Y] \end{array}\right]
\end{array}
\right]
$$

Now consider the following mating:

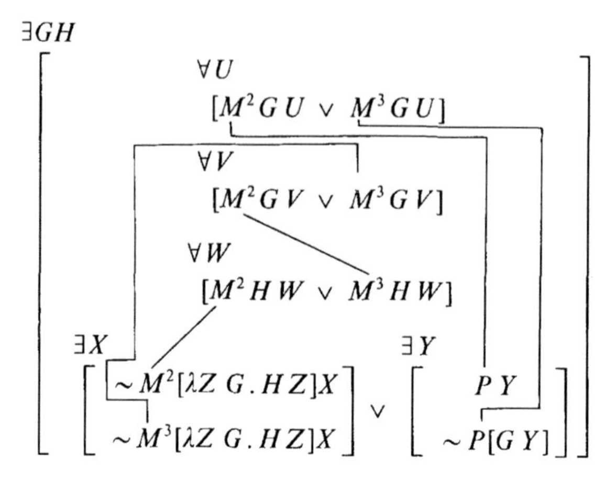

Let us first make the substitutions for M^2 and M^3 dictated by the mating above.

$$\exists G H \left[\begin{array}{l} \forall U \; [[\lambda F \lambda Z. \sim P Z] G U \vee [\lambda F \lambda Z\; P . F Z] G U] \\ \forall V \; [[\lambda F \lambda Z. \sim P Z] G V \vee [\lambda F \lambda Z\; P . F Z] G V] \\ \forall W \; [[\lambda F \lambda Z. \sim P Z] H W \vee [\lambda F \lambda Z\; P . F Z] H W] \\ \exists X \left[\begin{array}{l} \sim [\lambda F \lambda Z. \sim P Z][\lambda Z\; G . H Z] X \\ \sim [\lambda F \lambda Z\; P . F Z][\lambda Z\; G . H Z] X \end{array}\right] \vee \exists Y \left[\begin{array}{c} P Y \\ \sim P[G Y] \end{array}\right] \end{array}\right]$$

Note that the terms substituted for M^2 and M^3 contained no free variables, since P is a constant. Normalize and select G, H, X, and Y:

$$\left[\begin{array}{l} \forall U \; [\sim P U \vee P[G U]] \\ \forall V \; [\sim P V \vee P[G V]] \\ \forall W \; [\sim P W \vee P[H W]] \\ \left[\begin{array}{c} P X \\ \sim P[G . H X] \end{array}\right] \vee \left[\begin{array}{c} P Y \\ \sim P[G Y] \end{array}\right] \end{array}\right]$$

Now we expand on the universally quantified variables as dictated by the mating.

$$\begin{bmatrix} \sim P\,Y \lor P[G\,Y] \\ \sim P[H\,X] \lor P[G\,.\,H\,X] \\ \sim P\,X \lor P[H\,X] \\ \begin{bmatrix} P\,X \\ \sim P[G\,.\,H\,X] \end{bmatrix} \lor \begin{bmatrix} P\,Y \\ \sim P[G\,Y] \end{bmatrix} \end{bmatrix}$$

This wff is contradictory.

You might wonder why we list projections as primitive substitutions. Do we really need them? Let's have an example to show that projections are indeed needed.

THM104 is:

$$\forall U_\alpha \forall V_\alpha\,.\,\text{UNITSET } U = \text{UNITSET } V \supset U = V$$

The definition of UNITSET is $[\lambda W_\alpha \lambda Z_\alpha\,.\,Z = W]$. Thus [UNITSET W_α], which might be abbreviated as $\{W_\alpha\}$, is $[\lambda Z_\alpha\,.\,Z = W]$, which denotes the set of all Z such that $Z = W$. The definition of $=$ is $[\lambda X_\alpha \lambda Y_\alpha \forall Q_{o\alpha}\,.\,QX \supset QY]$. The definition of UNITSET with the definition of $=$ instantiated is $[\lambda W_\alpha \lambda Z_\alpha \forall Q_{o\alpha}\,.\,QZ \supset QW]$.

Nnf of the negation of THM104:

$$\exists UV \begin{bmatrix} \text{UNITSET } U = \text{UNITSET } V \\ \sim U = V \end{bmatrix}$$

Instantiate the definition of $=$:

$$\exists UV \begin{bmatrix} \forall P \\ [\sim P[\text{UNITSET } U] \lor P[\text{UNITSET } V]] \\ \exists Q \begin{bmatrix} QU \\ \sim Q\,V \end{bmatrix} \end{bmatrix}$$

Instantiate the definition of UNITSET:

$$\exists UV \begin{bmatrix} \forall P \\ [\sim P[\lambda Z\,.\,Z = U] \lor P[\lambda Z\,.\,Z = V]] \\ \exists Q \begin{bmatrix} Q\,U \\ \sim Q\,V \end{bmatrix} \end{bmatrix}$$

Instantiate the definition of $=$:

$$
\begin{array}{l}
\exists U V \\
\left[\begin{array}{l}
\forall P \\
\quad [\sim P[\lambda Z\ \forall R . R Z \supset R U] \vee P[\lambda Z \forall S . S Z \supset S V]] \\
\qquad\qquad\qquad \exists Q \\
\qquad\qquad\qquad \left[\begin{array}{c} Q U \\ \sim Q V \end{array}\right]
\end{array}\right]
\end{array}
$$

Select U, V, and Q:

$$
\left[\begin{array}{l}
\forall P \\
\quad [\sim P[\lambda Z\ \forall R . R Z \supset R U] \vee P[\lambda Z \forall S . S Z \supset S V]] \\
\qquad\qquad\qquad Q U \\
\qquad\qquad\qquad \sim Q V
\end{array}\right]
$$

Since U, V, and Q are constants, there is nothing to do but expand on $P_{o(o\alpha)}$. Clearly, we cannot derive a contradiction from the top line alone (since it just says that [UNITSET U] $=$ [UNITSET V]), so we must apply some operation Φ to the top line which will create some useful interaction between the top line and the two bottom literals. Note that modulo alphabetic changes of bound variables, the top line has the form $[\sim LU \vee LV]$. Clearly $\Phi[\sim LU \vee LV]$ must contain the literals $\sim QU$ and QV. Since $[\sim LU \vee LU]$ and $[\sim LV \vee LV]$ are useless tautologies, the result of applying Φ to either of these cannot be the same as $\Phi[\sim LU \vee LV]$. Therefore, the occurrences of U and V in the top line must actually contribute to the creation of at least one of the literals $\sim QU$ and QV. This cannot be accomplished by applying the primitive substitutions in Figure 10.4 which introduce quantifiers and connectives (possibly with iterations) and then applying the substitution induced by some mating. Thus, something else must be done. We apply the projection in Figure 10.4, and obtain the wff

$$
\begin{array}{l}
[\sim [\lambda S_{o\alpha} S . P^{8}_{\alpha(o\alpha)} S][\lambda Z_{\alpha}\ \forall R_{o\alpha} . R Z \supset R U_{\alpha}] \\
\quad \vee\ [\lambda S\ S . P^{8} S] . \lambda Z\ \forall S_{o\alpha} . S Z \supset S V_{\alpha}] \wedge . Q_{o\alpha} U \wedge \sim Q V
\end{array}
$$

which we can λ-normalize to

$$
\begin{array}{l}
[\sim \forall R_{o\alpha}[R[P^{8}_{\alpha(o\alpha)} . \lambda Z_{\alpha} \forall R . R Z \supset R U_{\alpha}] \supset R U] \\
\quad \vee\ \forall S_{o\alpha} . S[P^{8} . \lambda Z \forall S . S Z \supset S V_{\alpha}] \supset S V] \wedge . Q_{o\alpha} U \wedge \sim Q V
\end{array}
$$

To abbreviate this formula we give the names M and N to the arguments of P^8. Note that these have no free variables, since U and V are constants. We now write the formula as

$$
[\sim \forall R_{o\alpha}[R[P^{8}_{\alpha(o\alpha)} M_{o\alpha}] \supset R U_{\alpha}] \vee \forall S_{o\alpha} . S[P^{8} N_{o\alpha}] \supset S V_{\alpha}] \wedge . Q_{o\alpha} U \wedge \sim Q V
$$

The nnf of this wff can be displayed as:

$$\left[\begin{array}{l} \exists R \\ \left[\begin{array}{c} R[P^8 M] \\ \sim R\,U \end{array}\right] \vee \begin{array}{c} \forall S \\ [\sim S[P^8 N] \vee S\,V] \end{array} \\ \qquad\qquad Q\,U \\ \qquad\qquad \sim Q\,V \end{array}\right]$$

Select R, and consider the following mating:

$$\left[\begin{array}{l} \left[\begin{array}{c} R[P^8 M] \\ \sim R\,U \end{array}\right] \vee \begin{array}{c} \forall S \\ [\sim S[P^8 N] \vee S\,V] \end{array} \\ \qquad\qquad Q\,U \\ \qquad\qquad \sim Q\,V \end{array}\right]$$

From here on the mating dictates what is to be done. Substitute $\lambda H_{o\alpha} U_\alpha$ for P^8.

$$\left[\begin{array}{l} \left[\begin{array}{c} R[\lambda H\,U]M] \\ \sim R\,U \end{array}\right] \vee \begin{array}{c} \forall S \\ [\sim S[[\lambda H\,U]N] \vee S\,V] \end{array} \\ \qquad\qquad Q\,U \\ \qquad\qquad \sim Q\,V \end{array}\right]$$

λ-normalize:

$$\left[\begin{array}{l} \left[\begin{array}{c} R\,U \\ \sim R\,U \end{array}\right] \vee \begin{array}{c} \forall S \\ [\sim S\,U \vee S\,V] \end{array} \\ \qquad\qquad Q\,U \\ \qquad\qquad \sim Q\,V \end{array}\right]$$

Instantiate S:

$$\left[\begin{array}{l} \left[\begin{array}{c} R\,U \\ \sim R\,U \end{array}\right] \vee [\sim Q\,U \vee Q\,V] \\ \qquad\qquad Q\,U \\ \qquad\qquad \sim Q\,V \end{array}\right]$$

This wff is contradictory. Examination of the proof above shows that the expansion term finally substituted for $P_{o(o\alpha)}$ is $[\lambda S_{o\alpha} . S U_\alpha]$. Thus, the idea underlying the proof can be expressed as follows: suppose [UNITSET U] = [UNITSET V]; therefore [UNITSET U]U implies that [UNITSET V]U; the conclusion that $U = V$ follows easily using the definition of UNITSET.

In Huet's paper [22] on constrained resolution, Huet introduces 'splitting rules' which clearly introduce connectives and quantifiers, but one might wonder how constrained resolution handles THM104, where a projection must be introduced if one uses the method above. Actually, when using constrained resolution one generates a clause with the constraint that the literal $P_{o(o\alpha)}\,[\lambda Z_\alpha \forall S_{o\alpha} . SZ \supset SV_\alpha]$ must be unified with an expression of the form $\forall M_{o\alpha}[H_{o(o\alpha)}M]$, and the unifying substitution provides the projection for P.

Of course, generating expansion terms incrementally by using primitive substitutions as illustrated above is a rather primitive procedure, and additional methods are clearly needed. The most significant work on this problem which has been published so far in Bledsoe's work on set variables [16, 17]. His approach is based on the very sensible idea that it is often possible to construct expansion terms in a systematic way from expressions which are already present in the theorem to be proved. He gives basic rules and combining rules for doing this in certain contexts. With the aid of these rules his theorem prover is able to perform quite impressively on certain examples. However, much research remains to be done on this problem.

Let's think some·more about how to use primitive substitutions.

Of course, one can choose the connectives and quantifiers to be introduced by primitive substitutions in a variety of ways. If one uses a minimal complete set such as $\{\sim, \wedge, \forall\}$, the process of generating expansion terms incrementally may be unduly ardous, so it may be better to use $\{\sim, \wedge, \vee, \forall, \exists\}$. As Sunil Issar has pointed out, in most contexts we may let the bodies of expansion terms be in nnf, so negations need be introduced only sparingly. Indeed, the matingsearch process can decide which literals require negations, so negations which will have only atomic scope in the wff need not be introduced by primitive substitutions. However, we cannot always assume that expansion terms can be in nnf; if a variable which will be replaced by an expansion term in the deep formula occurs as an argument of some constant symbol, then that expansion term may need to have whatever form will make the literal containing it complementary to some other literal. In such cases we would hope to generate the expansion term by direct unification of the atoms rather than by primitive substitutions.

In general, there will be infinitely many primitive substitutions for a variable, since some of the substitution terms will contain expressions of the form $\forall \mathbf{x}_\alpha \mathbf{B}_o$, where α is an arbitrary type symbol. The problem of how to choose the type symbol α is deep and significant. Of course, a natural thing to do is to use type variables for type symbols and determine what to substitute for these later. This is a problem which invites extensive research. At the very least, it suggests the need for a unification algorithm for expressions of typed λ-calculus containing type variables. For now we shall simply assume that the infinitely many primitive substitutions, and other expressions generated from them, will be represented in some finite way. In actual mathematical practice it is rather rare to use quantifiers on variables whose types are much more complicated than those in the theorem being proved, so until we are much more ambitious it should suffice to simply use a fairly small set of simple types in place of the infinite set of all types.

Of course, it is desirable to limit the use of primitive substitutions as much as possible. Actually, they may usefully be applied only to certain variables in a wff, which we shall call *eligible* (for primitive substitutions). We shall call the first proper symbol of the λ-normal form of an atom the *head* of that atom, and call a variable which is the head of some atom of a wff a *head variable* of that wff. Only head variables need be eligible. Also, if a variable occurs only at the heads of atoms and these atoms are all positive or all negative, we can make all these atoms reduce to tautologies or contradictions by a simple substitution, so there is no need for such a variable to be eligible.

Usually an eligible variable will occur as head of both positively and negatively occurring atoms of the wff. In this case both conjunctive and disjunctive primitive substitutions for that variable will enlarge the number of paths through the wff, and an analysis of the costs and benefits to the matingsearch process of this enlargement may be useful.

The fact that it may be appropriate to apply several different primitive substitutions to the same variable, at least until the matingsearch process determines which substitutions are useful, makes it necessary to consider how this should be done in an expansion tree. We may assume that the processors which expand the formula work with a copy of the formula to which no substitutions are applied except those involved in the expansion process, and that the substitutions generated by the matingsearch process are not actually applied until all necessary expansions have been made. We may also assume that variables introduced as expansion terms, or as free variables in substitution terms, are all new variables which are distinct from one another. Therefore, it is easy to see that no variable will occur free in more than one expansion term. The result of applying the primitive substitutions $\theta^1, \ldots, \theta^n$ to the eligible variable $\mathbf{v}$ in an expansion tree may be described as follows. Let $\mathbf{t}$ be the expansion term in which v occurs free, and N be the expansion node from which the arc labeled with $\mathbf{t}$ descends. Create n variants $\mathbf{t}^1, \ldots, \mathbf{t}^n$ of $\mathbf{t}$ by substituting new and distinct variables for the free variables of $\mathbf{t}$ other than $\mathbf{v}$, so that these variants share no free variables except for $\mathbf{v}$. Now add n new arcs descending from node N labeled with expansion terms $(\theta^1\mathbf{t}^1), \ldots,$ and $(\theta^n\mathbf{t}^n)$, respectively. Note that the original arc labeled with $\mathbf{t}$ remains in the tree; we have no way of knowing whether it is appropriate to apply primitive substitutions to $\mathbf{v}$ or not, so we keep one copy of $\mathbf{v}$ without change.

We visualize a search procedure in which variables are instantiated only by primitive substitutions for head variables and by the unifier associated with a mating. It's natural to ask whether such a procedure is complete in an appropriate sense.

In principle, we don't need to backtrack if we make inappropriate expansions of the wff, since this simply enlarges the search space. This is similar to deriving redundant clauses in resolution, though perhaps more costly. However, if the wff is repeatedly expanded in a blind and unmotivated way, one will quickly generate an unmanageably large search space for the matingsearch process. Of course, such situations are well known to researchers in this field. In general, we would like the matingsearch process to motivate the particular expansions of the wff which are made. The development of heuristics to guide the process of expanding the wff is a major area for research. It is important to understand how appropriate expansions can be used to create the literals needed for the connections of an acceptable mating.

One quite simple expansion procedure is based on the idea that one should search first for expansion proofs in which the formula has not been expanded very much. Thus, the matingsearch processors should work on copies of the expansion tree to which only one expansion option has been applied until all such options have been explored, then on copies to which only two expansion options have been applied, etc. (More generally, one could assign weights to combinations of expansion options, and minimize these weights.) The names of the nodes in these copies can always be taken from the master expansion tree maintained by the expansion processors, so consistent contributions to the failure record can be made by these processors working on different fragments of the master tree.

11. Conclusion

Many problems arise in connection with the automation of higher-order logic. By formulating these precisely and dealing with them systematically, we can hope to eventually make progress in this important realm. There's lots of room for good research here.

References

1. Andrews, Peter B., 'Resolution in Type Theory', *Journal of Symbolic Logic* **36**, 414–432 (1971).
2. Andrews, Peter B., 'Resolution and the Consistency of Analysis', *Notre Dame Journal of Formal Logic* **15**, 73–84 (1974).
3. Andrews, Peter B., 'Refutations by Matings', *IEEE Transactions on Computers* **C-25**, 801–807 (1976).
4. Andrews, Peter B., 'Transforming Matings into Natural Deduction Proofs,' in *5th Conference on Automated Deduction*, edited by W. Bibel and R. Kowalski, Les Arcs, France, Lecture Notes in Computer Science 87, Springer-Verlag, 1980, pp. 281–292.
5. Andrews, Peter B., 'Theorem Proving via General Matings', *Journal of the ACM* **28**, 193–214 (1981).
6. Andrews, Peter B., Miller, Dale A., Cohen, Eve Longini, and Pfenning, Frank, 'Automating Higher-Order Logic', in *Automated Theorem Proving: After 25 Years*, edited by W. W. Bledsoe and D. W. Loveland, Contemporary Mathematics series, vol. 29, American Mathematical Society, 1984, pp. 169–192.
7. Andrews, Peter B., *An Introduction to Mathematical Logic and Type Theory: To Truth Through Proof*, Academic Press, 1986.
8. Andrews, Peter B., 'Connections and Higher-Order Logic', in *8th International Conference on Automated Deduction*, edited by Jorg H. Siekmann, Oxford, England, Lecture Notes in Computer Science 230, Springer-Verlag, 1986, pp. 1–4.
9. Bibel, Wolfgang, and Schreiber, J., 'Proof search in a Gentzen-like system of first-order logic', in *International Computing Symposium 1975*, edited by E. Gelenbe and D. Potier, North-Holland, Amsterdam, 1975, pp. 205–212.
10. Bibel, Wolfgang, 'Tautology Testing with a Generalized Matrix Reduction Method', *Theoretical Computer Science* **8**, 31–44 (1979).

11. Bibel, Wolfgang, 'On Matrices with Connections', *Journal of the ACM* **28**, 633–645 (1981).
12. Bibel, Wolfgang, *Automated Theorem Proving*, Vieweg, Braunschweig, 1982.
13. Bibel, Wolfgang, 'A Comparative Study of Several Proof Procedures', *Artificial Intelligence* **18**, 269–293 (1982).
14. Bibel, Wolfgang, 'Matings in Matrices', *Communications of the ACM* **26**, 844–852 (1983).
15. Bibel, Wolfgang, and Buchberger, Bruno, 'Towards a Connection Machine for Logical Inference', *Future Generations Computer Systems* **1**, 177–185 (1984–1985).
16. Bledsoe, W. W., 'A Maximal Method for Set Variables in Automatic Theorem Proving', in *Machine Intelligence 9*, Ellis Harwood Ltd., Chichester, and John Wiley & Sons, 1979, pp. 53–100.
17. Bledsoe, W. W., 'Using Examples to Generate Instantiations of Set Variables', in *Proceedings of IJCAI-83, Karlsruhe, Germany*, Aug 8–12, 1983, pp. 892–901.
18. Church, Alonzo, 'A Formulation of the Simple Theory of Types', *Journal of Symbolic Logic* **5**, 56–68 (1940).
19. Davis, Martin, 'Eliminating the Irrelevant from Mechanical Proofs', in *Experimental Arithmetic, High Speed Computing and Mathematics*, Proceedings of Symposia in Applied Mathematics XV, American Mathematical Society, 1963, pp. 15–30.
20. Dreben, Burton, Andrews, Peter, and Aanderaa, Stal. 'False Lemmas in Herbrand', *Bulletin of the American Mathematical Society* **69**, 699–706 (1963).
21. Dreben, Burton and Denton, John, 'A Supplement to Herbrand', *Journal of Symbolic Logic* **31**, 393–398 (1966).
22. Huet, Gérard P., 'A Mechanization of Type Theory', in *Proceedings of the Third International Joint Conference on Artificial Intelligence, IJCAI*, 1973, pp. 139–146.
23. Huet, Gérard P., 'A Unification Algorithm for Typed λ-Calculus', *Theoretical Computer Science* **1**, 27–57 (1975).
24. Jensen, D. C., and Pietrzykowski, T., 'Mechanizing ω-Order Type Theory Through Unification', *Theoretical Computer Science* **3**, 123–171 (1976).
25. Miller, Dale A., 'Proofs in Higher-Order Logic', Ph.D. Thesis, Carnegie Mellon University, 1983, 81 pp.
26. Miller, Dale A., 'Expansion Tree Proofs and Their Conversion to Natural Deduction Proofs', in *7th International Conference on Automated Deduction*, edited by R. E. Shostak, Napa, California, USA, Lecture Notes in Computer Science 170, Springer-Verlag, 1984, pp. 375–393.
27. Miller, Dale A., 'A Compact Representation of Proofs', *Studia Logica* **46**, 347–370 (1987).
28. Pfenning, Frank, 'Analytic and Non-analytic Proofs', in *7th International Conference on Automated Deduction*, edited by R. E. Shostak, Napa, California, USA, Lecture Notes in Computer Science 170, Springer-Verlag, 1984, pp. 394–413.
29. Pfenning, Frank, 'Proof Transformations in Higher-Order Logic', Ph.D. Thesis, Carnegie Mellon University, 1987, 156 pp.
30. Prawitz, Dag, 'Advances and Problems in Mechanical Proof Procedures', in *Machine Intelligence 4*, Edinburgh University Press, 1969, pp. 59–71.
31. Prawitz, Dag, 'A proof procedure with matrix reduction', in *Symposium on Automatic Demonstration, Versailles, France*, edited by M. Laudet, D. Lacombe, L. Nolin, and M. Schutzenberger, Lecture Notes in Mathematics 125, Springer-Verlag, 1970, pp. 207–214.
32. Shostak, Robert E., 'Refutation Graphs', *Artificial Intelligence* **7**, 51–64 (1976).

Andrews' Type Theory with Undefinedness

WILLIAM M. FARMER

1 Introduction

In 1940 Alonzo Church introduced in [4] a version of simple type theory with lambda-notation now known as *Church's type theory*. Church's students Leon Henkin and Peter B. Andrews extensively studied and refined Church's type theory. Henkin proved that Church's type theory is complete with respect to a semantics based on general models [14] and showed that Church's type theory can be reformulated so that it is based on only the primitive notions of function application, function abstraction, equality, and (definite) description [15]. Andrews devised a simple and elegant proof system for Henkin's reformulation of Church's type theory [1]. He also formulated a version of Church's type theory called $\mathcal{Q}_0$ that employs the ideas developed by Church, Henkin, and himself. $\mathcal{Q}_0$ is meticulously described and analyzed in [2] and is the logic of the TPS Theorem Proving System [3].

Church's type theory has had a profound impact on many areas of computer science, especially programming languages, automated reasoning, formal methods, type theory, and formalized mathematics. It is the fountainhead of a long stream of typed lambda calculi that includes systems such as *System F* [12], *Martin-Löf type theory* [17], and the *Calculus of Constructions* [5]. Several computer theorem proving systems are based on versions of Church's type theory including HOL [13], IMPS [10, 11], Isabelle [19], ProofPower [16], PVS [18], and TPS.

One of the principal virtues of Church's type theory is that it has great expressivity, both theoretical and practical. However, like other traditional logics, Church's type theory assumes that terms are always defined. Despite the fact that undefined terms are commonplace in mathematics (and computer science), undefined terms cannot be directly expressed in Church's type theory—as they are in mathematical practice.

A term is *undefined* if it has no prescribed meaning or if it denotes a value that does not exist.[1] There are two main sources of undefinedness

[1]Some of the text in this section concerning undefinedness is taken from [9].

in mathematics. The first source is terms that denote an application of a function. A function f usually has both a *domain of definition* D_f consisting of the values at which it is defined and a *domain of application* D^*_f consisting of the values to which it may be applied. (The domain of definition of a function is usually called simply the *domain* of the function.) These two domains are not always the same, but obviously $D_f \subseteq D^*_f$. A *function application* is a term $f(a)$ that denotes the application of a function f to an argument $a \in D^*_f$. $f(a)$ is *undefined* if $a \not\in D_f$. We will say that a function is *partial* if $D_f \neq D^*_f$ and *total* if $D_f = D^*_f$.

The second source of undefinedness is terms that are intended to uniquely describe a value. A *definite description* is a term t of the form "the x that has property P". t is *undefined* if there is no unique x (i.e., none or more than one) that has property P. Definite descriptions are quite common in mathematics but often occur in a disguised form. For example, "the limit of $\sin \frac{1}{x}$ as x approaches 0" is a definite description—which is undefined since the limit does not exist.

There is a *traditional approach to undefinedness* that is widely practiced in mathematics and even taught to some extent to students in secondary school. This approach treats undefined terms as legitimate, nondenoting terms that can be components of meaningful statements. The traditional approach is based on three principles:

1. Atomic terms (i.e., variables and constants) are always defined—they always denote something.

2. Compound terms may be undefined. A function application $f(a)$ is undefined if f is undefined, a is undefined, or $a \not\in D_f$. A definite description "the x that has property P" is undefined if there is no x that has property P or there is more than one x that has property P.

3. Formulas are always true or false, and hence, are always defined. To ensure the definedness of formulas, a function application $p(a)$ formed by applying a predicate p to an argument a is *false* if p is undefined, a is undefined, or $a \not\in D_p$.

A logic that formalizes the traditional approach to undefinedness has two advantages over a traditional logic that does not. First, the use of the traditional approach in informal mathematics can be directly formalized, yielding a result that is close to mathematical practice. Second, statements involving partial functions and undefined terms can be expressed very concisely. In particular, assumptions about the definedness of terms and functions often do not have to be made explicit. Concise informal mathematical statements involving partial functions or undefinedness can usually only be expressed in a traditional logic by verbose statements in which definedness assumptions are explicit. For evidence and further discussion of these assertions, see [9].

We presented in [6] a version of Church's type system named PF that formalizes the traditional approach to undefinedness. PF is the basis for LUTINS [7, 8], the logic of the IMPS theorem proving system [10, 11]. The paper [6] includes a proof that PF is complete with respect to a Henkin-style general models semantics. The proof, however, contains a mistake: the tautology theorem does not hold in PF as claimed. This mistake can be corrected by adding modus ponens and a technical axiom schema involving equality to PF's proof system. In [9] we introduced a version of Church's type system with undefinedness called STTwU which is simpler than PF. The proof system of STTwU is claimed to be complete, but a proof of completeness is not given in [9].

The purpose of this paper is to carefully show what changes have to be made to Church's type theory in order to formalize the traditional approach to undefinedness. We do this by presenting a modification of Andrews' type theory $\mathcal{Q}_0$ called $\mathcal{Q}_0^{\mathrm{u}}$. Our goal is to keep $\mathcal{Q}_0^{\mathrm{u}}$ as close to $\mathcal{Q}_0$ as possible, changing as few of the definitions in [2] concerning $\mathcal{Q}_0$ as possible. We present the syntax, semantics and proof system of $\mathcal{Q}_0^{\mathrm{u}}$ and prove that the proof system is sound and complete with respect to its semantics. A series of notes indicates precisely where and how $\mathcal{Q}_0$ and $\mathcal{Q}_0^{\mathrm{u}}$ diverge from each other.

Our presentation of $\mathcal{Q}_0^{\mathrm{u}}$ differs from the presentation of PF in [6] in the following ways:

1. The notation and terminology for $\mathcal{Q}_0^{\mathrm{u}}$ is almost identical to the notation and terminology for $\mathcal{Q}_0$ given in [2] unlike the notation and terminology for PF.

2. The semantics of $\mathcal{Q}_0^{\mathrm{u}}$ is simpler than the semantics of PF.

3. The proof system of $\mathcal{Q}_0^{\mathrm{u}}$ is complete unlike the proof system of PF.

4. The proof of the completeness theorem for $\mathcal{Q}_0^{\mathrm{u}}$ is presented in greater detail than the (erroneous) proof of the completeness theorem for PF.

The paper is organized as follows. The syntax of $\mathcal{Q}_0^{\mathrm{u}}$ is defined in section 2. A Henkin-style general models semantics for $\mathcal{Q}_0^{\mathrm{u}}$ is presented in section 3. Section 4 introduces several important defined logical constants and abbreviations. Section 5 gives the proof system of $\mathcal{Q}_0^{\mathrm{u}}$. Some metatheorems of $\mathcal{Q}_0^{\mathrm{u}}$ and the soundness and completeness theorems for $\mathcal{Q}_0^{\mathrm{u}}$ are proved in sections 6 and 7, respectively. The paper ends with a conclusion in section 8.

The great majority of the definitions for $\mathcal{Q}_0^{\mathrm{u}}$ are exactly the same as those for $\mathcal{Q}_0$ given in [2]. We repeat only the most important and least obvious definitions for $\mathcal{Q}_0$; for the others the reader is referred to [2].

2 Syntax of $\mathcal{Q}_0^{\mathrm{u}}$

The syntax of $\mathcal{Q}_0^{\mathrm{u}}$ is almost exactly the same as that of $\mathcal{Q}_0$. The only difference is that just one iota constant is primitive in $\mathcal{Q}_0$, while infinitely many iota constants are primitive in $\mathcal{Q}_0^{\mathrm{u}}$.

A *type symbol* of $\mathcal{Q}_0^{\mathrm{u}}$ is defined inductively as follows:

1. $\imath$ is a type symbol.
2. o is a type symbol.
3. If α and β are type symbols, then $(\alpha\beta)$ is a type symbol.

Let $\mathcal{T}$ denote the set of type symbols. $\alpha, \beta, \gamma, \ldots$ are syntactic variables ranging over type symbols. When there is no loss of meaning, matching pairs of parentheses in type symbols may be omitted. We assume that type combination associates to the left so that a type of the form $((\alpha\beta)\gamma)$ may be written as $\alpha\beta\gamma$.

The *primitive symbols* of $\mathcal{Q}_0^{\mathrm{u}}$ are the following:

1. *Improper symbols*: [,], λ.
2. A denumerable set of *variables* of type α for each $\alpha \in \mathcal{T}$: f_α, g_α, h_α, x_α, y_α, z_α, f_α^1, g_α^1, h_α^1, x_α^1, y_α^1, z_α^1, $\ldots$.
3. *Logical constants*: $\mathsf{Q}_{((o\alpha)\alpha)}$ for each $\alpha \in \mathcal{T}$ and $\iota_{(\alpha(o\alpha))}$ for each $\alpha \in \mathcal{T}$ with $\alpha \neq o$.
4. An unspecified set of *nonlogical constants* of various types.

$\mathbf{x}_\alpha, \mathbf{y}_\alpha, \mathbf{z}_\alpha, \mathbf{f}_\alpha, \mathbf{g}_\alpha, \mathbf{h}_\alpha, \ldots$ are syntactic variables ranging over variables of type α.

Note 1 (Iota Constants). Only $\iota_{\imath(o\imath)}$ is a primitive logical constant in $\mathcal{Q}_0$; each other $\iota_{\alpha(o\alpha)}$ is a nonprimitive logical constant in $\mathcal{Q}_0$ defined according to an inductive scheme presented by Church in [4] (see [2, pp. 233–4]). We will see in the next section that the iota constants have a different semantics in $\mathcal{Q}_0^{\mathrm{u}}$ than in $\mathcal{Q}_0$. As a result, it is not possible to define the iota constants in $\mathcal{Q}_0^{\mathrm{u}}$ as they are defined in $\mathcal{Q}_0$, and thus they must be primitive in $\mathcal{Q}_0^{\mathrm{u}}$. Notice that $\iota_{o(oo)}$ is not a primitive logical constant of $\mathcal{Q}_0^{\mathrm{u}}$. It has been left out because it serves no useful purpose. It can be defined as a nonprimitive logical constant as in [2, p. 233] if desired. ■

A *wff of type* α (*wff*$_\alpha$) is defined inductively as follows:

1. A variable or constant of type α is a wff$_\alpha$.

2. $[\mathbf{A}_{\alpha\beta}\mathbf{B}_\beta]$ is a wff$_\alpha$.

3. $[\lambda \mathbf{x}_\beta \mathbf{A}_\alpha]$ is a wff$_{\alpha\beta}$.

A wff of the form $[\mathbf{A}_{\alpha\beta}\mathbf{B}_\beta]$ is called a *function application* and a wff of the form $[\lambda \mathbf{x}_\beta \mathbf{A}_\alpha]$ is called a *function abstraction*. $\mathbf{A}_\alpha, \mathbf{B}_\alpha, \mathbf{C}_\alpha, \ldots$ are syntactic variables ranging over wffs of type α. When there is no loss of meaning, matching pairs of square brackets in wffs may be omitted. We assume that wff combination of the form $[\mathbf{A}_{\alpha\beta}\mathbf{B}_\beta]$ associates to the left so that a wff $[[\mathbf{C}_{\gamma\beta\alpha}\mathbf{A}_\alpha]\mathbf{B}_\beta]$ may be written as $\mathbf{C}_{\gamma\beta\alpha}\mathbf{A}_\alpha\mathbf{B}_\beta$.

3 Semantics of $\mathcal{Q}_0^{\mathrm{u}}$

The traditional approach to definedness is formalized in $\mathcal{Q}_0^{\mathrm{u}}$ by modifying the semantics of $\mathcal{Q}_0$. Two principal changes are made to the $\mathcal{Q}_0$ semantics: (1) The notion of a general model is redefined to include partial functions as well as total functions. (2) The valuation function for wffs is made into a partial function that assigns a value to a wff iff the wff is defined according to the traditional approach.

A *frame* is a collection $\{\mathcal{D}_\alpha \mid \alpha \in \mathcal{T}\}$ of nonempty domains such that:

1. $\mathcal{D}_o = \{\mathsf{T}, \mathsf{F}\}$.

2. For $\alpha, \beta \in \mathcal{T}$, $\mathcal{D}_{\alpha\beta}$ is some set of *total* functions from $\mathcal{D}_\beta$ to $\mathcal{D}_\alpha$ if $\alpha = o$ and is some set of *partial and total* functions from $\mathcal{D}_\beta$ to $\mathcal{D}_\alpha$ if $\alpha \neq o$.[2]

$\mathcal{D}_o$ is the *domain of truth values*, $\mathcal{D}_\iota$ is the *domain of individuals*, and for $\alpha, \beta \in \mathcal{T}$, $\mathcal{D}_{\alpha\beta}$ is a *function domain*. For all $\alpha \in \mathcal{T}$, the *identity relation* on $\mathcal{D}_\alpha$ is the total function $q \in \mathcal{D}_{o\alpha\alpha}$ such that, for all $x, y \in \mathcal{D}_\alpha$, $q(x)(y) = \mathsf{T}$ iff $x = y$. For all $\alpha \in \mathcal{T}$ with $\alpha \neq o$, the *unique member selector* on $\mathcal{D}_\alpha$ is the partial function $f \in \mathcal{D}_{\alpha(o\alpha)}$ such that, for all $s \in \mathcal{D}_{o\alpha}$, if the predicate

[2]The condition that a domain $D_{o\beta}$ contains only total functions is needed to ensure that the law of extensionality holds for predicates. This condition is weaker than the condition used in the semantics for PF and its extended versions PF* [7] and LUTINS. In these logics, a domain D_γ contains only total functions iff γ has the form $o\beta_1 \cdots \beta_n$ where $n \geq 1$. The weaker condition, which is due to Aaron Stump [20], yields a semantics that is simpler.

s represents a singleton $\{x\} \subseteq \mathcal{D}_\alpha$, then $f(s) = x$, and otherwise $f(s)$ is undefined.

Note 2 (Function Domains). In a $\mathcal{Q}_0$ frame a function domain $\mathcal{D}_{\alpha\beta}$ contains only total functions, while in a $\mathcal{Q}_0^{\mathrm{u}}$ frame a function domain $\mathcal{D}_{o\beta}$ contains only total functions but a function domain $\mathcal{D}_{\alpha\beta}$ with $\alpha \neq o$ contains partial functions as well as total functions. ■

An *interpretation* $\langle\{\mathcal{D}_\alpha \mid \alpha \in \mathcal{T}\}, \mathcal{J}\rangle$ of $\mathcal{Q}_0^{\mathrm{u}}$ consists of a frame and a function $\mathcal{J}$ that maps each constant of $\mathcal{Q}_0^{\mathrm{u}}$ of type α to an element of $\mathcal{D}_\alpha$ such that $\mathcal{J}(\mathsf{Q}_{o\alpha\alpha})$ is the identity relation on $\mathcal{D}_\alpha$ for each $\alpha \in \mathcal{T}$ and $\mathcal{J}(\iota_{\alpha(o\alpha)})$ is the unique member selector on $\mathcal{D}_\alpha$ for each $\alpha \in \mathcal{T}$ with $\alpha \neq o$.

Note 3 (Definite Description Operators). The $\iota_{\alpha(o\alpha)}$ in $\mathcal{Q}_0$ are *description operators*: if $\mathbf{A}_{o\alpha}$ denotes a singleton, then the value of $\iota_{\alpha(o\alpha)}\mathbf{A}_{o\alpha}$ is the unique member of the singleton, and otherwise the value of $\iota_{\alpha(o\alpha)}\mathbf{A}_{o\alpha}$ is *unspecified.* In contrast, the $\iota_{\alpha(o\alpha)}$ in $\mathcal{Q}_0^{\mathrm{u}}$ are *definite description operators*: if $\mathbf{A}_{o\alpha}$ denotes a singleton, then the value of $\iota_{\alpha(o\alpha)}\mathbf{A}_{o\alpha}$ is the unique member of the singleton, and otherwise the value of $\iota_{\alpha(o\alpha)}\mathbf{A}_{o\alpha}$ is *undefined.* ■

An *assignment* into a frame $\{\mathcal{D}_\alpha \mid \alpha \in \mathcal{T}\}$ is a function φ whose domain is the set of variables of $\mathcal{Q}_0^{\mathrm{u}}$ such that, for each variable $\mathbf{x}_\alpha$, $\varphi(\mathbf{x}_\alpha) \in \mathcal{D}_\alpha$. Given an assignment φ, a variable $\mathbf{x}_\alpha$, and $d \in \mathcal{D}_\alpha$, let $(\varphi : \mathbf{x}_\alpha/d)$ be the assignment ψ such that $\psi(\mathbf{x}_\alpha) = d$ and $\psi(\mathbf{y}_\beta) = \varphi(\mathbf{y}_\beta)$ for all variables $\mathbf{y}_\beta \neq \mathbf{x}_\alpha$.

An interpretation $\mathcal{M} = \langle\{\mathcal{D}_\alpha \mid \alpha \in \mathcal{T}\}, \mathcal{J}\rangle$ is a *general model* for $\mathcal{Q}_0^{\mathrm{u}}$ if there is a binary function $\mathcal{V}^{\mathcal{M}}$ such that for each assignment φ and wff $\mathbf{C}_\gamma$, either $\mathcal{V}_\varphi^{\mathcal{M}}(\mathbf{C}_\gamma) \in \mathcal{D}_\gamma$ or $\mathcal{V}_\varphi^{\mathcal{M}}(\mathbf{C}_\gamma)$ is undefined and the following conditions are satisfied for all assignments φ and all wffs $\mathbf{C}_\gamma$:

(a) Let $\mathbf{C}_\gamma$ be a variable of $\mathcal{Q}_0^{\mathrm{u}}$. Then $\mathcal{V}_\varphi^{\mathcal{M}}(\mathbf{C}_\gamma) = \varphi(\mathbf{C}_\gamma)$.

(b) Let $\mathbf{C}_\gamma$ be a constant of $\mathcal{Q}_0^{\mathrm{u}}$. Then $\mathcal{V}_\varphi^{\mathcal{M}}(\mathbf{C}_\gamma) = \mathcal{J}(\mathbf{C}_\gamma)$.

(c) Let $\mathbf{C}_\gamma$ be $[\mathbf{A}_{\alpha\beta}\mathbf{B}_\beta]$. If $\mathcal{V}_\varphi^{\mathcal{M}}(\mathbf{A}_{\alpha\beta})$ is defined, $\mathcal{V}_\varphi^{\mathcal{M}}(\mathbf{B}_\beta)$ is defined, and the function $\mathcal{V}_\varphi^{\mathcal{M}}(\mathbf{A}_{\alpha\beta})$ is defined at the argument $\mathcal{V}_\varphi^{\mathcal{M}}(\mathbf{B}_\beta)$, then

$$\mathcal{V}_\varphi^{\mathcal{M}}(\mathbf{C}_\gamma) = \mathcal{V}_\varphi^{\mathcal{M}}(\mathbf{A}_{\alpha\beta})(\mathcal{V}_\varphi^{\mathcal{M}}(\mathbf{B}_\beta)),$$

the value of the function $\mathcal{V}_\varphi^{\mathcal{M}}(\mathbf{A}_{\alpha\beta})$ at the argument $\mathcal{V}_\varphi^{\mathcal{M}}(\mathbf{B}_\beta)$. Otherwise, $\mathcal{V}_\varphi^{\mathcal{M}}(\mathbf{C}_\gamma) = \mathsf{F}$ if $\alpha = o$ and $\mathcal{V}_\varphi^{\mathcal{M}}(\mathbf{C}_\gamma)$ is undefined if $\alpha \neq o$.

(d) Let $\mathbf{C}_\gamma$ be $[\lambda \mathbf{x}_\alpha \mathbf{B}_\beta]$. Then $\mathcal{V}_\varphi^\mathcal{M}(\mathbf{C}_\gamma)$ is the (partial or total) function $f : \mathcal{D}_\alpha \rightarrow \mathcal{D}_\beta$ such that, for each $d \in \mathcal{D}_\alpha$, $f(d) = \mathcal{V}_{(\varphi:\mathbf{x}_\alpha/d)}^\mathcal{M}(\mathbf{B}_\beta)$ if $\mathcal{V}_{(\varphi:\mathbf{x}_\alpha/d)}^\mathcal{M}(\mathbf{B}_\beta)$ is defined and $f(d)$ is undefined if $\mathcal{V}_{(\varphi:\mathbf{x}_\alpha/d)}^\mathcal{M}(\mathbf{B}_\beta)$ is undefined.

Note 4 (Valuation Function). In $\mathcal{Q}_0$, if $\mathcal{M}$ is a general model, then $\mathcal{V}^\mathcal{M}$ is total and the value of $\mathcal{V}^\mathcal{M}$ on a function abstraction is always a total function. In $\mathcal{Q}_0^\mathrm{u}$, if $\mathcal{M}$ is a general model, then $\mathcal{V}^\mathcal{M}$ is partial and the value of $\mathcal{V}^\mathcal{M}$ on a function abstraction can be either a partial or a total function. ■

PROPOSITION 1. *Let $\mathcal{M}$ be a general model for $\mathcal{Q}_0^\mathrm{u}$. Then $\mathcal{V}^\mathcal{M}$ is defined on all variables, constants, function abstractions, and function applications of type o and is defined on a proper subset of function applications of type $\alpha \neq o$.*

Note 5 (Traditional Approach). $\mathcal{Q}_0^\mathrm{u}$ clearly satisfies the three principles of the traditional approach to undefinedness. Like other traditional logics, $\mathcal{Q}_0$ only satisfies the first principle. ■

Let $\mathcal{H}$ be a set of wffs$_o$ and $\mathcal{M}$ be a general model for $\mathcal{Q}_0^\mathrm{u}$. $\mathbf{A}_o$ is *valid* in $\mathcal{M}$, written $\mathcal{M} \models \mathbf{A}_o$, if $\mathcal{V}_\varphi^\mathcal{M}(\mathbf{A}_o) = \mathsf{T}$ for all assignments φ. $\mathcal{M}$ is a *general model* for $\mathcal{H}$ if $\mathcal{M} \models \mathbf{B}_o$ for all $\mathbf{B}_o \in \mathcal{H}$. $\mathbf{A}_o$ is *valid (in the general sense)* in $\mathcal{H}$, written $\mathcal{H} \models \mathbf{A}_o$, if $\mathcal{M} \models \mathbf{A}_o$ for every general model $\mathcal{M}$ for $\mathcal{H}$. $\mathbf{A}_o$ is *valid (in the general sense)* in $\mathcal{Q}_0^\mathrm{u}$, written $\models \mathbf{A}_o$, if $\emptyset \models \mathbf{A}_o$.

Note 6 (Mutual Interpretability). $\mathcal{Q}_0^\mathrm{u}$ can be interpreted in $\mathcal{Q}_0$ by viewing a function of type $\alpha\beta$ in $\mathcal{Q}_0^\mathrm{u}$ as a function (predicate) of type $o\alpha\beta$ in $\mathcal{Q}_0$. $\mathcal{Q}_0$ can be interpreted in $\mathcal{Q}_0^\mathrm{u}$ by viewing a function of type $\alpha\beta$ in $\mathcal{Q}_0$ as a total function of type $\alpha\beta$ in $\mathcal{Q}_0^\mathrm{u}$. Thus $\mathcal{Q}_0$ and $\mathcal{Q}_0^\mathrm{u}$ are equivalent in the sense of being mutually interpretable. ■

4 Definitions and Abbreviations

As Andrews does in [2, p. 212], we will introduce several defined logical constants and notational abbreviations. The former includes constants for true and false, the propositional connectives, and a canonical undefined wff. The latter includes notation for equality, the propositional connectives, universal and existential quantification, defined and undefined wffs, quasi-equality, and definite description.

$[\mathbf{A}_\alpha = \mathbf{B}_\alpha]$	stands for	$[\mathsf{Q}_{o\alpha\alpha}\mathbf{A}_\alpha\mathbf{B}_\alpha]$.
T_o	stands for	$[\mathsf{Q}_{ooo} = \mathsf{Q}_{ooo}]$.

F_o	stands for	$[\lambda x_o T_o] = [\lambda x_o x_o]$.
$[\forall \mathbf{x}_\alpha \mathbf{A}_o]$	stands for	$[\lambda y_\alpha T_o] = [\lambda \mathbf{x}_\alpha \mathbf{A}_o]$.
$\wedge_{ooo}$	stands for	$[\lambda x_o y_o[[\lambda g_{ooo}[g_{ooo} T_o T_o]] = [\lambda g_{ooo}[g_{ooo} x_o y_o]]]]$.
$[\mathbf{A}_o \wedge \mathbf{B}_o]$	stands for	$[\wedge_{ooo} \mathbf{A}_o \mathbf{B}_o]$.
$\supset_{ooo}$	stands for	$[\lambda x_o y_o[x_o = [x_o \wedge y_o]]]$.
$[\mathbf{A}_o \supset \mathbf{B}_o]$	stands for	$[\supset_{ooo} \mathbf{A}_o \mathbf{B}_o]$.
$\sim_{oo}$	stands for	$[\mathsf{Q}_{ooo} F_o]$.
$[\sim \mathbf{A}_o]$	stands for	$[\sim_{oo} \mathbf{A}_o]$.
$\vee_{ooo}$	stands for	$[\lambda x_o y_o \sim[[\sim x_o] \wedge [\sim y_o]]]$.
$[\mathbf{A}_o \vee \mathbf{B}_o]$	stands for	$[\vee_{ooo} \mathbf{A}_o \mathbf{B}_o]$.
$[\exists \mathbf{x}_\alpha \mathbf{A}_o]$	stands for	$[\sim[\forall \mathbf{x}_\alpha \sim \mathbf{A}_o]]$.
$[\exists_1 \mathbf{x}_\alpha \mathbf{A}_o]$	stands for	$[\exists y_\alpha[[\lambda \mathbf{x}_\alpha \mathbf{A}_o] = \mathsf{Q}_{o\alpha\alpha} y_\alpha]]$ where y_α does not occur in $\mathbf{A}_o$.
$[\mathbf{A}_\alpha \neq \mathbf{B}_\alpha]$	stands for	$[\sim[\mathbf{A}_\alpha = \mathbf{B}_\alpha]]$.
$[\mathbf{A}_\alpha\downarrow]$	stands for	$[\exists x_\alpha[x_\alpha = \mathbf{A}_\alpha]]$ where x_α does not occur in $\mathbf{A}_\alpha$.
$[\mathbf{A}_\alpha\uparrow]$	stands for	$[\sim[\mathbf{A}_\alpha\downarrow]]$
$[\mathbf{A}_\alpha \simeq \mathbf{B}_\alpha]$	stands for	$[\mathbf{A}_\alpha\downarrow \vee \mathbf{B}_\alpha\downarrow] \supset [\mathbf{A}_\alpha = \mathbf{B}_\alpha]$.
$[\mathrm{I} \mathbf{x}_\alpha \mathbf{A}_o]$	stands for	$[\iota_{\alpha(o\alpha)}[\lambda \mathbf{x}_\alpha \mathbf{A}_o]]$.
$\bot_\alpha$	stands for	$[\mathrm{I} x_\alpha[x_\alpha \neq x_\alpha]]$.

$[\exists_1 \mathbf{x}_\alpha \mathbf{A}_o]$ asserts that there is a unique $\mathbf{x}_\alpha$ that satisfies $\mathbf{A}_o$.

$[\mathrm{I} \mathbf{x}_\alpha \mathbf{A}_o]$ is called a *definite description*. It denotes the unique $\mathbf{x}_\alpha$ that satisfies $\mathbf{A}_o$. If there is no or more than one such $\mathbf{x}_\alpha$, it is undefined. Following Bertrand Russell and Church, Andrews denotes this definite description operator as an inverted lower case iota ($\imath$). We represent this operator by an (inverted) capital iota (I).

$[\mathbf{A}_\alpha\downarrow]$ says that $\mathbf{A}_\alpha$ is defined, and similarly, $[\mathbf{A}_\alpha\uparrow]$ says that $\mathbf{A}_\alpha$ is undefined. $[\mathbf{A}_\alpha \simeq \mathbf{B}_\alpha]$ says that $\mathbf{A}_\alpha$ and $\mathbf{B}_\alpha$ are *quasi-equal*, i.e., that $\mathbf{A}_\alpha$ and $\mathbf{B}_\alpha$ are either both defined and equal or both undefined. $\bot_\alpha$ is a canonical undefined wff of type α.

Note 7 (Definedness Notation). In $\mathcal{Q}_0$, $[\mathbf{A}_\alpha\downarrow]$ is always true, $[\mathbf{A}_\alpha\uparrow]$ is always false, $[\mathbf{A}_\alpha \simeq \mathbf{B}_\alpha]$ is always equal to $[\mathbf{A}_\alpha = \mathbf{B}_\alpha]$, and $\bot_\alpha$ denotes an unspecified value. ■

5 Proof System of $\mathcal{Q}_0^{\mathrm{u}}$

In this section we present the proof system of $\mathcal{Q}_0^{\mathrm{u}}$ which is derived from the proof system of $\mathcal{Q}_0$. The issue of definedness makes the proof system of $\mathcal{Q}_0^{\mathrm{u}}$ moderately more complicated than the proof system for $\mathcal{Q}_0$. While $\mathcal{Q}_0$

has only five axiom schemas and one rule of inference, $\mathcal{Q}_0^{\mathrm{u}}$ has the following thirteen axiom schemas and two rules of inference:

A1 (Truth Values)

$$[g_{oo}T_o \wedge g_{oo}F_o] = \forall x_o[g_{oo}x_o].$$

A2 (Leibniz' Law)

$$[x_\alpha = y_\alpha] \supset [h_{o\alpha}x_\alpha = h_{o\alpha}y_\alpha].$$

A3 (Extensionality)

$$[f_{\alpha\beta} = g_{\alpha\beta}] = \forall x_\beta[f_{\alpha\beta}x_\beta \simeq g_{\alpha\beta}x_\beta].$$

A4 (Beta-Reduction)

$$\mathbf{A}_\alpha\!\downarrow \supset [[\lambda \mathbf{x}_\alpha \mathbf{B}_\beta]\mathbf{A}_\alpha \simeq \mathsf{S}^{\mathbf{x}_\alpha}_{\mathbf{A}_\alpha}\mathbf{B}_\beta]$$

provided $\mathbf{A}_\alpha$ is free for $\mathbf{x}_\alpha$ in $\mathbf{B}_\beta$.

A5 (Variables are Defined)

$$\mathbf{x}_\alpha\!\downarrow.$$

A6 (Constants are Defined)

$\mathbf{c}_\alpha\!\downarrow$ where $\mathbf{c}_\alpha$ is a constant of $\mathcal{Q}_0^{\mathrm{u}}$.

A7 (Function Abstractions are Defined)

$$[\lambda \mathbf{x}_\alpha \mathbf{B}_\beta]\!\downarrow.$$

A8 (Function Applications of Type o are Defined)

$$\mathbf{A}_{o\beta}\mathbf{B}_\beta\!\downarrow.$$

A9 (Improper Function Application of Type o)

$$[\mathbf{A}_{o\beta}\!\uparrow \vee \mathbf{B}_\beta\!\uparrow] \supset {\sim}\mathbf{A}_{o\beta}\mathbf{B}_\beta.$$

A10 (Improper Function Application of Type $\alpha \neq o$)

$$[\mathbf{A}_{\alpha\beta}\uparrow \vee \mathbf{B}_{\beta}\uparrow] \supset \mathbf{A}_{\alpha\beta}\mathbf{B}_{\beta}\uparrow \quad \text{where } \alpha \neq o.$$

A11 (Equality and Quasi-Quality)

$$\mathbf{A}_{\alpha}\downarrow \supset [\mathbf{B}_{\alpha}\downarrow \supset [[\mathbf{A}_{\alpha} \simeq \mathbf{B}_{\alpha}] \simeq [\mathbf{A}_{\alpha} = \mathbf{B}_{\alpha}]]].$$

A12 (Proper Definite Description)

$$\exists_1 \mathbf{x}_{\alpha}\mathbf{A}_{o} \supset [[\mathrm{I}\mathbf{x}_{\alpha}\mathbf{A}_{o}]\downarrow \wedge \mathsf{S}^{\mathbf{x}_{\alpha}}_{[\mathrm{I}\mathbf{x}_{\alpha}\mathbf{A}_{o}]}\mathbf{A}_{o}] \quad \text{where } \alpha \neq o$$

and provided $\mathrm{I}\mathbf{x}_{\alpha}\mathbf{A}_{o}$ is free for $\mathbf{x}_{\alpha}$ in $\mathbf{A}_{o}$.

A13 (Improper Definite Description)

$$\sim[\exists_1 \mathbf{x}_{\alpha}\mathbf{A}_{o}] \supset [\mathrm{I}\mathbf{x}_{\alpha}\mathbf{A}_{o}]\uparrow \quad \text{where } \alpha \neq o.$$

R1 (Quasi-Equality Substitution) From $\mathbf{A}_{\alpha} \simeq \mathbf{B}_{\alpha}$ and $\mathbf{C}_{o}$ infer the result of replacing one occurrence of $\mathbf{A}_{\alpha}$ in $\mathbf{C}_{o}$ by an occurrence of $\mathbf{B}_{\alpha}$, provided that the occurrence of $\mathbf{A}_{\alpha}$ in $\mathbf{C}_{o}$ is not (an occurrence of a variable) immediately preceded by λ.

R2 (Modus Ponens) From $\mathbf{A}_{o}$ and $\mathbf{A}_{o} \supset \mathbf{B}_{o}$ infer $\mathbf{B}_{o}$.

Note 8 (Axiom Schemas). The axiom schemas A1, A2, A3, A4, and A12 of $\mathcal{Q}_0^{\mathrm{u}}$ correspond to the five axiom schemas of $\mathcal{Q}_0$. A1 and A2 are exactly the same as the first and second axiom schemas of $\mathcal{Q}_0$. A3 and A4 are modifications of the third and fourth axiom schemas of $\mathcal{Q}_0$. A3 is the axiom of extensionality for partial and total functions, and A4 is beta-reduction for functions that may be partial and arguments that may be undefined.

The seven axiom schemas A5–A11 of $\mathcal{Q}_0^{\mathrm{u}}$ deal with the definedness of wffs. A5 and A6 address the first principle of the traditional approach to undefinedness, A10 addresses the second principle, and A8 and A9 address the third principle. A7 states that a function abstraction always denotes some function, either partial or total. And A11 is a technical axiom schema for identifying equality with quasi-equality when applied to defined wffs.

The last two axiom schemas of $\mathcal{Q}_0^{\mathrm{u}}$ state the properties of definite descriptions. A12 states that proper definite descriptions are defined and denote the unique value satisfying the description; it corresponds to the fifth axiom schema of $\mathcal{Q}_0$. A13 states that improper definite descriptions are undefined. The proof system of $\mathcal{Q}_0$ leaves improper definite descriptions unspecified. ■

Note 9 (Rules of Inference). $\mathcal{Q}_0^{\mathrm{u}}$'s R1 rule of inference, Quasi-Equality Substitution, corresponds to $\mathcal{Q}_0$'s single rule of inference, which is equality substitution. These rules are exactly the same except that the $\mathcal{Q}_0^{\mathrm{u}}$ rule requires only *quasi-equality* ($\simeq$) between the target wff and the substitution wff, while the $\mathcal{Q}_0$ rule requires *equality* ($=$).

$\mathcal{Q}_0^{\mathrm{u}}$'s R2 rule of inference, Modus Ponens, is a primitive rule of inference, but modus ponens is a derived rule of inference in $\mathcal{Q}_0$. Modus ponens must be primitive in $\mathcal{Q}_0^{\mathrm{u}}$ since it is needed to discharge the definedness conditions on instances of A4, the schema for beta-reduction, and A11. ■

A *proof* of a wff$_o$ $\mathbf{A}_o$ in $\mathcal{Q}_0^{\mathrm{u}}$ is a finite sequence of wffs$_o$, ending with $\mathbf{A}_o$, such that each member in the sequence is an instance of an axiom schema of $\mathcal{Q}_0^{\mathrm{u}}$ or is inferred from preceding members in the sequence by a rule of inference of $\mathcal{Q}_0^{\mathrm{u}}$. A *theorem* of $\mathcal{Q}_0^{\mathrm{u}}$ is a wff$_o$ for which there is a proof in $\mathcal{Q}_0^{\mathrm{u}}$.

Let $\mathcal{H}$ be a set of wffs$_o$. A *proof of $\mathbf{A}_o$ from $\mathcal{H}$* in $\mathcal{Q}_0^{\mathrm{u}}$ consists of two finite sequences $\mathcal{S}_1$ and $\mathcal{S}_2$ of wffs$_o$ such that $\mathcal{S}_1$ is a proof in $\mathcal{Q}_0^{\mathrm{u}}$, $\mathbf{A}_o$ is the last member of $\mathcal{S}_2$, and each member $\mathbf{D}_o$ of $\mathcal{S}_2$ satisfies at least one of the following conditions:

1. $\mathbf{D}_o \in \mathcal{H}$.

2. $\mathbf{D}_o$ is a member of $\mathcal{S}_1$ (and hence a theorem of $\mathcal{Q}_0^{\mathrm{u}}$).

3. $\mathbf{D}_o$ is inferred from two preceding members $\mathbf{A}_\alpha \simeq \mathbf{B}_\alpha$ and $\mathbf{C}_o$ of $\mathcal{S}_2$ by R1, provided that the occurrence of $\mathbf{A}_\alpha$ in $\mathbf{C}_o$ is not in a well-formed part $\lambda \mathbf{x}_\beta \mathbf{E}_\gamma$ of $\mathbf{C}_o$ where $\mathbf{x}_\beta$ is free in a member of $\mathcal{H}$ and free in $\mathbf{A}_\alpha \simeq \mathbf{B}_\beta$.

4. $\mathbf{D}_o$ is inferred from two preceding members of $\mathcal{S}_2$ by R2.

We write $\mathcal{H} \vdash \mathbf{A}_o$ to mean there is a proof of $\mathbf{A}_o$ from $\mathcal{H}$ in $\mathcal{Q}_0^{\mathrm{u}}$. $\vdash \mathbf{A}_o$ is written instead of $\emptyset \vdash \mathbf{A}_o$. Clearly, $\mathbf{A}_o$ is a theorem of $\mathcal{Q}_0^{\mathrm{u}}$ iff $\vdash \mathbf{A}_o$.

The next two theorems follow immediately from the definition above.

THEOREM 2 (R1$'$). *If $\mathcal{H} \vdash \boldsymbol{A}_\alpha \simeq \boldsymbol{B}_\alpha$ and $\mathcal{H} \vdash \boldsymbol{C}_o$, then $\mathcal{H} \vdash \boldsymbol{D}_o$, where $\boldsymbol{D}_o$ is the result of replacing one occurrence of $\boldsymbol{A}_\alpha$ in $\boldsymbol{C}_o$ by an occurrence of $\boldsymbol{B}_\alpha$, provided that the occurrence of $\boldsymbol{A}_\alpha$ in $\boldsymbol{C}_o$ is not immediately preceded by λ or in a well-formed part $\lambda \boldsymbol{x}_\beta \boldsymbol{E}_\gamma$ of $\boldsymbol{C}_o$ where $\boldsymbol{x}_\beta$ is free in a member of $\mathcal{H}$ and free in $\boldsymbol{A}_\alpha \simeq \boldsymbol{B}_\alpha$.*

THEOREM 3 (R2$'$). *If $\mathcal{H} \vdash \boldsymbol{A}_o$ and $\mathcal{H} \vdash \boldsymbol{A}_o \supset \boldsymbol{B}_o$, then $\mathcal{H} \vdash \boldsymbol{B}_o$.*

6 Some Metatheorems

In this section we prove some metatheorems of $\mathcal{Q}_0^{\mathrm{u}}$ that are needed to prove the soundness and completeness of the proof system of $\mathcal{Q}_0^{\mathrm{u}}$.

PROPOSITION 4 (Wffs of type o are defined). $\vdash \boldsymbol{A}_o\!\downarrow$ *for all wffs* $\boldsymbol{A}_o$.

Proof. Directly implied by axiom schemas A5, A6, and A8. ■

THEOREM 5 (Beta-Reduction Rule). *If* $\mathcal{H} \vdash \boldsymbol{A}_\alpha\!\downarrow$ *and* $\mathcal{H} \vdash \boldsymbol{C}_o$*, then* $\mathcal{H} \vdash \boldsymbol{D}_o$*, where* $\boldsymbol{D}_o$ *is the result of replacing one occurrence of* $[\lambda \boldsymbol{x}_\alpha \boldsymbol{B}_\beta]\boldsymbol{A}_\alpha$ *in* $\boldsymbol{C}_o$ *by an occurrence of* $\mathsf{S}^{\mathbf{x}_\alpha}_{\mathbf{A}_\alpha}\boldsymbol{B}_\beta$*, provided* $\boldsymbol{A}_\alpha$ *is free for* $\boldsymbol{x}_\alpha$ *in* $\boldsymbol{B}_\beta$ *and the occurrence of* $[\lambda \boldsymbol{x}_\alpha \boldsymbol{B}_\beta]\boldsymbol{A}_\alpha$ *is not in a well-formed part* $\lambda \boldsymbol{y}_\gamma \boldsymbol{E}_\delta$ *of* $\boldsymbol{C}_o$ *where* $\boldsymbol{y}_\gamma$ *is free in a member of* $\mathcal{H}$ *and free in* $[\lambda \boldsymbol{x}_\alpha \boldsymbol{B}_\beta]\boldsymbol{A}_\alpha$.

Proof. Follows immediately from A4, R1′, and R2′. ■

LEMMA 6. $\vdash \boldsymbol{A}_\alpha \simeq \boldsymbol{A}_\alpha$.

Proof. Let $\mathbf{x}_\alpha$ be a variable that does not occur in $\mathbf{A}_\alpha$. Then $\mathbf{x}_\alpha\!\downarrow$ is an instance of A5, and $\mathbf{x}_\alpha\!\downarrow \supset [[\lambda \mathbf{x}_\alpha \mathbf{A}_\alpha]\mathbf{x}_\alpha \simeq \mathbf{A}_\alpha]$ is an instance of A4. By applying R2′ to these two wffs we obtain $\vdash [[\lambda \mathbf{x}_\alpha \mathbf{A}_\alpha]\mathbf{x}_\alpha \simeq \mathbf{A}_\alpha]$. The conclusion of the lemma then follows by the Beta-Reduction Rule. ■

LEMMA 7. *If* $\mathcal{H} \vdash \boldsymbol{A}_\alpha\!\downarrow$ *and* $\mathcal{H} \vdash \boldsymbol{B}_\alpha\!\downarrow$*, then* $\mathcal{H} \vdash \boldsymbol{A}_\alpha \simeq \boldsymbol{B}_\alpha$ *iff* $\mathcal{H} \vdash \boldsymbol{A}_\alpha = \boldsymbol{B}_\alpha$.

Proof.

($\Rightarrow$): Follows immediately from A11, R1′, and R2′.

($\Leftarrow$): $\mathcal{H} \vdash [\mathbf{A}_\alpha \simeq \mathbf{B}_\alpha] \simeq [\mathbf{A}_\alpha = \mathbf{B}_\alpha]$ by the first two hypotheses, A11, and R2′. $\vdash [\mathbf{A}_\alpha \simeq \mathbf{B}_\alpha] \simeq [\mathbf{A}_\alpha \simeq \mathbf{B}_\alpha]$ by Lemma 6. We obtain $\mathcal{H} \vdash [\mathbf{A}_\alpha = \mathbf{B}_\alpha] \simeq [\mathbf{A}_\alpha \simeq \mathbf{B}_\alpha]$ by applying R1′ to these two statements. The conclusion of the lemma then follows by applying R1′ to this last statement and $\mathcal{H} \vdash \mathbf{A}_\alpha = \mathbf{B}_\alpha$. ■

COROLLARY 8. *If* $\vdash \boldsymbol{A}_\alpha\!\downarrow$*, then* $\vdash \boldsymbol{A}_\alpha = \boldsymbol{A}_\alpha$.

Proof. By Lemmas 6 and 7. ■

LEMMA 9. *If* $\vdash \boldsymbol{A}_\alpha\!\downarrow$ *and* $\vdash \boldsymbol{B}_\beta \simeq \boldsymbol{C}_\beta$*, then* $\vdash \mathsf{S}^{\mathbf{x}_\alpha}_{\mathbf{A}_\alpha}[\boldsymbol{B}_\beta \simeq \boldsymbol{C}_\beta]$*, provided* $\boldsymbol{A}_\alpha$ *is free for* $\boldsymbol{x}_\alpha$ *in* $\boldsymbol{B}_\beta \simeq \boldsymbol{C}_\beta$.

Proof. Follows from Lemma 6 and the Beta-Reduction Rule in a way that is similar to the proof of Theorem 5209 in [2]. ■

COROLLARY 10. *If* $\vdash \mathbf{A}_\alpha\downarrow$ *and* $\vdash \mathbf{B}_o = \mathbf{C}_o$, *then* $\vdash \mathsf{S}^{\mathbf{x}_\alpha}_{\mathbf{A}_\alpha}[\mathbf{B}_o = \mathbf{C}_o]$, *provided* $\mathbf{A}_\alpha$ *is free for* $\mathbf{x}_\alpha$ *in* $\mathbf{B}_o = \mathbf{C}_o$.

Proof. By Proposition 4, Lemma 7, and Lemma 9. ∎

LEMMA 11. *If* $\vdash \mathbf{B}_\beta\downarrow$, *then* $\vdash T_o = [\mathbf{B}_\beta = \mathbf{B}_\beta]$.

Proof. The proof of $\vdash T_o = [\mathbf{B}_\beta \simeq \mathbf{B}_\beta]$ is similar to the proof of Theorem 5210 in [2] with Corollaries 8 and 10 used in place of Theorems 5200 and 5209, respectively. The lemma then follows from A11, R1, and R2. ∎

LEMMA 12. *If* $\vdash \mathbf{A}_o = \mathbf{B}_o$ *and* $\vdash \mathbf{C}_o = \mathbf{D}_o$, *then* $\vdash [\mathbf{A}_o = \mathbf{B}_o] \wedge [\mathbf{C}_o = \mathbf{D}_o]$.

Proof. Similar to the proof of Theorem 5213 in [2] with Lemma 11 used in place of Theorem 5210. ∎

The proofs of the next four theorems are similar to the proofs of Theorems 5215, 5220, 5234, and 5240 except that:

1. Rule R1 and Lemma 7 are used in place of rule R.
2. Rule R1′ and Lemma 7 are used in place of rule R′.
3. The Beta-Reduction Rule is used in place of the β-Contraction rule.
4. Corollary 8 is used in place of Theorem 5200.
5. Axiom schema A4 and Lemma 7 are used in place of Theorem 5207.
6. Corollary 10 is used in place of Theorem 5209.
7. Lemma 11 is used in place of Theorem 5210.
8. Lemma 12 is used in place of Theorem 5213.
9. Rule R2′ is used in place of Theorem 5224 (MP).
10. Axiom schemas A5–A8 are used to discharge definedness conditions.

THEOREM 13 (Universal Instantiation). *If* $\mathcal{H} \vdash \mathbf{A}_\alpha\downarrow$ *and* $\mathcal{H} \vdash \forall \mathbf{x}_\alpha \mathbf{B}_o$, *then* $\mathcal{H} \vdash \mathsf{S}^{\mathbf{x}_\alpha}_{\mathbf{A}_\alpha}\mathbf{B}_o$, *provided* $\mathbf{A}_\alpha$ *is free for* $\mathbf{x}_\alpha$ *in* $\mathbf{B}_o$.

Proof. Similar to the proof of Theorem 5215 ($\forall$I) in [2]. See the comment above. ∎

THEOREM 14 (Universal Generalization). *If* $\mathcal{H} \vdash \mathbf{A}_o$, *then* $\mathcal{H} \vdash \forall \mathbf{x}_\alpha \mathbf{A}_o$, *provided* $\mathbf{x}_\alpha$ *is not free in any wff in* $\mathcal{H}$.

Proof. Similar to the proof of Theorem 5220 (Gen) in [2]. See the comment above. ■

THEOREM 15 (Tautology Theorem). *If* $\mathcal{H} \vdash \boldsymbol{A}_o^1, \ldots, \mathcal{H} \vdash \boldsymbol{A}_o^n$ *and* $[\boldsymbol{A}_o^1 \wedge \cdots \wedge \boldsymbol{A}_o^n] \supset \boldsymbol{B}_o$ *is tautologous for* $n \geq 1$, *then* $\mathcal{H} \vdash \boldsymbol{B}_o$. *Also, if* $\boldsymbol{B}_o$ *is tautologous, then* $\mathcal{H} \vdash \boldsymbol{B}_o$.

Proof. Similar to the proof of Theorem 5234 (Rule P) in [2]. See the comment above. ■

PROPOSITION 16. $\vdash [\boldsymbol{A}_\alpha = \boldsymbol{B}_\alpha] \supset [\boldsymbol{A}_\alpha \simeq \boldsymbol{B}_\alpha]$.

Proof. Follows from the definition of $\simeq$ and the Tautology Theorem. ■

THEOREM 17 (Deduction Theorem). *If* $\mathcal{H} \cup \{\boldsymbol{H}_o\} \vdash \boldsymbol{P}_o$, *then* $\mathcal{H} \vdash \boldsymbol{H}_o \supset \boldsymbol{P}_o$.

Proof. Similar to the proof of Theorem 5240 in [2]. See the comment above. ■

7 Soundness and Completeness

In this section, let $\mathcal{H}$ be a set of wffs$_o$. $\mathcal{H}$ is *consistent* if there is no proof of F_o from $\mathcal{H}$ in $\mathcal{Q}_0^{\mathrm{u}}$.

THEOREM 18 (Soundness Theorem). *If* $\mathcal{H} \vdash \boldsymbol{A}_o$, *then* $\mathcal{H} \models \boldsymbol{A}_o$.

Proof. A straightforward verification shows that (1) each instance of each axiom schema of $\mathcal{Q}_0^{\mathrm{u}}$ is valid and (2) the rules of inference of $\mathcal{Q}_0^{\mathrm{u}}$, R1 and R2, preserve validity in every general model for $\mathcal{Q}_0^{\mathrm{u}}$. This shows that if $\vdash \mathbf{A}_o$, then $\models \mathbf{A}_o$.

Suppose $\mathcal{H} \vdash \mathbf{A}_o$ and $\mathcal{M}$ is a model for $\mathcal{H}$. Then there is a finite subset $\{\mathbf{H}_o^1, \ldots, \mathbf{H}_o^n\}$ of $\mathcal{H}$ such that $\{\mathbf{H}_o^1, \ldots, \mathbf{H}_o^n\} \vdash \mathbf{A}_o$. By the Deduction Theorem, this implies $\vdash \mathbf{H}_o^1 \supset \cdots \supset \mathbf{H}_o^n \supset \mathbf{A}_o$. By the result just above, $\mathcal{M} \models \mathbf{H}_o^1 \supset \cdots \supset \mathbf{H}_o^n \supset \mathbf{A}_o$. But $\mathcal{M} \models \mathbf{H}_o^i$ for all i with $1 \leq i \leq n$ since $\mathcal{M}$ is a model for $\mathcal{H}$. Therefore $\mathcal{M} \models \mathbf{A}_o$, and so $\mathcal{H} \models \mathbf{A}_o$. ■

THEOREM 19 (Consistency Theorem). *If* $\mathcal{H}$ *has a general model, then* $\mathcal{H}$ *is consistent.*

Proof. Let $\mathcal{M}$ be a general model for $\mathcal{H}$. Assume that $\mathcal{H}$ is inconsistent, i.e., that $\mathcal{H} \vdash F_o$. Then, by the Soundness Theorem, $\mathcal{H} \models F_o$ and hence $\mathcal{M} \models F_o$. This means that $\mathcal{V}_\varphi^{\mathcal{M}}(F_o) = \mathsf{T}$ (for any assignment φ), which contradicts the definition of a general model. ■

A *cwff* [*cwff*$_\alpha$] is a closed wff [closed wff$_\alpha$]. A *sentence* is a cwff$_o$. Let $\mathcal{H}$ be a set of sentences. $\mathcal{H}$ is *complete* in $\mathcal{Q}_0^{\mathrm{u}}$ if, for every sentence $\mathbf{A}_o$, either $\mathcal{H} \vdash \mathbf{A}_o$ or $\mathcal{H} \vdash \sim\mathbf{A}_o$. $\mathcal{H}$ is *extensionally complete* in $\mathcal{Q}_0^{\mathrm{u}}$ if, for every sentence of the form $\mathbf{A}_{\alpha\beta} = \mathbf{B}_{\alpha\beta}$, there is a cwff $\mathbf{C}_\beta$ such that:

1. $\mathcal{H} \vdash \mathbf{C}_\beta\!\downarrow$.

2. $\mathcal{H} \vdash [\mathbf{A}_{\alpha\beta}\!\downarrow \wedge \mathbf{B}_{\alpha\beta}\!\downarrow \wedge [\mathbf{A}_{\alpha\beta}\mathbf{C}_\beta \simeq \mathbf{B}_{\alpha\beta}\mathbf{C}_\beta]] \supset [\mathbf{A}_{\alpha\beta} = \mathbf{B}_{\alpha\beta}]$.

Let $\mathcal{L}(\mathcal{Q}_0^{\mathrm{u}})$ be the set of wffs of $\mathcal{Q}_0^{\mathrm{u}}$.

LEMMA 20 (Extension Lemma). *Let $\mathcal{G}$ be a consistent set of sentences of $\mathcal{Q}_0^{\mathrm{u}}$. Then there is an expansion $\overline{\mathcal{Q}_0^{\mathrm{u}}}$ of $\mathcal{Q}_0^{\mathrm{u}}$ and a set $\mathcal{H}$ of sentences of $\overline{\mathcal{Q}_0^{\mathrm{u}}}$ such that:*

1. *$\mathcal{G} \subseteq \mathcal{H}$.*

2. *$\mathcal{H}$ is consistent.*

3. *$\mathcal{H}$ is complete in $\overline{\mathcal{Q}_0^{\mathrm{u}}}$.*

4. *$\mathcal{H}$ is extensionally complete in $\overline{\mathcal{Q}_0^{\mathrm{u}}}$.*

5. *$card(\mathcal{L}(\overline{\mathcal{Q}_0^{\mathrm{u}}})) = card(\mathcal{L}(\mathcal{Q}_0^{\mathrm{u}}))$.*

Proof. The proof is very close to the proof of Theorem 5500 in [2]. The crucial difference is that, in case (c) of the definition of $\mathcal{G}_{\tau+1}$,

$$\mathcal{G}_{\tau+1} = \mathcal{G}_\tau \cup \{\sim[\mathbf{A}_{\alpha\beta}\!\downarrow \wedge \mathbf{B}_{\alpha\beta}\!\downarrow \wedge [\mathbf{A}_{\alpha\beta}\mathbf{c}_\beta \simeq \mathbf{B}_{\alpha\beta}\mathbf{c}_\beta]]\}$$

where $\mathbf{c}_\beta$ is the first constant in $\mathcal{C}_\beta$ that does not occur in $\mathcal{G}_\tau$ or $\mathbf{A}_{\alpha\beta} = \mathbf{B}_{\alpha\beta}$. (Notice that $\vdash \mathbf{c}_\beta$ by A6.)

To prove that $\mathcal{G}_{\tau+1}$ is consistent assuming $\mathcal{G}_\tau$ is consistent when $\mathcal{G}_{\tau+1}$ is obtained by case (c) , it is necessary to show that, if

$$\mathcal{G}_\tau \vdash \mathbf{A}_{\alpha\beta}\!\downarrow \wedge \mathbf{B}_{\alpha\beta}\!\downarrow \wedge [\mathbf{A}_{\alpha\beta}\mathbf{c}_\beta \simeq \mathbf{B}_{\alpha\beta}\mathbf{c}_\beta],$$

then $\mathcal{G}_\tau \vdash \mathbf{A}_{\alpha\beta} = \mathbf{B}_{\alpha\beta}$. Assume the hypothesis of this statement. Let $\mathcal{P}$ be a proof of

$$\mathbf{A}_{\alpha\beta}\!\downarrow \wedge \mathbf{B}_{\alpha\beta}\!\downarrow \wedge [\mathbf{A}_{\alpha\beta}\mathbf{c}_\beta \simeq \mathbf{B}_{\alpha\beta}\mathbf{c}_\beta]$$

from a finite subset $\mathcal{S}$ of $\mathcal{G}_\tau$, and let $\mathbf{x}_\beta$ be a variable that does not occur in $\mathcal{P}$ or $\mathcal{S}$. Since $\mathbf{c}_\beta$ does not occur in $\mathcal{G}_\tau$, $\mathbf{A}_{\alpha\beta}$, or $\mathbf{B}_{\alpha\beta}$, $\mathrm{S}^{\mathbf{c}_\beta}_{\mathbf{x}_\beta}\mathcal{P}$ is a proof of

$$\mathbf{A}_{\alpha\beta}\!\downarrow \wedge \mathbf{B}_{\alpha\beta}\!\downarrow \wedge [\mathbf{A}_{\alpha\beta}\mathbf{x}_\beta \simeq \mathbf{B}_{\alpha\beta}\mathbf{x}_\beta]$$

from $\mathcal{S}$. Therefore,

$$\mathcal{S} \vdash \mathbf{A}_{\alpha\beta}\downarrow \wedge \mathbf{B}_{\alpha\beta}\downarrow \wedge [\mathbf{A}_{\alpha\beta}\mathbf{x}_\beta \simeq \mathbf{B}_{\alpha\beta}\mathbf{x}_\beta].$$

This implies

$$\mathcal{S} \vdash \forall \mathbf{x}_\beta[\mathbf{A}_{\alpha\beta}\mathbf{x}_\beta \simeq \mathbf{B}_{\alpha\beta}\mathbf{x}_\beta]$$

by Universal Generalization since $\mathbf{x}_\beta$ does not occur in $\mathcal{S}$. From this, $\mathcal{S} \vdash \mathbf{A}_{\alpha\beta}\downarrow$, $\mathcal{S} \vdash \mathbf{B}_{\alpha\beta}\downarrow$, and the fact $\mathbf{x}_\beta$ does not occur in $\mathbf{A}_{\alpha\beta}$ or $\mathbf{B}_{\alpha\beta}$, it follows that $\mathcal{G}_\mathcal{T} \vdash \mathbf{A}_{\alpha\beta} = \mathbf{B}_{\alpha\beta}$ by A3, Universal Generalization, Universal Instantiation, Proposition 16, Lemma 6, R1$'$, and R2$'$.

The rest of the proof is essentially the same as the proof of Theorem 5500. ■

A general model $\langle\{\mathcal{D}_\alpha \mid \alpha \in \mathcal{T}\}, \mathcal{J}\rangle$ for $\mathcal{Q}_0^{\mathrm{u}}$ is *frugal* if $\mathrm{card}(\mathcal{D}_\alpha) \leq \mathrm{card}(\mathcal{L}(\mathcal{Q}_0^{\mathrm{u}}))$ for all $\alpha \in \mathcal{T}$.

THEOREM 21 (Henkin's Theorem for $\mathcal{Q}_0^{\mathrm{u}}$). *Every consistent set of sentences of $\mathcal{Q}_0^{\mathrm{u}}$ has a frugal general model.*

Proof. Let $\mathcal{G}$ be a consistent set of sentences of $\mathcal{Q}_0^{\mathrm{u}}$, and let $\mathcal{H}$ and $\overline{\mathcal{Q}_0^{\mathrm{u}}}$ be as described in the Extension Lemma. We define simultaneously, by induction on $\gamma \in \mathcal{T}$, a frame $\{\mathcal{D}_\alpha \mid \alpha \in \mathcal{T}\}$ and a partial function $\mathcal{V}$ whose domain is the set of cwffs of $\overline{\mathcal{Q}_0^{\mathrm{u}}}$ so that the following conditions hold for all $\gamma \in \mathcal{T}$:

(1^γ) $\mathcal{D}_\gamma = \{\mathcal{V}(\mathbf{A}_\gamma) \mid \mathbf{A}_\gamma \text{ is a cwff}_\gamma \text{ and } \mathcal{H} \vdash \mathbf{A}_\gamma\downarrow\}$.

(2^γ) $\mathcal{V}(\mathbf{A}_\gamma)$ is defined iff $\mathcal{H} \vdash \mathbf{A}_\gamma\downarrow$ for all cwffs $\mathbf{A}_\gamma$.

(3^γ) $\mathcal{V}(\mathbf{A}_\gamma) = \mathcal{V}(\mathbf{B}_\gamma)$ iff $\mathcal{H} \vdash \mathbf{A}_\gamma = \mathbf{B}_\gamma$ for all cwffs $\mathbf{A}_\gamma$ and $\mathbf{B}_\gamma$.

Let $\mathcal{V}(x) \simeq \mathcal{V}(y)$ mean either $\mathcal{V}(x)$ and $\mathcal{V}(y)$ are both defined and equal or $\mathcal{V}(x)$ and $\mathcal{V}(y)$ are both undefined.

For each cwff $\mathbf{A}_o$, if $\mathcal{H} \vdash \mathbf{A}_o$, let $\mathcal{V}(\mathbf{A}_o) = \mathsf{T}$; otherwise let $\mathcal{V}(\mathbf{A}_o) = \mathsf{F}$. Also, let $\mathcal{D}_o = \{\mathsf{T}, \mathsf{F}\}$. ($1^o$) is obviously satisfied; (2°) is satisfied by Proposition 4 and the consistency of $\mathcal{H}$; and (3°) is satisfied by the consistency and completeness of $\mathcal{H}$.

For each cwff $\mathbf{A}_\iota$, if $\mathcal{H} \vdash \mathbf{A}_\iota\downarrow$, let

$$\mathcal{V}(\mathbf{A}_\iota) = \{\mathbf{B}_\iota \mid \mathbf{B}_\iota \text{ is a cwff}_\iota \text{ and } \mathcal{H} \vdash \mathbf{A}_\iota = \mathbf{B}_\iota\};$$

otherwise $\mathcal{V}(\mathbf{A}_\iota)$ is undefined. Also, let

$$\mathcal{D}_\iota = \{\mathcal{V}(\mathbf{A}_\iota) \mid \mathbf{A}_\iota \text{ is a cwff}_\iota \text{ and } \mathcal{H} \vdash \mathbf{A}_\iota\downarrow\}.$$

($1^{\imath}$) and ($2^{\imath}$) is are clearly satisfied; ($3^{\imath}$) is satisfied by the consistency and completeness of $\mathcal{H}$.

Now suppose that $\mathcal{D}_\alpha$ and $\mathcal{D}_\beta$ are defined and that the conditions hold for α and β. For each cwff $\mathbf{A}_{\alpha\beta}$, if $\mathcal{H} \vdash \mathbf{A}_{\alpha\beta}\downarrow$, let $\mathcal{V}(\mathbf{A}_{\alpha\beta})$ be the (partial or total) function from $\mathcal{D}_\beta$ to $\mathcal{D}_\alpha$ whose value, for any argument $\mathcal{V}(\mathbf{B}_\beta) \in \mathcal{D}_\beta$, is $\mathcal{V}(\mathbf{A}_{\alpha\beta}\mathbf{B}_\beta)$ if $\mathcal{V}(\mathbf{A}_{\alpha\beta}\mathbf{B}_\beta)$ is defined and is undefined if $\mathcal{V}(\mathbf{A}_{\alpha\beta}\mathbf{B}_\beta)$ is undefined; otherwise let $\mathcal{V}(\mathbf{A}_{\alpha\beta})$ be undefined. We must show that this definition is independent of the particular cwff $\mathbf{B}_\beta$ used to represent the argument. So suppose $\mathcal{V}(\mathbf{B}_\beta) = \mathcal{V}(\mathbf{C}_\beta)$; then $\mathcal{H} \vdash \mathbf{B}_\beta = \mathbf{C}_\beta$ by (3^β), so $\mathcal{H} \vdash \mathbf{A}_{\alpha\beta}\mathbf{B}_\beta \simeq \mathbf{A}_{\alpha\beta}\mathbf{C}_\beta$ by Lemmas 6 and 7 and R1′, and so $\mathcal{V}(\mathbf{A}_{\alpha\beta}\mathbf{B}_\beta) \simeq \mathcal{V}(\mathbf{A}_{\alpha\beta}\mathbf{C}_\beta)$ by (2^β) and (3^β), Finally, let

$$\mathcal{D}_{\alpha\beta} = \{\mathcal{V}(\mathbf{A}_{\alpha\beta}) \mid \mathbf{A}_{\alpha\beta} \text{ is a cwff}_{\alpha\beta} \text{ and } \mathcal{H} \vdash \mathbf{A}_{\alpha\beta}\downarrow\}.$$

($1^{\alpha\beta}$) and ($2^{\alpha\beta}$) are clearly satisfied; we must show that ($3^{\alpha\beta}$) is satisfied. Suppose $\mathcal{V}(\mathbf{A}_{\alpha\beta}) = \mathcal{V}(\mathbf{B}_{\alpha\beta})$. Then $\mathcal{H} \vdash \mathbf{A}_{\alpha\beta}\downarrow$ and $\mathcal{H} \vdash \mathbf{B}_{\alpha\beta}\downarrow$. Since $\mathcal{H}$ is extensionally complete, there is a $\mathbf{C}_\beta$ such that $\mathcal{H} \vdash \mathbf{C}_\beta\downarrow$ and

$$\mathcal{H} \vdash [\mathbf{A}_{\alpha\beta}\downarrow \wedge \mathbf{B}_{\alpha\beta}\downarrow \wedge [\mathbf{A}_{\alpha\beta}\mathbf{C}_\beta \simeq \mathbf{B}_{\alpha\beta}\mathbf{C}_\beta]] \supset [\mathbf{A}_{\alpha\beta} = \mathbf{B}_{\alpha\beta}].$$

Then $\mathcal{V}(\mathbf{A}_{\alpha\beta}\mathbf{C}_\beta) \simeq \mathcal{V}(\mathbf{A}_{\alpha\beta})(\mathcal{V}(\mathbf{C}_\beta)) \simeq \mathcal{V}(\mathbf{B}_{\alpha\beta})(\mathcal{V}(\mathbf{C}_\beta)) \simeq \mathcal{V}(\mathbf{B}_{\alpha\beta}\mathbf{C}_\beta)$, so $\mathcal{H} \vdash \mathbf{A}_{\alpha\beta}\mathbf{C}_\beta \simeq \mathbf{B}_{\alpha\beta}\mathbf{C}_\beta$, and so $\mathcal{H} \vdash \mathbf{A}_{\alpha\beta} = \mathbf{B}_{\alpha\beta}$. Now suppose $\mathcal{H} \vdash \mathbf{A}_{\alpha\beta} = \mathbf{B}_{\alpha\beta}$. Then, for all cwffs $\mathbf{C}_\beta \in \mathcal{D}_\beta$, $\mathcal{H} \vdash \mathbf{A}_{\alpha\beta}\mathbf{C}_\beta \simeq \mathbf{B}_{\alpha\beta}\mathbf{C}_\beta$ by Lemmas 6 and 7 and R1′, and so $\mathcal{V}(\mathbf{A}_{\alpha\beta})(\mathcal{V}(\mathbf{C}_\beta)) \simeq \mathcal{V}(\mathbf{A}_{\alpha\beta}\mathbf{C}_\beta) \simeq \mathcal{V}(\mathbf{B}_{\alpha\beta}\mathbf{C}_\beta) \simeq \mathcal{V}(\mathbf{B}_{\alpha\beta})(\mathcal{V}(\mathbf{C}_\beta))$. Hence $\mathcal{V}(\mathbf{A}_{\alpha\beta}) = \mathcal{V}(\mathbf{B}_{\alpha\beta})$.

We claim that $\mathcal{M} = \langle\{\mathcal{D}_\alpha \mid \alpha \in \mathcal{T}\}, \mathcal{V}\rangle$ is an interpretation. For each constant $\mathbf{c}_\gamma$ of $\overline{\mathcal{Q}_0^{\mathrm{u}}}$, $\mathcal{H} \vdash \mathbf{c}_\gamma$ by A6, and thus $\mathcal{V}$ maps each constant of $\overline{\mathcal{Q}_0^{\mathrm{u}}}$ of type γ into $\mathcal{D}_\gamma$ by (1^γ) and (2^γ).

We must show that $\mathcal{V}(\mathsf{Q}_{o\alpha\alpha})$ is the identity relation on $\mathcal{D}_\alpha$. Let $\mathcal{V}(\mathbf{A}_\alpha)$ and $\mathcal{V}(\mathbf{B}_\alpha)$ be arbitrary members of $\mathcal{D}_\alpha$. Then $\mathcal{V}(\mathbf{A}_\alpha) = \mathcal{V}(\mathbf{B}_\alpha)$ iff $\mathcal{H} \vdash \mathbf{A}_\alpha = \mathbf{B}_\alpha$ iff $\mathcal{H} \vdash \mathsf{Q}_{o\alpha\alpha}\mathbf{A}_\alpha\mathbf{B}_\alpha$ iff $\mathsf{T} = \mathcal{V}(\mathsf{Q}_{o\alpha\alpha}\mathbf{A}_\alpha\mathbf{B}_\alpha) = \mathcal{V}(\mathsf{Q}_{o\alpha\alpha})(\mathcal{V}(\mathbf{A}_\alpha))(\mathcal{V}(\mathbf{B}_\alpha))$. Thus $\mathcal{V}(\mathsf{Q}_{o\alpha\alpha})$ is the identity relation on $\mathcal{D}_\alpha$.

We must show that, for $\alpha \neq o$, $\mathcal{V}(\iota_{\alpha(o\alpha)})$ is the unique member selector on $\mathcal{D}_\alpha$. For $\alpha \neq o$, let $\mathbf{A}_{o\alpha}$ be an arbitrary member of $\mathcal{D}_{o\alpha}$, $\mathbf{B}_\alpha$ be an arbitrary member of $\mathcal{D}_\alpha$, and $\mathbf{x}_\alpha$ be a variable that does not occur in $\mathbf{A}_{o\alpha}$. Using A12 and A13, $\mathcal{V}(\mathbf{A}_{o\alpha}) = \mathcal{V}(\mathsf{Q}_{o\alpha\alpha}\mathbf{B}_\alpha)$ iff $\mathcal{H} \vdash \mathbf{A}_{o\alpha} = \mathsf{Q}_{o\alpha\alpha}\mathbf{B}_\alpha$ iff $\mathcal{H} \vdash \iota_{\alpha(o\alpha)}\mathbf{A}_{o\alpha} = \mathbf{B}_\alpha$ iff $\mathcal{V}(\iota_{\alpha(o\alpha)}\mathbf{A}_{o\alpha}) = \mathcal{V}(\mathbf{B}_\alpha)$ iff $\mathcal{V}(\iota_{\alpha(o\alpha)})(\mathcal{V}(\mathbf{A}_{o\alpha})) = \mathcal{V}(\mathbf{B}_\alpha)$. Similarly, using A12 and A13, $\mathcal{V}(\sim\exists_1\mathbf{x}_\alpha[\mathbf{A}_{o\alpha}\mathbf{x}_\alpha]) = \mathsf{T}$ iff $\mathcal{H} \vdash \sim\exists_1\mathbf{x}_\alpha[\mathbf{A}_{o\alpha}\mathbf{x}_\alpha]$ iff $\mathcal{H} \vdash \iota_{\alpha(o\alpha)}\mathbf{A}_{o\alpha}\uparrow$ iff $\mathcal{V}(\iota_{\alpha(o\alpha)}\mathbf{A}_{o\alpha})$ is undefined iff $\mathcal{V}(\iota_{\alpha(o\alpha)})(\mathcal{V}(\mathbf{A}_{o\alpha}))$ is undefined. Thus $\mathcal{V}(\iota_{\alpha(o\alpha)})$ is the unique member selector on $\mathcal{D}_\alpha$.

Thus $\mathcal{M}$ is an interpretation. We claim further that $\mathcal{M}$ is a general model for $\overline{\mathcal{Q}_0^{\mathrm{u}}}$. For each assignment φ into $\mathcal{M}$ and wff $\mathbf{C}_\gamma$, let

$$\mathbf{C}_\gamma^\varphi = \mathsf{S}^{\mathbf{x}^1_{\delta_1}\cdots\mathbf{x}^n_{\delta_n}}_{\mathbf{E}^1_{\delta_1}\cdots\mathbf{E}^n_{\delta_n}}$$

where $\mathbf{x}^1_{\delta_1} \cdots \mathbf{x}^n_{\delta_n}$ are the free variables of $\mathbf{C}_\gamma$ and $\mathbf{E}^i_{\delta_i}$ is the first cwff (in some fixed enumeration) of $\overline{\mathcal{Q}^{\mathrm{u}}_0}$ such that $\varphi(\mathbf{x}^i_{\delta_i}) = \mathcal{V}(\mathbf{E}^i_{\delta_i})$ for all i with $1 \le i \le n$. Let $\mathcal{V}_\varphi(\mathbf{C}_\gamma) \simeq \mathcal{V}(\mathbf{C}^\varphi_\gamma)$. $\mathbf{C}^\varphi_\gamma$ is clearly a cwff$_\gamma$, so $\mathcal{V}_\varphi(\mathbf{C}_\gamma) \in \mathcal{D}_\gamma$ if $\mathcal{V}_\varphi(\mathbf{C}_\gamma)$ is defined.

(a) Let $\mathbf{C}_\gamma$ be a variable $\mathbf{x}_\delta$. Choose $\mathbf{E}_\delta$ so that $\varphi(\mathbf{x}_\delta) = \mathcal{V}(\mathbf{E}_\delta)$ as above. Then $\mathcal{V}_\varphi(\mathbf{C}_\gamma) = \mathcal{V}_\varphi(\mathbf{x}_\delta) = \mathcal{V}(\mathbf{x}^\varphi_\delta) = \mathcal{V}(\mathbf{E}_\delta) = \varphi(\mathbf{x}_\delta)$.

(b) Let $\mathbf{C}_\gamma$ be a constant. Then $\mathcal{V}_\varphi(\mathbf{C}_\gamma) = \mathcal{V}(\mathbf{C}^\varphi_\gamma) = \mathcal{V}(\mathbf{C}_\gamma)$.

(c) Let $\mathbf{C}_\gamma$ be $[\mathbf{A}_{\alpha\beta}\mathbf{B}_\beta]$. If $\mathcal{V}_\varphi(\mathbf{A}_{\alpha\beta})$ is defined, $\mathcal{V}_\varphi(\mathbf{B}_\beta)$ is defined, and $\mathcal{V}_\varphi(\mathbf{A}_{\alpha\beta})$ is defined at $\mathcal{V}_\varphi(\mathbf{B}_\beta)$, then $\mathcal{V}_\varphi(\mathbf{C}_\gamma) = \mathcal{V}_\varphi(\mathbf{A}_{\alpha\beta}\mathbf{B}_\beta) = \mathcal{V}(\mathbf{A}^\varphi_{\alpha\beta}\mathbf{B}^\varphi_\beta) = \mathcal{V}(\mathbf{A}^\varphi_{\alpha\beta})(\mathcal{V}(\mathbf{B}^\varphi_\beta)) = \mathcal{V}_\varphi(\mathbf{A}_{\alpha\beta})(\mathcal{V}_\varphi(\mathbf{B}_\beta))$. Now assume $\mathcal{V}_\varphi(\mathbf{A}_{\alpha\beta})$ is undefined, $\mathcal{V}_\varphi(\mathbf{B}_\beta)$ is undefined, or $\mathcal{V}_\varphi(\mathbf{A}_{\alpha\beta})$ is not defined at $\mathcal{V}_\varphi(\mathbf{B}_\beta)$. Then $\mathcal{H} \vdash \mathbf{A}^\varphi_{\alpha\beta}\uparrow$, $\mathcal{H} \vdash \mathbf{B}^\varphi_\beta\uparrow$, or $\mathcal{V}(\mathbf{A}^\varphi_{\alpha\beta}\mathbf{B}^\varphi_\beta)$ is undefined. If $\alpha = o$, then $\mathcal{H} \vdash \mathbf{A}^\varphi_{\alpha\beta}\uparrow$ or $\mathcal{H} \vdash \mathbf{B}^\varphi_\beta\uparrow$, which implies $\mathcal{H} \vdash {\sim}\mathbf{A}^\varphi_{\alpha\beta}\mathbf{B}^\varphi_\beta$ by A9, so $\mathcal{H} \vdash \mathbf{A}^\varphi_{\alpha\beta}\mathbf{B}^\varphi_\beta = F_o$, so $\mathcal{V}(\mathbf{A}^\varphi_{\alpha\beta}\mathbf{B}^\varphi_\beta) = \mathcal{V}(F_o)$, so $\mathcal{V}_\varphi(\mathbf{A}_{\alpha\beta}\mathbf{B}_\beta) = \mathcal{V}(F_o)$, and so $\mathcal{V}_\varphi(\mathbf{C}_\gamma) = \mathsf{F}$. If $\alpha \neq o$ and $\mathcal{H} \vdash \mathbf{A}^\varphi_{\alpha\beta}\uparrow$ or $\mathcal{H} \vdash \mathbf{B}^\varphi_\beta\uparrow$, then $\mathcal{H} \vdash [\mathbf{A}^\varphi_{\alpha\beta}\mathbf{B}^\varphi_\beta]\uparrow$ by A10, and so $\mathcal{V}(\mathbf{A}^\varphi_{\alpha\beta}\mathbf{B}^\varphi_\beta)$ is undefined. Hence, if $\alpha \neq o$, $\mathcal{V}_\varphi(\mathbf{C}_\gamma) \simeq \mathcal{V}_\varphi(\mathbf{A}_{\alpha\beta}\mathbf{B}_\beta) \simeq \mathcal{V}(\mathbf{A}^\varphi_{\alpha\beta}\mathbf{B}^\varphi_\beta)$ is undefined.

(d) Let $\mathbf{C}_\gamma$ be $[\lambda\mathbf{x}_\alpha\mathbf{B}_\beta]$. Let $\mathcal{V}(\mathbf{E}_\alpha)$ be an arbitrary member of $\mathcal{D}_\alpha$, and so $\mathbf{E}_\alpha$ is a cwff and $\mathcal{H} \vdash \mathbf{E}_\alpha\downarrow$. Given an assignment φ, let $\psi = (\varphi : \mathbf{x}_\alpha/\mathcal{V}(\mathbf{E}_\alpha)$. From A4 it follows that $\mathcal{H} \vdash [\lambda\mathbf{x}_\alpha\mathbf{B}_\beta]^\varphi\mathbf{E}_\alpha \simeq \mathbf{B}^\psi_\beta$. Then $\mathcal{V}_\varphi(\mathbf{C}_\gamma)(\mathcal{V}(\mathbf{E}_\alpha)) \simeq \mathcal{V}([\lambda\mathbf{x}_\alpha\mathbf{B}_\beta]^\varphi)(\mathcal{V}(\mathbf{E}_\alpha)) \simeq \mathcal{V}(\mathbf{B}^\psi_\beta) \simeq \mathcal{V}_\psi(\mathbf{B}_\beta)$. Thus $\mathcal{V}_\varphi(\mathbf{C}_\gamma)$ satisfies condition (d) in the definition of a general model.

Thus $\mathcal{M}$ is a general model for $\overline{\mathcal{Q}^{\mathrm{u}}_0}$ (and hence for $\mathcal{Q}^{\mathrm{u}}_0$). Also, if $\mathbf{A}_o \in \mathcal{G}$, then $\mathbf{A}_o \in \mathcal{H}$, so $\mathcal{H} \vdash \mathbf{A}_o$, so $\mathcal{V}(\mathbf{A}_o) = \mathsf{T}$ and $\mathcal{M} \models \mathbf{A}_o$, so $\mathcal{M}$ is a general model for $\mathcal{G}$. Clearly, (1) card$(\mathcal{D}_\alpha) \le$ card$(\mathcal{L}(\mathcal{Q}^{\mathrm{u}}_0))$ since $\mathcal{V}$ maps a subset of the cwffs$_\alpha$ of $\overline{\mathcal{Q}^{\mathrm{u}}_0}$ onto $\mathcal{D}_\alpha$ and (2) card$(\mathcal{L}(\overline{\mathcal{Q}^{\mathrm{u}}_0}))$ = card$(\mathcal{L}(\mathcal{Q}^{\mathrm{u}}_0))$, and so $\mathcal{M}$ is frugal. ∎

THEOREM 22 (Henkin's Completeness Theorem for $\mathcal{Q}^{\mathrm{u}}_0$). *Let $\mathcal{H}$ be a set of sentences of $\mathcal{Q}^{\mathrm{u}}_0$. If $\mathcal{H} \models \mathbf{A}_o$, then $\mathcal{H} \vdash \mathbf{A}_o$.*

Proof. Assume $\mathcal{H} \models \mathbf{A}_o$, and let $\mathbf{B}_o$ be the universal closure of $\mathbf{A}_o$. Then $\mathcal{H} \models \mathbf{B}_o$. Suppose $\mathcal{H} \cup \{{\sim}\mathbf{B}_o\}$ is consistent. Then, by Henkin's Theorem, there is a general model $\mathcal{M}_0$ for $\mathcal{H} \cup \{{\sim}\mathbf{B}_o\}$, and so $\mathcal{M}_0 \models {\sim}\mathbf{B}_o$. Since $\mathcal{M}_0$ is also a general model for $\mathcal{H}$, $\mathcal{M}_0 \models \mathbf{B}_o$. From this contradiction it follows that $\mathcal{H} \cup \{{\sim}\mathbf{B}_o\}$ is inconsistent. Hence $\mathcal{H} \vdash \mathbf{B}_o$ by the Deduction Theorem and the Tautology Theorem. Therefore, $\mathcal{H} \vdash \mathbf{A}_o$ by Universal Instantiation and A5. ∎

8 Conclusion

$\mathcal{Q}_0^{\mathrm{u}}$ is a version of Church's type theory that directly formalizes the traditional approach to undefinedness. In this paper we have presented the syntax, semantics, and proof system of $\mathcal{Q}_0^{\mathrm{u}}$. The semantics is based on Henkin-style general models. We have also proved that $\mathcal{Q}_0^{\mathrm{u}}$ is sound and complete with respect to its semantics.

$\mathcal{Q}_0^{\mathrm{u}}$ is a modification of $\mathcal{Q}_0$. Its syntax is essentially identical to the syntax of $\mathcal{Q}_0$. Its semantics is based on general models that include partial functions as well as total functions and in which terms may be nondenoting. Its proof system is derived from the proof system of $\mathcal{Q}_0$; the axiom schemas and rules of inference of $\mathcal{Q}_0$ have been modified to accommodate partial functions and undefined terms and to axiomatize definite description.

Our presentation of $\mathcal{Q}_0^{\mathrm{u}}$ is intended to show as clearly as possible what must be changed in Church's type theory in order to formalize the traditional approach to undefinedness. Our development of $\mathcal{Q}_0^{\mathrm{u}}$ closely follows Andrews' development of $\mathcal{Q}_0$. Notes indicate where and how $\mathcal{Q}_0$ and $\mathcal{Q}_0^{\mathrm{u}}$ differ from each other. And the proofs of the soundness and completeness theorems for $\mathcal{Q}_0^{\mathrm{u}}$ follow very closely the proofs of these theorems for $\mathcal{Q}_0$.

$\mathcal{Q}_0$ and $\mathcal{Q}_0^{\mathrm{u}}$ have the same *theoretical expressivity* (see Note 6). However, with its formalization of the traditional approach, $\mathcal{Q}_0^{\mathrm{u}}$ has significantly greater *practical expressivity* than $\mathcal{Q}_0$. Statements involving partial functions and undefined terms can be expressed in $\mathcal{Q}_0^{\mathrm{u}}$ more naturally and concisely than in $\mathcal{Q}_0$ (see [9]). All the standard laws of predicate logic hold in $\mathcal{Q}_0^{\mathrm{u}}$ except those involving equality and substitution, but these do hold for defined terms. In summary, $\mathcal{Q}_0^{\mathrm{u}}$ has the benefit of greater practical expressivity at the cost of a modest departure from standard predicate logic.

The benefits of a practical logic like $\mathcal{Q}_0^{\mathrm{u}}$ would be best realized by a computer implementation of the logic. $\mathcal{Q}_0^{\mathrm{u}}$ has not been implemented, but the related logic LUTINS [6, 7, 8] has been implemented in the IMPS theorem proving system [10, 11] and successfully used to prove hundreds of theorems in traditional mathematics, especially in mathematical analysis. LUTINS is essentially just a more sophisticated version of $\mathcal{Q}_0^{\mathrm{u}}$ with subtypes and additional expression constructors. An implemented logic that formalizes the traditional approach to undefinedness can reap the benefits of a proven approach developed in mathematical practice over hundreds of years.

9 Acknowledgments

Peter Andrews deserves special thanks for writing *An Introduction to Mathematical Logic and Type Theory: To Truth through Proof* [2]. The ideas embodied in $\mathcal{Q}_0^{\mathrm{u}}$ heavily depend on the presentation of $\mathcal{Q}_0$ given in this superb textbook.

BIBLIOGRAPHY

[1] P. B. Andrews. A reduction of the axioms for the theory of propositional types. *Fundamenta Mathematicae*, 52:345–350, 1963.

[2] P. B. Andrews. *An Introduction to Mathematical Logic and Type Theory: To Truth through Proof, Second Edition.* Kluwer, 2002.

[3] P. B. Andrews, M. Bishop, S. Issar, D. Nesmith, F. Pfennig, and H. Xi. TPS: A Theorem Proving System for Classical Type Theory. *Journal of Automated Reasoning*, 16:321–353, 1996.

[4] A. Church. A Formulation of the Simple Theory of Types. *Journal of Symbolic Logic*, 5:56–68, 1940.

[5] T. Coquand and G. Huet. The Calculus of Constructions. *Information and Computation*, 76:95–120, 1988.

[6] W. M. Farmer. A Partial Functions Version of Church's Simple Theory of Types. *Journal of Symbolic Logic*, 55:1269–91, 1990.

[7] W. M. Farmer. A Simple Type Theory with Partial Functions and Subtypes. *Annals of Pure and Applied Logic*, 64:211–240, 1993.

[8] W. M. Farmer. Theory Interpretation in Simple Type Theory. In J. Heering et al., editor, *Higher-Order Algebra, Logic, and Term Rewriting*, volume 816 of *Lecture Notes in Computer Science*, pages 96–123. Springer-Verlag, 1994.

[9] W. M. Farmer. Formalizing Undefinedness Arising in Calculus. In D. Basin and M. Rusinowitch, editors, *Automated Reasoning—IJCAR 2004*, volume 3097 of *Lecture Notes in Computer Science*, pages 475–489. Springer-Verlag, 2004.

[10] W. M. Farmer, J. D. Guttman, and F. J. Thayer. IMPS: An Interactive Mathematical Proof System. *Journal of Automated Reasoning*, 11:213–248, 1993.

[11] W. M. Farmer, J. D. Guttman, and F. J. Thayer Fábrega. IMPS: An Updated System Description. In M. McRobbie and J. Slaney, editors, *Automated Deduction—CADE-13*, volume 1104 of *Lecture Notes in Computer Science*, pages 298–302. Springer-Verlag, 1996.

[12] J.-Y. Girard, Y. Lafont, and P. Taylor. *Proofs and Types*, volume 7 of *Cambridge Tracts in Theoretical Computer Science.* Cambridge University Press, 1989.

[13] M. J. C. Gordon and T. F. Melham. *Introduction to HOL: A Theorem Proving Environment for Higher Order Logic.* Cambridge University Press, 1993.

[14] L. Henkin. Completeness in the Theory of Types. *Journal of Symbolic Logic*, 15:81–91, 1950.

[15] L. Henkin. A theory of propositional types. *Fundamenta Mathematicae*, 52:323–344, 1963.

[16] Lemma 1 Ltd. *ProofPower: Description*, 2000. Available at http://www.lemma-one.com/ProofPower/doc/doc.html.

[17] P. Martin-Löf. *Intuitionistic Type Theory.* Bibliopolis, 1984.

[18] S. Owre, S. Rajan, J. M. Rushby, N. Shankar, and M. Srivas. PVS: Combining Specification, Proof Checking, and Model Checking. In R. Alur and T. A. Henzinger, editors, *Computer Aided Verification: 8th International Conference, CAV '96*, volume 1102 of *Lecture Notes in Computer Science*, pages 411–414. Springer-Verlag, 1996.

[19] L. C. Paulson. *Isabelle: A Generic Theorem Prover*, volume 828 of *Lecture Notes in Computer Science.* Springer-Verlag, 1994.

[20] A. Stump. Subset Types and Partial Functions. In F. Baader, editor, *Automated Deduction—CADE-19*, volume 2741 of *Lecture Notes in Computer Science*, pages 151–165. Springer-Verlag, 2003.

A Finite Axiomatization of Propositional Type Theory in Pure Lambda Calculus

MARK KAMINSKI, GERT SMOLKA

ABSTRACT. We consider simply typed lambda terms obtained with a single base type B and two constants $\bot$ and $\to$, where B is interpreted as the set of the two truth values, $\bot$ as falsity, and $\to$ as implication. We show that every value of the full set-theoretic type hierarchy can be described by a closed term and that every valid equation can be derived from three axioms with β and η. In contrast to the established approach, we employ a pure lambda calculus where constants appear as a derived notion.

1 Introduction

Propositional type theory employs simply typed lambda terms obtained with a single base type B, which is interpreted as the set of the two truth values. A concrete propositional type theory fixes some logical constants and a deduction system. We require denotational and deductive completeness. Denotational completeness means that every value of the full set-theoretic type hierarchy can be denoted by a closed term. Deductive completeness means that every valid formula can be deduced. An example of a valid formula is $f(f(fx)) \equiv fx$ where $f : BB$ and $x : B$ are variables. Deciding the validity of formulas obtained with higher-order quantification requires nonelementary time [10].

The first propositional type theory was devised by Henkin [6]. As constants Henkin takes all identity predicates, which can easily express the propositional connectives and the quantifiers. His deductive system is a Hilbert system whose inference rules are β and replacement of equals with equals. Henkin proves denotational and deductive completeness. Henkin's deductive system has been simplified and generalized to general type theory by Andrews [2, 3].

Altenkirch and Uustalu [1] have devised a different propositional type theory. They take the simply typed lambda calculus with a single base type B, the constants *false* : B and *true* : B, and the polymorphic conditional $if_\sigma : B\sigma\sigma$ known from programming languages. They obtain deductive completeness with β, η, and four axiom schemes. Their completeness proof is based on Berger and Schwichtenberg's [4] normalization-as-evaluation technique.

Like Altenkirch and Uustalu, we base our propositional type theory on the simply typed lambda calculus with a single base type B. However, we omit the polymorphic conditional and employ only two constants, falsity and implication. We show that falsity and implication suffice for denotational completeness and obtain deductive completeness with β, η, and three axioms. Our proof of deductive completeness uses ideas from Henkin's completeness proof. We are the first ones to obtain a complete propositional type theory with only finitely many constants (the polymorphic conditional counts as infinitely many constants).

As it comes to the lambda calculus, we work with a pure system where constants appear only as a derived notion. Besides β and η we only have conversion steps of the form $us = ut$ where $s = t$ is an axiom. This simplifies the standard approach [9], where axioms can be applied with capture below binders (ξ-rule). Application with capture amounts to an implicit universal quantification of the free variables of the axioms. Implicit quantification is essential for algebraic systems, but is unnecessary for higher-order systems, which can express universal quantification with functional equality. For instance, $\forall x.\ x + 0 = x$ can be expressed as $\lambda x.\, x + 0 = \lambda x.x$.

The paper is organized as follows. We start with the pure lambda calculus and basic conversion proofs. We then provide a sequent system and show that it yields the same theorems as the conversion system. The rules of the sequent system formulate important properties of the conversion system. We then obtain a propositional type theory by committing to one base type, two constants, and three axioms. Next we formulate propositional sequent rules that are admissible for the general sequent system and that subsume the usual natural deduction rules. We then prove denotational and deductive completeness.

2 Terms and Conversion

We assume familiarity with the simply typed lambda calculus (see, e.g., [7]). *Types* (σ, τ, ρ) are obtained from *base types* (α) according to $\sigma ::= \alpha \mid \sigma\sigma$. Think of $\sigma\tau$ as the type of functions from σ to τ. *Terms* (s, t, u) are obtained from *names* (x, y, z, f, g) according to $s ::= x \mid \lambda x.s \mid ss$. We assume a *typing relation* $s : \sigma$ satisfying the following properties:

1. For every term s there is at most one type σ such that $s : \sigma$.
2. For every type σ there are infinitely many names x such that $x : \sigma$.

3. For all x, s, σ, τ: $\lambda x.s : \sigma\tau \iff x : \sigma \wedge s : \tau$.
4. For all s, t, σ: $st : \sigma \iff \exists\tau\colon\ s : \tau\sigma \wedge t : \tau$.

A term σ is *well-typed* if there is a type σ such that $s : \sigma$. We will only consider well-typed terms. We use Λ to denote the set of all well-typed terms. We omit parentheses according to $\sigma\tau\rho \rightsquigarrow \sigma(\tau\rho)$ and $stu \rightsquigarrow (st)u$.

An *equation* is a pair $s = t$ of two terms s, t of the same type. We use e as a metavariable for single equations and A as a metavariable for finite sets of equations. Since equations are pairs, A is a binary relation on Λ. We use $\mathcal{N}s$, $\mathcal{N}e$ and $\mathcal{N}A$ to denote the sets of names that *occur free* in s, e and A, respectively. *Contexts* are obtained according to $C ::= [] \mid \lambda x.C \mid C\,s \mid s\,C$. The notation $C[s]$ describes the term obtained by replacing the hole $[]$ of C with s (capturing is ok). We assume a *substitution operation* s^x_t that yields for s, x, t a term that can be obtained from s by *capture-free substitution* of t for x, possibly after renaming of local names.

We define three *reduction relations*:

$$\begin{aligned}
\rightarrow_\beta &:= \{\, (C[(\lambda x.s)t],\, C[s^x_t]) \mid C[(\lambda x.s)t] \in \Lambda \,\} \\
\rightarrow_\eta &:= \{\, (C[\lambda x.sx],\, C[s]) \mid C[\lambda x.sx] \in \Lambda \ \wedge\ x \notin \mathcal{N}s \,\} \\
\rightarrow_A &:= \{\, (us, ut) \mid us \in \Lambda \ \wedge\ (s,t) \in A \,\}
\end{aligned}$$

The elements of the reduction relations are called *reduction steps*. We define *A-conversion* $\sim_A$ as the least equivalence relation on Λ that contains $\rightarrow_\beta$, $\rightarrow_\eta$, and $\rightarrow_A$. *λ-conversion* $\sim_\lambda$ is obtained as $\sim_\emptyset$. We say that s, t are *A-convertible* [*λ-convertible*] if $s \sim_A t$ [$s \sim_\lambda t$]. A *β-conversion step* is an equation $s = t$ such that $s \rightarrow_\beta t$ or $t \rightarrow_\beta s$. The definition of *η- and A-conversion steps* is analogous. A *basic conversion proof of $s = t$ with A* is a tuple $(s_1, \ldots, s_n)$ such that $s_1 = s$, $s_n = t$, and $s_i = s_{i+1}$ is a β-, η- or A-conversion step for all $i \in \{1, \ldots, n-1\}$.

PROPOSITION 1.1. *$s \sim_A t$ iff there exists a basic conversion proof of $s = t$ with A.*

PROPOSITION 1.2 (α-equivalence). *$\lambda x.s \sim_\lambda \lambda y.s^x_y$ if $\lambda x.s \in \Lambda$, $y \notin \mathcal{N}(\lambda x.s)$, and x and y have the same type.*

Proof. Follows with the basic conversion proof $(\lambda x.s,\ \lambda y.(\lambda x.s)y,\ \lambda y.s^x_y)$ consisting of an η- and a β-step. ■

Our definition of $\rightarrow_A$ is non-standard. At first glance, $\rightarrow_A$ may seem too weak to yield first-order rewriting. However, if we quantify the equational axioms with λ, we get all we need. This can be seen from the following basic conversion proof of $f(y+0) = fy$ with $\{\lambda x.x{+}0 = \lambda x.x\}$:

$$\begin{aligned} f(y+0) &= f((\lambda x.x{+}0)y) && \beta \\ &= (\lambda g.f(gy))(\lambda x.x{+}0) && \beta \\ &= (\lambda g.f(gy))(\lambda x.x) && \lambda x.x{+}0 = \lambda x.x \\ &= f((\lambda x.x)y) && \beta \\ &= fy && \beta \end{aligned}$$

Our pure lambda calculus with A-conversion has the same expressive power as the lambda calculus with constants commonly used in programming language theory [9]. We prefer the pure calculus since it is technically simpler. In particular, it does not provide implicit quantification of the free variables of equational axioms. In the pure system, the distinction between variables and constants is not hard-wired but is obtained as a derived notion, as will be seen in § 5.

A context C *captures* a name x if the hole of C is in the scope of a binder λx. A context C is *admissible for* A if it does not capture any name in $\mathcal{N}A$.

PROPOSITION 1.3 (Compatibility). *If $s \sim_A t$ and C is admissible for A, then $C[s] \sim_A C[t]$.*

Proof. It suffices to show the property for single reduction steps. It is easy to verify that it holds for β- and η-steps. Let $us \to_A ut$ and C be admissible for A. Then we can choose a name x such that $(\lambda x.C[ux])s \to_\beta C[us]$ and $(\lambda x.C[ux])t \to_\beta C[ut]$. Hence $C[us] \sim_A C[ut]$. ■

A *generalized A-conversion step* is an equation $s = t$ such that either $s \sim_\lambda t$ or there exist an equation $(u{=}v) \in A$ and an A-admissible context C such that $s \sim_\lambda C[u]$ and $C[v] \sim_\lambda t$. A *generalized conversion proof of $s = t$ with A* is a tuple $(s_1, \ldots, s_n)$ such that $s_1 = s$, $s_n = t$, and $s_i = s_{i+1}$ is a generalized A-conversion step for all $i \in \{1, \ldots, n-1\}$.

PROPOSITION 1.4. *$s \sim_A t$ iff there exists a generalized conversion proof of $s = t$ with A.*

Generalized conversion proofs can be much shorter than basic conversion proofs. For instance, there is a generalized conversion proof of $f(y{+}0) = fy$ with $\{\lambda x.x{+}0 = \lambda x.x\}$ consisting of a single step. Both basic and generalized conversion proofs have their uses. The existence of basic conversion proofs is helpful when we show general properties of A-conversion.

$$\text{Triv}\quad \frac{}{A,\, e \vdash e} \qquad \text{Weak}\quad \frac{A' \vdash e}{A \vdash e}\; A' \subseteq A \qquad \text{Cut}\quad \frac{A \vdash e' \qquad A, e' \vdash e}{A \vdash e}$$

$$\text{Sub}\quad \frac{A \vdash e}{A^x_s \vdash e^x_s}$$

$$\text{Ref}\quad \frac{}{A \vdash s = s} \qquad \text{Sym}\quad \frac{A \vdash s = t}{A \vdash t = s} \qquad \text{Trans}\quad \frac{A \vdash s = t \qquad A \vdash t = u}{A \vdash s = u}$$

$$\text{Inst}\quad \frac{A \vdash s_1 = s_2}{A \vdash t_1 = t_2} \quad t_1 \sim_\lambda C[s_1] \text{ and } C[s_2] \sim_\lambda t_2 \text{ and } C \text{ admissible for } A$$

Figure 1.1. A Sequent-Based Proof System

3 A Sequent System

The rules in Figure 1.1 define a sequent-based proof system for equations. We will show that $A \vdash e$ iff there is a conversion proof of e with A. Thus the sequent rules show how conversion proofs can be combined into more complex conversion proofs. In fact, the rules formulate important properties of A-conversion. The rules Triv, Weak, Cut, and Sub express basic properties that are common for sequent systems. Note that Cut boosts conversion proofs since it allows conversion with respect to lemmas (i.e., provable equations). The rules Ref, Sym, and Trans account for the fact that conversion is an equivalence relation. The rule Inst (Instantiation) is the workhorse of the system. It incorporates λ-equivalence and closure under admissible contexts. Here is an instance of Inst that shows its power:

$$\frac{A \vdash \lambda xy.\, x + y = \lambda xy.\, y + x}{A \vdash s + t = t + s}$$

Inst also subsumes the rules

$$\zeta \quad \frac{A \vdash sx = tx}{A \vdash t = s} \quad x \notin \mathcal{N}(A \cup \{s{=}t\}) \qquad \xi \quad \frac{A \vdash s = t}{A \vdash \lambda x.s = \lambda x.t} \quad x \notin \mathcal{N}A$$

that generalize well-known properties of λ-conversion.

PROPOSITION 1.5. $s \sim_A t \iff A \vdash s = t$

Proof. The direction $\Rightarrow$ is straightforward since Ref, Sym, Trans and Inst can simulate basic conversion proofs. The other direction requires more work. For each of the rules one shows that one can obtain a generalized conversion proof for the conclusion if one has basic conversion proofs for the premises. Proposition 1.3 is helpful for Cut and Inst. ∎

4 Interpretations

An *interpretation* is a function $\mathcal{I}$ that maps every type to a nonempty set and every name $x : \sigma$ to an element of $\mathcal{I}\sigma$. We will only consider interpretations that map a functional type $\sigma\tau$ to the set of all total functions from $\mathcal{I}\sigma$ to $\mathcal{I}\tau$. Every interpretation $\mathcal{I}$ can be extended uniquely to a function $\hat{\mathcal{I}}$ which maps every term $s : \sigma$ to an element of $\mathcal{I}\sigma$ and treats applications and abstractions as one would expect. An interpretation $\mathcal{I}$ *satisfies an equation* $s = t$ if $\mathcal{I}s = \mathcal{I}t$. An interpretation *satisfies* A if it satisfies every equation in A. We write $s \approx_A t$ if every interpretation that satisfies A also satisfies $s = t$.

PROPOSITION 1.6 (Soundness). $s \sim_A t \Longrightarrow s \approx_A t$

5 Propositional Type Theory

Figure 1.2 shows our axiomatization of propositional type theory. Technically, we fix a base type B and five names $\bot$, $\rightarrow$, x, y, and f and define P as the set containing the following three equations:

$$\begin{array}{rcll} \lambda x.\, \top \rightarrow x & = & \lambda x.\, x & \qquad \text{I}\top \\ \lambda fx.\, f\bot \rightarrow f\top \rightarrow fx & = & \lambda fx.\, \top & \qquad \text{BCA} \\ \lambda xy.\, (x \equiv y) \rightarrow x & = & \lambda xy.\, (x \equiv y) \rightarrow y & \qquad \text{Rep} \end{array}$$

Note that only $\bot$ and $\rightarrow$ occur free in P. We call $\bot$ and $\rightarrow$ *constants* and all other names *variables*. The variable constant distinction is exploited in Figure 1.2, where the axioms appear with implicit quantification. Note that in our system implicit quantification is a notational device and not a syntactic feature.

Convention. In the rest of the paper we will only consider types, terms, and equations that can be obtained with the single base type B. We use $\mathcal{V}s$ to denote the set of all variables that occur free in s. A term s is *closed* if $\mathcal{V}s = \emptyset$, and *open* otherwise.

PROPOSITION 1.7 (I$\bot$). $P \vdash \bot \rightarrow x = \top$

Proof. Here is a (generalized) conversion proof:

$$\begin{array}{rll} \bot \rightarrow x & = \bot \rightarrow \top \rightarrow x & \qquad \text{I}\top \\ & = \top & \qquad \text{BCA} \end{array}$$

∎

Base Type B

Constants $\bot : B;\quad \to : BBB$

Variables $x, y : B;\quad f : BB$

Notations

$$\begin{array}{rclcrcl} \top & := & \bot \to \bot & \qquad & s \vee t & := & (s \to t) \to t \\ \neg s & := & s \to \bot & & s \wedge t & := & \neg(\neg s \vee \neg t) \\ & & & & s \equiv t & := & (s \to t) \wedge (t \to s) \end{array}$$

Precedence $=,\ \equiv,\ \to,\ \vee,\ \wedge,\ \neg$

Parentheses $s \to t \to u \quad \leadsto \quad s \to (t \to u)$

Axioms

$$\begin{array}{rcll} \top \to x & = & x & \qquad \mathrm{I}\top \\ f\bot \to f\top \to fx & = & \top & \qquad \mathrm{BCA} \\ (x \equiv y) \to x & = & (x \equiv y) \to y & \qquad \mathrm{Rep} \end{array}$$

Figure 1.2. Axiomatization of Propositional Type Theory

A *propositional interpretation* is an interpretation $\mathcal{I}$ such that $\mathcal{I}B = \{0,1\}$, $\mathcal{I}\bot = 0$ and $\mathcal{I}(\to)ab = \mathsf{if}\ a{=}0\ \mathsf{then}\ 1\ \mathsf{else}\ b$ for all $a, b \in \{0,1\}$.

PROPOSITION 1.8. *Every propositional interpretation satisfies P.*

An equation $s = t$ is *propositionally valid* if $\hat{\mathcal{I}}s = \hat{\mathcal{I}}t$ for every propositional interpretation $\mathcal{I}$. We write $s \approx_2 t$ if $s = t$ is propositionally valid. An example of a propositionally valid equation is $f(f(fx)) = fx$ where $f : BB$ and $x : B$ (to verify validity, consider all 4 functions $\{0,1\} \to \{0,1\}$).

PROPOSITION 1.9 (Semantic Completeness). $s \approx_P t \iff s \approx_2 t$

Proof. The direction $\Rightarrow$ holds by Proposition 1.8. For the other direction we assume that $\mathcal{I}$ is an interpretation that satisfies P. We show $\mathcal{I}s = \mathcal{I}t$. Case Analysis.

Let $\mathcal{I}\bot = \mathcal{I}\top$. Then $\mathcal{I}(\lambda x.x) = \mathcal{I}(\lambda x.\top)$ by I$\bot$ (Proposition 1.7) and Axiom I$\top$. Hence $\mathcal{I}B$ is a singleton. Thus all types denote singletons and hence $\mathcal{I}s = \mathcal{I}t$.

Let $\mathcal{I}\bot \neq \mathcal{I}\top$. Because of I$\bot$ and I$\top$ it suffices to show that $\mathcal{I}B \subseteq \{\mathcal{I}\bot, \mathcal{I}\top\}$. Suppose for contradiction $a \in \mathcal{I}B - \{\mathcal{I}\bot, \mathcal{I}\top\}$, and let b be a function from $\mathcal{I}B \to \mathcal{I}B$ such that $b(\mathcal{I}\bot) = b(\mathcal{I}\top) = \mathcal{I}\top$ and $ba = a$. Instantiating BCA and I$\top$ with a and b yields $a = \mathcal{I}\top$, thus contradicting the assumption. ■

Note that so far we have only used the axioms I$\top$ and BCA. The axiom Rep is only needed for deductive completeness. Brown [5] has constructed a general model that shows that P without Rep is deductively weaker than P. An alternative to Rep that yields the same deductive power is $\lambda xy.\, x\vee y = \lambda xy.\, y\vee x$.

A *propositional term* is a term that can be obtained according to $s ::= x \mid s \to s$ where $x : B$ is a name. A *propositional equation* is an equation $s = t$ where s, t are propositional terms. A *tautology* is a propositional equation $s = t$ such that $s \approx_2 t$. Note that the axioms I$\top$ and Rep are tautologies.

LEMMA 1.10. *Let s be a closed propositional term. Then:*

1. $s \sim_P \bot$ *or* $s \sim_P \top$.
2. *If* $s \approx_2 \top$, *then* $s \sim_P \top$.

Proof. A closed propositional term contains no other names but $\bot$ and $\to$. We show (1) by induction on the size of s. If $s = \bot$, then $s \sim_P \bot$ by Ref. If $s = s_1 \to s_2$, then $s_1 \sim_P \bot$ or $s_1 \sim_P \top$ and $s_2 \sim_P \bot$ or $s_2 \sim_P \top$ by induction. Now the claim follows with I$\top$ and I$\bot$. Claim (2) follows by (1) and Soundness. ■

A term s is *β-normal* if there exists no term t such that $s \to_\beta t$. It is well-known [7] that for every well-typed term s there exists a β-normal term t such that $s \sim_\lambda t$ and $\mathcal{N}t \subseteq \mathcal{N}s$.

PROPOSITION 1.11.
Every closed and β-normal term $s : B$ is propositional.

Proof. By induction on the size of s. Let $s : B$ be β-normal and closed. Then $s = xs_1 \ldots s_n$ where $s_1, \ldots, s_n$ are closed and β-normal. Since s is closed, either $x = \bot$ or $x = \to$. If $x = \bot$, then $n = 0$ and hence s is propositional. If $x = \to$, then $n = 2$ and s_1 and s_2 are propositional by induction. Hence s is propositional. ■

Let E be a set of equations. We say that P is *complete for E* if $s \approx_2 t$ implies $s \sim_P t$ for every equation $s = t$ in E. Eventually, we will show that P is complete for all equations.

PROPOSITION 1.12. *P is complete for all closed equations $s = \top$.*

Proof. Let $s \approx_2 \top$. There exists a β-normal and closed term $t : B$ such that $s \sim_\lambda t$. By Soundness and Semantic Completeness we have $t \approx_2 \top$. By Proposition 1.11 we know that t is propositional. Hence $t \sim_P \top$ by Lemma 1.10 (2). Thus $s \sim_P \top$. ■

$$\text{MP}\quad \frac{A \vdash s \to t = \top \qquad A \vdash s = \top}{A \vdash t = \top}\quad P \subseteq A$$

$$\text{BE}\quad \frac{A \vdash s \equiv t = \top}{A \vdash s = t}\quad P \subseteq A \qquad\qquad \text{BE}^{-}\quad \frac{A \vdash s = t}{A \vdash s \equiv t = \top}\quad P \subseteq A$$

$$\text{Taut}\quad \frac{}{A \vdash e}\quad P \subseteq A \text{ and } e \text{ tautology}$$

$$\text{CA}\quad \frac{A \vdash e^x_\bot \qquad A \vdash e^x_\top}{A \vdash e}\quad P \subseteq A \qquad\qquad \text{Ded}\quad \frac{A, s = \top \vdash t = \top}{A \vdash s \to t = \top}\quad P \subseteq A$$

$$\text{DE}\quad \frac{A,\, s = \top \vdash t = \top \qquad A,\, t = \top \vdash s = \top}{A \vdash s = t}\quad P \subseteq A$$

Figure 1.3. Admissible Rules for $\vdash$

6 Propositional Sequent Rules

Figure 1.3 collects some sequent rules that are admissible for the relation $\vdash$ defined in § 3. Admissibility of a rule for $\vdash$ means that the conclusion of the rule is satisfied by $\vdash$ if the premises of the rule are satisfied by $\vdash$. MP stands for modus ponens, BE for Boolean equality, CA for case analysis, Ded for deductivity, and DE for deductivity and Boolean equality.

The rules formulate important properties of the conversion relations $\sim_A$ with $P \subseteq A$. By Proposition 1.5 a rule is admissible for $\vdash$ iff there is a conversion proof for the conclusion of the rule if there are conversion proofs for the premises of the rule. Hence the rules tell us how we can construct complex conversion proofs from simpler ones. The rules will be crucial for the completeness proof to come. Seen semantically, the rules express well-known properties of propositional logic.

PROPOSITION 1.13. MP *is admissible for* $\vdash$.

Proof. Let $P \subseteq A$. By Cut it suffices to give a conversion proof of $t = \top$ with $A \cup \{s \to t = \top,\ s = \top\}$. Here it is:

$$\begin{aligned} t &= \top \to t && \text{I}\top \\ &= s \to t && s = \top \\ &= \top && s \to t = \top \end{aligned}$$

■

PROPOSITION 1.14. BE *is admissible for* $\vdash$.

Proof. Let $P \subseteq A$. Here is a conversion proof of $s = t$ with $A \cup \{s \equiv t = \top\}$:

$$
\begin{array}{ll}
s = \top \to s & \text{IT} \\
\quad = (s \equiv t) \to s & s \equiv t = \top \\
\quad = (s \equiv t) \to t & \text{Rep} \\
\quad = \top \to t & s \equiv t = \top \\
\quad = t & \text{IT}
\end{array}
$$

■

Axiom Rep is used for the first time in the above proof. In fact, our axiomatization contains Rep so that we can show that BE is admissible.

LEMMA 1.15. *If* $s = \top$ *is a tautology, then* $P \vdash s = \top$.

Proof. By induction on $|\mathcal{V}s|$. If s is closed, the claim follows by Lemma 1.10 (2). Otherwise, let $x \in \mathcal{V}s$. Then $s^x_\bot = \top$ and $s^x_\top = \top$ are tautologies. By induction $P \vdash s^x_\bot = \top$ and $P \vdash s^x_\top = \top$. Hence $P \vdash s = \top$ follows with a conversion proof:

$$
\begin{array}{ll}
s = \top \to s & \text{IT} \\
\quad = \top \to \top \to s & \text{IT} \\
\quad = s^x_\bot \to \top \to s & s^x_\bot = \top \\
\quad = s^x_\bot \to s^x_\top \to s & s^x_\top = \top \\
\quad = \top & \text{BCA}
\end{array}
$$

■

PROPOSITION 1.16. BE^- *is admissible for* $\vdash$.

Proof. Let $P \subseteq A$. Since $x \equiv x = \top$ is a tautology, it suffices by Lemma 1.15 to give a conversion proof of $s \equiv t = \top$ with $A \cup \{s = t,\ x \equiv x = \top\}$. This is straightforward. ■

PROPOSITION 1.17. Taut *is admissible for* $\vdash$.

Proof. By BE and Lemma 1.15. ■

PROPOSITION 1.18. CA *is admissible for* $\vdash$.

Proof. By BE and BE^- it suffices to show the claim for $e = (s = \top)$. This can be done with BCA, Inst, and MP. ■

LEMMA 1.19. $P \vdash x \to fx = x \to f\top$

Proof. Follows by CA and conversion proofs with I$\bot$. ■

LEMMA 1.20. *Let $P \subseteq A$ and $t_1 = t_2$ be an $A \cup \{s = \top\}$-conversion step. Then $A \vdash s{\to}t_1 = s{\to}t_2$.*

Proof. If $t_1 = t_2$ is a A-conversion step, the claim follows by Inst. If $t_1 = t_2$ is a $\{s = \top\}$-conversion step, we have $t_1 = us$ and $t_2 = u\top$ for some u. Hence $P \vdash s{\to}t_1 = s{\to}t_2$ by Lemma 1.19. The claim follows by Weak. ■

PROPOSITION 1.21. Ded *is admissible for* $\vdash$.

Proof. Let $P \subseteq A$ and $A, s = \top \vdash t = \top$. Then there exists a basic conversion proof of $t = \top$ with $A \cup \{s = \top\}$. Hence $A \vdash s{\to}t = s{\to}\top$ by Lemma 1.20. Since $A \vdash x{\to}\top = \top$ by Taut, we have $A \vdash s{\to}t = \top$. ■

PROPOSITION 1.22. DE *is admissible for* $\vdash$.

Proof. By Ded and BE it suffices to give a conversion proof of $s{\equiv}t = \top$ with $s{\to}t = \top$, $t{\to}s = \top$, and tautologies. This is straightforward. ■

7 Denotational Completeness

We fix a propositional interpretation $\mathcal{B}$. Then we have $\mathcal{I}\sigma = \mathcal{B}\sigma$ for every propositional interpretation $\mathcal{I}$ and every type σ. Moreover, we have $\hat{\mathcal{I}}s = \hat{\mathcal{B}}s$ for every closed term s and every propositional interpretation $\mathcal{I}$. We will show that P is *denotationally complete*, that is, for every type σ and every value $a \in \mathcal{B}\sigma$ there is a closed term $s : \sigma$ such that $\hat{\mathcal{B}}s = a$.

We will define a family of *quote functions* $\downarrow^\sigma : \mathcal{B}\sigma \to \Lambda_0^\sigma$ by recursion on types. Λ_0^σ is the set of all closed terms of type σ. The quote functions will satisfy $\hat{\mathcal{B}}(\downarrow^\sigma a) = a$ for all $a \in \mathcal{B}\sigma$ and all types σ.

The definition of the basic quote function $\downarrow^B$ is straightforward. To explain the definition of the other quote functions, we consider the special case $\downarrow^{\sigma B}$. We start with

$$\downarrow^{\sigma B}(a) \;=\; \lambda x.\; \bigvee_{\substack{b \in \mathcal{B}\sigma \\ ab=1}} x \doteq_\sigma (\downarrow^\sigma b)$$

It remains to define a closed term $\doteq_\sigma$ that denotes the identity predicate for $\mathcal{B}\sigma$. If $\sigma = B$, $\lambda xy.\, x \equiv y$ does the job. If $\sigma = \sigma_1\sigma_2$, we rely on recursion and define

$$\doteq_\sigma \;=\; \lambda fg.\; \bigwedge_{a \in \mathcal{B}\sigma_1} f(\downarrow^{\sigma_1} a) \doteq_{\sigma_2} g(\downarrow^{\sigma_1} a)$$

$$\downarrow^{\sigma_1 \ldots \sigma_n B} a \; := \; \lambda x_1 \ldots x_n . \bigvee_{\substack{\langle b_i \in \mathcal{B}\sigma_i \rangle \\ a b_1 \ldots b_n = 1}} \; \bigwedge_{1 \le j \le n} x_j \doteq_{\sigma_j} (\downarrow^{\sigma_j} b_j) \qquad \text{D}\downarrow$$

$$\forall_\sigma \; := \; \lambda f. \bigwedge_{a \in \mathcal{B}\sigma} f(\downarrow^\sigma a) \qquad \text{D}\forall$$

$$\doteq_B \; := \; \lambda xy.\; x \equiv y \qquad \text{D}\doteq$$

$$\doteq_{\sigma\tau} \; := \; \lambda fg.\; \forall_\sigma(\lambda x.\; fx \doteq_\tau gx) \qquad \text{D}\doteq$$

$\langle b_i \in \mathcal{B}\sigma_i \rangle$ abbreviates $(b_1, \ldots, b_n) \in \mathcal{B}\sigma_1 \times \cdots \times \mathcal{B}\sigma_n$

Figure 1.4. Quote Functions $\downarrow^\sigma$ and Terms $\forall_\sigma$ and $\doteq_\sigma$

Figure 1.4 shows the full definition of the quote functions. A disjunction with an empty index set denotes $\bot$, and a conjunction with an empty index set denotes $\top$. The notations $\doteq_\sigma$ and $\forall_\sigma$ will be used in the following. We will write $\forall_\sigma x.\, s$ for $\forall_\sigma(\lambda x.s)$. The notational operator $\doteq_\sigma$ will be used with a precedence higher than $\neg$ (i.e, $\neg\, s \doteq_\sigma t \leadsto \neg(s \doteq_\sigma t)$).

PROPOSITION 1.23. *The terms $\downarrow^\sigma a$, $\forall_\sigma$ and $\doteq_\sigma$ are closed.*

PROPOSITION 1.24. *Let σ be a type, $a, b \in \mathcal{B}\sigma$, and $\varphi \in \mathcal{B}(\sigma B)$. Then:*

1. $\hat{\mathcal{B}}(\downarrow^\sigma a) = a$
2. $\hat{\mathcal{B}}(\forall_\sigma)\varphi = 1 \iff \forall c \in \mathcal{B}\sigma\colon\; \varphi c = 1$
3. $\hat{\mathcal{B}}(\doteq_\sigma)ab = 1 \iff a = b$

THEOREM 1.25 (Denotational Completeness). *Let σ be a type and $a \in \mathcal{B}\sigma$. Then there is a closed term s such that $\hat{\mathcal{B}}s = a$.*

Proof. Follows from Proposition 1.24 (1). ■

8 Deductive Completeness

PROPOSITION 1.26. *P is complete for all equations if it is complete for all equations of the form $s = \top$.*

Proof. Assume P is complete for all equations of the form $s = \top$. Let $s \approx_2 t$. We show $s \sim_P t$. We choose distinct variables $\bar{x} = x_1 \ldots x_n$ that do not occur in $s = t$ such that $s\bar{x} : B$. We have $s\bar{x} \approx_2 t\bar{x}$ and hence $s\bar{x} \equiv t\bar{x} \approx_2 \top$. By the assumption we have $s\bar{x} \equiv t\bar{x} \sim_P \top$. By BE we have $P \vdash s\bar{x} = t\bar{x}$. Thus $P \vdash s = t$ by Inst (or repeated use of ζ). ■

LEMMA 1.27. *P is complete for all equations if for all types σ there are variables f, x such that $P \vdash \forall_\sigma f \rightarrow fx = \top$.*

Proof. Assume that for all types σ there are variables f, x such that $P \vdash \forall_\sigma f \rightarrow fx = \top$. Let $s \approx_2 \top$. By Proposition 1.26 it suffices to show that $s \sim_P \top$. There exist variables $\bar{x} = x_1 \dots x_n$ such that $\forall \bar{x}.s$ is closed. By the second assumption we have $\forall \bar{x}.s \approx_2 \top$. By Proposition 1.12 we have $P \vdash \forall \bar{x}.s = \top$. Hence $P \vdash s = \top$ by the first assumption, Sub and MP. ∎

LEMMA 1.28. *$P \vdash x \doteq_\sigma x = \top$*

Proof. By induction on σ, exploiting Taut. ∎

LEMMA 1.29. *If $a, b \in \mathcal{B}\sigma$ are distinct, then $P \vdash (\downarrow^\sigma a) \doteq_\sigma (\downarrow^\sigma b) = \bot$.*

Proof. Let $a, b \in \mathcal{B}\sigma$ be distinct. Then:

$\neg\, (\downarrow^\sigma a) \doteq_\sigma (\downarrow^\sigma b) \approx_2 \top$	Proposition 1.24
$P \vdash\ \neg\, (\downarrow^\sigma a) \doteq_\sigma (\downarrow^\sigma b) = \top$	Proposition 1.12
$P \vdash\ \neg\neg\, (\downarrow^\sigma a) \doteq_\sigma (\downarrow^\sigma b) = \neg\top$	Inst
$P \vdash\ (\downarrow^\sigma a) \doteq_\sigma (\downarrow^\sigma b) = \bot$	Taut

∎

LEMMA 1.30. *If $\sigma = \rho\tau$, then for all $a \in \mathcal{B}\sigma$ and $b \in \mathcal{B}\rho$:*
$P \vdash (\downarrow^\sigma a)(\downarrow^\rho b) = \downarrow^\tau (ab)$.

Proof. Let $\sigma = \rho\tau$ and $\tau = \tau_1 \dots \tau_n B$. Let $a \in \mathcal{B}\sigma$ and $b \in \mathcal{B}\rho$. By D$\downarrow$ and β, we have

$$(\downarrow^{\rho\tau_1 \dots \tau_n B} a)(\downarrow^\rho b)$$

$$= \left(\lambda x y_1 \dots y_n. \bigvee_{\substack{c \in \mathcal{B}\rho \\ \langle d_i \in \mathcal{B}\tau_i \rangle \\ a c d_1 \dots d_n = 1}} x \doteq_\rho (\downarrow^\rho c) \wedge \bigwedge_{1 \le j \le n} y_j \doteq_{\tau_j} (\downarrow^{\tau_j} d_j) \right) (\downarrow^\rho b)$$

$$= \lambda y_1 \dots y_n. \bigvee_{\substack{c \in \mathcal{B}\rho \\ \langle d_i \in \mathcal{B}\tau_i \rangle \\ a c d_1 \dots d_n = 1}} (\downarrow^\rho b) \doteq_\rho (\downarrow^\rho c) \wedge \bigwedge_{1 \le j \le n} y_j \doteq_{\tau_j} (\downarrow^{\tau_j} d_j)$$

By Lemma 1.28, Lemma 1.29, and Taut we obtain

$$= \lambda y_1 \dots y_n. \bigvee_{\substack{\langle d_i \in \mathcal{B}\tau_i \rangle \\ a b d_1 \dots d_n = 1}} \bigwedge_{1 \le j \le n} y_j \doteq_{\tau_j} (\downarrow^{\tau_j} d_j)$$

$$= \downarrow^{\tau_1 \dots \tau_n} (ab)$$

∎

LEMMA 1.31. *Let I and J be finite sets and $x_{i,j} : B$ be a variable for all $i \in I$, $j \in J$. Moreover, let $[I \to J]$ be the set of all total functions $I \to J$. Then the equation*

$$\bigwedge_{i \in I} \bigvee_{j \in J} x_{i,j} = \bigvee_{\varphi \in [I \to J]} \bigwedge_{i \in I} x_{i,\varphi i}$$

is a tautology and hence is provable with P.

Proof. Let s and t be the left and the right term of the equation, and let $\mathcal{I}$ be a propositional interpretation. We have to show that $\hat{\mathcal{I}}s = 1$ iff $\hat{\mathcal{I}}t = 1$. Let $\hat{\mathcal{I}}s = 1$. Then for every $i \in I$ there exists a $j \in J$ such that $\mathcal{I}(x_{i,j}) = 1$. Hence there exists a function $\varphi \in [I \to J]$ such that $\mathcal{I}(x_{i,\varphi i}) = 1$ for every $i \in I$. Hence $\hat{\mathcal{I}}t = 1$. The other direction follows analogously. ■

LEMMA 1.32. $P \vdash \bigvee_{a \in \mathcal{B}\sigma} x \doteq_\sigma (\downarrow^\sigma a) = \top$

Proof. By induction on σ. If $\sigma = B$, the claim follows with Taut. Otherwise, let $\sigma = \sigma_1\sigma_2$. We show the claim with a conversion proof.

$$\begin{aligned}
& \bigvee_{a \in \mathcal{B}\sigma} x \doteq_\sigma (\downarrow^\sigma a) \\
= & \bigvee_{a \in \mathcal{B}\sigma} \forall_{\sigma_1} y.\ xy \doteq_{\sigma_2} (\downarrow^\sigma a)y && \text{D}\doteq \\
= & \bigvee_{a \in \mathcal{B}\sigma} \bigwedge_{b \in \mathcal{B}\sigma_1} x(\downarrow^{\sigma_1} b) \doteq_{\sigma_2} (\downarrow^\sigma a)(\downarrow^{\sigma_1} b) && \text{D}\forall \\
= & \bigvee_{a \in \mathcal{B}\sigma} \bigwedge_{b \in \mathcal{B}\sigma_1} x(\downarrow^{\sigma_1} b) \doteq_{\sigma_2} (\downarrow^{\sigma_2}(ab)) && \text{Lemma 1.30} \\
= & \bigwedge_{b \in \mathcal{B}\sigma_1} \bigvee_{c \in \mathcal{B}\sigma_2} x(\downarrow^{\sigma_1} b) \doteq_{\sigma_2} (\downarrow^{\sigma_2} c) && \text{Lemma 1.31, } \mathcal{B}\sigma = [\mathcal{B}\sigma_1 \to \mathcal{B}\sigma_2] \\
= & \bigwedge_{b \in \mathcal{B}\sigma_1} \top && \text{induction for } \sigma_2 \\
= & \top && \text{Taut}
\end{aligned}$$

■

LEMMA 1.33. *If $P, x \doteq_\sigma y = \top \vdash x = y$, then $P \vdash \forall_\sigma f \to fx = \top$.*

Proof. Let $P, x \doteq_\sigma y = \top \vdash x = y$. Then we obtain with DE and Ded

$$P \vdash x \doteq_\sigma y \to fx = x \doteq_\sigma y \to fy \qquad (1.1)$$

Now we show the claim with a conversion proof.

$$\begin{aligned}
& \forall_\sigma f \to fx \\
= {} & \top \to \forall_\sigma f \to fx && \text{Taut} \\
= {} & \left(\bigvee_{a \in \mathcal{B}\sigma} x \doteq_\sigma (\downarrow^\sigma a) \right) \to \forall_\sigma f \to fx && \text{Lemma 1.32} \\
= {} & \bigwedge_{a \in \mathcal{B}\sigma} x \doteq_\sigma (\downarrow^\sigma a) \to \forall_\sigma f \to fx && \text{Taut} \\
= {} & \bigwedge_{a \in \mathcal{B}\sigma} x \doteq_\sigma (\downarrow^\sigma a) \to \forall_\sigma f \to f(\downarrow^\sigma a) && (1.1) \\
= {} & \bigwedge_{a \in \mathcal{B}\sigma} x \doteq_\sigma (\downarrow^\sigma a) \to \left(\bigwedge_{b \in \mathcal{B}\sigma} f(\downarrow^\sigma b) \right) \to f(\downarrow^\sigma a) && \text{D}\forall \\
= {} & \top && \text{Taut}
\end{aligned}$$

■

LEMMA 1.34. $P,\ x \doteq_\sigma y = \top \ \vdash \ x = y$

Proof. By induction on σ. If $\sigma = B$, the claim follows with BE. Otherwise, let $\sigma = \sigma_1\sigma_2$. By D$\doteq$ we have

$$P,\ x \doteq_\sigma y = \top \ \vdash \ \forall_{\sigma_1} z.\ xz \doteq_{\sigma_2} yz = \top$$

for some variable z. By induction for σ_1 and Lemma 1.33 we have

$$P \ \vdash \ (\forall_{\sigma_1} z.\ xz \doteq_{\sigma_2} yz) \to xz \doteq_{\sigma_2} yz = \top$$

By MP we have

$$P,\ x \doteq_\sigma y = \top \ \vdash \ xz \doteq_{\sigma_2} yz = \top$$

Now induction for σ_2 and Cut yield

$$P,\ x \doteq_\sigma y = \top \ \vdash \ xz = yz$$

which yields the claim with Inst. ■

THEOREM 1.35 (Deductive Completeness).
P is complete for all equations.

Proof. Follows by Lemma 1.34, Lemma 1.33 and Lemma 1.27. ■

9 Final Remarks

We have shown that the pure simply typed lambda calculus furnished with three axioms is a complete deduction system for propositional type theory with falsity and implication. This yields the most minimal set-up of propositional type theory known so far. Our results carry over to propositional type theories with quantifiers and identity predicates [8]. The motivation of our research is to better understand the deductive power of the lambda-calculus when applied to higher-order logic. The idea to capture higher-order logic as a lambda theory appears in Mitchell [9] (Example 4.4.7).

Acknowledgments. We benefited from discussions with Chad E. Brown and Jan Schwinghammer. Jan pointed us to [1].

BIBLIOGRAPHY

[1] Thorsten Altenkirch and Tarmo Uustalu. Normalization by evaluation for $\lambda^{\to 2}$. In Yukiyoshi Kameyama and Peter J. Stuckey, editors, *Symposium on Functional and Logic Programming*, volume 2998 of *LNCS*, pages 260–275. Springer, 2004.

[2] Peter B. Andrews. A Reduction of the Axioms for the Theory of Propositional Types. *Fundamenta Mathematicae*, 52:345–350, 1963.

[3] Peter B. Andrews. *An Introduction to Mathematical Logic and Type Theory: To Truth Through Proof*, volume 27 of *Applied Logic Series*. Kluwer Academic Publishers, 2nd edition, 2002.

[4] Ulrich Berger and Helmut Schwichtenberg. An Inverse of the Evaluation Functional for Typed Lambda-Calculus. In *Proc. 6th Annual IEEE Symposium on Logic in Computer Science (LICS'91)*, pages 203–211. IEEE Computer Society Press, 1991.

[5] Chad E. Brown. Personal communication, January 2007.

[6] Leon Henkin. A Theory of Propositional Types. *Fundamenta Mathematicae*, 52:323–344, 1963.

[7] J. Roger Hindley and Jonathan P. Seldin. *Introduction to Combinators and λ-calculus.* Cambridge University Press, 1986.

[8] Mark Kaminski. *Completeness Results for Higher-Order Equational Logic.* Master's Thesis, Saarland University, 2006.

[9] John C. Mitchell. *Foundations for Programming Languages.* Foundations of Computing. The MIT Press, 1996.

[10] S. Vorobyov. The most nonelementary theory. *Information and Computation*, 190(2):196–219, 2004.

Skolemization in Simple Type Theory: the Logical and the Theoretical Points of View

GILLES DOWEK

Peter Andrews has proposed, in 1971, the problem of finding an analog of the Skolem theorem for Simple Type Theory. A first idea lead to a naive rule that worked only for Simple Type Theory with the axiom of choice and the general case has only been solved, more than ten years later, by Dale Miller [9, 10]. More recently, we have proposed with Thérèse Hardin and Claude Kirchner [7] a new way to prove analogs of the Miller theorem for different, but equivalent, formulations of Simple Type Theory.

In this paper, that does not contain new technical results, I try to show that the history of the skolemization problem and of its various solutions is an illustration of a tension between two points of view on Simple Type Theory: the *logical* and the *theoretical* points of view.

1 Skolemization

1.1 The Skolem theorem

Let $\mathcal{T}$ be a theory in first-order predicate logic containing an axiom of the form

$$\forall x_1 ... \forall x_n \exists y \; A$$

and $\mathcal{T}'$ be the theory obtained by replacing this axiom by

$$\forall x_1 ... \forall x_n \; ((f(x_1, ..., x_n)/y)A)$$

where f is a function symbol not used in $\mathcal{T}$. Then, the Skolem theorem asserts that the theory $\mathcal{T}'$ is a conservative extension of $\mathcal{T}$ and, in particular, that one theory is contradictory if and only if the other is.

1.2 Extending Skolem theorem to Simple Type Theory

The Skolem theorem plays a key role in automated theorem proving because it permits to eliminate quantifier alternation in the proposition to be proved, or refuted, and this alternation is often delicate to manage. This explains why, in 1971, seeking for a generalization of the Resolution method to Simple

Type Theory [2], Peter Andrews has proposed the problem of finding an analog of the Skolem theorem for Simple Type Theory.

Following the Skolem theorem for first-order predicate logic, we can try to replace an axiom of the form

$$\forall x_1 ... \forall x_n \exists y \ A$$

where x_1, ..., x_n and y are variables of type T_1, ..., T_n and U, by

$$\forall x_1 ... \forall x_n \ (((f \ x_1 \ ... \ x_n)/y)A)$$

where f is now a new constant of type $T_1 \to ... \to T_n \to U$. Unfortunately, the theory $\mathcal{T}'$ obtained this way is not always a conservative extension of the theory $\mathcal{T}$. For instance, it is not possible to prove the proposition $\exists g \ \forall x \ (P \ x \ (g \ x))$ from the axiom $\forall x \ \exists y \ (P \ x \ y)$, because, as proved again by Peter Andrews in 1972 [3], the axiom of choice

$$(\forall x \ \exists y \ (P \ x \ y)) \Rightarrow (\exists g \ \forall x \ (P \ x \ (g \ x)))$$

is not provable in Simple Type Theory. But this proposition is obviously provable from the skolemized form of this axiom: $\forall x \ (P \ x \ (f \ x))$. Thus, although this naive skolemization can be used in Simple Type Theory extended with the axiom of choice, it cannot be used in the usual formulation of Simple Type Theory, without the axiom of choice.

A more restricted form of skolemization has been proposed in 1983 by Dale Miller. In the Miller theorem, the Skolem symbol f is not only given the type $T_1 \to ... \to T_n \to U$ but also the arity $\langle T_1, ..., T_n, U \rangle$ and, unlike the usual symbols of a functional type, the Skolem symbols are not terms *per se*. To form a term with a Skolem symbol f of arity $\langle T_1, ..., T_n, U \rangle$, it is necessary to apply it to terms t_1, ..., t_n of type T_1, ..., T_n, called the *necessary arguments* of the symbol f. This reflects the intuition that, in a model of the negation of the axiom of choice, we may have for each n-uple t_1, ..., t_n the object $(f \ t_1 \ ... \ t_n)$, without having the function f itself as an object.

As it is not a term, the symbol f cannot be used as a witness to prove the proposition $\exists g \ \forall x \ (P \ x \ (g \ x))$. However, this restriction is not sufficient, because, although the symbol f cannot be used as a witness, the term $\lambda z \ (f \ z)$, can, yielding, after normalization, the same result: $\forall x \ (P \ x \ (f \ x))$. This explains why Miller has introduced a second restriction: that the variables free in the necessary arguments of a Skolem symbol cannot be bound by a λ-abstraction, higher in a term substituted for a variable in a quantifier rule.

With these two restrictions, Miller has been able to give a syntactic proof of an analogous of the Skolem theorem: any proof using the axiom

$$\forall x_1 ... \forall x_n \ (((f \ x_1 \ ... \ x_n)/y)A)$$

and whose conclusion does not use the symbol f can be transformed into a proof using the axiom

$$\forall x_1 ... \forall x_n \exists y \ A$$

2 The logical and the theoretical point of view

Simple Type Theory is otherwise known as *higher-order logic.* This duality of names reveals that this formalism is sometimes seen as a theory and sometimes as a logic. As a theory, it should be compared to other theories, such as arithmetic or set theory. As a logic, to other logics, and in particular to first-order predicate logic, that, historically, is a restriction of it.

From the logical point of view, it is natural to try to *extend* the theorems and algorithms known for first-order predicate logic, such as the Gödel completeness theorem, the Skolem theorem, the Resolution method, ... to higher-order predicate logic. But, from the theoretical point of view, it is more natural to try to express Simple Type Theory as an axiomatic theory in first-order predicate logic and *apply* these theorems and algorithms to this particular theory.

This tension between the logical and the theoretical point of view is illustrated in a 1968 discussion between J. Alan Robinson and Martin Davis [11, 5]. Robinson calls first-order predicate logic: "the restricted predicate calculus" and higher-order logic: "the full predicate calculus", while Davis, replying that it is a simple matter to express Simple Type Theory as a theory in first-order predicate logic, calls such a theory expressed in first-order predicate logic a "theory with standard formulation".

The logical point of view has dominated the history of Simple Type Theory: Henkin models, Higher-order resolution, ... have been designed specifically for Simple Type Theory and not for a class of theories formulated in first-order predicate logic, among which Simple Type Theory is one instance. It is only recently that presentations of Simple Type Theory in first-order predicate logic have shed a new light on Henkin models, as anticipated by Martin Davis [5], on higher-order unification, on proof search algorithms, on cut elimination theorems, on functional interpretations of constructive proofs, ... and on the Miller theorem.

Indeed, as shown in [7], expressing Simple Type Theory as a theory in first-order predicate logic allows to apply the Skolem theorem and this way to reconstruct the Miller theorem.

3 Simple Type Theory as a theory in first-order predicate logic

We shall express Simple Type Theory as a theory in many-sorted first-order predicate logic with equality. If a single-sorted formulation were needed, the usual relativization method could be applied.

3.1 Sorts

As, in Simple Type Theory, functions and predicates are objects, the sorts of the theory are not only the base types ι and o of Simple Type Theory but all its types: ι, o, $\iota \to \iota$, $\iota \to o$, $o \to o$, ...

As usual in many-sorted first-order predicate logic, terms are assigned a sort and function and predicate symbols are assigned a tuple of sort called an *arity* or a *rank*. If f is a function symbol of arity $\langle T_1, ..., T_n, U\rangle$ and t_1, ..., t_n are terms of sort T_1, ..., T_n, then $f(t_1, ..., t_n)$ is a term of sort U and if P is a predicate symbol of arity $\langle T_1, ..., T_n\rangle$ and t_1, ..., t_n are terms of sort T_1, ..., T_n, then $P(t_1, ..., t_n)$ is an atomic proposition.

3.2 Symbols

It is well-known that making a predicate P an object requires to introduce a copula $\in$ and write $a \in P$ what was previously written $P(a)$. In the same way making a function f an object requires to introduce an application symbol α and write $\alpha(f, a)$ what was previously written $f(a)$.

In Simple Type Theory, we need a function symbol $\alpha_{T,U}$ of arity $\langle T \to U, T, U\rangle$ for each pair of sorts T, U and a single predicate symbol ε of arity $\langle o\rangle$ to promote a term t of sort o to a proposition $\varepsilon(t)$.

For instance, if P is a term of sort $T \to o$ and t a term of sort T, then the proposition usually written $P(t)$ is not written $t \in P$, like in set theory, but $\varepsilon(\alpha_{T,o}(P, t))$ where P is first applied to t using the function symbol $\alpha_{T,o}$ to build a term of sort o, that is then promoted to a proposition using the predicate symbol ε. In the same way, the proposition usually written $\forall P\ (P \Rightarrow P)$ is written $\forall p\ (\varepsilon(p) \Rightarrow \varepsilon(p))$.

Then, we need symbols to construct terms expressing functions and predicates. Introducing the binding symbol λ is not possible in first-order predicate logic and we have to use a first-order encoding of λ-calculus. A simple solution is to use the combinators S and K. Thus, for each triple of sorts T, U, V, we introduce a constant $S_{T,U,V}$ of sort $(T \to U \to V) \to (T \to U) \to T \to V$ and for each pair of sorts T, U, we introduce a constant $K_{T,U}$ of sort $T \to U \to T$. Finally, we need similar combinators to build terms of sort o. Thus, we introduce constants $\dot{=}_T$ of sort $T \to T \to o$, $\dot{\top}$ and $\dot{\bot}$ of sorts o, $\dot{\neg}$ of sort $o \to o$, $\dot{\wedge}$, $\dot{\vee}$ and $\dot{\Rightarrow}$ of sort $o \to o \to o$, $\dot{\forall}_T$ and $\dot{\exists}_T$ of sort $(T \to o) \to o$. Of course, some of these symbols are redundant and could be defined from others using de Morgan's law. It is also possible to define all these symbols from equality $\dot{=}_T$, following the idea of Leon Henkin and Peter Andrews [8, 1].

Indices may be omitted when they can be reconstructed from the context.

3.3 Axioms

Finally, we need axioms expressing the meaning of these symbols. Besides the axioms of equality, we take the axioms

$$\forall x \forall y \forall z\ (\alpha(\alpha(\alpha(S, x), y), z) = \alpha(\alpha(x, z), \alpha(y, z)))$$

$$\forall x \forall y\ (\alpha(\alpha(K, x), y) = x)$$

$$\forall x \forall y\ (\varepsilon(\alpha(\alpha(\dot{=}, x), y)) \Leftrightarrow (x = y))$$

$$\varepsilon(\dot{\top}) \Leftrightarrow \top$$

$$\varepsilon(\dot{\bot}) \Leftrightarrow \bot$$

$$\forall x\ (\varepsilon(\alpha(\dot{\neg}, x)) \Leftrightarrow \neg\varepsilon(x))$$

$$\forall x \forall y\ (\varepsilon(\alpha(\alpha(\dot{\wedge}, x), y)) \Leftrightarrow (\varepsilon(x) \wedge \varepsilon(y)))$$

$$\forall x \forall y\ (\varepsilon(\alpha(\alpha(\dot{\vee}, x), y)) \Leftrightarrow (\varepsilon(x) \vee \varepsilon(y)))$$

$$\forall x \forall y\ (\varepsilon(\alpha(\alpha(\dot{\Rightarrow}, x), y)) \Leftrightarrow (\varepsilon(x) \Rightarrow \varepsilon(y)))$$

$$\forall x\ (\varepsilon(\alpha(\dot{\forall}, x)) \Leftrightarrow \forall y\ \varepsilon(\alpha(x, y)))$$

$$\forall x\ (\varepsilon(\alpha(\dot{\exists}, x)) \Leftrightarrow \exists y\ \varepsilon(\alpha(x, y)))$$

To these axioms, we may add, as usual, the extensionality axioms, the axiom of infinity, the description axiom and the axiom of choice.

To recall the choice we have made to express terms with combinators, we call this theory *HOL-SK*.

3.4 Properties

An easy induction on the structure of t permits to prove that for each term t there exists a term u such that the proposition

$$\alpha(u, x) = t$$

is provable. The term u is often written $\hat{\lambda} x\ t$. In the same way, an easy induction on the structure of P permits to prove that for each proposition P there exists a term u such that the proposition

$$\varepsilon(u) \Leftrightarrow P$$

is provable.

This way, we can translate all the terms of the usual formulation of Simple Type Theory with λ-calculus to terms of HOL-SK and prove that if t is a term of sort o and t' its translation, then t is provable in the usual formulation of Simple Type theory if and only if $\varepsilon(t')$ is provable in HOL-SK. However, this theorem requires that the extensionality axioms are added to both theories, because the combinators S and K do not simulate λ-calculus exactly, but only up to extensionality.

4 Skolemization

As a consequence of the Skolem theorem, we get that if $\mathcal{T}$ is a theory containing an axiom of the form

$$\forall x_1 ... \forall x_n \exists y \; A$$

where x_1, ..., x_n and y are variables of sorts T_1, ..., T_n and U, and $\mathcal{T}'$ is the theory obtained by replacing this axiom by

$$\forall x_1 ... \forall x_n \; ((f(x_1, ..., x_n)/y)A)$$

then HOL-SK $\cup \, \mathcal{T}'$ is a conservative extension of HOL-SK $\cup \, \mathcal{T}$.

The symbol f is a function symbol of arity $\langle T_1, ..., T_n, U \rangle$ and not a constant of sort $T_1 \to ... \to T_n \to U$, first because the Skolem theorem for first-order predicate logic introduces a function symbol and not a constant and then because it ignores the internal structure of sorts. Thus, the symbol f alone is not a term, and the term obtained by applying the symbol f to the terms t_1, ..., t_n is the term $f(t_1, ..., t_n)$ and not the ill-formed term $\alpha(...\alpha(f, t_1), ..., t_n)$. In contrast, when the sort U has the form $V \to W$, the term $f(t_1, ..., t_n)$ can be further applied to a term t_{n+1} using the application symbol: $\alpha(f(t_1, ..., t_n), t_{n+1})$.

In short, the Skolem symbols are not at the level of the symbols S or K, but at the level of the symbols $\alpha_{T,U}$.

We prove this way an analogous of the Miller theorem for HOL-SK. For this formulation Miller's second condition vanishes as there is no binder λ. All that remains is the first condition: the fact that Skolem symbols must be applied. Notice however that the term $\hat{\lambda} x \; t$ cannot be defined for all terms t containing the symbol f, but only those that do not have an occurrence of the variable x in an argument of the symbol f.

An advantage of expressing Simple Type Theory as a theory in first-order predicate logic is that the proof of the Miller theorem is simplified as it is then proved as a consequence of the Skolem theorem. Moreover this shows that arities are not a feature of the Skolem symbols only, but that all function symbols have arities, in particular the symbols $\alpha_{T,U}$ and the Skolem symbols.

5 Miller's second condition

We may wonder if it is possible to go one step further and reconstruct Miller's second condition as a consequence of the Skolem theorem for first-order predicate logic. As we shall see, this is possible but this requires to use a more precise first-order encoding of the λ-calculus: the λ-calculus with nameless dummies introduced by Nicolaas de Bruijn [6].

5.1 De Bruijn indices

In a λ-term, we may add, to each occurrence of a bound variable, a natural number expressing the number of abstractions separating the occurrence from its binder. For instance, adding indices to the term

$$\lambda x \lambda y\ (x\ \lambda z\ (y\ x\ z))$$

yields

$$\lambda x \lambda y\ (x^2\ \lambda z\ (y^2\ x^3\ z^1))$$

Indeed, the index of the occurrence of the variable z is 1 because the binder λz is just one level above in the term, while the index of the second occurrence of the variable x is 3 as the the binder λx is three levels above. Once the indices are added this way, the names of the bound variables are immaterial and can be dropped. We thus get the term

$$\lambda\lambda\ (\_^2\ \lambda\ (\_^2\ \_^3\ \_^1))$$

Notice that the term $\lambda x \lambda y\ (x\ \lambda z\ (y\ x\ z))$ is closed but that its subterm $\lambda y\ (x\ \lambda z\ (y\ x\ z))$ contains a free variable x. In the same way, in the term $\lambda\lambda\ (\_^2\ \lambda\ (\_^2\ \_^3\ \_^1))$ all the de Bruijn indices refer to a binder, but its subterm $\lambda\ (\_^2\ \lambda\ (\_^2\ \_^3\ \_^1))$ contain two "free indices", that exceeded the number of binders above them, and correspond to the former occurrences of the variable x.

The λ-calculus with de Bruijn indices is a first-order language as, once variables names have been dropped, the symbol λ is not a binder anymore. This language is formed with an infinite number of constants $\_^1, \_^2, \_^3, ...$, a binary function symbol α for application and a unary function symbol λ. Closed terms, such as $\lambda\ (\_^2\ \lambda\ (\_^2\ \_^3\ \_^1))$ or $\_^1$, may contain free indices, exceeding the number of binders above them.

5.2 Sorts

When type-checking an open λ-term *à la* Church, *i.e.* with explicitly typed bound variables, for instance

$$\lambda y_{((\iota \to \iota) \to \iota) \to \iota \to \iota} \ (x \ \lambda z_\iota \ (y \ x \ z))$$

it is necessary to have a context defining the type of the free variables such as x. Indeed, if x is assigned the type $(\iota \to \iota) \to \iota$, then the term is well-typed, but not if it is assigned, for instance, the type ι.

In the same way, when typing a term that may contains free de Bruijn indices, even if this term is closed (*i.e.* does not contain named variables), we need a context defining the types of the de Bruijn indices exceeding the number of binders above them. This context is a finite list of types: the type of the indices exceeding by 1 the number of binders above them, that of the indices exceeding by 2 the number of binders above them, ... For instance the term

$$\lambda_{((\iota \to \iota) \to \iota) \to \iota \to \iota} \ (\_^2 \ \lambda_\iota \ (\_^2 \ \_^3 \ \_^1))$$

has type $(((\iota \to \iota) \to \iota) \to \iota \to \iota) \to \iota$ in the context $[(\iota \to \iota) \to \iota]$.

Thus, even if it is closed, a term t can be assigned a type T, only relatively to a context Γ. In other words, it can be absolutely assigned an ordered pair formed with a context Γ and a type T. We write such a pair $\Gamma \vdash T$.

When we express Simple Type Theory as a first-order theory, using the λ-calculus with de Bruijn indices as a first-order encoding of the λ-calculus, the sort are not just the simple types, like in HOL-SK, but such pairs $\Gamma \vdash T$ formed with a list Γ of simple types and a simple type T.

In this formulation, the quantified variables, in contrast to the λ-abstracted ones, are not replaced by de Bruijn indices but, as in all theories expressed in first-order predicate logic, they are kept as standard named variables and they are assigned a sort of the form $\vdash T$, where the context is the empty list.

The last step of the construction of this theory would be to introduce explicit substitutions, hence its name *HOL-$\lambda\sigma$*. I do not want to go into these details here and the interested reader can refer to [7]. But, I want to insist on two points. First, the fact that this theory is *intentionally* equivalent to the usual presentation of Simple Type Theory with λ-calculus, *i.e.* even if the extensionality axioms are not assumed. Second, that, as explained above, the sorts are pairs $\Gamma \vdash T$, *i.e.* that they contain scoping information. In particular terms of a sort $\vdash T$, with an empty context, are de Bruijn-closed, *i.e.* they do not contain indices that may be bound higher in the term by a λ-abstraction.

5.3 Skolemization

As a consequence of the Skolem theorem, we get that if $\mathcal{T}$ is a theory containing an axiom of the form

$$\forall x_1 ... \forall x_n \exists y\ A$$

where x_1, ..., x_n and y are variables of sort $\vdash T_1$, ..., $\vdash T_n$ and $\vdash U$, and $\mathcal{T}'$ is the theory obtained by replacing this axiom by

$$\forall x_1 ... \forall x_n\ ((f(x_1, ..., x_n)/y)A)$$

then HOL-$\lambda\sigma \cup \mathcal{T}'$ is a conservative extension of HOL-$\lambda\sigma \cup \mathcal{T}$. The symbol f is a function symbol of arity $\langle \vdash T_1, ..., \vdash T_n, \vdash U \rangle$. Thus, not only f is a function symbol, but a function symbol whose arguments must not contain indices that may be bound by a λ-abstraction higher in a term. We get this way exactly Miller's conditions. The first is rephrased as the fact that f is a function symbol and the second as the fact that the arguments of these function symbol must have a sort with an empty context.

We obtain this way an alternative proof of the Miller theorem for HOL-$\lambda\sigma$, as consequence of the Skolem theorem.

A difficult point in the Miller theorem is the discrepancy between the general formation rules for the terms and the propositions and the more restricted ones for the terms substituted for variables in quantifier rules. As we have seen, the variables free in the necessary arguments of the Skolem symbols cannot be bound in a term substituted for a variable. But this restriction does not apply to the terms and propositions in general, because, if it did, the skolemized axiom itself would not be well-formed, as the arguments of the Skolem symbol in this skolemized axiom are universally bound variables.

In HOL-$\lambda\sigma$ the situation is slightly different. The formation rules for the terms substituted for variables in quantifier rules are the same as those of the other terms. But the λ-bound variables and the quantified variables are treated differently. As we have seen, the λ-bound variables are replaced by de Bruijn indices, but the quantified variables are kept as named variables, as in all theories expressed in first-order predicate logic. Thus the λ-bound variables cannot appear in the arguments of a Skolem symbol but the quantified variables can and, this way, the skolemized axiom is well-formed.

6 From the theoretical point of view

With this example of skolemization, we have seen two advantages of the theoretical point of view on Simple Type Theory. First the proofs of the theorems are simplified, because we can take advantage of having already proved similar theorems for first-order predicate logic. Then some difficult points of these theorems are explained.

We can mention several other theorems and algorithms that have been simplified and explained this way. The Henkin completeness theorem, already mentioned in 1968 by Martin Davis, independence results, in particular of the extensionality axioms, cut elimination theorems, both model-based ones and reduction-based ones, proof search algorithms and the functional interpretation of constructive proofs.

This succession of points of view on Simple Type Theory is itself an example of a common back and forth movement in the development of science: first, new objects and new results are discovered, breaking with the old framework, they are supposed not to fit in. Then, after a possible evolution of the general framework, these new objects are integrated back. We may try to understand what kind of evolution of first-order predicate logic, the integration of Simple Type Theory requires or suggests.

The first evolution is the shift from the single-sorted first-order predicate logic to the many-sorted one. Although it is always possible to relativize a many-sorted theory to a single-sorted one, the natural framework to express Simple Type Theory is many-sorted first-order predicate logic. It is interesting to see that early papers on many-sorted first-order predicate logic, *e.g.* [12], already motivate the introduction of many-sorted first-order predicate logic by the will to express Simple Type Theory in it.

A second evolution is to take into account the possibility to consider terms and propositions up to reduction. The axioms of HOL-SK (and those of HOL-$\lambda\sigma$) can easily be transformed into rewrite rules:

$$\alpha(\alpha(\alpha(S,x),y),z) \longrightarrow \alpha(\alpha(x,z),\alpha(y,z))$$

$$\alpha(\alpha(K,x),y) \longrightarrow x$$

$$\varepsilon(\alpha(\alpha(\dot{=},x),y)) \longrightarrow x = y$$

$$\varepsilon(\dot{\top}) \longrightarrow \top$$

$$\varepsilon(\dot{\bot}) \longrightarrow \bot$$

$$\varepsilon(\alpha(\dot{\neg},x)) \longrightarrow \neg\varepsilon(x)$$

etc. and identifying equivalent propositions is more natural than keeping these axioms as such, exactly like in the usual formulation of Simple Type Theory, β-normalizing terms and propositions is more natural than keeping β-conversion as an axiom. This idea has lead to the development of *Deduction modulo*, that again was initially motivated by the will to express Simple Type Theory in it.

A third evolution is the introduction of binders in first-order predicate logic. Although this has been a hot topic recently, it is fair to say that we have not yet a completely satisfactory extension of first-order predicate logic with such binders.

To conclude, I want to mention the influence of Peter Andrews' work, course and book [4], on the emergence of the theoretical point of view on Simple Type Theory (although I do not attribute any point of view to anyone except myself). First, it is striking that in his course, Peter Andrews compared Simple Type Theory not only to first-order predicate logic but also to set theory in particular that he insisted on the fact that both theories are restriction of the naive, inconsistent, set theory with full comprehension. More technically, the alternative characterization of Henkin models, given by Peter Andrews [3] showed the way out of Henkin mysterious conditions that all λ-terms must have a denotation to a standard condition that the axioms S and K must be valid and thus to the idea that Henkin models were just models of some theory expressed in first-order predicate logic.

7 Acknowledgements

The author wants to thank Claude Kirchner and Dale Miller for helpful comments on a previous version of this paper.

BIBLIOGRAPHY

[1] Peter B. Andrews. A Reduction of the Axioms of the Theory of Propositional Types. *Fund. Math.*, 52:345–350, 1963.

[2] Peter B. Andrews. Resolution in Type Theory. *The Journal of Symbolic Logic*, 36(3):414–432, 1971.

[3] Peter B. Andrews. General Models, Descriptions, and Choice in Type Theory. *The Journal of Symbolic Logic*, 37(2):385–394, 1972.

[4] Peter B. Andrews. *An Introduction to Mathematical Logic and Type Theory: to Truth through Proofs*. Academic Press, 1986.

[5] Martin Davis. Invited Commentray to [11]. In *Information Processing 68, Proceedings of the IFIP Congress 1968*, pages 67–68. North-Holland, 1969.

[6] Nicolaas G. de Bruijn. Lambda Calculus Notation with Nameless Dummies, a Tool for Automatic Formula Manipulation, with Application to the Church-Rosser Theorem. *Indagationes Mathematicae*, 34(5):381–392, 1972.

[7] Gilles Dowek, Thérèse Hardin, and Claude Kirchner. HOL-lambda-sigma: an Intentional First-order Expression of Higher-order Logic. *Mathematical Structures in Computer Science*, 11:1–25, 2001.

[8] Leon Henkin. A theory of propositional types. *Fund. Math.*, 52:323–344, 1963.

[9] Dale Miller. *Proofs in Higher-order logic*. PhD thesis, Carnegie Mellon University, 1983.

[10] Dale A. Miller. A Compact Representation of Proofs. *Studia Logica*, 46(4):347–370, 1987.

[11] J. Alan Robinson. New Directions In Mechanical Theorem Proving. In *Information Processing 68, Proceedings of the IFIP Congress 1968*, pages 63–67. North-Holland, 1969.

[12] Hao Wang. Logic of Many-sorted Theories. *The Journal of Symbolic Logic*, 17(2):105–116, 1952.

The Lambda-Calculus is Nominal Algebraic

MURDOCH J. GABBAY[1] AND AAD MATHIJSSEN[2,3]

1 Introduction

The λ-calculus is fundamental in the study of logic and computation. Partly this is because it is a tool to study functions and functions are an important object of study in this field. Partly this is because the λ-calculus seems to be, for *homo sapiens*, an ergonomic formal syntax.

DEFINITION 1. λ-terms g, h, k are inductively defined by

$$g \quad ::= \quad a \mid \lambda a.g \mid gg \mid \mathsf{c}.$$

In this paper we will write -$[a \mapsto$ -$]$ as shorthand for $(\lambda a.\text{-})\text{-}$. Thus $g[a \mapsto h]$ stands for $(\lambda a.g)h$ and *not* for the term resulting from 'substituting h for a in g' (we write that as $g[h/a]$, see Definition 44).

The λ-calculus represents functions in programming languages [25, 31], logic [7, 20], theorem-provers [3, 24], higher-order rewriting [5], and much more besides. However, the 'λ' in the λ-calculus has proved resistent to a treatment in universal algebra [8]. For example the property that

$$\text{“}(\lambda a.g)[b \mapsto h] = \lambda a.(g[b \mapsto h]) \text{ when } a \text{ does not occur free in } h\text{”}$$

cannot be represented in an algebraic framework, at least not obviously so, because of the *freshness condition* 'a does not occur free in h' which is necessary to avoid 'accidental capture' by λ. Similarly for the property "$\lambda a.(ga) = g$ when a does not occur free in g".

Nominal algebra is a form of universal algebra enriched with primitive constructs to handle names, binding, and freshness conditions — just like those that appear in informal specifications of the λ-calculus and other

[1]Homepage: http://www.gabbay.org.uk

[2]Email: a.h.j.mathijssen@tue.nl

[3]We are grateful to a dilligent anonymous referee of a previous paper for making a suggestion which put us on the path to write this paper, and to Pablo Nogueira and Chad Brown for useful advice on improving the exposition.

$$
\begin{array}{rrrcl}
(\mathbf{var}{\mapsto}) & \vdash & a[a \mapsto X] & = & X \\
(\#{\mapsto}) & a\#Z \vdash & Z[a \mapsto X] & = & Z \\
(\mathbf{app}{\mapsto}) & \vdash & (Z'Z)[a \mapsto X] & = & (Z'[a \mapsto X])(Z[a \mapsto X]) \\
(\mathbf{abs}{\mapsto}) & b\#X \vdash & (\lambda b.Z)[a \mapsto X] & = & \lambda b.(Z[a \mapsto X]) \\
(\mathbf{id}{\mapsto}) & \vdash & Z[a \mapsto a] & = & Z
\end{array}
$$

Figure 1. Axioms of ULAM

languages with binders. Nominal algebra has the feature that, thanks to the enriched constructs, it allows fully formal algebraic reasoning which is pleasingly close to informal practice, including explicit reasoning on α-renaming and freshness side-conditions.

In this paper we introduce ULAM (Figure 1), a nominal algebra theory for the untyped λ-calculus. The axioms of ULAM make fundamental use of characteristic 'nominal' features of nominal algebra:

- We use *nominal unknowns* Z, Z', and X to represent unknown elements. Instantiation of nominal unknowns does not avoid capture; see Definition 11.
- *Freshness conditions* $a\#Z$, $b\#X$, and $b\#Z$ are a framework to prevent 'accidental capture' of names by binders.

We shall prove that ULAM is sound and complete with respect to a model constructed out of λ-terms quotiented by $\alpha\beta$-equivalence; the rest of the paper makes these observations formal.

Nominal techniques subscribe to a mathematical view according to which names are first-class entities in the denotation. This was used, for example, to develop the Gabbay-Pitts И quantifier and the Gabbay-Pitts model of α-abstraction [17]. A traditional view is that names arise as a syntax for talking about inputs to functions, and therefore they range over elements of the underlying domain.[1] The λ-calculus expresses this latter idea. With ULAM our nominal algebraic axiomatisation of the λ-calculus we make a novel connection between the two worlds; the axioms of ULAM express the properties that must be added to convert a nominal-style atom into a λ-calculus style variable, and a nominal-style abstraction into a λ-calculus binding.

[1] In [17], names had no functional content at all; they were used just to build datatypes of abstract syntax trees with binding. Higher-order abstract syntax [26] is a way to do the same thing using the 'names as arguments to functions' philosophy.

Map of the paper. Section 2 introduces nominal algebra, giving basic definitions and results about syntax, freshness conditions, equality, and nominal algebra theories. Section 3 introduces the syntax and operational semantics of the untyped λ-calculus. Section 4 proves that ULAM is sound and complete (Subsections 4.1 and 4.2), and that it is conservative over the native nominal terms theory of α-equivalence (Subsection 4.3). The most technical material in this paper is concentrated in the proofs in Subsection 4.2. Finally, Section 5 discusses related and future work.

2 Nominal algebra

2.1 The syntax of nominal terms

We define a syntax of nominal terms. It is tailored to our application to the untyped λ-calculus; see elsewhere for general treatments [32, 13, 14, 23].

DEFINITION 2. Fix the following:

- A countably infinite set of **atoms** $\mathbb{A}$. We let $a, b, c, \ldots$ range over atoms. These model λ-calculus variables.

  We use a **permutative convention** that a and b range *permutatively* over atoms unless stated otherwise. For example in (**#ab**) from Figure 3, and in (**perm**) from Figure 4, a and b range over any two *distinct* atoms.

- A countably infinite collection of **unknowns**. We let $X, Y, Z, T, U, \ldots$ range over unknowns. These represent unknown elements in nominal algebra axioms.

  We also use a permutative convention that X and Y range permutatively over unknowns. In Figure 1 we make a fixed but arbitrary choice of unknown. That is, ULAM contains five axioms — not infinitely many for every possible a, b, X, Z, and Z' — but the choice is immaterial for all practical purposes as we shall see in (**ax**) of Figure 4 and in Definition 25.

- A possibly infinite collection of **constant symbols** $\mathsf{c} \in \mathsf{C}$.

Unknowns, atoms, and other distinct syntactic classes, are assumed disjoint.

REMARK 3. For the reader's convenience we provide Figure 2: a 'cheat-sheet' linking in a single list concepts from informal practice to some of the main definitions and lemmas which will soon follow. This list, by its nature, contains forward references.

We can now set about building the machinery of nominal algebra.

- a is an atom. It represents an object-level variable symbol.
- X is an unknown. It represents a meta-variable.
- $\pi \cdot t$ is a permutative renaming of the atoms in t. We use this to provide a 'naturally capture-avoiding' theory of α-equivalence. $\pi \cdot X$ has the intuition of 'permute π in whatever X is instantiated to'.
- $t\sigma$ is t with meta-variables substituted, we can think of this as *instantiation.* If $\pi \cdot X$ appears in t then π acts on $\sigma(X)$ and the result is included in $t\sigma$.
- $a\#t$ asserts a freshness. It has the intuition 'a cannot be free in t'. $a\#X$ has the intuition of 'a is fresh for whatever X is instantiated to'.

Figure 2. Cheat-sheet for intuitive reading of notation

DEFINITION 4. A **permutation** π of atoms is a bijection on atoms with **finite support** meaning that for some finite set of atoms $\pi(a) \not\equiv a$, and for all other atoms $\pi(a) = a$. In words: For 'most' atoms π is the identity.

DEFINITION 5. Let **terms** t, u, v be inductively defined by:

$$t \quad ::= \quad a \mid \pi \cdot X \mid \lambda a.t \mid tt \mid \mathsf{c}.$$

We write **syntactic identity** of terms t, u as $t \equiv u$ to distinguish it from '=' the derivable equality-in-freshness-context we construct in Subsection 2.3. Note that if $\pi = \pi'$ then $\pi \cdot X \equiv \pi' \cdot X$, since permutations are represented by themselves. Also note that we do *not* quotient terms in any way.

We give some intuition of terms:

- An atom a represents a λ-calculus variable symbol.
- We call $\pi \cdot X$ a **moderated unknown**. This represents an unknown term, on which a permutation of atoms will be performed when X is instantiated. The use of permutations provides primitive support for *α-equivalence.*
- $t't$ represents the usual λ-calculus application.
- $\lambda a.t$ represents the usual λ-abstraction. Recall from the Introduction that we write $t[a \mapsto u]$ as shorthand for $(\lambda a.t)u$, for example in Figure 1.

A typed version of this syntax is possible; the interaction between atoms, unknowns, permutations, and λ-abstraction raises subtle and unexpected issues which have been investigated independently in the general framework of nominal rewriting [12]. Types would cause no essential difficulties for the results which follow.

2.2 Permutation, substitution and freshness

We need some more notation to talk about permutations (Definition 4).

DEFINITION 6. As usual we write id for the **identity** permutation, π^{-1} for the **inverse** of π, and $\pi \circ \pi'$ for the **composition** of π and π', i.e. $(\pi \circ \pi')(a) = \pi(\pi'(a))$. id is also the identity of composition: $id \circ \pi = \pi$ and $\pi \circ id = \pi$. Importantly, we shall write $(a\ b)$ for the permutation that **swaps** a and b, i.e. the permutation that maps a to b and vice versa, and maps all other c to themselves; note that this is the same permutation as $(b\ a)$. We may drop $\circ$ between swappings, writing for example $(a\ b) \circ (b\ c)$ as $(a\ b)(b\ c)$; this is a standard notation in the theory of permutations. We may write X as shorthand for $id \cdot X$.

DEFINITION 7. We write $a \in \pi$ when $\pi(a) \neq a$. We extend this inductively to $a \in t$ as follows:

$$\frac{}{a \in a} \qquad \frac{a \in \pi}{a \in \pi \cdot X} \qquad \frac{a \in t'}{a \in t't} \qquad \frac{a \in t}{a \in t't} \qquad \frac{}{a \in \lambda a.t} \qquad \frac{a \in t}{a \in \lambda b.t}$$

If $a \in t$ is not derivable write $a \notin t$. We read '$a \in$' as 'a occurs in'.

We also write $X \in t$ when X occurs anywhere in t, and $X \notin t$ otherwise. Occurrence is literal, for example $a \in \lambda a.a$ and $a \in (a\ b) \cdot X$.

DEFINITION 8. Define a **permutation action** $\pi \cdot t$ inductively by:

$$\pi \cdot a \equiv \pi(a) \qquad \pi \cdot (\pi' \cdot X) \equiv (\pi \circ \pi') \cdot X \qquad \pi \cdot \lambda a.t \equiv \lambda(\pi(a)).(\pi \cdot t)$$
$$\pi \cdot (t't) \equiv (\pi \cdot t')(\pi \cdot t) \qquad \pi \cdot \mathsf{c} \equiv \mathsf{c}.$$

Intuitively π propagates through the structure of t until it reaches an atom or a moderated unknown. Note that in the clause for λ, π acts also on the 'a'. Following Cheney (verbal communication) we say that the permutation action is *inherently capture-avoiding*, meaning that it can act uniformly and will not require special behaviour for binders. For example:

$$(a\ b) \cdot \lambda a.X \equiv \lambda b.(a\ b) \cdot X$$

LEMMA 9. $\pi \cdot (\pi' \cdot t) \equiv (\pi \circ \pi') \cdot t$ *and* $id \cdot t \equiv t$.

Proof. By an easy induction on the structure of t using the fact that composition of permutations is just composition of functions and so is associative. ∎

DEFINITION 10. We call a **substitution** σ a function from unknowns to terms. We write $[t_1/X_1, \ldots, t_n/X_n]$ for the substitution mapping X_i to t_i for $1 \leq i \leq n$, and mapping Y to Y for all other Y.

DEFINITION 11. Define a **substitution action** $t\sigma$ inductively by:

$$a\sigma \equiv a \qquad (\pi \cdot X)\sigma \equiv \pi \cdot \sigma(X) \qquad (\lambda a.t)\sigma \equiv \lambda a.(t\sigma)$$
$$(t't)\sigma \equiv (t'\sigma)(t\sigma) \qquad \mathsf{c}\sigma \equiv \mathsf{c}.$$

We may call $t\sigma$ an **instance** of t.

For example:

$$(\lambda a.X)[a/X] \equiv \lambda a.(X[a/X]) \equiv \lambda a.a$$

$$\begin{aligned}(\lambda b.(a\ b) \cdot X)[a/X] &\equiv \lambda b.(((a\ b) \cdot X)[a/X]) \\ &\equiv \lambda b.((a\ b) \cdot (X[a/X])) \equiv \lambda b.(a\ b) \cdot a \equiv \lambda b.b\end{aligned}$$

$$X\sigma \equiv \sigma(X)$$

The final example is direct from the definition and Lemma 9; $\sigma(X)$ is 'the function σ applied to X', whereas $X\sigma$ is shorthand for $(id \cdot X)\sigma$.

Intuitively, σ propagates through the structure of t until it reaches an atom or a moderated unknown. σ 'evaporates' on an atom a, and on $\pi \cdot X$ it instantiates X to $\sigma(X)$ — then the permutation acts on $\sigma(X)$.

Substitution does not avoid capture; $(\lambda a.X)[a/X] \equiv \lambda a.a$. There is an deliberate analogy here with context substitution, which is the substitution used when we write 'let - be a in $\lambda a.$-', or 'suppose t is a in $\lambda a.t$'; we obtain $\lambda a.a$. Moderated unknowns behave exactly like the 'hole' - in syntactic contexts — except that an unknown can occur multiple times in a nominal term; unknowns occur in terms with a moderating permutation because this is needed, along with freshness contexts, to manage α-equivalence.

Recall from Definition 11 that $(\pi \cdot X)\sigma \equiv \pi \cdot \sigma(X)$. This extends easily to all terms:

LEMMA 12. $(\pi \cdot t)\sigma \equiv \pi \cdot (t\sigma)$.

Proof. By induction on t. The only interesting case if when $t \equiv \pi' \cdot X$. Then unpacking definitions

$$(\pi \cdot (\pi' \cdot X))\sigma \equiv (\pi \circ \pi') \cdot \sigma(X) \quad \text{and} \quad \pi \cdot ((\pi' \cdot X)\sigma) \equiv \pi \cdot (\pi' \cdot \sigma(X)).$$

$$\frac{}{a\#b}(\#\mathbf{ab}) \qquad \frac{\pi^{-1}(a)\#X}{a\#\pi\cdot X}(\#\mathbf{X}) \qquad \frac{}{a\#\lambda a.t}(\#\lambda\mathbf{a}) \qquad \frac{a\#t}{a\#\lambda b.t}(\#\lambda\mathbf{b})$$

$$\frac{a\#t' \; a\#t}{a\#t't}(\#\mathbf{app}) \qquad \frac{}{a\#\mathsf{c}}(\#\mathsf{c})$$

Figure 3. Freshness derivation rules for nominal terms

The result follows by Lemma 9. ∎

DEFINITION 13. A **freshness (assertion)** is a pair $a\#t$ of an atom and a term. We call a freshness of the form $a\#X$ (so $t \equiv X$) **primitive**. We write Δ and ∇ for (finite, and possibly empty) sets of *primitive* freshnesses and call them **freshness contexts**.

We may drop set brackets in freshness contexts, e.g. writing $a\#X, b\#Y$ for $\{a\#X, b\#Y\}$. Also, we may write $a, b\#X$ for $a\#X, b\#X$. We write $a \in \Delta$ when $a\#X \in \Delta$ for some X, and $X \in \Delta$ when $a\#X \in \Delta$ for some a.

DEFINITION 14. We define **derivability on freshnesses** in natural deduction style [19] by the rules in Figure 3. In accordance with our permutative convention, a and b range over distinct atoms.

REMARK 15. A sequent style presentation of Figure 3 is also possible; for example (#**X**) becomes $\frac{\Delta \vdash \pi^{-1}(a)\#X}{\Delta \vdash a\#\pi\cdot X}$ and we need a new rule $\frac{[a\#X \in \Delta]}{\Delta \vdash a\#X}$ where the 'X' in the bottom line is shorthand for $id{\cdot}X$ and square brackets indicate a side-condition.

DEFINITION 16. We write $\Delta \vdash a\#t$ when a derivation of $a\#t$ exists using the rules of Figure 3, such that the assumptions are elements of Δ. In words, we say "$a\#t$ is **derivable** from Δ". We usually write $\emptyset \vdash a\#t$ as $\vdash a\#t$.

For example $\vdash a\#\lambda b.b$, $\vdash a\#\lambda a.a$, and $a\#X \vdash a\#X(\lambda a.Y)$:

$$\frac{\frac{}{a\#b}(\#\mathbf{ab})}{a\#\lambda b.b}(\#\lambda\mathbf{b}) \qquad \frac{}{a\#\lambda a.a}(\#\lambda\mathbf{a}) \qquad \frac{a\#X \quad \frac{}{a\#\lambda a.Y}(\#\lambda\mathbf{a})}{a\#X(\lambda a.Y)}(\#\mathbf{app})$$

The example of $\vdash a\#\lambda a.a$ demonstrates that $a\#t$, whose intuitive reading is 'is not free in', differs from $a \notin t$, whose intuitive reading is 'does not occur in'.

The derivation rules are completely syntax-directed. Therefore:

LEMMA 17.

1. $\Delta \vdash a\#b$ *always.*
2. *If* $\Delta \vdash a\#X$ *then* $a\#X \in \Delta$.
3. *If* $\Delta \vdash a\#\pi \cdot X$ *then* $\Delta \vdash \pi^{-1}(a)\#X$.
4. $\Delta \vdash a\#\lambda a.t$ *always.*
5. *If* $\Delta \vdash a\#\lambda b.t$ *then* $\Delta \vdash a\#t$.
6. *If* $\Delta \vdash a\#t't$ *then* $\Delta \vdash a\#t'$ *and* $\Delta \vdash a\#t$.
7. $\Delta \vdash a\#\mathsf{c}$ *always.*

Proof. By an easy induction on the structure of the derivation rules in Figure 3. ■

We can *strengthen* the freshness context Δ in freshness derivations:

THEOREM 18. *If* $c\#Z, \Delta \vdash a\#t$ *and* $c \notin t$ *then* $\Delta \vdash a\#t$.

Proof. We transform a derivation of $c\#Z, \Delta \vdash a\#t$ into a derivation of $\Delta \vdash a\#t$:

- If $c\#Z, \Delta \vdash a\#X$ by assumption then $a\#X \in c\#Z, \Delta$. Since $a \neq c$, we know $a\#X \in \Delta$, and we conclude $\Delta \vdash a\#X$ by assumption.

  The case of $a\#Z$ is similar.
- (**#X**): Suppose $c\#Z, \Delta \vdash a\#\pi \cdot X$ is derived using (**#X**). Then $c\#Z, \Delta \vdash \pi^{-1}(a)\#X$. By assumption $c \notin \pi \cdot X$, so $\pi(c) = c$. Since also $a \neq c$, we know $\pi^{-1}(a) \neq c$. By the inductive hypothesis and the simple fact that $c \notin X$ we obtain $\Delta \vdash \pi^{-1}(a)\#X$. We conclude $\Delta \vdash a\#\pi \cdot X$ using (**#X**).

  The case of $a\#\pi \cdot Z$ is similar.
- (**#ab**), (**#λa**) and (**#c**) carry over directly.
- (**#λb**) and (**#app**) are straightforward using the inductive hypothesis and the following facts: if $a \notin \lambda b.t$ then $a \notin t$, and if $a \notin t't$ then $a \notin t'$ and $a \notin t$.

■

Freshness context *weakening* also holds and is part of a more general result; see Theorem 21.

Equivariance is a characteristic property of nominal techniques. Equivariance arises from the use nominal techniques make of permutations as opposed to renamings (possibly non-bijective functions on atoms). Intuitively, equivariance states that if something is true, then it should remain true if we permute atoms. Formally we write:

LEMMA 19. *If* $\Delta \vdash a\#t$ *then* $\Delta \vdash \pi(a)\#\pi \cdot t$.

Proof. By induction on the structure of derivations of $a\#t$ from Δ. The only non-trivial case is (#**X**). Suppose $a\#\pi' \cdot X$ is derived from $\pi'^{-1}(a)\#X$ using (#**X**). We must show that $\pi(a)\#(\pi \circ \pi') \cdot X$. This follows from $(\pi \circ \pi')^{-1}(\pi(a))\#X$. It is a fact that $(\pi \circ \pi')^{-1}(\pi(a)) = \pi'^{-1}(a)$. The result follows. ■

Lemma 19 is part of a collection of equivariance properties and these are responsible for much of the technical convenience of the nominal treatment of names. See for example [32, Lemma 2.7], [13, Lemma 20], [16, Appendix A], and [17, Lemma 4.7].

We can extend the substitution action to freshness contexts:

DEFINITION 20. We write

$$\Delta\sigma \quad \text{for} \quad \{a\#\sigma(X) \mid a\#X \in \Delta\}.$$

Intuitively read this as 'apply σ to every X in Δ'.

We write $\Delta' \vdash \Delta\sigma$ when $\Delta' \vdash a\#\sigma(X)$ for every $a\#X \in \Delta$.

Note that $\Delta\sigma$ need not be a freshness context, because it might contain $a\#t$ for t not an unknown.

THEOREM 21. *For any* Δ', Δ, σ, *if* $\Delta \vdash a\#t$ *and* $\Delta' \vdash \Delta\sigma$ *then* $\Delta' \vdash a\#t\sigma$.

As a corollary, if $\Delta \vdash a\#t$ *and* $\Delta \subseteq \Delta'$ *then* $\Delta' \vdash a\#t$.

Proof. The structure of natural deduction derivations is such that the conclusion of one derivation may be 'plugged in' to an assumption in another derivation, if assumption and conclusion are syntactically identical. The structure of all the rules except for (#**X**) is such that if unknowns are instantiated by σ nothing need change.

For the case of (#**X**) we use Lemma 19.

The corollary follows taking $\sigma(X) \equiv id \cdot X$ for all X. ■

2.3 Equality, axioms and theories

DEFINITION 22. We call a pair $t = u$ an **equality (assertion)**. We call the pair $\nabla \vdash t = u$ of a freshness context ∇ and an equality $t = u$ an **axiom**. We may write $\emptyset \vdash t = u$ as $\vdash t = u$.

We call a set of axioms T a **theory**. The theories needed in this paper are:

- CORE: the empty set of axioms.
- ULAM: the axioms from Figure 1.

REMARK 23. Theory ULAM axiomatises a non-extensional λ-calculus. An extensional version is obtained if we add the following axiom to Figure 1:

$$(\eta) \qquad a\#Z \vdash \lambda a.(Za) = Z$$

We see no difficulties with extending the results of this paper to the extensional case.

REMARK 24. A word on the history of the ideas behind ULAM: ULAM grew out of SUB [15], which was based on a nominal rewrite system for the λ-calculus [13, Example 43], which was itself based on an example signature used in nominal unification [32, Example 2.2]. The axioms of ULAM directly and deliberately identify nominal abstraction $[a]t$ (notation in the style of [32, 15]) with λ-abstraction $\lambda a.t$, and nominal substitution $\mathsf{sub}(t, u)$ with λ-calculus application tu. In [32, 13, 15] term-formers such as λ-abstraction and application exist separately and a sort-system ensures for example that the t in $\mathsf{sub}(t, u)$ is an abstraction:

- λ is a unary term-former. It takes a nominal abstraction as an argument and returns a term of base sort. In the style of [15, 32] we write that it has arity $([\mathbb{A}]\mathbb{T})\mathbb{T}$.
- Substitution is a binary term-former sub with arity $([\mathbb{A}]\mathbb{T}, \mathbb{T})\mathbb{T}$.

The direct identification of nominal abstraction with λ-abstraction, and substitution with application, is possible because terms of ULAM are unsorted (i.e. all terms have base sort, including abstractions) and because nominal algebra allows us to assert the relevant equalities in a logical framework.

DEFINITION 25. Define **derivability on equalities** in natural deduction style by the rules in Figure 4. We write $\Delta \vdash_{\mathsf{T}} t = u$ when a derivation of $t = u$ exists using these rules such that:

- for each instance of $(\mathbf{ax}_{\nabla\vdash\mathbf{t=u}})$, $\nabla \vdash t = u$ is an axiom from T.
- in the derivations of freshnesses (introduced by instances of $(\mathbf{ax}_{\nabla\vdash\mathbf{t=u}})$ and $(\mathbf{perm})$) the freshness assumptions used are from Δ only.

We write $\emptyset \vdash_{\mathsf{T}} t = u$ as $\vdash_{\mathsf{T}} t = u$.

We briefly discuss the most interesting rules of Figure 4:

- $(\mathbf{perm})$. This rule provides us with a concise way to express α-equivalence. See Lemma 28 and Theorem 49 for formal expressions of this intuition.

$$\frac{}{t = t}\ (\mathbf{refl}) \qquad \frac{t = u}{u = t}\ (\mathbf{symm}) \qquad \frac{t = u \quad u = v}{t = v}\ (\mathbf{tran})$$

$$\frac{t = u}{\lambda a.t = \lambda a.u}\ (\mathbf{cong}\lambda) \qquad \frac{t' = u' \quad t = u}{t't = u'u}\ (\mathbf{congapp})$$

$$\frac{a\#t \quad b\#t}{(a\ b) \cdot t = t}\ (\mathbf{perm}) \qquad \frac{\nabla\sigma}{\pi \cdot t\sigma = \pi \cdot u\sigma}\ (\mathbf{ax}_{\nabla \vdash \mathbf{t=u}})$$

$$\begin{array}{c} [a\#X] \quad \Delta \\ \vdots \\ \dfrac{t = u}{t = u}\ (\mathbf{fr}) \quad (a \notin t, u) \end{array}$$

Figure 4. Derivation rules for nominal equality

- $(\mathbf{ax}_{\nabla \vdash \mathbf{t=u}})$. This rule formally expresses how we obtain instances of axioms: instantiate unknowns by terms (using substitutions) and rename atoms (using permutations) such that the hypotheses are corresponding instances of freshness conditions of the axiom. In this rule σ ranges over substitutions, π ranges over permutations, and $\nabla\sigma$ stands for the hypotheses $\{a\#\sigma(X) \mid a\#X \in \nabla\}$.

  The reader might expect the premise of the axiom rule to be $\pi \cdot \nabla\sigma$ instead of $\nabla\sigma$. It turns out that both versions are correct, because of Lemma 19: $\Delta \vdash \nabla\sigma$ if and only if $\Delta \vdash \pi \cdot \nabla\sigma$ for any Δ.

- $(\mathbf{fr})$. This rule allows us to introduce a fresh atom into the derivation. In $(\mathbf{fr})$ the square brackets denote *discharge* in the sense of natural deduction, for example as in implication introduction [19]; Δ denotes the other assumptions of the derivation of $t = u$. In sequent style $(\mathbf{fr})$ would be $\frac{a\#X, \Delta \vdash t = u}{\Delta \vdash t = u}\ (a \notin t, u)$.

  We will *always* be able to find a fresh atom, no matter how unknowns are instantiated, since all our syntax is finite and can never mention more than finitely many atoms.

To provide some intuition for these rules, we give a number of examples.

EXAMPLE 26. The $(\mathbf{perm})$ rule allows us to show standard α-equivalence properties such as $\vdash_{\mathsf{CORE}} \lambda a.a = \lambda b.b$ and $\vdash_{\mathsf{CORE}} \lambda a.\lambda b.(ab) = \lambda b.\lambda a.(ba)$:

$$\dfrac{\dfrac{\dfrac{}{a\#b}(\mathbf{\#ab})}{a\#\lambda b.b}(\mathbf{\#\lambda b}) \quad \dfrac{}{b\#\lambda b.b}(\mathbf{\#\lambda a})}{\lambda a.a = \lambda b.b}(\mathbf{perm}) \qquad \dfrac{\dfrac{\dfrac{}{a\#\lambda a.(ba)}(\mathbf{\#\lambda a})}{a\#\lambda b.\lambda a.(ba)}(\mathbf{\#\lambda b}) \quad \dfrac{}{b\#\lambda b.\lambda a.(ba)}(\mathbf{\#\lambda a})}{\lambda a.\lambda b.(ab) = \lambda b.\lambda a.(ba)}(\mathbf{perm})$$

Note that we use $\lambda a.a \equiv (a\ b) \cdot \lambda b.b$ and $\lambda a.\lambda b.(ab) \equiv (a\ b) \cdot \lambda b.\lambda a(ba)$ in the conclusions of the derivations.

EXAMPLE 27. We give a full derivation of $\vdash_{\mathsf{ULAM}} (\lambda b.(\lambda a.b))a = \lambda c.a$. We use the shorthand from the Introduction to write for example $(\lambda b.(\lambda a.b))a$ as $(\lambda a.b)[b \mapsto a]$.

$$\Pi \quad = \quad \dfrac{\dfrac{\dfrac{\dfrac{\dfrac{}{a\#b}(\mathbf{\#ab})}{a\#\lambda c.b}(\mathbf{\#\lambda b}) \quad \dfrac{}{c\#\lambda c.b}(\mathbf{\#\lambda a})}{\lambda a.b = \lambda c.b}(\mathbf{perm})}{\lambda b.\lambda a.b = \lambda b.\lambda c.b}(\mathbf{cong\lambda}) \qquad \dfrac{}{a = a}(\mathbf{refl})}{(\lambda a.b)[b \mapsto a] = (\lambda c.b)[b \mapsto a]}(\mathbf{congapp})$$

$$\Pi' \quad = \quad \dfrac{\dfrac{\dfrac{}{c\#a}(\mathbf{\#ab})}{(\lambda c.b)[b \mapsto a] = \lambda c.(b[b \mapsto a])}(\mathbf{ax_{abs\mapsto}}) \qquad \dfrac{\dfrac{}{b[b \mapsto a] = a}(\mathbf{ax_{var\mapsto}})}{\lambda c.(b[b \mapsto a]) = \lambda c.a}(\mathbf{cong\lambda})}{(\lambda c.b)[b \mapsto a] = \lambda c.a}(\mathbf{tran})$$

$$\dfrac{\Pi \qquad \Pi'}{(\lambda a.b)[b \mapsto a] = \lambda c.a}(\mathbf{tran})$$

The examples above do not showcase the full power of nominal terms, because we are not reasoning on terms containing *unknowns*. The raison d'étre of nominal terms is their unknowns and how their interaction with permutations and freshness allow us to manage α-conversion. We now consider examples of derivations with α-renaming and freshness in the presence of unknowns:

LEMMA 28. $b\#X \vdash_{\mathsf{CORE}} X[a \mapsto T] = ((b\ a) \cdot X)[b \mapsto T]$

Proof. De-sugaring we must derive $(\lambda a.X)T = (\lambda b.(b\ a) \cdot X)T$ from $b\#X$:

$$\dfrac{\dfrac{\dfrac{\dfrac{b\#X}{a\#(b\ a)\cdot X}(\mathbf{\#X})}{a\#\lambda b.(b\ a)\cdot X}(\mathbf{\#\lambda b}) \qquad \dfrac{}{b\#\lambda b.(b\ a)\cdot X}(\mathbf{\#\lambda a})}{\lambda a.X = \lambda b.(b\ a)\cdot X}(\mathbf{perm}) \qquad \dfrac{}{T=T}(\mathbf{refl})}{(\lambda a.X)T = (\lambda b.(b\ a)\cdot X)T}(\mathbf{congapp})$$

The instance of (**perm**) relies on the fact that $(b\ a) \cdot \lambda a.X \equiv \lambda b.(b\ a) \cdot X$. ∎

LEMMA 29 (Substitution Lemma).

$$a\#Y \vdash_{\mathsf{ULAM}} Z[a \mapsto X][b \mapsto Y] = Z[b \mapsto Y][a \mapsto X[b \mapsto Y]].$$

The usual proof of the substitution lemma is by induction on Z but now Z is just a formal symbol and part of the syntax — but instead, the axioms of ULAM capture this behaviour.

Proof. By (**tran**) it suffices to derive

$$((\lambda a.Z)X)[b \mapsto Y] = ((\lambda a.Z)[b \mapsto Y])(X[b \mapsto Y])$$

and

$$((\lambda a.Z)[b \mapsto Y])(X[b \mapsto Y]) = (\lambda a.(Z[b \mapsto Y]))(X[b \mapsto Y])$$

from $a\#Y$. The first part follows by axiom ($\mathbf{app}{\mapsto}$). The second part can be derived as follows:

$$\dfrac{\dfrac{a\#Y}{(\lambda a.Z)[b \mapsto Y] = \lambda a.(Z[b \mapsto Y])}(\mathbf{ax_{abs\mapsto}}) \qquad \dfrac{}{X[b \mapsto Y] = X[b \mapsto Y]}(\mathbf{refl})}{((\lambda a.Z)[b \mapsto Y])(X[b \mapsto Y]) = (\lambda a.(Z[b \mapsto Y]))(X[b \mapsto Y])}(\mathbf{congapp})$$

∎

The following lemma formally connects the usual use of substitution to handle α-renaming, with the unusual treatment of α-renaming which is primitive to nominal terms, based on permutations.

LEMMA 30. $b\#Z \vdash_{\mathsf{ULAM}} Z[a \mapsto b] = (b\ a) \cdot Z$.

Proof. We sketch the derivation:

$$\dfrac{\dfrac{\begin{array}{c} b\#Z \\ \vdots \\ \text{As for Lemma 28}\end{array}}{Z[a\mapsto b] = ((b\ a)\cdot Z)[b\mapsto b]} \qquad \dfrac{}{((b\ a)\cdot Z)[b\mapsto b] = (b\ a)\cdot Z}\,(\mathbf{ax_{id\mapsto}})}{Z[a\mapsto b] = (b\ a)\cdot Z}\,(\mathbf{tran})$$

■

REMARK 31. Axiom $(\mathbf{id}\mapsto)$ is *equivalent* to Lemma 30. That is, we can derive $Z[a\mapsto a] = Z$ using Lemma 30:

$$\dfrac{\dfrac{\dfrac{\begin{array}{c} [b\#Z]^1 \\ \vdots \\ \text{As for Lemma 28}\end{array}}{Z[a\mapsto a] = ((b\ a)\cdot Z)[b\mapsto a]} \qquad \dfrac{\dfrac{[b\#Z]^1}{a\#(b\ a)\cdot Z}\,(\#\mathbf{X})}{((b\ a)\cdot Z)[b\mapsto a] = Z}\,(\text{Lemma 30})}{Z[a\mapsto a] = Z}\,(\mathbf{tran})}{Z[a\mapsto a] = Z}\,(\mathbf{fr})^1$$

In this derivation the superscript number one ^{1} is an annotation associating the instance of the rule $(\mathbf{fr})$ with the assumption it discharges in the derivation. Note how we use $(\mathbf{fr})$ to *generate* a 'fresh atom'.

LEMMA 32. *If $a\#X$ for every unknown in t, and $a \notin t$, then $\vdash a\#t$.*

Proof. By an easy induction on t using the rules in Figure 3. ■

DEFINITION 33. We write $\mathrm{ds}(\pi, \pi')$ for the set $\{a \mid \pi(a) \neq \pi'(a)\}$, the **difference set** of permutations π and π'. We write $\Delta \vdash \mathrm{ds}(\pi, \pi')\#X$ for a set of proof-obligations $\Delta \vdash a\#X$, one for each $a \in \mathrm{ds}(\pi, \pi')$.

REMARK 34. Figure 4 contains some redundancy; $(\mathbf{refl})$ may be emulated using $(\mathbf{fr})$ and $(\mathbf{tran})$ as follows:

$$\dfrac{\dfrac{\dfrac{\begin{array}{c}\vdots \\ a\#(a\ b)\cdot t\end{array} \quad \begin{array}{c}\vdots \\ b\#(a\ b)\cdot t\end{array}}{t = (a\ b)\cdot t}\,(\mathbf{perm}) \qquad \dfrac{\begin{array}{c}\vdots \\ a\#t\end{array} \quad \begin{array}{c}\vdots \\ b\#t\end{array}}{(a\ b)\cdot t = t}\,(\mathbf{perm})}{t = t}\,(\mathbf{tran})}{t = t}\,(\mathbf{fr})\ \text{for all unknowns in } t$$

Here a, b are chosen fresh (so $a \notin t$ and $b \notin t$). The vertical dots elide derivations described in Lemma 32.

It is convenient in a logic of equality to be able to derive that $t = t$ without 'going round the houses' with (**tran**) and (**fr**), so we include both (**perm**) and (**refl**).

(**perm**) and (**refl**) are both instances of the following rule:

$$\frac{\mathrm{ds}(\pi, \pi')\#t}{\pi \cdot t = \pi' \cdot t}\ (\mathrm{ds}\mathbf{refl})$$

Here $\mathrm{ds}(\pi, \pi')\#t$ is shorthand for the set of $a\#t$ for all $a \in \mathrm{ds}(\pi, \pi')$, if any. This rule looks complicated, so we do not use it in Figure 4.

We can derive syntactic criteria for determining equality in CORE. These will be useful later:

THEOREM 35. *$\Delta \vdash_{\mathsf{CORE}} t = u$ precisely when one of the following hold:*

1. *$t \equiv a$ and $u \equiv a$.*
2. *$t \equiv \pi \cdot X$ and $u = \pi' \cdot X$ and $\Delta \vdash \mathrm{ds}(\pi, \pi')\#X$.*
3. *$t \equiv \lambda a.t'$ and $u \equiv \lambda a.u'$ and $\Delta \vdash_{\mathsf{CORE}} t' = u'$.*
4. *$t \equiv \lambda a.t'$ and $u \equiv \lambda b.u'$ and $\Delta \vdash b\#t'$ and $\Delta \vdash_{\mathsf{CORE}} (b\ a) \cdot t' = u'$.*
5. *$t \equiv t't$ and $u \equiv u'u$ and $\Delta \vdash_{\mathsf{CORE}} t' = u'$ and $\Delta \vdash_{\mathsf{CORE}} t = u$.*
6. *$t \equiv \mathsf{c}$ and $u \equiv \mathsf{c}$.*

Proof. See [15, Corollary 2.32] or [23, Corollary 2.5.4]. ∎

COROLLARY 36 (Consistency). *For all Δ there are t and u such that $\Delta \nvdash_{\mathsf{CORE}} t = u$.*

Proof. By Theorem 35, $\Delta \vdash_{\mathsf{CORE}} a = b$ is never derivable. ∎

COROLLARY 37 (Decidability). *It is decidable whether $\Delta \vdash_{\mathsf{CORE}} t = u$ is true or not.*

Proof. By an easy induction on t, using the syntactic criteria of Theorem 35. ∎

A number of properties on freshnesses also hold for equalities of any theory T. For instance we can strengthen the freshnesses context Δ in equational derivations:

LEMMA 38. *If $c\#Z, \Delta \vdash_{\mathsf{T}} t = u$ and $c \notin t$ and $c \notin u$, then $\Delta \vdash_{\mathsf{T}} t = u$.*

Proof. We extend the derivation with (**fr**). ∎

Also we may permute atoms and instantiate unknowns in equational derivations:

LEMMA 39. *If* $\Delta \vdash_{\mathsf{T}} t = u$ *then* $\Delta \vdash_{\mathsf{T}} \pi \cdot t = \pi \cdot u$.

Proof. By induction on the structure of the rules of Figure 4. ∎

THEOREM 40. *For any* T, Δ', Δ, σ, *if* $\Delta \vdash_{\mathsf{T}} t = u$ *and* $\Delta' \vdash \Delta\sigma$ *then* $\Delta' \vdash_{\mathsf{T}} t\sigma = u\sigma$.

As a corollary, if $\Delta \vdash_{\mathsf{T}} t = u$ *and* $\Delta \subseteq \Delta'$ *then* $\Delta' \vdash_{\mathsf{T}} t = u$.

Proof. Analogous to the proof of Theorem 21.

In the case that σ mentions a 'fresh' atom used in an instance of (**fr**), we rename that atom to be 'fresher'. The inductive hypothesis is valid also for the 'freshened' derivation because of the mathematical principle of ZFA equivariance ([16, Appendix A] or [17]); induction on a measure of the depth of derivations is also possible, subject to uninteresting lemmas that renaming atoms does not change depth. ∎

3 Untyped λ-terms

DEFINITION 41. We call a term **ground** when it mentions no unknowns. Ground terms are inductively characterised by the grammar in Definition 1.

It is no coincidence that ground terms are characterised by Definition 1; as discussed in Subsection 2.1 the syntax of nominal terms used in this paper is specialised to the intended application.

The rest of this section sketches a formal development of the syntax and operational semantics of λ-terms and $\alpha\beta$-reduction, and links it to the 'nominal' exposition. For more detailed treatments of λ-terms and $\alpha\beta$-reduction see elsewhere [2, 6].

DEFINITION 42. Define the **free atoms** $fa(g)$ inductively by:

$$fa(a) = \{a\} \quad fa(\lambda a.g) = fa(g) \setminus \{a\} \quad fa(gg') = fa(g) \cup fa(g') \quad fa(\mathsf{c}) = \{\}.$$

This is standard.

LEMMA 43. $a \notin fa(g)$ *if and only if* $\vdash a\#g$.

Also, if $a \notin g$ *(Definition 7) then* $\vdash a\#g$.

Proof. By routine inductions on the structure of g. ∎

DEFINITION 44. Define the **size** of a ground term inductively by:

$$|a| = 1 \qquad |\lambda a.g| = |g| + 1 \qquad |g'g| = |g'| + |g| + 1 \qquad |\mathsf{c}| = 1.$$

We define a **capture-avoiding substitution** action $g[h/a]$ inductively on the size of g by:

$$a[h/a] \equiv h \qquad b[h/a] \equiv b \qquad (\lambda a.g)[h/a] \equiv \lambda a.g$$
$$(\lambda b.g)[h/a] \equiv \lambda b.(g[h/a]) \quad (b \notin fa(h))$$
$$(\lambda b.g)[h/a] \equiv \lambda c.(g[c/b][h/a]) \quad (b \in fa(h),\ c \text{ fresh})$$
$$(g'g)[h/a] \equiv (g'[h/a])(g[h/a]) \qquad \mathsf{c}[h/a] = \mathsf{c}.$$

In the clause for $(\lambda b.g)[h/a]$ we make some fixed and arbitrary choice of fresh c (the 'c fresh'), for each b, g, h, a.

A basic property is useful in the proofs of the results which follow:

LEMMA 45. *Capture-avoiding substitution of atoms for atoms preserves size. More formally,* $|g[a/a]| = |g|$ *and* $|g[b/a]| = |g|$.

Proof. By an easy induction on $|g|$. ■

LEMMA 46. $fa(g[h/a]) \subseteq (fa(g)\backslash\{a\}) \cup fa(h)$.
Also, if $\vdash a\#h$ *then* $\vdash a\#g[h/a]$.

Proof. The first part is by a routine induction on size. For the second part we prove the contrapositive. By Lemma 43 it suffices to show that $a \in fa(g[h/a])$ implies $a \in fa(h)$. This also follows by a routine induction on the size of g. ■

We introduce the usual notion of α-equivalence.

DEFINITION 47. We write $=_\alpha$ for the **α-equivalence** relation, which is obtained by extending syntactic equivalence with the following rule to rename bound variables:

$$\lambda a.g =_\alpha \lambda b.h \quad \text{when} \quad g[c/a] =_\alpha h[c/b] \quad \text{for some fresh atom } c.$$

The following lemma shows how permutations interact with $=_\alpha$:

LEMMA 48.

1. *If* $a, b \notin fa(g)$ *then* $(a\ b) \cdot g =_\alpha g$.
2. *If* $b \notin fa(g)$ *then* $g[b/a] =_\alpha (b\ a) \cdot g$.

Proof. For the first part, we observe that all a and b in g must occur in the scope of λa and λb. We traverse the structure of g bottom-up and rename these to fresh atoms (for example $\lambda a'$ and $\lambda b'$ which do not occur anywhere in g). Call the resulting term g'. Now $(a\ b) \cdot g' \equiv g'$ because $a, b \notin g'$. Equality is symmetric, so we reverse the process to return to g.

The second part follows by an induction on $|g|$. ■

THEOREM 49. *Derivable equality in* CORE *coincides with* $=_\alpha$ *on ground terms.*

Proof. See [15, Theorem 3.9] or [23, Theorem 4.3.13]. ■

REMARK 50. We do not quotient terms by α-conversion and we do not use a syntax based on a nominal-style datatype of syntax-with-binding [17]. This is because the proof of Theorem 57 involves keeping careful track of what atoms do and do not appear in terms, and if we quotient by α-equivalence now we lose information — names of abstracted atoms — which is useful for expressing that proof.

DEFINITION 51. Let β-**reduction** $g \rightarrow_\beta h$ be inductively defined by:

$$\frac{}{g[a \mapsto h] \rightarrow_\beta g[h/a]} \qquad \frac{g \rightarrow_\beta g'}{\lambda a.g \rightarrow_\beta \lambda a.g'} \qquad \frac{g \rightarrow_\beta g' \quad h \rightarrow_\beta h'}{gh \rightarrow_\beta g'h'}$$

We call g a β-**normal form** when there is no g' such that $g \rightarrow_\beta g'$.

We write $g \rightarrow_{\alpha\beta} h$ when there exist g' and h' such that

$$g =_\alpha g', \qquad g' \rightarrow_\beta h', \quad \text{and} \quad h' =_\alpha h.$$

We write $=_{\alpha\beta}$ for the transitive reflexive symmetric closure of $\rightarrow_{\alpha\beta}$.

THEOREM 52. $\rightarrow_{\alpha\beta}$ *is confluent.*

Proof. See elsewhere [6]. ■

4 Soundness, completeness, conservativity

4.1 Soundness

DEFINITION 53. We call a substitution ς **ground** for a set of unknowns $\mathcal{X}$ when $\varsigma(X)$ is ground for every $X \in \mathcal{X}$. We call ς ground for Δ, t, u when ς is ground for the set of unknowns appearing anywhere in Δ, t, or u.

LEMMA 54. $fa(\pi \cdot g) = \{\pi(a) \mid a \in fa(g)\}$.

Proof. By Lemmas 43 and 19. The result also follows directly by a routine induction on the syntax of g. ■

It is now easy to state and prove a notion of *soundness* for ULAM:

THEOREM 55 (Soundness). *Suppose that ς is a ground substitution for Δ, t, and u. Suppose further that $a \notin fa(\varsigma(X))$ for every $a\#X \in \Delta$. Then:*

- $\Delta \vdash a\#t$ *implies* $a \notin fa(t\varsigma)$.
- $\Delta \vdash_{\mathsf{ULAM}} t = u$ *implies* $t\varsigma =_{\alpha\beta} u\varsigma$.

Proof. We proceed by induction on ULAM derivations.
We consider the rules in Figure 3 in turn:

- By assumption. We must show that if $a\#X \in \Delta$ then $a \notin fa(\varsigma(X))$, which follows by our assumptions on ς.
- The case (**#ab**). It is a fact of Definition 42 that $a \notin fa(b)$.
- The case (**#λa**). It is a fact that $a \notin fa(\lambda a.(t\varsigma))$.
- The cases of (**#λb**), (**#app**), and (**#c**) are no harder.
- The case (**#X**). By Lemma 54, $a \in fa(\pi \cdot g)$ if and only if $\pi^{-1}(a) \in fa(g)$.

We consider the rules in Figure 4 in turn:

- The cases (**refl**), (**symm**), (**tran**), (**congλ**) and (**congapp**) follow by induction using the fact that $=_{\alpha\beta}$ is an equivalence relation and a congruence.
- The case (**perm**). Suppose that $a, b \notin fa(g)$. Then $(a\ b) \cdot g =_{\alpha\beta} g$ follows by Lemma 48, since $=_\alpha$ implies $=_{\alpha\beta}$.
- The case (**ax**). It remains to check the validity of the axioms of ULAM. It suffices to verify that:
  - (**var**$\mapsto$). We must show
    $$\pi(a)[\pi(a) \mapsto (\pi \cdot \sigma(X))\varsigma] =_{\alpha\beta} (\pi \cdot \sigma(X))\varsigma$$
    for any permutation π and substitution σ. This follows from the property that
    $$b[b \mapsto h] =_{\alpha\beta} h$$
    always (i.e. for any atom b and ground term h), which is a fact about $\alpha\beta$-equivalence.
  - (**#**$\mapsto$). Suppose that $(\pi \cdot \sigma(Z))[\pi(a) \mapsto \pi \cdot \sigma(X)] = \pi \cdot \sigma(Z)$ is derived from $\pi(a)\#\pi \cdot \sigma(Z)$ using the assumptions from Δ. By inductive hypothesis, we know $\pi(a) \notin fa((\pi \cdot \sigma(Z))\varsigma)$. We must show $((\pi \cdot \sigma(Z))\varsigma)[\pi(a) \mapsto (\pi \cdot \sigma(X))\varsigma] =_{\alpha\beta} (\pi \cdot \sigma(Z))\varsigma$. This follows from the basic fact about $\alpha\beta$-equivalence that
    $$b \notin fa(g) \quad \text{implies} \quad g[b \mapsto h] =_{\alpha\beta} g$$
    always.

    Other cases are similar:

- **(app$\mapsto$)**. $(g'g)[b \mapsto h] =_{\alpha\beta} (g'[b \mapsto h])(g[b \mapsto h])$.
- **(abs$\mapsto$)**. If $c \notin fa(h)$ then $(\lambda c.g)[b \mapsto h] =_{\alpha\beta} \lambda c.(g[b \mapsto h])$.
- **(id$\mapsto$)**. $g[b \mapsto b] =_{\alpha\beta} g$.

- The case **(fr)**. If necessary, we rename the fresh atom in **(fr)** to a 'fresher' atom. By ZFA equivariance ([16, Appendix A]), the inductive hypothesis also holds for the 'freshened' derivation, and the result easily follows.

■

4.2 Completeness

We start with a weak form of completeness:

LEMMA 56. *For ground terms g and h, if $g =_{\alpha\beta} h$ then $\vdash_{\mathsf{ULAM}} g = h$.*

It is not hard to prove this by concrete calculations, but it also follows as a corollary of the following more powerful result:

THEOREM 57 (Completeness). *Suppose that $t\varsigma =_{\alpha\beta} u\varsigma$ for every substitution ς such that*

- *ς is ground for Δ, t, u and*
- *$a \notin fa(\varsigma(X))$ for every $a\#X \in \Delta$.*

Then $\Delta \vdash_{\mathsf{ULAM}} t = u$ is derivable.

The proof occupies the rest of this section. Fix some freshness context Δ and two terms t and u.

DEFINITION 58. Let $\mathcal{A}$ be the atoms mentioned anywhere in Δ, t, or u. Let $\mathcal{X}$ be the unknowns mentioned anywhere in Δ, t, or u. For each $X \in \mathcal{X}$ fix the following data:

- An order $a_{X1}, \ldots, a_{Xk_X}$ on the atoms in $\mathcal{A}$ such that $a\#X \notin \Delta$.
- Some entirely fresh atom c_X.

We write $\mathcal{C}$ for the set $\{c_X \mid X \in \mathcal{X}\}$.

DEFINITION 59. Specify ς a ground substitution for $\mathcal{X}$ by:

- $\varsigma(X) \equiv c_X a_{X1} \ldots a_{Xk_X}$ if $X \in \mathcal{X}$.
- $\varsigma(Y) \equiv Y$ for $Y \notin \mathcal{X}$.

LEMMA 60. *If $a\#X \in \Delta$ then $a \notin fa(\varsigma(X))$.*

Proof. By construction of $\varsigma(X)$. ■

By assumption $t\varsigma =_{\alpha\beta} u\varsigma$. The untyped λ-calculus is confluent (Theorem 52) so $t\varsigma$ and $u\varsigma$ rewrite to a common term, call it p:

DEFINITION 61. Fix two chains

$$\begin{array}{ll} t\varsigma & \equiv g_1 =_\alpha g_2 \to_\beta g_3 =_\alpha g_4 \to_\beta \ldots \to_\beta g_{m-1} =_\alpha g_m \equiv p \\ u\varsigma & \equiv h_1 =_\alpha h_2 \to_\beta h_3 =_\alpha h_4 \to_\beta \ldots \to_\beta h_{n-1} =_\alpha h_n \equiv p \end{array}$$

Without loss of generality we assume these α-conversions and β-reductions do not introduce abstractions by atoms from $\mathcal{C}$.

DEFINITION 62. Let $\mathcal{A}^+$ be the set of *all* atoms mentioned anywhere in the chains of Definition 61, extended with a set of fresh atoms

$$\mathcal{B} = \{b_{Xi} \mid X, i \text{ such that } a_{Xi} \in \mathcal{A}\}.$$

in bijection with $\mathcal{A}$. (So $\mathcal{B}$ is disjoint from $\mathcal{C}$ and from all atoms mentioned in Δ, t, u, $g_1, \ldots, g_m$, $h_1, \ldots, h_n$.) Let Δ^+ be Δ enriched with freshness assumptions $a\#X$ for every $a \in \mathcal{A}^+ \setminus \mathcal{A}$ and every $X \in \mathcal{X}$.

DEFINITION 63. We call a ground term g **accurate** when:

- g mentions only atoms in $\mathcal{A}^+ \setminus \mathcal{B}$.
- $c_X \in \mathcal{C}$ never occurs abstracted. That is, no term contains 'λc_X'.
- $c_X \in \mathcal{C}$ appears, if it appears at all, in head position applied to a list of terms in a subterm of the form '$c_X g_1 \ldots g_{k_X}$'.

Note that $g_1, \ldots, g_m$ and $h_1, \ldots, h_n$ are accurate by construction of $t\varsigma$ and $u\varsigma$ and by the nature of α-conversions and β-reductions.

DEFINITION 64. Define an **inverse translation** from accurate terms to (possibly non-ground) terms inductively by:

$$a^{-1} \equiv a \quad (a \notin \mathcal{C}) \qquad (\lambda a.g)^{-1} \equiv \lambda a.(g^{-1}) \qquad (gh)^{-1} \equiv (g^{-1})(h^{-1}) \qquad \mathsf{c}^{-1} \equiv \mathsf{c}$$
$$(c_X)^{-1} \equiv \lambda b_{X1}.\cdots\lambda b_{Xk_X}.(b_{X1}\ a_{X1})\cdots(b_{Xk_X}\ a_{Xk_X})\cdot X \quad (c_X \in \mathcal{C})$$

The inverse translation $\_^{-1}$ is an inverse of ς in the following sense:

LEMMA 65. $\Delta^+ \vdash_{\mathsf{ULAM}} (t\varsigma)^{-1} = t$, *and* $\Delta^+ \vdash_{\mathsf{ULAM}} (u\varsigma)^{-1} = u$.

Proof. We prove by induction that if v is a subterm of t or u then $\Delta^+ \vdash_{\mathsf{ULAM}} (v\varsigma)^{-1} = v$.

The only interesting case is when $v \equiv \pi \cdot X$. We must show

$$\Delta^+ \vdash_{\mathsf{ULAM}} (\pi_1 \cdot X)[b_{Xk_X} \mapsto \pi(a_{Xk_X})] \cdots [b_{X1} \mapsto \pi(a_{X1})] = \pi \cdot X$$

where $\pi_1 = (b_{X1}\ a_{X1}) \cdots (b_{Xk_X}\ a_{Xk_X})$.

Take $\pi_2 = (b_{X1}\ \pi(a_{X1})) \cdots (b_{Xk_X}\ \pi(a_{Xk_X}))$. By transitivity it suffices to show:

1. $\Delta^+ \vdash_{\mathsf{ULAM}} (\pi_1 \cdot X)[b_{Xk_X} \mapsto \pi(a_{Xk_X})] \cdots [b_{X1} \mapsto \pi(a_{X1})] = (\pi_2 \circ \pi_1) \cdot X$.
2. $\Delta^+ \vdash_{\mathsf{ULAM}} (\pi_2 \circ \pi_1) \cdot X = \pi \cdot X$.

The first part follows using Lemma 30 when, for $1 \leq i \leq k_X$,

$$\Delta^+ \vdash \pi(a_{Xi})\#(\pi_1 \cdot X)[b_{Xk_X} \mapsto \pi(a_{Xk_X})] \cdots [b_{Xi+1} \mapsto \pi(a_{Xi+1})].$$

By the rules for freshness, this follows from $\Delta^+ \vdash \pi(a_{Xi})\#\pi_1 \cdot X$ since the $\pi(a_{X1}), \ldots, \pi(a_{Xk_X})$ are all disjoint. We conclude using a case distinction on $\pi(a_{Xi})$:

- $\pi(a_{Xi}) \neq a_{Xj}$ for all j: then $\pi(a_{Xi})\#X \in \Delta^+$ since $\pi(a_{Xi})\#X \in \Delta$.
- $\pi(a_{Xi}) = a_{Xj}$ for some j: then $b_{Xj}\#X \in \Delta^+$ by definition.

We still have to show that $\Delta^+ \vdash_{\mathsf{ULAM}} (\pi_2 \circ \pi_1) \cdot X = \pi \cdot X$. It is convenient to show the stronger property $\Delta^+ \vdash_{\mathsf{CORE}} (\pi_2 \circ \pi_1) \cdot X = \pi \cdot X$. By Theorem 35 we need only show that $\Delta^+ \vdash \mathrm{ds}(\pi_2 \circ \pi_1, \pi)\#X$. That is, we must show that $\Delta^+ \vdash a\#X$ for every a such that $(\pi_2 \circ \pi_1)(a) \neq \pi(a)$. We consider every possible a (every $a \in \pi_2 \circ \pi_1$ and $a \in \pi$):

- $a = b_{Xi}$: then $b_{Xi}\#X \in \Delta^+$ by definition, and the result follows.
- $a = a_{Xi}$: then $(\pi_2 \circ \pi_1)(a_{Xi}) = \pi(a_{Xi})$ and there is nothing to prove.
- $a = \pi(a_{Xi})$: then we distinguish two cases:
  - if $\pi(a_{Xi}) = a_{Xj}$ for some j, the result follows by the case of a_{Xi};
  - if $\pi(a_{Xi}) \neq a_{Xj}$ for all j, then $\pi(a_{Xi})\#X \in \Delta^+$ by definition.
- $a \in \pi$, but $a \neq a_{Xj}$ for all j, then $a\#X \in \Delta^+$ by definition.

■

REMARK 66. The reader might wonder why the inverse mapping of the c_X renames a_{Xi} to the fresh b_{Xi}. Consider for example $(a_{X1}\ a_{X2}) \cdot X$ in the empty freshness context $\emptyset$, so we do not know $a_{X1}\#X$ or $a_{X2}\#X$. Then

$$((a_{X1}\ a_{X2}) \cdot X)\varsigma^{-1} \equiv ((b_{X1}\ a_{X1})(b_{X2}\ a_{X2}) \cdot X)[b_{X2} \mapsto a_{X1}][b_{X1} \mapsto a_{X2}].$$

By calculations we can verify Lemma 65:

$$\emptyset^+ \vdash_{\mathsf{ULAM}} ((a_{X1}\ a_{X2}) \cdot X)\varsigma^{-1} = (a_{X1}\ a_{X2}) \cdot X$$

where $\emptyset^+ = \{b_{X1}\#X, b_{X2}\#X, c_X\#X\}$. Had we left out the renaming to fresh atoms then $((a_{X1}\ a_{X2}) \cdot X)\varsigma^{-1}$ would be $X[a_{X2} \mapsto a_{X1}][a_{X1} \mapsto a_{X2}]$, which is not equal to $(a_{X1}\ a_{X2}) \cdot X$, since for example

$$((a_{X1}\ a_{X2}) \cdot X)[a_{X2}/X] = a_{X1} \text{ but } X[a_{X2} \mapsto a_{X1}][a_{X1} \mapsto a_{X2}][a_{X2}/X] = a_{X2}.$$

A technical lemma about freshness will be useful.

LEMMA 67. *Suppose that g is accurate. For any $a \in \mathcal{A}^+$, if $a \notin fa(g)$ then $\Delta^+ \vdash a\#g^{\text{-}1}$.*

Proof. By induction on g. The only non-trivial case is when $g \equiv c_X$. Suppose $a \notin fa(c_X)$. Then $a \neq c_X$ and we must show

$$\Delta^+ \vdash a\#\lambda b_{X1}.\cdots\lambda b_{Xk_X}.(\pi \cdot X),$$

where $\pi = (b_{X1}\ a_{X1})\cdots(b_{Xk_X}\ a_{Xk_X})$. We distinguish two cases:

- $a = b_{Xk_j}$ for some j: then $b_{Xk_j}\#\lambda b_{Xk_j}.\cdots\lambda b_{Xk_X}(\pi \cdot X)$ by $(\#\lambda\mathbf{a})$, and the result follows by the rules of freshness using the inductive hypothesis.
- $a \neq b_{Xk_j}$ for all j: then $\pi^{\text{-}1}(a) \neq a_{Xk_j}$ for all j, so $\pi^{\text{-}1}(a)\#X \in \Delta^+$ by definition. The result follows using the rules for freshness and the inductive hypothesis.

■

We need a technical lemma:

LEMMA 68. *Suppose that g is accurate. Suppose that π is a permutation such that $\pi(a) = a$ for all $a \notin \mathcal{A}^+ \setminus (\mathcal{B} \cup \mathcal{C})$. Then $\Delta^+ \vdash_{\mathsf{CORE}} (\pi\cdot g)^{\text{-}1} = \pi\cdot(g^{\text{-}1})$.*

Proof. By a routine induction on g. In the case of $g \equiv c_X$ we use the fact that $\pi(a) = a$ for all $a \in \mathcal{B} \cup \mathcal{C}$. ■

LEMMA 69. *Suppose that g and h are accurate. Then if $g =_\alpha h$ then $\Delta^+ \vdash_{\mathsf{CORE}} g^{\text{-}1} = h^{\text{-}1}$.*

Proof. By Theorem 49 $g =_\alpha h$ coincides with $\vdash_{\mathsf{CORE}} g = h$. We therefore work by induction on the structure of g using the syntactic criteria of Theorem 35.

The only non-trivial case is when

$$g \equiv \lambda a.g', \quad h \equiv \lambda b.h', \quad \vdash b\#g', \quad \text{and} \quad \vdash_{\mathsf{CORE}} (b\ a)\cdot g' = h'.$$

By assumption $a, b \in \mathcal{A}^+ \setminus (\mathcal{B} \cup \mathcal{C})$. By Lemma 67 we have $\Delta^+ \vdash b\#g'^{\text{-}1}$. By inductive hypothesis $\Delta^+ \vdash_{\mathsf{CORE}} ((b\ a)\cdot g')^{\text{-}1} = h'^{\text{-}1}$. By Lemma 68 we have $\Delta^+ \vdash_{\mathsf{CORE}} ((b\ a)\cdot g')^{\text{-}1} = (b\ a)\cdot(g'^{\text{-}1})$. From the rules for freshness and equality $\Delta^+ \vdash_{\mathsf{CORE}} \lambda a.(g'^{\text{-}1}) = \lambda b.(h'^{\text{-}1})$ follows. ■

LEMMA 70. *Suppose that $a \in \mathcal{A}^+ \setminus \mathcal{C}$. Suppose that g, h, and $g[h/a]$ are accurate. Then $\Delta^+ \vdash_{\mathsf{ULAM}} g^{\text{-}1}[a \mapsto h^{\text{-}1}] = (g[h/a])^{\text{-}1}$.*

Proof. We work by induction on the size of g:

- $a[h/a]$. $\Delta^+ \vdash_{\mathsf{ULAM}} a[a \mapsto h^{-1}] = h^{-1}$ by $(\mathbf{var}\mapsto)$.
- $b[h/a]$. $\Delta^+ \vdash_{\mathsf{ULAM}} b[a \mapsto h^{-1}] = b$ by $(\#\mapsto)$ since $\Delta^+ \vdash a\#b$. The cases of $(\lambda a.g)[h/a]$ and $\mathsf{c}[h/a]$ are similar.
- $(\lambda b.g)[h/a]$ where $b \notin fa(h)$. We must show

  $$\Delta^+ \vdash_{\mathsf{ULAM}} (\lambda b.(g^{-1}))[a \mapsto h^{-1}] = \lambda b.((g[h/a])^{-1}).$$

  By the inductive hypothesis and the rules for equality, it suffices to show

  $$\Delta^+ \vdash_{\mathsf{ULAM}} (\lambda b.(g^{-1}))[a \mapsto h^{-1}] = \lambda b.(g^{-1}[a \mapsto h^{-1}]).$$

  By Lemma 67 we know $\Delta^+ \vdash b\#h^{-1}$. The result follows from $(\mathbf{abs}\mapsto)$.
- $(\lambda b.g)[h/a]$ where $b \in fa(h)$. By assumption $\lambda b.g$ is accurate, therefore $b \notin \mathcal{B} \cup \mathcal{C}$.

  Recall from Definition 44 that $(\lambda b.g)[h/a] \equiv \lambda c.(g[c/b][h/a])$ for some choice of fresh c (so $c \notin fa(g)$ and $c \notin fa(h)$). Now by assumption $\lambda c.(g[c/b][h/a])$ is accurate, so $c \notin \mathcal{B} \cup \mathcal{C}$ and $g[c/b][h/a]$ is accurate.

  We must show

  $$\Delta^+ \vdash_{\mathsf{ULAM}} (\lambda b.(g^{-1}))[a \mapsto h^{-1}] = \lambda c.((g[c/b][h/a])^{-1}).$$

  Note that by Lemma 67, $\Delta^+ \vdash c\#g^{-1}$ and $\Delta^+ \vdash c\#h^{-1}$, and therefore $\Delta^+ \vdash c\#\lambda b.(g^{-1})$ by $(\#\lambda\mathbf{b})$. Also $\Delta^+ \vdash b\#\lambda b.(g^{-1})$ is immediate by $(\#\lambda\mathbf{a})$. We present the rest of the proof in a calculational style:

  $$\begin{array}{ll} & \lambda c.((g[c/b][h/a])^{-1}) \\ = & \quad \{\ g[c/b] =_\alpha (c\ b) \cdot g \text{ by Lemma 48 since } c \notin fa(g)\ \} \\ & \lambda c.(((c\ b) \cdot g)[h/a])^{-1} \\ = & \quad \{\ \text{inductive hypothesis, since } (c\ b) \cdot g \text{ is accurate}\ \} \\ & \lambda c.(((c\ b) \cdot g)^{-1}[a \mapsto h^{-1}]) \\ = & \quad \{\ \text{Lemma 68}\ \} \\ & \lambda c.(((c\ b) \cdot (g^{-1}))[a \mapsto h^{-1}]) \\ = & \quad \{\ (\mathbf{abs}\mapsto), \text{ since } \Delta^+ \vdash c\#h^{-1}\ \} \\ & (\lambda c.((c\ b) \cdot (g^{-1})))[a \mapsto h^{-1}] \\ = & \quad \{\ (\mathbf{perm}) \text{ since } \Delta^+ \vdash b\#\lambda b.(g^{-1}) \text{ and } \Delta^+ \vdash c\#\lambda b.(g^{-1})\ \} \\ & (\lambda b.(g^{-1}))[a \mapsto h^{-1}] \end{array}$$

The result follows by transitivity.

- $(gg')[h/a]$. By the inductive hypothesis and the rules for equality, $\Delta^+ \vdash_{\mathsf{ULAM}} ((g^{-1})(g'^{-1}))[a \mapsto h^{-1}] = ((gg')[h/a])^{-1}$ follows from

  $$\Delta^+ \vdash_{\mathsf{ULAM}} ((g^{-1})(g'^{-1}))[a \mapsto h^{-1}] = (g^{-1}[a \mapsto h^{-1}])(g'^{-1}[a \mapsto h^{-1}]).$$

  We conclude using axiom (**app**$\mapsto$).

- $c_X[h/a]$ where $c_X \in \mathcal{C}$. By assumption $a \neq c_X$, so we must show

  $$\Delta^+ \vdash_{\mathsf{ULAM}} (\lambda b_{X1}.\cdots\lambda b_{Xk_X}.(\pi \cdot X))[a \mapsto h^{-1}] = \lambda b_{X1}.\cdots\lambda b_{Xk_X}.(\pi \cdot X),$$

  where $\pi = (b_{X1}\ a_{X1})\cdots(b_{Xk_X}\ a_{Xk_X})$.

  By assumption $b_{Xi} \notin h$ for all $1 \leq i \leq k_X$, therefore also $b_{Xi} \notin fa(h)$. By Lemma 67 also $\Delta^+ \vdash b_{Xi}\#h^{-1}$. Then we can show by a number of applications of axiom (**abs**$\mapsto$) that $(\lambda b_{X1}.\cdots\lambda b_{Xk_X}.(\pi \cdot X))[a \mapsto h^{-1}]$ is equal to $\lambda b_{X1}.\cdots\lambda b_{Xk_X}.(\pi \cdot X)[a \mapsto h^{-1}]$. By the rules for equality, it suffices to show

  $$\Delta^+ \vdash_{\mathsf{ULAM}} (\pi \cdot X)[a \mapsto h^{-1}] = \pi \cdot X.$$

  By axiom ($\#\mapsto$) this follows if $\pi^{-1}(a)\#X \in \Delta^+$. There are two possibilities:

  - $a = a_{Xj}$ for some j: then $\pi^{-1}(a) = b_{Xj}$, and $b_{Xj}\#X \in \Delta^+$ by definition.
  - $a \neq a_{Xj}$ for all j: then $\pi^{-1}(a) = a$, and $a\#X \in \Delta^+$ by definition.

The result follows. ∎

COROLLARY 71. *Suppose that g and h are accurate. If $g \to_\beta h$ then $\Delta^+ \vdash_{\mathsf{ULAM}} g^{-1} = h^{-1}$.*

Proof. By induction on the derivation rules for $\to_\beta$ from Definition 51. It suffices to show the following (here g, g', h, h', and $g[h/a]$ are accurate and $a \in \mathcal{A}^+ \setminus \mathcal{C}$):

1. $\Delta^+ \vdash_{\mathsf{ULAM}} g^{-1}[a \mapsto h^{-1}] = (g[h/a])^{-1}$.
2. If $\Delta^+ \vdash_{\mathsf{ULAM}} g^{-1} = g'^{-1}$ then $\Delta^+ \vdash_{\mathsf{ULAM}} \lambda a.(g^{-1}) = \lambda a.(g'^{-1})$.
3. If $\Delta^+ \vdash_{\mathsf{ULAM}} g^{-1} = g'^{-1}$ and $\Delta^+ \vdash_{\mathsf{ULAM}} h^{-1} = h'^{-1}$ then $\Delta^+ \vdash_{\mathsf{ULAM}} (g^{-1})(h^{-1}) = (g'^{-1})(h'^{-1})$.

The first part is Lemma 70. The second and third parts follow by (**cong**λ) and (**congapp**). ∎

We are now ready to prove Theorem 57:

Proof. Recall from Definition 61 the chains

$$t\varsigma \equiv g_1 =_\alpha g_2 \to_\beta g_3 =_\alpha g_4 \to_\beta \dots \to_\beta g_{m-1} =_\alpha g_m \equiv p$$
$$u\varsigma \equiv h_1 =_\alpha h_2 \to_\beta h_3 =_\alpha h_4 \to_\beta \dots \to_\beta h_{n-1} =_\alpha h_n \equiv p.$$

By Lemma 69 and Corollary 71

$$\Delta^+ \vdash_{\mathsf{ULAM}} (t\varsigma)^{\text{-}1} \equiv g_1^{\text{-}1} = g_2^{\text{-}1} = \dots = g_m^{\text{-}1} \equiv p^{\text{-}1}$$
$$\Delta^+ \vdash_{\mathsf{ULAM}} (u\varsigma)^{\text{-}1} \equiv h_1^{\text{-}1} = h_2^{\text{-}1} = \dots = h_n^{\text{-}1} \equiv p^{\text{-}1}.$$

By transitivity

$$\Delta^+ \vdash_{\mathsf{ULAM}} (t\varsigma)^{\text{-}1} = p^{\text{-}1} \quad \text{and} \quad \Delta^+ \vdash_{\mathsf{ULAM}} (u\varsigma)^{\text{-}1} = p^{\text{-}1}$$

so by symmetry and transitivity $\Delta^+ \vdash_{\mathsf{ULAM}} (t\varsigma)^{\text{-}1} = (u\varsigma)^{\text{-}1}$. By Lemma 65 then also $\Delta^+ \vdash_{\mathsf{ULAM}} t = u$. Since Δ^+ extends Δ with atoms that are not mentioned in t and u by Lemma 38 we conclude $\Delta \vdash_{\mathsf{ULAM}} t = u$ as required. ■

4.3 Conservativity over CORE

We can exploit the ς from Subsection 4.2 to prove conservativity of ULAM over CORE.

LEMMA 72. *Fix Δ. Suppose that t and u contain no subterm of the form $v[a \mapsto w]$. Then for ς the ground substitution constructed in Subsection 4.2, $t\varsigma$ and $u\varsigma$ are β-normal forms.*

Proof. $\varsigma(X) \equiv c_X a_{X1} \dots a_{Xk_X}$ for every X appearing in Δ, t, or u. Applying this substitution to t and u cannot introduce subterms of the form $v[a \mapsto w]$. ■

THEOREM 73 (Conservativity). *Suppose that t and u contain no subterm of the form $v[a \mapsto w]$. Then*

$$\Delta \vdash_{\mathsf{ULAM}} t = u \qquad \textit{if and only if} \qquad \Delta \vdash_{\mathsf{CORE}} t = u.$$

Proof. A derivation in CORE is also a derivation in ULAM so the right-to-left implication is immediate.

Now suppose that $\Delta \vdash_{\mathsf{ULAM}} t = u$. We construct ς as in Subsection 4.2. By Theorem 55, $t\varsigma =_{\alpha\beta} u\varsigma$. By Lemma 72 we know that $t\varsigma$ and $u\varsigma$ are β-normal forms. By confluence of the λ-calculus (Theorem 52), $t\varsigma =_\alpha u\varsigma$.

We now prove $\Delta \vdash_{\mathsf{CORE}} t = u$ by induction on t. The calculations are detailed but entirely routine. We consider just one case, the hardest one:

Suppose $t \equiv \pi \cdot X$. Then $t\varsigma \equiv c_X \pi(a_{X1}) \dots \pi(a_{Xk_X})$. By Theorem 35, if $t\varsigma =_\alpha u\varsigma$ it *must* be that $u\varsigma \equiv c_X \pi(a_{X1}) \dots \pi(a_{Xk_X})$.

By the construction of $u\varsigma$ and the way we chose $a_{X1}, \dots, a_{Xk_X}$ to be the atoms mentioned in Δ, t, or u which are *not* provably fresh for X in Δ, it follows that u must have been equal to $\pi' \cdot X$, for some π', such that $\Delta \vdash \mathrm{ds}(\pi, \pi')\#X$. It follows that $\Delta \vdash_{\mathsf{CORE}} t = u$ as required. ■

5 Conclusions

5.1 Related work not using nominal techniques

$\alpha\beta$-equivalence on λ-calculus syntax is a good idea; we find it realised in different ways in different systems. If the λ-calculus syntax in question serves as the language of a logic, then $\alpha\beta$-equivalence may have the status of axioms. For example Andrews's logic $\mathcal{Q}_0$ [1, §51] contains five axioms $(\mathbf{4_1})$, $(\mathbf{4_2})$, $(\mathbf{4_3})$, $(\mathbf{4_4})$, and $(\mathbf{4_5})$ ([1, page 164]). In fact they are axiom *schemes*, containing meta-variables $\mathbf{A}$ and $\mathbf{B}$ in the informal meta-level ranging over terms (and also meta-variables x, y ranging permutatively over variable-symbols). The relationship which Figure 1 bears to them is clear but here, axioms feature in the formal framework of Nominal Algebra — a formal logic, not an informal meta-level. We claim that Nominal Algebra captures part of the 'informal meta-level' in which researchers routinely work — and that within that, ULAM captures the (untyped) λ-calculus.

Salibra's Lambda Abstraction Algebras [28] axiomatise the λ-calculus using universal algebra. The method is cylindric in the sense of cylindric algebras [18]; abstraction is represented by infinitely many term-formers (there is a term-former 'λa' for every a) and freshness is encoded in the structure of the terms in the following sense: consider Salibra's rule (β_4) from [28, page 6]:

$$(\beta_4) \qquad (\lambda x.(\lambda x.\xi))\mu = \lambda x.\xi.$$

We can rewrite this in our notation as $(\lambda a.Z)[a \mapsto X] = \lambda a.Z$ and this is a version of $(\#{\mapsto})$, where the freshness condition $a\#Z$ has been built into the structure of the term by replacing Z with $\lambda a.Z$. Similarly Salibra's rule (α)

$$(\alpha) \qquad \lambda x.(\lambda y.\xi)z = \lambda y.(\lambda x.(\lambda y.\xi)z)y$$

can be rewritten in our notation as $\lambda a.(Z[b \mapsto c]) = \lambda b.(Z[b \mapsto c][a \mapsto b])$ and this plays an analogous rôle to our Lemma 28. The notion of *dimension set* [28, Definition 4] corresponds to freshness. The proof of a main result, Theorem 13 of [28] (first carried out by Salibra, improved by Selinger), uses definitions reminiscent of, though not identical to, Definitions 58 and 59.

The 'nominal' techniques give a clear separation of the parts having to do with names and abstraction, and the parts having to do with λ and application, which is not possible with Lambda Abstraction Algebras. However, Salibra's results [28, 22] show what can be achieved using algebraic methods.

Curry discovered *combinatory algebra* [10]. The signature contains a binary term-former *application* and two constants S and K. Axioms are $Kxy = x$ and $Sxyz = (xz)(yz)$. This syntax is parsimonious and the axioms are compact, but it is not natural or ergonomic to program in; that most ergonomic feature of the λ-calculus which makes it so very useful in practice, the λ, is missing. There is also a mathematical issue: the natural encoding of closed λ-terms into combinatory algebra syntax ([6, Section 7] or [29, Subsection 1.4]) does not map $\alpha\beta$-equivalent λ-terms ($\lambda z.(\lambda x.x)z$ and $\lambda z.z$) to provably equal terms in combinatory algebra. This is resolved if we strengthen combinatory algebra to *lambda algebra* by adding five axioms, due to Curry [29, Proposition 5]. However the translation is still not sound, in the sense that there exist λ-terms M and N such that the translation of M is derivably equal to the translation of N, but the translation of $\lambda x.M$ is not derivably equal to the translation of $\lambda x.N$. To ensure soundness, we must add the Meyer-Scott axiom [29, Proposition 20] (Selinger calls it 'the notorious rule').

In short, combinators do not capture the model of functions expressed by the λ-calculus. Selinger [29] identifies the problem with the interpretation of variables and argues for denotations with fresh 'indeterminates'; this reminds us of nominal techniques with its set of atoms in the denotation and well-developed theories of freshness.

λ-calculi of explicit substitution decompose the substitution used in β-reduction into many explicit reduction steps [21]. This is similar to the way ULAM breaks down the calculation of an equality into many explicit algebraic equalities. A calculus of explicit substitutions is a calculus, not a logic; all reasoning using calculi occurs informally in natural language. ULAM on the other hand is intended to support algebraic reasoning on the λ-calculus within a formal framework, while remaining very close to informal practice.

5.2 Related work using nominal techniques

The first application of nominal techniques was to datatypes of syntax with binding [17]. The syntax of the untyped λ-calculus is often used as a paradigmatic example of such a datatype. This paper is *not* another such study. 'The λ-calculus' studied in some other publications — using nominal sets [17], nominal logic [27], and also using higher-order abstract syntax [26], de Bruijn terms [11], and so on — is a collection of (syntax) trees. This paper studies functions.

This paper is part of a broader research programme developing nominal techniques in general and nominal algebra in particular. A rewrite system for the λ-calculus appeared already in [13] but without any statement or proof of completeness (indeed, the system considered there was not complete). The authors have axiomatised first-order logic as a theory FOL [16] and also substitution as a theory SUB [15]. Since β-reduction is based on substitution, this paper shares technical results with the study of SUB [15]. The completeness result we give for ULAM is for a model based on untyped λ-terms quotiented by $\alpha\beta$-equality. The completeness result for SUB is for a model based on trees with an explicit substitution. Derivable equality in ULAM includes the full power of the λ-calculus and is undecidable; equality in SUB is equality of syntax with a substitution action and is decidable. For example

$$\vdash_{\mathsf{ULAM}} (ba)[b \mapsto \lambda a.a] = a$$

is derivable in ULAM, but (in notation from [15]) only

$$\vdash_{\mathsf{SUB}} \mathsf{app}(b, a)[b \mapsto \mathsf{lam}([a]a)] = \mathsf{app}(\mathsf{lam}([a]a), a)$$

is derivable in SUB. The proofs in this paper have been improved and simplified. Technical issues have been avoided and we do not rely on a strong normalisation property which the proofs of [15] required (probably unnecessarily). Note that the treatment of SUB is parametric over a range of signatures and would be the appropriate theory where substitution is precisely what we require — for example as part of an axiomatisation of quantifiers in first-order logic.

A version of ULAM for a *typed* λ-calculus should be possible. It would sit 'in between' ULAM and SUB, as one might expect, but to make this formal a typing system for nominal terms is required (such that atoms are assigned types by a typing context). This has been investigated to some extent [12]. Investigations in nominal algebra are for future work.

Nominal algebra has a cousin, nominal equational logic (**NEL**) [9], which was derived from nominal algebra but making different design decisions. NEL satisfies a completeness result for a generic class of models in nominal sets [9], as does nominal algebra [14, 23]. The completeness result of this paper is much stronger than the generic results, because it is completeness for a single elementary model; similarly for the authors' treatments of substitution [15] and first-order logic [16]. We know of no like treatments of substitution, logic, and the λ-calculus in NEL. If and when this is done it will be interesting to compare the results.

5.3 Future work

The nominal algebraic framework proved itself capable of translating, with remarkable accuracy and uniformity, informal mathematical specification into formal nominal algebra axioms, almost symbol for symbol. This paper is both a study of the algebraisation of functions using nominal algebra, and a case study in nominal techniques applied to the λ-calculus. We hope that future study of nominal techniques may benefit from the 'off-the-shelf' axiomatisation provided in this paper. We have proved a soundness result (Theorem 55) and a strong completeness result (Theorem 57). We hope that as the theory of nominal algebra itself is improved, this will be of direct benefit to λ-calculus theory.

It would be interesting to consider direct 'nominal' versions of models of the λ-calculus, such as graph models or domain models of the λ-calculus [6, 30]. It would also be interesting to consider work using the language of categories (for example [29, 4]) using categories based on nominal sets.

It remains to develop the theory of nominal algebra itself, such as to prove the HSP theorem [8]; we would then be able to apply it to ULAM to study the λ-calculus, and likewise for other nominal algebraic theories.

It is also possible to investigate how well nominal algebra, or a system like it, can serve as the basis of a theorem-prover. Theorem-provers based on the λ-calculus [3] are the state of the art, and the Isabelle theorem-prover demonstrates how a weak meta-logic (such as Isabelle/Pure) can encode powerful object-logics (such as Isabelle/HOL) [24]. Is it possible that an elaboration of Nominal Algebra could serve as the foundation of a generic theorem-prover in the spirit of Isabelle, offering a new set of reasoning-principles 'ϵ away from informal practice'? The construction of ULAM in nominal algebra is a useful preliminary step.

5.4 Conclusions

The λ-calculus is a fundamental model of functions in logic and computation. We have given axioms ULAM for the untyped λ-calculus in *nominal algebra*, a recently-developed logical framework based on nominal techniques. This gives a logical theory pleasingly close to informal practice, while remaining mathematically completely rigorous. ULAM completes a trio of papers on nominal algebra and first-order logic [16], substitution [15], and with this paper, the λ-calculus. Researchers using nominal techniques might find ULAM and its completeness result a useful off-the-shelf component in later and larger works.

ULAM formally connects 'nominal atoms' and 'λ-calculus variables'. Discussions about this connection — usually based on not-entirely-explicit criteria of practical usefulness — have continued since nominal techniques were introduced [17] and they may continue into the forseeable future. ULAM makes a nice mathematical contribution to this discussion; the axioms of ULAM are an algebraic measure of the distance we must travel from nominal-style atoms and atoms-abstraction, to λ-calculus style variables and λ-binding. This is gives new sense of how nominal techniques fit into a long tradition of functions in logic and computation.

BIBLIOGRAPHY

[1] Peter B. Andrews. *An introduction to mathematical logic and type theory: to truth through proof.* Academic Press, 1986.

[2] Peter B. Andrews. Classical type theory. In *Handbook of Automated Reasoning*, volume 2, chapter 15, pages 965–1007. Elsevier Science, 2001.

[3] Peter B. Andrews, Matthew Bishop, Sunil Issar, Dan Nesmith, Frank Pfenning, and Hongwei Xi. TPS: A Theorem Proving System for Classical Type Theory. *Journal of Automated Reasoning*, 16(3):321–353, 1996.

[4] Giulio Manzonetto Antonio Bucciarelli, Thomas Ehrhard. Not Enough Points Is Enough. In *Computer Science Logic*, pages 298–312, 2007.

[5] Franz Baader and Tobias Nipkow. *Term rewriting and all that.* Cambridge University Press, Great Britain, 1998.

[6] H. P. Barendregt. *The Lambda Calculus: its Syntax and Semantics (revised ed.).* North-Holland, 1984.

[7] J. Barwise. An Introduction to First-Order Logic. In J. Barwise, editor, *Handbook of Mathematical Logic*, pages 5–46. North Holland, 1977.

[8] S. Burris and H. Sankappanavar. *A Course in Universal Algebra.* Graduate texts in mathematics. Springer, 1981.

[9] Ranald A. Clouston and Andrew M. Pitts. Nominal Equational Logic. *Electronic Notes in Theoretical Computer Science*, 172:223–257, 2007.

[10] Haskell B. Curry and R. Feys. *Combinatory Logic*, volume 1. North Holland, 1958.

[11] N. G. de Bruijn. Lambda Calculus Notation With Nameless Dummies, A Tool For Automatic Formula Manipulation, With Application To The Church-Rosser Theorem. *Indagationes Mathematicae*, 5(34):381–392, 1972.

[12] Maribel Fernández and Murdoch J. Gabbay. Curry-style types for nominal terms. In *Types for Proofs and Programs (proceedings of TYPES'06)*, volume 4502 of *Lecture Notes in Computer Science*, pages 125–139. Springer, 2007.

[13] Maribel Fernández and Murdoch J. Gabbay. Nominal rewriting (journal version). *Information and Computation*, 205(6):917–965, 2007.

[14] Murdoch J. Gabbay and Aad Mathijssen. A formal calculus for informal equality with binding. In *WoLLIC'07: 14th Workshop on Logic, Language, Information and Computation*, volume 4576 of *Lecture Notes in Computer Science*, pages 162–176, 2007.

[15] Murdoch J. Gabbay and Aad Mathijssen. Capture-Avoiding Substitution as a Nominal Algebra. *Formal Aspects of Computing*, 20(4-5):451–479, August 2008.

[16] Murdoch J. Gabbay and Aad Mathijssen. One-and-a-halfth-order Logic. *Journal of Logic and Computation*, 18(4):521–562, August 2008.

[17] Murdoch J. Gabbay and A. M. Pitts. A New Approach to Abstract Syntax with Variable Binding (journal version). *Formal Aspects of Computing*, 13(3–5):341–363, 2001.

[18] L. Henkin, J. D. Monk, and A. Tarski. *Cylindric Algebras.* North Holland, 1971 and 1985. Parts I and II.

[19] Wilfrid Hodges. Elementary predicate logic. In D.M. Gabbay and F. Guenthner, editors, *Handbook of Philosophical Logic, 2nd Edition*, volume 1, pages 1–131. Kluwer, 2001.

[20] Daniel Leivant. Higher order logic. In D. Gabbay, C.J. Hogger, and J.A. Robinson, editors, *Handbook of Logic in Artificial Intelligence and Logic Programming*, volume 2, pages 229–322. Oxford University Press, 1994.

[21] Pierre Lescanne. From Lambda-sigma to Lambda-upsilon: a Journey Through Calculi of Explicit Substitutions. In *POPL*, pages 60–69. ACM, 1994.

[22] G. Manzonetto and A. Salibra. Boolean Algebras for Lambda Calculus. In *21th IEEE Symposium on Logic in Computer Science (LICS 2006)*, pages 317–326. IEEE Computer Society, 2006.

[23] Aad Mathijssen. *Logical Calculi for Reasoning with Binding*. PhD thesis, Technische Universiteit Eindhoven, 2007.

[24] Lawrence C. Paulson. The Foundation of a Generic Theorem Prover. *Journal of Automated Reasoning*, 5(3):363–397, 1989.

[25] Lawrence C. Paulson. *ML for the working programmer (2nd ed.)*. Cambridge University Press, 1996.

[26] F. Pfenning and C. Elliot. Higher-Order Abstract Syntax. In *PLDI (Programming Language design and Implementation)*, pages 199–208. ACM Press, 1988.

[27] A. M. Pitts. Nominal Logic, A First Order Theory of Names and Binding. *Information and Computation*, 186(2):165–193, 2003.

[28] Antonino Salibra. On the algebraic models of lambda calculus. *Theoretical Computer Science*, 249(1):197–240, 2000.

[29] Peter Selinger. The lambda calculus is algebraic. *Journal of Functional Programming*, 12(6):549–566, 2002.

[30] Joseph E. Stoy. *Denotational Semantics: The Scott-Strachey Approach to Programming Language Theory*. MIT Press, Cambridge, MA, USA, 1977.

[31] Simon Thompson. *Haskell: The Craft of Functional Programming*. Addison Wesley, 1996.

[32] Christian Urban, Andrew M. Pitts, and Murdoch J. Gabbay. Nominal Unification. *Theoretical Computer Science*, 323(1–3):473–497, 2004.

Church and Curry: Combining Intrinsic and Extrinsic Typing

FRANK PFENNING

1 Introduction

Church's formulation of the simple theory of types [1] has the pleasing property that every well-formed term has a unique type. The type is thus an intrinsic attribute of a term. Furthermore, we can restrict attention to well-formed terms as the only meaningful ones, because the property of being well-formed is evidently and easily decidable.

Curry's formulation of combinatory logic [3], later adapted to the λ-calculus [9], assigns types to terms extrinsically and terms may have many different types. In other words, types capture properties of terms which have meaning independent of the types we might assign. This formulation very easily supports both finitary and infinitary polymorphism. Especially the former, expressed as *intersection types* [2], seems to be incompatible with uniqueness of types.

In this paper we show that it can be very fruitful to consider a two-layer approach. In the first layer, we have an intrinsically typed λ-calculus in the tradition of Church. A second layer of types, constructed in the tradition of Curry, captures properties of terms, but only those already well-formed according to the first layer. In order to avoid confusion, in the remainder of the paper we call types from the second layer *sorts* and use *types* only to refer to the first layer.

The resulting system combines the strengths of the two approaches. We can restrict attention to well-formed terms, and we exploit this when defining substitution and related operations. Moreover, we can easily define finitary sort polymorphism, and the system is quite precise in the properties it can assign to terms without losing decidability of sort checking.

The outline of the paper is as follows. We first define an intrinsically typed λ-calculus in the tradition of Church and consider canonical forms, which are β-normal and η-long. The theory of canonical forms includes a brief study of hereditary substitution and iterated expansion. We then define a system of sorts, including subsorting and intersection sorts, as a second layer. Of particular interest for this layer are the preservation of sorts under hereditary substitution, its converse, and iterated expansion. Finally we extend subsorting to higher types and conclude with some remarks on related and future work.

Not much in this paper is new: hereditary substitution and iterated expansion go back to Watkins [13], who devised them to deal with the complexities of a dependent type theory with linearity and a monad. Much of the work on sorts and subsorting at higher types is joint work with Lovas [10] in the context of a dependent type theory or with Dunfield [6] in the context of functional programming. My goal in writing this paper was to distill these ideas to their purest form to make them as accessible as I could. In the process I rediscovered some of the virtues of Church's presentation of type theory. I also came to understand much better that the construction of sorts as a refinement of types represents a synthesis of Church's and Curry's approaches to type theory which are often seen as antithetical.

2 An Intrinsically Typed λ-Calculus

Church's formulation of type theory has two base types: o for truth values and ι for individuals. Since we are interested in the underlying λ-calculus and not the logic, we just consider ι as the only base type. We use a modern notation for function types; Church wrote $(\beta\,\alpha)$ for $\alpha \to \beta$.

DEFINITION 1 (Types). We define the set of types inductively.

1. ι is a type.

2. If α and β are types, then $\alpha \to \beta$ is a type.

We assume that for each type α we have an infinite supply of variables x^α. We also have typed constants c^α. Well-formed terms (henceforth just called *terms*) satisfy the following inductive definition.

DEFINITION 2 (Well-formed terms). We define the set of well-formed terms M^α of type α.

1. Any variable x^α or constant c^α is a term.

2. If x^α is a variable and M^β a term then $(\lambda x.\, M)^{\alpha\to\beta}$ is a term.

3. If $M_1^{\alpha\to\beta}$ and M_2^α are terms, then $(M_1\, M_2)^\beta$ is a term.

It is an easy inductive property that the type of a term is unique, and that we can decide if a term is well-formed according to these rules. We elide the definitions defining the usual notions of free and bound variables and α-conversion, β-conversion, and η-conversion. We write $\text{fv}(M)$ for the free variables in a term M. As is common practice, we will mostly omit type superscripts, since they can usually be determined from context.

> **Convention.** *In the remainder of this paper we assume that all terms are well-formed according to the above definition. Moreover, we will tacitly apply α-conversion to satisfy conditions on bound variables.*

3 Canonical Forms

In the area of logical frameworks, the notion of *canonical form* is particularly important. The representation methodology of logical frameworks such as LF [8] or hereditary Harrop formulas [11] puts the canonical forms of a given type in one-to-one correspondence with the objects to be modelled. To achieve this kind of bijection, canonical forms should be β-normal and η-long. The following mutually inductive definition of canonical and atomic terms captures this.

DEFINITION 3 (Canonical and atomic terms).

1. If x^α is a variable and N^β is canonical then $(\lambda x.\, N)^{\alpha\to\beta}$ is canonical.
2. If R^ι is atomic then R^ι is canonical.
3. If $R^{\beta\to\alpha}$ is atomic and N^β is canonical then $(R\,N)^\alpha$ is atomic.
4. A variable x^α or constant c^α is atomic.

We have made the intrinisic types of the terms explicit, but except for clause 2 where we must specify the type ι, the types are entirely redundant.

We can see from the definition that a canonical term cannot contain a β-redex, because the left-hand side of an application $R\,N$ must be atomic, which can only be a variable or another application. It is also fully η-expanded because the body of a λ-term is either another λ-term (in which case we do not care about its type) or an atomic term, in which case it must be of type ι.

> **Convention.** *Unless explicitly noted otherwise, in the remainder of this paper M and N stand for canonical terms and R stands for atomic terms.*

4 Hereditary Substitution

Once α-conversion is understood, the crucial operation in the λ-calculus is substitution, written as $[M/x]N$. Church only considered it if the type of M matches the type of the variable x, and if the bound variables of N are distinct from x and the free variables of M so as to avoid variable capture. The condition on the bound variables of N can always be achieved by renaming of bound variables, so substitution is a total operation modulo α-conversion.

Unfortunately, canonical forms are not closed under substitution. A simple example is $[(\lambda x^\iota . x)/y](y^{\iota\to\iota}\, z^\iota)$ which yields $(\lambda x^\iota . x)\, z$. If we are to develop a complete theory of canonical forms, we need a different operation. The central insight is that instead of creating the redex $(\lambda x^\iota . x)\, z$ we can spawn another substitution operation, $[z/x^\iota]x$. This new substitution *is at a smaller type* than the original substitution. In this example, we reduced the type from $\iota \to \iota$ to ι. This observation holds in general, as we show below.

We define *hereditary substitution* using three related operations. The principal operation is the substitution of a canonical term into a canonical term, yielding a canonical term. We write $[M^\alpha/x^\alpha]^n(N^\beta) = N'^\beta$. When substituting for x in an atomic term R we need to distinguish two cases: either the variable at the head of R is x, or else it is different from x. In the former case we may need to substitute hereditarily so as to avoid creating a redex. In the latter case we just proceed compositionally since the head of the result will remain unchanged.

We write $[M/x^\alpha]^{rn}(R^\beta) = N^\beta$ if the head of R is x and $[M^\alpha/x^\alpha]^{rr}(R^\beta) = R'^\beta$ if the head of R is different from x. The superscript indicates if we are mapping an atomic term (r) to another atomic term (r) or to a canonical term (n).

DEFINITION 4 (Head). We define the head $\mathrm{hd}(R)$ of an atomic term R with

$$\begin{array}{lcl} \mathrm{hd}(x) & = & x \\ \mathrm{hd}(c) & = & c \\ \mathrm{hd}(R\,N) & = & \mathrm{hd}(R) \end{array}$$

DEFINITION 5 (Hereditary substitution). We define three forms of hereditary substitution

1. $[M^\alpha/x^\alpha]^n(N^\beta) = N'^\beta$,

2. $[M^\alpha/x^\alpha]^{rr}(R^\beta) = R'^\beta$ for $\mathrm{hd}(R) \neq x$, and

3. $[M^\alpha/x^\alpha]^{rn}(R^\beta) = N^\beta$ for $\mathrm{hd}(R) = x$

by the following equations. They constitute an inductive definition, first on the structure of α and second on the structure of N and R, as confirmed by the subsequent theorem.

$$
\begin{array}{lll}
[M/x^\alpha]^n(\lambda y.\, N) & = \lambda y.\, [M/x^\alpha]^n N & \text{provided } y \neq x \text{ and } y \notin \mathrm{fv}(M) \\
[M/x^\alpha]^n(R) & = [M/x^\alpha]^{rr}(R) & \text{if } \mathrm{hd}(R) \neq x \\
[M/x^\alpha]^n(R) & = [M/x^\alpha]^{rn}(R) & \text{if } \mathrm{hd}(R) = x
\end{array}
$$

$$
\begin{array}{lll}
[M/x^\alpha]^{rr}(R_1\, N_2) & = ([M/x^\alpha]^{rr} R_1)\, ([M/x^\alpha]^n N_2) & \\
[M/x^\alpha]^{rr}(y) & = y & \text{for } x \neq y \\
[M/x^\alpha]^{rr}(c) & = c &
\end{array}
$$

$$
\begin{array}{lll}
[M/x^\alpha]^{rn}(R_1\, N_2) & = [N_2'/y_2^{\beta_2}]^n(N_1) & \text{where } [M/x^\alpha]^{rn}(R_1) = \lambda y_2^{\beta_2}.\, N_1^{\beta_1} \\
& & \text{and } [M/x^\alpha]^n(N_2) = N_2' \\
[M/x^\alpha]^{rn}(x) & = M^\alpha &
\end{array}
$$

Hereditary substitution is always defined. The key observation is that if $[M/x^\alpha]^{rn}(R^\beta) = N^\beta$ then β is a subexpression of α. We refer to this property as *type reduction*. We write $\alpha \geq \beta$ if β is a subexpression of α, and $\alpha > \beta$ if β is a strict subexpression of α.[1] In the crucial last case of the hereditary substitution property below, we have $\alpha > \beta_2$ so the hereditary substitution does indeed take place at a strictly smaller type.

THEOREM 6 (Hereditary substitution).

1. $[M^\alpha/x^\alpha]^n(N^\beta) = N'^\beta$ *for a unique canonical* N'.

2. $[M^\alpha/x^\alpha]^{rr}(R^\beta) = R'^\beta$ *for a unique atomic* R' *if* $\mathrm{hd}(R) \neq x$.

3. $[M^\alpha/x^\alpha]^{rn}(R^\beta) = N'^\beta$ *for a unique canonical* N' *if* $\mathrm{hd}(R) = x$. *Furthermore, in this case,* $\alpha \geq \beta$.

Proof. Uniqueness is straightforward, since the clauses in the definition of hereditary substitution do not overlap.

We prove existence by nested induction, first on the type α and second the terms N and R. Furthermore, the type and term may remain the same when (1) appeals to (2) or (3), but must become strictly smaller when (2) or (3) appeal to (1).

Case(1): $N = \lambda y_2.\, N_1$ with $y_2 \neq x$ and $y_2 \notin \mathrm{fv}(M)$. Then

[1] This should not be confused with a subtyping as familiar from programming languages. We introduce a corresponding notion of *subsorting* later in this paper.

$[M/x^\alpha]^n N_1 = N_1'$ for some N_1' — by i.h.(1) on α and N_1

$[M/x^\alpha]^n(\lambda y_2.\, N_1) = \lambda y_2.\, N_1'$ — by defn. of $[\,]^n$.

Case(1): $N = R^\iota$ with $\mathrm{hd}(R) \neq x$. Then

$[M/x^\alpha]^{rr}(R) = R'$ for some R' — by i.h.(2) on α and R

$[M/x^\alpha]^n(R) = R'$ — by defn. of $[\,]^n$

Case(1): $N = R^\iota$ with $\mathrm{hd}(R) = x$. Then

$[M/x^\alpha]^{rn}(R) = N'$ for some N' — by i.h.(3) on α and R

$[M/x^\alpha]^n(R) = N'$ — by defn. of $[\,]^n$

Case(2): $R = x$. Impossible, since we assumed $\mathrm{hd}(R) \neq x$.

Case(2): $R = y$ with $y \neq x$. Then

$[M/x^\alpha]^{rr}(y) = y$ — by defn. of $[\,]^{rr}$

Case(2): $R = c$. As in the previous case.

Case(2): $R = R_1\, N_2$. Then

$[M/x^\alpha]^{rr}(R_1) = R_1'$ for some R_1' — by i.h.(2) on α and R_1

$[M/x^\alpha]^n(N_2) = N_2'$ for some N_2' — by i.h.(1) on α and N_2

$[M/x^\alpha]^{rr}(R_1\, N_2) = R_1'\, N_2'$ — by defn. of $[\,]^{rr}$

Case(3): $R = x$. Then

$[M/x^\alpha]^{rn}(x) = M^\alpha$ — by defn. of $[\,]^{rn}$

$\alpha \geq \alpha$ — by defn. of $\geq$

Case(3): $R = y$ with $y \neq x$ or $R = c$. Impossible, since we assumed $\mathrm{hd}(R) = x$.

Case(3): $R^\beta = R_1^{\beta_2 \to \beta_1}\, N_2^{\beta_2}$ with $\beta = \beta_1$. Then

$[M/x^\alpha]^{rn}(R_1) = N_1'$ for some $N_1'^{\beta_2 \to \beta_1}$ and

$\alpha \geq \beta_2 \to \beta_1$ — by i.h.(3) on α and R_1

$N_1'^{\beta_2 \to \beta_1} = \lambda y_2.\, N_1$ for some $y_2^{\beta_2}$ and $N_1^{\beta_1}$ — since N_1' is canonical

$[M/x^\alpha]^n(N_2) = N_2'$ — by i.h.(1) on α and N_2

$\alpha > \beta_2$ — by defn. of $>$

$[N_2'/y_2^{\beta_2}]^n(N_1) = N'$ — by i.h.(1) on β_2 and N_1

$[M/x^\alpha]^n(R_1\, N_2) = N'$ — by defn. of $[\,]^{rn}$

$\alpha > \beta_1$ — by defn. of $>$

■

The hereditary substitution theorem expresses that we can substitute canonical terms for variables (which are atomic) in a canonical term and obtain a canonical term.

Conversely, if we have an atomic term, we can convert it to a canonical term by a process analogous to several η-expansions.

DEFINITION 7 (Iterated expansion). We define $\eta^\alpha(R^\alpha) = N^\alpha$ by induction on α.

$$\begin{array}{lcll} \eta^\iota(R) & = & R & \\ \eta^{\alpha\to\beta}(R) & = & \lambda x^\alpha.\, \eta^\beta(R\, \eta^\alpha(x)) & \text{choosing } x \notin \text{fv}(R) \end{array}$$

Again, this is easily seen to be well-founded and to return a canonical term.

THEOREM 8 (Iterated expansion). $\eta^\alpha(R^\alpha) = N^\alpha$ *for some canonical* N.

Proof. By induction on the structure of α.

Case: $\alpha = \iota$. Then

$\eta^\iota(R) = R$	by defn. of η
R^ι canonical	by defn. of canonical

Case: $\alpha = \alpha_2 \to \alpha_1$. Then

Let $x_2^{\alpha_2}$ be a variable not in fv(R)	
$\eta^{\alpha_2}(x_2) = N_2$ for some canonical N_2	by i.h. on α_2
$\eta^{\alpha_1}(R\, N_2) = N_1$ for some canonical N_1	by i.h. on α_1
$\eta^\alpha(R) = \lambda x_2.\, N_1$	by defn. of η
$\lambda x_2.\, N_1$ canonical	by defn. of canonical

■

We have been especially detailed in the analysis of the definitions of hereditary substitution and iterated expansion because the inductive patterns of these definitions recur multiple times in our development below.

5 Composition and Identity

With ordinary substitution, we usually need some simple lemmas that show composition and identity properties. For example, $[M_1/x_1][M_2/x_2]N =$

$[[M_1/x_1]M_2/x_2]([M_1/x_1]N)$ and $[x/x]N = N$. The corresponding properties for hereditary substitution are bit more complex to state because we need to obey the discipline of canonical and atomic terms which entails that there are several forms of hereditary substitution, as we have seen in the previous section.

Also, proofs become a bit more tedious. In the cases of ordinary substitution above, they are straightforward by induction on the structure of N. Hereditary substitutions are defined by nested induction on a type and a term, so the proofs of the composition properties employ a corresponding nested induction.

When all is said and done, though, these proofs are really not much more difficult since the structure of the definitions guides every single step of the development.

We need one more preparatory lemma.

THEOREM 9 (Vacuous Substitution).

1. $[M/x]^n(N) = N$ *if* $x \notin \mathrm{fv}(N)$
2. $[M/x]^{rr}(R) = R$ *if* $x \notin \mathrm{fv}(R)$

Proof. By straightforward induction on the structure of N and R. Note that $[M/x]^{rn}(R)$ is never needed when $x \notin \mathrm{fv}(R)$ because $\mathrm{hd}(R)$ can not be x. ■

THEOREM 10 (Composition of hereditary substitutions). *Assume* $x_1 \neq x_2$ *and* $x_2 \notin \mathrm{fv}(M_1)$. *Then*

$$[M_1/x_1]^n([M_2/x_2]^n(N)) = [[M_1/x_1]^n(M_2)/x_2]^n([M_1/x]^n(N))$$

Proof. We generalize to the following statements, assuming $x_1 \neq x_2$ and $x_2 \notin \mathrm{fv}(M_1)$.

1. $[M_1/x_1]^n([M_2/x_2]^n(N)) = [[M_1/x_1]^n(M_2)/x_2]^n([M_1/x]^n(N))$
2. $[M_1/x_1]^{rr}([M_2/x_2]^{rr}(R)) = [[M_1/x_1]^n(M_2)/x_2]^{rr}([M_1/x_1]^{rr}(R))$ if $\mathrm{hd}(R) \neq x_1$ and $\mathrm{hd}(R) \neq x_2$
3. $[M_1/x_1]^n([M_2/x_2]^{rn}(R)) = [[M_1/x_1]^n(M_2)/x_2]^{rn}([M_1/x_1]^{rr}(R))$ if $\mathrm{hd}(R) = x_2$
4. $[M_1/x_1]^{rn}([M_2/x_2]^{rr}(R)) = [[M_1/x_1]^n(M_2)/x_2]^n([M_1/x_1]^{rn}(R))$ if $\mathrm{hd}(R) = x_1$

Let $x_1^{\alpha_1}$ and $x_2^{\alpha_2}$. Then the proof proceeds by nested induction, first on α_1 and α_2, and second on the terms N and R. Also, part (1) may appeal to parts (2), (3), and (4) with the same types and terms, but any other appeal has to strictly decrease the induction measure. For the outer induction, when α_1 decreases then α_2 stays the same and vice versa, and in the first case of part (4) they change roles, so one could take the unordered pair of the two types or the sum of their sizes as the induction measure.

Case(1): $N = \lambda y_2. N_1$.

$$\begin{array}{ll} [M_1/x_1]^n([M_2/x_2]^n(\lambda y_2. N_1)) & \\ = [M_1/x_1]^n(\lambda y_2. [M_2/x_2]^n(N_1)) & \text{by defn. of } []^n \\ = \lambda y_2. [M_1/x_1]^n([M_2/x_2]^n(N_1)) & \text{by defn. of } []^n \\ = \lambda y_2. [[M_1/x_1]^n(M_2)/x_2]^n([M_1/x_1]^n(N_1)) & \text{by i.h.(1) on } \alpha_1, \alpha_2, N_1 \\ = [[M_1/x_1](M_2)/x_2]^n([M_1/x_1](\lambda y_2. N_1)) & \text{by defn. of } []^n \text{ (twice)} \end{array}$$

Case(1): $N = R$ with $\mathrm{hd}(R_1) \neq x_1$ and $\mathrm{hd}(R_1) \neq x_2$.

$$\begin{array}{ll} [M_1/x_1]^n([M_2/x_2]^n(R)) & \\ = [M_1/x_1]^{rr}([M_2/x_2]^{rr}(R)) & \text{by defn. of } []^n \text{ (twice)} \\ = [[M_1/x_1]^n(M_2)/x_2]^{rr}([M_1/x_1]^{rr}(R)) & \text{by i.h.(2) on } \alpha_1, \alpha_2, R \\ = [[M_1/x_1]^n(M_2)/x_2]^n([M_1/x_1]^n(R)) & \text{by defn. of } []^n \text{ (twice)} \end{array}$$

Case(1): $N = R$ with $\mathrm{hd}(R) = x_2$.

$$\begin{array}{ll} [M_1/x_1]^n([M_2/x_2]^n(R)) & \\ = [M_1/x_1]^n([M_2/x_2]^{rn}(R)) & \text{by defn. of } []^n \\ = [[M_1/x_1]^n(M_2)/x_2]^{rn}([M_1/x_1]^{rr}(R)) & \text{by i.h.(3) on } \alpha_1, \alpha_2, R \\ = [[M_1/x_1]^n(M_2)/x_2]^n([M_1/x_1]^n(R)) & \text{defn. of } []^n \text{ (twice)} \end{array}$$

Case(1): $N = R$ with $\mathrm{hd}(R) = x_1$.

$$\begin{array}{ll} [M_1/x_1]^n([M_2/x_2]^n(R)) & \\ = [M_1/x_1]^{rn}([M_2/x_2]^{rr}(R)) & \text{by defn. of } []^n \text{ (twice)} \\ = [[M_1/x_1]^n(M_2)/x_2]^n([M_1/x_1]^{rn}(R)) & \text{by i.h.(4) on } \alpha_1, \alpha_2, R \\ = [[M_1/x_1]^n(M_2)/x_2]^n([M_1/x_1]^n(R)) & \text{by defn. of } []^n \end{array}$$

Case(2): $R = R_1\, N_2$ with $\mathrm{hd}(R) \neq x_1$ and $\mathrm{hd}(R) \neq x_2$.

$$\begin{array}{l} [M_1/x_1]^{rr}([M_2/x_2]^{rr}(R_1\, N_2)) \\ = [M_1/x_1]^{rr}([M_2/x_2]^{rr}(R_1))\ [M_1/x_1]^n([M_2/x_2]^n(N_2)) \end{array}$$

$$\text{by defn. of } [\,]^{rr} \text{ (twice)}$$
$$= [[M_1/x_1]^n(M_2)/x_2]^{rr}([M_1/x_1]^{rr}(R_1))\ [M_1/x_1]^n([M_2/x_2]^n(N_2))$$
$$\text{by i.h.(2) on } \alpha_1,\ \alpha_2,\ R_1$$
$$= [[M_1/x_1]^n(M_2)/x_2]^{rr}([M_1/x_1]^{rr}(R_1))$$
$$[[M_1/x_1]^n(M_2)/x_2]^n([M_1/x_1]^n(N_2)) \quad \text{by i.h.(1) on } \alpha_1,\ \alpha_2,\ N_2$$
$$= [[M_1/x_1]^n(M_2)/x_2]^{rr}([M_1/x_1]^{rr}(R_1\, N_2)) \quad \text{by defn. of } [\,]^{rr} \text{ (twice)}$$

Case(2): $R = y$ with $y \neq x_1$ and $y \neq x_2$.

$$[M_1/x_1]^{rr}([M_2/x_2]^{rr}(y))$$
$$= y \qquad \text{by defn. of } [\,]^{rr} \text{ (twice)}$$
$$= [[M_1/x_1]^{rr}(M_2)/x_2]([M_1/x_1]^{rr}(y)) \qquad \text{by defn. of } [\,]^{rr} \text{ (twice)}$$

Case(2): $R = c$. Like the previous case.

Case(3): $R = R_1\, N_2$ with $\mathrm{hd}(R_1) = x_2$.

$$[M_1/x_1]^n([M_2/x_2]^{rn}(R_1\, N_2))$$
$$= [M_1/x_1]^n([[M_2/x_2]^n(N_2)/y_2]^n(N_1)) \qquad \text{by defn. of } [\,]^{rn} \text{ and } (*)$$
$$= [[M_1/x_1]^n([M_2/x_2]^n(N_2))/y_2]^n([M_1/x_1]^n(N_1))$$
$$\text{by i.h.(1) on } \alpha_1 \text{ and } \beta_2$$
$$= [[M_1/x_1]^n(M_2)/x_2]^{rn}([M_1/x_1]^{rr}(R_1)\,[M_1/x_1]^n(N_2))$$
$$\text{by defn. of } [\,]^{rn} \text{ and } (**) \text{ and } (***)$$
$$= [[M_1/x_1]^n(M_2)/x_2]^{rn}([M_1/x_1]^{rr}(R_1\, N_2)) \qquad \text{by defn. of } [\,]^{rr}$$

where (*)

$[M_2/x_2]^{rn}(R_1) = \lambda y_2.\, N_1$ for some $y_2^{\beta_2}$ and $N_1^{\beta_1}$ — by hered. subst.
and $\alpha_2 \geq \beta_2 \to \beta_1$ — by type reduction of $[\,]^{rn}$

and (**)

$$[[M_1/x_1]^n(M_2)/x_2]^{rn}([M_1/x_1]^{rr}(R_1))$$
$$= [M_1/x_1]^n([M_2/x_2]^{rn}(R_1)) \qquad \text{by i.h.(3) on } \alpha_1,\ \alpha_2,\ R_1$$
$$= [M_1/x_1]^n(\lambda y_2.\, N_1) \qquad \text{equality } (*)$$
$$= \lambda y_2.\, [M_1/x_1]^n(N_1) \qquad \text{by defn. of } [\,]^n$$

and (***)

$$[[M_1/x_1]^n(M_2)/x_2]^n([M_1/x_1]^n(N_2))$$
$$= [M_1/x_1]^n([M_2/x_2]^n(N_2)) \qquad \text{by i.h.(1) on } \alpha_1,\ \alpha_2,\ N_2$$

Case(3): $R = x_2$.

$$
\begin{array}{ll}
[M_1/x_1]^n([M_2/x_2]^{rn}(x_2)) & \\
= [M_1/x_1]^n(M_2) & \text{by defn. of } [\,]^{rn} \\
= [[M_1/x_1]^n(M_2)/x_2]^{rn}(x_2) & \text{by defn. of } [\,]^{rn} \\
= [[M_1/x_1]^n(M_2)/x_2]^{rn}([M_1/x_1]^{rr}(x_2)) & \text{by defn. of } [\,]^{rr}
\end{array}
$$

Case(4): $R = R_1\, N_2$ with $\mathrm{hd}(R_1) = x_1$.

$$
\begin{array}{l}
[M_1/x_1]^{rn}([M_2/x_2]^{rr}(R_1\, N_2)) \\
= [M_1/x_1]^{rn}([M_2/x_2]^{rr}(R_1)\, [M_2/x_2]^n(N_2)) \qquad \text{by defn. of } [\,]^{rr} \\
= [[[M_1/x_1]^n(M_2)/x_2]^n([M_1/x_1]^n(N_2))/y_2]^n([[M_1/x_1](M_2)/x_2](N_1)) \\
\hfill \text{by defn. of } [\,]^{rn}, (**) \text{ and } (***) \\
= [[M_1/x_1]^n(M_2)/x_2]^n([[M_1/x_1]^n(N_2)/y_2]^n(N_1)) \\
\hfill \text{by i.h.(1) on } \alpha_2, \beta_2 \\
= [[M_1/x_1]^n(M_2)/x_2]^n([M_1/x_1]^{rn}(R_1\, N_2)) \quad \text{by defn. of } [\,]^{rn} \text{ and } (*)
\end{array}
$$

where (*)

$$
\begin{array}{ll}
[M_1/x_1]^{rn}(R_1) = \lambda y_2.\, N_1 \text{ for some } y_2^{\beta_2} \text{ and } N_1^{\beta_1} & \text{by hered. subst.} \\
\text{and } \alpha_1 \geq \beta_2 \rightarrow \beta_1 & \text{by type reduction of } [\,]^{rn}
\end{array}
$$

and (**)

$$
\begin{array}{ll}
[M_1/x_1]^{rn}([M_2/x_2]^{rr}(R_1)) & \\
= [[M_1/x_1]^n(M_2)/x_2]^n([M_1/x_1]^{rn}(R_1)) & \text{by i.h.(4) on } \alpha_1, \alpha_2 \text{ and } R_1 \\
= [[M_1/x_1]^n(M_2)/x_2]^n(\lambda y_2.\, N_1) & \text{by equality } (*) \\
= \lambda y_2.\, [[M_1/x_1]^n(M_2)/x_2]^n(N_1) & \text{by defn. of } [\,]^n
\end{array}
$$

and (***)

$$
\begin{array}{ll}
[M_1/x_1]^n([M_2/x_2]^n(N_2)) & \\
= [[M_1/x_1]^n(M_2)/x_2]^n([M_1/x_1]^n(N_2)) & \text{by i.h.(1) on } \alpha_1, \alpha_2, N_2
\end{array}
$$

Case(4): $R = x_1$.

$$
\begin{array}{ll}
[M_1/x_1]^{rn}([M_2/x_2]^{rr}(x_1)) & \\
= [M_1/x_1]^{rn}(x_1) & \text{by defn. of } [\,]^{rr} \\
= M_1 & \text{by defn. of } [\,]^{rn} \\
= [[M_1/x_1]^n(M_2)/x_2]^n(M_1) & \text{by vacuous substitution} \\
= [[M_1/x_1]^n(M_2)/x_2]^n([M_1/x_1]^{rn}(x_1)) & \text{by defn. of } [\,]^{rn}
\end{array}
$$

■

The identity property of the ordinary substitution $[x/x]N = N$ is almost trivial, and the reverse $[N/x]x = N$ is part of the definition of substitution.

Here, the identity properties reveal an interplay between iterated expansion and hereditary substitution. We cannot substitute $[x/x]$ because x may not be canonical. Instead, we substitute the iterated expansion of x, so the *left identity* property shows that the iterated expansion of a variable behaves like the variable itself under substitution. Conversely, the *right identity* shows that substituting into the expansion of a variable amounts to substituting into the variable itself.

THEOREM 11 (Identity of iterated expansion).

1. *(Left identity)* $[\eta^\alpha(x)/x]^n(N) = N$
2. *(Right identity)* $[N/x]^n(\eta^\alpha(x)) = N$

Proof. We generalize to the following five properties.

1. $[\eta^\alpha(x)/x]^n(N) = N$
2. $[\eta^\alpha(x)/x]^{rr}(R) = R$ if $\mathrm{hd}(R) \neq x$
3. $[\eta^\alpha(x)/x]^{rn}(R) = \eta^\beta(R^\beta)$ if $\mathrm{hd}(R) = x$
4. $[N/x]^n(\eta^\alpha(R)) = \eta^\alpha([N/x]^{rr}(R))$ if $\mathrm{hd}(R) \neq x$
5. $[N/x]^n(\eta^\alpha(R)) = [N/x]^{rn}(R)$ if $\mathrm{hd}(R) = x$

Right identity follows from part (5) using $R = x$.

The proof is by nested induction, first on α and second on N and R. Also, (1) may appeal to (2) and (3) with unchanged type or term, but appeals from (2) and (3) to any part will strictly decrease the measure of the parameters.

Case(1): $N = \lambda y_2.\, N_1$ for $y_2 \neq x$.

$$
\begin{array}{lr}
[\eta^\alpha(x)/x]^n(\lambda y_2.\, N_1) & \\
= \lambda y_2.\, [\eta^\alpha(x)/x]^n(N_1) & \text{by defn. of } [\,]^n \\
= \lambda y_2.\, N_1 & \text{by i.h.(1) on } \alpha \text{ and } N_1
\end{array}
$$

Case(1): $N = R^\iota$ with $\mathrm{hd}(R) \neq x$.

$$
\begin{array}{lr}
[\eta^\alpha(x)/x]^n(R) & \\
= [\eta^\alpha(x)/x]^{rr}(R) & \text{by defn. of } [\,]^n \\
= R & \text{by i.h.(2) on } \alpha \text{ and } R
\end{array}
$$

Case(1): $N = R^\iota$ with $\mathrm{hd}(R) = x$.

$[\eta^\alpha(x)/x]^n(R)$	
$= [\eta^\alpha(x)/x]^{rn}(R)$	by defn. of $[\,]^n$
$= \eta^\iota(R)$	by i.h.(3) on α and R
$= R$	by defn. of η

Case(2): $R = R_1\,N_2$.

$[\eta^\alpha(x)/x]^{rr}(R_1\,N_2)$	
$= [\eta^\alpha(x)/x]^{rr}(R_1)\,[\eta^\alpha(x)/x]^n(N_2)$	by defn. of $[\,]^{rr}$
$= R_1\,[\eta^\alpha(x)/x]^n(N_2)$	by i.h.(2) on α and R_1
$= R_1\,N_2$	by i.h.(1) on α and N_2

Case(2): $R = y$ for $y \neq x$.

$[\eta^\alpha(x)/x]^{rr}(y)$	
$= y$	by defn. of $[\,]^{rr}$

Case(2): $R = c$. Like the previous case.

Case(3): $R = R_1^{\beta_2 \to \beta_1}\,N_2^{\beta_2}$ where $\mathrm{hd}(R_1) = x$.

$[\eta^\alpha(x)/x]^{rn}(R_1\,N_2)$	
$= [N_2/y_2]^n(\eta^{\beta_1}(R_1\,\eta^{\beta_2}(y_2)))$	by defn. of $[\,]^{rn}$, (*) and (**)
$= \eta^{\beta_1}([N_2/y_2]^{rr}(R_1\,\eta^{\beta_2}(y_2)))$	by i.h.(4) on β_1
$= \eta^{\beta_1}([N_2/y_2]^{rr}(R_1)\,[N_2/y_2]^n(\eta^{\beta_2}(y_2)))$	by defn. of $[\,]^{rr}$
$= \eta^{\beta_1}(R_1\,[N_2/y_2]^n(\eta^{\beta_2}(y_2)))$	by vacuous substitution
$= \eta^{\beta_1}(R_1\,[N_2/y_2]^{rn}(y_2))$	by i.h.(5) on β_2
$= \eta^{\beta_1}(R_1\,N_2)$	by defn. of $[\,]^{rn}$

where (*)

$[\eta^\alpha(x)/x]^{rn}(R_1)$	
$= \eta^{\beta_2 \to \beta_1}(R_1)$	by i.h.(3) on α and R_1
$= \lambda y_2.\,\eta^{\beta_1}(R_1\,\eta^{\beta_2}(y_2))$	by defn. of η
$\alpha \geq \beta_2 \to \beta_1$	by type reduction of $[\,]^{rn}$

and (**)

$[\eta^\alpha(x)/x]^n(N_2)$	
$= N_2$	by i.h.(1) on α and N_2

Case(3): $R = x$.

$$
\begin{aligned}
&[\eta^{\alpha}(x)/x]^{rn}(x) \\
&= \eta^{\alpha}(x) && \text{by defn. of } [\,]^{rn}
\end{aligned}
$$

Case(4): $\alpha = \alpha_2 \to \alpha_1$ and $\mathrm{hd}(R) \neq x$.

$$
\begin{aligned}
&[N/x]^n(\eta^{\alpha_2\to\alpha_1}(R)) \\
&= [N/x]^n(\lambda y_2.\, \eta^{\alpha_1}(R\,\eta^{\alpha_2}(y_2))) && \text{by defn. of } \eta \\
&= \lambda y_2.\, [N/x]^n(\eta^{\alpha_1}(R\,\eta^{\alpha_2}(y_2))) && \text{by defn. of } [\,]^n \\
&= \lambda y_2.\, \eta^{\alpha_1}([N/x]^{rr}(R\,\eta^{\alpha_2}(y_2))) && \text{by i.h.(4) on } \alpha_1 \\
&= \lambda y_2.\, \eta^{\alpha_1}([N/x]^{rr}(R)\,[N/x]^n(\eta^{\alpha_2}(y_2))) && \text{by defn. of } [\,]^{rr} \\
&= \lambda y_2.\, \eta^{\alpha_1}([N/x]^{rr}(R)\,\eta^{\alpha_2}([N/x]^{rr}(y_2))) && \text{by i.h.(4) on } \alpha_2 \\
&= \lambda y_2.\, \eta^{\alpha_1}([N/x]^{rr}(R)\,\eta^{\alpha_2}(y_2)) && \text{by defn. of } [\,]^{rr} \\
&= \eta^{\alpha_2\to\alpha_1}([N/x]^{rr}(R)) && \text{by defn. of } \eta
\end{aligned}
$$

Case(4): $\alpha = \iota$ and $\mathrm{hd}(R) \neq x$.

$$
\begin{aligned}
&[N/x]^n(\eta^{\iota}(R)) \\
&= [N/x]^n(R) && \text{by defn. of } \eta \\
&= [N/x]^{rr}(R) && \text{by defn. of } [\,]^n \\
&= \eta^{\iota}([N/x]^{rr}(R)) && \text{by defn. of } \eta
\end{aligned}
$$

Case(5): $\alpha = \alpha_2 \to \alpha_1$ and $\mathrm{hd}(R) = x$.

$$
\begin{aligned}
&[N/x]^n(\eta^{\alpha_2\to\alpha_1}(R)) \\
&= [N/x]^n(\lambda y_2.\, \eta^{\alpha_1}(R\,\eta^{\alpha_2}(y_2))),\ y_2 \notin \mathrm{fv}(N) \cup \mathrm{fv}(R) && \text{by defn. of } \eta \\
&= \lambda y_2.\, [N/x]^n(\eta^{\alpha_1}(R\,\eta^{\alpha_2}(y_2))) && \text{by defn. of } [\,]^n \\
&= \lambda y_2.\, [N/x]^{rn}(R\,\eta^{\alpha_2}(y_2)) && \text{by i.h.(5) on } \alpha_1 \\
&= \lambda y_2.\, [\eta^{\alpha_2}(y_2)/y_2]^n(N_1) && \text{by defn. of } [\,]^{rn}\text{, (*) and (**)} \\
&= \lambda y_2.\, N_1 && \text{by i.h.(1) on } \alpha_2 \\
&= [N/x]^{rn}(R) && \text{by equality (*)}
\end{aligned}
$$

where (*)

$$
\begin{aligned}
&[N/x]^{rn}(R) \\
&= \lambda y_2.\, N_1 \text{ for some } N_1^{\alpha_1} && \text{by hereditary substitution} \\
& && \text{and renaming, since } y_2 \notin \mathrm{fv}(N) \cup \mathrm{fv}(R)
\end{aligned}
$$

and (**)

$$
\begin{aligned}
&[N/x]^n(\eta^{\alpha_2}(y_2)) \\
&= \eta^{\alpha_2}([N/x]^{rr}(y_2)) && \text{by i.h.(4) on } \alpha_2 \\
&= \eta^{\alpha_2}(y_2) && \text{by defn. of } [\,]^{rr}
\end{aligned}
$$

Case(5): $\alpha = \iota$ and $\mathrm{hd}(R) = x$.

$$\begin{array}{ll} [N/x]^n(\eta^\iota(R)) & \\ = [N/x]^n(R) & \text{by defn. of } \eta \\ = [N/x]^{rn}(R) & \text{by defn. of } [\,]^n \end{array}$$

■

6 Simple Sorts

Now that we have defined an intrinsically typed λ-calculus in the tradition of Church, we can define a second layer of *sorts* in the tradition of Curry [3]. In this and the following section we temporarily allow general well-formed terms, and not just canonical forms.

We begin by considering the type ι. For Church's purposes, a single type of individuals was sufficient: other "types" could be explicitly defined using the higher-order features of his type theory, or one could define new constants and represent sorts as predicates. For example, if we wanted to introduce natural numbers, we might declare constants z^ι, $\mathsf{s}^{\iota\to\iota}$, and $\mathsf{nat}^{\iota\to o}$ and the assumptions

$$\begin{array}{l} \mathsf{nat}(\mathsf{z}) \\ \forall x^\iota.\, \mathsf{nat}(x) \supset \mathsf{nat}(\mathsf{s}(x)) \end{array}$$

However, reasoning with natural numbers now requires a good deal of explicit inferences with the nat predicate, while from a modern perspective it seems that static type-checking should be able ascertain when a given term represents a natural number.

In order to explore this, we allow declarations of sorts such as nat. In addition we can declare new constants and give them sorts. For example:

$$\begin{array}{lcl} \mathsf{nat}^\iota \ \text{sort} & & \\ \mathsf{z}^\iota & : & \mathsf{nat} \\ \mathsf{s}^{\iota\to\iota} & : & \mathsf{nat} \to \mathsf{nat} \end{array}$$

The first declaration states that nat is a new sort constant that represents a subset of the terms of type ι. We say that nat *refines* ι. We use Q for sort constants, also called base sorts.

From the last declaration we can see that for this new device to be useful, the language of sorts needs to include *function sorts*, and they must match the types in a consistent way. Declarations for base sorts and constants are collected in a signature Σ. We call these simple sorts, since we extend them later to include intersections, and because they match simple types.

DEFINITION 12 (Simple sorts). We define simple sorts inductively with respect to a signature Σ.

1. A base sort Q^ι declared in Σ is a simple sort.

2. If S^α and T^β are simple sorts, then $(S \to T)^{\alpha\to\beta}$ is a simple sort.

The general form of a constant declaration is $c^\alpha : S^\alpha$, that is, the sort assigned to a constant must refine its type. This is essential in this two-layer approach so we do not lose the good properties of the underlying intrinsically typed λ-calculus.

We follow a similar restriction when we assign sorts to variables $x^\alpha : S^\alpha$. Since sorts are extrinsic to variables, just as they are extrinsic to terms, we must now explicitly track the sort we would like to assign to a variable. We collect the sort assignments to variables in a *context* Γ which is a collection of variable declarations as above. We write $\mathrm{dv}(\Gamma)$ for the set of variables declared in a context.

More generally, type assignment is a relation between a term and a sort in a context. We write $\Gamma \vdash M^\alpha : S^\alpha$. In modern parlance we refer to this as a *judgment* and define it by a set of inference rules, but it could just as well be written out as an inductive definition. Interestingly, Curry's original paper also used an axiomatic formulation to establish that a term has a sort by logical reasoning, while Church's original paper gave typing as an explicitly inductive definition. We assume all sort and term constants are declared in a fixed signature Σ.

$$\frac{x{:}S \in \Gamma}{\Gamma \vdash x : S} \qquad \frac{c{:}S \in \Sigma}{\Gamma \vdash c : S}$$

$$\frac{\Gamma, x{:}S \vdash M : T \quad (x \notin \mathrm{dv}(\Gamma))}{\Gamma \vdash \lambda x.\, M : S \to T} \quad \frac{\Gamma \vdash M : S \to T \quad \Gamma \vdash N : S}{\Gamma \vdash M\, N : T}$$

We have omitted all the types, but it is essential to remember that we consider $M : S$ only when M has type α and S refines α, that is, $M^\alpha : S^\alpha$. If we ignore this restriction, then the rules above define a Curry-Howard style type assignment system for the simply-typed λ-calculus [9] which is essentially isomorphic to Church's system. We do not precisely formulate or prove such a theorem here in order to concentrate on richer sort systems.

7 Subsorts

The first generalization will be to allow subsorts. Continuing the previous example, we can distinguish the sort with just the constant z (zero) and the sort of all positive terms (pos). We also express that they are *subsorts* of the sort of natural numbers (nat).

$$\begin{array}{lcl} \mathsf{zero} & \leq & \mathsf{nat} \\ \mathsf{pos} & \leq & \mathsf{nat} \\ \mathsf{z} & : & \mathsf{zero} \\ \mathsf{s} & : & \mathsf{nat} \rightarrow \mathsf{pos} \end{array}$$

We now also allow multiple sort declarations for constants. Since a sort represents a property of terms, this is perfectly natural. We also need a rule of *subsumption*, allowing us to exploit the knowledge of subsorting. Remarkably, the remainder of the system does not need to be changed.

$$\frac{}{Q \leq Q} \qquad \frac{Q_1 \leq Q_2 \quad Q_2 \leq Q_3}{Q_1 \leq Q_3}$$

$$\frac{\Gamma \vdash M : Q \quad Q \leq Q'}{\Gamma \vdash M : Q'}$$

We can now prove, for example, $\mathsf{s}\,\mathsf{z} : \mathsf{pos}$ or $\lambda x.\, \mathsf{s}\, x : \mathsf{zero} \rightarrow \mathsf{pos}$.

However, this system is unfortunately very weak. For example, $\lambda x.\, x$ has the property expressed by the sort $(\mathsf{nat} \rightarrow \mathsf{zero}) \rightarrow (\mathsf{zero} \rightarrow \mathsf{nat})$. In words: any function mapping an arbitrary natural number to 0 will also map 0 to some natural number. But this typing cannot be established in the system above, because subtyping is too weak. Fortunately, this problem can be solved by strengthening the subtyping relation and raising it to higher type, as we will do in Section 12.

8 Sort Checking Canonical Forms

We now restrict our attention again to canonical forms. Careful study of a sorting system for canonical forms can be transferred to a calculus that permits arbitrary well-formed terms, although we do not pursue this transfer in this paper.

The rules for sort assignment in the previous section do not immediately describe an algorithm for performing sort checking. If we try to infer sorts for terms, then the rule for λ-abstraction presents a problem in that, even if we can infer the sort of the body, the sort of the bound variable has to be guessed. Conversely, if we instead try to check a term against a known sort, then the rule for application creates problems because we do not know the sort of the argument. This is in contrast to Church's system where we can systematically construct the unique type of a term (or fail, if it is not well-formed). One avenue now would be to introduce unification into the sort-checking process, but this will be difficult to extend to more expressive sort systems. Instead, we observe that on *canonical forms* sort-checking can be performed very elegantly.

We distinguish two judgments, $\Gamma \vdash N \Leftarrow S$ and $\Gamma \vdash R \Rightarrow S$. For the first one, we envision Γ, N, and S to be given, and we want to verify if N has sort S. For the second, we assume that Γ and R are given, and we want to synthesize a sort S for R. At the place where checking meets synthesis (or: where an atomic term is viewed as a canonical one) we must have a base sort that refines ι.

$$\frac{x{:}S \in \Gamma}{\Gamma \vdash x \Rightarrow S} \qquad \frac{c{:}S \in \Sigma}{\Gamma \vdash c \Rightarrow S}$$

$$\frac{\Gamma, x{:}S \vdash N \Leftarrow T \quad (x \notin \mathrm{dv}(\Gamma))}{\Gamma \vdash \lambda x.\, N \Leftarrow S \to T} \qquad \frac{\Gamma \vdash R \Rightarrow S \to T \quad \Gamma \vdash N \Leftarrow S}{\Gamma \vdash R\, N \Rightarrow T}$$

$$\frac{\Gamma \vdash R \Rightarrow Q' \quad Q' \leq Q}{\Gamma \vdash R \Leftarrow Q}$$

This system can easily be recognized as a decision procedure for sorting of terms, based on the intuition given above, as long as basic subsorting is also decidable. Since $Q' \leq Q$ is just the reflexive and transitive closure of the subsorting declarations in the signature, this also holds.

We hold off on a formal statement of this property, since we prove a more general statement in the next section. We close this section by reconsidering the example showing incompleteness of the rules from the previous section. We noted that

$$\not\vdash \lambda x.\, x : (\mathsf{nat} \to \mathsf{zero}) \to (\mathsf{zero} \to \mathsf{nat})$$

This term, $\lambda x^{\iota \to \iota}.\, x$, is not canonical. If we expand it to its canonical form we obtain

$$\lambda x^{\iota \to \iota}.\, \eta^{\iota \to \iota}(x) = \lambda x.\, \lambda y.\, \eta^{\iota}(x\, \eta^{\iota}(y)) = \lambda x.\, \lambda y.\, x\, y$$

Now we can indeed verify that

$$\vdash \lambda x.\, \lambda y.\, x\, y \Leftarrow (\mathsf{nat} \to \mathsf{zero}) \to (\mathsf{zero} \to \mathsf{nat})$$

This derivation uses $\mathsf{zero} \leq \mathsf{nat}$ twice: once for y and once for $x\, y$.

9 Intersection Sorts

The system so far, on canonical forms, has a number of desirable properties, but it lacks the basic ability to ascribe more than one property to a term.

So we can prove, for example,

$$\begin{array}{l} \vdash \lambda x^{\iota}.\, x \Leftarrow \mathsf{nat} \to \mathsf{nat} \\ \vdash \lambda x^{\iota}.\, x \Leftarrow \mathsf{zero} \to \mathsf{zero} \\ \vdash \lambda x^{\iota}.\, x \Leftarrow \mathsf{pos} \to \mathsf{pos} \end{array}$$

but we cannot combine these pieces of information into a single judgment. At first one might think of a form of parametric polymorphism to solve this problem, for example, $\forall q^{\iota}.\, q \to q$ as a common sort. To see that this would not be sufficient, consider the additional sorts even and odd.

$$\begin{array}{lcl} \mathsf{even} & \leq & \mathsf{nat} \\ \mathsf{odd} & \leq & \mathsf{nat} \\ \mathsf{z} & : & \mathsf{even} \\ \mathsf{s} & : & \mathsf{even} \to \mathsf{odd} \\ \mathsf{s} & : & \mathsf{odd} \to \mathsf{even} \end{array}$$

Now we have, for example,

$$\begin{array}{l} \vdash \lambda x^{\iota}.\, \mathsf{s}(\mathsf{s}(\mathsf{s}\, x)) \Leftarrow \mathsf{even} \to \mathsf{odd} \\ \vdash \lambda x^{\iota}.\, \mathsf{s}(\mathsf{s}(\mathsf{s}\, x)) \Leftarrow \mathsf{odd} \to \mathsf{even} \end{array}$$

but we cannot combine them using universal quantification.

To increase the expressive power we introduce finitary polymorphism in the form of *intersection sorts*, $S_1 \wedge S_2$. Because we construct a Curry-style sort assignment system on top of Church-style typing, a single term M can only be assigned two sorts S_1 and S_2 if $M^{\alpha} : S_1^{\alpha}$ and $M^{\alpha} : S_2^{\alpha}$. Therefore, to form the intersection, both sorts must refine the same type $S_1^{\alpha} \wedge S_2^{\alpha}$, and the intersection will again refine α.

The rules to introduce and eliminate intersection are quite straightforward: a term M, which need not be canonical, has sort $S_1 \wedge S_2$ if and only if it has both sorts.

$$\frac{\Gamma \vdash M : S_1 \quad \Gamma \vdash M : S_2}{\Gamma \vdash M : S_1 \wedge S_2} \qquad \frac{\Gamma \vdash M : S_1 \wedge S_2}{\Gamma \vdash M : S_1} \qquad \frac{\Gamma \vdash M : S_1 \wedge S_2}{\Gamma \vdash M : S_2}$$

To integrate them into the canonical forms system, we have to think about the direction they should apply.

The first consideration is the availability of information. In the checking rules, we know the sort we are checking against, and we must make sure we also know the sort in the premises. This means that the intersection introduction rule on the left should be a checking rule. In the synthesis rules we may assume that the premise yields a type, but we must make sure we can synthesize the type in the conclusion. This means that the two elimination rules on the right should be synthesis rules.

$$\frac{\Gamma \vdash N \Leftarrow S_1 \quad \Gamma \vdash N \Leftarrow S_2}{\Gamma \vdash N \Leftarrow S_1 \wedge S_2} \qquad \frac{\Gamma \vdash R \Rightarrow S_1 \wedge S_2}{\Gamma \vdash R \Rightarrow S_1} \quad \frac{\Gamma \vdash R \Rightarrow S_1 \wedge S_2}{\Gamma \vdash R \Rightarrow S_2}$$

We can also think about it from the logical perspective: an introduction rule, read bottom-up, will decompose the formula (here: the sort $S_1 \wedge S_2$). It should therefore become a checking rule, which also decomposes the sort bottom-up. Conversely, an elimination rule, read top-down, will decompose a formula (here: the sort $S_1 \wedge S_2$). It should therefore become a synthesis rule which also decomposes the sort top-down.

We introduce one more sort, $\top$, which is the unit of intersection and can be thought of as a conjunction of zero conjuncts. The introduction rule (checking), therefore, has zero premises which must be satisfied. And instead of two eliminations rules (synthesis) we have none.

$$\frac{}{\Gamma \vdash N \Leftarrow \top} \qquad \text{No } \top \text{ elimination rule}$$

This rule may at first appear troublesome, but due to our two-layer construction it does not mean that every term is well-sorted, just those that are well-formed according to the typing discipline. So $\top$ is just an uninformative property of well-formed terms.

We consolidate our language of sorts into a definition, extending the previous definition of simple sorts.

DEFINITION 13 (Sorts). We define sorts inductively with respect to a signature Σ.

1. A base sort Q^ι declared in Σ is a sort.

2. If S^α and T^β are sorts, then $(S \to T)^{\alpha \to \beta}$ is a sort.

3. If S^α and T^α are sorts then $(S \wedge T)^\alpha$ is a sort.

4. $\top^\alpha$ is a sort for each type α.

The complete rules for sort assignment can be found in Figure 1. Now we can verify that, indeed, at least for canonical forms sort checking is decidable.

THEOREM 14 (Decidability of sort checking). *Assume a signature Σ with sort declarations is fixed.*

Subsorting $Q' \leq Q$.

$$\frac{}{Q \leq Q} \qquad \frac{Q_1 \leq Q_2 \quad Q_2 \leq Q_3}{Q_1 \leq Q_3} \qquad \frac{Q' \leq Q \in \Sigma}{Q' \leq Q}$$

Sort checking $\Gamma \vdash N^\alpha \Leftarrow S^\alpha$ and
sort synthesis $\Gamma \vdash R^\alpha \Rightarrow T^\alpha$.

$$\frac{x{:}S \in \Gamma}{\Gamma \vdash x \Rightarrow S} \qquad \frac{c{:}S \in \Sigma}{\Gamma \vdash c \Rightarrow S}$$

$$\frac{\Gamma, x{:}S \vdash N \Leftarrow T \quad (x \notin \mathrm{dv}(\Gamma))}{\Gamma \vdash \lambda x.\, N \Leftarrow S \to T} \qquad \frac{\Gamma \vdash R \Rightarrow S \to T \quad \Gamma \vdash N \Leftarrow S}{\Gamma \vdash R\,N \Rightarrow T}$$

$$\frac{\Gamma \vdash R \Rightarrow Q' \quad Q' \leq Q}{\Gamma \vdash R \Leftarrow Q}$$

$$\frac{\Gamma \vdash N \Leftarrow S_1 \quad \Gamma \vdash N \Leftarrow S_2}{\Gamma \vdash N \Leftarrow S_1 \wedge S_2} \qquad \frac{\Gamma \vdash R \Rightarrow S_1 \wedge S_2}{\Gamma \vdash R \Rightarrow S_1} \qquad \frac{\Gamma \vdash R \Rightarrow S_1 \wedge S_2}{\Gamma \vdash R \Rightarrow S_2}$$

$$\frac{}{\Gamma \vdash N \Leftarrow \top} \qquad \text{No } \top \text{ elimination rule}$$

Figure 1. Sort assignment rules for canonical and atomic terms

1. It is decidable whether $Q' \leq Q$.

2. It is decidable whether $\Gamma \vdash N \Leftarrow S$.

3. There is a finite number of sorts T such that $\Gamma \vdash R \Rightarrow T$.

We assume that all judgments are well-formed in the manner explained before. For example N must be of some type α and S must refine α.

Proof. Decidability of subsorting follows since it is just the reflexive and transitive closure of the given subsort declarations.

Parts (2) and (3) follow by nested induction, first on the terms N and R. Second, (2) may appeal to (3) with the same term, but (3) may appeal to (2) only on a strictly smaller term. Finally within each part we may break down the structure of S and T if the term stays the same.

Case(2) $S = \top$. Then $\Gamma \vdash N \Leftarrow \top$ by rule.

Case(2) $S = S_1 \wedge S_2$. Then $\Gamma \vdash N \Leftarrow S_1 \wedge S_2$ iff $\Gamma \vdash N \Leftarrow S_1$ and $\Gamma \vdash N \Leftarrow S_2$ by inversion. Both of these are decidable by induction hypothesis (2) on N and S_1 and S_2, respectively, and so $\Gamma \vdash N \Leftarrow S_1 \wedge S_2$ is decidable.

Case(2) $S = S_1 \to S_2$. Then $\Gamma \vdash N \Leftarrow S_1 \to S_2$ iff $N = \lambda x_1. N_2$ and $\Gamma, x_1{:}S_1 \vdash N_2 \Leftarrow S_2$. This is decidable by induction hypothesis on N_2 and S_2.

Case(2) $S = Q$. Then $\Gamma \vdash N \Leftarrow Q$ iff $N = R$ and there exists a Q' such that $\Gamma \vdash R \Rightarrow Q'$ and $Q' \leq Q$. By induction hypothesis (3) on R, there is a finite set of Q' such that $\Gamma \vdash R \Rightarrow Q'$. By part (1) we can test each one in turn to see if at least one of them is a subsort of Q or not.

Case(3) $R = x$. Then there is a unique T such that $x{:}T \in \Gamma$, so we start with $\Gamma \vdash x \Rightarrow T$. We can saturate the set of judgments $\Gamma \vdash x \Rightarrow T'$ by applying the two intersection elimination rules, the only ones that do not change x. Since there are only finitely many conjuncts in T, this will terminate with a finite set.

Case(3) $R = c$. As in the case for variables, except we may already start with a finite set of T with $\Gamma \vdash c \Rightarrow T_i$ since there may be finitely many declarations for c.

Case(3) $R = R_1\, N_2$. By induction hypothesis on R_1, there are finitely many T' such that $\Gamma \vdash R_1 \Rightarrow T'$. For each T' of the form $T_2 \to T_1$

for some T_2 and T_1 we can test, by induction hypothesis, whether $\Gamma \vdash N_2 \Leftarrow T_2$. Then for each matching T_1 we get $\Gamma \vdash R_1\, N_2 \Rightarrow T_1$. We complete the set of all sorts synthesized by $R_1\, N_2$ by saturating with respect to the two intersection elimination rules.

∎

10 Sort Preservation

The decidability of sort checking is a useful property for an implementation, but we should verify that the system is well-constructed. On canonical forms, this means the preservation of sorts under hereditary substitution and iterated expansions, which are generalizations of the same properties for types. We constructed the rules systematically, which makes the proof of these preservation properties quite straightforward.

We need some properties of typing derivations, specifically that the order of the typing assumptions does not matter (*order irrelevance*), and that we can adjoin additional unused assumptions (*weakening*) without changing the structure of the typing derivation. We elide the straightforward formal statement and proof of these properties. Recall that we assume that all judgments are well-formed with respect to the underlying layer of types.

THEOREM 15 (Sort preservation under hereditary substitution).
Assume $\Gamma \vdash M \Leftarrow S$.

1. *If* $\Gamma, x{:}S \vdash N \Leftarrow T$ *then* $\Gamma \vdash [M/x]^n(N) \Leftarrow T$.

2. *If* $\Gamma, x{:}S \vdash R \Rightarrow T$ *and* $\mathrm{hd}(R) \neq x$ *then* $\Gamma \vdash [M/x]^{rr}(R) \Rightarrow T$.

3. *If* $\Gamma, x{:}S \vdash R \Rightarrow T$ *and* $\mathrm{hd}(R) = x$ *then* $\Gamma \vdash [M/x]^{rn}(R) \Leftarrow T$.

Proof.

There are two forms of induction we can use for this property. One is a nested induction on the structure of the type α of M^α first, and the structure of the given sort derivations second. This is very similar to the proof given for composition of substitutions earlier.

An alternative is to use a nested induction, first over the structure of the computation of $[M/x]^n(N)$, $[M/x]^{rr}(R)$, and $[M/x]^{rn}(R)$, and second on the sort derivation $\mathcal{D}$ for N and R. For this, we consider the equations defining hereditary substitutions as rules of computation, reading them left to right, where we can apply the induction hypothesis to any subcomputation. Since hereditary substitution is well-founded, this is a well-founded form of induction.

We use this alternative method to illustrate it. When we write "*by i.h. on* $[M/x]^n(N)$" we mean the computation starting at this term, which should be a subcomputation of the given one.

Case(1): $[M/x]^n(N)$ is arbitrary and

$$\mathcal{D} = \frac{\begin{array}{cc}\mathcal{D}_1 & \mathcal{D}_2 \\ \Gamma, x{:}S \vdash N \Leftarrow T_1 & \Gamma, x{:}S \vdash N \Leftarrow T_2\end{array}}{\Gamma, x{:}S \vdash N \Leftarrow T_1 \wedge T_2}$$

$\Gamma \vdash [M/x]^n(N) \Leftarrow T_1$ — by i.h.(1) on $[M/x]^n(N)$ and $\mathcal{D}_1$
$\Gamma \vdash [M/x]^n(N) \Leftarrow T_2$ — by i.h.(2) on $[M/x]^n(N)$ and $\mathcal{D}_2$
$\Gamma \vdash [M/x]^n(N) \Leftarrow T_1 \wedge T_2$ — by rule

Case(1): $[M/x]^n(N)$ is arbitrary and

$$\mathcal{D} = \frac{}{\Gamma, x{:}S \vdash N \Leftarrow \top}$$

$\Gamma \vdash [M/x]^n(N) \Leftarrow \top$ — by rule

Case(1): $[M/x]^n(\lambda y_2.\, N_1) = \lambda y_2.\, [M/x]^n(N_1)$ where $y_2 \neq x$ and $y_2 \notin \mathrm{fv}(M)$ and

$$\mathcal{D} = \frac{\begin{array}{c}\mathcal{D}_1 \\ \Gamma, x{:}S, y_2{:}T_2 \vdash N_1 \Leftarrow T_1\end{array}}{\Gamma, x{:}S \vdash \lambda y_2.\, N_1 \Leftarrow T_2 \to T_1}$$

$\Gamma, y_2{:}T_2, x{:}S \vdash N_1 \Leftarrow T_1$ — by order irrelevance
$\Gamma, y_2{:}T_2 \vdash M \Leftarrow S$ — by weakening
$\Gamma, y_2{:}T_2 \vdash [M/x]^n(N_1) \Leftarrow T_1$ — by i.h.(1) on $[M/x]^n(N_1)$ and $\mathcal{D}_1$
$\Gamma \vdash \lambda y_2.\, [M/x]^n(N_1) \Leftarrow T_2 \to T_1$ — by rule

Case(1): $[M/x]^n(R) = [M/x]^{rr}(R)$ with $\mathrm{hd}(R) \neq x$ and

$$\mathcal{D} = \frac{\begin{array}{cc}\mathcal{D}' & \mathcal{Q} \\ \Gamma, x{:}S \vdash R \Rightarrow Q' & Q' \leq Q\end{array}}{\Gamma, x{:}S \vdash R \Leftarrow Q}$$

$\Gamma \vdash [M/x]^{rr}(R) \Rightarrow Q'$ — by i.h.(2) on $[M/x]^{rr}(R)$ and $\mathcal{D}'$
$\Gamma \vdash [M/x]^n(R) \Leftarrow Q$ — by rule

Case(1): $[M/x]^n(R) = [M/x]^{rn}(R)$ with $\mathrm{hd}(R) = x$ and

$$\mathcal{D} = \cfrac{\begin{array}{cc} \mathcal{D}' & \mathcal{Q} \\ \Gamma, x{:}S \vdash R \Rightarrow Q' & Q' \leq Q \end{array}}{\Gamma, x{:}S \vdash R \Leftarrow Q}$$

$\Gamma \vdash [M/x]^{rn}(R) \Leftarrow Q'$	by i.h.(3) on $\mathcal{D}'$
$\Gamma \vdash [M/x]^{rn}(R) \Rightarrow Q''$ and $Q'' \leq Q'$	by inversion
$Q'' \leq Q$	by transitivity
$\Gamma \vdash [M/x]^{rn}(R) \Leftarrow Q$	by rule

Case(2): $[M/x]^{rr}(R)$ arbitrary and

$$\mathcal{D} = \cfrac{\begin{array}{c} \mathcal{D}_{12} \\ \Gamma, x{:}S \vdash R \Rightarrow T_1 \wedge T_2 \end{array}}{\Gamma, x{:}S \vdash R \Rightarrow T_1}$$

$\Gamma \vdash [M/x]^{rr}(R) \Rightarrow T_1 \wedge T_2$	by i.h.(2) on $[M/x]^{rr}(R)$ and $\mathcal{D}_{12}$
$\Gamma \vdash [M/x]^{rr}(R) \Rightarrow T_1$	by rule

Case(2): $[M/x]^{rr}(R)$ is arbitrary and $\mathcal{D}$ ends in the second intersection elimination rule. Symmetric to the previous case.

Case(2): $[M/x]^{rr}(R_1\, N_2) = ([M/x]^{rr} R_1)\, ([M/x]^n N_2)$ and

$$\mathcal{D} = \cfrac{\begin{array}{cc} \mathcal{D}_1 & \mathcal{D}_2 \\ \Gamma, x{:}S \vdash R_1 \Rightarrow T_2 \to T_1 & \Gamma, x{:}S \vdash N_2 \Leftarrow T_2 \end{array}}{\Gamma, x{:}S \vdash R_1\, N_2 \Rightarrow T_1}$$

$\Gamma \vdash [M/x]^{rr} R_1 \Rightarrow T_2 \to T_1$	By i.h.(2) on $[M/x]^{rr} R_1$ and $\mathcal{D}_1$
$\Gamma \vdash [M/x]^n N_2 \Leftarrow T_2$	By i.h.(1) on $[M/x]^n N_2$ and $\mathcal{D}_2$
$\Gamma \vdash ([M/x]^{rr} R_1)\, ([M/x]^n N_2) \Rightarrow T_1$	by rule

Case(2): $[M/x]^{rr}(y) = y$ with $y \neq x$ and

$$\mathcal{D} = \cfrac{y{:}T \in \Gamma, x{:}S}{\Gamma, x{:}S \vdash y \Rightarrow T}$$

$\Gamma \vdash y \Rightarrow T$	by rule

Case(2): $[M/x]^{rr}(c) = c$ and $\mathcal{D}$ ends in the sorting rules for constants. As in the previous case.

Case(3): $[M/x]^{rn}(R)$ is arbitrary and

$$\mathcal{D} = \dfrac{\begin{array}{c}\mathcal{D}_{12}\\ \Gamma, x{:}S \vdash R \Rightarrow T_1 \wedge T_2\end{array}}{\Gamma, x{:}S \vdash R \Rightarrow T_1}$$

$\Gamma \vdash [M/x]^{rn}(R) \Leftarrow T_1 \wedge T_2$	by i.h.(3) in $[M/x]^{rn}(R)$ and $\mathcal{D}_{12}$
$\Gamma \vdash [M/x]^{rn}(R) \Leftarrow T_1$	by inversion

Case(3): $[M/x]^{rn}(R)$ is arbitrary and $\mathcal{D}$ ends in the second intersection elimination rule. Symmetric to the previous case.

Case(3): $[M/x]^{rn}(R_1\, N_2) = [N_2'/y_2]^n(N_1)$ where $[M/x]^{rn}(R_1) = \lambda y_2.\, N_1$ and $[M/x]^n(N_2) = N_2'$ and

$$\mathcal{D} = \dfrac{\begin{array}{c}\mathcal{D}_1\\ \Gamma, x{:}S \vdash R_1 \Rightarrow T_2 \rightarrow T_1\end{array} \quad \begin{array}{c}\mathcal{D}_2\\ \Gamma, x{:}S \vdash N_2 \Leftarrow T_2\end{array}}{\Gamma, x{:}S \vdash R_1\, N_2 \Rightarrow T_1}$$

$\Gamma \vdash [M/x]^{rn}(R_1) \Leftarrow T_2 \rightarrow T_1$	by i.h.(3) on $[M/x]^{rn}(R_1)$ and $\mathcal{D}_1$
$\Gamma \vdash \lambda y_2.\, N_1 \Leftarrow T_2 \rightarrow T_1$	by equality
$\Gamma, y_2{:}T_2 \vdash N_1 \Leftarrow T_1$	by inversion
$\Gamma \vdash [M/x]^n(N_2) \Leftarrow T_2$	by i.h.(1) on $[M/x]^n(N_2)$ and $\mathcal{D}_2$
$\Gamma \vdash N_2' \Leftarrow T_2$	by equality
$\Gamma \vdash [N_2'/y_2]^n(N_1) \Leftarrow T_1$	by i.h.(1) on $[N_2'/y_2]^n(N_1)$

Case(3): $[M/x]^{rn}(x) = M$ and

$$\mathcal{D} = \dfrac{x{:}S \in (\Gamma, x{:}S)}{\Gamma, x{:}S \vdash x \Rightarrow S}$$

$\Gamma \vdash M \Leftarrow S$	by assumption

■

A second property is sort preservation under iterated expansion. In some sense this is dual to the previous property: hereditary substitution allows us to replace a hypothesis $x{:}S$ which should be read as $x{\Rightarrow}S$ with a term $M \Leftarrow S$ and thus provides a way to go from checking to synthesis. Conversely, iterated expansion takes us from $x \Rightarrow S$ to $\eta^\alpha(x) \Leftarrow S$, that is, from synthesis to checking.

THEOREM 16 (Sort preservation under iterated expansion).
If $\Gamma \vdash R^\alpha \Rightarrow T$ *then* $\Gamma \vdash \eta^\alpha(R) \Leftarrow T$.

Proof. By induction on the structure of T.

Case: $T = T_1 \wedge T_2$. Then

$$
\begin{array}{ll}
\Gamma \vdash R \Rightarrow T_1 \wedge T_2 & \text{assumption} \\
\Gamma \vdash R \Rightarrow T_1 & \text{by rule} \\
\Gamma \vdash \eta^{\alpha}(R) \Leftarrow T_1 & \text{by i.h. on } T_1 \\
\Gamma \vdash R \Rightarrow T_2 & \text{by rule} \\
\Gamma \vdash \eta^{\alpha}(R) \Leftarrow T_2 & \text{by i.h. on } T_2 \\
\Gamma \vdash \eta^{\alpha}(R) \Leftarrow T_1 \wedge T_2 & \text{by rule}
\end{array}
$$

Case: $T = \top$. Then

$$
\begin{array}{ll}
\Gamma \vdash \eta^{\alpha}(R) \Leftarrow \top & \text{by rule}
\end{array}
$$

Case: $T = T_2 \to T_1$ and $\alpha = \alpha_2 \to \alpha_1$.

Let x_2 be a variable not in $\text{fv}(R)$.

$$
\begin{array}{ll}
\Gamma, x_2{:}T_2 \vdash x_2 \Rightarrow T_2 & \text{by rule} \\
\Gamma, x_2{:}T_2 \vdash \eta^{\alpha_2}(x_2) \Leftarrow T_2 & \text{by i.h. on } T_2 \\
\Gamma, x_2{:}T_2 \vdash R \Rightarrow T_2 \to T_1 & \text{by weakening} \\
\Gamma, x_2{:}T_2 \vdash R\,\eta^{\alpha_2}(x_2) \Rightarrow T_1 & \text{by rule} \\
\Gamma, x_2{:}T_2 \vdash \eta^{\alpha_1}(R\,\eta^{\alpha_2}(x_2)) \Leftarrow T_1 & \text{by i.h. on } T_1 \\
\Gamma \vdash \lambda x_2.\, \eta^{\alpha_1}(R\,\eta^{\alpha_2}(x_2)) \Leftarrow T_2 \to T_2 & \text{by rule}
\end{array}
$$

Case: $T = Q$ and $\alpha = \iota$.

$$
\begin{array}{ll}
\Gamma \vdash R \Rightarrow Q & \text{assumption} \\
Q \leq Q & \text{by reflexivity} \\
\Gamma \vdash R \Leftarrow Q & \text{by rule}
\end{array}
$$

■

11 Sort Preservation under Converse Substitution

One of the surprising properties of intersection types that was realized early on in their development [2] is that the converse of β-reduction preserves types. This means if we can type a normal form, we can type every term that reduces to it. Here, in the setting of canonical forms, we don't have redexes, but we can ask a corresponding question about the converse of hereditary substitution. We will show in this section that the converse of hereditary substitution preserves sorts.

We first prove a few useful lemmas.

LEMMA 17 (Strengthening).

1. *If* $\Gamma, x{:}S \vdash N \Leftarrow T$ *and* $x \notin \mathrm{fv}(N)$ *then* $\Gamma \vdash N \Leftarrow T$.

2. *If* $\Gamma, x{:}S \vdash R \Rightarrow T$ *and* $x \notin \mathrm{fv}(R)$ *then* $\Gamma \vdash R \Rightarrow T$.

Proof. By straightforward mutual induction on N and R. ■

We write $[R^\alpha/x^\alpha]N$ and $[R^\alpha/x^\alpha]R'$ for the *direct substitution* of R for x in N and R', which should be capture-avoiding as usual. This substitution is *not* hereditary. Intuitively, this is because we are replacing a variable, which is atomic, with an atomic term of the same sort.

An easy induction verifies that direct substitution preserves types and, once we know this, also sorts.

LEMMA 18 (Sort preservation under direct substitution).
Assume $\Gamma \vdash R \Rightarrow S$.

1. *If* $\Gamma, x{:}S \vdash N \Leftarrow T$ *then* $\Gamma \vdash [R/x]N \Leftarrow T$.

2. *If* $\Gamma, x{:}S \vdash R' \Rightarrow T$ *then* $\Gamma \vdash [R/x]R' \Rightarrow T$

Proof. By straightforward mutual induction on the typing derivations for N and R'. ■

LEMMA 19 (Hypothesis intersection).

1. *If* $\Gamma, x{:}S_1 \vdash N \Leftarrow T$ *then* $\Gamma, x{:}S_1 \wedge S_2 \vdash N \Leftarrow T$.

2. *If* $\Gamma, x{:}S_2 \vdash N \Leftarrow T$ *then* $\Gamma, x{:}S_1 \wedge S_2 \vdash N \Leftarrow T$.

3. *If* $\Gamma, x{:}S_1 \vdash R \Rightarrow T$ *then* $\Gamma, x{:}S_1 \wedge S_2 \vdash R \Rightarrow T$.

4. *If* $\Gamma, x{:}S_2 \vdash R \Rightarrow T$ *then* $\Gamma, x{:}S_1 \wedge S_2 \vdash R \Rightarrow T$.

Proof. We show the proof of part (1). The others are analogous.

$\Gamma, x{:}S_1 \vdash N \Leftarrow T$	Given
$\Gamma, y{:}S_1 \wedge S_2, x{:}S_1 \vdash N \Leftarrow T$	by weakening
$\Gamma, y{:}S_1 \wedge S_2 \vdash y \Rightarrow S_1 \wedge S_2$	by rule
$\Gamma, y{:}S_1 \wedge S_2 \vdash y \Rightarrow S_1$	by rule
$\Gamma, y{:}S_1 \wedge S_2 \vdash [y/x]N \Leftarrow T$	by direct substitution
$\Gamma, x{:}S_1 \wedge S_2 \vdash N \Leftarrow T$	by renaming

■

THEOREM 20 (Sort preservation under converse hereditary substitution).

1. *If* $\Gamma \vdash [M/x]^n(N) \Leftarrow T$ *then there exists an* S *such that* $\Gamma \vdash M \Leftarrow S$ *and* $\Gamma, x{:}S \vdash N \Leftarrow T$.

2. *If* $\Gamma \vdash [M/x]^{rr}(R) \Rightarrow T$ *then there exists an* S *such that* $\Gamma \vdash M \Leftarrow S$ *and* $\Gamma, x{:}S \vdash R \Rightarrow T$.

3. *If* $\Gamma \vdash [M/x]^{rn}(R) \Leftarrow T$ *then there exists an* S *such that* $\Gamma \vdash M \Leftarrow S$ *and* $\Gamma, x{:}S \vdash R \Rightarrow T$.

Proof. Again, we have a choice: we can prove this by a nested induction on a type and a sorting derivation, or we can use computation induction on the definition of hereditary substitution. Choosing the latter, the proof is by nested induction, first on the computation of $[M/x]^n(N)$, $[M/x]^{rr}(R)$ or $[M/x]^{rn}(R)$ and second on the given typing derivation $\mathcal{D}$.

Case(1): $[M/x]^n(N)$ is arbitrary and

$$\mathcal{D} = \cfrac{\begin{array}{cc} \mathcal{D}_1 & \mathcal{D}_2 \\ \Gamma \vdash [M/x]^n(N) \Leftarrow T_1 & \Gamma \vdash [M/x]^n(N) \Leftarrow T_2 \end{array}}{\Gamma \vdash [M/x]^n(N) \Leftarrow T_1 \wedge T_2}$$

There exists S_1 such that	
$\Gamma \vdash M \Leftarrow S_1$ and	
$\Gamma, x{:}S_1 \vdash N \Leftarrow T_1$	by i.h.(1) on $[M/x]^n(N)$ and $\mathcal{D}_1$
There exists S_2 such that	
$\Gamma \vdash M \Leftarrow S_2$ and	
$\Gamma, x{:}S_2 \vdash N \Leftarrow T_2$	by i.h.(1) on $[M/x]^n(N)$ and $\mathcal{D}_2$
$\Gamma, x{:}S_1 \wedge S_2 \vdash N \Leftarrow T_1$	by hypothesis intersection
$\Gamma, x{:}S_1 \wedge S_2 \vdash N \Leftarrow T_2$	by hypothesis intersection
$\Gamma, x{:}S_1 \wedge S_1 \vdash N \Leftarrow T_1 \wedge T_2$	by rule
$\Gamma \vdash M \Leftarrow S_1 \wedge S_2$	by rule

Case(1): $[M/x]^n(N)$ is arbitrary and

$$\mathcal{D} = \cfrac{}{\Gamma \vdash [M/x]^n(N) \Leftarrow \top}$$

$\Gamma \vdash M \Leftarrow \top$	by rule
$\Gamma, x{:}\top \vdash N \Leftarrow \top$	by rule

Case(1): $[M/x]^n(\lambda y_2.\, N_1) = \lambda y_2.\, [M/x]^n(N_1)$ where $y_2 \neq x$ and $y_2 \notin \mathrm{fv}(M)$ and

$$\mathcal{D} = \frac{\begin{array}{c}\mathcal{D}_1\\ \Gamma, y_2{:}T_2 \vdash [M/x]^n(N_1) \Leftarrow T_1\end{array}}{\Gamma \vdash \lambda y_2.\, [M/x]^n(N_1) \Leftarrow T_2 \to T_1}$$

There exists S such that
$\Gamma, y_2{:}T_2 \vdash M \Leftarrow S$ and
$\Gamma, y_2{:}T_2, x{:}S \vdash N_1 \Leftarrow T_1$ — by i.h.(1) on $[M/x]^n(N_1)$ and $\mathcal{D}_1$
$\Gamma, x{:}S, y_2{:}T_2 \vdash N_1 \Leftarrow T_1$ — by order irrelevance
$\Gamma, x{:}S \vdash \lambda y_2{:}T_2.\, N_1 \Leftarrow T_2 \to T_1$ — by rule
$\Gamma \vdash M \Leftarrow S$ — by strengthening since $y_2 \notin \mathrm{fv}(M)$

Case(1): $[M/x]^n(R) = [M/x]^{rr}(R)$ where $\mathrm{hd}(R) \neq x$ and

$$\mathcal{D} = \frac{\begin{array}{c}\mathcal{D}'\\ \Gamma \vdash [M/x]^{rr}R \Rightarrow Q' \quad Q' \leq Q\end{array}}{\Gamma \vdash [M/x]^{rr}R \Leftarrow Q}$$

There exists S such that
$\Gamma \vdash M \Leftarrow S$ and
$\Gamma, x{:}S \vdash R \Rightarrow Q'$ — by i.h.(2) on $[M/x]^{rr}(R)$ and $\mathcal{D}'$
$\Gamma, x{:}S \vdash R \Leftarrow Q$ — by rule

Case(1): $[M/x]^n(R) = [M/x]^{rn}(R)$ where $\mathrm{hd}(R) = x$ and

$$\mathcal{D} = \frac{\begin{array}{c}\mathcal{D}'\\ \Gamma \vdash [M/x]^{rn}(R) \Rightarrow Q' \quad Q' \leq Q\end{array}}{\Gamma \vdash [M/x]^{rn}(R) \Leftarrow Q}$$

There exists S such that
$\Gamma \vdash M \Leftarrow S$ and
$\Gamma, x{:}S \vdash R \Rightarrow Q'$ — by i.h.(3) on $[M/x]^{rn}(R)$ and $\mathcal{D}'$
$\Gamma, x{:}S \vdash R \Leftarrow Q$ — by rule

Case(2): $[M/x]^{rr}(R)$ is arbitrary and

$$\mathcal{D} = \frac{\begin{array}{c}\mathcal{D}_{12}\\ \Gamma \vdash [M/x]^{rr}(R) \Rightarrow T_1 \wedge T_2\end{array}}{\Gamma \vdash [M/x]^{rr}(R) \Rightarrow T_1}$$

There exists S such that
$\Gamma \vdash M \Leftarrow S$ and
$\Gamma, x{:}S \vdash R \Rightarrow T_1 \wedge T_2$ by i.h.(2) on $\mathcal{D}_{12}$
$\Gamma, x{:}S \vdash R \Rightarrow T_1$ by rule

Case(2): $[M/x]^{rr}(R)$ is arbitrary $\mathcal{D}$ ends in the second intersection elimination. Symmetric to the previous case.

Case(2): $[M/x]^{rr}(R_1\, N_2) = [M/x]^{rr}(R_1)\, [M/x]^{n}(N_2)$ and

$$\mathcal{D} = \cfrac{\begin{array}{cc}\mathcal{D}_1 & \mathcal{D}_2 \\ \Gamma \vdash [M/x]^{rr}(R_1) \Rightarrow T_2 \to T_1 & \Gamma \vdash [M/x]^{n}(N_2) \Leftarrow T_2\end{array}}{\Gamma \vdash [M/x]^{rr}(R_1)\, [M/x]^{n}(N_2) \Rightarrow T_1}$$

There exists S_1 such that
$\Gamma \vdash M \Leftarrow S_1$ and
$\Gamma, x{:}S_1 \vdash R_1 \Rightarrow T_2 \to T_1$ by i.h.(2) on $[M/x]^{rr}(R_1)$ and $\mathcal{D}_1$
There exists S_2 such that
$\Gamma \vdash M \Leftarrow S_2$ and
$\Gamma, x{:}S_2 \vdash N_2 \Leftarrow T_2$ by i.h.(1) on $[M/x]^{n}(N_2)$ and $\mathcal{D}_2$
$\Gamma \vdash M \Leftarrow S_1 \wedge S_2$ by rule
$\Gamma, x{:}S_1 \wedge S_2 \vdash R_1 \Rightarrow T_2 \to T_1$ by hypothesis intersection
$\Gamma, x{:}S_1 \wedge S_2 \vdash N_2 \Leftarrow T_2$ by hypothesis intersection
$\Gamma, x{:}S_1 \wedge S_2 \vdash R_1\, N_2 \Rightarrow T_1$ by rule

Case(2): $[M/x]^{rr}(y) = y$ for $y \neq x$ and

$$\mathcal{D} = \cfrac{y{:}T \in \Gamma}{\Gamma \vdash y \Rightarrow T}$$

$\Gamma \vdash M \Leftarrow \top$ by rule
$\Gamma, x{:}\top \vdash y \Rightarrow T$ by rule

Case(2): $[M/x]^{rr}(c) = c$ and $\mathcal{D}$ ends in the rule for constants. Proceed as in the previous case.

Case(3): $[M/x]^{rn}(R_1\, N_2) = [N_2'/y_2]^{n}(N_1)$ where $[M/x]^{rn}(R_1) = \lambda y_2.\, N_1$ and $[M/x]^{n}(N_2) = N_2'$ and

$$\begin{array}{c}\mathcal{D} \\ \Gamma \vdash [N_2'/y_2]^{n}(N_1) \Leftarrow T\end{array}$$

is arbitrary.

There exists S_2 such that
$\Gamma \vdash N_2' \Leftarrow S_2$ and
$\Gamma, y_2{:}S_2 \vdash N_1 \Leftarrow T$ — by i.h.(1) on $[N_2'/y_2]^n(N_1)$ and $\mathcal{D}$
$\Gamma \vdash [M/x]^n(N_2) \Leftarrow S_2$ — by equality
There exists S such that
$\Gamma \vdash M \Leftarrow S$ and
$\Gamma, x{:}S \vdash N_2 \Leftarrow S_2$ — by i.h.(1) on $[M/x]^n(N_2)$
$\Gamma \vdash \lambda y_2.\, N_1 \Leftarrow S_2 \rightarrow T$ — by rule
$\Gamma \vdash [M/x]^{rn}(R_1) \Leftarrow S_2 \rightarrow T$ — by equality
There exists S' such that
$\Gamma \vdash M \Leftarrow S'$ and
$\Gamma, x{:}S' \vdash R_1 \Rightarrow S_2 \rightarrow T$ — by i.h.(3) on $[M/x]^{rn}(R_1)$
$\Gamma \vdash M \Leftarrow S \wedge S'$ — by rule
$\Gamma, x{:}S \wedge S' \vdash R_1 \Rightarrow S_2 \rightarrow T$ — by hypothesis intersection
$\Gamma, x{:}S \wedge S' \vdash N_2 \Leftarrow S_2$ — by hypothesis intersection
$\Gamma, x{:}S \wedge S' \vdash R_1\, N_2 \Rightarrow T$ — by rule

Case(3): $[M/x]^{rn}(x) = M$ and

$$\begin{array}{c}\mathcal{D}\\ \Gamma \vdash M \Leftarrow T\end{array}$$

is arbitrary.

$\Gamma \vdash M \Leftarrow T$ — by $\mathcal{D}$
$\Gamma, x{:}T \vdash x \Rightarrow T$ — by rule

■

12 Subsorting at Higher Types

Recall that iterated expansion preserves sorts: if $R^\alpha \Rightarrow S$ then $\eta^\alpha(R) \Leftarrow S$. Since the substitution property has a (perhaps surprising) converse, we might conjecture that the identity property also has a converse.

Alas, it is not the case that $\eta^\alpha(R) \Leftarrow S$ implies $R \Rightarrow S$. A simple counterexample is $x{:}\mathsf{even} \vdash x \Leftarrow \mathsf{nat}$, but $x{:}\mathsf{even} \not\vdash x \Rightarrow \mathsf{nat}$. Various slightly extended conjectures also fail: synthesis is inherently very weak when compared to checking.

Nevertheless, there are still other interesting developments to consider regarding iterated expansion and the identity theorem. So far, subsorting has been confined to base sorts and this has been sufficient, essentially because we only consider η-long terms. But we can introduce a derived notion of subsorting beyond base sorts and show that we have a system which is both sound and complete with respect to other characterizations of what subtyping might be. Finally, although beyond the scope of the present paper, we can use it to complete the general system of sort assignment sketched earlier which does not rely on canonical forms, but has a general subsumption rule.

One desirable property would be that S is a subsort of T if and only if every term of sort S is also a term of sort T. This means subsorting is essentially inclusion among sets of terms. However, this quantifies over all canonical terms, and is thus not an a priori decidable condition and of limited immediate utility.

The basic idea on how to reduce subsorting to a property of sorting and iterated expansion is to check whether $x{:}S \vdash \eta^{\alpha}(x) \Leftarrow T$. If so, S should be a subsort of T because $\eta^{\alpha}(x)$ is the identity. Because sort-checking is decidable, this is an effective criterion.

DEFINITION 21 (Subsorting at higher types). For two sorts S^{α} and T^{α} we write $S \leq T$ iff $x{:}S \vdash \eta^{\alpha}(x) \Leftarrow T$.

We are justified in reusing the symbol $\leq$, because on base sorts the prior judgment $Q' \leq Q$ coincides with the extended one: the derivation

$$\frac{\dfrac{}{x{:}Q' \vdash x \Rightarrow Q'} \quad Q' \leq Q}{x{:}Q' \vdash \eta^{\iota}(x) \Leftarrow Q}$$

is entirely forced, noting that $\eta^{\iota}(x) = x$.

We can now show that subsorting via sort checking of identities is sound and complete with respect to inclusion among sets of canonical forms.

THEOREM 22 (Alternative characterization of subtyping).
$S \leq T$ *if and only if forall all* Γ *and* N, $\Gamma \vdash N \Leftarrow S$ *implies* $\Gamma \vdash N \Leftarrow T$.

Proof. Direct in each direction.

($\Longrightarrow$)

$S \leq T$	assumption
$\Gamma \vdash N \Leftarrow S$	assumption
$x{:}S \vdash \eta^{\alpha}(x) \Leftarrow T$	by defn. of $\leq$
$\Gamma, x{:}S \vdash \eta^{\alpha}(x) \Leftarrow T$	by weakening
$\Gamma \vdash [N/x]^{n}(\eta^{\alpha}(x)) \Leftarrow T$	by sort preservation under substitution
$\Gamma \vdash N \Leftarrow T$	by right identity

($\Longleftarrow$)

$x{:}S \vdash x \Rightarrow S$	by rule
$x{:}S \vdash \eta^\alpha(x) \Leftarrow S$	by sort preservation under expansion
$x{:}S \vdash \eta^\alpha(x) \Leftarrow T$	by assumption
$S \leq T$	by defn. of $\leq$

■

13 Conclusion

We have developed a calculus of canonical forms for a λ-calculus formulated in the style of Church [1] where every well-formed term intrinsically has a unique simple type. The critical ingredient is the operation of *hereditary substitution* which returns a canonical form when substituting a canonical form for a variable in a canonical form. Its counterpart is *iterated expansion* which returns a canonical form when given an atomic one. By an argument relying crucially on the intrinsic types, we could see that these operations are always properly defined. Moreover, hereditary substitutions can be composed with expected results and iterated expansion of variables returns both a left and right identity for hereditary substitutions.

As a second layer we defined a system in the style of Curry [3] where sorts are assigned to terms already known to be well-typed. Sorts thus stand for properties of terms, which is the essence of Curry's approach. An interesting possibility we explored was to introduce finitary polymorphism through which we can explicitly state multiple properties of a single term. We presented an elegant formulation of subsorting and intersection sorts, which can immediately be seen to be decidable. In addition, sorts are preserved under hereditary substitution and iterated expansion, which means that Curry-style sort assignment harmoniously coexists with Church-style typing. Interestingly, the presence of finitary intersection allows us to validate the converse of substitution. In practice this means that we can sort-check a term not in canonical form by converting it to one and sort-checking the result, although the corresponding theorem is beyond the scope of this paper. We cannot directly validate the converse of iterated expansion because sort synthesis of atomic terms is too weak. But we introduce a notion of subsorting based on iterated expansion which characterizes, precisely, inclusion among sets of canonical forms and is directly decidable.

A natural next step is to give a sort assignment system for well-formed terms that are not necessarily in canonical form and prove (a) normalization via hereditary substitution and iterated expansion, (b) that sorting can be decided by conversion to canonical form and application of the algorithms in this paper. We conjecture that this should be possible and relatively straightforward based on the presented results.

We have carried out a related analysis of canonical forms and sorts for a dependently typed λ-calculus [10] and for types for a modal λ-calculus with metavariables [12]. These results illustrate the robustness of the ideas presented here. *Refinement types* [7, 4, 5] for functional programs represent another line of development that shares many ideas with this paper, although there are significant technical differences between a call-by-value functional language with data types, recursion, and effects, and Church's simply-typed λ-calculus. A promising idea toward a synthesis and unification of these threads is to generalize canonical forms via focusing in polarized logic [14] which is closely related to the concurrent logical framework [13]. Perhaps it is no accident that the latter originated the idea of hereditary substitutions.

Acknowledgments. I would like to dedicate this paper to my advisor, Peter Andrews, who introduced me to Church's type theory and taught me the beauty, precision, and utility of mathematical logic. I am greatly indebted to him. I would also like to thank Kevin Watkins, who first discovered hereditary substitutions and patiently explained them to me, and William Lovas and Noam Zeilberger for many insightful discussions regarding canonical forms, subtyping, and intersection types. Chad Brown and William Lovas deserve special thanks for their careful proof-reading and corrections to an earlier draft of this paper. This work has been supported by NSF grants NSF-0702381 *Integrating Types and Verification* and CCR-0306313 *Formal Digital Library*.

BIBLIOGRAPHY

[1] Alonzo Church. A Formulation of the Simple Theory of Types. *Journal of Symbolic Logic*, 5:56–68, 1940.

[2] Mario Coppo, Maria Dezani-Ciancaglini, and Betti Venneri. Functional Character of Solvable Terms. *Zeitschrift für mathematische Logic und Grundlagen der Mathematik*, 27:45–58, 1981.

[3] H. B. Curry. Functionality in Combinatory Logic. *Proceedings of the National Academy of Sciences, U.S.A.*, 20:584–590, 1934.

[4] Rowan Davies. *Practical Refinement-Type Checking.* PhD thesis, Carnegie Mellon University, May 2005. Available as Technical Report CMU-CS-05-110.

[5] Joshua Dunfield. *A Unified System of Type Refinements.* PhD thesis, Carnegie Mellon University, August 2007. Available as Technical Report CMU-CS-07-129.

[6] Joshua Dunfield and Frank Pfenning. Type Assignment for Intersections and Unions in Call-by-Value Languages. In A.D. Gordon, editor, *Proceedings of the 6th International Conference on Foundations of Software Science and Computation Structures (FOSSACS'03)*, pages 250–266, Warsaw, Poland, April 2003. Springer-Verlag LNCS 2620.

[7] Tim Freeman and Frank Pfenning. Refinement Types for ML. In *Proceedings of the Symposium on Programming Language Design and Implementation (PLDI'91)*, pages 268–277, Toronto, Ontario, June 1991. ACM Press.

[8] Robert Harper, Furio Honsell, and Gordon Plotkin. A Framework for Defining Logics. *Journal of the Association for Computing Machinery*, 40(1):143–184, January 1993.

[9] W. A. Howard. The formulae-as-types notion of construction. In J. P. Seldin and J. R. Hindley, editors, *To H. B. Curry: Essays on Combinatory Logic, Lambda Calculus and Formalism*, pages 479–490. Academic Press, 1980. Hitherto unpublished note of 1969, rearranged, corrected, and annotated by Howard.

[10] William Lovas and Frank Pfenning. A Bidirectional Refinement Type System for LF. In B. Pientka and C. Schürmann, editors, *Proceedings of the Second International Workshop on Logical Frameworks and Meta-Languages: Theory and Practice*, pages 113–128, Bremen, Germany, July 2007. Electronic Notes in Theoretical Computer Science (ENTCS), vol 196.

[11] Dale Miller, Gopalan Nadathur, Frank Pfenning, and Andre Scedrov. Uniform Proofs as a Foundation for Logic Programming. *Annals of Pure and Applied Logic*, 51:125–157, 1991.

[12] Aleksandar Nanevski, Frank Pfenning, and Brigitte Pientka. Contextual Modal Type Theory. *Transactions on Computational Logic*, 2008. To appear.

[13] Kevin Watkins, Iliano Cervesato, Frank Pfenning, and David Walker. A Concurrent Logical Framework I: Judgments and Properties. Technical Report CMU-CS-02-101, Department of Computer Science, Carnegie Mellon University, 2002. Revised May 2003.

[14] Noam Zeilberger. On the Unity of Duality. *Annals of Pure and Applied Logic*, 153(1–3):66–96, April 2008. Special issue on *Classical Logic and Computation*.

Strategic Computation and Deduction

CLAUDE KIRCHNER, FLORENT KIRCHNER,
HÉLÈNE KIRCHNER

> I'd like to conclude by emphasizing what a wonderful field this is to work in. Logical reasoning plays such a fundamental role in the spectrum of intellectual activities that advances in automating logic will inevitably have a profound impact in many intellectual disciplines. Of course, these things take time. We tend to be impatient, but we need some historical perspective. The study of logic has a very long history, going back at least as far as Aristotle. During some of this time not very much progress was made. It's gratifying to realize how much has been accomplished in the less than fifty years since serious efforts to mechanize logic began.
>
> Peter B. Andrews [2]

1 Introduction

Strategies, tactics, tacticals, proof plans, are terms widely used in artificial intelligence, in automated or interactive reasoning, in semantics of programming languages as well as in every day life. But what do they mean? What are these concepts used for and why is there so many different points of view? We do not claim to address all these questions here, but will contribute to define in a uniform way what strategies are for computation and deduction.

The complementarity between deduction and computation, as emphasized in particular in deduction modulo [19], allows us to now envision a completely new generation of proof assistants where customized deductions are performed modulo appropriate and user definable computations [11, 12]. This has in particular the advantage to allow for a uniform implementation of higher-order and first-order logics [16, 13] making possible the safe use of existing dedicated proof environments [36, 20, 9]. This generalizes the approaches typical in first-order theorem proving [45], as well as higher-order ones like PVS [49], TPS [3, 4], Omega [8, 53] or Coq [21], to mention just a few.

To encompass this view of computation and deduction uniformly, we start from a rule-based view point. Rule-based reasoning is present in many domains of computer science: in formal specifications, rewriting is used for prototyping specifications; in theorem proving, for dealing with equality, simplifying the formulas and pruning the search space; in programming languages, the rule object can be explicit like in PROLOG, OBJ or ML, or hidden in the operational semantics; expert systems use rules to describe actions to perform; in constraint logic programming, solvers are described via rules transforming constraint systems. XML document transformations, access-control policies or bio-chemical reactions are a few examples of application domains.

But deterministic rule-based computations or deductions are often not sufficient to capture every computation or proof development. A formal mechanism is needed, for instance, to sequentialize the search for different solutions, to check context conditions, to request user input to instantiate variables, to process subgoals in a particular order, etc. This is the place where the notion of strategy comes in.

Reduction strategies in term rewriting study which expressions should be selected for evaluation and which rules should be applied. These choices usually increase efficiency of evaluation but may affect fundamental properties of computations such as confluence or (non-)termination. Programming languages like ELAN, Maude and Stratego allow for the explicit definition of the evaluation strategy, whereas languages like Clean, Curry, and Haskell allow for its modification.

In theorem proving environments, including automated theorem provers, proof checkers, and logical frameworks, strategies (also called tacticals in some contexts) are used for various purposes, such as proof search and proof planning, restriction of search spaces, specification of control components, combination of different proof techniques and computation paradigms, or meta-level programming in reasoning systems.

Strategies are thus ubiquitous in automated deduction and reasoning systems, yet only recently have they been studied in their own right. In the two communities of automated deduction and rewriting, workshops have been launched to make progress towards a deeper understanding of the nature of strategies, their descriptions, their properties, and their usage.

In this paper we are contributing to the theoretical foundations of strategies and to the convergence of different points of view, namely rewriting-based computations on one hand, rule-based deduction and proof-search on the other hand. We will rely on previous works and strategy languages that have been recently designed and studied [32]. In rewriting, from elementary strategies expressions directly issued from a term rewrite system R, more elaborated strategies expressions can be built using a strategy language like in ELAN [34, 10], Stratego [56], TOM [6] or more recently Maude [44]. The semantics of such a language is naturally described in the rewriting calculus [14, 15]. A similar mechanism also exists in proof systems, where a set of core strategies, historically derived from the language LCF, is used to program sophisticated proof search patterns. The semantics of such a language appears in [18, 30, 35].

Building upon the definition of abstract reduction systems (recalled in Section 2), the main contributions of this paper are as follows: we propose in Section 3 a notion of abstract strategies together with adequate properties of termination, confluence and normalization under strategy. Thanks to this abstract concept and to the rewriting calculus (recalled in Section 4), we are able to draw a parallel between strategies for computation and strategies for deduction. While strategies for computation, developed in Section 5, essentially rely on the largely explored and well-known domain of term reduction by rewriting or narrowing, strategies for deduction, developed in Section 6, require to introduce an original point of view: we define deduction rules as rewrite rules, a deduction step as a rewriting step, a deduction system as an abstract reduction system. Strategic deductions are there of interest for developing complete proof trees. The same vision allows us to introduce proof construction as narrowing derivation. Computation, deduction and proof search are then captured as the foundational concept of abstract strategy.

2 Abstract reduction systems

When abstracting the notion of strategies, one important preliminary remark is that we need to start from an appropriate notion of *abstract reduction system* (ARS) based on the notion of graph instead of relation. This is due to the fact that, speaking of derivations, we need to make a difference between "being in relation" and "being connected". Typically modeling ARS as relations as in [5] allows us to say that, e.g, a and b are in relation but not that there may be several different ways to derive b from a. Consequently, we need to use the more expressive approach of [54, 38], based on a notion of oriented graph instead of that of a relation.

DEFINITION 1. An *abstract reduction system* (ARS) is a labelled oriented graph $(\mathcal{O}, \mathcal{S})$. The nodes in $\mathcal{O}$ are called *objects*, the oriented edges in $\mathcal{S}$ are called *steps*.

EXAMPLE 2. We use the standard graphical representation of binary oriented graph.

1. A first standard example: $\mathcal{A}_{lc} = $

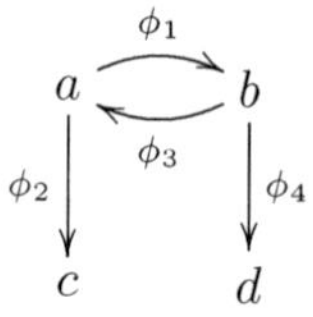

$$\begin{array}{ccc} a & \underset{\phi_3}{\overset{\phi_1}{\rightleftarrows}} & b \\ \phi_2 \downarrow & & \downarrow \phi_4 \\ c & & d \end{array}$$

2. The interest of using graph instead of binary relations is exemplified by $\mathcal{A}_c = a$ with two loops ϕ_1 and ϕ_2 on a.

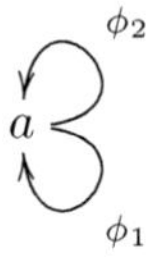

The next definitions can be seen as a renaming of usual ones in graph theory. Their interest is to allow us to define uniformly term rewriting derivations and strategies. If the concepts are not original, their presentation is.

DEFINITION 3 (Derivation). For a given ARS $\mathcal{A}$:

1. A *reduction step* is a labelled edge ϕ together with its source a and target b. This is written $a \rightarrow_{\mathcal{A}}^{\phi} b$, or simply $a \rightarrow^{\phi} b$ when unambiguous.

2. An $\mathcal{A}$-*derivation* or $\mathcal{A}$-*reduction sequence* is a path π in the graph $\mathcal{A}$.

3. When it is finite, an $\mathcal{A}$-derivation π can be written $a_0 \rightarrow^{\phi_0} a_1 \rightarrow^{\phi_1} a_2 \ldots \rightarrow^{\phi_{n-1}} a_n$ and we say that a_0 reduces to a_n by the derivation $\pi = \phi_0\phi_1 \ldots \phi_{n-1}$; this is also denoted $a_0 \rightarrow^{\phi_0\phi_1\ldots\phi_{n-1}} a_n$ or simply $a_0 \rightarrow^{\pi} a_n$. We call n the *length* of π.

   (a) The *source* of π is the object a_0 and its domain is defined as the singleton $dom(\pi) = \{a_0\}$.

   (b) The *target* of π is the object a_n and the application of the derivation π to a_0 is the singleton denoted $(\pi\ a_0) = \{a_n\}$. This is also denoted simply πa_0 when there is no syntactic ambiguity.

4. The set of all derivations is denoted $\mathcal{D}(\mathcal{A})$.

5. A derivation is *empty* when its length is zero, in which case its source and target are the same. The empty derivation issued from a is denoted id_a.

6. The *concatenation* of two derivations π_1 and π_2 is defined when $dom(\pi_1) = \{a\}$ and $\pi_1 a = dom(\pi_2)$. Then $\pi_1; \pi_2$ denotes the new $\mathcal{A}$-derivation $a \rightarrow_{\mathcal{A}}^{\pi_1} b \rightarrow_{\mathcal{A}}^{\pi_2} c$ and $((\pi_1; \pi_2)\ a) = (\pi_2(\pi_1 a)) = \{c\}$.

Note that an $\mathcal{A}$-derivation is the concatenation of its reduction steps.

EXAMPLE 4. Following the previous examples, we have:

1. $\mathcal{D}(\mathcal{A}_{lc}) \supset \{id_a, \phi_1, \phi_1\phi_3, \phi_1\phi_4, \phi_1\phi_3\phi_1, (\phi_1\phi_3)^n, (\phi_1\phi_3)^\omega, \ldots\}$, where ϕ^n denotes the n-steps iteration of ϕ and ϕ^ω denotes the infinite iteration of ϕ;

2. $\mathcal{D}(\mathcal{A}_c) \supset \{\phi_1, \phi_2, \phi_1\phi_2, \ldots, (\phi_1)^\omega, (\phi_2)^\omega, \ldots\}$.

The following definitions state general properties of an ARS.

DEFINITION 5 (Termination). For a given ARS $\mathcal{A} = (\mathcal{O}, \mathcal{S})$:

- $\mathcal{A}$ is *terminating* (or *strongly normalizing*) if all its derivations are of finite length;
- An object a in $\mathcal{O}$ is *normalized* when the empty derivation is the only one with source a (e.g., a is the source of no edge);
- A derivation is *normalizing* when its target is normalized;
- An ARS is *weakly terminating* if every object a is the source of a normalizing derivation.

DEFINITION 6 (Confluence). An ARS $\mathcal{A} = (\mathcal{O}, \mathcal{S})$ is *confluent* if for all objects a, b, c in $\mathcal{O}$, and all $\mathcal{A}$-derivations π_1 and π_2, when $a \twoheadrightarrow^{\pi_1} b$ and $a \twoheadrightarrow^{\pi_2} c$, there exist d in $\mathcal{O}$ and two $\mathcal{A}$-derivations π_3, π_4 such that $c \twoheadrightarrow^{\pi_3} d$ and $b \twoheadrightarrow^{\pi_4} d$.

3 Abstract strategies

We use a general definition, compliant with [34] and slightly different from the one used in, e.g., [54].

DEFINITION 7 (Abstract Strategy). For a given ARS $\mathcal{A}$:

1. An *abstract strategy* ζ is a subset of the set of all derivations (finite or not) of $\mathcal{A}$.

2. Applying the strategy ζ on an object a is denoted ζa. It denotes the set of all objects that can be reached from a using a derivation in ζ:

$$\zeta a = \{b \mid \exists \pi \in \zeta \text{ such that } a \twoheadrightarrow^{\pi} b\} = \{\pi a \mid \pi \in \zeta\}$$

   When no derivation in ζ has source a, we say that the strategy application on a fails.

3. Applying the strategy ζ on a set of objects consists in applying ζ to each element a of the set. The result is the union of ζa for all a in the set of objects.

4. The *domain* of a strategy is the set of objects that are source of a derivation in ζ:

$$dom(\zeta) = \bigcup_{\delta \in \zeta} dom(\delta)$$

5. The strategy that contains all the empty derivations is denoted Id:

$$Id = \{id_a \mid a \in \mathcal{O}\}$$

With this definition, a strategy that contains only infinite derivations from source a is not defined on a. This has to be distinguished from the case where a strategy contains no derivations from source a and returns the empty set $\emptyset$, i.e., the strategy application fails. The empty set of derivations is a strategy called *Fail*; its application always fails. Notice finally that instead of returning *sets* of results, we might define strategies to return *multisets* or even *list* or any appropriate data structure.

EXAMPLE 8 (Example 2 continued). For $\mathcal{A}_{lc}$, let us define and examine a few strategies:

1. $\zeta_1 = \mathcal{D}(\mathcal{A}_{lc})$, i.e., all the derivations. So we have for example: $\zeta_1 a = \{a, b, c, d\}$.
   If multisets of results are considered, $\zeta_1 a = \{a, a, ..., b, c, d\}$ in which a occurs for each finite derivation which approximates $(\phi_1\phi_3)^\omega$.

2. $\zeta_2 = \emptyset(= \textit{Fail})$: failure, no applicable derivation, i.e., for all x in $\mathcal{O}_{lc}$, $\zeta_2 x = \emptyset$.

3. $\zeta_3 = \{(\phi_1\phi_3)^*\phi_4\}$,
   a always converges to d: $\zeta_3 a = \{d\}$;
   b is not transformed (as well as c and d): $\zeta_3 b = \emptyset$.

4. The result of $((\phi_1\phi_3)^\omega\ a)$ is the empty set.

The standard notions of ARS termination and confluence must be carefully extended to abstract strategies.

DEFINITION 9 (Termination under strategy). For a given ARS $\mathcal{A} = (\mathcal{O}, \mathcal{S})$ and strategy ζ:

- $\mathcal{A}$ is *ζ-terminating* if all derivations in ζ are of finite length;

- An object a in $\mathcal{O}$ is ζ-normalized when the empty derivation is the only one in ζ with source a;

- A derivation is *ζ-normalizing* when its target is ζ-normalized;

- An ARS is *weakly ζ-terminating* if every object a is the source of a ζ-normalizing derivation.

EXAMPLE 10 (Example 2 continued). Given $\mathcal{A}_{lc}$ and the strategy ζ defined as $a \twoheadrightarrow^{\phi_1} b \twoheadrightarrow^{\phi_4} d$, b is ζ-normalized since there is no derivation in ζ with source b.

It could be tempting to generalize Definition 6 as follows:
An ARS $\mathcal{A} = (\mathcal{O}, \mathcal{S})$ is confluent *under strategy ζ if for all objects a, b, c in $\mathcal{O}$, and all $\mathcal{A}$-derivations π_1 and π_2 in ζ, when $a \twoheadrightarrow^{\pi_1} b$ and $a \twoheadrightarrow^{\pi_2} c$ there exists d in $\mathcal{O}$ and two $\mathcal{A}$-derivations π_3, π_4 in ζ such that $c \twoheadrightarrow^{\pi_3} d$ and $b \twoheadrightarrow^{\pi_4} d$.*
However this generalization is not correct since nothing ensures that the derivations $c \twoheadrightarrow^{\pi_3} d$ and $b \twoheadrightarrow^{\pi_4} d$ belong to the strategy. This leads to two possible definitions of respectively weak and strong confluence under strategy.

DEFINITION 11 (Weak Confluence under strategy). An ARS $\mathcal{A} = (\mathcal{O}, \mathcal{S})$ is *weakly confluent* under strategy ζ if for all objects a, b, c in $\mathcal{O}$, and all $\mathcal{A}$-derivations π_1 and π_2 in ζ, when $a \twoheadrightarrow^{\pi_1} b$ and $a \twoheadrightarrow^{\pi_2} c$ there exists d in $\mathcal{O}$ and two $\mathcal{A}$-derivations π_3', π_4' in ζ such that $\pi_3' : a \twoheadrightarrow b \twoheadrightarrow d$ and $\pi_4' : a \twoheadrightarrow c \twoheadrightarrow d$.

DEFINITION 12 (Strong Confluence under strategy). An ARS $\mathcal{A} = (\mathcal{O}, \mathcal{S})$ is *strongly confluent* under strategy ζ if for all objects a, b, c in $\mathcal{O}$, and all $\mathcal{A}$-derivations π_1 and π_2 in ζ, when $a \twoheadrightarrow^{\pi_1} b$ and $a \twoheadrightarrow^{\pi_2} c$ there exists d in $\mathcal{O}$ and two $\mathcal{A}$-derivations π_3, π_4 in ζ such that:

1. $b \twoheadrightarrow^{\pi_3} d$ and $c \twoheadrightarrow^{\pi_4} d$;

2. $\pi_1; \pi_3$ and $\pi_2; \pi_4$ belong to ζ.

EXAMPLE 13 (Example 2 continued). Let us again consider $\mathcal{A}_{lc}$ and the following various strategies:

1. $\zeta_1 = \mathcal{D}(\mathcal{A}_{lc})$, i.e., all the derivations. $\mathcal{A}_{lc}$ is neither weakly nor strongly confluent under ζ_1: just consider $\pi_1 : a \twoheadrightarrow^{\phi_1} b \twoheadrightarrow^{\phi_4} d$ and $\pi_2 : a \twoheadrightarrow^{\phi_2} c$.

2. $\zeta_2 = \emptyset (= \mathit{Fail})$: $\mathcal{A}_{lc}$ is trivially both weakly and strongly confluent under ζ_2.

3. $\zeta_3 = \{(\phi_1\phi_3)^*\phi_4\}$: $\mathcal{A}_{lc}$ is also weakly and strongly confluent under ζ_3.

4. For a different reason, this is also the case for $\zeta_4 = (\phi_1\phi_3)^\omega$ whose result is the empty set.

To understand the difference between weak and strong confluence let us consider $\mathcal{O} = \{a, b, c, d\}$ and reduction steps $\phi_1, \phi_2, \phi_3, \phi_4, \phi'_1, \phi'_2, \phi'_3, \phi'_4$. This ARS $\mathcal{A}_{lc}$ is weakly and strongly confluent under the strategy $\zeta = \{a \twoheadrightarrow^{\phi_1} b, a \twoheadrightarrow^{\phi_2} c, b \twoheadrightarrow^{\phi_3} d, c \twoheadrightarrow^{\phi_4} d, a \twoheadrightarrow^{\phi_1} b \twoheadrightarrow^{\phi_3} d, a \twoheadrightarrow^{\phi_2} c \twoheadrightarrow^{\phi_4} d\}$, but is not under $\zeta = \{a \twoheadrightarrow^{\phi_1} b, a \twoheadrightarrow^{\phi_2} c, b \twoheadrightarrow^{\phi_3} d, c \twoheadrightarrow^{\phi_4} d\}$. A is weakly but not strongly confluent under the strategy $\zeta = \{a \twoheadrightarrow^{\phi_1} b, a \twoheadrightarrow^{\phi_2} c, b \twoheadrightarrow^{\phi_3} d, c \twoheadrightarrow^{\phi_4} d, a \twoheadrightarrow^{\phi'_1} b \twoheadrightarrow^{\phi'_3} d, a \twoheadrightarrow^{\phi'_2} c \twoheadrightarrow^{\phi'_4} d\}$.

4 The rewriting calculus

The rewriting calculus or ρ-calculus generalizes term rewriting and lambda-calculus. It has been introduced in [14]. We recall here the main syntactic ingredients necessary to set-up the framework developed in this paper, using a syntax that restricts patterns to be only algebraic, a case simpler than the general one where pattern may contain abstractions (see [14]).

Definition 1 (Syntax). We consider the symbols "$\_ \twoheadrightarrow \_$" (abstraction operator), "$\_ \wr \_$" (structure operator), the (hidden) application operator, a set $\mathcal{X}$ of variables and a set $\mathcal{K}$ of constants, each of them having an arity denoted $ar(\mathcal{K})$. The syntax of the basic rewriting calculus is the following:

$$\mathcal{T} ::= \mathcal{X} \mid \mathcal{K} \mid \mathcal{P} \twoheadrightarrow \mathcal{T} \mid \mathcal{T}\,\mathcal{T} \mid \mathcal{T} \wr \mathcal{T} \qquad \text{Terms}$$

$$\mathcal{P} ::= \mathcal{X} \mid (\ldots((\mathcal{K}\,\mathcal{P}_1)\,\mathcal{P}_2)\ldots\mathcal{P}_{ar(\mathcal{K})} \qquad \text{Algebraic Patterns}$$

A *linear* pattern is a pattern where every variable occurs at most once.

We assume that the application operator associates to the left, while the other operators associate to the right. The priority of the application is higher than that of "$\_ \twoheadrightarrow \_$" which is, in turn, of higher priority than the "$\_ \wr \_$". In the following, the symbols $A, B, C, \ldots$ range over the set $\mathcal{T}$ of terms, the symbols $x, y, z, \ldots$ range over the set $\mathcal{X}$ of variables ($\mathcal{X} \subseteq \mathcal{T}$), the symbols $a, b, c, \ldots, f, g, h, \ldots$ and strings built from them range over a set $\mathcal{K}$ of constants ($\mathcal{K} \subseteq \mathcal{T}$). Finally, the symbols P, Q range over the set $\mathcal{P}$ of patterns, ($\mathcal{X} \subseteq \mathcal{P} \subseteq \mathcal{T}$). Vectors of terms $(A_1, \ldots, A_n)$ are denoted by $\overline{A}$. We usually denote a term of the form $(\ldots((f\ A_1)\ A_2)\ldots\ A_n)$ by $f(A_1, A_2, \ldots, A_n)$. Identity of terms is denoted by $\equiv$.

A term of the form $P \twoheadrightarrow B$ is an *abstraction* (or *rule*) with pattern P and body B; intuitively, the free variables of P are bound in B. The term $A \wr B$ is called a *structure*.

As for the lambda-calculus for instance, the reduction relation $\mapsto\!\!\!\twoheadrightarrow$ of the rewriting calculus is defined as the smallest congruence generated by the following top-level reductions:

$$(P \twoheadrightarrow A)\, B \quad \rightharpoondown_{\rho} \quad A\theta \qquad\qquad \text{if } \exists\theta.\ P\theta \equiv B$$

$$(A \wr B)\, C \quad \rightharpoondown_{\delta} \quad A\, C \wr B\, C$$

The one step reduction relation is denoted by $\mapsto_{\rho}$ or $\mapsto_{\delta}$ according to the reduction rule which is used. Notice that in the main reduction $(\rightharpoondown_{\rho})$, the condition consists in checking the existence of a matching substitution θ. When it exists, this substitution is applied to A, taking care of free and bounded variables as usual in a higher-order setting (remember that $\twoheadrightarrow$ is an abstractor).

5 Strategic reduction

Let us now come back to the more classical context of first-order terms and term rewriting to see how it fits to the general notions presented above. Basic definitions on term rewriting can be found in [33, 37, 5]. The following standard notations will be used in the following. $\mathcal{T}(\mathcal{F},\mathcal{X})$ is the set of first-order terms built from a given finite set $\mathcal{F}$ of function symbols and a denumerable set $\mathcal{X}$ of variables. The set of variables occurring in a term t is denoted by $\mathcal{V}ar(t)$. If $\mathcal{V}ar(t)$ is empty, t is called a *ground term* and $\mathcal{T}(\mathcal{F})$ is the set of ground terms. A substitution σ is an assignment from $\mathcal{X}$ to $\mathcal{T}(\mathcal{F},\mathcal{X})$, with a finite domain $\{x_1,\dots,x_k\}$ and is written $\sigma = \{x_1 \mapsto t_1,\dots,x_k \mapsto t_k\}$.

5.1 Term rewriting

A rewrite rule is an ordered pair of terms $l, r \in \mathcal{T}(\mathcal{F},\mathcal{X})$, denoted $l \rightarrow r$, where it is often required that l is not a variable and $\mathcal{V}ar(r) \subseteq \mathcal{V}ar(l)$. We will drop these restrictions later. The terms l and r are respectively called the left-hand side and the right-hand side of the rule. A *rewrite system* is a (finite or infinite) set of rewrite rules. Rules can be labelled to easily talk about them.

DEFINITION 14 (One step rewriting). Given a rewrite system R, an algebraic term t in $\mathcal{T}(\mathcal{F},\mathcal{X})$ rewrites to an algebraic term t' if there exists a rewrite rule $l \rightarrow r$ of R and a position ω in t, such that in the ρ-calculus, applying the ρ-term $(l \twoheadrightarrow r)$ to t at position ω evaluates to t', which is denoted by

$$t[(l \twoheadrightarrow r)t_{|\omega}]_{\omega} \mapsto_{\rho} t'$$

This is denoted $t \longrightarrow_{\omega}^{l\rightarrow r} t'$ or $t \longrightarrow_{\omega}^{R} t'$ when we do not need to make precise which rewrite rule is used.

Indeed this means that there exists a substitution σ such that $t_{|\omega} = \sigma(l)$ and $t' = t[\sigma(r)]_\omega$. A subterm $t_{|\omega}$ where the rewriting step is applied is called *redex*. A term that has no redex is said to be irreducible for R or to be in *R-normal form*.

To a set of rewrite rules corresponds directly a unique abstract reduction system that can be seen as a generic way to describe the set of all derivations.

DEFINITION 15 (Reduction system). Given a set of terms $\mathcal{T}(\mathcal{F},\mathcal{X})$ and a set of rewrite rules R, a reduction system is an abstract reduction system $\mathcal{R} = (\mathcal{O}_R, \mathcal{S}_R)$ with:

- $\mathcal{O}_R = \mathcal{T}(\mathcal{F},\mathcal{X})$, and
- $\mathcal{S}_R = \{t \to t' | t \longrightarrow^R_\omega t' \text{ for } \omega \text{ a position in } t\}$.

It is now possible to give the definition of strategic rewriting:

DEFINITION 16 (Strategic rewriting).

Given an abstract reduction system $\mathcal{R} = (\mathcal{O}_R, \mathcal{S}_R)$ generated by a term rewrite system R, and a strategy ζ of $\mathcal{R}$, a strategic rewriting derivation (or rewriting derivation under strategy ζ) is an element of ζ. A strategic rewriting step under ζ is a rewriting step $t \longrightarrow^R_\omega t'$ that occurs in a derivation of ζ.

Notice that the previous definitions extend immediatly to the case where matching is performed modulo an equational theory like associativity-commutativity of some fonction symbols. Strategic rewriting can therefore be defined for rewriting modulo relations as defined in [51, 31].

Besides the rewriting relation, another reduction relation, called narrowing, is worth considering to develop an ARS. This well-known process, introduced in [22, 29], is quite similar to rewriting but there, matching is replaced by unification. Let us recall its usual definition:

DEFINITION 17 (One step narrowing). Given a term rewrite system R, an algebraic term t in $\mathcal{T}(\mathcal{F},\mathcal{X})$ narrows to an algebraic term t' if there exists a rewrite rule $l \to r$ of R and a position ω in t, such that $t|_\omega$ and l are unifiable with a most general unifier σ. Then $t' = \sigma(t[r]_\omega)$. This is denoted $t \leadsto^{l\to r}_{\omega,\sigma} t'$ where either ω or σ may be omitted, or simply $t \leadsto^R t'$ when we do not need to make precise which rewrite rule is used.

For instance, using the rewrite rule $f(x,x) \to x$ the term $f(g(y),z)$ narrows to the term $g(y)$ with the substitution $\sigma = \{x \mapsto g(y), z \mapsto g(y)\}$.

Then, we can also consider that a rewrite system R generates in a similar way an abstract reduction system $\mathcal{N} = (\mathcal{O}_R, \mathcal{S}_R)$ with:

- $\mathcal{O}_R = \mathcal{T}(\mathcal{F},\mathcal{X})$, and

- $\mathcal{S}_R = \{t \rightsquigarrow t' | t \rightsquigarrow_{\omega}^{l \to r} t'$ for ω a position in $t\}$.

As for the rewriting relation, we will therefore use the definitions given in the abstract setting and we can define strategic narrowing:

DEFINITION 18 (Strategic narrowing).

Given an abstract reduction system $\mathcal{N} = (\mathcal{O}_R, \mathcal{S}_R)$ generated by a term rewrite system R, and a strategy ζ of $\mathcal{N}$, a strategic narrowing derivation (or narrowing derivation under strategy ζ) is an element of ζ. A strategic narrowing step under ζ is a narrowing step $t \rightsquigarrow_{\omega}^{R} t'$ that occurs in a derivation of ζ.

5.2 Term rewriting strategies

In the classical setting of first-order term rewriting, strategies have been used to determine at each step of a derivation which is the next redex. Thus they have often been defined as functions on the set of terms like in [54]. In our general setting of abstract strategies, the definitions of these standard strategies express a property that a derivation must satisfy to belong to the strategy of interest. Let us illustrate this point of view by a few examples of strategies that have been primarily provided to describe the operational semantics of functional programming languages and the related notions of call by value, call by name, call by need.

Leftmost-innermost and outermost reduction

Let R be a rewrite system on $\mathcal{T}(\mathcal{F}, \mathcal{X})$. A rewriting derivation under the innermost (resp. outermost) strategy verifies : for a given term $t \in \mathcal{T}(\mathcal{F}, \mathcal{X})$, $t \to_{\omega}^{R} t'$ and the rewriting position ω in t is such that there is no suffix (resp. prefix) position ω' of ω such that t rewrites at position ω'.

Lazy reduction

Lazy reduction, as performed in Haskell for instance, or described in [23], combines innermost and outermost strategies, in order to avoid redundant rewrite steps. For that, operators of the signature have labels specifying which arguments are lazy or eager. Positions in terms are then annotated as lazy or eager, and the strategy consists in reducing the eager subterms only when their reduction allows a reduction step higher in the term [48].

Lazy reduction can also be performed with other mechanisms, such as local strategies or context-sensitive rewriting.

Reduction under local strategies

Local strategies on operators, used for instance in the OBJ-like languages [25], introduce a function LS from $\mathcal{F}$ to the set $\mathcal{L}(\mathbb{N})$ of lists of integers to specify which and in which order the arguments of each function symbol have to be reduced. At each step of the derivation, for the current term t, the next redex is chosen according to $LS(f)$ where f is the top operator of the term t.

Context-sensitive reduction

Local strategies are close to context-sensitive rewriting [41, 42, 1], where rewriting is also allowed only at some specified positions in the terms. The former specify an ordering on these rewriting positions, so they are more specific than context-sensitive rewriting where a redex is chosen in a set of positions. More precisley, here, a replacement map is a mapping $\mu : \mathcal{F} \mapsto \mathcal{P}(\mathbb{N})$ satisfying $\mu(f) \subseteq \{1, \ldots, k\}$, for each k-ary symbol f of a signature $\mathcal{F}$ [40]. Replacement maps discriminate the argument positions on which the rewriting steps are allowed.

Needed and standard reduction

In [27, 28], the notions of needed and strongly needed redexes are defined for orthogonal rewrite systems. The main idea here is to find the optimal way, when it exists, to reach the normal form a term. A redex is needed when there is no way to avoid reducing it to reach the normal form. Reducing only needed redexes is clearly the optimal reduction strategy, as soon as needed redexes can be decided, which is not the case in general.

5.3 Strategy language for reduction

This section is devoted to expressing rewriting strategies; exactly the same principles could be used to express other strategies as we will see later.

In the 1990s, generalizing OBJ's concept of local strategies, the idea has emerged to better formalize the control on rewriting which was performed implicitly in interpreters or compilers of rule-based programming languages. Then instead of describing a strategy by a property of its derivations, the idea is now to provide a *strategy language* to specify which derivations we are interested in.

Various approaches have followed, yielding different strategy languages such as ELAN [34, 10], APS [39], Stratego [56], TOM [6] or more recently Maude [44]. All these languages share the concern to provide abstract ways to express control of rule applications, by using reflexivity and the meta-level for Maude, or a strategy language for ELAN, Stratego or ASF+SDF. Strategies such as bottom-up, top-down or leftmost-innermost are high level ways to describe how rewrite rules should be applied. TOM, ELAN, Maude and Stratego provide flexible and expressive strategy languages where high-level strategies are defined by combining low-level primitives. In this section, we choose TOM [7][1] to illustrate the construction of strategy languages of this family and we describe below the main elements of the language. The

[1] `http://tom.loria.fr`

semantics of the TOM strategy language as well as others are naturally described in the rewriting calculus [14, 15].

We can distinguish two classes of constructs in the strategy language: the first one allows construction of derivations from the basic elements, namely the rewrite rules. The second one corresponds to constructs that express the control, especially deterministic and undeterministic choice. Moreover, the capablility of expressing recursion in the language brings even more expressive power.

Elementary strategies

An elementary strategy is either *Identity* which corresponds to the set Id of all empty derivations, *Fail* which denotes the empty set of derivations *Fail*, or a set of rewrite rules R which represents one-step derivations with rules in R at the root position. $Sequence(s_1, s_2)$, also denoted $s_2; s_1$ is the concatenation of s_1 and s_2 whenever it exists: $Sequence(s_1, s_2)t = s_2(s_1 t)$.

Control strategies

A few constructions are needed to build derivations, branching and take into account the structure of elements (terms).

choice $Choice(s_1, s_2)$ selects the first strategy that does not fail; it fails if both fail:

$$Choice(s_1, s_2)t = s_1 t$$

if $s_1 t$ does not fail else $s_2 t$.

all subterms On a term t, $All(s)$ applies the strategy s on all immediate subterms:

$$All(s)f(t_1, ..., t_n) = f(t'_1, ..., t'_n)$$

if $st_1 = t'_1, ..., st_n = t'_n$; it fails if there exists i such that st_i fails.

one subterm On a term t, $One(s)$ applies the strategy s on the first immediate subterm where s does not fail:

$$One(s)f(t_1, ..., t_n) = f(t_1, ..., t'_i, ..., t_n)$$

if for all $j < i$, st_j fails, and $st_i = t'_i$; it fails if for all i, st_i fails.

fixpoint The μ recursion operator (comparable to *rec* in OCaml) is introduced to allow the recursive definition of strategies. $\mu x.s$ applies the derivation in s with the variable x instantiated with $\mu x.s$:

$$\mu x.s = s[x \leftarrow \mu x.s]$$

These strategies are then composed to build more elaborated ones. For instance $Try(s) = choice(s, Identity)$: $Try(s)$ applies s if it can, and performs the *Identity* otherwise. The *All* and *One* combinators are used in combination with the fixpoint operator to define tree traversals. For example, we have $TopDown(s) = \mu x.Sequence(s, All(x))$: the strategy s is first applied on top of the considered term, then the strategy $TopDown(s)$ is recursively called on all immediate subterms of the term.

6 Strategic deduction

In the previous section, we have seen how the abstract strategy concept instantiates to control rule-based computation. Let us now explore the deduction side.

In proof assistants such as Coq [21] and PVS [49], the interaction language (or *proof language*) contains two kinds of *proof commands*: *tactics*, which construct the proof tree by applying logical rules, and *strategies* which provide proof search control[2]. In this section, we provide innovative definitions of these two concepts, starting with tactics.

6.1 Proof rewriting

We describe the proof construction mechanism using the abstract reduction system framework. Our purpose in this section is not to formalize any particular instance of a deduction system, such as classical natural deduction [52] or intuitionistic sequent calculus [24]: our model is set-theoretic. Roughly, sequents/typing judgements are composed of a finite set of labelled hypotheses, a finite set of labelled conclusions, and a proof term.

DEFINITION 19 (Labelled proposition). Given a proof term language $\mathcal{T}$ and a logical language $\mathcal{L}$, a labelled proposition is a pair consisting of a proof term $\pi \in \mathcal{T}$ and a logical proposition $A \in \mathcal{L}$, noted $\pi : A$.

DEFINITION 20 (Sequent). Given a proof term language $\mathcal{T}$ and a logical language $\mathcal{L}$, a sequent is a triple $s = (\pi, \Gamma, \Delta)$, where π is a proof term of $\mathcal{T}$, and where Γ and Δ (respectively the *antecedent* and the *consequent*) are finite sets of labelled propositions of $\mathcal{L}$. In the following, this triple will be denoted: $\pi \therefore (\Gamma \vdash \Delta)$, or in short $\pi \therefore s$. The (infinite) set of sequents will be denoted $\mathcal{S}(\mathcal{T}, \mathcal{L})$.

This representation of sequents differs from the usual definition, given in natural deduction, where all proof terms are attached to a proposition in the antecedent or in the consequent, and a sequent is just the pair $\Gamma \vdash \Delta$.

[2]Various other terminologies exist in this field: for instance, PVS tactics are called *proof rules*, while Coq strategies are names *tacticals*. We use the words "tactic" and "strategy" because they appropriately convey a notion of locality.

The usual representation can be encoded in our formalism using a reserved formula label '$*$', as illustrated by the following example:

EXAMPLE 21. Let us take Λ, the set of simply typed lambda-terms, for $\mathcal{T}$, and FOL, the set of all first-order propositions, for $\mathcal{L}$. The sequent stating that under the assumption $(t : A)$, one of the statements $(\lambda x.t : B \Rightarrow A)$ or $(u : C)$ is provable, is denoted in natural deduction:

$$t : A \vdash \lambda x.t : B \Rightarrow A, u : C$$

Assume a proof of this sequent begins with the deconstruction of the λ-proof term $\lambda x.t$, using the "$\Rightarrow$-intro" deduction rule. Since it is the central element of the next deduction step, this proof term has a particular status, and can be qualified as *pivotal*. In order to express the aforementioned sequent within the framework set by Definition 20, the pivotal proof term is separated from its proposition, and the '$*$' placeholder put in its place to label the proposition. Hence the sequent is written:

$$\lambda x.t \therefore (t : A \vdash * : B \Rightarrow A, u : C)$$

Since a pivotal proof term can always be pinpointed, such a transformation can be made on any sequent of a proof in natural deduction.

On the other hand, separating the proof term from the antecedent and consequent allows us to capture more involved logical settings, where the pivotal proof terms might appear, not only at the level of formulas, but also at the level of sequents. For instance, in the case of sequent calculus *à la* Curien-Herbelin [17], the sequents $\Gamma \vdash \lambda x.v : A \Rightarrow B; \Delta$ and $\langle v|e\rangle : (\Gamma \vdash \Delta)$, where $\lambda x.v$ and $\langle v|e\rangle$ are pivotal $\bar{\lambda}\mu\tilde{\mu}$-proof terms, are well-formed. In our formalism, while the first sequent is expressed using the aforementioned '$*$' label using the same process as in Example 21, the second sequent is trivially represented by the isomorphic triple $\langle v|e\rangle \therefore (\Gamma \vdash \Delta)$, in which no '$*$' label appear. The same holds for Urban's sequent calculus [55], which places *all* proof terms on the same level as the antecedents and consequents. Finally we need to add a notion of *list of sequents*: therefore we introduce the usual list constructors *cons* and *nil*, noted ';' and '$\varnothing$'. For typesetting purposes and when unambiguous, we abbreviate the list $\phi; \varnothing$ to ϕ, and we omit all list constructors for unary lists. Finally, the notion of substitution carried by terms and propositions is easily extended through simple subterm propagation to range over sequents and lists of sequents. The set of lists of sequents is denoted $\mathcal{LS}(\mathcal{T}, \mathcal{L})$.

A deduction system can be seen as a rewrite system operating on typed proof terms, i.e., sequents. The following definitions lay the groundwork and lead to the expression of deduction systems as instances of the abstract reduction system framework.

DEFINITION 22 (Deduction rule). A rewrite rule, also called *deduction rule*, connects a single sequent to a list of sequents:

$$l \therefore (\Gamma_l \vdash \Delta_l) \rightarrow r_1 \therefore (\Gamma_{r_1} \vdash \Delta_{r_1}); \ldots; r_n \therefore (\Gamma_{r_n} \vdash \Delta_{r_n})$$

In order to capture the expressivity of logical rules, rewrite rules in this context may have a variable as left-hand side and may introduce in the right-hand side some variables that do not occur in the left-hand side. Indeed, relaxing these syntactic restrictions makes even more crucial the need of controlling rule application thanks to strategies.

DEFINITION 23 (One step deduction). We say that the sequent list $\phi = \pi_1 \therefore s_1; \ldots; \pi_m \therefore s_m$ reduces in one step to the sequent list ϕ' by the deduction rule $l \therefore s_l \rightarrow r_1 \therefore s_1; \ldots; r_n \therefore s_n$ if and only if there exists a substitution σ such that $\pi_i \therefore s_i = \sigma(l \therefore s_l)$ and

$$\phi' = \pi_1 \therefore s_1; \ldots; \pi_{i-1} \therefore s_{i-1}; \\ \sigma(r_1 \therefore s_1; \ldots; r_n \therefore s_n); \\ \pi_{i+1} \therefore s_{i+1}; \ldots; \pi_m \therefore s_m$$

When appropriate, the matching step used in the previous definition can be applied modulo an equational theory, like in the following example.

EXAMPLE 24. First-order minimal natural deduction is perhaps the most well-known logical framework to support the proof-as-terms morphism. It considers first-order propositions as types for lambda-terms, and has typing judgements of the form $\Gamma \vdash t : A$. In our formalism, we have $\mathcal{T} = \Lambda$, $\mathcal{L} = \text{FOL}$ and sequents of the form $t \therefore (\Gamma \vdash * : A)$. The deduction rules for minimal natural deduction can be expressed as the following rewrite rules, where Γ is a variable representing a multiset of propositions (i.e., the separator "," is associative and commutative), t, u are proof term variables, A, B are proposition variables.

$$\begin{array}{rcll} I \therefore (\Gamma \vdash * : \top) & \rightarrow & \varnothing & (D_1) \\ t \therefore (\Gamma, t : A \vdash * : A) & \rightarrow & \varnothing & (D_2) \\ \lambda x.t \therefore (\Gamma \vdash * : A \Rightarrow B) & \rightarrow & t \therefore (\Gamma, x : A \vdash * : B) & (D_3) \\ tu \therefore (\Gamma \vdash * : B) & \rightarrow & t \therefore (\Gamma \vdash * : A \Rightarrow B); u \therefore (\Gamma \vdash * : A) & (D_4) \\ \lambda x.t \therefore (\Gamma \vdash * : \forall x B) & \rightarrow & t \therefore (\Gamma \vdash * : B) & (D_5) \\ tu \therefore (\Gamma \vdash * : B[u]_\omega) & \rightarrow & t \therefore (\Gamma \vdash * : \forall x B[x]_\omega) & (D_6) \end{array}$$

where I is a proof term constant. Notice the use of a free variable A (*resp.* x) in the right hand-side of rule (D_4) (*resp.* (D_6)). Based on this set of rewrite rules, a series of four reduction steps could be:

$$\begin{aligned}
&\lambda x.\lambda y.\,(yx) \therefore (\vdash * : A \Rightarrow (A \Rightarrow B) \Rightarrow B) \\
&\qquad \rightarrow^{(D_3)} \lambda y.\,(yx) \therefore (x : A \vdash * : (A \Rightarrow B) \Rightarrow B) \\
&\qquad \rightarrow^{(D_3)} (yx) \therefore (x : A, y : A \Rightarrow B \vdash * : B) \\
&\qquad \rightarrow^{(D_4)} y \therefore (x : A, y : A \Rightarrow B \vdash * : A \Rightarrow B); \\
&\qquad\qquad x \therefore (x : A, y : A \Rightarrow B \vdash * : A) \\
&\qquad \rightarrow^{(D_2)} \varnothing; \varnothing
\end{aligned}$$

Note that in the previous example, the rule (D_2) can be applied because we assume the symbol "," to be associative and commutative. Another possibility could be to add a commutation rule, with the risk to inherit of the explicit treatment of commutativity at the proof design level.

DEFINITION 25 (Deduction system). Given a logical language $\mathcal{L}$ and a term algebra $\mathcal{T}$, a deduction system is an abstract reduction system $\mathcal{R} = (\mathcal{O}_R, \mathcal{S}_R)$ where:

- $\mathcal{O}_R$ contains elements of $\mathcal{LS}(\mathcal{T}, \mathcal{L})$, and
- $\mathcal{S}_R = \{\phi \rightarrow \phi' \mid \phi \longrightarrow_{\omega}^{\phi_l \rightarrow \phi_r} \phi'\}$ where ω is a position in ϕ (i.e., one of its sequents) and $\phi_l \rightarrow \phi_r$ is a deduction rule.

As for term rewriting, we will hereafter use all the definitions given in the abstract case for abstract reduction systems.

For a given deduction system $\mathcal{DS}$, a *completed proof* of a sequent ϕ is a derivation issued from ϕ and reaching the list $\varnothing; \ldots; \varnothing$ using the rules in $\mathcal{DS}$.

It is important to notice that the notion of derivation considered here concerns a full proof tree and not only one of its branches. This is due to the fact that we want to define strategies able to consistently develop complete proof trees and not only part of them. Therefore, the notions of deduction systems and of strategies are defined accordingly on *lists* of sequents representing proof trees.

Following these definitions, completed proofs are caracterized as reduction sequences, or equivalently as paths into the ARS formed by a deduction system. However, this leaves the question of the *construction* of a proof, i.e., of a proof term, unanswered. Yet this is a fundamental mechanism in the design of proof assistants. In order to address this question, various authors [43, 47, 30] have shown that the syntax of proof terms and sequents can be extended with *proof metavariables*, that represent the still unknown parts of a proof.

DEFINITION 26 (Proof metavariables). Consider the syntax of proof terms extended with a special set of variables $\mathcal{M}$, called *metavariables*, noted $\mathsf{X}, \mathsf{Y}, \mathsf{Z}\ldots$. We note $\mathcal{T}_{\mathcal{M}}$ the algebra of proof terms with metavariables, and $\mathrm{meta}(\pi)$ the set of metavariables appearing in π. We call *open sequent* a triple $\pi \therefore (\Gamma \vdash \Delta)$ where π contains at least one metavariable.

As shown in [47], the use of metavariables in logical settings featuring dependent types or polymorphism requires the addition of *constraints* to the representation of sequents. Since this extension is more elaborated, interesting but largely orthogonal to our current purpose, it will not be further detailed in this paper.

DEFINITION 27 (Proof grafting rule). A proof grafting rule is an ordered pair of sequents, denoted $\mathsf{L} \therefore s \to r \therefore s$, where L is a proof metavariable and r may or may not contain metavariables.

A proof construction step can then be simply defined as the application of a proof grafting rule to a list of sequents, similarly to Definition 23. The sequent featuring the newly introduced proof term is then reduced using deduction rules, until the proof ends or a new grafting step is necessary.

EXAMPLE 28. Recall the minimal natural deduction formalism presented in Example 24. Assume the following rules for metavariable instantiation:

$$\begin{aligned}
\mathsf{X} \therefore (\Gamma \vdash * : \top) &\to I \therefore (\Gamma \vdash * : \top) && (C_1)\\
\mathsf{X} \therefore (\Gamma, x : A \vdash * : A) &\to x \therefore (\Gamma, x : A \vdash A) && (C_2)\\
\mathsf{X} \therefore (\Gamma \vdash * : A \Rightarrow B) &\to \lambda x^A.\mathsf{Y} \therefore (\Gamma \vdash * : A \Rightarrow B) && (C_3)\\
\mathsf{X} \therefore (\Gamma \vdash * : A) &\to (\mathsf{YZ}) \therefore (\Gamma \vdash * : A) && (C_4)\\
\mathsf{X} \therefore (\Gamma \vdash * : \forall x B) &\to \lambda x.t \therefore (\Gamma \vdash * : \forall x B) && (C_5)\\
\mathsf{X} \therefore (\Gamma \vdash * : B[u]_\omega) &\to tu \therefore (\Gamma \vdash * : B[u]_\omega) && (C_6)
\end{aligned}$$

While the proof grafting rules can be arbitrarily chosen, these six rules each describe the construction of a λ-proof term head symbol. Furthermore, it is easy to verify that they build proofs terms that can be typechecked using the deduction rules for natural deduction. Based on this set of rewrite rules, the construction of the proof term of example 24 is conducted by an alternation of deduction and grafting rules:

$$
\begin{array}{l}
\mathsf{X}_1 \therefore (\vdash * : A \Rightarrow (A \Rightarrow B) \Rightarrow B) \\
\quad \rightarrow^{(C_3)} \lambda x^A.\mathsf{X}_2 \therefore (\vdash * : A \Rightarrow (A \Rightarrow B) \Rightarrow B) \\
\quad \rightarrow^{(D_3)} \mathsf{X}_2 \therefore (x : A \vdash * : (A \Rightarrow B) \Rightarrow B) \\
\quad \rightarrow^{(C_3)} \lambda y^{A \Rightarrow B}.\mathsf{X}_3 \therefore (x : A \vdash * : A \Rightarrow (A \Rightarrow B) \Rightarrow B) \\
\quad \rightarrow^{(D_3)} \mathsf{X}_3 \therefore (x : A, y : A \Rightarrow B \vdash * : A \Rightarrow (A \Rightarrow B) \Rightarrow B) \\
\quad \rightarrow^{(C_4)} (\mathsf{X}_4 \mathsf{X}_5) \therefore x : A, y : A \Rightarrow B \vdash A \Rightarrow (A \Rightarrow B) \Rightarrow B \\
\quad \rightarrow^{(D_4)} \mathsf{X}_4 \therefore (x : A, y : A \Rightarrow B \vdash * : A \Rightarrow B); \\
\qquad\qquad \mathsf{X}_5 \therefore (x : A, y : A \Rightarrow B \vdash * : A) \\
\quad \rightarrow^{(C_2)} \mathsf{y} \therefore (x : A, y : A \Rightarrow B \vdash * : A \Rightarrow B); \\
\qquad\qquad \mathsf{X}_5 \therefore (x : A, y : A \Rightarrow B \vdash * : A) \\
\quad \rightarrow^{(D_2)} \varnothing; \mathsf{X}_5 \therefore (x : A, y : A \Rightarrow B \vdash * : A) \\
\quad \rightarrow^{(C_2)} \varnothing; \mathsf{x} \therefore (x : A, y : A \Rightarrow B \vdash * : A) \\
\quad \rightarrow^{(D_2)} \varnothing; \varnothing
\end{array}
$$

The synthesis of the grafting step instances then yields the final proof term. The alternate $C_i; D_i$ reductions correspond to the LCF definition of a tactic, i.e., a grafting step that instantiates a proof term metavariable followed by a series of deduction steps that infers new goals from the resulting sequent.

Yet this approach using metavariables and grafting can be significantly condensed by using a mechanism specific to the term rewriting domain. Indeed, the instantiation of a metavariable by a term, such that the latter becomes reducible by a rule of the deduction system, is exactly a narrowing step, as defined in Definition 17.

EXAMPLE 29. The deduction system of Example 24 can be used to *find* a proof of the proposition $A \Rightarrow (A \Rightarrow B) \Rightarrow B$ by narrowing the initial query in the following way.

$$
\begin{array}{rl}
\mathsf{U} \therefore (\vdash * : A \Rightarrow (A \Rightarrow B) \Rightarrow B) & \rightsquigarrow^{(D_3)}_{\mathsf{U} \mapsto \lambda x.\mathsf{X}} \mathsf{X} \therefore (x : A \vdash * : (A \Rightarrow B) \Rightarrow B) \\
& \rightsquigarrow^{(D_3)}_{\mathsf{X} \mapsto \lambda y.\mathsf{Y}} \mathsf{Y} \therefore (x : A; y : A \Rightarrow B \vdash * : B) \\
& \rightsquigarrow^{(D_4)}_{\mathsf{Y} \mapsto \mathsf{TZ}} \mathsf{T} \therefore (x : A; y : A \Rightarrow B \vdash * : \alpha \Rightarrow B); \\
& \qquad\quad \mathsf{Z} \therefore (x : A; y : A \Rightarrow B \vdash * : \alpha) \\
& \rightsquigarrow^{(D_2)}_{\alpha \mapsto A, \mathsf{T} \mapsto y, \mathsf{Z} \mapsto x} \varnothing; \varnothing
\end{array}
$$

The proof grafting rules can be recovered by annotating the unification assignments with the appropriate sequents:

$$
\begin{aligned}
\mathsf{X} \therefore (\Gamma \vdash * : A \Rightarrow B) &\rightarrow \lambda x^A.\mathsf{Y} \therefore (\Gamma \vdash * : A \Rightarrow B) \\
\mathsf{Y} \therefore (\Gamma \vdash * : A) &\rightarrow (\mathsf{T}\mathsf{Z}) \therefore (\Gamma \vdash * : A) \\
\mathsf{T} \therefore (\Gamma, x : A \vdash * : A) &\rightarrow x \therefore (\Gamma, x : A \vdash A)
\end{aligned}
$$

and the computed proof term is simply obtained by composing the narrowing substitutions:

$$
\begin{aligned}
\mathsf{U} &= \{\alpha \mapsto A, \mathsf{T} \mapsto y, \mathsf{Z} \mapsto x\}(\{\mathsf{Y} \mapsto (\mathsf{T}\mathsf{Z})\}(\{\mathsf{X} \mapsto \lambda y.\mathsf{Y}\}(\lambda x.\mathsf{X}))) \\
&= \lambda x.\lambda y.\ (yx)
\end{aligned}
$$

Notice in the previous example the use made of proof term variables (i.e., $\mathsf{U}, \mathsf{X}, \mathsf{Y}, \mathsf{Z}, \mathsf{T}$) and the use of one proposition variable, i.e., α.

Hence, while traditionally a proof construction system was defined as an alternation of metavariable grafting and deduction steps, what emerges here is a more concise definition, based on the narrowing relation derived from a deduction system.

DEFINITION 30 (Proof construction system). Given a logical language $\mathcal{L}$, a term algebra $\mathcal{T}$ and a deduction system $\mathcal{DS}$ defined from a set D of proof rewrite rules, a *proof construction system* is an abstract reduction system $\mathcal{N} = (\mathcal{O}_D, \mathcal{S}_D)$ where:

- $\mathcal{O}_D$ contains elements of $\mathcal{LS}(\mathcal{T}, \mathcal{L})$, and
- $\mathcal{S}_D = \{\phi \rightsquigarrow \phi' \mid \phi \rightsquigarrow_\omega^D \phi' \text{ for } \omega \text{ a position in } \phi\}$.

This definition provides an explicit link between the deduction rules, which are usually well-known and documented as part of the underlying logical framework of a proof construction system, and the metavariable instantiation steps, which often are left implicit. As such, this formalism can be seen as an innovative contribution to the semantics of proof systems.

6.2 Proof rewriting strategies

The need for the creation of a mechanism for proof strategies was acknowledged along with the first theorem provers implementations. In [26], the authors remark that their initial attempts at using LCF [46], one of the first deduction software, were

> "limited by the fixed, and rather primitive, nature of its repertoire of commands",

and compared the experience to using a top-level interactive assembly language. What was needed, they analyzed, was a way to specify "recipes for proofs", i.e., strategies.

Interestingly, expressing a proof construction and deduction system as an instance of an abstract reduction system, gives us the possibility to extend the notion of abstract strategy (as defined in Section 3) to deduction systems. The result is the following definition of a proof strategy.

DEFINITION 31 (Proof strategy). A proof strategy is a subset of the set of all possible proof derivations.

The application and domain of a proof strategy, as well as the strategy containing all the empty derivations, are defined accordingly to Definition 7. However, this definition in extension does not provide much guidance regarding the actual implementation of proof strategies in a theorem prover.

In modern LCF, the strategies are introduced as combinators whose behavior depends on the *state* of the proof after the application of its arguments. The state of the proof usually contains non-logical information that signals whether the previous tactic application has, e.g., solved the goal, or failed. In an interactive proof development, this information is fed back to the user, who uses it to further his understanding of the proof. An example of LCF strategy can be found in the `orelse` combinator, which when provided two tactics t_1 and t_2, applies t_2 only if the application of t_1 has either failed, or did not modify the proof. All major implementations of proof strategy languages are based on a core set of commands inherited from the LCF prover [50], namely: `then`, `thenl`, `orelse`, `idtac`, `fail`, and `repeat`.

The natural question to ask is whether it is possible to bridge the divide between a LCF-style proof strategy language and the concept of a proof strategy as given by Definition 31. More generally, because of the uniform approach proposed in this paper, in the upcoming section we investigate the possibility of having an *abstract strategy language*, i.e., a language for building strategies for abstract reduction systems that can be instantiated into rewriting and proof strategy languages. The LCF proof strategy language will be a starting point for the design of this abstract strategy language.

6.3 Strategy language for deduction

The strategy language of modern procedural theorem provers is based on the language of modern LCF, as summarized in Section 6. With time, this core has been extended to include a variety of commands, either to enrich the interaction with the proof systems (postpone goals, undo changes, etc.), or to build more complex and automated deduction rules (including unification, narrowing, etc.) by combining simple rules using strategic constructs.

In the following, we propose a classification of the different strategic commands into the two aforementioned branches: programming and interaction. As a guideline through this taxonomy, we pay particular attention to the place of strategies in the *representation* (that is, the final format) of proofs and in the role they play with regard to the proof *construction* mechanism.

Strategies as programming constructs

Strategies are used to build other commands, i.e., they act as a programming language at the level of the proof language. Unlike deduction and proof construction rules, they are not used directly to construct proofs per se, but rather to generate more complex rules (build the proof builders). In general, strategies are not necessary to represent the proofs. Indeed, because they are in essence programming constructs, only the *results* of the strategic programs are to be retained. As an example, take the following proof script:

```
orelse (apply lemma_1) (apply lemma_2)
```

that attempts to perform a deduction by applying the result of a first lemma, and that applies a second lemma if the first one is unsuccessful. Whichever branch of the `orelse` strategy gets selected should be recorded in the final representation of the proof. However, operationally, there is no point in storing the whole script. We call these *programming strategies.*

Strategies as proof structuring commands

There are two strategic constructs that, although used to build other commands, are important to distinguish from the programming strategies. Indeed, they constitute exceptions to the non representativity of the commands in the previous category. On the contrary they are *fundamental* in the representation of proofs:

`thens`, which combines commands in a tree structure, isomorphic to the structure of the proof. It is the "glue" that holds the building bricks of the proof together, and thus it is necessary to the representation of non-trivial proofs.

`idtac`, that "does nothing" when applied to a sequent. It is used to represent incomplete proofs, and in the tree of commands, fills the place where later tactics might be employed.

Together with tactics, these two strategies are all that are needed to represent proofs. In reference to their tree-structuring functionality, we call these special programming strategies *bark strategies.*

Strategies as interaction controls

Building proofs has become a largely interactive process, with goals being presented one by one, postponed, changes undone, etc.. The second group of strategic commands are the ones that control such interactions. They do not contribute directly to the construction of the proof, neither should they appear in a finished proof script. Examples of such commands include the `postpone` construct evoked in the introduction, PVS's `(hide)` / `(reveal)` or Coq's `focus`. Also, Coq's unassuming '`.`' command, that (literally) punctuates command applications, should be seen as both an evaluation trigger and an interactive control, returning the first subgoal of the active subtree once the evaluation is complete. We call these controls *interactive commands.*

7 Conclusion and further work

We have given in this paper the definition of abstract strategies and formalized the properties of termination, confluence and normalization under strategy. We then provided instantiations of this framework both in the field of computation and deduction systems. Finally, we exposed and classified the notions of strategies used in these two fields.

We believe that our description of strategies for computation and deduction, combined with our use of a similar framework for expressing their underlying derivation system, makes them ripe for a comparison, and that their similarities are manifest. As a nice bonus, in the process, we have proposed a new definition of the concept of proof construction system, which sheds new light on the semantics of the mechanisms involved.

Further work involves two directions: first, the definition of abstract strategy has to be refined according to the objects on which they are applied: this would lead to more operational definitions and properties for strategies on terms. Second, a proposal for an abstract strategy language should emerge from this work and could be the basis either for future program and proof assistants, or for cooperation between existing ones.

Acknowledgements

This paper beneficiate of many interactions we had, within the Protheo team in Nancy since 15 years, with Thérèse Hardin and Gilles Dowek, and from our previous works on Coq, PVS, OBJ, ELAN and TOM. Many thanks also to Dan Dougherty for constructive interaction on strategies and to Christoph Benzmüller for his precise reading and comments.

BIBLIOGRAPHY

[1] María Alpuente, Santiago Escobar, and Salvador Lucas. Correct and Complete (Positive) Strategy Annotations for OBJ. In *Proceedings of the 5th International Workshop on Rewriting Logic and its Applications (RTA)*, volume 71 of *Elecronic Notes In Theoretical Computer Science*, pages 70–89, 2004.

[2] Peter B. Andrews. Herbrand Award Acceptance Speech. *Journal of automated deduction*, 31:169–187, 2003.

[3] Peter B. Andrews, Matthew Bishop, Sunil Issar, Dan Nesmith, Frank Pfenning, and Hongwei Xi. TPS: A Theorem Proving System for Classical Type Theory. *Journal of Automated Reasoning*, 16(3):321–353, June 1996.

[4] Peter B. Andrews and Chad E. Brown. TPS: A hybrid automatic-interactive system for developing proofs. *J. Applied Logic*, 4(4):367–395, 2006.

[5] Franz Baader and Tobias Nipkow. *Term Rewriting and all That*. Cambridge University Press, 1998.

[6] Emilie Balland, Paul Brauner, Radu Kopetz, Pierre-Etienne Moreau, and Antoine Reilles. Tom Manual. LORIA, Nancy (France), version 2.4 edition, October 2006.

[7] Emilie Balland, Paul Brauner, Radu Kopetz, Pierre-Etienne Moreau, and Antoine Reilles. Tom: Piggybacking Rewriting on Java. In *Proceedings of the 18th Conference on Rewriting Techniques and Applications*, volume 4533 of *Lecture Notes in Computer Science*, pages 36–47. Springer-Verlag, 2007.

[8] Christoph Benzmüller, Matthew Bishop, and Volker Sorge. Integrating TPS and Omega. *J. UCS*, 5(3):188–207, 1999.

[9] Frédéric Blanqui, Jean-Pierre Jouannaud, and Pierre-Yves Strub. Building Decision Procedures in the Calculus of Inductive Constructions. In Jacques Duparc and Thomas Henziger, editors, *16th Annual Conference on Computer Science and Logic - CSL 2007*, volume 4646 of *Lecture Notes in Computer Science*, Lausanne, Suisse, 2007. Springer Verlag.

[10] Peter Borovanský, Claude Kirchner, Helene Kirchner, and Christophe Ringeissen. Rewriting with strategies in ELAN: a functional semantics. *International Journal of Foundations of Computer Science*, 12(1):69–98, February 2001.

[11] Paul Brauner, Clément Houtmann, and Claude Kirchner. Principle of superdeduction. In Luke Ong, editor, *Proceedings of LICS*, pages 41–50, jul 2007.

[12] Paul Brauner, Clément Houtmann, and Claude Kirchner. Superdeduction at work. In Hubert Comon, Claude Kirchner, and Hélène Kirchner, editors, *Rewriting, Computation and Proof. Essays Dedicated to Jean-Pierre Jouannaud on the Occasion of His 60th Birthday*, volume 4600. Springer, jun 2007.

[13] Guillaume Burel. Superdeduction as a Logical Framework. submitted, jan 2008.

[14] Horatiu Cirstea and Claude Kirchner. The rewriting calculus — Part I *and* II. *Logic Journal of the Interest Group in Pure and Applied Logics*, 9:427–498, May 2001.

[15] Horatiu Cirstea, Claude Kirchner, Luigi Liquori, and Benjamin Wack. Rewrite strategies in the rewriting calculus. In Bernhard Gramlich and Salvador Lucas, editors, *Electronic Notes in Theoretical Computer Science*, volume 86. Elsevier, 2003.

[16] Denis Cousineau and Gilles Dowek. Embedding Pure Type Systems in the lambda-Pi-calculus modulo. In Simona Ronchi Della Rocca, editor, *TLCA*, volume 4583 of *Lecture Notes in Computer Science*, pages 102–117. Springer-Verlag, 2007.

[17] Pierre-Louis Curien and Hugo Herbelin. The Duality of Computation. *ACM sigplan notices*, 35(9), 2000.

[18] David Delahaye. A Tactic Language for the System Coq. In Michel Parigot and Andrei Voronkov, editors, *Proc. 7th Int. Conf. on Logic for Programming and Automated Reasoning*, volume 1955 of *Lecture Notes in computer Science*, pages 85–95. Springer-Verlag, November 2000.

[19] Gilles Dowek, Thérèse Hardin, and Claude Kirchner. Theorem Proving Modulo. *Journal of Automated Reasoning*, 31(1):33–72, Nov 2003.

[20] Gilles Dowek and Benjamin Werner. Arithmetic as a Theory Modulo. In Jürgen Giesl, editor, *RTA*, volume 3467 of *Lecture Notes in Computer Science*, pages 423–437. Springer-Verlag, 2005.

[21] Bruno Barras et al. *The Coq Proof Assistant Reference Manual*, 2006.

[22] Michael Fay. First Order Unification in Equational Theories. In *Proceedings 4th Workshop on Automated Deduction, Austin (Tex., USA)*, pages 161–167, 1979.

[23] Wan Fokkink, Jasper Kamperman, and Pum Walters. Lazy rewriting on eager machinery. *ACM Transactions on Programming Languages and Systems*, 22(1):45–86, 2000.

[24] Gerhard Gentzen. Untersuchungen über das Logisches Schließen. *Mathematische Zeitschrift*, 1:176–210, 1935.

[25] Joseph Goguen, Claude Kirchner, Hélène Kirchner, Aristide Mégrelis, José Meseguer, and Tim Winkler. An Introduction to OBJ-3. In J.-P. Jouannaud and S. Kaplan, editors, *Proceedings 1st International Workshop on Conditional Term Rewriting Systems, Orsay (France)*, volume 308 of *Lecture Notes in Computer Science*, pages 258–263. Springer-Verlag, July 1987. Also as internal report CRIN: 88-R-001.

[26] Michael Gordon, Robin Milner, Lockwood Morris, Malcolm Newey, and Christopher Wadsworth. A Metalanguage for Interactive Proof in LCF. In *Proc. 5th ACM Symp. on Principles of Programming Languages*, pages 119–130. ACM, ACM Press, January 1978.

[27] Gérard Huet and Jean-Jacques Lévy. Computations in Non-ambiguous Linear Term Rewriting Systems. Technical Report, INRIA Laboria, 1979.

[28] Gérard Huet and Jean-Jacques Lévy. *Computational Logic – Essays in Honor of Alan Robinson*, chapter Computations in orthogonal rewriting systems, Part I+II, pages 349–405. MITPress, 1991.

[29] Jean-Marie Hullot. Canonical Forms and Unification. In *Proceedings 5th International Conference on Automated Deduction, Les Arcs (France)*, pages 318–334, July 1980.

[30] Gueorgui Jojgov. Holes with Binding Power. In *Proc. 2002 Int. Workshop on Proofs and Programs*. Springer-Verlag, 2003.

[31] Jean-Pierre Jouannaud and Hélène Kirchner. Completion of a set of rules modulo a set of Equations. *SIAM Journal of Computing*, 15(4):1155–1194, 1986.

[32] Claude Kirchner. Strategic Rewriting. *Electr. Notes Theor. Comput. Sci. Proceedings of the 4th International Workshop on Reduction Strategies in Rewriting and Programming - WRS'2004, Aachen, Germany*, 124(2):3–9, 2005.

[33] Claude Kirchner and Hélène Kirchner. Rewriting, Solving, Proving. A preliminary version of a book available at `http://www.loria.fr/~ckirchne/=rsp/rsp.pdf`, 1999.

[34] Claude Kirchner, Hélène Kirchner, and Marian Vittek. Designing Constraint Logic Programming Languages using Computational Systems. In P. Van Hentenryck and V. Saraswat, editors, *Principles and Practice of Constraint Programming. The Newport Papers.*, chapter 8, pages 131–158. The MIT press, 1995.

[35] Florent Kirchner. *Interoperable proof systems.* PhD thesis, École Polytechnique, 2007.

[36] Florent Kirchner and Claudio Sacerdoti Coen. The Fellowship proof manager. `www.lix.polytechnique.fr/Labo/Florent.Kirchner/fellowship/`, 2007.

[37] Jan Klop. Term Rewriting Systems. In S. Abramsky, D. Gabbay, and T. Maibaum, editors, *Handbook of Logic in Computer Science*, volume 1, chapter 6. Oxford University Press, 1990.

[38] Jan Willem Klop, Vincent van Oostrom, and Femke van Raamsdonk. Reduction Strategies and Acyclicity. In Hubert Comon-Lundh, Claude Kirchner, and Hélène Kirchner, editors, *Rewriting, Computation and Proof*, volume 4600. Springer, jun 2007.

[39] Alexander Letichevsky. Development of Rewriting Strategies. In Maurice Bruynooghe and Jaan Penjam, editors, *PLILP*, volume 714 of *Lecture Notes in Computer Science*, pages 378–390. Springer, 1993.

[40] Salvador Lucas. Context-sensitive computations in functional and functional logic programs. *Journal of Functional and Logic Programming*, 1:1–61, January 1998.

[41] Salvador Lucas. Termination of on-demand rewriting and termination of OBJ programs. In H. Sondergaard, editor, *Proceedings of the 3rd International ACM SIGPLAN Conference on Principles and Practice of Declarative Programming, PPDP'01*, pages 82–93, Firenze, Italy, September 2001. ACM Press, New York.

[42] Salvador Lucas. Termination of Rewriting With Strategy Annotations. In A. Voronkov and R. Nieuwenhuis, editors, *Proceedings of the 8th International Conference on Logic for Programming, Artificial Intelligence and Reasoning, LPAR'01*, volume 2250 of *Lecture Notes in Artificial Intelligence*, pages 669–684, La Habana, Cuba, December 2001. Springer-Verlag, Berlin.

[43] Lena Magnusson. *The Implementation of ALF: A Proof Editor Based on Martin-Löf's Monomorphic Type Theory with Explicit Substitution.* PhD thesis, Chalmers University of Technology and Göteborg University, January 1995.

[44] Narciso Martí-Oliet, José Meseguer, and Alberto Verdejo. Towards a Strategy Language for Maude. In Narciso Martí-Oliet, editor, *Proceedings Fifth International Workshop on Rewriting Logic and its Applications, WRLA 2004, Barcelona, Spain, March 27 - April 4, 2004*, volume 117 of *Electronic Notes in Theoretical Computer Science*, pages 417–441. Elsevier, 2005.

[45] William McCune. Semantic Guidance for Saturation Provers. *Artificial Intelligence and Symbolic Computation*, pages 18–24, 2006.

[46] Robin Milner. Implementation and applications of Scott's logic for computable functions. *ACM SIGPLAN Notices*, 7(1):1–6, January 1972.

[47] César Muñoz. Proof-term Synthesis on Dependent-Type Systems via Explicit Substitutions. *Theoretical Computer Science*, 266(1-2):407–440, 2001.

[48] Quang-Huy Nguyen. Compact Normalisation Trace via Lazy Rewriting. In S. Lucas and B. Gramlich, editors, *Proceedings of the 1st International Workshop on Reduction Strategies in Rewriting and Programming (WRS 2001)*, volume 57 of *Elecronic Notes In Theoretical Computer Science.* Elsevier, 2001.

[49] Sam Owre, John Rushby, and Natarajan Shankar. PVS: A Prototype Verification System. In Deepak Kapur, editor, *Proc. 11th Int. Conf. on Automated Deduction*, volume 607 of *Lecture Notes in Artificial Intelligence*, pages 748–752. Springer-Verlag, June 1992.

[50] Lawrence Paulson. *Logic and Computation : Interactive proof with Cambridge LCF*, volume 2 of *Cambridge Tracts in Theoretical Computer Science.* Cambridge University Press, 1987.

[51] Gerald E. Peterson and Mark E. Stickel. Complete Sets of Reductions for Some Equational Theories. *Journal of the ACM*, 28:233–264, 1981.

[52] Dag Prawitz. *Natural Deduction: a Proof-Theoretical Study*, volume 3 of *Stockholm Studies in Philosophy.* Almqvist & Wiksell, 1965.

[53] Jörg H. Siekmann, Christoph Benzmüller, and Serge Autexier. Computer supported mathematics with Omega. *J. Applied Logic*, 4(4):533–559, 2006.

[54] Terese. *Term Rewriting Systems.* Cambridge University Press, 2003. M. Bezem, J. W. Klop and R. de Vrijer, eds.

[55] Christian Urban and Gavin Bierman. Strong Normalisation of Cut-elimination in Classical Logic. In *Proc. 4th Int. Conf. on Typed Lambda Calculi and Applications*, pages 365–380. Springer-Verlag, 1999.

[56] Eelco Visser. Stratego: A Language for Program Transformation based on Rewriting Strategies. System Description of Stratego 0.5. In A. Middeldorp, editor, *Rewriting Techniques and Applications (RTA'01)*, volume 2051 of *Lecture Notes in Computer Science*, pages 357–361. Springer-Verlag, May 2001.

Part IV

Higher-Order Automated Reasoning in Practice

Journal of Automated Reasoning **16:** 321–353, 1996.

TPS: A Theorem-Proving System for Classical Type Theory*

PETER B. ANDREWS [a], MATTHEW BISHOP [a], SUNIL ISSAR [b], DAN NESMITH [c], FRANK PFENNING [d] and HONGWEI XI [a]
[a] *Mathematics Department, Carnegie Mellon University, Pittsburgh, PA 15213, U.S.A. e-mail: Andrews+@cmu.edu; mbishop@cs.cmu.edu; hwxi@cs.cmu.edu*
[b] *School of Computer Science, Carnegie Mellon University, Pittsburgh, PA 15213, U.S.A. e-mail: si@cmu.edu*
[c] *e-mail: nesmith@infinity.ccsi.com*
[d] *Department of Computer Science, Carnegie Mellon University, Pittsburgh, PA 15213, U.S.A. e-mail: fp+@cs.cmu.edu*

(Received: 13 June 1994; in final form: 23 November 1994)

Abstract. This is description of **TPS**, a theorem-proving system for classical type theory (Church's typed λ-calculus). **TPS** has been designed to be a general research tool for manipulating wffs of first- and higher-order logic, and searching for proofs of such wffs interactively or automatically, or in a combination of these modes. An important feature of **TPS** is the ability to translate between expansion proofs and natural deduction proofs. Examples of theorems that **TPS** can prove completely automatically are given to illustrate certain aspects of **TPS**'s behavior and problems of theorem proving in higher-order logic.

AMS Subject Classification: 03-04, 68T15, 03B35, 03B15, 03B10.

Key words: higher-order logic, type theory, mating, connection, expansion proof, natural deduction.

1. Introduction

TPS is a theorem-proving system for classical type theory** (Church's typed λ-calculus [20]) which has been under development at Carnegie Mellon University for a number years. This paper gives a general description of **TPS**, serves as a report on our implementations of ideas that were discussed in previous papers, and illuminates what can be accomplished using these ideas. Many of the ideas underlying **TPS** are summarized in [8], with which we shall assume familiarity.

We start with a brief history of the **TPS** project.

---

* This material is based upon work supported by the National Science Foundation under grant CCR-9201893 and previous grants.

** Type theory was introduced by Bertrand Russell [51, 52], and was used extensively in [57]. In [20] Church introduced an improved and simplified formulation of type theory that incorporated the notation of the λ-calculus. Since then, other type theories have been developed, so we refer to Church's formulation as classical type theory.

TPS is based on an approach to automated theorem proving called the *mating method* [5], which is essentially the same as the *connection method* developed independently by Bibel [13]. The mating method arose from reflections [3] on what a proof by resolution [50] reveals about the logical structure of the theorem being proved, but a distinguishing characteristic of the mating method is that it does not require reduction to clausal form.

Matings provide significant insight into the logical structure of theorems, but it is not always easy for people to grasp them intuitively or to relate them to other approaches to theorem proving, so a procedure for automatically transforming acceptable matings into proofs in natural deduction style was developed [4]. The ideas in [4, 5], and [32] were implemented in a theorem-proving system which we now call **TPS1**. This was described in [37] and [6]. It automatically proved certain theorems of type theory (higher-order logic) as well as first-order logic, and embodied a proof procedure which was in principle complete for first-order logic, though not for type theory.

As a start toward extending the mating method to a complete method for proving theorems of higher-order logic, it was shown in [6] that a sentence is a theorem of elementary type theory (the system of [1] and [2]) if and only if it has a tautologous *development*, where a development is the analogue of a Herbrand expansion of a sentence of first-order logic. Once one has found a tautologous development for a theorem, one can construct a proof of it in natural deduction style without further search. Thus, the problem of finding proofs for theorems of higher-order logic can be reduced to the problem of finding tautologous developments for them, and the search can be carried on in a context where one can hope to analyze the essential logical structure of the theorem.

Dale Miller explored this subject more deeply, proved an analogue of the metatheorem mentioned above in which the notion of a development was replaced by that of an *expansion proof*, gave the details of an explicit algorithm for converting expansion proofs into natural deduction proofs, and proved it correct. An expansion proof is an elegant and concise representation of a theorem of type theory, its tautologous development, and the relation between them. Matings are naturally embedded in expansion proofs, and Miller's work [38–40] provided a firm theoretical foundation for the extension of the matings approach to theorem proving from first-order logic to higher-order logic.

In [45] and [46] it was shown how to translate natural deduction proofs into expansion proofs, and an improved method of translating expansion proofs into natural deduction proofs was given. The papers [46] and [47] also contain discussions of equality and extensionality, and methods of generating more elegant natural deduction proofs.

About 1985 work started on the design and implementation of a completely new version of **TPS** to accommodate the transition from general matings to expansion proofs, take advantage of new versions of Lisp and new computers, and incorporate various improvements in the program. The initial work was done in MacLisp, yielding a program called **TPS2**, and this was later translated into Common Lisp to create the current version of **TPS**, which is called **TPS3**. **TPS3** owes a great deal to the work Dale Miller did on **TPS1** in addition to his theoretical contributions. Carl Klapper also contributed to the development of **TPS3**. Henceforth, we refer to **TPS3** simply as **TPS**, since the previous versions of **TPS** are now obsolete.

The desirability of finding a search procedure in which expansions of the formula are motivated by the needs of the mating search process has long been evident. In [33] and [34] a mating search procedure is described in which quantifier replications are localized to vertical paths (thus reducing the enormous growth in the number of paths which accompanies replications), and the replications for each path are generated as needed to permit the construction of a mating that spans that path. The search space grows and contracts as different vertical paths are considered. This procedure has been implemented in **TPS** and has improved its speed very significantly.

TPS combines ideas from two fields which, regrettably, have not achieved much cross-fertilization. On the one hand, there is the 'traditional' work in first-order theorem proving using such methods as resolution, model elimination, or connection graphs. On the other hand, we find 'avant-garde' proof checkers and theorem provers for type theories of a variety of flavors, mostly centered on interactive proof construction with the aid of tactics.

In traditional theorem provers for first-order logic, relatively little attention has been paid to issues of human-computer interaction, but much attention has been paid to finding complete strategies that can be implemented very efficiently using, for example, advanced indexing schemes. While the use of first-order logic produces simplicity and efficiency in basic syntax and certain processes, many theorems of mathematics and other disciplines can be expressed very simply in type theory, but only in a rather complex way in first-order logic. This complexity can enormously enlarge the search space one confronts when one tries to proves these theorems.

On the other hand, tactic-based theorem provers, beginning with LCF [27] and including systems such HOL [28, 29], Nuprl [21], the Calculus of Constructions [22], Isabelle [44] and IMPS [23], have paid considerable attention to user interaction and to the problem of formulating and supporting expressive languages for the formalization of mathematics. Techniques developed for first-order theorem proving, however, have been essentially ignored with the exception of unification, which now plays an important role in a number of these systems. Another point to note is that some of these systems chose to work in constructive logics for a variety of reasons, and that classical theorem-proving techniques do not immediately apply in these circumstances. (Only recently has this gap between classical and constructive theorem-proving techniques begun to close [43, 56]).

TPS unifies important ideas and concepts from both of these lines of research into a single system. It is based on classical higher-order logic, in which much of mathematics can be formalized very directly. It provides a natural deduction interface which can take advantage of the underlying theorem proving engine (the mating search procedure). It employs higher-order unification, finds instantiations for quantifiers on higher-order variables, and uses a machine-oriented representation of the wff for the search process. It is the combination of these features that makes **TPS** unique. Of course, the systems described in [11] and [17] also find proofs automatically by using techniques that have essential relevance to higher-order logic. **TPS** is far from comprehensive, and the systems mentioned above have many other features that are not available in **TPS**. Perhaps closest in spirit to **TPS** is the work by Helmink and Ahn [30], who have also proven significant theorems in type theory (such as Cantor's theorem) completely automatically.

2. An Overview of TPS

Our experience has shown that even if one is primarily interested in the problem of proving theorems automatically, one needs good interactive tools in order to efficiently investigate examples that illuminate the fundamental logical problems of finding proofs. **TPS** has been designed to be a general research tool for manipulating wffs of first- and higher-order logic, and searching for proofs of such wffs interactively or automatically, or in a combination of these modes.

TPS handles two sorts of proofs:

1. *Natural deduction proofs* (*natural deductions*). These are human-readable (though at present boringly detailed) formal proofs. Examples are given in the figures below. In these examples we use Church's convention* that a dot in a wff stands for a left bracket whose mate is as far to the right as is consistent with the pairing of brackets already present and the well-formedness of the formula. See [4] or [7] for more details about this formulation of natural deduction.
2. *Expansion proofs*. These are described briefly in [8] and studied in [38–40, 45–47]. The structure of an expansion proof is closely and directly related to the structure of the theorem it establishes, and provides a context for search that facilitates concentrating on the essential logical structure of the theorem. At the same time, it abstracts from many details of concrete deductions. This balance between preservation of formula structure (compared with resolution refutations, for example) and abstraction of proof structure (compared with sequent derivations, for example) makes expansion proofs universal structures for cut-free (or normal) proofs. Wallen [56] provides further evidence for this by showing how expansion proofs can be adapted naturally to nonclassical logics. Despite their many advantages, expansion proofs have

* As noted on p. 75 of [19], the use of dots to replace brackets was introduced by Peano and was adopted by Whitehead and Russell in [57]. Church's convention is a modification of theirs.

a severe deficiency in that they are distant from formats that can be used effectively by humans.

TPS has facilities for searching for expansion proofs automatically or interactively, translating these into natural deduction proofs, constructing natural deduction proofs interactively, translating natural deduction proofs that are in normal form into expansion proofs, and solving unification problems in higher-order logic, as well as a variety of utilities designed to facilitate research and efficient interaction with the program.

The ability to translate between expansion proofs and natural deduction proofs is one of the important and attractive features of **TPS**. It permits both humans and computers to work in contexts that are appropriate to them. Also, we are much more confident that **TPS** has correctly proved a theorem when it presents us with a proof in natural deduction style than we would be if it simply indicated that it had found an expansion proof.

TPS has a number of top levels, each with its own commands. The main top level is for constructing natural deduction proofs, and there are others devoted to matings and expansion proofs and to higher-order unification problems. Another top level is a formula editor which facilitates constructing new wffs from others already known to **TPS**. There are editor commands for λ-conversion, Skolemization, transforming to normal forms, expanding abbreviations, counting vertical and horizontal paths, and many other manipulations of wffs. When one enters the editor, windows display the formula being edited and the particular part of the wff one is focused on.

Many aspects of the program's behavior can be controlled by setting flags, and there are over 300 of these flags. **TPS** has a top level called Review for examining and changing the settings of flags, and for defining and reusing groups of flag settings called *modes*.

Still another top level is a library facility for saving and displaying wffs, definitions, modes, and disagreement pairs for higher-order unification problems. Definitions can be polymorphic (i.e., contain type variables) and can contain other definitions to any level of nesting. When **TPS** retrieves a definition or theorem, it retrieves all the necessary subsidiary definitions. When **TPS** finishes proving a theorem, information about the heuristics used and statistics about the search can be stored automatically with the theorem in the library.

Yet another top level, called Test, permits one to set up experiments in which **TPS** will automatically try to prove a theorem a number of times, with different modes for each run, and record which mode produces the quickest proof. This top level is still rather unsophisticated.

TPS uses a type-inference mechanism based on an algorithm by Milner [41] as modified by Dan Leivant. For example, we supply the following description of a wff: "a subset b and $f(\mathrm{DB})[gx(\mathrm{G})]$ in a implies b union $c = d$". The notation $f(\mathrm{DB})$ means that f has type $(\delta\beta)$, i.e., that it is a function mapping objects of type β to objects of type δ. Similarly, the notation x(G) means that x has type γ. From this information **TPS** determines the types of the function g and the sets a, b, c, and d, and (if one is using the X11 window system on one's workstation) prints the wff as

$$a_{o\delta} \subseteq b_{o\delta} \wedge f_{\delta\beta}[g_{\beta\gamma} x_\gamma] \in a \supset b \cup c_{o\delta} = d_{o\delta}.$$

TPS understands various conventions for omitting brackets, but by changing a suitable flag one can make **TPS** print the wff above as

$$[[a_{o\delta} \subseteq b_{o\delta}] \wedge [[f_{\delta\beta}[g_{\beta\gamma} x_\gamma]] \in a]] \supset [[b \cup c_{o\delta}] = d_{o\delta}].$$

The display of type symbols can be suppressed. (This is particularly appropriate when displaying wffs of first-order logic.)

When proving a theorem automatically, one presents the theorem to **TPS** in the readable form illustrated above (and in the examples in Section 6), and all the processing necessary to put the theorem into the form used by the search process is done automatically.

TPS can display wffs in the two-dimensional format (called a vpform) which was introduced in [5] to help one visualize the vertical paths through the wff. Examples are given in Figures 5.2 and 5.3 below.

Proofs in natural deduction style can be printed in files that are processed by Scribe or Tex, so that familiar notations of logic appear in printed proofs as well as in wffs displayed on the screen.

When one is working to construct a proof interactively, the proof can be displayed in a window called a proofwindow, which is updated automatically whenever a command that changes the proof is executed. Another proofwindow displays only the 'active lines' of the proof, facilitating concentration on the essentials of the problem. One can work forwards, backwards, or in a combination of these modes, and one can easily rearrange proofs and delete parts of proofs. Complete or incomplete proofs can be saved in files, and read in at another time for continued work. The entire sequence of commands which have been executed can also be saved, and re-executed later.

When trying to understand someone else's natural deduction proof intuitively, it is sometimes more informative to watch the proof being constructed (working backwards and forwards at appropriate times) than to read the finished proof. **TPS** can translate an expansion proof into a natural deduction one step at a time, each prompted by the user, who can watch the natural deduction growing in the proofwindows at leisure.

Online help is available for all commands, as well as for their arguments. Considerable documentation [9, 10, 14, 35, 42, 48] has been written, though more is needed. The Facilities Guides are produced automatically.

For a number of years the purely interactive facilities of **TPS** have been used under the name **ETPS** (Educational Theorem Proving System) by students in logic courses at Carnegie Mellon to construct natural deduction proofs. Students generally learn to use **ETPS** fairly quickly just by reading the manual (which contains some complete examples) and playing with the system. **ETPS** permits students to concentrate on basic decisions about applying rules of inference while constructing formal proofs, gives them immediate feedback for both correct and incorrect actions, and relieves them of many of the trivial and burdensome details of writing proofs. After reviewing eight programs that support the teaching of logic, the authors of [26] (which was partially reprinted in [25]) concluded: "For elementary and advanced courses in mathematical logic for students with a formal background, we choose ETPS, a powerful tool that is also easy to learn."

If the teacher of a course using **ETPS** wishes to use a set of rules of inference that is different from the set that comes with the program, it is quite easy to do this by using what is called the RULES module of **TPS**. One simply describes the new rules in a simple lisp meta-language (using the existing rules as examples), and uses commands in **TPS** to create the lisp code for executing these commands. (Of course, much more work would be required if one also wished to be able to automatically translate expansion proofs into natural deductions using different rules.)

TPS is a large program whose uncompiled source code contains more than 114,000 lines (including comments). The compressed tar file for distribution of **TPS**, which contains documentation as well as the source code, occupies about 3.2 megabytes. **TPS** is portable, runs in a variety of implementations of Common Lisp, and has been distributed to a number of researchers.*

3. Tactics and Proof Translations

The basic tools in **TPS** for automatically applying rules of inference to construct natural deductions are *tactics*, which can be combined using *tacticals* [27]. A tactic examines a given goal situation (the problem of deriving a conclusion from a set of assumptions) and reduces it to the problem of solving a number of subgoals. This is done by applying rules of inference (forward or backward) to derive new proof lines or to justify certain lines of the proof while introducing other lines which may still require justification.

One can use tactics to speed the process of constructing proofs interactively. **TPS** has a command called GO2 that calls a number of tactics to apply mundane rules of inference to construct the easy parts of the proof, and quickly bring one to the point where some judgment and insight are needed. The user can choose whether or not to be prompted for approval before each of these tactics is applied.

The main use of tactics in **TPS** is to translate an expansion proof into a natural deduction by the methods of [46]; in this context, the tactics can consult the

* Information about distribution of **TPS** can be obtained from the **TPS** World-Wide Web home page at `http://www.cs.cmu.edu/afs/cs/user/andrews/www/tps.html`, or by sending e-mail to Andrews+@cmu.edu.

expansion proof for useful information through a number of predefined functions. We give two paradigmatic examples of such functions.

When proving a goal $[\mathbf{A} \vee \mathbf{B}]$ from some assumptions, we may need to consult the expansion proof to determine if **A** by itself already follows from the assumptions. If this is the case, we can apply disjunction introduction on the left. If not, the 'disj-left' tactic fails, i.e., it does not apply. If it and its dual 'disj-right' both fail, we probably want to defer a decision on the goal and try to reason forward from the assumptions, or try an indirect proof. An example where an indirect proof would be appropriate is $[\sim \mathbf{A} \vee \mathbf{A}]$. When proving $[\mathbf{A} \vee \mathbf{B}]$ from $[\mathbf{B} \vee \mathbf{A}]$, we need to distinguish the two cases (either **B** or **A**), before we can proceed with the disjunction introduction in the two subproofs. Expansion proofs are crucial in determining such information, as in general it is undecidable whether **A** or **B** follows directly. Other proof formats may also contain enough information to answer such questions, but in many cases it is obscured by preprocessing or other idiosyncrasies of various data structures devised for search.

Another example is a goal of the form $\exists \mathbf{x} \mathbf{A}$. In this case, we can determine from the expansion proof whether there is a substitution term **t** for **x** such that $[\mathbf{t}/\mathbf{x}]\mathbf{A}$ is provable from the current set of assumptions. If so, we can derive the goal by existential generalization; otherwise, we might have to postpone application of this rule. In some cases, the expansion proof might even indicate that only the rule of indirect proof will make progress.

Basic tactics may check conditions on the expansion proof and apply appropriate inference rules. They can be combined with other tactics in different ways which can lead to different styles of proof construction. The current **TPS** system contains a number of basic styles which can additionally be modified through flags. The styles differ in their preference for certain inference rules and in the granularity of the rules applied. For example, one tactic applies Rule P [7], which uses arbitrary propositional tautologies, while another uses only simple rules of inference. Proofs constructed using the latter tactic are more appropriate for students of logic early in their education, while those constructed using the former are often more appropriate for mathematical arguments. Using the former tactic, **TPS** produces a one-line proof of $[[P_o \equiv Q_o] \equiv R_o] \equiv [P \equiv [Q \equiv R]]$, but when it uses the latter, it produces a proof 170 lines long.

The currently implemented tactics almost always produce natural deductions in normal form. This is the primary limitation of the current system, but it is clearly a consequence of the basic analytic structure of expansion proofs (and machine-generated proofs in general). The only exception is the application of *symmetric simplification* [47] to introduce simple variations of the law of excluded middle into the deduction.

TPS also partially implements a translation in the other direction, mapping normal natural deductions into expansion proofs. The utility of this translation is severely limited by the current restriction to normal deductions and has not been fully explored.

Line	Hyps		Formula	Justification
(1)	1	$\vdash$	$\sim\exists y\ \forall x\ .P\ y \supset P\ x$	Assume negation
(2)	1	$\vdash$	$\forall y\ .\sim\forall x\ .P\ y \supset P\ x$	Neg: 1
(3)	1	$\vdash$	$\sim\forall x\ .P\ y^1 \supset P\ x$	UI: y^1 2
(4)	1	$\vdash$	$\exists x\ .\sim.P\ y^1 \supset P\ x$	Neg: 3
(5)	1	$\vdash$	$\sim\forall x\ .P\ y^2 \supset P\ x$	UI: y^2 2
(6)	1,6	$\vdash$	$\sim.P\ y^1 \supset P\ y^2$	Choose: y^2
(7)	1	$\vdash$	$\exists x\ .\sim.P\ y^2 \supset P\ x$	Neg: 5
(8)	1,6,8	$\vdash$	$\sim.P\ y^2 \supset P\ y^3$	Choose: y^3
(9)	1,6	$\vdash$	$P\ y^1 \land \sim P\ y^2$	Neg: 6
(10)	1,6	$\vdash$	$P\ y^1$	Conj: 9
(11)	1,6	$\vdash$	$\sim P\ y^2$	Conj: 9
(12)	1,6,8	$\vdash$	$P\ y^2 \land \sim P\ y^3$	Neg: 8
(13)	1,6,8	$\vdash$	$P\ y^2$	Conj: 12
(14)	1,6,8	$\vdash$	$\sim P\ y^3$	Conj: 12
(15)	1,6,8	$\vdash$	$\bot$	RuleP: 11 13
(16)	1,6	$\vdash$	$\bot$	RuleC: 7 15
(17)	1	$\vdash$	$\bot$	RuleC: 4 16
(18)		$\vdash$	$\exists y\ \forall x\ .P\ y \supset P\ x$	Indirect: 17

Fig. 3.1. Original proof of X2119.

Line	Hyps		Formula	Justification
(1)		$\vdash$	$\forall x\ P\ x \lor \sim\forall x\ P\ x$	RuleP
(2)	2	$\vdash$	$\forall x\ P\ x$	Case 1: 1
(3)	2	$\vdash$	$P\ w$	UI: w 2
(4)	2	$\vdash$	$P\ y^1 \supset P\ w$	Deduct: 3
(5)	2	$\vdash$	$\forall w\ .P\ y^1 \supset P\ w$	UGen: w 4
(6)	2	$\vdash$	$\forall x\ .P\ y^1 \supset P\ x$	AB: 5
(7)	2	$\vdash$	$\exists y\ \forall x\ .P\ y \supset P\ x$	EGen: y^1 6
(8)	8	$\vdash$	$\sim\forall x\ P\ x$	Case 2: 1
(9)	9	$\vdash$	$\sim\exists y\ \forall x\ .P\ y \supset P\ x$	Assume negation
(10)	10	$\vdash$	$\sim P\ y^2$	Assume negation
(11)	10	$\vdash$	$P\ y^2 \supset P\ x$	RuleP: 10
(12)	10	$\vdash$	$\forall x\ .P\ y^2 \supset P\ x$	UGen: x 11
(13)	10	$\vdash$	$\exists y\ \forall x\ .P\ y \supset P\ x$	EGen: y^2 12
(14)	9,10	$\vdash$	$\bot$	NegElim: 9 13
(15)	9	$\vdash$	$P\ y^2$	Indirect: 14
(16)	9	$\vdash$	$\forall y^2\ P\ y^2$	UGen: y^2 15
(17)	9	$\vdash$	$\forall x\ P\ x$	AB: 16
(18)	8,9	$\vdash$	$\bot$	NegElim: 8 17
(19)	8	$\vdash$	$\exists y\ \forall x\ .P\ y \supset P\ x$	Indirect: 18
(20)		$\vdash$	$\exists y\ \forall x\ .P\ y \supset P\ x$	Cases: 1 7 19

Fig. 3.2. Transformed proof of X2119.

While considering how to translate expansion proofs into natural deductions, we have come to consider various aspects of the question: "How can we take advantage of the information in an expansion proof for a theorem **A** when constructing a natural deduction for **A**?" It is the *form* of the natural deduction that is of primary concern; we are not content with constructing an arbitrary one from a given expansion proof.

Translating back and forth between natural deductions and expansion proofs can be used as a mechanism for intelligent restructuring of natural deductions. This mechanism can transform the structure of a natural deduction rather drastically. For example, the proof in Figure 3.1 was translated into an expansion

```
(1)  1 ⊢  U X_α = U Z_α                                         Hyp
(2)  1 ⊢  λy_α [X_α = y] = λy .Z_α = y                   EquivWffs: 1
(3)  1 ⊢  ∀q_o(oα) .q [λy_α .X_α = y] ⊃ q .λy .Z_α = y    Equality: 2
(5)  1 ⊢  [λw_oα .w X_α] [λy_α .X = y] ⊃ [λw .w X] .λy .Z_α = y
                                                  UI: [λw_oα .w X_α] 3
(6)  1 ⊢  X_α = X ⊃ Z_α = X                                 Lambda: 5
(7)     ⊢  q_oα X_α ⊃ q X                                       RuleP
(8)     ⊢  ∀q_oα .q X_α ⊃ q X                            UGen: q_oα 7
(9)     ⊢  X_α = X                                        Equality: 8
(10) 1 ⊢  Z_α = X_α                                             MP: 9 6
```

Fig. 3.3. Main part of proof of THM104 with REMOVE–LEIBNIZ set to NIL.

```
(1)  1 ⊢  U X_α = U Z_α                                         Hyp
(2)  1 ⊢  λy_α [X_α = y] = λy .Z_α = y                   EquivWffs: 1
(3)     ⊢  X_α = X                                      Assert REFL=
(4)     ⊢  [λy_α .X_α = y] X                                Lambda: 3
(5)  1 ⊢  [λy_α .Z_α = y] X_α                             Subst=: 4 2
(6)  1 ⊢  Z_α = X_α                                         Lambda: 5
```

Fig. 3.4. Main part of proof of THM104 with REMOVE–LEIBNIZ set to T.

proof, and then back to a natural deduction using symmetric simplification to produce the proof in Figure 3.2. The original proof consisted of a rather unintuitive, brute force, indirect proof, while the transformed proof identifies the crucial case distinction which should be made: either 'P' is true everywhere or not. An even simpler proof could have been obtained using the lemma $[\forall x P x \vee \exists x \sim Px]$, which is beyond the scope of our current methods. (See [47] for further discussion.)

The equality relation is ubiquitous in mathematical reasoning, and special mechanisms for dealing with equality, such as those in [24] and [53], are clearly needed. At present, **TPS** has no such mechanisms, and simply defines equality by the Leibniz definition $[\lambda x_\alpha \lambda y_\alpha \forall q_{o\alpha} \cdot qx \supset qy]$ or by the extensional definition $[\lambda x_{\alpha\beta} \lambda y_{\alpha\beta} \forall z_\beta \cdot xz = yz]$ for equality between functions or sets, and uses ordinary laws of logic to prove results involving equality. An example of this is in Figure 3.3, where we show the main part of a proof constructed automatically for THM104. (This theorem and the definition of U are discussed further in Section 6). In [46] it was shown how an expansion proof for a theorem involving equality can be translated into a natural deduction containing traditional equality inferences, and this has been implemented in **TPS**. The application of this feature is controlled by the flag REMOVE–LEIBNIZ. Thus, when we change the value of this flag from NIL to T, the same expansion proof that generated the natural deduction in Figure 3.3 generates the one in Figure 3.4.

4. Automatic Search

When one asks **TPS** to find a proof for a theorem automatically, it starts out by searching for an expansion proof of the theorem, and then translates this into a natural deduction proof. It sets up an expansion tree to represent the wff, and searches for an acceptable mating [5] of its literals. (An expansion tree that is appropriately expanded and has an acceptable mating is an expansion proof.)

The search process is controlled by a number of flags. Ideally, **TPS** would have heuristics to decide how to set these flags, but at present the user does this interactively before starting the automatic search. The flags provide a convenient way to explore many different aspects of the problem of searching for proofs.

When one is seeking an expansion proof for a theorem of higher-order logic, not all necessary substitution terms can be generated by unification of formulas already present, so certain *expansion options* [8] are applied to the expansion tree that represents the theorem. In [8] we discussed expansion options that consisted of applying *primitive substitutions* to predicate variables. These substitutions introduce a single quantifier or connective, and contain variables for which additional substitutions can be made at a later stage. However, we do not yet have good heuristics to guide the process of applying primitive substitutions incrementally, so we currently use a procedure that introduces substitution terms containing more logical structure, and then searches for a mating without making further substitutions for the variables in the substitution terms except as dictated by the unifier associated with the mating. In order to limit the number of forms which must be considered for the substitution terms, we often use (in addition to projections) terms whose bodies are in prenex normal form with matrix in conjunctive or disjunctive normal form. We call these substitutions *gensubs* (general substitutions). Primitive substitutions are special cases of gensubs. The logical complexity of the gensubs which **TPS** will attempt to apply is limited by a flag called MAX-PRIM-DEPTH. Examples of gensubs for $r_{o\beta(o\beta)}$ are given in Figure 4.1. The gensubs for a variable are determined (up to renaming of auxiliary variables) by the type of the variable. Since a gensub substitutes a term for just one variable, we shall sometimes use the name of the gensub to denote the substitution term. The types of quantified variables in gensubs (such as β in the figure) are chosen from a small fixed set of types that is specified as the value of the flag PRIM-BDTYPES. One can permit **TPS** to apply a rather naive algorithm for choosing this set of types by setting another flag, called PRIM-BDTYPES-AUTO.

Different sets of expansion options are applied to create different expansion trees that are all subtrees of a master expansion tree. The sets of expansion options are generated in a systematic and exhaustive way whose details are determined by certain flags. Thus a potentially infinite list of subtrees is generated; smaller subtrees are explored before larger ones in an attempt to keep the search space manageable. Of course, this blind generation of sets of expansion options is rather

Gensubs with MAX-PRIM-DEPTH 2:

P0 $\lambda w^1_{o\beta}\ \lambda w^2_{\beta} . r^1_{o\beta(o\beta)}\ w^1\ w^2 \wedge r^2_{o\beta(o\beta)}\ w^1\ w^2$

P1 $\lambda w^1_{o\beta}\ \lambda w^2_{\beta} . r^3_{o\beta(o\beta)}\ w^1\ w^2 \vee r^4_{o\beta(o\beta)}\ w^1\ w^2$

P2 $\lambda w^1_{o\beta}\ \lambda w^2_{\beta}\ w^1 . r^5_{\beta\beta(o\beta)}\ w^1\ w^2$

P3 $\lambda w^1_{o\beta}\ \lambda w^2_{\beta}\ \exists w^3_{\beta}\ r^6_{o\beta\beta(o\beta)}\ w^1\ w^2\ w^3$

P4 $\lambda w^1_{o\beta}\ \lambda w^2_{\beta}\ \forall w^3_{\beta}\ r^7_{o\beta\beta(o\beta)}\ w^1\ w^2\ w^3$

P5 $\lambda w^1_{o\beta}\ \lambda w^2_{\beta}\ \exists w^4_{\beta} . r^8_{o\beta\beta(o\beta)}\ w^1\ w^2\ w^4 \vee r^9_{o\beta\beta(o\beta)}\ w^1\ w^2\ w^4$

P6 $\lambda w^1_{o\beta}\ \lambda w^2_{\beta}\ \exists w^4_{\beta} . r^{10}_{o\beta\beta(o\beta)}\ w^1\ w^2\ w^4 \wedge r^{11}_{o\beta\beta(o\beta)}\ w^1\ w^2\ w^4$

P7 $\lambda w^1_{o\beta}\ \lambda w^2_{\beta}\ \forall w^4_{\beta} . r^{12}_{o\beta\beta(o\beta)}\ w^1\ w^2\ w^4 \vee r^{13}_{o\beta\beta(o\beta)}\ w^1\ w^2\ w^4$

P8 $\lambda w^1_{o\beta}\ \lambda w^2_{\beta}\ \forall w^4_{\beta} . r^{14}_{o\beta\beta(o\beta)}\ w^1\ w^2\ w^4 \wedge r^{15}_{o\beta\beta(o\beta)}\ w^1\ w^2\ w^4$

Examples of additional gensubs with MAX-PRIM-DEPTH 3:

P13 $\lambda w^1_{o\beta}\ \lambda w^2_{\beta}\ \exists w^5_{\beta}\ \forall w^6_{\beta} .$

$[r^{40}_{o\beta\beta\beta(o\beta)}\ w^1\ w^2\ w^5\ w^6 \wedge r^{41}_{o\beta\beta\beta(o\beta)}\ w^1\ w^2\ w^5\ w^6]$

$\vee\ [r^{42}_{o\beta\beta\beta(o\beta)}\ w^1\ w^2\ w^5\ w^6 \wedge r^{43}_{o\beta\beta\beta(o\beta)}\ w^1\ w^2\ w^5\ w^6]$

$\vee\ [r^{44}_{o\beta\beta\beta(o\beta)}\ w^1\ w^2\ w^5\ w^6 \wedge r^{45}_{o\beta\beta\beta(o\beta)}\ w^1\ w^2\ w^5\ w^6]$

P16 $\lambda w^1_{o\beta}\ \lambda w^2_{\beta}\ \forall w^5_{\beta}\ \forall w^6_{\beta} .$

$[r^{58}_{o\beta\beta\beta(o\beta)}\ w^1\ w^2\ w^5\ w^6 \vee r^{59}_{o\beta\beta\beta(o\beta)}\ w^1\ w^2\ w^5\ w^6]$

$\wedge\ [r^{60}_{o\beta\beta\beta(o\beta)}\ w^1\ w^2\ w^5\ w^6 \vee r^{61}_{o\beta\beta\beta(o\beta)}\ w^1\ w^2\ w^5\ w^6]$

$\wedge\ [r^{62}_{o\beta\beta\beta(o\beta)}\ w^1\ w^2\ w^5\ w^6 \vee r^{63}_{o\beta\beta\beta(o\beta)}\ w^1\ w^2\ w^5\ w^6]$

Fig. 4.1. Gensubs for $r_{o\beta(o\beta)}$.

crude. We look forward to the development of heuristics and metatheorems to improve the sophistication of this process by guiding or restricting it.

Before searching for an acceptable mating of the literals in a subtree, **TPS** converts the subtree into an alternative representation called a *jform* (junctive form). If the connectives truth and falsehood occur, they are eliminated by applying the

```
(1)  1 ⊢  ∀r_oβ(oβ) .∀p_oβ ∃x_β r p x ⊃ ∃j_β(oβ) ∀p r p .j p                Hyp
(2)  1 ⊢    ∀p_oβ ∃x_β [λp λy_β .∃x p x ⊃ p y] p x
                ⊃ ∃j_β(oβ) ∀p [λp λy .∃x p x ⊃ p y] p .j p
                              UI: [λp_oβ λy_β.∃x_β p x ⊃ p y] 1
(3)  1 ⊢  ∀p_oβ ∃x_β [∃x p x ⊃ p x] ⊃ ∃j_β(oβ) ∀p .∃x p x ⊃ p.j p
                                                            Lambda: 2
              ...
(17) 1 ⊢  ∃j_β(oβ) ∀p_oβ .∃x_β p x ⊃ p .j p                           PLAN3
(18)   ⊢    ∀r_oβ(oβ) [∀p_oβ ∃x_β r p x ⊃ ∃j_β(oβ) ∀p r p .j p]
                ⊃ ∃j ∀p .∃x p x ⊃ p .j p                     Deduct: 17
(19)   ⊢    ∀r_oβ(oβ) [∀x_oβ ∃y_β r x y ⊃ ∃f_β(oβ) ∀x r x .f x]
                ⊃ ∃j_β(oβ) ∀p_oβ .∃x_β p x ⊃ p .j p               AB: 18
```

Fig. 4.2. Outline of proof of X5310.

relevant laws of propositional calculus. Then **TPS** tries to build up a mating that spans every vertical path through the jform by progressively adding links to span the paths. When working on higher-order theorems, it uses Huet's higher-order unification algorithm [32] to check the compatibility of the connections in the partial mating. A unification tree is associated with the partial mating, and when a new connection is added to the mating, the associated disagreement pair is added to all the leaves of the unification tree. When incompatibilities are encountered, the process backtracks. Since higher-order unification may not terminate, **TPS** is not permitted to generate nodes of the unification tree with depth greater than the value of the flag MAX-SEARCH-DEPTH. (If this flag is set to a high value, **TPS** will often spend an enormous amount of time generating higher-order unification search trees.) The search for an acceptable mating within a given jform may not terminate, so when the time spent on it reaches the value of the flag SEARCH-TIME-LIMIT, **TPS** temporarily abandons this jform and tries another one. It may return to this jform later, but there is also a limit (specified by the flag MAX-SEARCH-LIMIT) on the total amount of time that can be spent on searching for a proof in any jform. Once an acceptable mating is found, it is converted into an expansion proof, which is simplified by a process called *merging*, and then translated into a natural deduction proof.

Now that we have outlined the general procedure **TPS** uses to find a proof, let us illustrate the use of gensubs by discussing how **TPS** proves theorem X5310, which may be found at the end of Section 6. In order to understand X5310, the reader is advised to first look at the companion theorem X5308. The outline of a simple proof of X5310 is in Figure 4.2. The theorem to be proved is in line (19), but the alphabetic variant of it in (18) is easier to work with. The problem is to derive (17) (which is the Axiom of Choice) from the hypothesis in (1). The key step is to instantiate the quantifier $\forall r_{o\beta(o\beta)}$ in (1) with the term $[\lambda p_{o\beta}\, \lambda y_\beta .\, \exists x_\beta p x \supset p y]$, which we shall call N. Doing this yields (3), whose antecedent is easily provable, and whose consequent is (17).

```
Original gensub:
```

P7 $[\lambda w^1_{o\beta}\ \lambda w^2_\beta\ \forall w^4_\beta . r^{12}_{o\beta\beta(o\beta)}\ w^1\ w^2\ w^4 \lor r^{13}_{o\beta\beta(o\beta)}\ w^1\ w^2\ w^4]$

```
Result of substituting for the free variables of P7:
```

M $[\lambda w^1_{o\beta}\ \lambda w^2_\beta\ \forall w^4_\beta . w^1\ w^2 \lor \sim w^1\ w^4]$

```
Alphabetic variant of M:
```

M1 $[\lambda p_{o\beta}\ \lambda y_\beta\ \forall x_\beta . p\ y \lor \sim p\ x]$

```
Crucial substitution term in Figure 4-2:
```

N $[\lambda p_{o\beta}\ \lambda y_\beta . \exists x_\beta\ p\ x \supset p\ y]$

Fig. 4.3. Substitution terms for $r_{o\beta(o\beta)}$ for proof of X5310.

In the search for a proof, **TPS** eventually considers the set of expansion options that simply substitutes gensub P7 of Figure 4.1 (which is also displayed in Fig. 4.3) for $r_{o\beta(o\beta)}$. **TPS** finds an acceptable mating of the associated jform, and applies the unifier associated with the mating to the free variables $r^{12}_{o\beta\beta(o\beta)}$ and $r^{13}_{o\beta\beta(o\beta)}$ of P7 to obtain the wff M of Figure 4.3; an alphabetic variant of this is the term with which **TPS** instantiates $\forall r_{o\beta(o\beta)}$ in the natural deduction proof. For convenience we display the alphabetic variant M1 of M, which can be transformed to the wff N mentioned above by applying elementary logical equivalences.

Note that while the quantified variable in P7 occurs in both the left and the right scopes of the disjunction, this is not the case for M1. Thus the gensub P7 is truly a general wff which can take several forms one of which is M.

While the use of prenex normal forms in this context dramatically reduces the number of sets of expansion options that must be considered, and seems not to complicate the search for an acceptable mating as long as the quantifiers thus introduced need not be duplicated, using a prenex formula such as M instead of a miniscope formula such as N may produce a rather awkward natural deduction proof. The natural deduction proof for X5310 that **TPS** constructs using M is 57 lines long and is rather clumsy. In order to remedy this, if the flag MIN-QUANT-ETREE is set to T, after the search is completed and the actual terms needed to instantiate quantifiers on higher-order variables are known, these terms are put into miniscope form, an expansion proof using these terms is constructed, and it is translated into a natural deduction proof. The proof thus constructed for X5310 is 19 lines long and very similar to that outlined in Figure 4.2.

As noted in [8], restricting expansion terms to some normal form may (depending on other details of the implementation) entail loss of completeness of a proof procedure, but for the present we are content to explore the benefits of using gensubs. Questions related to combining this method of instantiating quantifiers on higher-order variables with other methods, such as those of [11, 16], and [17], need further study. Some of the examples found in these papers are discussed below.

TPS can duplicate quantifiers during the search for a mating by using outermost-quantifier duplication [5] or by using path-focused duplication [33]. It can also generate sets of expansion options in several ways, and there are several implementations of both first- and higher-order unification algorithms. These basic facilities are combined into a number of search procedures in **TPS**. Much remains to be done in exploring the relative merits of various search techniques in various situations, and strategies for systematically incrementing the flags that control the size of the multidimensional search space.

Some of the search procedures in **TPS** take into account the fact for any subformula of the form $[\mathbf{A} \vee \mathbf{B}]$, any path that passes through **A** has a variant that passes through **B**, and both of these paths must be spanned by an acceptable mating. Therefore, when the matingsearch procedure comes to **B**, if **A** has no mate, then no mate for **B** will be sought. Also, if **A** has a mate, but no mate for **B** can be found, then the search will backtrack, throwing out the mate for **A** and all links that were subsequently added to the mating.

Since the search procedures used by **TPS** treat the paths very systematically (and unimaginatively) in their natural order, **TPS** requires very little space to keep track of what it has done. (This may be contrasted, for example, with resolution systems which store vast numbers of clauses.) The higher-order unification procedure does introduce many auxiliary variables, but they are used only briefly, and it was found that by simply uninterning them a great deal of space could be reclaimed. Consequently, **TPS** can run for weeks without running out of space (particularly on problems of first-order logic). However, for many theorems one clearly needs to use more flexible heuristically guided search procedures. It is a significant problem to find a good balance between applying sophisticated search heuristics and limiting the amount of memory required to keep track of the process of exploring the search space.

Of course, many methods could be used to find matings, including resolution [31]. The use of resolution in this context is complicated by the facts that higher-order unification may not terminate and that most general unifiers may not exist.

Following (and slightly extending) terminology introduced by Huet [32], we refer to literals whose atoms are headed by predicate variables as *flexible*, and to literals whose atoms are headed by predicate constants as *rigid*. In first-order logic all predicate symbols may be regarded as constants, so flexible literals need not be considered. However, they do frequently occur in higher-order logic, for example when the Leibniz definition of equality is instantiated. Applications of gensubs to the head variables of flexible literals create even more such literals. Since flexible literals can be mated with arbitrary literals, the search space associated with finding an acceptable mating for a wff that contains many such literals can be extremely large. Two heuristics, each controlled by a flag, are available in **TPS** for trying to cope with this problem. First, the user can set the flag MAX-MATES to limit the number of mates any literal-occurrence may have. (It is fairly rare to encounter examples of acceptable matings in which literals have many mates.) Since the algorithm for constructing matings in **TPS** tries to span each path by mating the first available pair of literals, and a flexible literal typically occurs on many paths, this limit quickly excludes from consideration many matings that the unification process would find incompatible only after considerable work. Second, one can rearrange the jform before the search for a mating commences. If the user sets the flag ORDER-COMPONENTS to the value PREFER-RIGID1, **TPS** applies an algorithm to rearrange the jform (using the commutativity and associativity of conjunction and disjunction) so that rigid literals tend to be encountered by the search process before flexible literals; this postpones finding mates for flexible literals until constraints introduced by mating other literals have been introduced.

Additional complexity arises when one mates a pair of flexible literals. Mating such a pair can cause either literal to become the negation of the other (after substitution, λ-reduction, and elimination of double negations). Both possibilities must be considered, since the variables in these literals may occur in other literals too. When **TPS** considers adding such a pair to the mating, it puts a disagreement pair corresponding to it into the leaves of the unification tree, and proceeds with the unification process. If this process encounters a disagreement pair of the form $\langle \mathbf{A}, \sim \mathbf{B} \rangle$, where **A** starts with a constant but **B** does not, it replaces this pair with $\langle \sim \mathbf{A}, \mathbf{B} \rangle$ and continues. Thus, in a very economical fashion **TPS** finds which way of mating these literals is ultimately compatible with the unification problem for the entire mating.

If one is working on a theorem that is too hard for **TPS** to prove all by itself, one can still use the automatic facilities of **TPS** to provide assistance. One can use the interactive facilities of **TPS** (supplemented by GO2) to develop the general outline of the proof in natural deduction style, and ask **TPS** to help by automatically proving certain lines of the proof from other specified lines.

5. An Example

One of the most interesting theorems that **TPS** has proved automatically is THM15B:

$$\forall \mathtt{f}_{\iota\iota} \cdot \exists \mathtt{g}_{\iota\iota} [\mathtt{ITERATE+}\ \mathtt{fg} \wedge \exists \mathtt{x}_{\iota} \cdot \mathtt{gx} = \mathtt{x} \wedge \forall \mathtt{z}_{\iota} \cdot \mathtt{gz} = \mathtt{z} \supset \mathtt{z} = \mathtt{x}]$$
$$\supset \exists \mathtt{y}_{\iota} \cdot \mathtt{fy} = \mathtt{y}.$$

(1) 1 $\vdash \exists g_{\iota\iota} .\ \mathrm{ITERATE+}\ f_{\iota\iota}\ g$
$\quad \land \exists x_\iota .g\ x = x \land \forall z_\iota .g\ z = z \supset z = x$ Hyp

(2) 1,2 $\vdash \mathrm{ITERATE+}\ f_{\iota\iota}\ g_{\iota\iota} \land \exists x_\iota .g\ x = x \land \forall z_\iota .g\ z = z \supset z = x$
Choose: $g_{\iota\iota}$ 1

(3) 1,2 $\vdash \mathrm{ITERATE+}\ f_{\iota\iota}\ g_{\iota\iota}$ RuleP: 2

(4) 1,2 $\vdash \exists x_\iota .g_{\iota\iota}\ x = x \land \forall z_\iota .g\ z = z \supset z = x$ RuleP: 2

(5) 1,2 $\vdash \forall p_{o(\iota\iota)} .p\ f_{\iota\iota} \land \forall j_{\iota\iota}\ [p\ j \supset p .\lambda x_\iota\ f .j\ x] \supset p\ g_{\iota\iota}$
EquivWffs: 3

(6) 1,2,6 $\vdash g_{\iota\iota}\ x_\iota = x \land \forall z_\iota .g\ z = z \supset z = x$ Choose: x_ι 4

(7) 1,2,6 $\vdash g_{\iota\iota}\ x_\iota = x$ RuleP: 6

(8) 1,2,6 $\vdash \forall z_\iota .g_{\iota\iota}\ z = z \supset z = x_\iota$ RuleP: 6

(9) 1,2,6 $\vdash g_{\iota\iota}\ [f_{\iota\iota}\ x_\iota] = f\ x \supset f\ x = x$ UI: $[f_{\iota\iota}\ x_\iota]$ 8

(10) 1,2 $\vdash \quad [\lambda j_{\iota\iota}\ \forall P_{o\iota} .P\ [f_{\iota\iota} .j\ x_\iota] \lor \sim P .j .f\ x]\ f$
$\quad \land \forall j\ [\quad [\lambda j\ \forall P .P\ [f .j\ x] \lor \sim P .j .f\ x]\ j$
$\quad\quad \supset [\lambda j\ \forall P .P\ [f .j\ x] \lor \sim P .j .f\ x] .\lambda x\ f .j\ x]$
$\supset [\lambda j\ \forall P .P\ [f .j\ x] \lor \sim P .j .f\ x]\ g_{\iota\iota}$
UI: $[\lambda j_{\iota\iota}\ \forall P_{o\iota}.P\ [f_{\iota\iota}.j\ x_\iota] \lor \sim P.j.f\ x]$ 5

(11) 1,2 $\vdash \quad \forall P_{o\iota}\ [P\ [f_{\iota\iota} .f\ x_\iota] \lor \sim P .f .f\ x]$
$\quad \land \forall j_{\iota\iota}\ [\quad \forall P\ [P\ [f .j\ x] \lor \sim P .j .f\ x]$
$\quad\quad \supset \forall P .P\ [f .f .j\ x] \lor \sim P .f .j .f\ x]$
$\supset \forall P .P\ [f .g_{\iota\iota}\ x] \lor \sim P .g .f\ x$ Lambda: 10

(12) $\vdash P_{o\iota}\ [f_{\iota\iota} .f\ x_\iota] \lor \sim P .f .f\ x$ RuleP

(13) $\vdash \forall P_{o\iota} .P\ [f_{\iota\iota} .f\ x_\iota] \lor \sim P .f .f\ x$ UGen: $P_{o\iota}$ 12

(14) 14 $\vdash \forall P_{o\iota} .P\ [f_{\iota\iota} .j_{\iota\iota}\ x_\iota] \lor \sim P .j .f\ x$ Hyp

(15) 14 $\vdash [\lambda w_\iota\ W^9_{o\iota} .f_{\iota\iota}\ w]\ [f .j_{\iota\iota}\ x_\iota] \lor \sim [\lambda w\ W^9 .f\ w] .j .f\ x$
UI: $[\lambda w_\iota\ W^9_{o\iota}.f_{\iota\iota}\ w]$ 14

(16) 14 $\vdash W^9_{o\iota}\ [f_{\iota\iota} .f .j_{\iota\iota}\ x_\iota] \lor \sim W^9 .f .j .f\ x$ Lambda: 15

(17) 14 $\vdash \forall W^9_{o\iota} .W^9\ [f_{\iota\iota} .f .j_{\iota\iota}\ x_\iota] \lor \sim W^9 .f .j .f\ x$ UGen: $W^9_{o\iota}$ 16

(18) 14 $\vdash \forall P_{o\iota} .P\ [f_{\iota\iota} .f .j_{\iota\iota}\ x_\iota] \lor \sim P .f .j .f\ x$ AB: 17

(19) $\vdash \quad \forall P_{o\iota}\ [P\ [f_{\iota\iota} .j_{\iota\iota}\ x_\iota] \lor \sim P .j .f\ x]$
$\supset \forall P .P\ [f .f .j\ x] \lor \sim P .f .j .f\ x$ Deduct: 18

(20) $\vdash \forall j_{\iota\iota} .\quad \forall P_{o\iota}\ [P\ [f_{\iota\iota} .j\ x_\iota] \lor \sim P .j .f\ x]$
$\quad \supset \forall P .P\ [f .f .j\ x] \lor \sim P .f .j .f\ x$ UGen: $j_{\iota\iota}$ 19

(21) $\vdash \quad \forall P_{o\iota}\ [P\ [f_{\iota\iota} .f\ x_\iota] \lor \sim P .f .f\ x]$
$\land \forall j_{\iota\iota} .\quad \forall P\ [P\ [f .j\ x] \lor \sim P .j .f\ x]$
$\quad\quad \supset \forall P .P\ [f .f .j\ x] \lor \sim P .f .j .f\ x$
RuleP: 13 20

(22) 1,2 $\vdash \forall P_{o\iota} .P\ [f_{\iota\iota} .g_{\iota\iota}\ x_\iota] \lor \sim P .g .f\ x$ MP: 21 11

(23) 1,2 $\vdash g_{\iota\iota}\ [f_{\iota\iota}\ x_\iota] = f\ [g\ x] \lor \sim .g\ [f\ x] = g .f\ x$
UI: $[=.g_{\iota\iota}.f_{\iota\iota}\ x_\iota]$ 22

(24) 1,2,24 $\vdash g_{\iota\iota}\ [f_{\iota\iota}\ x_\iota] = f .g\ x$ Case 1: 23

(25) 1,2,25 $\vdash \sim .g_{\iota\iota}\ [f_{\iota\iota}\ x_\iota] = g .f\ x$ Case 2: 23

(26) 1,2,25 $\vdash \sim \mathbf{T}$ Refl=: 25

(27) 1,2,25 $\vdash g_{\iota\iota}\ [f_{\iota\iota}\ x_\iota] = f .g\ x$ RuleP: 26

(28) 1,2 $\vdash g_{\iota\iota}\ [f_{\iota\iota}\ x_\iota] = f .g\ x$ Cases: 23 24 27

(29) 1,2,6 $\vdash g_{\iota\iota}\ [f_{\iota\iota}\ x_\iota] = f\ x$ Subst=: 28 7

(30) 1,2,6 $\vdash f_{\iota\iota}\ x_\iota = x$ MP: 29 9

(31) 1,2,6 $\vdash \exists y_\iota .f_{\iota\iota}\ y = y$ EGen: x_ι 30

(32) 1,2 $\vdash \exists y_\iota .f_{\iota\iota}\ y = y$ RuleC: 4 31

(33) 1 $\vdash \exists y_\iota .f_{\iota\iota}\ y = y$ RuleC: 1 32

(34) $\vdash \quad \exists g_{\iota\iota}\ [\quad \mathrm{ITERATE+}\ f_{\iota\iota}\ g$
$\quad\quad \land \exists x_\iota .g\ x = x \land \forall z_\iota .g\ z = z \supset z = x]$
$\supset \exists y_\iota .f\ y = y$ Deduct: 33

(35) $\vdash \forall f_{\iota\iota} .\quad \exists g_{\iota\iota}\ [\quad \mathrm{ITERATE+}\ f\ g$
$\quad\quad \land \exists x_\iota .g\ x = x \land \forall z_\iota .g\ z = z \supset z = x]$
$\quad \supset \exists y_\iota .f\ y = y$ UGen: $f_{\iota\iota}$ 34

Fig. 5.1. Proof of THM15B.

TPS takes about 2.5 hours to prove this theorem,* which asserts that if some iterate of a function f has a unique fixed point, then f has a fixed point. (The definition of ITERATE+ is in Section 6.) The theorem appears in [36], where it is noted that the theorem can be used in solving integral equations of the second kind; it justifies showing that a fixed point equation has a solution by showing that the iterated equation has a unique solution. This was posed as a problem for theorem provers in [1], and it is gratifying that we are finally able to prove it automatically. It is a hard theorem for **TPS** because so many flexible literals are created when the definition of equality is instantiated.

The very natural, though overly detailed, proof that **TPS** finds is shown in Figure 5.1. Let us summarize it by providing comments for some of the lines of the proof. **TPS** starts out by assuming that g is an iterate of f,

$$(3) \quad 1,2 \quad \vdash \text{ITERATE+}\, f_{\iota\iota} g_{\iota\iota} \qquad \text{RuleP: } 2$$

that x is a fixed point of g,

$$(7) \quad 1,2,6 \quad \vdash g_{\iota\iota} x_\iota = x \qquad \text{RuleP: } 6$$

and that x is the only fixed point of g.

$$(8) \quad 1,2,6 \quad \vdash \forall z_\iota \cdot g_{\iota\iota} z = z \supset z = x_\iota \qquad \text{RuleP: } 6$$

In lines (5) and (10)–(28) **TPS** then gives an inductive proof, based on the definition in (3), of the fact that

$$(28) \quad 1,2 \quad \vdash g_{\iota\iota}[f_{\iota\iota} x_\iota] = f \cdot gx \qquad \text{Cases: } 23\ 24\ 27$$

Of course, (28) follows from the fact that f commutes with all its iterates. No knowledge or heuristics about induction, commutativity, or iterates are built into **TPS** except the definition of ITERATE+. **TPS** decides to prove (28), and to prove it by induction, simply by applying general logical principles in its search for an expansion proof of the theorem. From (7) and (28) **TPS** concludes that

$$(29) \quad 1,2,6 \quad \vdash g_{\iota\iota}[f_{\iota\iota} x_\iota] = fx \qquad \text{Subst} =: 28\ 7$$

Line (29) shows that fx is also a fixed point of g, so it must be the same as x:

$$(9) \quad 1,2,6 \quad \vdash g_{\iota\iota}[f_{\iota\iota} x_\iota] = fx \supset fx = x \qquad \text{UI: } [f_{\iota\iota} x_\iota]\ 8$$

$$(30) \quad 1,2,6 \quad \vdash f_{\iota\iota} x_\iota = x \qquad \text{MP: } 29\ 9$$

Thus x is a fixed point of f.

* Note added in proof: Improvements in **TPS** subsequent to the acceptance of this paper have reduced this time to 5.7 minutes.

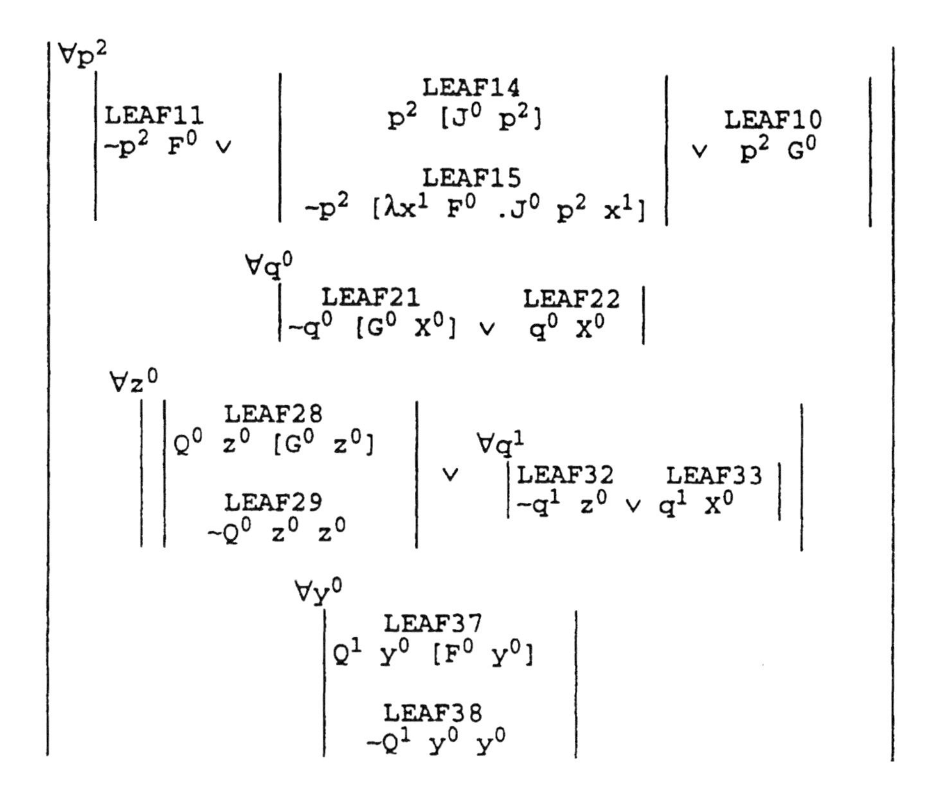

Fig. 5.2. Vpform for THM15B.

It is natural to wonder how such a proof can be found, so let us see how **TPS** does this. In the preprocessing stage, **TPS** negates the theorem and eliminates the definitions obtaining

$$\begin{aligned}&\exists f_{\iota\iota}.\exists g_{\iota\iota}\ [\forall p_{o(\iota\iota)}[pf \wedge \forall j_{\iota\iota}[pj \supset p\,.\,\lambda x_\iota f.\,jx] \supset pg]\\ &\wedge\ \exists x\,.\,gx = x\\ &\wedge\ \forall z_\iota.\sim [gz = z] \vee z = x]\\ &\wedge\ \forall y_\iota.\sim .\,fy = y\end{aligned}$$

It then expands the equality formulas using the Leibniz definition, and puts the wff into negation normal form, obtaining

$$\begin{aligned}&\exists f_{\iota\iota}.\exists g_{\iota\iota}[\forall p_{o(\iota\iota)}[\sim pf \vee\ \exists j_{\iota\iota}[pj\ \wedge \sim p.\,\lambda x_\iota f\,.\,jx] \vee pg]\\ &\wedge\ \exists x\,.\ \ \forall q_{o\iota}[\ \sim q[gx] \vee qx]\\ &\wedge\ \forall z_\iota.\ \ \exists q[q[gz]\ \wedge \sim qz] \vee\ \ \forall q.\sim qz \vee qx]\\ &\wedge\ \forall y_\iota\ \exists q.\,q[fy]\ \wedge \sim qy\end{aligned}$$

It then Skolemizes the formula (using capital letters as Skolem constants) and displays the formula (with type symbols deleted) in a vpform that is shown in Figure 5.2.

TPS then starts the search for an expansion proof. It considers, in turn, the following substitutions for the predicate variable $p_{o(\iota\iota)}$ (shown as p^2 in Figure 5.2) which was introduced in the definition of ITERATE+:

$$
\begin{array}{ll}
\texttt{oset}-0: & \texttt{None.} \\
\texttt{oset}-1: & \lambda w^2_{\iota\iota}.\, p^1_{o(\iota\iota)} w^2 \wedge p^2_{o(\iota\iota)} w^2 \\
\texttt{oset}-2: & \lambda w^3_{\iota\iota}.\, p^3_{o(\iota\iota)} w^3 \vee p^4_{o(\iota\iota)} w^3 \\
\texttt{oset}-3: & \lambda w^4_{\iota\iota}\ \exists w^3_{o\iota} p^5_{o(o\iota)(\iota\iota)} w^4 w^3 \\
\texttt{oset}-4: & \lambda w^5_{\iota\iota}\ \forall w^4_{o\iota} p^6_{o(o\iota)(o\iota)} w^5 w^4 \\
\texttt{oset}-5: & \lambda w^6_{\iota\iota}\ \exists w^5_{o\iota}.\, p^7_{o(o\iota)(o\iota)} w^6 w^5 \vee p^8_{o(o\iota)(o\iota)} w^6 w^5 \\
\texttt{oset}-6: & \lambda w^7_{\iota\iota}\ \exists w^6_{o\iota}.\, p^9_{o(o\iota)(o\iota)} w^7 w^6 \wedge p^{10}_{o(o\iota)(o\iota)} w^7 w^6
\end{array}
$$

It spends about 19 minutes exploring each of these expansion options. Finally it comes to

$$\texttt{oset}-7:\ \lambda w^8_{\iota\iota}\ \forall w^7_{o\iota} \cdot p^{11}_{o(o\iota)(o\iota)} w^8 w^7 \vee p^{12}_{o(o\iota)(\iota\iota)} w^8 w^7$$

It applies the substitution and preprocesses to obtain a jform that is displayed in Figure 5.3.

TPS searches on this jform for an acceptable mating and finds the following:

(LEAF28 . LEAF162)	(LEAF38 . LEAF33)	(LEAF37 . LEAF32)
(LEAF29 . LEAF22)	(LEAF21 . LEAF161)	(LEAF159 . LEAF156)
(LEAF158 . LEAF155)	(LEAF150 . LEAF149)	

The substitution associated with the mating is

$$
\begin{array}{l}
q^0_{o\iota} \rightarrow \lambda w^{122}_{\iota} Q^0_{o\iota\iota}[F^0_{\iota\iota} X^0_{\iota}].\, F^0 w^{122} \\
z^0_{\iota} \rightarrow F^0_{\iota\iota} X^0_{\iota} \\
p^{11}_{o(o\iota)(o\iota)} \rightarrow \lambda w^{123}_{\iota\iota} \lambda w^{124}_{o\iota} w^{124}.\, F^0_{\iota\iota}.\, w^{123} X^0_{\iota} \\
w^{17}_{o\iota} \rightarrow \lambda w^{125}_{\iota} W^3_{o\iota(o(\iota\iota))} \left[\lambda w^8_{\iota\iota} \Pi_{o(o(o\iota))}.\, \lambda w^7_{o\iota} w^7 [F^0_{\iota\iota}.\, w^8 X^0_{\iota}]\right. \\
\qquad\qquad\qquad \left. \vee \sim w^7.\, w^8.\, F^0 X^0\right].\, F^0 w^{125} \\
w^{18}_{o\iota} \rightarrow \lambda w^{122}_{\iota} Q^0_{o\iota\iota}[F^0_{\iota\iota} X^0_{\iota}] w^{122} \\
p^{12}_{o(o\iota)(\iota\iota)} \rightarrow \lambda w^{120}_{\iota\iota} \lambda w^{121}_{o\iota}.\sim w^{121}.\, w^{120}.\, F^0_{\iota\iota} X^0_{\iota} \\
q^1_{o\iota} \rightarrow \lambda w^{122}_{\iota} Q^1_{o\iota\iota} X^0_{\iota} w^{122} \\
y^0_{\iota} \rightarrow X^0_{\iota}
\end{array}
$$

Let us examine the computation of the important compound substitution for the predicate variable $p_{o(\iota\iota)}$. The substitution term from oset-7 is

$$\lambda w^8_{\iota\iota}\ \forall w^7_{o\iota}.\, p^{11}_{o(o\iota)(o\iota)} w^8 w^7 \vee p^{12}_{o(o\iota)(\iota\iota)} w^8 w^7$$

```
|∀p11p12
||                      |∀w17                                                                  ||
||                      ||LEAF155   LEAF156 |  |                                               ||
||  | LEAF149   |       ||p11Jw17  v  p12Jw17|  |                                               ||
||  | ~p11F0Wa  |       |                       |     ∀w18                                      ||
||  |           |  v    |       LEAF158         |  v  |LEAF161          LEAF162 |               ||
||  | LEAF150   |       | ~p11[λx1F0.Jx1]Wb     |     |p11G0w18    v   p12G0w18 |               ||
||  | ~p12F0Wa  |       |                       |                                               ||
||                      |       LEAF159         |                                               ||
||                      | ~p12[λx1F0.Jx1]Wb     |                                               ||
|                        ∀q0
|                         |   LEAF21        LEAF22  |
|                         |-q0  [G0 X0]  v  q0 X0   |
|               ∀z0
|               ||     LEAF28          |        ∀q1
|               ||Q0  z0  [G0  z0]     |   v    |LEAF32    LEAF33 |  |
|               ||     LEAF29          |        |-q1  z0  v q1 X0 |  |
|               || -Q0  z0  z0         |
|                           ∀y0
|                           |    LEAF37          |
|                           |Q1  y0  [F0  y0]    |
|                           |    LEAF38          |
|                           |  ~Q1  y0  y0       |                                             |

where
J  is [J0  .λw8  ∀w7  .p11  w8  w7  v p12  w8  w7]
Wa is [W2  .λw8  ∀w7  .p11  w8  w7  v p12  w8  w7]
Wb is [W3  .λw8  ∀w7  .p11  w8  w7  v p12  w8  w7]
```

Fig. 5.3. Vpform for expanded form of THM15B.

Applying the substitutions for $p^{11}_{o(o\iota)(\iota\iota)}$ and $p^{12}_{o(o\iota)(\iota\iota)}$ produces

$$\lambda w^{8}_{\iota\iota}\ \ \forall w^{7}_{o\iota}.\ [\lambda w^{123}_{\iota\iota}\lambda w^{124}_{o\iota} w^{124}.\, F^{0}_{\iota\iota}.\, w^{123} X^{0}_{\iota}] w^{8} w^{7}$$
$$\vee\ \ [\lambda w^{120}_{\iota\iota}\lambda w^{121}_{o\iota}.\sim w^{121}.\, w^{120}.\, F^{0} X^{0}] w^{8} w^{7}$$

λ-normalizing transforms this to

$$\lambda w^{8}_{\iota\iota}\ \ \forall w^{7}_{o\iota}.\, w^{7}[F^{0}_{\iota\iota}.\, w^{8} X^{0}_{1}]\ \vee \sim w^{7}.\, w^{8}.\, F^{0} X^{0}$$

TPS makes alphabetic changes of the variables to convert this to

$$\lambda j_{\iota\iota}\ \ \forall P_{o\iota}.\, P[F^{0}_{\iota\iota}.\, j X^{0}_{\iota}]\ \vee \sim P.\, j.\, F^{0} X^{0}$$

TPS then constructs an expansion proof from this mating, merges it, and constructs the natural deduction proof in Figure 5.1, guided by the information in the expansion tree. To see how this works, let us look at two stages in this process.

By applying rules of inference in both forward and backward directions in a rather natural way, and using the substitutions for z^0_ι and y^0_ι, **TPS** constructs the

```
(1-4)
(5)   1,2    ⊢ ∀p_o(ιι) .p f_ιι ∧ ∀j_ιι [p j ⊃ p .λx_ι f .j x] ⊃ p g_ιι
                                                          EquivWffs: 3
(6)
(7)   1,2,6  ⊢ g_ιι x_ι = x                                  RuleP: 6
(8)
(9)   1,2,6  ⊢ g_ιι [f_ιι x_ι] = f x ⊃ f x = x      UI: [f_ιι x_ι] 8
              ...
(30)  1,2,6  ⊢ f_ιι x_ι = x                                   PLAN21
(31-35)
```

Fig. 5.4. Early stage in construction of proof of THM15B.

```
(1-8)
(9)   1,2,6  ⊢ g_ιι [f_ιι x_ι] = f x ⊃ f x = x      UI: [f_ιι x_ι] 8
(10)
(11)  1,2    ⊢       ∀P_oι [P [f_ιι .f x_ι] ∨ ~P .f .f x]
                   ∧ ∀j_ιι [  ∀P [P [f .j x] ∨ ~P .j .f x]
                           ⊃ ∀P .P [f .f .j x] ∨ ~P .f .j .f x]
                 ⊃ ∀P .P [f .g_ιι x] ∨ ~P .g .f x          Lambda: 10
              ...
(28)  1,2    ⊢ g_ιι [f_ιι x_ι] = f .g x                       PLAN28
(29)  1,2,6  ⊢ g_ιι [f_ιι x_ι] = f x                    Subst=: 28 7
(30)  1,2,6  ⊢ f_ιι x_ι = x                                   PLAN21
(31-35)
```

Fig. 5.5. Later stage in construction of proof of THM15B.

partial proof displayed in Figure 5.4. At this stage the proof contains only lines (1)–(9) and (30)–(35). **TPS** is planning to prove (30), and it knows that only (5), (7), and (9) need be actively used to do this. The other lines are *inactive* and will not be used again in the process of constructing the proof. This *status information* is represented simply as (30 9 7 5); the first entry is the number of the line to be proved, and the other entries are numbers of lines that may now be used to prove that line. In the figure we display the active lines and only the numbers of the inactive lines that are now present in the proof.

TPS next derives (10) by applying universal instantiation to (5), using the substitution for $p_{o(\iota\iota)}$ discussed above. This makes (5) inactive, so the status information is now (30 10 9 7).

With a few more inferences the proof reaches the form in Figure 5.6. Now the status information is (28 11 9)(30 29 11 9), which means that both (28) and (30) are to be proven.

It is hard to explain exactly how **TPS** decided to infer (29) from (7) and (28) without a detailed discussion of the tactics for dealing with equality that were invoked by setting the flag REMOVE-LEIBNIZ to T for this proof. Suffice it to say that (7) is descended from LEAF21 and LEAF22 in Figure 5.3, (28) is descended from LEAF161, and LEAF21 is mated to LEAF161 in the expansion proof. (The process is easier to understand when REMOVE-LEIBNIZ is NIL, but the proof thus obtained is not so elegant.)

Similarly, the antecedent of (9) is descended from LEAF28 and LEAF29, and (29) is descended from LEAF22, so the mating between LEAF29 and LEAF22 guides the derivation of (30) by modus ponens (MP) from (9) and (29).

Since the consequent of (11) is essentially the assertion in (28) (modulo the Leibniz definition of equality and the symmetry of disjunction), it can be seen that the same general methods suffice to complete the construction of the proof.

6. Theorems Proved Automatically

While **TPS** is still in a rudimentary state as a system for automatically proving serious theorems of type theory, it is a well developed platform for experimenting with these theorems and developing ideas about the basic issues involved. In this section we discuss some examples of theorems that **TPS** has proved completely automatically.

Naturally, **TPS** can be used to prove theorems of first-order logic, but we focus mainly on examples from higher-order logic. (Apart from the development of path-focused duplication, a relatively small part of the development effort for **TPS** has been devoted thus far to certain basic issues of search that are important for first-order logic.)

For ease of reference, we list the theorems in the order of their labels; these simply reflect the way examples have been put into our library over the years. Theorems (such as X2129) whose names start with an 'X' are exercises in [7] (or will be exercises in the next edition) and are available in **TPS** and **ETPS** whether or not one has any library files.

When **TPS** proves a theorem in automatic mode, it records the time used to do a number of things, including searching for an acceptable mating (*search*), merging the expansion proof (*merge*), translating the expansion proof into a natural deduction proof (*translate*), and printing the proof on the screen (*print*). It also records the total time used to do all these things and produce a natural deduction proof of the theorem (*total*). For each example below we give the internal runtime, minus garbage-collect time, used by **TPS** for some or all of these processes while running on a Hewlett Packard Apollo 9000 model 735 workstation equipped with 208 megabytes of RAM. Times are in seconds (*sec*), minutes (*min*), or hours (*h*), as seems most convenient. These numbers are useful only for their approximate magnitudes; they are quite dependent on how various flags are set, and in many cases probably do not represent optimal settings of the flags. They fluctuate both up and down as changes are made in **TPS**. It should also be noted that the time to produce output on the screen is not negligible. In one run for THM47 which is reported below, the total runtime was 25.08 sec. However, when we ran this again in a mode that minimized output to the screen, the total runtime was 22.67 sec.

It will be noted that in many cases the process of translating an expansion proof into a natural deduction takes a surprisingly large amount of time, even though it involves no deep search or backtracking. This is because the conditions checked on expansion proofs in some steps of the translation are computationally expensive in order to arrive at the most natural proof possible with the current tactics. Furthermore, no attempt has been made to optimize this part of the program.

Definitions

The definitions below, which are used in various theorems we shall discuss, are built into **TPS** or stored in the **TPS** library, and the user can easily add more definitions to the library. The way **TPS** handles definitions during the search process is determined by the settings of certain flags, such as REWRITE-DEFNS and REWRITE-EQUALITIES. When these flags are set to T, **TPS** simply eliminates these definitions from the theorem while preparing the expansion tree for the search process. In some cases, this expands the search space in a very undesirable way, and more sophisticated ways of deciding when to instantiate definitions in the search process are clearly needed. (Discussions of this issue may be found in [12], [15] (where 'peeking' is discussed), and [58].) Once **TPS** finds an expansion proof, the translation tactics cause the definitions to be handled rather naturally in the final natural deduction proof.

We remark that a wff of the form $[\lambda x_\alpha \mathbf{B}]$, where $\mathbf{B}$ is a statement about $\mathbf{x}_\alpha$, denotes the set $\{x_\alpha \mid \mathbf{B}\}$. Also, $[\mathbf{P}_{o\alpha}\mathbf{x}_\alpha]$ means $[\mathbf{x}_\alpha \,\varepsilon\, \mathbf{P}_{o\alpha}]$. Binary operators are often written in infix position.

ε (set membership): $[\lambda x_\alpha \, \lambda p_{o\alpha} px]$.

$\subseteq$ (subset): $[\lambda p_{o\alpha} \, \lambda r_{o\alpha} \, \forall x_\alpha \cdot px \supset rx]$.

$\cup$ (union): $[\lambda p_{o\alpha} \, \lambda r_{o\alpha} \, \lambda z_\alpha \cdot pz \vee rz]$.

$\bigcup$ (union of a collection of sets): $[\lambda w_{o(o\alpha)} \, \lambda x_\alpha \, \exists s_{o\alpha} \cdot ws \wedge sx]$.

$\bigcap$ (intersection of a collection of sets): $[\lambda w_{o(o\alpha)} \, \lambda x_\alpha \, \forall s_{o\alpha} \cdot ws \supset sx]$.

$\circ$ (composition of functions): $[\lambda f_{\alpha\beta} \, \lambda g_{\beta\chi} \, \lambda x_\chi f \cdot gx$.

INJECTIVE: $[\lambda f_{\alpha\beta} \, \forall x_\beta \, \forall y_\beta \cdot fx = fy \supset x = y]$

\# (image): $[\lambda f_{\alpha\beta} \, \lambda x_{o\beta} \, \lambda z_\alpha \, \exists t_\beta \cdot xt \wedge z = ft]$

$[\# f_{\alpha\beta} x_{o\beta}]$ is the image of the set $x_{o\beta}$ under the function $f_{\alpha\beta}$.

U (unit set): $[\lambda x_\alpha \, \lambda y_\alpha \cdot x = y]$

$[Ux]$ is customarily written as $\{x\}$.

IND: $\forall p_{o\iota} \cdot p 0_\iota \wedge \forall x_\iota [px \supset p \cdot S_{\iota\iota} x] \supset \forall x \, px$

IND express a simple induction axiom for the natural numbers. 0_ι is zero, and $S_{\iota\iota}$ is the successor function.

INTERATE+: $[\lambda f_{\alpha\alpha}\, \lambda g_{\alpha\alpha}\, \forall p_{o(\alpha\alpha)} \cdot pf \wedge \forall j_{\alpha\alpha}[pj \supset p \cdot f \circ j] \supset pg]$

[ITERATE+ f g] means that g is a composition of one or more copies of f. Note how easy it is to express this inductive definition in type theory.

ITERATE: $[\lambda f_{\alpha\alpha}\, \lambda g_{\alpha\alpha}\, \forall p_{o(\alpha\alpha)} \cdot p[\lambda u_{\alpha} u] \wedge \forall j_{\alpha\alpha}[pj \supset p \cdot f \circ j] \supset pg]$

[ITERATE f g] means that g is a composition of zero or more copies of f.

HOMOM2: $[\lambda h_{\alpha\beta}\, \lambda f_{\beta\beta\beta}\, \lambda g_{\alpha\alpha\alpha}\, \forall x_{\beta} \forall y_{\beta} \cdot h[fxy] = g[hx] \cdot hy]$

[HOMOM2 h f g] means that h is a homomorphism from objects of type β to objects of type α, where f and g are binary operators on the types β and α, respectively.

MAPS: $[\lambda h_{\alpha\beta}\, \lambda u_{o\beta}\, \lambda v_{o\alpha}\, \forall x_{\beta} \cdot ux \supset v \cdot hx]$

[MAPS h u v] means that the function h maps the set u into the set v.

–CLOSED: $[\lambda h_{\alpha\alpha}\, \lambda u_{o\alpha}\, \text{MAPS}_{o(o\alpha)(o\alpha)(\alpha\alpha)} huu]$

[h–CLOSED u] means that the set u is closed under the function h.

HOM: $\lambda h_{\alpha\beta}\, \lambda r_{o\beta}\, \lambda f_{\beta\beta}\, \lambda s_{o\alpha}\, \lambda g_{\alpha\alpha} \cdot f\text{–CLOSED}\, r \wedge g\text{–CLOSED}\, s \wedge \text{MAPS}_{o(o\alpha)(o\beta)(\alpha\beta)} hrs \wedge \forall x_{\beta} \cdot rx \supset h[fx] = g \cdot hx$

[HOM h rfs g] means that h is a homomorphism from $\langle r, f\rangle$ to $\langle s, g\rangle$, where r and s are sets, f is a unary operator on r, and g is a unary operator on s.

Theorems

THM15B: $\forall f_{\iota\iota} \cdot\ \exists g_{\iota\iota}[\text{ITERATE+}\, fg \wedge\ \exists x_{\iota} \cdot gx = x \wedge\ \forall z_{\iota} \cdot gz = z \supset z = x] \supset\ \exists y_{\iota} \cdot fy = y$

(search: 2.5 h   total: 2.5 h)

This theorem was discussed in Section 5.

THM30: $R_{o\alpha} \subseteq S_{o\alpha} \equiv\ \forall F_{\alpha\alpha} \cdot \#FR \subseteq \#FS$

(search: 0.56 sec   translate: 1.19 sec   total: 2.52 sec)

THM47: $\forall X_{\iota}\, \forall Y_{\iota} \cdot\ \forall Q_{o\iota}[QX \supset QY] \equiv\ \forall R_{o\iota\iota} \cdot\ \forall Z_{\iota} RZZ \supset RXY$

Run with MATING-VERBOSE MAX and UNIFY-VERBOSE MAX:

(search: 22.82 sec   total: 25.08 sec)

Run with MATING-VERBOSE SILENT and UNIFY-VERBOSE SILENT:

(search: 20.54 sec   total: 22.67 sec)

THM47 shows the equivalence of two ways of defining equality in type theory: the Leibniz definition, and the intersection of all reflexive relations.

THM48: $\forall F_{\alpha\beta}\, \forall G_{\beta\chi} \cdot \text{INJECTIVE } F \wedge \text{INJECTIVE } G \supset \text{INJECTIVE} \cdot F \circ G$

Trial with REWRITE-EQUALITIES set to T:
(search: 89.74 sec total: 91.43 sec)
Trial with REWRITE-EQUALITIES set to NIL:
(search: 0.04 sec total: 0.81 sec)
THM48 asserts that the composition of injective functions is injective. The definition of equality which is contained in the definition of INJECTIVE is actually not needed in order to prove this theorem, and the time required to prove the theorem is dramatically affected by whether the equalities are instantiated or not.

THM67: $\forall S_{o\alpha}\, \forall T_{o\alpha}[S \subseteq T \supset F_{o\alpha(o\alpha)}T \subseteq FS] \wedge [\, \forall S[S \subseteq F[G_{o\alpha(o\alpha)}S]] \wedge \forall S[S \subseteq G[FS]]] \supset\ \forall S\, \forall x_\alpha[F[G[FS]]x \equiv FSx]$
(search: 11.79 sec total: 12.75 sec)

Next we have two examples which were discussed in [8] as examples of theorems that require instantiations for set-variables that cannot be obtained by unification from literals in the theorem.

THM104: $\forall X_\alpha\, \forall Z_\alpha \cdot UX = UZ \supset X = Z$
(search: 9.67 sec total: 10.55 sec)
The proof referred to here was obtained using the Leibniz definition of equality, and uses a projection as an expansion term. However, if we change the value of the flag REWRITE-EQUAL-EXT so that **TPS** uses the extensional definition of equality between sets, no expansion option is needed, and the times for the proof are (search: 0.14 sec, total: 1.22 sec)

THM112: $\forall P_{o\iota}\, \exists M_{o(\iota\iota)}\, \forall G_{\iota\iota}\, \forall H_{\iota\iota} \cdot MG \wedge MH \supset M[G \circ H] \wedge\ \forall Y_\iota \cdot PY \supset P \cdot GY$
(search: 6.13 sec total: 6.13 sec)
THM112 asserts that for any set P, there is a set M of functions mapping P into P which is closed under composition. **TPS** quickly finds a trivial proof where the set M is $[\lambda w_{\iota\iota} \cdot \sim h_0 \wedge h]$, i.e., the empty set of functions. To make the problem slightly less trivial, we excluded this solution in the statement of THM112A below.

THM112A: $\forall P_{o\iota}\, \exists M_{o(\iota\iota)} \cdot M[\lambda x_\iota x] \wedge\ \forall G_{\iota\iota}\, \forall H_{\iota\iota} \cdot MG \wedge MH \supset M[G \circ H] \wedge\ \forall Y_\iota \cdot PY \supset P \cdot GY$
(search: 4.4 min total: 4.4 min)
For the proof of THM112A, **TPS** finds that it suffices to let M be $[\lambda f_{\iota\iota}\, \forall u_\iota \cdot P_{o\iota}[fu] \vee \sim Pu]$. This can be rewritten as $[\lambda f_{\iota\iota}\, \forall u_\iota \cdot P_{o\iota}u \supset P \cdot fu]$, and denotes the set of functions that map P into itself.

THM117C: $\forall x_{o\iota} \forall z_{\iota}[z \,\varepsilon\, x \supset \exists y_{\iota} \cdot y \,\varepsilon\, x \wedge \forall w_{\iota} \cdot R_{o\iota\iota}\, y\, w \supset\sim w \,\varepsilon\, x] \wedge \forall x1_{\iota}$
$[\forall y1_{\iota}[y1 \,\varepsilon\, s_{o\iota} \wedge R\; x1\; y1 \supset P_{o\iota}\; y1] \supset P\; x1] \supset$
$\forall x2_{\iota} \cdot x2 \,\varepsilon\, s \supset P\; x2$
(search: 0.19 sec total: 1.63 sec)
This is the TRANSFINITE INDUCTION theorem of [11] (page 396) expressed in the language of type theory. Think of Ryw as saying that $y > w$. The theorem asserts that if R is a well-founded relation and P is an inductive property over R restricted to the set s, then everything in s has property P.

THM129: $\text{IND} \wedge \forall x_{\iota} +_{o\iota\iota\iota} 0_{\iota}xx \wedge \forall x \forall y_{\iota} \forall z_{\iota}[+yxz \supset +[S_{\iota\iota}y]x \cdot Sz] \supset$
$\forall y \forall x \exists z + yxz$
(search: 0.57 sec total: 2.33 sec)

THM130: $\text{IND} \wedge r_{o\iota\iota}0_{\iota}0 \wedge \forall x_{\iota} \forall y_{\iota}[rxy \supset r[S_{\iota\iota}x] \cdot Sy] \supset \forall x \exists y\; rxy$
(search: 0.51 sec total: 1.18 sec)
This is a theorem in which the conclusion is weaker than the statement that must be proved by induction. From the hypotheses **TPS** proves $\forall x_{\iota}\; r_{o\iota\iota}xx$ by induction, and from this derives the desired conclusion $\forall x_{\iota}\; \exists y_{\iota}\; r_{o\iota\iota}xy$. No special mechanism for deciding what to prove by induction is built into **TPS**; it falls naturally out of a purely logical analysis of the structure of THM130.

THM131: $\forall h1_{\beta\gamma} \forall h2_{\alpha\beta} \forall s1_{o\gamma} \forall f1_{\gamma\gamma} \forall s2_{o\beta} \forall f2_{\beta\beta} \forall s3_{o\alpha} \forall f3_{\alpha\alpha} \cdot$
$\text{HOM}\, h1\; s1\; f1\; s2\; f2 \wedge \text{HOM}\, h2\; s2\; f2\; s3\; f3 \supset$
$\text{HOM}\, [h2 \circ h1] s1\; f1\; s3\; f3$
(search: 30.5 min merging: 6.2 min translate: 3.99 sec total: 36.8 min)
This example, which asserts that the composition of homomorphisms is a homomorphism, was suggested in [18], though the formulation of the theorem given there in terms of the primitives of axiomatic set theory makes it much harder to prove. It may be enlightening to compare this with THM133 below, which is much easier for **TPS** since the closure of the sets under the appropriate functions is dealt with implicitly through the use of types in THM133.

THM133: $\forall h1_{\beta\gamma} \forall h2_{\alpha\beta} \forall f1_{\gamma\gamma\gamma} \forall f2_{\beta\beta\beta} \forall f3_{\alpha\alpha\alpha} \cdot \text{HOMOM2}\; h1\; f1\; f2 \wedge$
$\text{HOMOM2}\; h2\; f2\; f3 \supset \text{HOMOM2}[h2 \circ h1] f1\; f3$
(search: 3.38 sec total: 5.15 sec)

THM134: $\forall z_{\iota} \forall g_{\iota\iota} \cdot \text{ITERATE} + [\lambda x_{\iota}\; z] g \supset \forall x \cdot gx = z$
(search: 0.05 sec total: 1.02 sec)
THM134 can be paraphrased as saying that the only positive iterate of a constant function is that function.

THM135: $\forall f_{\alpha\alpha}\ \forall g1_{\alpha\alpha}\ \forall g2_{\alpha\alpha} \cdot \text{ITERATE}\ f\ g1 \wedge \text{ITERATE}\ f\ g2 \supset$
$\text{ITERATE}\ f \cdot g1 \circ g2$
(search: 3.4 min   total: 3.5 min)
This theorem asserts that the composition of iterates of a function is an iterate of that function.

THM141: $\forall f_{\iota\iota} \cdot\ \exists g_{\iota\iota}[f \circ g = g \circ f \wedge\ \exists x_\iota \cdot gx = x \wedge\ \forall z_\iota \cdot gz = z \supset$
$z = x] \supset \exists y_\iota \cdot fy = y$
(search: 8.62 sec   translate: 5.69 sec   total: 16.54 sec)
This theorem, which is inspired by the proof of THM15B, asserts that if some function that commutes with f has a unique fixed point, then f has a fixed point.

THM142: $\forall a_{o\alpha\beta}\ \forall y_\alpha\ \forall r_{o\beta} \cdot r =\ \lambda j_\beta[y \,\varepsilon\, aj] \supset$
$\exists p_{o(o\beta)(o\beta)}\ \forall s_{o\beta} \cdot y \,\varepsilon \bigcap [\#\ a\ s] \equiv p\ r\ s$
(search: 11.6 min   total: 11.8 min)
This theorem concerns the formulation of a lemma that can be used as part of a proof of the Principle of Inclusion and Exclusion of combinatorics. Think of $a_{o\alpha\beta}$ as an indexed family of sets and $s_{o\beta}$ as a set of indices; a typical set in the family is $a_{o\alpha\beta}j_\beta$, which would simply be written as a_j in conventional mathematical notation. **TPS** finds that $[p\ r\ s]$ can be $[s \subseteq r]$.

In the next two theorems [DOUBLE $u\ v$] means that $2u = v$, and [HALF $u\ v$] means that the greatest integer in $u/2$ is v.

THM300A: $\forall u_\iota\ \forall v_\iota[\text{HALF}_{o\iota\iota} u\ v \equiv \forall Q_{o\iota\iota} \cdot Q0_\iota 0 \wedge Q[S_{\iota\iota}0]0 \wedge\ \forall x_\iota\ \forall y_\iota[Qxy \supset$
$Q[S \cdot Sx] \cdot Sy] \supset Quv] \wedge \text{DOUBLE}_{o\iota\iota}\ 0\ 0$
$\wedge\ \forall x\ \forall y[\text{DOUBLE}\ x\ y \supset \text{DOUBLE}\ [Sx] \cdot S \cdot Sy] \supset$
$\forall u\ \forall v \cdot \text{HALF}\, u\ v \supset \text{DOUBLE}\, v\ u \vee \text{DOUBLE}\ [Sv] \cdot Su$
(search: 83.56 sec   total: 88.13 sec)

THM301A: $\forall u_\iota\ \forall v_\iota[\text{DOUBLE}_{o\iota\iota}\ u\ v \equiv\ \forall Q_{o\iota\iota} \cdot Q0_\iota 0 \wedge\ \forall x_\iota\ \forall y_\iota[Qxy$
$\supset Q[S_{\iota\iota}x] \cdot S \cdot Sy] \supset Quv] \wedge \text{HALF}_{o\iota\iota}\ 0\ 0 \wedge \text{HALF}\ [S0]\ 0 \wedge$
$\forall x\ \forall y[\text{HALF}\ x\ y \supset \text{HALF}\ [S \cdot Sx] \cdot Sy] \supset$
$\forall u\ \forall v \cdot \text{DOUBLE}\ u\ v \supset \text{HALF}\ v\ u$
(search: 56.90 sec   total: 62.79 sec*)

THM303: $\text{EVEN}_{o\iota}0_\iota \wedge\ \forall n_\iota[\text{EVEN}\, n \supset \text{EVEN} \cdot S_{\iota\iota} \cdot Sn] \wedge [\text{ODD}_{o\iota}[S0] \wedge$
$\forall n \cdot \text{ODD}\, n \supset \text{ODD} \cdot S \cdot Sn] \wedge \text{IND} \wedge\ \forall n[\text{NUMBER}_{o\iota}n \equiv$
$\text{EVEN}\ n \vee \text{ODD}\ n] \supset\ \forall n\ \text{NUMBER}\, n$
(search: 33.8 min   total: 33.9 min)

* Note added in proof: Improvements in **TPS** subsequent to the acceptance of this paper have reduced this time to 9.58 sec.

After assuming the antecedent, **TPS** proves $\forall x_\iota$ [NUMBER$_{o\iota}x \wedge$ NUMBER $\cdot S_{\iota\iota}x$] by induction (using IND), and from this derives $\forall n_\iota$ NUMBER$_{o\iota}$n. Note that a direct inductive proof of $\forall n_\iota$ NUMBER$_{o\iota}n$ does not work.

BLEDSOE-FENG-SV-I1: $\forall A_{o\iota}[AO_\iota \wedge \forall x_\iota[Ax \supset A \cdot 1{+}_{\iota\iota}x] \supset An_\iota]$
$\wedge P_{o\iota}O \wedge \forall x[Px \supset P \cdot 1{+}x] \supset Pn$
(search: 0.14 sec total: 0.61 sec)
This is Example I1 from [17].

BLEDSOE-FENG-SV-I2: $\forall A_{o\iota\iota}[AO_\iota O \wedge \forall x_\iota \forall y_\iota[Axy \supset A[s_{\iota\iota}x] \cdot sy]$
$\supset An_\iota m_\iota] \wedge P_{o\iota}n \supset Pm$
(search: 8.15 sec total: 9.91 sec)
This is Example I2 from [17].

NUM016-1: $\sim \cdot \forall X_\iota[\sim \text{less}_{o\iota\iota}XX] \wedge$
$\forall X \forall Y_\iota[\sim \text{less}\, XY \vee \sim \text{less}\, YX] \wedge \forall X$
$\text{divides}_{o\iota\iota}XX \wedge \forall X\ \forall Y\ \forall Z_\iota\ [\sim\text{divides}\ XY \vee \sim\text{divides}\ YZ \vee$ $\text{divides}\ XZ] \wedge \forall X\ \forall Y[\sim\text{divides}XY \vee \sim \text{less}\ YX] \wedge \forall X$ less X[factorial_plus_one$_{\iota\iota}X$]$\wedge \forall X\ \forall Y[\sim$ divides X[factorial_plus_one $Y] \vee$ less $YX] \wedge \forall X$ [prime$_{o\iota}X \vee$ divides [prime_divisor$_{\iota\iota}X]X] \wedge$ $\forall X$[prime $X \vee$ prime·prime_divisor $X] \wedge \forall X$[prime $X \vee$ less [prime_divisor $X]X] \wedge$ prime $a_\iota \wedge \forall X\cdot \sim$prime $X \vee \sim$less a $X \vee$ less [factorial_plus_one a] X
(search: 7.38 sec merge: 5.73 sec translate: 6.23 sec total: 20.79 sec)
This is example NUM016-1 from the TPTP Problem Library [54]. It is more commonly referred to as LS17.

SYN031-1: $\sim \cdot \forall A_\iota[g_{o\iota\iota}Aa_\iota \vee g[f_{\iota\iota}A]A] \wedge \forall A[gAa \vee gA \cdot fA] \wedge \forall A\ \forall B_\iota$
$[\sim gAB \vee g[fB]B] \wedge$
$\forall A\ \forall B[\sim gAB \vee gB \cdot fB] \wedge \forall A\ \forall B \cdot \sim gAB \vee \sim gBa$
(search: 64.08 sec total: 65.79 sec)
This is example SYN031-1 from the TPTP Problem Library. It is more commonly referred to as MQW.

X2115: $\forall x_\iota[\ \exists y_\iota P_{o\iota\iota}xy \supset \forall z_\iota Pzz] \wedge \forall u_\iota\ \exists v_\iota[Puv \vee M_{o\iota}u \wedge Q_{o\iota} \cdot f_{\iota\iota\iota}uv] \wedge$
$\forall w_\iota[Qw \supset\sim M \cdot g_{\iota\iota}w] \supset \forall u\ \exists v \cdot P[gu]v \wedge Puu$
(search: 0.12 sec merge: 0.72 sec translate: 2.57 sec total: 4.73 sec)

X2116: $\forall x_\iota\ \exists y_\iota[P_{o\iota}x \supset R_{o\iota\iota}x[g_{\iota\iota} \cdot h_{\iota\iota}y] \wedge Py] \wedge \forall w_\iota[Pw \supset P[gw] \wedge$
$P \cdot hw] \supset \forall x \cdot Px \supset \exists y \cdot Rxy \wedge Py$
(search: 0.42 sec total: 0.93 sec)

X2129: $\exists x\, \forall y[Px \equiv Py] \equiv [\, \exists x Qx \equiv\ \forall y Py] \equiv \cdot\, \exists x\, \forall y[Qx \equiv Qy]$
$\equiv \cdot\, \exists x Px \equiv\ \forall y Qy$
(search: 0.14 sec merge: 44.05 sec translate: 70.87 sec print: 5.67 sec total: 122.48 sec)

This was presented as a challenge problem by Andrews at the Fourth Workshop on Automated Deduction in 1979. Other researchers (see references cited in [34] and [49]) have found ways to deal with this problem, often by making reductions of it in the preprocessing stage. **TPS** found a refutation for this problem using path-focused duplication. The natural deduction proof is 584 lines long.

X5200: $x \cup y = \bigcup \cdot \lambda v \cdot v = x \vee v = y$
(search: 13.45 sec total: 16.62 sec)

X5205: $\# f_{\alpha\beta}[\bigcap w_{o(o\beta)}] \subseteq \bigcap \cdot \#[\# f] w$
(search: 4.44 min total: 6.31 min)

X5304: $\sim \exists g_{o\alpha\alpha}\, \forall f_{o\alpha}\, \exists j_{\alpha} \cdot gj = f$
(search: 0.09 sec total: 0.43 sec)
This is the Simple Cantor Theorem for Sets, which **TPS1** could prove [6]. It is stated here for comparison with X5305 below, which is harder to prove.

X5305: $\forall s_{o\alpha} \cdot \sim\ \exists g_{o\alpha\alpha}\, \forall f_{o\alpha} \cdot f \subseteq s \supset\ \exists j_{\alpha} \cdot sj \wedge gj = f$
(search: 0.37 sec total: 1.42 sec)
We call this the General Cantor Theorem for Sets. It says that there is no mapping g from a set s onto its power set.

X5308: $\exists j_{\beta(o\beta)}\, \forall p_{o\beta}[\, \exists x_{\beta}\ p\ x \supset p \cdot j\ p] \supset \cdot\ \forall x_{\alpha}\, \exists y_{\beta}\ r_{o\beta\alpha}\ x\ y \equiv$
$\exists f_{\beta\alpha}\, \forall x\ r\ x \cdot fx$
(search: 0.31 sec total: 1.21 sec)
The Axiom of Choice (for type β) is $[\, \exists j_{\beta(o\beta)}\, \forall p_{o\beta} \cdot\ \exists x_{\beta}\ p\ x \supset p \cdot j\ p]$; it asserts that there is a choice function $j_{\beta(o\beta)}$ that chooses an element $j_{\beta(o\beta)}p_{o\beta}$ from every nonempty set $p_{o\beta}$. X5308 shows a consequence of this axiom.

X5310: $\forall r_{o\beta(o\beta)}[\forall x_{o\beta}\, \exists y_{\beta}\ r\ x\ y \supset \exists f_{\beta(o\beta)}\, \forall x\ r\ x \cdot fx] \supset \exists j_{\beta(o\beta)}$
$\forall p_{o\beta} \cdot\ \exists x_{\beta}\ p\ x \supset p \cdot j\ p$
(search: 21.7 min total: 21.8 min)
X5310 implies the converse of X5308 (suitably generalized) when α is $(o\beta)$. The proof of this theorem was discussed in Section 4.

7. Conclusion

While **TPS** is still under development, it already provides a rich environment for exploring the complexities of theorem proving in higher-order logic. Experience with **TPS** leads to the following conclusions:

- In many contexts, the use of higher-order logic is very advantageous.
- When verifying theorems that are too hard to prove completely automatically, it is extremely valuable to be able to work in a mixture of automatic and interactive modes, whereby the user builds up the outline of the proof in natural deduction style and inserts lemmas into the proof, and calls on the automatic procedures to fill gaps in the proof.
- Searching for expansion proofs and then translating these into natural deduction proofs is a good way of constructing natural deduction proofs automatically. Expansion proofs provide a good context for search, as well as for the development of new ideas related to higher-order theorem proving.
- The use of primitive substitutions and gensubs is an effective mechanism for generating substitution terms for higher-order variables. We arbitrarily guess that certain connectives and quantifiers may be needed in these terms, and the details of the structure of the terms are determined by the search process and the unification algorithm.

There is much to be done in the development of methods for higher-order theorem proving and the improvement of **TPS**. Some major areas where work is needed are

- the basic mechanisms of searching for matings;
- the efficiency of higher-order unification;
- the treatment of equality;
- the introduction of rewrite rules;
- methods of instantiating quantifiers on higher-order variables;
- methods of deciding when to instantiate definitions;
- improved methods of transforming proofs from one format to another, and improving the style of the proofs that are obtained.

With the aid of **TPS**, one can think in a concrete way about these problems.

References

1. Andrews, P. B.: Resolution in type theory, *J. Symbolic Logic* **36** (1971), 414–342.
2. Andrews, P. B.: Provability in elementary type theory, *Z. Math. Logic und Grundlagen der Mathematik* **20** (1974), 411–418.
3. Andrews, P. B.: Refutations by matings, *IEEE Trans. Computers* **C-25** (1976), 801–807.
4. Andrews, P. B.: Transforming matings into natural deduction proofs, in W. Bibel and R. Kowalski (eds), *5th Conf. Automated Deduction, Les Arcs, France*, Lecture Notes in Computer Sci. 87, Springer-Verlag, 1980, pp. 281–292.
5. Andrews, P. B.: Theorem proving via general matings, *J. ACM* **28** (1981), 193–214.

6. Andrews, P. B., Miller, D. A., Cohen, E. L., and Pfenning, F.: Automating higher-order logic, in W. W. Bledsoe and D. W. Loveland (eds), *Automated Theorem Proving: After 25 Years*, Contemporary Mathematics Series, Vol. 29, Amer. Math. Soc., 1984, pp. 169–192.
7. Andrews, P. B.: *An Introduction to Mathematical Logic and Type Theory: To Truth through Proof*, Academic Press, New York, 1986.
8. Andrews, P. B.: On connections and higher-order logic, *J. Automated Reasoning* **5** (1989), 257–291.
9. Andrews, P. B., Issar, S., Nesmith, D., Pfenning, F., Xi, H., and Bishop, M.: *TPS3 Facilities Guide for Programmers and Users*, 1994.
10. Andrews, P. B., Issar, S., Nesmith, D., Pfenning, F., Xi, H., and Bishop, M.: *TPS3 Facilities Guide for Users*, 1994.
11. Bailin, S. C. and Barker-Plummer, D.: Z-match: An inference rule for incrementally elaborating set instantiations, *J. Automated Reasoning* **11** (1993), 391–428. Errata: JAR **12** (1994), 411–412.
12. Barker-Plummer, D.: Gazing: An approach to the problem of definition and lemma use, *J. Automated Reasoning* **8** (1992), 311–344.
13. Bibel, W.: *Automated Theorem Proving*, Vieweg, Braunschweig, 1987.
14. Bishop, M., Nesmith, D., Pfenning, F., Issar, S., Andrews, P. B., and Xi, H.: *TPS3 Programmer's Guide*, 1994.
15. Bledsoe, W. W. and Bruell, P.: A man-machine theorem-proving system, *Artificial Intelligence* **5** (1974), 51–72.
16. Bledsoe, W. W.: A maximal method for set variables in automatic theorem proving, in J. E. Hayes, D. Michie, and L. I. Mikulich (eds), *Machine Intelligence*, Vol. 9, Ellis Harwood Ltd., Chichester, and Wiley, New York, 1979, pp. 53–100.
17. Bledsoe, W. W. and Feng, G.: Set-var, *J. Automated Reasoning* **11** (1993), 293–314.
18. Boyer, R., Lusk, E., McCune, W., Overbeek, R., Stickel, M., and Wos, L.: Set theory in first-order logic: Clauses for Gödel's axioms, *J. Automated Reasoning* **2** (1986), 287–327.
19. Church, A.: *Introduction to Mathematical Logic*, Princeton University Press, Princeton, NJ, 1956.
20. Church, A.: A formulation of the simple theory of types, *J. Symbolic Logic* **5** (1940), 56–68.
21. Constable, R. L. et al.: *Implementing Mathematics with the Nuprl Proof Development System*, Prentice Hall, 1986.
22. Coquand, T. and Huet, G.: The calculus of constructions, *Information and Computation* **76** (1988), 95–120.
23. Farmer, W. M., Guttman, J. D., and Thayer, J.: IMPS: An interactive mathematical proof system, *J. Automated Reasoning* **11** (1993), 213–248.
24. Gallier, J. H., Narendran, P., Raatz, S., and Snyder, W.: Theorem proving using equational matings and rigid E-unification, *J. ACM* **39** (1992), 377–429.
25. Goldson, D. and Reeves, S.: Using programs to teach logic to computer scientists, *Notices Amer. Math. Soc.* **40** (1993), 143–148.
26. Goldson, D., Reeves, S., and Bornat, R.: A review of several programs for the teaching of logic, *Computer J.* **36** (1993), 373–386.
27. Gordon, M. J. C., Milner, A. J., and Wadsworth, C. P.: Edinburgh LCF, *Lecture Notes Computer Sci.* **78**, Springer-Verlag, 1979.
28. Gordon, M. J. C.: HOL: A proof generating system for higher-order logic, in G. Birtwistle and P. A. Subrahmanyam (eds), *VLSI Specification, Verification, and Synthesis*, Kluwer, Dordrecht, 1988, pp. 73–128.
29. Gordon, M. J. C. and Melham, T. F.(eds): *Introduction to HOL: A Theorem-Proving Environment for Higher-Order Logic*, Cambridge University Press, 1993.
30. Helmink, L. and Ahn, R.: Goal directed proof construction in type theory, in G. Huet and G. Plotkin (eds), *Logical Frameworks*, Cambridge University Press, 1991, pp. 120–148.
31. Huet, G. P.: A mechanization of type theory, in *Proc. 3rd Int. Joint Conf. Artificial Intelligence*, IJCAI, 1973, pp. 139–146.
32. Huet, G. P.: A unification algorithm for typed λ-calculus, *Theor. Comp. Sci.* **1** (1975), 27–57.

33. Issar, S.: Path-focused duplication: A search procedure for general matings, in *AAAI-90 Proc. 8th Nat. Conf. Artificial Intelligence*, AAAI Press/The MIT Press, 1990, pp. 221–226.
34. Issar, S.: Operational issues in automated theorem proving using matings, PhD Thesis, Carnegie Mellon University, 1991.
35. Issar, S., Andrews, P. B., Pfenning, F., and Nesmith, D.: *GRADER Manual*, 1992.
36. Kolodner, I. I.: Fixed points, *Amer. Math. Monthly* **71** (1964), 906.
37. Miller, D. A., Cohen, E. L., and Andrews, P. B.: A look at TPS, in D. W. Loveland (ed.), *6th Conf. Automated Deduction*, Lecture Notes in Computer Sci. 138, Springer-Verlag, 1982, pp. 50–69.
38. Miller, D. A.: Proofs in higher-order logic, PhD Thesis, Carnegie Mellon University, 1983.
39. Miller, D. A.: Expansion tree proofs and their conversion to natural deduction proofs, in R. E. Shostak (ed.), *7th Int. Conf. Automated Deduction*, Lecture Notes in Computer Sci. 170, Springer-Verlag, 1984, pp. 375–393.
40. Miller, D. A.: A compact representation of proofs, *Stud. Logica* **46** (1987), 347–370.
41. Milner, R.: A theory of type polymorphism in programming, *J. Computer System Sci.* **17** (1978), 348–375.
42. Nesmith, D., Bishop, M., Andrews, P. B., Issar, S., Pfenning, F., and Xi, H.: *TPS User's Manual*, 1994.
43. Ohlbach, H. J.: A resolution calculus for modal logics, PhD Thesis, Department of Computer Science, University of Kaiserslautern, 1988.
44. Paulson, L. C.: The foundation of a generic theorem prover, *J. Automated Reasoning* **5** (1989), 363–397.
45. Pfenning, F.: Analytic and non-analytic proofs, in R. E. Shostak (ed.), *7th Int. Conf. Automated Deduction*, Lecture Notes Computer Sci. 170, Springer-Verlag, 1984, pp. 394–413.
46. Pfenning, F.: Proof transformations in higher-order logic, PhD Thesis, Carnegie Mellon University, 1987.
47. Pfenning, F. and Nesmith, D.: Presenting intuitive deductions via symmetric simplification, in M. E. Stickel (ed.), *10th Int. Conf. Automated Deduction*, Lecture Notes in Artificial Intelligence 449, Springer-Verlag, 1990, pp. 336–350.
48. Pfenning, F., Issar, S., Nesmith, D., Andrews, P. B., Xi, H., and Bishop, M.: *ETPS User's Manual*, 1994.
49. Quaife, A.: Andrews' challenge problem revisited, *AAR Newslet.* **15** (1990), 3–7.
50. Robinson, J. A.: A machine-oriented logic based on the resolution principle, *J. ACM* **12** (1965), 23–41.
51. Russell, B.: *The Principles of Mathematics*, Cambridge University Press, 1903.
52. Russell, B.: Mathematical logic as based on the theory of types, *Amer. J. Math.* **30** (1908), 222–262. Reprinted in [55], pp. 150–182.
53. Snyder, W.: Higher-order E-unification, in M. E. Stickel (ed.), *10th Int. Conf. Automated Deduction, Kaiserslautern, FRG*, Lecture Notes in Artificial Intelligence 449, Springer-Verlag, 1990, pp. 573–587.
54. Sutcliffe, G., Suttner, Ch., and Yemenis, T.: The TPTP problem library, in A. Bundy (ed.), *Automated Deduction – CADE-12, 12th Int. Conf. Automated Deduction, Nancy, France*, Lecture Notes in Artificial Intelligence 814, Springer-Verlag, 1994, pp. 252–266.
55. van Heijenoort, J.: *From Frege to Gödel*, Harvard University Press, Cambridge, MA, 1967.
56. Wallen, L. A.: *Automated Deduction in Nonclassical Logics*, The MIT Press, 1990.
57. Whitehead, A. N. and Russell, B.: *Principia Mathematica*, 3 vols, Cambridge University Press, Cambridge, England, 1910–1913.
58. Wos, L.: The problem of definition expansion and contraction, *J. Automated Reasoning* **3** (1987), 433–435.

Exploring Properties of Normal Multimodal Logics in Simple Type Theory with LEO-II[1]

CHRISTOPH BENZMÜLLER AND LAWRENCE C. PAULSON

To Peter B. Andrews

1 Introduction

There are two well investigated approaches to automate reasoning in modal logics: the direct approach and the translational approach. The direct approach [6, 7, 14, 27] develops specific calculi and tools for the task; the translational approach [29, 30] transforms modal logic formulas into first-order logic and applies standard first-order tools. Embeddings of modal logics into higher-order logic, however, have not yet been widely studied, although multimodal logic can be regarded as a natural fragment of simple type theory. Gallin [15] appears to mention the idea first. He presents an embedding of modal logic into a 2-sorted type theory. This idea is picked up by Gamut [16] and a related embedding has recently been studied by Hardt and Smolka [17]. Carpenter [12] proposes to use lifted connectives, an idea that is also underlying the embeddings presented by Merz [26], Brown [11], Harrison [18, Chap. 20], and Kaminski and Smolka [22].

In this paper we pick up and extend the embedding of multimodal logics in simple type theory as studied by Brown [11]. The starting point is a characterization of multimodal logic formulas as particular λ-terms in simple type theory. A distinctive characteristic of the encoding is that the definiens of the $\Box_R$ operator λ-abstracts over the accessibility relation R. We illustrate that this supports the formulation of meta properties of encoded multimodal logics such as the correspondence between certain axioms and properties of the accessibility relation R. We show that some of these meta properties can even be efficiently automated within our higher-order theorem prover LEO-II [9] via cooperation with the first-order automated

[1]This work was supported by EPSRC grant EP/D070511/1 (LEO II: An Effective Higher-Order Theorem Prover) at Cambridge University, UK.

theorem prover E [32]. We also discuss some challenges to higher-order reasoning implied by this application direction. Moreover, we extend the presented embedding to first-order and higher-order quantified multimodal logics.

2 Simple Type Theory

Classical higher-order logic or simple type theory [4, 13] is a formalism built on top of the simply typed λ-calculus. The set $\mathcal{T}$ of simple types is usually freely generated from a set of basic types $\mathcal{BT} = \{o, \iota\}$ using the function type constructor $\rightarrow$. Here we allow an arbitrary but fixed number of additional base types to be specified.

For formulae, we start with a set of typed variables $X_\alpha, Y, Z, X^1_\beta, X^2_\gamma \ldots$ and a set of typed constants $c_\alpha, f_{\alpha\rightarrow\beta}, \ldots$. The set of constants includes the primitive logical connectives $\neg_{o\rightarrow o}$, $\vee_{o\rightarrow o\rightarrow o}$ and $\Pi_{(\alpha\rightarrow o)\rightarrow o}$ (abbreviated Π^α) and $=_{\alpha\rightarrow\alpha\rightarrow o}$ (abbreviated $=^\alpha$) for each type α.[1] Other logical connectives can be defined in terms of the these primitive ones in the usual way. In the remainder of this paper, we assume that any signature Σ we consider at least contains the primitive logical connectives $\neg_{o\rightarrow o}$, $\vee_{o\rightarrow o\rightarrow o}$ and $\Pi_{(\alpha\rightarrow o)\rightarrow o}$.

Terms and formulae are constructed from typed variables and constants using application and λ-abstraction. We use Church's dot notation so that $\centerdot$ stands for an implicit left bracket whose mate is as far to the right as possible (consistent with explicit brackets). We use infix notation $A \vee B$ for $((\vee A)B)$ and binder notation $\forall X_\alpha \centerdot A$ for $(\Pi^\alpha(\lambda X_\alpha \centerdot A_o))$.

Standard and Henkin semantics of simple type theory is well understood and thoroughly documented in the literature [1, 2, 8, 19].

One particular challenge for the complete automation of higher-order logic is *primitive substitution* [3] or splitting [20]. Primitive substitutions blindly guess some logical structure for free predicate or set variables in a clause that cannot be synthesized otherwise, and they introduce new free variables in order to delay some further decisions. The instantiation of the new and the remaining free variables is ideally supported by higher-order pre-unification. Generally, however, the primitive substitution process has to be iterated which leads to very challenging search space for clause sets containing many free variables. Some of the proof problems discussed in this paper require simple primitive substitutions.

[1] Note that there are infinitely many different Π^α and $=^\alpha$ introduced here .

3 Encoding Multimodal Logics in Simple Type Theory

Simple type theory is an expressive logic and it is thus no surprise that modal logic can be encoded in several ways in it. Harrison [18], for instance, presents a 'deep embedding' of modal logics by formalizing standard Kripke semantics and a 'shallow embedding' of the temporal logic LTL. The latter encoding more naturally exploits the expressiveness of higher-order logic. Harrison's shallow embedding is an instance of the encoding due to Brown [11]. Here we adapt and further extend Brown's suggestion and show that this approach is well suited for reasoning within and about modal logics.

The idea of the encoding is simple: Choose a base type — we choose ι — to denote the set of all possible worlds. Certain formulas of type $\iota \to o$ then correspond to multimodal logic expressions. The modal operators $\neg$, $\vee$, and $\Box_R$ become λ-terms of types $(\iota \to o) \to (\iota \to o)$, $(\iota \to o) \to (\iota \to o) \to (\iota \to o)$, and $(\iota \to \iota \to o) \to (\iota \to o) \to (\iota \to o)$ respectively. Note that $\neg$ forms the complement of a set of worlds, while $\vee$ forms the union of two such sets.

DEFINITION 1 (Propositional Multimodal Logic Λ_0^{mm}). Given a signature Σ, we define the the set Λ_0^{mm} of propositional multimodal logic propositions as follows.

1. We introduce the logical constants of Λ_0^{mm} as abbreviations for the following λ-terms:

$$\begin{aligned}
\neg_{(\iota\to o)\to(\iota\to o)} &= \lambda A_{\iota\to o}\text{.}\lambda X_\iota\text{.}\neg A\,X \\
\vee_{(\iota\to o)\to(\iota\to o)\to(\iota\to o)} &= \lambda A_{\iota\to o}\text{.}\lambda B_{\iota\to o}\text{.}\lambda X_\iota\text{.}A\,X \vee B\,X \\
\Box_{R\,(\iota\to\iota\to o)\to(\iota\to o)\to(\iota\to o)} &= \\
&\lambda R_{\iota\to\iota\to o}\text{.}\lambda A_{\iota\to o}\text{.}\lambda X_\iota\text{.}\forall Y_\iota\text{.}R\,X\,Y \Rightarrow A\,Y
\end{aligned}$$

2. We define the set of Λ_0^{mm}-propositions as the smallest set of simply typed λ-terms for which the following hold:

   - Each constant $p_{\iota\to o} \in \Sigma$ is an atomic Λ_0^{mm}-proposition.
   - If ϕ and ψ are Λ_0^{mm}-propositions, then so are $\neg\,\phi$, $\phi \vee \psi$ and $\Box_R\,\phi$, where $\neg$, $\vee$, and $\Box_R$ are defined as above and where R is a term of type $\iota \to \iota \to o$.

3. The propositional multimodal logic operators $\Rightarrow$, $\Leftrightarrow$, $\Diamond_r$, etc. can be defined in terms of $\neg$, $\vee$ and $\Box_R$ in the usual way.

Note that the encoding of the modal operator $\Box_R$ depends explicitly on an accessibility relation R of type $\iota \to \iota \to o$ given as its first argument. Hence, we basically introduce a generic framework for modeling multimodal logics. This idea is where Brown [11] differs from the LTL encoding of Harrison. The latter chooses the interpreted type num of numerals and then uses the predefined relation $\leq$ over numerals as a fixed accessibility relation in the definitions of $\Box$ and $\Diamond$.

By making the dependency of $\Box_R$ and $\Diamond_R$ on the accessibility relation R explicit, we can formalize and automatically prove some properties of multimodal logics in simple type theory, as we will later illustrate.

Given a signature Σ containing some constants $a_{\iota\to o}, b_{\iota\to o}$ (basic modal propositions) and $r_{\iota\to\iota\to o}, s_{\iota\to\iota\to o}$ (accessibility relation constants) we can formulate statements in Λ_0^{mm}, such as these:

$$\Box_r \top \qquad \Box_r a \Rightarrow \Box_r a \qquad \Box_r a \Rightarrow \Box_s a \qquad \Box_s (\Box_r a \Rightarrow \Box_r a)$$

We assume that $\Box_r$ binds more strongly than the propositional connectives and, hence, $\Box_r a \Rightarrow \Box_r a$ stands for $(\Box_r a) \Rightarrow (\Box_r a)$.

Next, we define validity of modal logic expressions $A_{\iota\to o} \in \Lambda_0^{mm}$: formula A is valid iff for all possible worlds W_ι we have $W \in A$, that is, iff $A\,W$ holds.

DEFINITION 2 (Validity, Satisfiability, Falsifiable). We define the following notions as simply typed λ-terms:

$$\begin{aligned}
\text{valid} &:= \lambda A_{\iota\to o}.\forall W_\iota.A\,W \\
\text{satisfiable} &:= \lambda A_{\iota\to o}.\exists W_\iota.A\,W \\
\text{falsifiable} &:= \lambda A_{\iota\to o}.\exists W_\iota.\neg A\,W \\
\text{unsatisfiable} &:= \lambda A_{\iota\to o}.\forall W_\iota.\neg A\,W
\end{aligned}$$

4 Solving Simple Λ_0^{mm}-problems with LEO-II

We can now exploit this framework to automatically analyze modal logic formulas within a higher-order theorem prover such as our LEO-II. Table 1 presents performance results for LEO-II on some example problems for the multimodal logic K. Since LEO-II cooperates with a first-order theorem prover, we actually present two times in the table. The first one is the total reasoning time used by both cooperating systems. The second time is that used by LEO-II alone. (Hence their difference is the time spent in the first-order theorem prover.) The times are given in seconds. All experiments reported in this paper were conducted with version v0.9 of LEO-II on a notebook computer with a Intel Pentium 1.60GHz processor with 1GB memory running Linux.

K^c cannot be solved by LEO-II, which is fortunate: this statement is clearly not valid without imposing restrictions on r and s. The other state-

Table 1. Runtimes for Proving Modal Logic Theorems

	problem	LEO-II+E / LEO-II
K^a	$\text{valid}(\Box_r \top)$	0.024/0.008
K^b	$\text{valid}(\Box_r a \Rightarrow \Box_r a)$	0.032/0.012
K^c	$\text{valid}(\Box_r a \Rightarrow \Box_s a)$	–
K^d	$\text{valid}(\Box_s (\Box_r a \Rightarrow \Box_r a))$	0.033/0.016
K^e	$\text{valid}(\Box_r (a \wedge b) \Leftrightarrow (\Box_r a \wedge \Box_r b))$	0.051/0.020
K^f	$\text{valid}(\Diamond_r (a \Rightarrow b) \Rightarrow \Box_r a \Rightarrow \Diamond_r b)$	0.038/0.016
K^g	$\text{valid}(\neg \Diamond_r a \Rightarrow \Box_r (a \Rightarrow b))$	0.031/0.012
K^h	$\text{valid}(\Box_r b \Rightarrow \Box_r (a \Rightarrow b))$	0.032/0.012
K^i	$\text{valid}((\Diamond_r a \Rightarrow \Box_r b) \Rightarrow \Box_r (a \Rightarrow b)$	0.032/0.012
K^j	$\text{valid}((\Diamond_r a \Rightarrow \Box_r b) \Rightarrow (\Box_r a \Rightarrow \Box_r b))$	0.035/0.016
K^k	$\text{valid}((\Diamond_r a \Rightarrow \Box_r b) \Rightarrow (\Diamond_r a \Rightarrow \Diamond_r b))$	0.035/0.016

ments expand into trivially refutable problems and they are quickly solved by the LEO-II+E cooperation.

We explain how the cooperation between LEO-II and E solves problem K^d. When the problem is initially input to LEO-II, the prover creates the following clause:

$$\mathcal{C}_1 : \quad [\text{valid}(\Box_s (\Box_r a \Rightarrow \Box_r a))]^F$$

This clause consists of one negated literal, as indicated by the superscript F. The atom of the literal, enclosed in square brackets, consists of the original problem statement. LEO-II unfolds the definitions and thereby rewrites this clause into

$$\begin{aligned}\mathcal{C}_2 : \quad & [\forall X^0_\iota . \forall X^1_\iota . : \neg(s\, X^0\, X^1) \\ & \vee ((\neg(\forall X^2_\iota . \neg(r\, X^1\, X^2) \vee (a\, X^2))) \vee (\forall X^2_\iota . \neg(r\, X^1\, X^2) \vee (a\, X^2)))]^F.\end{aligned}$$

This clause is obviously not in normal form yet. Clause normalization is the next step performed in LEO-II. This produces the following four normal form clauses, where the sk^j are Skolem constants and the V^1 is a variable:

$$\begin{aligned}\mathcal{C}_3 : \quad & [s\, sk^1\, sk^2]^T \\ \mathcal{C}_4 : \quad & [a\, sk^3]^F \\ \mathcal{C}_5 : \quad & [r\, sk^2\, sk^3]^T \\ \mathcal{C}_6 : \quad & [a\, V^1]^T \vee [r\, sk^2\, V^1]^F\end{aligned}$$

Subsequent to these pre-processing steps, LEO-II calls the first-order theorem prover E. The translation from higher-order to first-order logic is based on the ideas of Kerber [23].[2] This translation introduces a family of application operators @ and encodes type information into their names. For example, the operator $@_{(\beta\alpha)\_\beta}$ encodes the information that its first argument has function type $\beta \rightarrow \alpha$ and its second argument type β. Each clause in the domain of the translation is passed after executing the transformation to the prover E. In our example, four first-order clauses are generated and passed to E:

$$
\begin{aligned}
\mathcal{C}_{3'} :&\quad @_{(io)\_i}(@_{(i(io))\_i}(s, sk^1), sk^2)\\
\mathcal{C}_{4'} :&\quad @_{(\iota o)\_\iota}(a, sk^3)\\
\mathcal{C}_{5'} :&\quad @_{(io)\_i}(@_{(i(io))\_i}(r, sk^2), sk^3)\\
\mathcal{C}_{6'} :&\quad @_{(\iota o)\_\iota}(a, V^1) \vee (r, sk^2), V^1)
\end{aligned}
$$

E immediately finds a refutation for this trivial set of clauses and signals success back to LEO-II. This in turn triggers LEO-II to stop its proof search and to report success to the user. The proof protocol generated by LEO-II for this problem is given in Appendix A.

LEO-II alone, not cooperating with E, would also quickly refute this trivial set of clauses, but here E signals success before LEO-II even starts its own main proof loop. Generally the subset of clauses in LEO-II's search space that can be passed to E usually becomes much bigger and these are the cases where the cooperation between LEO-II and the first-order prover really pays off: the first-order specialist prover may quickly find the refutation while LEO-II alone gets stuck in its much more challenging search space. In the refutation proof of problem K^e, for example, LEO-II passes 24 first-order clauses to E, which then generates 322 further clauses before signaling that a refutation has been found. While the search space here is still trivial for both, LEO-II and E, we will see examples in Section 5 where E generates more than 20000 clauses before it finds a refutation for the subproblem passed to it by LEO-II.

More information of the design of LEO-II is provided in Benzmüller et al. [9]. In general terms, the first-order specialist provers support LEO-II by periodically trying to detect subsets of clauses in LEO-II's search space that can be quickly refuted by them.

This simple idea can be further generalized

(i) by realizing analogous cooperations with specialist reasoners working for other interesting and efficiently mechanizable fragments of higher-

[2] Further translations have been discussed in the literature; see for example [21, 24, 25].

order logic such as propositional logic, guarded fragment, monadic second-order logic, etc.,

(ii) by allowing LEO-II and these reasoners to run in parallel, and

(iii) by supporting alternative translations to first-order logic and other fragments of higher-order logic.

In fact, LEO-II already provides a link to the first-order theorem prover SPASS as an alternative to E. And it also already supports an alternative translation to first-order logic. In order to achieve the distribution goals we plan to adapt the OANTS system [10] to LEO-II.

We could now go on to study increasingly complex examples of the kind as presented in the table above and they would clearly provide nice exercises for LEO-II+E. However, we do not expect that LEO-II+E can compete with specialist modal logic provers on challenge problems. In order to gain a highly efficient mechanization of challenge problems, we may therefore want to develop a cooperation with fast modal logic provers as well and pass the problems before expansion of the definitions from LEO-II directly to them. Verification of their results within LEO-II could then be tackled afterwards. Verification of the cooperative proofs is an open issue also in our current translational approach since the refutations of the prover E cannot yet be translated back into refutations in LEO-II. This is also on our list of future work.

In the next Section will turn our focus to a more interesting question: using LEO-II to reason *about* some properties of different modal logics, rather than to reason *within* such modal logics.

5 Exploring Properties of Λ_0^{mm} with LEO-II

We now study study questions such as, *'What is the concrete modal logic we have introduced?'* First, we turn our attention to the weakest modal logic, namely K. The essential properties of K are the necessitation rule N and the distribution axiom D (for all accessibility relations R and modal propositions A, B):

N	*If A is a theorem of K, then so is $\Box_R A$*
D	$\Box_R (A \Rightarrow B) \Rightarrow (\Box_R A \Rightarrow \Box_R B)$

Our modal logic definitions in fact entail these principles and LEO-II can easily prove them: We formalize problem K_N, expressing that N is entailed by our definitions so far, as

$$\forall R \boldsymbol{.} \forall A \boldsymbol{.} \text{valid}(A) \Rightarrow \text{valid}(\Box_R A)$$

LEO-II+E can prove it easily in 0.027 seconds. Problem K_D, expressing that D is valid, is formalized as

$$\forall R \mathbf{.} \forall A \mathbf{.} \forall B \mathbf{.} \text{valid}(\Box_R (A \Rightarrow B) \Rightarrow (\Box_R A \Rightarrow \Box_R B))$$

LEO-II+E proves it in 0.029 seconds.

This provides some evidence that we have indeed correctly modeled the modal logic K in simple type theory. So let us call the system we developed so far $\boldsymbol{K}$. Modal logic $S4$ is obtained from modal logic K by adding the axioms T and 4. These axioms are well known to correspond to reflexivity and transitivity of the accessibility relation R.

$$\begin{array}{lll} T & \Box_R A \Rightarrow A & \textit{corresponds to}\text{: refl}(R) \\ 4 & \Box_R A \Rightarrow \Box_R \Box_R A & \textit{corresponds to}\text{: trans}(R) \end{array}$$

Reflexivity and transitivity are defined in the obvious manner:

$$\begin{aligned} \text{refl} &:= \lambda R \mathbf{.} \forall X \mathbf{.} R\, X\, X \\ \text{trans} &:= \lambda R \mathbf{.} \forall X \mathbf{.} \forall Y \mathbf{.} \forall Z \mathbf{.} R\, X\, Y \wedge R\, Y\, Z \Rightarrow R\, X\, Z. \end{aligned}$$

We will later also use irreflexivity and symmetry:

$$\begin{aligned} \text{irrefl} &:= \lambda R \mathbf{.} \forall X \mathbf{.} \neg (R\, X\, X) \\ \text{sym} &:= \lambda R \mathbf{.} \forall X \mathbf{.} \forall Y \mathbf{.} R\, X\, Y \Rightarrow R\, Y\, X. \end{aligned}$$

We now study the following obvious questions *about* our system $\boldsymbol{K}$ within LEO-II:

K_T^a: *Is axiom T is valid in $\boldsymbol{K}$?* As expected, LEO-II cannot prove

$$\forall R \mathbf{.} \forall A \mathbf{.} \text{valid}(\Box_R A \Rightarrow A).$$

K_T^b: *Is there a relation R such that for all modal propositions A, axiom T is valid in $\boldsymbol{K}$?* LEO-II+E can prove

$$\exists R \mathbf{.} \forall A \mathbf{.} \text{valid}(\Box_R A \Rightarrow A)$$

in 0.456 seconds; the prover E is unsuccessfully called three times in LEO-II's reasoning loop before the refutation is finally detected within LEO-II itself. A clever instantiation for relation R is actually needed to solve this problem (as we know from our undergraduate modal logic course, R obviously needs to be reflexive) and this instantiation can neither be synthesized by E's first-order unification nor by

LEO-II's higher-order pre-unification. In fact, such an instantiation needs to be guessed by primitive substitution. LEO-II applies primitive substitution after the initial pre-processing phase and it proposes, amongst others, to consider the equality relation as a candidate instantiation for R. Then, LEO-II quickly finds the refutation. To be more precise, LEO-II does not instantiate R with $\lambda X_\iota . \lambda Y_\iota . X = Y$ in the crucial primitive substitution step but with the more general term $\lambda X_\iota . \lambda Y_\iota . (V\ X\ Y) = (W\ X\ Y)$ where V and W are new free variables. By instantiating V and W with projections on the first and second argument the former term could be introduced. However, as the proof protocol generated by LEO-II for this example problem in the Appendix B illustrates, the second, more general term already leads to a refutation (ending in pre-unification constraint consisting only in a flex-flex unification pair) without further instantiation of V or W.

For the exploration of the properties of modal logic $\boldsymbol{K}$we have thus gained useful information: by considering the equality relation as an accessibility relation, which amongst other properties is reflexive, we are able to prove this statement. Thus, one way to proceed with the exploration is to investigate the connection between reflexivity and T in $\boldsymbol{K}$.

K_T^c: *Is axiom T indeed equivalent to reflexivity of R in* $\boldsymbol{K}$*?* LEO-II+E takes 0.068 seconds to prove

$$\forall R . (\forall A . \text{valid}(\Box_R A \Rightarrow A) \Leftrightarrow \text{refl}(R)).$$

K_4^a: *Is axiom 4 valid in* $\boldsymbol{K}$*?* As expected, LEO-II+E cannot prove

$$\forall R . \forall A . \text{valid}(\Box_R A \Rightarrow \Box_R \Box_R A).$$

K_4^b: *Is there a relation R and a modal proposition A for which axiom 4 is valid in* $\boldsymbol{K}$*?* LEO-II+E takes 0.055 seconds to prove

$$\exists R . \forall A . \text{valid}(\Box_R A \Rightarrow \Box_R \Box_R A).$$

Interestingly, the relation generated for R is $\lambda X_\iota . \lambda Y_\iota . (V\ X\ Y) \neq (W\ X\ Y)$ where V and W are new free variables; see the proof protocol generated by LEO-II in the Appendix C. A simpler relation would be the empty relation $\lambda X_\iota . \lambda Y_\iota . X \neq X$ which is obviously an instance of the above one.[3] We proceed with the exploration. Let us be naive this time and consider irreflexivity and symmetry as interesting properties first before investigating transitivity.

K_4^c: *Is axiom 4 equivalent to irreflexivity?* LEO-II+E cannot prove

$$\forall R\mathbin{.}(\forall A\mathbin{.}\text{valid}(\Box_R A \Rightarrow \Box_R \Box_R A)) \Leftrightarrow \text{irrefl}(R).$$

K_4^d: *Is axiom 4 equivalent to symmetry?* LEO-II+E cannot prove

$$\forall R\mathbin{.}(\forall A\mathbin{.}\text{valid}(\Box_R A \Rightarrow \Box_R \Box_R A)) \Leftrightarrow \text{sym}(R).$$

K_4^e: *Is axiom 4 equivalent to transitivity of R in **K**?* LEO-II+E takes 0.193 seconds to prove

$$\forall R\mathbin{.}(\forall A\mathbin{.}\text{valid}(\Box_R A \Rightarrow \Box_R \Box_R A)) \Leftrightarrow \text{trans}(R),$$

K_{T4}^a: *Are axioms T and 4 equivalent to reflexivity and transitivity of R in **K**?* LEO-II+E takes 2.329 seconds (of which LEO-II uses 0.164 seconds) to prove

$$\forall R\mathbin{.}(\forall A\mathbin{.}\text{valid}(\Box_R A \Rightarrow A) \wedge \text{valid}(\Box_R A \Rightarrow \Box_R \Box_R A)) \\ \Leftrightarrow (\text{refl}(R) \wedge \text{trans}(R))$$

E receives 70 clauses and generates 21769 before it finds the refutation. This example well illustrates the benefits of the cooperation between LEO-II and E, since the first-order refutation required in this example is already too challenging to be easily detected by LEO-II alone in its much more challenging search space.

It is much easier, however, to prove the two directions separately. LEO-II+E takes 0.040 seconds to prove

$$\forall R\mathbin{.}(\forall A\mathbin{.}\text{valid}(\Box_R A \Rightarrow A) \wedge \text{valid}(\Box_R A \Rightarrow \Box_R \Box_R A)) \\ \Rightarrow (\text{refl}(R) \wedge \text{trans}(R))$$

[3] The proof protocol in the Appendix C shows that from the initial problem statement LEO-II derives the clause 27 : $[V^1\ (sk^3 V^1)\ (sk^5 V^1)]^T$. It then instantiates V^1 with the term $t = \lambda V^9\mathbin{.}\lambda V^{10}\mathbin{.}((\lambda X\mathbin{.}\lambda Y\mathbin{.}\neg(X = Y))\ (V^{11}\ V^9\ V^{10})\ (V^{12}\ V^9\ V^{10}))$ by primitive substitution to obtain the clause 39 : $[\neg(V^{11}\ (sk^3\ t)\ (sk^5\ t)) = (V^{12}\ (sk^3\ t)\ (sk^5\ t))]^T$ and subsequently it generates the flex-flex unification constraint clause 59 : $[(V^{11}\ (sk^3\ t)\ (sk^5\ t)) = (V^{12}\ (sk^3\ t)\ (sk^5\ t))]^F$ by clause normalization. Flex-flex unification constraint clauses can always be refuted and thus proof search in LEO-II terminates here. For example, if we would instantiate V^{11} and V^{12} both with a term $\lambda X\mathbin{.}\lambda Y\mathbin{.}a$ for an arbitrary constant a then we would generate the the clause $[a = a]^F$ in which case the contradiction becomes obvious.

and 0.039 seconds for

$$\forall R\,.\,(\forall A\,.\,\text{valid}(\Box_R\,A \Rightarrow A) \wedge \text{valid}(\Box_R\,A \Rightarrow \Box_R\,\Box_R\,A))$$
$$\Leftarrow (\text{refl}(R) \wedge \text{trans}(R)).$$

Thus, we have successfully explored the properties of modal logic $S4$ with LEO-II+E. And we could go on to explore properties of other more specialized modal logics and multi modal logics in exactly the same way.

6 Quantified Multimodal Logics Λ_1^{mm} and Λ_ω^{mm}

We adapt the definition of propositional multimodal logic to the first-order and higher-order case.

DEFINITION 3 (First-order Quantified Multimodal Logic Λ_1^{mm}). In addition to base type ι we introduce a second base type, μ. The idea is that ι is reserved to denote the set of all possible worlds while μ denotes the set of individuals. Let Σ be a signature.

1. Λ_1^{mm}-terms are defined as the smallest set of simply typed λ-terms for which the following hold:

   - Each constant $c_\mu \in \Sigma$ and variable $X_\mu \in \Sigma$ is a Λ_1^{mm}-term.
   - If $t_\mu^1, \ldots, t_\mu^n$ are Λ_1^{mm}-terms and $f_{\mu\to\ldots\to\mu\to\mu} \in \Sigma$ is an n-ary (curried) function symbol, then $(\ldots(f\,t^1)\ldots t^n)_\mu$ is a Λ_1^{mm}-term.

   Note that Λ_1^{mm}-terms must not depend on worlds, that is, subterms of a Λ_1^{mm}-term are never of type ι.

2. The modal operators $\neg$, $\vee$, $\Box_R$ are defined as before.

3. Modal universal quantification $\forall X_\mu\,.\,\phi_{\iota\to o}$ is defined as the term

   $$\lambda w_\iota\,.\,\forall X_\mu\,.\,\phi\, w$$

   In fact we can employ the standard trick in simple type theory to avoid introducing a new binder for universal quantification. For this we introduce the logical constant $\mathbf{\Pi}_{\mu\to(\iota\to o)}$ and use this to encode modal universal quantification as follows: $\forall X_\mu\,.\,\phi_{\iota\to o}$ stands for

   $$\mathbf{\Pi}_{\mu\to(\iota\to o)}(\lambda X_\mu\,.\,\phi_{\iota\to o})$$

   and the modal operator $\mathbf{\Pi}$ is itself defined as

   $$\lambda\phi'_{\mu\to(\iota\to o)}\,.\,\lambda W_\iota\,.\,\forall X_\mu\,.\,\phi'\,X\,W.$$

4. Λ_1^{mm}-propositions are defined as the smallest set of simply typed terms for which the following hold:

   - If $t_\mu^1, \ldots, t_\mu^n$ are Λ_1^{mm}-terms and $p_{\mu\to\ldots\to\mu\to(\iota\to o)} \in \Sigma$ is an n-ary (curried) predicate symbol (for $n \geq 0$), then $(\ldots(p\,t^1)\ldots t^n)_{\iota\to o}$ is an atomic Λ_1^{mm}-proposition.
   - If ϕ and ψ are Λ_1^{mm}-propositions, then so are $\neg\,\phi$, $\phi \vee \psi$ and $\Box_R\,\phi$, where R is a term of type $\iota \to \iota \to o$.
   - If $X_\mu \in \Sigma$ is a variable of type μ and $\phi_{\iota\to o}$ is a Λ_1^{mm}-proposition, then $\forall X \centerdot \phi$ is a Λ_1^{mm}-proposition.

First-order quantified multimodal logic has been studied in Nguyen [28], who defines a basic quantified normal multimodal logic called $K(m)$; like K in the propositional case, it serves as the starting point for the introduction of further quantified normal multimodal logics.

Here we briefly investigate whether Λ_1^{mm} fulfill the characteristics of the normal multimodal logic $K(m)$.

CLAIM 4 (Λ_1^{mm} is a normal multimodal logic). Λ_1^{mm} fulfill the properties of the normal multimodal logic $K(m)$ of [28]. That is:

1. Λ_1^{mm} validates the axioms for classical predicate logic.
2. Λ_1^{mm} validates the K-axioms: $\Box_R\,(\phi \Rightarrow \psi) \Rightarrow (\Box_R\,\phi \Rightarrow \Box_R\,\psi)$.
3. Λ_1^{mm} validates the Barcan formula axioms: $(\forall X \centerdot \Box_R\,\phi) \Rightarrow \Box_R \forall X \centerdot \phi$.
4. Λ_1^{mm} validates the axioms defining $\Diamond_R$: $\Diamond_R\,\phi \Leftrightarrow \neg\Box_R\,\neg\phi$.
5. Λ_1^{mm} validates the modus ponens rule: if ϕ and $\phi \Rightarrow \psi$ are valid, then ψ is valid.
6. Λ_1^{mm} validates the generalization rule: if ϕ is valid, then $\forall X \centerdot \phi$ is valid.
7. Λ_1^{mm} validates the modal generalization rule: if ϕ is valid, then $\Box_R\,\phi$ is valid.
8. $K(M)$ validates the converse Barcan formula $(\forall X \centerdot \Box_R\,\phi) \Leftarrow \Box_R \forall X \centerdot \phi$.

We do not give a formal proof of this claim here and instead argue as follows:

1. The axioms for classical predicate logic are obviously valid in simple type theory (and LEO-II can handle them).

2. LEO-II+E can prove $\forall R \centerdot \forall \phi \centerdot \forall \psi \centerdot \text{valid}(\Box_R\,(\phi \Rightarrow \psi) \Rightarrow (\Box_R\,\phi \Rightarrow \Box_R\,\psi)$ in 0.059 seconds.

3. LEO-II+E can prove $\forall R \centerdot \forall \phi \centerdot \text{valid}(\forall X \centerdot \Box_R\,\phi \Rightarrow \Box_R\,\forall X \centerdot \phi)$ in 0.066 seconds.

4. LEO-II+E can prove $\forall R \centerdot \forall \phi \centerdot \text{valid}(\Diamond_R\,\phi \Leftrightarrow \neg\,\Box_R\,\neg\phi)$ in 0.078 seconds.

5. LEO-II+E can prove $\forall \phi \centerdot \forall \psi \centerdot \text{valid}(\phi) \wedge \text{valid}(\phi \Rightarrow \psi) \Rightarrow \text{valid}(\psi)$ in 0.062 seconds.

6. LEO-II+E can prove

$$\forall P_{\mu\to(\iota\to o)} \centerdot (\forall X_\mu \centerdot \text{valid}(P\,X)) \Rightarrow \text{valid}(\forall X_\mu \centerdot (P\,X)$$

in 0.061 seconds.

7. LEO-II+E can prove $\forall R \centerdot \forall \phi \centerdot \text{valid}(\phi) \Rightarrow \text{valid}(\Box_R\,\phi)$ in 0.056 seconds.

8. LEO-II+E can prove $\Box_R\,\forall X \centerdot \phi \Rightarrow \forall X \centerdot \Box_R\,\phi$, the converse Barcan formula, in 0.049 seconds.

Assuming that LEO-II+E is correct, this provides evidence that Λ_1^{mm} indeed adequately models the basic normal multimodal logic $K(m)$ of Nguyen [28]. Validity of the Barcan formula and its converse illustrate that quantification in Λ_1^{mm} is a fixed-domain quantification.

The first-order quantified multimodal logic Λ_1^{mm} can be further generalized. Doing so we obtain the following proposal for a higher order quantified multimodal logic Λ_ω^{mm}, again with fixed-domain quantification.

DEFINITION 5 (Higher-order Quantified Multimodal Logic Λ_ω^{mm}). The key idea of this definition is to exclude statements about possible worlds from the language Λ_ω^{mm}. The rest is as before.

1. Given a simply typed λ-term t_γ we define the function *obt* (occurring base types) as follows:

   - $obt(t = c_\delta \in \Sigma) = obt'(\delta)$
   - $obt(t = (t^1\,t^2)) = obt(t^1) \cup obt(t^2)$
   - $obt(t = (\lambda X \centerdot t^1)) = obt(X) \cup obt(t^1)$
   - $obt'(\delta \in \mathcal{BT}) = \{\delta\}$
   - $obt'(\beta \to \delta) = obt'(\beta) \cup obt'(\delta)$

Let $t_{\iota\to o}$ be a simply typed λ-term. If $t_{\iota\to o}$ is a constant or variable, then $t_{\iota\to o}$ is a Λ_{ω}^{mm}-term. If $t_{\iota\to o}$ is an abstraction $\lambda X_\iota . p_o$, then $t_{\iota\to o}$ is a Λ_{ω}^{mm}-term if $\iota \notin obt(p_o)$. If t is an application $(q_{\alpha\to(\iota\to o)}\ s_\alpha)$, then $t_{\iota\to o}$ is a Λ_{ω}^{mm}-term if $\iota \notin obt(s_\alpha)$.

2. The modal operators $\neg$, $\vee$, $\Box_R$ are defined as before.

3. Modal universal quantification $\forall X_\gamma . \phi_{\iota\to o}$ is now defined follows: Let $\phi_{\iota\to o}$ be an Λ_{ω}^{mm}-term and let $X_\gamma \in \Sigma$ be a variable such that $\iota \notin obt(X_\gamma)$. Then $\forall X_\gamma . \phi$ stands for $\lambda w_\iota . \forall X_\gamma . \phi\, w$.

   Again we can employ the standard trick in simple type theory to avoid introducing a new binder for universal quantification. This time we introduce the logical constants $\mathbf{\Pi}_{\gamma\to(\iota\to o)}$ for all types γ such that $\iota \notin obt'(\gamma)$. We use them to encode modal universal quantification as follows: $\forall X_\gamma . \phi_{\iota\to o}$ stands for $\mathbf{\Pi}_{\gamma\to(\iota\to o)}(\lambda X_\gamma . \phi_{\iota\to o})$ and the modal operator $\mathbf{\Pi}$ is itself defined as

$$\lambda \phi'_{\gamma\to(\iota\to o)} . \lambda W_\iota . \forall X_\gamma . \phi'\, X\, W.$$

4. Λ_{ω}^{mm}-propositions are defined as the smallest set of simply typed λ-terms for which the following hold:

   - Each Λ_{ω}^{mm}-term $t_{\iota\to o}$ is an atomic Λ_{ω}^{mm}-proposition.
   - If ϕ and ψ are Λ_{ω}^{mm}-propositions, then so are $\neg\,\phi$, $\phi \vee \psi$ and $\Box_R\,\phi$.
   - If $X_\gamma \in \Sigma$ is a variable of type μ and $\phi_{\iota\to o}$ is a Λ_{ω}^{mm}-proposition, then $\forall X . \phi$ is a Λ_{ω}^{mm}-proposition.

We could now proceed with a systematic exploration of the properties of Λ_{1}^{mm} and Λ_{ω}^{mm}. This, however, will be future work.

7 Discussion

In this paper we explore an interesting and promising research direction: the embedding of multimodal logic in simple type theory, the development of reasoning tools such as LEO-II or TPS [5] to support reasoning in and about multimodal logics in simple type theory, and the systematic computer supported exploration of properties of multimodal logic in simple type theory.

Interesting future work will be to employ the presented encoding of normal multimodal logics in order to attack the $100 modal logic challenge[4]

[4] http://www.cs.miami.edu/~tptp/HHDC/

with LEO-II. This challenge, originally proposed by John Halleck, calls for a computer program, which given the formalizations of any two modal logics determines their relationship. The approach presented in this paper should generally be applicable to this challenge when restricting it to *normal* multimodal logics.

We illustrate the idea by an example. Consider the statement

$$\exists R.\exists A.\exists B.(\neg\text{valid}(\Box_R A \Rightarrow A)) \vee (\neg\text{valid}(\Box_R B \Rightarrow \Box_R \Box_R B))$$

LEO-II alone (without E) can prove this statement in 17.305 seconds.[5] Using primitive substitution, LEO-II instantiates R in this proof with the "not equals" relation, which is non-reflexive. We can conclude from this proof that modal logic $S4$ is not entailed by modal logic K ($S4 \not\subseteq K$).

The time of 17.305 seconds required by LEO-II (without any tuning) is clearly a massive improvement over the 16 minutes pre-processing time plus 2710 seconds solving time reported by Rabe et al. [31] for the same problem. They employ a specifically tuned system based on standard first-order theorem provers and standard translations of modal logics into first-order logic.

However, in order to solve the challenge for all normal multimodal logics, a specific tuning of LEO-II seems also unavoidable. The problem is related to primitive substitution. It is well illustrated by the following statement, which expresses that modal logic T (which adds only axiom T to K) is not entailed in K:

$$\exists R.\exists B.(\neg\text{valid}(\Box_R B \Rightarrow \Box_R \Box_R B))$$

We would expect that this statement can be quickly proved when instantiating R with a non-transitive relation similar to primitive substitution steps required in examples K_T^b and K_4^b. For good reasons LEO-II fails to do so. In fact, LEO-II also fails to prove the related statement

$$\exists R.\neg\text{trans}(R)$$

The reason is that without further assumptions this statement is not a theorem. We have neither assumed the axiom of infinity nor that there exist at least two different possible worlds. Hence, our domain of possible worlds may well just consist of a single world w in which case a non-transitive accessibility relation cannot be provided.

[5]The proof in LEO-II+E takes 198.818 seconds because of several unsuccessful but time consuming calls to E. We have already mentioned that the current sequential architecture of LEO-II+E needs to be replaced by a distributed in which LEO-II cooperates with the first-order prover in a distributed and incremental manner.

LEO-II can in fact prove the existence of a non-transitive accessibility relation under the additional assumption

$$\exists X . \exists Y . X \neq Y$$

Like the special purpose approach used by Rabe et al. [31], we could probably successfully tune LEO-II to provide and employ a domain specific "menu" of the interesting accessibility relations in connection with interesting constraints on the domain of possible worlds in order to attack the $100 modal logic challenge.

Acknowledgment:
We thank Chad Brown, Franz Baader, Gert Smolka, and Mark Kaminski for their valuable comments to earlier versions of this paper and John Harrison for reminding us of this interesting application area. We also thank Catalin Hritcu for proofreading.

BIBLIOGRAPHY

[1] Peter B. Andrews. General Models and Extensionality. *Journal of Symbolic Logic*, 37:395–397, 1972.
[2] Peter B. Andrews. General Models, Descriptions, and Choice in Type Theory. *Journal of Symbolic Logic*, 37:385–394, 1972.
[3] Peter B. Andrews. On Connections and Higher-Order Logic. *Journal of Automated Reasoning*, 5:257–291, 1989.
[4] Peter B. Andrews. *An Introduction to Mathematical Logic and Type Theory: To Truth Through Proof.* Kluwer Academic Publishers, second edition, 2002.
[5] Peter B. Andrews and Chad E. Brown. TPS: A hybrid automatic-interactive system for developing proofs. *J. Applied Logic*, 4(4):367–395, 2006.
[6] Philippe Balbiani, Luis Fariñas del Cerro, and Andreas Herzig. Declarative Semantics for Modal Logic Programs. In *FGCS*, pages 507–514, 1988.
[7] Matteo Baldoni, Laura Giordano, and Alberto Martelli. A Framework for a Modal Logic Programming. In *Joint International Conference and Symposium on Logic Programming*, pages 52–66, 1996.
[8] Christoph Benzmüller, Chad E. Brown, and Michael Kohlhase. Higher Order Semantics and Extensionality. *Journal of Symbolic Logic*, 69:1027–1088, 2004.
[9] Christoph Benzmüller, Larry Paulson, Frank Theiss, and Arnaud Fietzke. Progress Report on LEO-II – An Automatic Theorem Prover for Higher-Order Logic. In *Emerging Trends at the 20th International Conference on Theorem Proving in Higher Order Logics.* University Kaiserslautern, Germany, 2007.
[10] Christoph Benzmüller, Volker Sorge, Mateja Jamnik, and Manfred Kerber. Combined Reasoning by Automated Cooperation. *Journal of Applied Logic*, 2007. To appear.
[11] Chad E. Brown. Encoding Hybrid Logic in Higher-Order Logic. Unpublished slides from an invited talk presented at Loria Nancy, France, April 2005. `http://mathgate.info/cebrown/papers/hybrid-hol.pdf`.
[12] Bob Carpenter. *Type-logical semantics.* MIT Press, Cambridge, MA, USA, 1998.
[13] Alonzo Church. A Formulation of the Simple Theory of Types. *Journal of Symbolic Logic*, 5:56–68, 1940.

[14] Luis Fariñas del Cerro. MOLOG: A System That Extends PROLOG with Modal Logic. *New Generation Comput.*, 4(1):35–50, 1986.
[15] Daniel Gallin. *Intensional and Higher-Order Modal Logic*, volume 19 of *North-Holland Mathematics Studies*. North-Holland, Amsterdam, 1975.
[16] L. T. F. Gamut. *Logic, Language, and Meaning. Volume II. Intensional Logic and Logical Grammar*, volume 2. The University of Chicago Press, 1991.
[17] Moritz Hardt and Gert Smolka. Higher-Order Syntax and Saturation Algorithms for Hybrid Logic. *Electr. Notes Theor. Comput. Sci.*, 174(6):15–27, 2007.
[18] John Harrison. *HOL Light Tutorial (for version 2.20)*. Intel JF1-13, September 2006. http://www.cl.cam.ac.uk/~jrh13/hol-light/tutorial_220.pdf.
[19] Leon Henkin. Completeness in the Theory of Types. *Journal of Symbolic Logic*, 15:81–91, 1950.
[20] Gérard P. Huet. A Mechanization of Type Theory. In *Proceedings of the Third International Joint Conference on Artificial Intelligence*, pages 139–146, Stanford University, California, USA, 1973. IJCAI.
[21] J. Hurd. An LCF-Style Interface between HOL and First-Order Logic. In *Automated Deduction — CADE-18*, volume 2392 of *LNAI*, pages 134–138. Springer, 2002.
[22] Mark Kaminski and Gert Smolka. Hybrid Tableax for the Difference Modality, 2007. Accepted to Methods for Modalities 5 (M4M-5), Cachan, France.
[23] Manfred Kerber. *On the Representation of Mathematical Concepts and their Translation into First-Order Logic*. PhD thesis, Fachbereich Informatik, Universität Kaiserslautern, Kaiserslautern, Germany, 1992.
[24] Jia Meng and Lawrence C. Paulson. Experiments on Supporting Interactive Proof Using Resolution. In *In Proc. of IJCAR 2004*, volume 3097 of *LNCS*, pages 372–384. Springer, 2004.
[25] Jia Meng, Claire Quigley, and Lawrence C. Paulson. Automation for interactive proof: first prototype. *Inf. Comput.*, 204(10):1575–1596, 2006.
[26] Stephan Merz. Yet another encoding of TLA in Isabelle. Available on the Internet: http://www.loria.fr/~merz/projects/isabelle-tla/doc/design.ps.gz, 1999.
[27] Linh Anh Nguyen. A Fixpoint Semantics and an SLD-Resolution Calculus for Modal Logic Programs. *Fundam. Inform.*, 55(1):63–100, 2003.
[28] Linh Anh Nguyen. Multimodal logic programming. *Theor. Comput. Sci.*, 360(1-3):247–288, 2006.
[29] Andreas Nonnengart. How to Use Modalities and Sorts in Prolog. In Craig MacNish, David Pearce, and Luís Moniz Pereira, editors, *JELIA*, volume 838 of *Lecture Notes in Computer Science*, pages 365–378. Springer, 1994.
[30] Hans Jürgen Ohlbach. A Resolution Calculus for Modal Logics. In Ewing L. Lusk and Ross A. Overbeek, editors, *CADE*, volume 310 of *Lecture Notes in Computer Science*, pages 500–516. Springer, 1988.
[31] Florian Rabe, Petr Pudlak, Geoff Sutcliffe, and Weina Shen. Solving the $100 modal logic challenge. *Journal of Applied Logic*, 2007.
[32] Stephan Schulz. E – A Brainiac Theorem Prover. *Journal of AI Communications*, 15(2/3):111–126, 2002.

A LEO-II Proof Protocol for Example K^d

```
**** Protocol for Problem: K^d.thf
**** Beginning of derivation protocol ****
1: (mvalid @ ((mbox @ s) @ ((mimpl @ ((mbox @ r) @ a)) @
 ((mbox @ r) @ a))))=$true
 ---  theorem(file(K^d.thf,[thm]))
2: (mvalid @ ((mbox @ s) @ ((mimpl @ ((mbox @ r) @ a)) @
 ((mbox @ r) @ a))))=$false
 ---  neg_input 1
```

```
3: (! [x0:$i,x1:$i] : ((~ ((s @ x0) @ x1)) | ((~ (! [x2:$i] :
 ((~ ((r @ x1) @ x2)) | (a @ x2)))) | (! [x2:$i] :
 ((~ ((r @ x1) @ x2)) | (a @ x2))))))=$false
 --- unfold_def 2
4: (! [x1:$i] : ((~ ((s @ sk1) @ x1)) | ((~ (! [x2:$i] :
 ((~ ((r @ x1) @ x2)) | (a @ x2)))) | (! [x2:$i] :
 ((~ ((r @ x1) @ x2)) | (a @ x2))))))=$false
 --- cnf 3
5: ((~ ((s @ sk1) @ sk2)) | ((~ (! [x2:$i] :
 ((~ ((r @ sk2) @ x2)) | (a @ x2)))) | (! [x2:$i] :
 ((~ ((r @ sk2) @ x2)) | (a @ x2)))))=$false
 --- cnf 4
6: ((~ (! [x2:$i] : ((~ ((r @ sk2) @ x2)) | (a @ x2)))) |
 (! [x2:$i] : ((~ ((r @ sk2) @ x2)) | (a @ x2))))=$false
 --- cnf 5
7: (~ ((s @ sk1) @ sk2))=$false
 --- cnf 5
8: ((s @ sk1) @ sk2)=$true
 --- cnf 7
9: (! [x2:$i] : ((~ ((r @ sk2) @ x2)) | (a @ x2)))=$false
 --- cnf 6
10: (~ (! [x2:$i] : ((~ ((r @ sk2) @ x2)) | (a @ x2))))=$false
 --- cnf 6
11: (! [x2:$i] : ((~ ((r @ sk2) @ x2)) | (a @ x2)))=$true
 --- cnf 10
12: ((~ ((r @ sk2) @ sk3)) | (a @ sk3))=$false
 --- cnf 9
13: ((~ ((r @ sk2) @ V1)) | (a @ V1))=$true
 --- cnf 11
14: (a @ sk3)=$false
 --- cnf 12
15: (~ ((r @ sk2) @ sk3))=$false
 --- cnf 12
16: (~ ((r @ sk2) @ V1))=$true | (a @ V1)=$true
 --- cnf 13
17: ((r @ sk2) @ sk3)=$true
 --- cnf 15
18: ((r @ sk2) @ V1)=$false | (a @ V1)=$true
 --- cnf 16
19: ($false)=$true
 --- fo-atp 8 14 17 18
**** End of derivation protocol ****
**** no. of clauses: 19 ****
```

B LEO-II Proof Protocol for Example K_T^b

```
**** Protocol for Problem: K_T^b.thf
**** Beginning of derivation protocol ****
1: (? [R:$i>($i>$o)] : (! [A:$i>$o] :
 (mvalid @ ((mimpl @ ((mbox @ R) @ A)) @ A))))=$true
 --- theorem(file(K_T^b.thf,[thm]))
2: (? [R:$i>($i>$o)] : (! [A:$i>$o] :
 (mvalid @ ((mimpl @ ((mbox @ R) @ A)) @ A))))=$false
```

```
 ---  neg_input 1
3: (~ (! [x0:$i>($i>$o)] : (~ (! [x1:$i>$o,x2:$i] :
 ((~ (! [x3:$i] : ((~ ((x0 @ x2) @ x3)) | (x1 @ x3)))) |
 (x1 @ x2))))))=$false
 ---  unfold_def 2
4: (! [x0:$i>($i>$o)] : (~ (! [x1:$i>$o,x2:$i] :
 ((~ (! [x3:$i] : ((~ ((x0 @ x2) @ x3)) | (x1 @ x3)))) |
 (x1 @ x2)))))=$true
 ---  cnf 3
5: (~ (! [x1:$i>$o,x2:$i] : ((~ (! [x3:$i] : ((~ ((V1 @ x2) @ x3)) |
 (x1 @ x3)))) | (x1 @ x2))))=$true
 ---  cnf 4
6: (! [x1:$i>$o,x2:$i] : ((~ (! [x3:$i] :
 ((~ ((V1 @ x2) @ x3)) | (x1 @ x3)))) | (x1 @ x2)))=$false
 ---  cnf 5
7: (! [x2:$i] : ((~ (! [x3:$i] : ((~ ((V1 @ x2) @ x3)) |
 ((sk1 @ V1) @ x3)))) | ((sk1 @ V1) @ x2)))=$false
 ---  cnf 6
8: ((~ (! [x3:$i] : ((~ ((V1 @ (sk2 @ V1)) @ x3)) |
 ((sk1 @ V1) @ x3)))) | ((sk1 @ V1) @ (sk2 @ V1)))=$false
 ---  cnf 7
9: ((sk1 @ V1) @ (sk2 @ V1))=$false
 ---  cnf 8
10: (~ (! [x3:$i] : ((~ ((V1 @ (sk2 @ V1)) @ x3)) |
 ((sk1 @ V1) @ x3))))=$false
 ---  cnf 8
11: (! [x3:$i] : ((~ ((V1 @ (sk2 @ V1)) @ x3)) |
 ((sk1 @ V1) @ x3)))=$true
 ---  cnf 10
12: ((~ ((V1 @ (sk2 @ V1)) @ V2)) | ((sk1 @ V1) @ V2))=$true
 ---  cnf 11
13: (~ ((V1 @ (sk2 @ V1)) @ V2))=$true | ((sk1 @ V1) @ V2)=$true
 ---  cnf 12
14: ((V1 @ (sk2 @ V1)) @ V2)=$false | ((sk1 @ V1) @ V2)=$true
 ---  cnf 13
19: (((V19 @ (sk2 @ (^ [x0:$i,x1:$i] :
 (((V19 @ x0) @ x1) = ((V20 @ x0) @ x1))))) @ V2) =
 ((V20 @ (sk2 @ (^ [x0:$i,x1:$i] : (((V19 @ x0) @ x1) =
 ((V20 @ x0) @ x1))))) @ V2))=$false |
 ((sk1 @ (^ [x0:$i,x1:$i] : (((V19 @ x0) @ x1) =
 ((V20 @ x0) @ x1)))) @ V2)=$true
 ---  prim-subst (V1-->lambda [V17]: lambda [V18]:
 (eq ((V19 V17) V18)) ((V20 V17) V18)) 14
31: ((sk1 @ (^ [x0:$i,x1:$i] : (((V19 @ x0) @ x1) =
 ((V20 @ x0) @ x1)))) @ V2)=$true |
 (((V19 @ (sk2 @ (^ [x0:$i,x1:$i] : (((V19 @ x0) @ x1) =
 ((V20 @ x0) @ x1))))) @ V2) = ((V20 @ (sk2 @ (^ [x0:$i,x1:$i] :
 (((V19 @ x0) @ x1) = ((V20 @ x0) @ x1))))) @ V2))=$false
 ---  uni () 19
32: ((sk1 @ V32) @ (sk2 @ V32))=$false
 ---  rename 9
94: (((sk1 @ (^ [x0:$i,x1:$i] : (((V19 @ x0) @ x1) =
```

```
 ((V20 @ x0) @ x1)))) @ V2) = ((sk1 @ V32) @ (sk2 @ V32)))=$false |
 (((V19 @ (sk2 @ (^ [x0:$i,x1:$i] : (((V19 @ x0) @ x1) =
 ((V20 @ x0) @ x1))))) @ V2) = ((V20 @ (sk2 @ (^ [x0:$i,x1:$i] :
 (((V19 @ x0) @ x1) = ((V20 @ x0) @ x1))))) @ V2))=$false
 ---  res 31 32
97: (((V64 @ ((((sk9 @ (sk2 @ (^ [x0:$i,x1:$i] : (((V64 @ x0) @ x1) =
 ((V65 @ x0) @ x1))))) @ (^ [x0:$i,x1:$i] : (((V64 @ x0) @ x1) =
 ((V65 @ x0) @ x1)))) @ V20) @ V19)) @ ((((sk10 @ V19) @ V20) @
 (sk2 @ (^ [x0:$i,x1:$i] : (((V64 @ x0) @ x1) = ((V65 @ x0) @ x1))))) @
 (^ [x0:$i,x1:$i] : (((V64 @ x0) @ x1) = ((V65 @ x0) @ x1))))) =
 ((V65 @ ((((sk9 @ (sk2 @ (^ [x0:$i,x1:$i] : (((V64 @ x0) @ x1) =
 ((V65 @ x0) @ x1))))) @ (^ [x0:$i,x1:$i] : (((V64 @ x0) @ x1) =
 ((V65 @ x0) @ x1)))) @ V20) @ V19)) @ ((((sk10 @ V19) @ V20) @
 (sk2 @ (^ [x0:$i,x1:$i] : (((V64 @ x0) @ x1) = ((V65 @ x0) @ x1))))) @
 (^ [x0:$i,x1:$i] : (((V64 @ x0) @ x1) = ((V65 @ x0) @ x1))))))=$false |
 (((V19 @ ((((sk9 @ (sk2 @ (^ [x0:$i,x1:$i] : (((V64 @ x0) @ x1) =
 ((V65 @ x0) @ x1))))) @ (^ [x0:$i,x1:$i] : (((V64 @ x0) @ x1) =
 ((V65 @ x0) @ x1)))) @ V20) @ V19)) @ ((((sk10 @ V19) @ V20) @
 (sk2 @ (^ [x0:$i,x1:$i] : (((V64 @ x0) @ x1) = ((V65 @ x0) @ x1))))) @
 (^ [x0:$i,x1:$i] : (((V64 @ x0) @ x1) = ((V65 @ x0) @ x1))))) =
 ((V20 @ ((((sk9 @ (sk2 @ (^ [x0:$i,x1:$i] : (((V64 @ x0) @ x1) =
 ((V65 @ x0) @ x1))))) @ (^ [x0:$i,x1:$i] : (((V64 @ x0) @ x1) =
 ((V65 @ x0) @ x1)))) @ V20) @ V19)) @ ((((sk10 @ V19) @ V20) @ (sk2 @
 (^ [x0:$i,x1:$i] : (((V64 @ x0) @ x1) = ((V65 @ x0) @ x1))))) @
 (^ [x0:$i,x1:$i] : (((V64 @ x0) @ x1) = ((V65 @ x0) @ x1))))))=$false |
 (((V19 @ (sk2 @ (^ [x0:$i,x1:$i] : (((V19 @ x0) @ x1) =
 ((V20 @ x0) @ x1))))) @ (sk2 @ (^ [x0:$i,x1:$i] : (((V64 @ x0) @ x1) =
 ((V65 @ x0) @ x1))))) = ((V20 @ (sk2 @ (^ [x0:$i,x1:$i] :
 (((V19 @ x0) @ x1) = ((V20 @ x0) @ x1))))) @ (sk2 @ (^ [x0:$i,x1:$i] :
 (((V64 @ x0) @ x1) = ((V65 @ x0) @ x1))))))=$false
 ---  uni ( V32/lambda [V62]: lambda [V63]: (eq ((V64 V62) V63))
 ((V65 V62) V63)  V2/sk2 (lambda [x0]: lambda [x1]: (eq ((V64 x0) x1))
 ((V65 x0) x1)) ) 94
98: ($false)=$true  ---  flexflex 97
**** End of derivation protocol ****
**** no. of clauses: 20 ****
```

C LEO-II Proof Protocol for Example K_4^b

```
**** Protocol for Problem: K_4^b.thf
**** Beginning of derivation protocol ****
2: (? [R:$i>($i>$o)] : (! [A:$i>$o] : (mvalid @ ((mimpl @ ((mbox @ R) @ A))
 @ ((mbox @ R) @ ((mbox @ R) @ A))))))=$true
 ---  theorem(file(K_4^b.thf,[thm]))
4: (? [R:$i>($i>$o)] : (! [A:$i>$o] : (mvalid @ ((mimpl @ ((mbox @ R) @ A))
 @ ((mbox @ R) @ ((mbox @ R) @ A))))))=$false
 ---  neg_input 2
5: (~ (! [x0:$i>($i>$o)] : (~ (! [x1:$i>$o,x2:$i] : ((~ (! [x3:$i] :
 ((~ ((x0 @ x2) @ x3)) | (x1 @ x3)))) | (! [x3:$i] : ((~ ((x0 @ x2) @ x3))
 | (! [x4:$i] : ((~ ((x0 @ x3) @ x4)) | (x1 @ x4))))))))))=$false
 ---  unfold_def 4
7: (! [x0:$i>($i>$o)] : (~ (! [x1:$i>$o,x2:$i] : ((~ (! [x3:$i] :
 ((~ ((x0 @ x2) @ x3)) | (x1 @ x3)))) | (! [x3:$i] : ((~ ((x0 @ x2) @ x3))
```

| (! [x4:$i] : ((~ ((x0 @ x3) @ x4)) | (x1 @ x4)))))))))=$true
--- cnf 5
9: (~ (! [x1:$i>$o,x2:$i] : ((~ (! [x3:$i] : ((~ ((V1 @ x2) @ x3)) |
(x1 @ x3)))) | (! [x3:$i] : ((~ ((V1 @ x2) @ x3)) | (! [x4:$i] :
((~ ((V1 @ x3) @ x4)) | (x1 @ x4))))))))=$true
--- cnf 7
11: (! [x1:$i>$o,x2:$i] : ((~ (! [x3:$i] : ((~ ((V1 @ x2) @ x3)) |
(x1 @ x3)))) | (! [x3:$i] : ((~ ((V1 @ x2) @ x3)) | (! [x4:$i] :
((~ ((V1 @ x3) @ x4)) | (x1 @ x4)))))))=$false
--- cnf 9
13: (! [x2:$i] : ((~ (! [x3:$i] : ((~ ((V1 @ x2) @ x3)) |
((sk2 @ V1) @ x3)))) | (! [x3:$i] : ((~ ((V1 @ x2) @ x3)) | (! [x4:$i] :
((~ ((V1 @ x3) @ x4)) | ((sk2 @ V1) @ x4)))))))=$false
--- cnf 11
15: ((~ (! [x3:$i] : ((~ ((V1 @ (sk3 @ V1)) @ x3)) | ((sk2 @ V1) @ x3)))) |
(! [x3:$i] : ((~ ((V1 @ (sk3 @ V1)) @ x3)) | (! [x4:$i] :
((~ ((V1 @ x3) @ x4)) | ((sk2 @ V1) @ x4))))))=$false
--- cnf 13
17: (! [x3:$i] : ((~ ((V1 @ (sk3 @ V1)) @ x3)) | (! [x4:$i] :
((~ ((V1 @ x3) @ x4)) | ((sk2 @ V1) @ x4)))))=$false
--- cnf 15
21: ((~ ((V1 @ (sk3 @ V1)) @ (sk5 @ V1))) | (! [x4:$i] :
((~ ((V1 @ (sk5 @ V1)) @ x4)) | ((sk2 @ V1) @ x4))))=$false
--- cnf 17
25: (~ ((V1 @ (sk3 @ V1)) @ (sk5 @ V1)))=$false
--- cnf 21
27: ((V1 @ (sk3 @ V1)) @ (sk5 @ V1))=$true
--- cnf 25
39: (~ (((V11 @ (sk3 @ (^ [x0:$i,x1:$i] : (~ (((V11 @ x0) @ x1) =
((V12 @ x0) @ x1)))))) @ (sk5 @ (^ [x0:$i,x1:$i] : (~ (((V11 @ x0) @ x1) =
((V12 @ x0) @ x1)))))) = ((V12 @ (sk3 @ (^ [x0:$i,x1:$i] :
(~ (((V11 @ x0) @ x1) = ((V12 @ x0) @ x1)))))) @ (sk5 @ (^ [x0:$i,x1:$i] :
(~ (((V11 @ x0) @ x1) = ((V12 @ x0) @ x1))))))))=$true
--- prim-subst (V1-->lambda [V9]: lambda [V10]: ((lambda [X]: lambda [Y]:
neg ((eq X) Y)) ((V11 V9) V10)) ((V12 V9) V10)) 27
59: (((V11 @ (sk3 @ (^ [x0:$i,x1:$i] : (~ (((V11 @ x0) @ x1) =
((V12 @ x0) @ x1)))))) @ (sk5 @ (^ [x0:$i,x1:$i] : (~ (((V11 @ x0) @ x1) =
((V12 @ x0) @ x1)))))) = ((V12 @ (sk3 @ (^ [x0:$i,x1:$i] :
(~ (((V11 @ x0) @ x1) = ((V12 @ x0) @ x1)))))) @ (sk5 @ (^ [x0:$i,x1:$i] :
(~ (((V11 @ x0) @ x1) = ((V12 @ x0) @ x1)))))))=$false
--- cnf 39
60: ($false)=$true --- flexflex 59
**** End of derivation protocol ****
**** no. of clauses: 15 ****

A Proof-Theoretic Approach to the Static Analysis of Logic Programs

DALE MILLER

Dedicated to Peter Andrews on the occasion of his 70th birthday.

1 Introduction

Static analysis of programs can provide useful information for programmers and compilers. Type checking, a common form of static analysis, can help identify errors during program compilation that might otherwise be found only when the program is executed, possibly by someone other than the programmer. The concise invariants that comes from static analysis can also provide valuable documentation about the meaning of code.

We describe a method that approximates a data structure by a collection of the elements it contains and statically determines whether or not the relations computed by a logic program satisfy certain relations over those approximations. More specifically, we shall use *multisets* and *sets* to *approximate* more structured data such as lists and binary trees. Consider, for example, a list sorting program that maintains duplicates of elements during sorting. Part of the correctness of a sort program includes the fact that if the atomic formula $(sort\ t\ s)$ is provable then s is a permutation of t that is in-order. The proof of such a property is likely to involve inductive arguments requiring the invention of invariants: in other words, this is not likely to be a property that can be inferred statically. On the other hand, if the lists t and s are approximated by multisets (that is, if we forget the order of items in lists), then it might be possible to prove automatically half of this property about the sorting program: namely, if the atomic formula $(sort\ t\ s)$ is provable then the multiset associated to t is equal to the multiset associated to s. If that is so, then it is immediate that the lists t and s are, in fact, permutations of one another (in other words, no elements were dropped, duplicated, or created during sorting). As we shall see, such properties based on using multisets to approximate lists can often be proved statically.

This paper, which is based on [21], exploits three aspects of proof theory to present a scheme for static analysis. First, logical formulas, even those comprising just first-order Horn clauses, are considered as part of a higher-order logic, such as the Simple Theory of Types [4, 7]. In such a setting, all constants, including predicate and function constants, can be abstracted and instantiated by other logical expressions: such abstractions and instantiations can be completely explained following the usual rules for the λ-calculus. Second, traces of logic program executions can be seen as cut-free sequent calculus proofs [22] and since sequent calculus proofs also support rich notions of abstraction and instantiation, it is possible to reason directly on logic program computations via standard proof-theoretic notions. Third, linear logic can be seen as the computational logic behind logic and via the instantiation mechanisms available for both formulas and proofs, linear logic can be put behind-the-scenes of Horn clause computation. In this background world, linear logic is used to perform basic computations with sets and multisets.

2 The undercurrents

There are various themes that underlie this approach to inferring properties of Horn clause programs. This section enumerates several such themes. The rest of this paper can be seen as a particular manifestation of these themes.

2.1 If typing is important, why use only one?

Types and other static properties of programming languages have proved important on a number of levels. Typing is useful to programmers since it offers important invariants and documentation for code. Static analysis can also be used by compilers to uncover useful structures that allow compilers to improve program execution. While compilers might make use of multiple static analyses, programmers do not have convenient access to multiple static analyzes. Sometimes a programming language definition provides for no static analysis, as is the case with Lisp and Prolog. Other programming languages offer exactly one typing discipline, such as the polymorphic typing disciplines of Standard ML and λProlog. Simple and fixed static checks are occasionally also part of a language definition, as is the case in SML where static checking is done to determine if a given function definition over concrete data structures covers all possible input values. It seems clear, however, that static analysis of code, if it can be done quickly and incrementally, should have significant benefits for programmers during the process of writing code. For example, a programmer might find it valuable to know that the recursive program that she has just written has linear or quadratic run-time complexity, or that a relation she just specified actually defines a function. The Ciao system pre-processor [14] provides for such functionality by allowing a programmer to write various properties about code that the pre-processor attempts to verify. Providing a flexible framework for the integration of static analysis into programming languages is an interesting direction of research in the design of programming languages. The paper provides one possible scheme for such integration.

2.2 Abstracting over programs and computation traces

A computational system can be seen as encoding symbolic systems on a number of levels: types, program expressions, static analysis expressions, and computation traces. All of these can benefit from representations that allow for natural notions of abstractions and instantiations. For example, we shall consider first-order Horn clauses as part of the Church's Simple Theory of Types [4, 7]. As is well understood in that setting, quantifier instantiation is completely described using the theory's underlying rules for λ-conversion. Similarly, traces of logic program computations can be seen as cut-free proofs and such proof objects also have rich notions of abstraction and application, given by the cut-elimination theorem for sequent calculus. The fact that proofs and programs can be related simply in a setting where substitution into both has well understood properties is certainly one of the strengths of the proof-theoretic foundations of logic programming (see, for example, [22]).

2.3 What good are atomic formulas?

In proof theory, there is an interesting problem of duality involving atomic formulas. The *initial rule* and the *cut rule*, given as

$$\frac{}{C \vdash C}\ \text{Initial} \quad \text{and} \quad \frac{\Gamma_1 \vdash C, \Delta_1 \quad \Gamma_2, C \vdash \Delta_2}{\Gamma_1, \Gamma_2 \vdash \Delta_1, \Delta_2}\ \text{Cut},$$

can be seen as being dual to each other [13]. In particular, the initial rule states that an occurrence of a formula on the left is stronger than the same occurrence on the right, whereas the cut rule states the dual: an occurrence of a formula on the right is strong enough to remove the same occurrence from the left. In most well designed proof systems, atomic and non-atomic occurrences of the cut-rule can be eliminated whereas only non-atomic initial rules (where C is non-atomic) can be eliminated. Atoms seem to spoil the elegant duality of the meta-theory of these inference rules.

While the logic programming world is most comfortable with the existence of atomic formulas, there have been a couple of recent proof-theoretic developments that try to eliminate them entirely. For example, in the work on *definitions* and *fixed points* described in [11, 17, 26], atoms are defined to be other formulas and the only primitive judgment involving terms is that of equality. Furthermore, if fixed points are *stratified* (no recursion through negations) and *noetherian* (no infinite descent in recursion), then *all* instances of cut and initial can be removed. Girard's *ludics* [12] is a more radical presentation of logic in which atomic formulas do not exist: formulas can be probed to arbitrary depth to uncover "subformulas". Another approach to atoms is to consider *all* constants as being variables: such an approach is possible in a higher-order logic by, say, replacing constants with universally quantified variables. On one hand this is a trivial position: if there are no constants (thus, no predicate constants) there are no atomic formulas (which are defined as formulas with non-logical constants at their head). On the other hand, adopting a point-of-view that constants can vary has some appeal. We describe this next.

2.4 Viewing constants and variables as one

The inference rule of $\forall$-generalization states that if B is provable then $\forall x.B$ is provable (with the appropriate proviso if the proof of B depends on hypotheses). In a first-order setting, only a free first-order variable, say x of B, can become bound in $\forall x.B$ by this inference rule. In a higher-order setting, any variable in any expression, even those that play the role of predicates or functions, can be quantified.

The difference between constants and variables can be seen as one of "scope," at least from a syntactic, proof-theoretic, and computational point of view. For example, variables are intended as syntactic objects that can "vary." During the computation of, say, the relation of appending lists, universal quantified variables surrounding Horn clauses change via substitution (during back-chaining and unification) but the constructors for lists as well as the symbol denoting the append relation do not change (are not instantiated) and, hence, can be seen as constants. But from a compiling and linking point-of-view, the append predicate might be considered something that varies: if append is in a module of Prolog that is separately compiled, the append symbol might denote a particular object in the compiled code that is later changed when the code is loaded and linked. In a similar fashion, we shall allow ourselves to instantiate constants with expressions during static analysis: that is, constants can be seen as varying over different approximations of their intended meaning.

Substituting for constants allows us to "split the atom": that is, by substituting for the predicate p in the atom $p(t_1, \ldots, t_n)$, we can replace that atom with a formula, which, in this paper, will be a linear logic formula that accounts for some resources related to the arguments $t_1, \ldots, t_n$.

2.5 Linear logic underlies computational logic

Linear logic [10] is able to explain the proof theory of usual Horn clause logic programming (and even richer logic programming languages [15]). It is also able to provide means to reason about resources, such as items in multisets and sets. Thus, linear logic will allow us to sit within one declarative framework to describe both usual logic programming as well as "sub-atomic" reasoning about the resources implicit in the arguments of predicates.

3 A primer for linear logic

Linear logic connectives can be divided into the following groups: the multiplicatives $\parr$, $\bot$, $\otimes$, $\mathbf{1}$; the additives $\oplus$, $\mathbf{0}$, $\&$, $\top$; the exponentials $!$, $?$; the implications $\multimap$ (where $B \multimap C$ can be defined as $B^\bot \parr C$) and $\Rightarrow$ (where $B \Rightarrow C$ can be defined as $(!\,B)^\bot \parr C$); and the quantifiers $\forall$ and $\exists$ (higher-order quantification is allowed). The equivalence of formulas in linear logic, $B \multimapboth C$, is defined as the formula $(B \multimap C) \;\&\; (C \multimap B)$. The quantifiers should be typed, say as $\forall_\tau$ and $\exists_\tau$, where τ is a simple type: in general, however, we will not write these type subscripts and assume that the reader can reconstruct them from context when their value is important.

First-order Horn clauses are formulas of the form

$$\forall x_1 \ldots \forall x_m [A_1 \wedge \ldots \wedge A_n \supset A_0] \qquad (n, m \geq 0)$$

where $\wedge$ and $\supset$ are intuitionistic or classical logic conjunction and implication and $x_1, \ldots, x_m$ are variables of primitive types. There are at least two natural mappings of Horn clauses into linear logic. The "multiplicative" mapping uses the $\otimes$ and $\multimap$ for the conjunction and implication: this encoding is used in, say, the linear logic programming settings, such as Lolli [15], where Horn clause programming can interact with the surrounding linear aspects of the full programming language. Here, we are not interested in linear logic programming *per se* but with using linear logic to help establish invariants about Horn clauses when these are interpreted in the usual, classical setting. As a result, we shall encode Horn clauses into linear logic using the "additive" conjunction $\&$ and the implication $\Rightarrow$: that is, we take Horn clauses to be formulas of the form

$$\forall x_1 \ldots \forall x_m [A_1 \,\&\, \ldots \,\&\, A_n \Rightarrow A_0]. \qquad (n, m \geq 0)$$

The usual proof search behavior of first-order Horn clauses in classical (and intuitionistic) logic is captured precisely when this style of linear logic encoding is used. An example of a Horn clause logic program is given in Figure 1.

4 A primer for proof theory

A *sequent* is a triple of the form $\Sigma; \Gamma \vdash \Delta$ where Σ, the signature, is a list of non-logical constants and eigenvariables paired with a simple type, and where both Γ and Δ are multisets of Σ-formulas (i.e., formulas all of whose

$$\forall \texttt{K}.\ [\texttt{append nil K K}]$$
$$\forall \texttt{X}, \texttt{L}, \texttt{K}, \texttt{M}.\ [\texttt{append L K M} \Rightarrow \texttt{append (cons X L) K (cons X M)}]$$
$$\forall \texttt{X}.\ [\texttt{split X nil nil nil}]$$
$$\forall \texttt{X}, \texttt{A}, \texttt{B}, \texttt{R}, \texttt{S}.[\texttt{le A X} \,\&\, \texttt{split X R S B} \Rightarrow \texttt{split X (cons A R) (cons A S) B}]$$
$$\forall \texttt{X}, \texttt{A}, \texttt{B}, \texttt{R}, \texttt{S}.[\texttt{gr A X} \,\&\, \texttt{split X R S B} \Rightarrow \texttt{split X (cons A R) S (cons A B)}]$$
$$\texttt{sort nil nil}$$
$$\forall \texttt{F}, \texttt{R}, \texttt{S}, \texttt{Sm}, \texttt{B}, \texttt{SS}, \texttt{BS}.\ [$$
$$\texttt{split F R Sm B} \,\&\, \texttt{sort Sm SS} \,\&\, \texttt{sort B BS} \,\&\, \texttt{append SS (cons F BS) S}$$
$$\Rightarrow \texttt{sort (cons F R) S}]$$

Figure 1. Some Horn clauses for specifying a sorting relation.

non-logical symbols are in Σ). The rules for linear logic are the standard ones [10], except here signatures have been added to sequents. The rules for quantifier introduction are the only rules that require the signature and they are reproduced here.

$$\frac{\Sigma, y{:}\,\tau; B[y/x], \Gamma \vdash \Delta}{\Sigma; \exists_\tau x.B, \Gamma \vdash \Delta}\,\exists L \qquad \frac{\Sigma \vdash t{:}\,\tau \quad \Sigma; \Gamma \vdash B[t/x], \Delta}{\Sigma; \Gamma \vdash \exists_\tau x.B, \Delta}\,\exists R$$

$$\frac{\Sigma \vdash t{:}\,\tau \quad \Sigma; B[t/x], \Gamma \vdash \Delta}{\Sigma; \forall_\tau x.B, \Gamma \vdash \Delta}\,\forall L \qquad \frac{\Sigma, y : \tau; \Gamma \vdash B[y/x], \Delta}{\Sigma; \Gamma \vdash \forall_\tau x.B, \Delta}\,\forall R$$

The premise $\Sigma \vdash t{:}\,\tau$ is the judgment that the term t has the (simple) type τ given the typing declaration contained in Σ.

We now outline three ways to do instantiation within the sequent calculus.

4.1 Substituting for types

Following Church [7], we shall assume that formulas are given the simple type o (omicron). Simple type expressions appear within sequent calculus proofs (in particular, within signatures and subscripts to quantifiers) without abstractions: that is, they are global and (in this setting) admit no bindings. As a result, it is an easy matter to show that if one replaces every occurrence of a type constant (different from o) in a proof with another type expressions, the result is another valid proof. We shall do this kind of substitution for type constants later when we replace a list by a multiset that approximates it: since we use linear logic formulas to encode multisets, we shall replace the type constant `list` with `o`.

4.2 Substituting for non-logical constants

Consider the linear logic sequent

$$\Sigma, p\colon \tau; !\, D_1, !\, D_2, !\,\Gamma \vdash p(t_1, \ldots, t_m),$$

where the type τ is a predicate type (that is, it is of the form $\tau_1 \rightarrow \cdots \rightarrow \tau_m \rightarrow o$) and where p appears in, say, D_1 and D_2 and in no formula of Γ. The linear logic exponential ! is used here to encode the fact that the formulas D_1 and D_2 are available for arbitrary reuse within a proof (the usual case for program clauses). Using the right introduction rules for implication and the universal quantifier, it follows that the sequent

$$\Sigma; !\,\Gamma \vdash \forall p[D_1 \Rightarrow D_2 \Rightarrow p(t_1, \ldots, t_m)]$$

is also provable. Using the rules for universal quantifiers, there must be proofs for all instances of this quantifier. Let θ be the substitution $[p \mapsto \lambda x_1 \ldots \lambda x_m . S]$, where S is a formula over the signature $\Sigma \cup \{x_1, \ldots, x_m\}$ of type o. A consequence of the proof theory of linear logic is that there is a proof also of

$$\Sigma; !\,\Gamma \vdash D_1\theta \Rightarrow D_2\theta \Rightarrow S[t_1/x_1, \ldots, t_m/x_m]$$

and of the sequent

$$\Sigma; !\, D_1\theta, !\, D_2\theta, !\,\Gamma \vdash S[t_1/x_1, \ldots, t_m/x_m].$$

As this example illustrates, it is possible to instantiate a predicate (here p) with an abstraction of a formula (here, $\lambda x_1 \ldots \lambda x_m.\ S$): such an instantiation carries a provable sequent to a provable sequent.

4.3 Substituting for assumptions

An instance of the cut-rule (mentioned earlier) is the following:

$$\frac{\Sigma; \Gamma_1 \vdash B \qquad \Sigma; B, \Gamma_2 \vdash C}{\Sigma; \Gamma_1, \Gamma_2 \vdash C}$$

This inference rule (especially when associated with the cut-elimination procedure) provides a way to instantiate a hypothetical use of a formula (here, B) with a proof of that formula. For example, consider the following situation. Given the example in the Section 4.2, assume that we can prove

$$\Sigma; !\,\Gamma \vdash !\, D_1\theta \quad \text{and} \quad \Sigma; !\,\Gamma \vdash !\, D_2\theta.$$

Using two instances of the cut rule and the proofs of these sequent, it is possible to obtain a proof of the sequent

$$\Sigma; !\,\Gamma \vdash S[t_1/x_1, \ldots, t_m/x_m]$$

(contraction on the left for !'ed formulas must be applied).

Thus, by a series of instantiations of proofs, it is possible to move from a proof of, say,

$$\Sigma, p\colon \tau; !\,D_1, !\,D_2, !\,\Gamma \vdash p(t_1, \ldots, t_m)$$

to a proof of

$$\Sigma; !\,\Gamma \vdash S[t_1/x_1, \ldots, t_m/x_m].$$

Such reasoning about proofs allows us to "split the atom" $p(t_1, \ldots, t_m)$ into a formula $S[t_1/x_1, \ldots, t_m/x_m]$ and to transform proofs involving that atom into proofs involving that formula. In what follows, the formula S will be a linear logic formula that provides an encoding of some judgment about the data structures encoded in the terms $t_1, \ldots, t_m$.

A few simple examples of using higher-order instantiations of logic programs in order to support reasoning about them appear in [20].

5 Encoding multisets as formulas

Linear logic can encode multisets and sets as well as simple judgments about them (such as inclusion and equality). We consider multisets first and tackle sets in Section 8. Let the token *item* be a linear logic predicate of one argument: the linear logic atomic formula *item* x will denote the multiset containing the element x occurring once. There are two natural encoding of multisets into formulas using this predicate. The *conjunctive* encoding uses **1** for the empty multiset and $\otimes$ to combine two multisets. For example, the multiset $\{1, 2, 2\}$ is encoded by the linear logic formula *item* $1 \otimes$ *item* $2 \otimes$ *item* 2. Proofs search using this style encoding places multisets on the left of the sequent arrow. This approach is favored when an intuitionistic subset of linear logic is used, such as in Lolli [15], LinearLF [6], and MSR [5]. The dual encoding, the *disjunctive* encoding, uses $\bot$ for the empty multiset and $⅋$ to combine two multisets. Proofs search using this style encoding places multisets on the right of the sequent arrow and, hence, multiple conclusion sequents are now required. Systems such as LO [2] and Forum [19] use this style of encoding. If negation is available, then the choice of which encoding one chooses is mostly a matter of style. We pick the disjunctive encoding for the rather shallow reason that the inclusion judgment for multisets and sets is encoded as an implication instead of a reverse implication, as we shall now see.

Let S and T be the two formulas

$$item\ s_1 ⅋ \cdots ⅋\ item\ s_n \text{ and } item\ t_1 ⅋ \cdots ⅋\ item\ t_m, \quad (n, m \geq 0)$$

respectively. Notice that $\vdash S \multimap T$ if and only if $\vdash T \multimap S$ if and only if the two multisets $\{s_1, \ldots, s_n\}$ and $\{t_1, \ldots, t_m\}$ are equal. Consider the following two ways for encoding the multiset inclusion $S \sqsubseteq T$.

- $S \parr 0 \multimap T$. This formula mixes multiplicative connectives with the additive connective 0: the latter allows items that are not matched between S and T to be deleted.

- $\exists q(S \parr q \multimap T)$. This formula mixes multiplicative connectives with a higher-order quantifier. While we can consider the instantiation for q to be the multiset difference of S from T, there is no easy way in the logic to enforce that interpretation of the quantifier.

As it turns out, these two approaches are equivalent in linear logic: in particular, $\vdash \mathbf{0} \multimapboth \forall p.p$ (linear logic absurdity) and

$$\vdash \forall S \forall T[(S \parr 0 \multimap T) \multimapboth \exists q(S \parr q \multimap T)].$$

Thus, below we can choose either one of these encodings for multiset inclusion.

6 Multisets approximations

A *multiset expression* is a formula in linear logic built from the predicate symbol *item* (denoting the singleton multiset), the linear logic multiplicative disjunction $\parr$ (for multiset union), and the unit $\bot$ for $\parr$ (used to denote the empty multiset). We shall also allow a propositional variable (a variable of type o) to be used to denote a (necessarily open) multiset expression. An example of an open multiset expression is $item\ f(X) \parr \bot \parr Y$, where Y is a variable of type o, X is a first-order variable, and f is some first-order term constructor.

Let S and T be two multiset expressions. The two *multiset judgments* that we wish to capture are multiset inclusion, written as $S \sqsubseteq T$, and equality, written as $S \stackrel{m}{=} T$. We shall use the syntactic variable ρ to range over these two judgments, which are formally binary relations of type $o \to o \to o$. A *multiset statement* is a closed formula of the form

$$\forall \bar{x}[S_1\ \rho_1\ T_1\ \&\ \cdots\ \&\ S_n\ \rho_n\ T_n \Rightarrow S_0\ \rho_0\ T_0],$$

where the quantified variables $\bar{x}$ are either first-order or of type o and formulas $S_0, T_0, \ldots, S_n, T_n$ are possibly open multiset expressions.

If S and T are closed multiset expressions, then we write $\models_m S \sqsubseteq T$ whenever the multiset (of closed first-order terms) denoted by S is contained in the multiset denoted by T, and we write $\models_m S \stackrel{m}{=} T$ whenever the multisets denoted by S and T are equal. Similarly, we write

$$\models_m \forall \bar{x}[S_1\ \rho_1\ T_1\ \&\ \cdots\ \&\ S_n\ \rho_n\ T_n \Rightarrow S_0\ \rho_0\ T_0]$$

if for all *multiset-valued* closed substitutions θ such that $\models_m S_i\theta \; \rho_i \; T_i\theta$ for all $i = 1, \ldots, n$, it is the case that $\models_m S_0\theta \; \rho_0 \; T_0\theta$. A *multiset-valued* substitution is one where variables of propositional type (type o) are mapped to multiset expressions.

The following proposition is central to our use of linear logic to establish multiset statements for Horn clause programs. The proof of this proposition is given in Section 9.2.

Proposition 1. Let $S_0, T_0, \ldots, S_n, T_n$ $(n \geq 0)$ be multiset expressions all of whose free variables are in the list of variables $\bar{x}$. For each judgment $s \; \rho \; t$ we write $s \; \hat{\rho} \; t$ to denote $s \mathbin{⅋} \mathbf{0} \multimap t$ if ρ is $\sqsubseteq$ and $t \multimapboth s$ if ρ is $\stackrel{m}{=}$. If

$$\forall \bar{x}[S_1 \; \hat{\rho}_1 \; T_1 \;\&\; \ldots \;\&\; S_n \; \hat{\rho}_n \; T_n \Rightarrow S_0 \; \hat{\rho}_0 \; T_0]$$

is provable in linear logic, then

$$\models_m \forall \bar{x}[S_1 \; \rho_1 \; T_1 \;\&\; \cdots \;\&\; S_n \; \rho_n \; T_n \Rightarrow S_0 \; \rho_0 \; T_0]$$

This proposition shows that linear logic can be used in a sound way to infer valid multiset statements. On the other hand, the converse (completeness) does not hold: the statement

$$\forall x \forall y.(x \sqsubseteq y) \;\&\; (y \sqsubseteq x) \Rightarrow (x \stackrel{m}{=} y)$$

is valid but its translation into linear logic is not provable.

To illustrate how deduction in linear logic can be used to establish a property of a logic program, consider the first-order Horn clause program in Figure 1. The signature for this collection of clauses can be given as follows:

```
nil     : list
cons    : int  -> list -> list
append  : list -> list -> list -> o
split   : int  -> list -> list -> list -> o
sort    : list -> list -> o
le, gr  : int  -> int  -> o
```

The first two declarations provide constructors for empty and non-empty lists; the next three are predicates defined by Horn clause in Figure 1; and the last two are order relations that are apparently defined elsewhere.

If we think of lists as collections of items, then we might want to check that the sort program as written does not drop, duplicate, or create any elements. That is, if the atom (`sort` t s) is provable then the multiset of items in the list denoted by t is equal to the multiset of items in the list denoted by s. If this property holds then s and t are lists that are permutations of each other: of course, this does not say that it is the correct permutation but this more simple fact is one that, as we show, can be inferred automatically.

Checking this property of our example logic programming follows the following three steps.

First, we provide an approximation of lists as being, in fact, multisets: more precisely, as linear logic *formulas* denoting multisets. The first step, therefore, must be to substitute `o` for `list` in the signature above. Now we can interpret the constructors for lists using the substitution

$$\texttt{nil} \mapsto \bot \qquad \texttt{cons} \mapsto \lambda x \lambda y.\ item\ x \invamp y.$$

Under such a mapping, the list (`cons 1 (cons 3 (cons 2 nil))`) is mapped to the multiset expression $item\ 1 \invamp item\ 3 \invamp item\ 2 \invamp \bot$.

Second, we associate with each predicate in Figure 1 a multiset judgment that encodes an invariant concerning the multisets denoted by the predicate's arguments. For example, if (`append` $r\ s\ t$) or (`split` $u\ t\ r\ s$) is provable then the multiset union of the items in r with those in s is equal to the multiset of items in t, and if (`sort` $t\ s$) is provable then the multisets of items in lists s and t are equal. This association of multiset judgments to atomic formulas can be achieved formally using the following substitutions for constants:

$$\texttt{append} \mapsto \lambda x \lambda y \lambda z.\ (x \invamp y) \multimapboth z \qquad \texttt{split} \mapsto \lambda u \lambda x \lambda y \lambda z.\ (y \invamp z) \multimapboth x$$
$$\texttt{sort} \mapsto \lambda x \lambda y.\ x \multimapboth y$$

The predicates `le` and `gr` (for the less-than-or-equal-to and greater-than relations) make no statement about collections of items: thus they can be mapped to the trivial tautology **1** (the multiplicative truth) via the substitution

$$\texttt{le} \mapsto \lambda x \lambda y.\ \mathbf{1} \qquad \texttt{gr} \mapsto \lambda x \lambda y.\ \mathbf{1}.$$

Figure 2 presents the result of applying these mappings to Figure 1.

Third, we must now attempt to prove each of the resulting formulas. In the case of Figure 2, all the displayed formulas are trivial theorems of linear logic.

Having taken these three steps, we now claim that we have proved the intended collection judgments associate to each of the logic programming predicates above: in particular, we have now shown that our particular sort program computes a permutation.

7 Formalizing the method

The formal correctness of this three stage approach is easily justified given the substitution properties we presented in Section 4 for the sequent calculus

$$\forall \mathtt{K}.\ [\bot \parr \mathtt{K} \multimapboth \mathtt{K}]$$
$$\forall \mathtt{X}, \mathtt{L}, \mathtt{K}, \mathtt{M}.\ [\mathtt{L} \parr \mathtt{K} \multimapboth \mathtt{M} \Rightarrow \mathit{item}\ \mathtt{X} \parr \mathtt{L} \parr \mathtt{K} \multimapboth \mathit{item}\ \mathtt{X} \parr \mathtt{M}]$$
$$\forall \mathtt{X}.\ [\bot \parr \bot \multimapboth \bot]$$
$$\forall \mathtt{X}, \mathtt{A}, \mathtt{B}, \mathtt{R}, \mathtt{S}.\ [(\mathtt{S} \parr \mathtt{B}) \multimapboth \mathtt{R} \Rightarrow \mathbf{1} \Rightarrow \mathit{item}\ \mathtt{A} \parr \mathtt{S} \parr \mathtt{B} \multimapboth \mathit{item}\ \mathtt{A} \parr \mathtt{R}$$
$$\forall \mathtt{X}, \mathtt{A}, \mathtt{B}, \mathtt{R}, \mathtt{S}.\ [(\mathtt{S} \parr \mathtt{B}) \multimapboth \mathtt{R} \Rightarrow \mathbf{1} \Rightarrow \mathtt{S} \parr \mathit{item}\ \mathtt{A} \parr \mathtt{B} \multimapboth \mathit{item}\ \mathtt{A} \parr \mathtt{R}$$
$$[\bot \multimapboth \bot]$$
$$\forall \mathtt{F}, \mathtt{R}, \mathtt{S}, \mathtt{Sm}, \mathtt{Bg}, \mathtt{SS}, \mathtt{BS}.\ [\ \mathtt{Sm} \parr \mathtt{B} \multimapboth \mathtt{R} \mathbin{\&} \mathtt{Sm} \multimapboth \mathtt{SS} \mathbin{\&} \mathtt{B} \multimapboth \mathtt{BS} \mathbin{\&} \mathtt{SS} \parr \mathit{item}\ \mathtt{F} \parr \mathtt{BS} \multimapboth \mathtt{S} \Rightarrow \mathit{item}\ \mathtt{F} \parr \mathtt{R} \multimapboth \mathtt{S}]$$

Figure 2. The linear logic formulas that result from instantiating the non-logical constants in the Horn clauses in Figure 1.

presentation of linear logic.

Let Γ denote a set of formulas displayed in Figure 1 and let Σ be the signature for Γ. Let θ denote the substitution described above for the type `list`, for the constructors `nil` and `cons`, and for the predicates in Figure 1. Finally, let Σ' be the signature of the range of θ (in this case, it just contains the constant *item*). Then, $\Gamma\theta$ is the set of formula in Figure 2.

Assume now that $\Sigma; \Gamma \vdash (\mathit{sort}\ t\ s)$ is provable. Given the discussion in Sections 4.1 and 4.2, we know that

$$\Sigma'; \Gamma\theta \vdash t\theta \multimapboth s\theta$$

is provable. Since the formulas in $\Gamma\theta$ are provable, we can use substitution into proofs (Section 4.3) to conclude that $\Sigma'; \vdash t\theta \multimapboth s\theta$ is provable. Given Proposition 1, we can conclude that $\models_m t\theta \overset{m}{=} s\theta$: that is, that $t\theta$ and $s\theta$ encode the same multiset.

Consider the following model theoretic argument for establishing similar properties of Horn clauses. Let $\mathcal{M}$ be the Herbrand model that captures the invariants that we have in mind. In particular, $\mathcal{M}$ contains the atoms ($\mathtt{append}\ r\ s\ t$) and ($\mathtt{split}\ u\ t\ r\ s$) if the items in the list r added to the items in list s are the same as the items in t. Furthermore, $\mathcal{M}$ contains all closed atoms of the form ($\mathtt{le}\ t\ s$) and ($\mathtt{gr}\ t\ s$), and closed atoms ($\mathtt{sort}\ t\ s$) where s and t are lists that are permutations of one another. One can now show that $\mathcal{M}$ satisfies all the Horn clauses in Figure 1. As a consequence of the soundness of first-order classical logic, any atom provable from the clauses in Figure 1, must be true in $\mathcal{M}$. By construction of $\mathcal{M}$, this means that the desired invariant holds for all atoms proved from the program.

The approach of this paper essentially replaces the construction of a model and the determination of truth in that model with deduction in linear logic.

8 Sets approximations

Linear logic can also be used to encode sets and the judgments of set equality and inclusion. In fact, the transition to sets from multisets is provided by the use of the linear logic exponential: since we are using the disjunctive encoding of collections (see the discussion in Section 5), we use the ? exponential (if we were using the conjunctive encoding, we would use the ! exponential).

The expression $?\,item\ t$ can be seen as describing the presence of an item for which the exact multiplicity does not matter: this formula represents the capacity to be used any number of times. Thus, the set $\{x_1, \ldots, x_n\}$ can be encoded as $?\,item\ x_1 \mathbin{⅋} \cdots \mathbin{⅋} ?\,item\ x_n$. Using logical equivalences of linear logic, this formula is also equivalent to the formula $?(item\ x_1 \oplus \cdots \oplus item\ x_n)$. This latter encoding is the one that we shall use for building our encoding of sets.

A *set expression* is a formula in linear logic built from the predicate symbol *item* (denoting the singleton set), the linear logic additive disjunction $\oplus$ (for set union), and the unit $\mathbf{0}$ for $\oplus$ (used to denote the empty set). We shall also allow a propositional variable (a variable of type o) to be used to denote a (necessarily open) set expression. An example of an open set expression is $item\ f(X) \oplus \mathbf{0} \oplus Y$, where Y is a variable of type o, X is a first-order variable, and f is some first-order term constructor.

Let S and T be two set expressions. The two *set judgments* that we wish to capture are set inclusion, written as $S \subseteq T$, and equality, written as $S \stackrel{s}{=} T$. We shall use the syntactic variable ρ to range over these two judgments, which are formally binary relations of type $o \to o \to o$. A *set statement* is a formula of the form

$$\forall \bar{x}[S_1\ \rho_1\ T_1 \mathbin{\&} \cdots \mathbin{\&} S_n\ \rho_n\ T_n \Rightarrow S_0\ \rho_0\ T_0]$$

where the quantified variables $\bar{x}$ are either first-order or of type o and formulas $T_0, S_0, \ldots, T_n, S_n$ are possibly open set expressions.

If S and T are closed set expressions, then we write $\models_s S \subseteq T$ whenever the set (of closed first-order terms) denoted by S is contained in the set denoted by T, and we write $\models_s S \stackrel{s}{=} T$ whenever the sets denoted by S and T are equal. Similarly, we write

$$\models_s \forall \bar{x}[S_1\ \rho_1\ T_1 \mathbin{\&} \cdots \mathbin{\&} S_n\ \rho_n\ T_n \Rightarrow S_0\ \rho_0\ T_0]$$

if for all *set-valued* closed substitutions θ such that $\models_s S_i\theta\ \rho_i\ T_i\theta$ for all $i = 1, \ldots, n$, it is the case that $\models_s S_0\theta\ \rho_0\ T_0\theta$. A *set-valued* substitution is one where variables of propositional type (type o) are mapped to set expressions.

The following proposition is central to our use of linear logic to establish set statements for Horn clause programs. The proof of this proposition in Section 9.1.

Proposition 2. Let $S_0, T_0, \ldots, S_n, T_n$ ($n \geq 0$) be set expressions all of whose free variables are in the list of variables $\bar{x}$. For each judgment $s\ \rho\ t$ we write $s\ \hat{\rho}\ t$ to denote $?\,s \multimap ?\,t$ if ρ is $\subseteq$ and $(?\,s \multimap ?\,t)\ \&\ (?\,t \multimap ?\,s)$ if ρ is $\stackrel{s}{=}$. If

$$\forall \bar{x}[S_1\ \hat{\rho}_1\ T_1\ \&\ \ldots\ \&\ S_n\ \hat{\rho}_n\ T_n \Rightarrow S_0\ \hat{\rho}_0\ T_0]$$

is provable in linear logic, then

$$\models_s \forall \bar{x}[S_1\ \rho_1\ T_1\ \&\ \cdots\ \&\ S_n\ \rho_n\ T_n \Rightarrow S_0\ \rho_0\ T_0]$$

Lists can be approximated by sets by using the following substitution:

$$\texttt{nil} \mapsto \mathbf{0} \qquad \texttt{cons} \mapsto \lambda x \lambda y.\ \mathit{item}\ x \oplus y.$$

Under such a mapping, the list (`cons 1 (cons 2 (cons 2 nil))`) is mapped to the set expression $\mathit{item}\ 1 \oplus \mathit{item}\ 2 \oplus \mathit{item}\ 2 \oplus \mathbf{0}$. This expression is equivalent ($\multimapboth$) to the set expression $\mathit{item}\ 1 \oplus \mathit{item}\ 2$.

For a simple example of using set approximations, consider modifying the sorting program provided before so that duplicates are not kept in the sorted list. Do this modification by replacing the previous definition for splitting a list with the clauses in Figure 3. That figure contains a new definition of splitting that contains three clauses for deciding whether or not the "pivot" `X` for the splitting is equal to, less than, or greater than the first member of the list being split. Using the following substitutions for predicates

$$\begin{aligned}
\texttt{append} &\mapsto \lambda x \lambda y \lambda z.\ ?(x \oplus y) \multimapboth ?\,z \\
\texttt{split} &\mapsto \lambda u \lambda x \lambda y \lambda z.\ ?(\mathit{item}\ u \oplus x) \multimapboth ?(\mathit{item}\ u \oplus y \oplus z) \\
\texttt{sort} &\mapsto \lambda x \lambda y.\ ?\,x \multimapboth ?\,y
\end{aligned}$$

(as well as the trivial substitution for `gr`), we can show that sort relates two lists only if those lists are approximated by the same set. Figure 4 contains the result of instantiating the `split` specification in Figure 3.

In the case of determining the validity of a set statement, the use of linear logic here appears to be rather weak when compared to the large body of results for solving set-based constraint systems [1, 25].

$$
\begin{array}{r}
\forall \texttt{X}.\ [\texttt{split X nil nil nil}] \\
\forall \texttt{X,B,R,S}.\ [\texttt{split X R S B} \Rightarrow \texttt{split X (cons X R) S B}] \\
\forall \texttt{X,A,B,R,S}.[\texttt{gr X A} \,\&\, \texttt{split X R S B} \Rightarrow \texttt{split X (cons A R) (cons A S) B}] \\
\forall \texttt{X,A,B,R,S}.[\texttt{gr A X} \,\&\, \texttt{split X R S B} \Rightarrow \texttt{split X (cons A R) S (cons A B)}]
\end{array}
$$

Figure 3. A specification of splitting lists that drops duplicates.

$$
\begin{array}{l}
\forall \texttt{X}.\ [(?(item\ \texttt{X} \oplus \mathbf{0}) \multimap\!\circ\ ?(item\ \texttt{X} \oplus \mathbf{0} \oplus \mathbf{0}))] \\
\forall \texttt{X,B,R,S}.\ [(?(item\ \texttt{X} \oplus \texttt{R}) \multimap\!\circ\ ?(item\ \texttt{X} \oplus \texttt{S} \oplus \texttt{B})) \Rightarrow \\
\qquad (?(item\ \texttt{X} \oplus item\ \texttt{X} \oplus \texttt{R}) \multimap\!\circ\ ?(item\ \texttt{X} \oplus \texttt{S} \oplus \texttt{B}))] \\
\forall \texttt{X,A,B,R,S}.\ [\mathbf{1}\&(?(item\ \texttt{X} \oplus \texttt{R}) \multimap\!\circ\ ?(item\ \texttt{X} \oplus \texttt{S} \oplus \texttt{B})) \Rightarrow \\
\qquad (?(item\ \texttt{X} \oplus item\ \texttt{A} \oplus \texttt{R}) \multimap\!\circ\ ?(item\ \texttt{X} \oplus item\ \texttt{A} \oplus \texttt{S} \oplus \texttt{B}))] \\
\forall \texttt{X,A,B,R,S}.\ [\mathbf{1}\&(?(item\ \texttt{X} \oplus \texttt{R}) \multimap\!\circ\ ?(item\ \texttt{X} \oplus \texttt{S} \oplus \texttt{B})) \Rightarrow \\
\qquad (?(item\ \texttt{X} \oplus item\ \texttt{A} \oplus \texttt{R}) \multimap\!\circ\ ?(item\ \texttt{X} \oplus \texttt{S} \oplus item\ \texttt{A} \oplus \texttt{B}))]
\end{array}
$$

Figure 4. The result of substituting set approximations into the `split` program.

9 Automation of deduction

We describe some proof theory results that can be used to automate deduction for the linear logic formulas that occur in Propositions 1 and 2. The key result of linear logic surrounding the search for cut-free proofs is given by the completeness of *focused proofs* [3]. Focused proofs are a normal form that significantly generalizes standard completeness results in logic programming, including the completeness of SLD-resolution and uniform proofs as well as various forms of bottom-up and top-down reasoning [16].

9.1 An proof system for additive connectives

We first analyze the nature of proof search for the linear logic translation of set statements. Note that when considering provability of set statements, there is no loss of generality if the only set judgment it contains is the subset judgment since set equality can be expressed as two inclusions. We now prove that the proof system in Figure 5 is sound and complete for proving set statements.

Proposition 3. Let $S_0, T_0, \ldots, S_n, T_n$ $(n \geq 0)$ be set expressions all of whose free variables are in the list of variables $\bar{x}$. The formula

$$\forall \bar{x}[(?\,S_1 \multimap ?\,T_1) \,\&\, \ldots \,\&\, (?\,S_n \multimap ?\,T_n) \Rightarrow (?\,S_0 \multimap ?\,T_0)]$$

$$\frac{}{\Gamma; A_i \vdash A_1 \oplus \cdots \oplus A_n} \oplus \mathrm{R} \qquad \frac{\Gamma; A_1 \vdash C \quad \ldots \quad \Gamma; A_n \vdash C}{\Gamma; A_1 \oplus \cdots \oplus A_n \vdash C} \oplus \mathrm{L}$$

$$\frac{\Gamma; B_1 \oplus \cdots \oplus B_m \vdash C}{\Gamma; A \vdash C} \mathrm{FC}$$

Figure 5. Specialized proof rules for proving set statements. Here, $n, m \geq 0$, $1 \leq i \leq n$, and in the FC (forward-chaining) inference rule, the formula $?(A_1 \oplus \cdots \oplus A_n) \multimap ?(B_1 \oplus \cdots \oplus B_m)$ must be a member of Γ and $A \in \{A_1, \ldots, A_n\}$.

is provable in linear logic if and only if the sequent

$$(?\, S_1 \multimap ?\, T_1), \ldots, (?\, S_n \multimap ?\, T_n); S_0 \vdash T_0$$

is provable using the proof system in Figure 5.

Proof. The soundness part of this proposition ("if") is easy to show. For completeness ("only if"), we use the completeness of focused proofs in [3]. In order to use that result, we need to give a polarity to all atomic formulas. We do this by assigning all atomic formulas (those of the form *item* $(\cdot)$ and those symbols in $\bar{x}$ of type o) positive polarity. Second, we need to translate the two-sided sequent $\Gamma; S \vdash T$ to $\Gamma^{\perp}, T; \Uparrow S^{\perp}$ when S is not atomic and to $\Gamma^{\perp}, T; S^{\perp} \Uparrow \cdot$ when S is a atom. Completeness then follows directly from the structure of focused proofs. ∎

Notice that when building proofs in a bottom-to-top fashion using the inference rules in Figure 5, the left-hand-side of sequents change until one reaches the top inference rule.

We can now provide a proof of Proposition 2. Assume that

$$\forall \bar{x}[S_1\ \rho_1\ T_1 \mathbin{\&} \cdots \mathbin{\&} S_n\ \rho_n\ T_n \Rightarrow S_0\ \rho_0\ T_0]$$

is provable in linear logic and let θ be a set-valued substitution for $\bar{x}$. Thus, the formula

$$S_1\theta\ \rho_1\ T_1\theta \mathbin{\&} \cdots \mathbin{\&} S_n\theta\ \rho_n\ T_n\theta \Rightarrow S_0\theta\ \rho_0\ T_0\theta$$

is provable. By Proposition 3, it follows by a simple induction on the structure of proofs in Figure 5 that

$$\models_s S_1\theta\ \rho_1\ T_1\theta \mathbin{\&} \cdots \mathbin{\&} S_n\theta\ \rho_n\ T_n\theta \Rightarrow S_0\theta\ \rho_0\ T_0\theta.$$

$$\frac{}{\Gamma; A_1 \wp \cdots \wp A_n \vdash A_1, \ldots, A_n} \wp\text{L}$$

$$\frac{}{\Gamma; A_1 \wp \cdots \wp A_n \wp \mathbf{0} \vdash A_1, \ldots, A_n, \Delta} \wp\mathbf{0}\text{L}$$

$$\frac{\Gamma; S \vdash T_1, T_2, \Delta}{\Gamma; S \vdash T_1 \wp T_2, \Delta} \wp\text{R} \qquad \frac{\Gamma; S \vdash T, \Delta}{\Gamma; S \vdash A_1, \ldots, A_n, \Delta} \text{BC}$$

Figure 6. Specialized proof rules for proving multiset statements. Here, $n \geq 0$ and $A_1, \ldots, A_n$ are atomic (in particular, they are not $\mathbf{0}$). In the BC inference rule, $T \multimap (A_1 \wp \cdots \wp A_m)$ must be a member of Γ.

9.2 A proof system for multiplicative connectives

The proof system in Figure 6 can be used to characterize the structure of proofs of the linear logic encoding of multiset statements. Let

$$\forall \bar{x}[S_1 \; \hat{\rho}_1 \; T_1 \;\&\; \ldots \;\&\; S_n \; \hat{\rho}_n \; T_n \Rightarrow S_0 \; \hat{\rho}_0 \; T_0]$$

be the translation of a multiset statement into linear logic. Provability of this formula can be reduced to attempting to prove $S_0 \; \hat{\rho}_0 \; T_0$ from assumptions of one of the following two forms:

$$(B_1 \wp \cdots \wp B_n) \multimap (A_1 \wp \cdots \wp A_m)$$

$$(B_1 \wp \cdots \wp B_n \wp \mathbf{0}) \multimap (A_1 \wp \cdots \wp A_m)$$

Here, $A_1, \ldots, A_m, B_1, \ldots, B_n$ are atomic formulas.

Proposition 4. Let S_0 and T_0 be multiset expressions all of whose free variables are in the list of variables $\bar{x}$ and let Γ be a set of (linear logic encodings of) multiset judgments. The formula $S_0 \multimap T_0$ is a linear logic consequence of Γ if and only if the sequent $\Gamma; S_0 \vdash T_0$ is provable using the inference rules in Figure 6. Similarly, the formula $S_0 \wp \mathbf{0} \multimap T_0$ is a linear logic consequence of Γ if and only if the sequent $\Gamma; S_0 \wp \mathbf{0} \vdash T_0$ is provable using the inference rules in Figure 6.

Proof. The soundness part of this proposition ("if") is easy to show. Completeness ("only if") is proved elsewhere, for example, in [18, Proposition 2]. It is also an easy consequence of the the completeness of focused proofs in [3]: fix the polarity to all atomic formulas to be negative. ∎

Notice that proofs using the rules in Figure 6 are straight-line proofs (no branching) and that they are goal-directed in the sense that the right-hand side (the "goal") changes during the bottom-to-top construction of a proof.

A proof of Proposition 1 follows from the Proposition 4 by a simple induction on the structure of proofs using the proof system in Figure 6.

9.3 Decidability and Practical Implementation

Determining whether or not the (additive) linear logic translation of a set statement is provable is decidable. In particular, notice that in a proof of the endsequent $\Gamma; S \vdash T$ using the inference rules in Figure 5, all sequents in the proof have the form $\Gamma; S' \vdash T$, where S is either an atomic formula or the conclusion of some implication in Γ. Thus, the search for a proof either succeeds (proof search ends by placing $\oplus$ R on top), or fails to find a proof, or it cycles, a case we can always detect since there is only a finite number of different formulas that can be S'.

Decidability for the proof system of Figure 6 is currently open. If all judgments in a multiset statement are equivalences ($\stackrel{m}{=}$) then deduction in the multiplicative proof system is an example of multiset rewriting and, as such, is a subset of the Petri net reachability problem, which is know to be decidable [9].

A simple prototype implementation of the proof systems in this section within the λProlog programming language [23] illustrates that a naive implementation of provability can be effective in finding proofs of provable linear logic statements generated by the examples in this paper. Also, when proofs existed, they existed under the assumption that any given assumed implication is used at most once. Exploiting such an observation allows one to search for short proofs first will a high chance of success.

10 Extensions

Various extensions of the basic scheme described here are natural to consider. In particular, it should be easy to consider approximating data structures that contain items of differing types: each of these types could be mapped into different $item_\alpha(\cdot)$ predicates, one for each type α.

It should also be simple to construct approximating mappings given the *polymorphic* typing of a given constructor's type. For example, if we are given the following declaration of binary trees (written here in λProlog syntax)

```
kind btree   type -> type.
type emp     btree A.
type bt      A -> btree A -> btree A -> btree A.
```

it should be possible to automatically construct the mapping

$$\begin{aligned} \texttt{btree} &\mapsto \lambda x.\texttt{o} \\ \texttt{emp} &\mapsto \bot \\ \texttt{bt} &\mapsto \lambda x \lambda y \lambda z.\ item_A(x) \parr y \parr z \end{aligned}$$

that can, for example, approximate a binary tree with the multiset of the labels for internal nodes.

Extending this work to do static analysis for higher-order Horn clauses [24] also seems most natural to consider. In the paper [21], collections based on lists and (functional) difference lists are also considered.

Abstract interpretation [8] associates to a program an approximation to its semantics. Such approximations can help to determine various kinds of properties of programs. It will be interesting to see how well the particular notions of collection analysis described here can be related to abstract interpretation. More challenging would be to see to what extent the general methodology described here – the substitution into proofs (computation traces) and use of linear logic – can be related to the general methodology of abstract interpretation.

BIBLIOGRAPHY

[1] Alexander Aiken. Set constraints: results, applications, and future directions. In *PPCP94: Principles and Practice of Constraint Programming*, LNCS 874, pages 171–179, 1994.

[2] J. M. Andreoli and R. Pareschi. Linear Objects: Logical Processes with Built-in Inheritance. *New Generation Computing*, 9(3-4):445–473, 1991.

[3] Jean-Marc Andreoli. Logic Programming with Focusing Proofs in Linear Logic. *J. of Logic and Computation*, 2(3):297–347, 1992.

[4] Peter B. Andrews. *An Introduction to Mathematical Logic and Type Theory*. Academic Press, 1986.

[5] Iliano Cervesato, Nancy A. Durgin, Patrick D. Lincoln, John C. Mitchell, and Andre Scedrov. A Meta-Notation for Protocol Analysis. In R. Gorrieri, editor, *Proceedings of the 12th IEEE Computer Security Foundations Workshop — CSFW'99*, pages 55–69, Mordano, Italy, 28–30 June 1999. IEEE Computer Society Press.

[6] Iliano Cervesato and Frank Pfenning. A Linear Logic Framework. In *Proceedings, Eleventh Annual IEEE Symposium on Logic in Computer Science*, pages 264–275, New Brunswick, New Jersey, July 1996. IEEE Computer Society Press.

[7] Alonzo Church. A Formulation of the Simple Theory of Types. *J. of Symbolic Logic*, 5:56–68, 1940.

[8] Patrick Cousot and Radhia Cousot. Abstract Interpretation: A Unified Lattice Model for Static Analysis of Programs by Construction or Approximation of Fixpoints. In *POPL*, pages 238–252, 1977.

[9] Javier Esparza and Mogens Nielsen. Decidability Issues for Petri Nets - a survey. *Bulletin of the EATCS*, 52:244–262, 1994.

[10] Jean-Yves Girard. Linear Logic. *Theoretical Computer Science*, 50:1–102, 1987.

[11] Jean-Yves Girard. A Fixpoint Theorem in Linear Logic. An email posting to the mailing list linear@cs.stanford.edu, February 1992.

[12] Jean-Yves Girard. Locus solum. *Mathematical Structures in Computer Science*, 11(3):301–506, June 2001.

[13] Jean-Yves Girard, Paul Taylor, and Yves Lafont. *Proofs and Types*. Cambridge University Press, 1989.

[14] Manuel V. Hermenegildo, Germán Puebla, Francisco Bueno, and Pedro López-García. Integrated program debugging, verification, and optimization using abstract interpretation (and the Ciao system preprocessor). *Sci. Comput. Program.*, 58(1-2):115–140, 2005.

[15] Joshua Hodas and Dale Miller. Logic Programming in a Fragment of Intuitionistic Linear Logic. *Information and Computation*, 110(2):327–365, 1994.

[16] Chuck Liang and Dale Miller. Focusing and polarization in intuitionistic logic. In J. Duparc and T. A. Henzinger, editors, *CSL 2007: Computer Science Logic*, LNCS 4646, pages 451–465. Springer-Verlag, 2007.

[17] Raymond McDowell and Dale Miller. Cut-elimination for a logic with definitions and induction. *Theoretical Computer Science*, 232:91–119, 2000.

[18] Dale Miller. The π-calculus as a theory in linear logic: Preliminary results. In E. Lamma and P. Mello, editors, *3rd Workshop on Extensions to Logic Programming*, LNCS 660, pages 242–265, Bologna, Italy, 1993. Springer-Verlag.

[19] Dale Miller. Forum: A Multiple-Conclusion Specification Logic. *Theoretical Computer Science*, 165(1):201–232, September 1996.

[20] Dale Miller. Higher-order quantification and proof search. In Hélène Kirchner and Christophe Ringeissen, editors, *Proceedings of AMAST 2002*, LNCS 2422, pages 60–74, 2002.

[21] Dale Miller. Collection analysis for Horn clause programs. In *Proceedings of PPDP 2006: 8th International ACM SIGPLAN Conference on Principles and Practice of Declarative Programming*, pages 179–188, July 2006.

[22] Dale Miller, Gopalan Nadathur, Frank Pfenning, and Andre Scedrov. Uniform Proofs as a Foundation for Logic Programming. *Annals of Pure and Applied Logic*, 51:125–157, 1991.

[23] Gopalan Nadathur and Dale Miller. An Overview of λProlog. In *Fifth International Logic Programming Conference*, pages 810–827, Seattle, August 1988. MIT Press.

[24] Gopalan Nadathur and Dale Miller. Higher-order Horn Clauses. *Journal of the ACM*, 37(4):777–814, October 1990.

[25] Leszek Pacholski and Andreas Podelski. Set Constraints: A Pearl in Research on Constraints. In *Principles and Practice of Constraint Programming - CP97*, LNCS 1330, pages 549–562. Springer, 1997.

[26] Peter Schroeder-Heister. Rules of Definitional Reflection. In M. Vardi, editor, *Eighth Annual Symposium on Logic in Computer Science*, pages 222–232. IEEE Computer Society Press, IEEE, June 1993.

ATS/LF: A Type System for Constructing Proofs as Total Functional Programs[1]

HONGWEI XI

1 Introduction

The development of *Applied Type System* ($\mathcal{ATS}$) [36, 31] stems from an earlier attempt to introduce dependent types into practical programming [38, 37]. While there is already a framework *Pure Type System* [4] ($\mathcal{PTS}$) that offers a simple and general approach to designing and formalizing type systems, it has been understood that there are some acute problems with $\mathcal{PTS}$ that make it difficult to support, especially, in the presence of dependent types various common programming features such as general recursion [10], recursive types [19], effects [17], exceptions [15] and input/output, etc. To address such limitations of $\mathcal{PTS}$, $\mathcal{ATS}$ is proposed to allow for designing and formalizing type systems that can readily accommodate common realistic programming features. The key salient feature of $\mathcal{ATS}$ lies in a complete separation of the statics, where types are formed and reasoned about, from the dynamics, where programs are constructed and evaluated. With this separation, it is no longer possible for a program to occur in a type as is otherwise allowed in $\mathcal{PTS}$.

We have now designed and implemented ATS, a programming language with its type system rooted in $\mathcal{ATS}$. The work we report here is primarily motivated by a need for combining programs with proofs in ATS. Before going into further details, we would like to present an example to clearly illustrate the motivation. In ATS, we can declare a function *append* (through a form of syntax rather similar to that of Standard ML [20]) in Figure 1. We use **list** as a type constructor. When applied to a type T and an integer I, $\mathbf{list}(T, I)$ forms a type for lists of length I in which each element is of type T. The two list constructors *nil* and *cons* are assigned the following

[1] *Partially supported by NSF grants no. CCR-0229480 and no. CCF-0702665

```
fun append {a:type} {m,n:nat}
  (xs: list (a, m), ys: list (a, n)): list (a, m+n) =
  case xs of
  | nil () => ys  // 1st clause
  | cons (x, xs) => cons (x, append (xs, ys)) // 2nd clause

dataprop MUL (int, int, int) =
  | {n:int} MULbas (0, n, 0)
  | {m,n,p:int | m >= 0} MULind (m+1, n, p+n) of MUL (m, n, p)
  | {m,n,p:int | m > 0} MULneg (~m, n, ~p) of MUL (m, n, p)

fun concat {a:type} {m,n:nat} (xxs: list (list (a, n), m))
  : [p:nat] (MUL (m, n, p) | list (a, p)) =
  case xxs of
  | nil () => (MULbas () | nil ())
  | cons (xs, xss) => let
      val (pf | res) = concat xss
    in
      (MULind pf | append (xs, res))
    end
```

Figure 1. An example of proof construction in ATS/LF

constant types (or c-types):

$$\begin{array}{lll} nil & : & \forall a : type.\ \mathbf{list}(a, 0) \\ cons & : & \forall a : type.\forall n : nat.\ (a, \mathbf{list}(a, n)) \Rightarrow \mathbf{list}(a, n+1) \end{array}$$

The header of the function *append* indicates that *append* is assigned the following type:

$$\forall a : type.\forall m : nat.\forall n : nat.\ (\mathbf{list}(a, m), \mathbf{list}(a, n)) \rightarrow \mathbf{list}(a, m+n)$$

which means that *append* returns a list of length $m+n$ when applied to two lists of length m and n, respectively. Note that *type* is a built-in sort in ATS, and a static term of sort *type* is a type (for dynamic terms). Also, *int* is a built-in sort for integers in ATS, and *nat* is the subset sort $\{a : int \mid a \geq 0\}$ for natural numbers. We will explain later that subset sorts are just of a form of syntactic sugar.

When type-checking the definition of *append*, we essentially generate the following two constraints:

$$\begin{array}{l} \forall m : nat.\forall n : nat.\ m = 0 \supset n = m + n \\ \forall m : nat.\forall n : nat.\forall m' : int.\ m = m' + 1 \supset (m' + n) + 1 = m + n \end{array}$$

The first constraint is generated when the first clause in the definition of *append* is type-checked; the constraint is needed for determining that the types $\mathbf{list}(a, n)$ and $\mathbf{list}(a, m+n)$ are equal under the assumption that $\mathbf{list}(a, m)$ equals $\mathbf{list}(a, 0)$. Similarly, the second constraint is generated when the second clause in the definition of *append* is type-checked; the constraint is needed for determining that for any integer m', the types $\mathbf{list}(a, (m'+n)+1)$ and $\mathbf{list}(a, m+n)$ are equal under the assumption that $\mathbf{list}(a, m)$ equals $\mathbf{list}(a, m'+1)$. Clearly, we need to impose certain restrictions on the form of constraints allowed in practice so that an effective means can be found to solve such constraints automatically. In ATS, we require that (arithmetic) constraints like those presented above be linear,[1] and we rely on a constraint solver based on the Fourier-Motzkin variable elimination method [12] to solve such constraints. While this is indeed a simple design, it is inherently *ad hoc* and can also be too restrictive, sometimes, in a situation where nonlinear constraints need to be handled. For instance, a function *concat* that concatenates m lists of length n may be given the following type:

$$\forall a : \mathit{type}.\forall m : \mathit{nat}.\forall n : \mathit{nat}.\ \mathbf{list}(\mathbf{list}(a, n), m) \rightarrow \mathbf{list}(a, m \times n)$$

Unfortunately, this type is not allowed in ATS as $m \times n$ is not a linear arithmetic expression and thus cannot be used as a type index. To address this issue, a recursive dependent *prop* constructor **MUL** is declared in Figure 1 for encoding the multiplication function on integers. In general, a prop is like a type, and the essential difference between them is that a prop can only be assigned to total terms (to be formally defined later), which we often refer to as *proof terms*. The concrete syntax used to declare **MUL** indicates that there are three (proof) value constructors associated with **MUL**, which are given the following constant props (or c-props):

$$\begin{array}{lcl} \mathit{MULbas} & : & \forall n : \mathit{int}.\ () \Rightarrow \mathbf{MUL}(0, n, 0) \\ \mathit{MULind} & : & \forall m : \mathit{int}.\forall n : \mathit{int}.\forall p : \mathit{int}.\ m \geq 0 \supset \\ & & (\mathbf{MUL}(m, n, p) \Rightarrow \mathbf{MUL}(m+1, n, p+n)) \\ \mathit{MULneg} & : & \forall m : \mathit{int}.\forall n : \mathit{int}.\forall p : \mathit{int}.\ m > 0 \supset \\ & & (\mathbf{MUL}(m, n, p) \Rightarrow \mathbf{MUL}(-m, n, -p)) \end{array}$$

Given integers I_1, I_2 and I_3, it is clear that $I_1 \times I_2 = I_3$ holds if and only if $\mathbf{MUL}(I_1, I_2, I_3)$ can be assigned to a closed (proof) value. In essence, *MULbas*, *MULind* and *MULneg* correspond to the following three equations in an inductive definition of the multiplication function on integers:

$$\begin{array}{rcl} 0 \times n & = & 0; \\ (m+1) \times n & = & m \times n + n \ \text{ if } m \geq 0; \\ (-m) \times n & = & -(m \times n) \ \text{ if } m > 0. \end{array}$$

[1]More precisely, each arithmetic constraint is required to be turned into a linear programming problem.

The function *concat* can now be given the following type:

$$\forall a : type.\forall m : nat.\forall n : nat.$$
$$\quad \mathbf{list}(\mathbf{list}(a, n), m) \rightarrow \exists p : nat.\ \mathbf{MUL}(m, n, p) * \mathbf{list}(a, p)$$

The code for implementing *concat* is given in Figure 1. We write $(\ldots)$ to form tuples. Also, we use the bar symbol $(|)$ as a separator to separate proofs from programs. Given an argument *xss* of type $\mathbf{list}(\mathbf{list}(T, I_2), I_1)$, the function *concat* returns a pair (pf, res) such that *pf* is a proof value of prop $\mathbf{MUL}(I_1, I_2, I_3)$ for some integer I_3 and *res* is a list of type $\mathbf{list}(T, I_3)$. Therefore, *pf* acts as a witness to certify that the length of *res* equals $I_1 \times I_2$.[2] Now suppose we would like to assign *concat* the following type:

$$\forall a : type.\forall m : nat.\forall n : nat.$$
$$\quad \mathbf{list}(\mathbf{list}(a, n), m) \rightarrow \exists p : nat.\ \mathbf{MUL}(n, m, p) * \mathbf{list}(a, p)$$

which is obtained from replacing $\mathbf{MUL}(m, n, p)$ with $\mathbf{MUL}(n, m, p)$ in the above type assigned to *concat*. Then we need to replace *MULbas*() and *MULind*(*pf*) in the definition of *concat* with *lemma0*() and *lemma1*(*pf*), respectively, where *lemma0* and *lemma1* are the following (proof) functions:

```
prfun lemma0 {n:nat} .<n>. (): MUL (n, 0, 0) =
  // [sif] forms a conditional with a static condition
  sif n > 0 then MULind (lemma0 {n-1} ()) else MULbas ()

prfun lemma1 {m,n:nat} {p:int} .<n>. // <n> is a termination metric
   (pf: MUL (n, m, p)): MUL (n, m+1, p+n) = case pf of
  | MULbas () => MULbas () | MULind pf' => MULind (lemma1 pf')
```

Note that the keyword *prfun* indicates the implementation of a proof function. We now choose *lemma1* for further explanation. The prop assigned to *lemma1* is

$$\forall m : nat.\forall n : nat.\forall p : int.\ \mathbf{MUL}(n, m, p) \rightarrow \mathbf{MUL}(n, m+1, p+n)$$

Essentially, *lemma1* represents an inductive proof of $n \times m = p \supset n \times (m + 1) = p + n$ for all natural numbers m, n and integers p, where the induction is on n. In particular, the following two linear arithmetic constraints, which can be easily verified, are generated when the two clauses in the body of *lemma1* are type-checked:

$$\forall n : nat.\forall p : int.\ n = 0 \supset (p = 0 \supset 0 = p + n)$$
$$\forall m : nat.\forall n : nat.\forall p : int.\forall n' : int.\forall p' : int.$$
$$\quad n = n' + 1 \supset (p = p' + m \supset p + n = (p' + n') + (m + 1))$$

[2]However, there is really no need for constructing proof values like *pf* at run-time, and this issue is already investigated elsewhere [6].

In order for *lemma1* to represent a proof, we need to show that *lemma1* is a total function, that is, given *pf* of prop $\mathbf{MUL}(I_2, I_1, I_3)$ for natural numbers I_1 and I_2 and integer I_3, *lemma1*(*pf*) is guaranteed to return a proof value of prop $\mathbf{MUL}(I_2, I_1+1, I_3+I_2)$. In this paper, we are to present a type system *ATS/LF* in which every well-typed function is guaranteed to be total. Generally speaking, when implementing a recursive function in *ATS/LF*, the programmer is required to provide a metric that can be used to verify the termination of the function. For instance, in the definition of *lemma1*, $\langle n \rangle$ is the provided metric for verifying that *lemma1* is terminating; when *lemma1* is applied to a value of prop $\mathbf{MUL}(I_2, I_1, I_3)$, the label $\langle I_2 \rangle$ is associated with this call; in the case where a recursive call to *lemma1* is made subsequently, the label associated with the recursive call is $\langle I_2 - 1 \rangle$ (since pf' in the definition of *lemma1* is given the type $\mathbf{MUL}(I_2 - 1, I_1, I_3 - I_1)$), which is strictly less than the label $\langle I_2 \rangle$ associated with the original call; as a label associated with *lemma1* is always a natural number, it is evident that *lemma1* is terminating. To show that *lemma1* is total, we also need to verify that pattern matching in the definition of *lemma1* can never fail, which is a topic already studied elsewhere [33, 35].

The primary motivation for developing *ATS/LF* is to support in ATS a programming paradigm that combines programs with proofs. For brevity, we, however, are unable to formally demonstrate in this paper as to how such a combination can take place, and we refer the interested reader to [6] for further details on this subject. Instead, we focus on the formalization of *ATS/LF*, establishing that every-well typed program in *ATS/LF* is total. A secondary motivation we have is to use *ATS/LF* as a logical framework for encoding deduction systems and their properties, and we are to present some examples in support of such an application of *ATS/LF*.

The rest of the paper is organized as follows. In Section 2, we first mention some closely related work. We then formalize *ATS/LF* in Section 3. In particular, we make use of the notion of reducibility [30] in proving that every well-typed program in *ATS/LF* is total. In support of using *ATS/LF* as a (meta) logical framework, we demonstrate how deduction systems can be encoded in *ATS/LF* by presenting some interesting examples in Section 5, which are all verified in ATS. We conclude in Section 6.

2 Related Work

The approach to termination verification in *ATS/LF* is essentially taken from an earlier work [34], where a notion of termination metrics is introduced into Dependent ML (DML) [38, 37] to support termination verification for programs in DML. In this paper, we give an account for this approach in a more general setting (e.g., functional type indexes, which are not allowed in DML, are supported in *ATS/LF*).

There is certainly a vast body of literature on termination verification. In type theory, a standard approach to proving termination (of functions or programs) derives from the notion of accessible predicate [1, 21], and a detailed study based on it can be found in [5].

When used as a logical framework, *ATS/LF* is probably most closely related to Twelf [27]. In particular, a dataprop declaration in *ATS/LF* corresponds to a declaration for a type constructor in Twelf plus the constants associated with the type constructor. The approach to termination verification (for logical programs) in Twelf [28] is similar to that of *ATS/LF* in the aspect that it requires a structural ordering (possibly on higher-order terms) to be provided by the user, though the justification for the correctness of this approach is not based on the notion of reducibility. In Twelf, a proof is really a meta concept and it can not be represented within Twelf while in *ATS/LF*, a proof is just a well-typed program. This is a fundamental difference between Twelf and *ATS/LF*, which greatly influences the manner in which deduction systems are encoded. Recently, Delphin, a functional programming language built on top of Twelf, is proposed [29] and termination proofs in Twelf are expected to be represented as total functional programs in Delphin.

Of course, the related work also includes various (interactive) theorem proving systems based on type theory such as NuPrl [9], Coq [14] and Isabelle [23]. In order to effectively reason about program properties within a type theory, the underlying functional language of a theorem proving system is often required to be relatively pure, making it difficult to support many realistic programming features (e.g., general recursion, references, exceptions). In general, it is inflexible as well as involved to construct programs in a theorem proving system, though significant progress has been made in this direction. In contrast, *ATS/LF* is primarily designed to be part of ATS, a programming language that can readily support realistic programming features (e.g., pointers and pointer arithmetic [39]). In this respect, the design of *ATS/LF* is unique, and it has not been seen elsewhere in the literature.

3 Formal Development

In this section, we formally present *ATS/LF*, a type system rooted in the framework $\mathcal{ATS}$ that can guarantee the totality of every-well typed program. There are two components in *ATS/LF*: the static component (statics) and the dynamic component (dynamics). We first give the syntax for the statics

$$\frac{\Sigma(a) = \sigma}{\Sigma \vdash a : \sigma} \qquad \frac{\vdash sc : (\sigma_1, \ldots, \sigma_n) \Rightarrow \sigma \quad \Sigma \vdash s_i : \sigma_i \ \text{ for } 1 \leq i \leq n}{\Sigma \vdash sc(s_1, \ldots, s_n) : \sigma}$$

$$\frac{\Sigma \vdash s_1 : \sigma_1 \quad \Sigma \vdash s_2 : \sigma_2}{\Sigma \vdash \langle s_1, s_2 \rangle : \sigma_1 * \sigma_2} \qquad \frac{\Sigma \vdash s : \sigma_1 * \sigma_2}{\Sigma \vdash \pi_1(s) : \sigma_1} \qquad \frac{\Sigma \vdash s : \sigma_1 * \sigma_2}{\Sigma \vdash \pi_2(s) : \sigma_2}$$

$$\frac{\Sigma, a : \sigma_1 \vdash s : \sigma_2}{\Sigma \vdash \lambda a : \sigma_1.s : \sigma_1 \to \sigma_2} \qquad \frac{\Sigma \vdash s_1 : \sigma_1 \to \sigma_2 \quad \Sigma \vdash s_2 : \sigma_1}{\Sigma \vdash s_1(s_2) : \sigma_2}$$

Figure 2. The rules for assigning sorts to static terms

as follows.

sorts	σ	$::=$	$b \mid \sigma_1 * \sigma_2 \mid \sigma_1 \to \sigma_2$
static terms	s	$::=$	$a \mid sc(\vec{s}) \mid$
			$\langle s_1, s_2 \rangle \mid \pi_1(s_1) \mid \pi_2(s_2) \mid$
			$\lambda a : \sigma.s \mid s_1(s_2)$
props	P	$::=$	$\delta(\vec{s}) \mid P_1 \to P_2 \mid B \supset P \mid$
			$\forall a : \sigma.\ P \mid B \wedge P \mid \exists a : \sigma.\ P$
static variable contexts	Σ	$::=$	$\emptyset \mid \Sigma, a : \sigma$
static substitutions	Θ	$::=$	$[] \mid \Theta[a \mapsto s]$

The statics itself is a simply typed language and a type in it is called *sort.* We use b for base sorts, σ for sorts, a for variables, and s for terms in the statics. There are certain constants sc in statics, which are either constructors scc or functions scf. Each sc is given a constant sort (c-sort) of the form $(\sigma_1, \ldots, \sigma_n) \Rightarrow \sigma$, indicating that $sc(s_1, \ldots, s_n)$ is a term of sort σ if s_i can be assigned sorts σ_i for $1 \leq i \leq n$, and we may write scc for $scc()$. Note that a c-sort is *not* considered a (regular) sort.

We use Σ for static variable contexts, which assign sorts to static variables, and $\mathbf{dom}(\Sigma)$ for the set of variables declared in Σ. We may write $\vec{a} : \vec{\sigma}$ for the static variable context $\emptyset, a_1 : \sigma_1, \ldots, a_n : \sigma_n$, where $\vec{a} = a_1, \ldots, a_n$ and $\vec{\sigma} = \sigma_1, \ldots, \sigma_n$. A sorting judgment is of the form $\Sigma \vdash s : \sigma$, which means that s can be assigned the sort σ under Σ. The rules for assigning sorts to terms are given in Figure 2. Also, we may write $\Sigma \vdash \vec{s} : \vec{\sigma}$ to mean that $\Sigma \vdash s_i : \sigma_i$ for $1 \leq i \leq n$, where $\vec{s} = s_1, \ldots, s_n$ and $\vec{\sigma} = \sigma_1, \ldots, \sigma_n$.

We use the names *static variable*, *static constant* and *static term* for a variable, a constant and a term in statics. A static substitution is a finite mapping from static variables to static terms. We use $[]$ for the empty mapping and $\Theta[a \mapsto s]$ for the mapping that extends Θ with a link from a to s. Also, we write $\bullet[\Theta]$ for the result of applying Θ to some syntax $\bullet$. Given Σ

$$\frac{\vdash \delta(\vec{\sigma})\ [\mathbf{prop}] \quad \Sigma \vdash \vec{s} : \vec{\sigma}}{\Sigma \vdash \delta(\vec{s})\ [\mathbf{prop}]} \qquad \frac{\Sigma \vdash P_1\ [\mathbf{prop}] \quad \Sigma \vdash P_2\ [\mathbf{prop}]}{\Sigma \vdash P_1 \to P_2\ [\mathbf{prop}]}$$

$$\frac{\Sigma \vdash B : bool \quad \Sigma \vdash P\ [\mathbf{prop}]}{\Sigma \vdash B \supset P\ [\mathbf{prop}]} \qquad \frac{\Sigma, a : \sigma \vdash P\ [\mathbf{prop}]}{\Sigma \vdash \forall a : \sigma.\ P\ [\mathbf{prop}]}$$

$$\frac{\Sigma \vdash B : bool \quad \Sigma \vdash P\ [\mathbf{prop}]}{\Sigma \vdash B \land P\ [\mathbf{prop}]} \qquad \frac{\Sigma, a : \sigma \vdash P\ [\mathbf{prop}]}{\Sigma \vdash \exists a : \sigma.\ P\ [\mathbf{prop}]}$$

Figure 3. The rules for forming props

and Θ, we write $\Theta : \Sigma$ to mean that $\mathbf{dom}(\Theta) = \mathbf{dom}(\Sigma)$ and $\emptyset \vdash \Theta(a) : \Sigma(a)$ holds for each $a \in \mathbf{dom}(\Theta)$. In general, we write $\Sigma_1 \vdash \Theta : \Sigma_2$ to mean that $\Sigma_1 \vdash \Theta(a) : \Sigma_2(a)$ holds for each $a \in \mathbf{dom}(\Theta) = \mathbf{dom}(\Sigma_2)$.

We assume the existence of a base sort *bool* and the static constants *tt* and *ff* that are given the c-sort $() \Rightarrow bool$. Also, we may write B for static terms of sort *bool*. For each sort σ, we have a binary relation $=_\sigma$ of c-sort $(\sigma, \sigma) \Rightarrow bool$. In practice, we have a sort *int* for integers, and we also provide syntax for the programmer to declare new sorts, which is to be shown in the examples we present.

The rule for forming props are given in Figure 3. We write $\vdash \delta(\vec{\sigma})$ [**prop**] to mean that δ is prop constructor that takes static terms $\vec{s}$ of sorts $\vec{\sigma}$ to form a prop. As a convenient notation, we may write $\forall\Sigma$ for a sequence of quantifiers: $\forall a_1 : \sigma_1 \ldots \forall a_n : \sigma_n$, where $\Sigma = \emptyset, a_1 : \sigma_1, \ldots, a_n : \sigma_n$. We call the forms $B \supset P$ and $B \land P$ guarded props and asserting props, respectively, and we often employ subset sorts, a form of syntactic sugar, in the concrete syntax of *ATS/LF* when forming such props. For example, when writing $\forall a : nat.\ P$ ($\exists a : nat.\ P$), we really mean $\forall a : int.\ a \geq 0 \supset P$ ($\exists a : int.\ a \geq 0 \land P$). We now introduce the notion of constraints as follows.

DEFINITION 1. A constraint relation is of the form $\Sigma; \overline{B} \models B_0$, where $\overline{B}$ stands for a sequence $B_1, \ldots, B_n$ such that $\Sigma \vdash B_i : bool$ holds for each $0 \leq i \leq n$. A constraint $\Sigma; \overline{B} \models B_0$ is satisfied if for each substitution $\Theta : \Sigma$, $\overline{B}[\Theta]$ contains *ff* or $B_0[\Theta]$ is *tt*. In addition, we write $\Sigma; \overline{B} \models \overline{B}_0$ to mean that $\Sigma; \overline{B} \models B_0$ holds for each B_0 in $\overline{B}_0$.

A proper approach to defining constraint relation is through the use of models for type theory [16, 2]. This is done, for instance, in [31], where some techniques in [3] are employed to construct a model for the statics of a generic applied type system. Also, a detailed construction of such a

$$\frac{\Sigma; \overline{B} \models s_i =_\sigma s'_i}{\Sigma; \overline{B} \models \delta(s_1, \ldots, s_n) \leq_{pr} \delta(s'_1, \ldots, s'_n)} \qquad \frac{\Sigma; \overline{B} \models P'_1 \leq_{pr} P_1 \quad \Sigma; \overline{B} \models P_2 \leq_{pr} P'_2}{\Sigma; \overline{B} \models P_1 \to P_2 \leq_{pr} P'_1 \to P'_2}$$

$$\frac{\Sigma; \overline{B}, B' \models B \quad \Sigma; \overline{B}, B' \models P \leq_{pr} P'}{\Sigma; \overline{B} \models B \supset P \leq_{pr} B' \supset P'} \qquad \frac{\Sigma, a : \sigma; \overline{B} \models P \leq_{pr} P'}{\Sigma; \overline{B} \models \forall a : \sigma.\ P \leq_{pr} \forall a : \sigma.\ P'}$$

$$\frac{\Sigma; \overline{B}, B \models B' \quad \Sigma; \overline{B}, B \models P \leq_{pr} P'}{\Sigma; \overline{B} \models B \land P \leq_{pr} B' \land P'} \qquad \frac{\Sigma, a : \sigma; \overline{B} \models P \leq_{pr} P'}{\Sigma; \overline{B} \models \exists a : \sigma.\ P \leq_{pr} \exists a : \sigma.\ P'}$$

Figure 4. The subprop rules for *ATS/LF*

$$\begin{array}{llll}
\text{dynamic terms} & d & ::= & x \mid f \mid dc[\vec{s}](\vec{d}) \mid \mathbf{lam}\, x.d \mid \mathbf{app}(d_1, d_2) \mid \\
 & & & \mathbf{fix}\, f[\vec{a}].m \Rightarrow d \mid \mathbf{sif}(s, d_1, d_2) \mid \\
 & & & \supset^+(d) \mid \supset^-(d) \mid \mathbf{lam}\, \vec{a}.d \mid \mathbf{app}(d, \vec{s}) \mid \\
 & & & \land(d) \mid \mathbf{let}\ \land(x) = d_1\ \mathbf{in}\ d_2 \mid \\
 & & & \langle s, d \rangle \mid \mathbf{let}\ \langle a, x \rangle = d_1\ \mathbf{in}\ d_2 \mid \\
\text{dynamic values} & v & ::= & x \mid dcc[\vec{s}](\vec{v}) \mid \mathbf{lam}\, x.d \mid \\
 & & & \supset^+(d) \mid \mathbf{lam}\, \vec{a}.d \mid \land(v) \mid \langle s, v \rangle \\
\text{dynamic var. ctx.} & \Pi & ::= & \emptyset \mid \Pi, x : P \mid \Pi, f : \forall \Sigma.\ m \Rightarrow P \\
\text{labeling} & \mu & ::= & \emptyset \mid \mu, f : s \\
\text{dynamic subst.} & \theta & ::= & [] \mid \theta[xf \mapsto d]
\end{array}$$

Figure 5. The syntax for the dynamics of *ATS/LF*

model can be found in [37]. In practice, we often impose certain restrictions on the constraint relation so that an effective means can be found to solve constraints automatically. For instance, in our current implementation of ATS [32], the imposed restrictions guarantee that each (arithmetic) constraint can be turned into a problem in linear integer programming.

Given two props P and P', we write $P \leq_{pr} P'$ to mean that P is a subprop of P'. A subprop judgment is of the form $\Sigma; \overline{B} \models P \leq_{pr} P'$, and the rules for deriving such judgments are given in Figure 4. Note that a subprop judgment $\Sigma; \overline{B} \models P \leq_{pr} P'$ is *conditional* in the sense that $P \leq_{pr} P'$ is decided under the conditions $\overline{B}$. This is a rather powerful notion in $\mathcal{ATS}$. In contrast, type equality in Martin-Löf's constructive type theory [18, 22] and related systems such as construction of calculus [11] are unconditional in the sense that two types are considered equal if and only if they are $\beta\eta$-equivalent.

The syntax for the dynamics of ATS/LF is given in Figure 5. We use x for a lam-variable and f for a fix-variable, and xf for a dynamic variable, which is either an x or an f. We use dc for dynamic constants, which are either dynamic constructors dcc or dynamic functions dcf. We assume that each dynamic constant dc is assigned a constant prop (c-prop) of the form $\forall\Sigma.\ \overline{B} \supset ((P_1, \ldots, P_n) \Rightarrow P)$. Note that a c-prop is not considered a (regular) prop. A dynamic variable context Π assigns props to lam-variables and decorated props, which are of the form $\forall\Sigma.\ m \Rightarrow P$, to fix-variables. The construct **sif** forms a conditional expression where the condition is a static term. A labeling μ associates static terms with fix-variables, and a dynamic substitution maps dynamic variables to dynamic terms. In ATS/LF, we often use the name *proof term* and *proof value* to refer to a dynamic term and a dynamic value, respectively.

We now introduce termination metrics as follows, which play a key rôle in guaranteeing that each well-typed program in ATS/LF is terminating.

DEFINITION 2. Given a sort σ, a binary transitive relation $<$ on static terms s of the sort σ is well-founded if there are no infinitely many s_i of the sort σ such that $s_{i+1} < s_i$ hold for all $i \geq 0$. For instance, a common well-founded ordering we use is the lexicographic ordering on tuples of natural numbers. Given a well-founded binary relation $<$ on static terms of sorts σ and a static term s of sort σ, $(<, s)$ is a metric. Note that s may contain free static variables. We use m for metrics.

A judgment for assigning a prop to a dynamic term in ATS/LF is of the form $\Sigma; \overline{B}; \Pi \vdash d : P \ll \mu$. The rules for assigning props to dynamic terms in ATS/LF are listed in Figure 6. Note that the obvious side conditions associated with certain rules are omitted. In the case where μ is empty, we may write $\Sigma; \overline{B}; \Pi \vdash d : P$ for $\Sigma; \overline{B}; \Pi \vdash d : P \ll \mu$. The rule **(pr-fix-var)** indicates that each occurrence of a fix-variable f in a dynamic term must be inside a term of the form $\mathbf{app}(f, \vec{s})$; if f is assigned a decorated prop $\forall \vec{a} : \vec{\sigma}.\ m \Rightarrow P_0$ for $m = (<, s_0)$, then we say that a label $s_0[\vec{a} \mapsto \vec{s}]$ is attached to this occurrence of f. A judgment of the form $\Sigma; \overline{B}; \Pi \vdash d : P \ll \mu$ basically means that d can be assigned the prop P under $\Sigma; \overline{B}; \Pi$; in addition, given any fix-variable f such that $\Sigma(f) = \forall \vec{a} : \vec{\sigma}.\ m \Rightarrow P_0$ for $m = (<, s_0)$ and $\mu(f) = s_1$ (i.e., $f : s_1$ occurs in μ), the label attached to each occurrence of f in d is strictly less than s_1 (according to the ordering $<$).

As usual, we have the following substitution lemma in ATS/LF.

LEMMA 3 (Substitution). *Assume that $\Sigma; \overline{B}; \Pi \vdash d : P \ll \mu$ is derivable.*

1. *If $\Sigma = \Sigma', \Sigma''$ and $\Sigma' \vdash \Theta : \Sigma''$ is derivable, then $\Sigma'; \overline{B}[\Theta]; \Pi[\Theta] \vdash d[\Theta] : P[\Theta] \ll \mu[\Theta]$ is also derivable.*

$$\frac{\Sigma; \overline{B}; \Pi \vdash d : P \quad \Sigma; \overline{B} \models P \leq_{pr} P'}{\Sigma; \overline{B}; \Pi \vdash d : P'} \textbf{(pr-sub)}$$

$$\frac{\Pi(x) = P}{\Sigma; \overline{B}; \Pi \vdash x : P \ll \mu} \textbf{(pr-lam-var)}$$

$$\frac{\Pi(f) = \forall \Sigma_0 .\ (<, s_0) \Rightarrow P \quad \Sigma \vdash \Theta : \Sigma_0 \quad \Sigma; \overline{B} \models s_0[\Theta] < \mu(f)}{\Sigma; \overline{B}; \Pi \vdash \mathbf{app}(f, \vec{s}) : P \ll \mu} \textbf{(pr-fix-var)}$$

$$\frac{\begin{array}{c}\vdash dc : \forall \vec{a} : \vec{\sigma}.\ \overline{B}_0 \supset ((P_1, \ldots, P_n) \Rightarrow P) \quad \Sigma \vdash \vec{s} : \vec{\sigma} \quad \Sigma \models \overline{B}_0[\vec{a} \mapsto \vec{s}] \\ \Sigma; \overline{B}; \Pi \vdash d_1 : P_1[\vec{a} \mapsto \vec{s}] \quad \ldots \quad \Sigma; \overline{B}; \Pi \vdash d_n : P_n[\vec{a} \mapsto \vec{s}]\end{array}}{\Sigma; \overline{B}; \Pi \vdash dc[\vec{s}](d_1, \ldots, d_n) : P[\vec{a} \mapsto \vec{s}]} \textbf{(pr-const)}$$

$$\frac{\Sigma; \overline{B}; \Pi, x : P_1 \vdash d : P_2 \ll \mu}{\Sigma; \overline{B}; \Pi \vdash \mathbf{lam}\, x.d : P_1 \to P_2 \ll \mu} \textbf{(pr-lam)}$$

$$\frac{\Sigma; \overline{B}; \Pi \vdash d_1 : P_1 \to P_2 \ll \mu \quad \Sigma; \overline{B}; \Pi \vdash d_2 : P_1 \ll \mu}{\Sigma; \overline{B}; \Pi \vdash \mathbf{app}(d_1, d_2) : P_2 \ll \mu} \textbf{(pr-app)}$$

$$\frac{\Sigma, \vec{a} : \vec{\sigma}; \overline{B}; \Pi, f : \forall \vec{a} : \vec{\sigma}.\ m \Rightarrow P \vdash d : P \ll \mu, f : s \quad m = (<, s)}{\Sigma; \overline{B}; \Pi \vdash \mathbf{fix}\, f[\vec{a}].m \Rightarrow d : \forall \vec{a} : \vec{\sigma}.\ P \ll \mu} \textbf{(pr-fix)}$$

$$\frac{\Sigma; \overline{B}, B; \Pi \vdash d : P \ll \mu}{\Sigma; \overline{B}; \Pi \vdash \supset^{+}(d) : B \supset P \ll \mu} \textbf{(pr-}\supset\textbf{-intro)}$$

$$\frac{\Sigma; \overline{B}; \Pi \vdash d : B \supset P \ll \mu \quad \Sigma; \overline{B} \models B}{\Sigma; \overline{B}; \Pi \vdash \supset^{-}(d) : P \ll \mu} \textbf{(pr-}\supset\textbf{-elim)}$$

$$\frac{\Sigma, \vec{a} : \vec{\sigma}; \overline{B}; \Pi \vdash d : P \ll \mu}{\Sigma; \overline{B}; \Pi \vdash \mathbf{lam}\, \vec{a}.d : \forall \vec{a} : \vec{\sigma}.\ P \ll \mu} \textbf{(pr-}\forall\textbf{-intro)}$$

$$\frac{\Sigma, a : \sigma; \overline{B}; \Pi \vdash d : \forall \vec{a} : \vec{\sigma}.\ P \ll \mu \quad \Sigma \vdash \vec{s} : \vec{\sigma}}{\Sigma; \overline{B}; \Pi \vdash \mathbf{app}(d, \vec{s}) : P[\vec{a} \mapsto \vec{s}] \ll \mu} \textbf{(pr-}\forall\textbf{-elim)}$$

$$\frac{\Sigma; \overline{B} \models B \quad \Sigma; \overline{B}; \Pi \vdash d : P \ll \mu}{\Sigma; \overline{B}; \Pi \vdash \wedge(d) : B \wedge P \ll \mu} \textbf{(pr-}\wedge\textbf{-intro)}$$

$$\frac{\Sigma; \overline{B}; \Pi \vdash d_1 : B \wedge P_1 \ll \mu \quad \Sigma; \overline{B}, B; \Pi, x : P_1 \vdash d_2 : P_2 \ll \mu}{\Sigma; \overline{B}; \Pi \vdash \mathbf{let}\ \wedge(x) = d_1\ \mathbf{in}\ d_2 : P_2 \ll \mu} \textbf{(pr-}\wedge\textbf{-elim)}$$

$$\frac{\Sigma \vdash s : \sigma \quad \Sigma; \overline{B}; \Pi \vdash d : P[a \mapsto s] \ll \mu}{\Sigma; \overline{B}; \Pi \vdash \langle s, d \rangle : \exists a : \sigma.\ P \ll \mu} \textbf{(pr-}\exists\textbf{-intro)}$$

$$\frac{\Sigma; \overline{B}; \Pi \vdash d_1 : \exists a : \sigma.\ P_1 \ll \mu \quad \Sigma, a : \sigma; \overline{B}; \Pi, x : P_1 \vdash d_2 : P_2 \ll \mu}{\Sigma; \overline{B}; \Pi \vdash \mathbf{let}\ \langle a, x \rangle = d_1\ \mathbf{in}\ d_2 : P_2 \ll \mu} \textbf{(pr-}\exists\textbf{-elim)}$$

$$\frac{\Sigma \vdash s : bool \quad \Sigma; \overline{B}, s; \Pi \vdash d_1 : P \ll \mu \quad \Sigma; \overline{B}, \neg s; \Pi \vdash d_2 : P \ll \mu}{\Sigma; \overline{B}; \Pi \vdash \mathbf{sif}(s, d_1, d_2) : P \ll \mu} \textbf{(pr-sta-if)}$$

Figure 6. The rules for assigning props to dynamic terms

2. *If* $\overline{B} = \overline{B}', \overline{B}''$ *and* $\Sigma; \overline{B}' \models \overline{B}''$ *holds, then* $\Sigma; \overline{B}'; \Pi \vdash d : P \ll \mu$ *is derivable.*

3. *If* $\Pi = \Pi', x : P'$ *and* $\Sigma; \overline{B}; \Pi' \vdash d' : P' \ll \mu$ *holds, then* $\Sigma; \overline{B}; \Pi' \vdash d[x \mapsto d'] : P \ll \mu$ *is derivable.*

4. *If* $\Pi = \Pi', f : \forall \Sigma_0.\ m \Rightarrow P'$ *and* $\mu = \mu', f : m$ *and* $\Sigma; \overline{B}; \Pi' \vdash d' : \forall \Sigma_0.\ P' \ll \mu'$ *holds, then* $\Sigma; \overline{B}; \Pi' \vdash d[x \mapsto d'] : P \ll \mu'$ *is derivable.*

Proof By standard structural induction. ■

We now introduce evaluation contexts as follows for assigning dynamic semantics to dynamic terms:

$$\begin{array}{ll} \text{eval. ctx.} & E \ ::= \\ & [\,] \mid dc[\vec{s}](v_1, \ldots, v_{i-1}, E, d_i, \ldots, d_n) \mid \mathbf{app}(E, d) \mid \mathbf{app}(v, E) \mid \\ & \supset^-(E) \mid \mathbf{app}(E, \vec{s}) \mid \mathbf{let} \wedge(x) = E \mathbf{~in~} d \mid \mathbf{let} \langle a, x \rangle = E \mathbf{~in~} d \end{array}$$

DEFINITION 4. We define redexes and their reducts as follows.

1. $dcf(v_1, \ldots, v_n)$ is a redex if it is defined, and its reduct is its defined value.

2. $\mathbf{app}(\mathbf{lam}\, x.d, v)$ is a redex, and its reduct is $d[x \mapsto v]$.

3. $\mathbf{fix}\, f[\vec{a}].m \Rightarrow d$ is a redex, and its reduct is:

$$\mathbf{lam}\, \vec{a}.d[f \mapsto \mathbf{fix}\, f[\vec{a}].m \Rightarrow d]$$

4. $\supset^-(\supset^+(d))$ is a redex and its reduct is d.

5. $\mathbf{app}(\mathbf{lam}\, \vec{a}.d, \vec{s})$ is a redex, and its reduct is $d[\vec{a} \mapsto \vec{s}]$.

6. $\mathbf{let} \wedge(x) = \wedge(v) \mathbf{~in~} d$ is a redex and its reduct is $d[x \mapsto v]$.

7. $\mathbf{let} \langle a, x \rangle = \langle s, v \rangle \mathbf{~in~} d$ is a redex and its reduct is $d[a \mapsto s][x \mapsto v]$.

8. $\mathbf{sif}(tt, d_1, d_2)$ is a redex, and its reduct is d_1.

9. $\mathbf{sif}(f\!f, d_1, d_2)$ is a redex, and its reduct is d_2.

Given $d_1 = E[d]$ and $d_2 = E[d']$, where d is a redex and d' is the reduct of d, we write $d_1 \rightarrow d_2$ and say that d_1 reduces to d_2 in one step. Let $\rightarrow^*$ be the reflexive and transitive closure of $\rightarrow$.

THEOREM 5 (Subject Reduction). *Assume that* $\Sigma; \overline{B}; \Pi \vdash d : P \ll \mu$ *is derivable and* $d \rightarrow d'$ *holds. Then* $\Sigma; \overline{B}; \Pi \vdash d' : P \ll \mu$ *is also derivable.*

Proof By structural induction on the derivation of $\Sigma; \overline{B}; \Pi \vdash d : P$. ■

THEOREM 6 (Progress). *Assume that* $\emptyset; \emptyset; \emptyset \vdash d : P$ *is derivable. Then* d *is either a value or* $d \rightarrow d'$ *holds for some* d'.

Proof By structural induction on the derivation of $\emptyset; \emptyset; \emptyset \vdash d : P$. ■

We next establish that every closed dynamic term d can be reduced to a value v if d can be assigned a prop in ATS/LF. The proof technique we employ is based on the notion of reducibility [30]. Given a dynamic term d, we write $d\downarrow$ to mean that there is *no* infinite reduction sequence from d: $d = d_0 \rightarrow d_1 \rightarrow d_2 \rightarrow \ldots$.

DEFINITION 7 (Reducibility). Suppose that d is a closed dynamic term of prop P, that is, $\emptyset; \emptyset; \emptyset \vdash d : P$ is derivable. We define that d is reducible of prop P by induction on the complexity of P, namely, the number of prop constructors $\rightarrow$, $\supset$, $\forall$, $\wedge$ and $\exists$ in P.

1. P is a base prop. Then d is reducible of prop P if $d\downarrow$ holds.

2. $P = P_1 \rightarrow P_2$. Then d is reducible of prop P if $d\downarrow$ holds and $d_0[x \mapsto v]$ is reducible of prop P_2 for any value v reducible of prop P_1 whenever $d \rightarrow^* \mathbf{lam}\, x.d_0$ holds for some d_0.

3. $P = B \supset P_0$. Then d is reducible of prop P if $d\downarrow$ holds and $\models B$ implies that d_0 is reducible of prop P_0 whenever $d \rightarrow^* \supset^+(d_0)$ holds for some d_0.

4. $P = B \wedge P_0$. Then d is reducible of prop P if $d\downarrow$ holds and v is reducible of prop P_0 whenever $d \rightarrow^* \wedge(v)$ holds.

5. $P = \forall \vec{a} : \vec{\sigma}.\ P_0$. Then d is reducible of prop P if $d\downarrow$ holds and $d_0[\vec{a} \mapsto \vec{s}]$ is reducible of prop $P_0[\vec{a} \mapsto \vec{s}]$ for any $\vec{s}$ of sorts $\vec{\sigma}$ whenever $d \rightarrow^* \mathbf{lam}\, \vec{a}.d_0$ holds for some d_0.

6. $P = \exists a : \sigma.\ P_0$. Then d is reducible of prop P if $d\downarrow$ holds and v is reducible of prop $P_0[a \mapsto s]$ whenever $d \rightarrow^* \langle s, v\rangle$ holds for some s and v.

For handling the fixed-point construct, we also introduce a notion of labeled reducibility as follows.

DEFINITION 8 (Labeled Reducibility). Assume that $\mathbf{fix}\, f[\vec{a}].m \Rightarrow d$ is a closed dynamic term of prop $\forall \vec{a} : \vec{\sigma}.\ P$. Given a label s_1, d is s_1-reducible of prop $\forall \vec{a} : \vec{\sigma}.\ P$ if $\mathbf{app}(\mathbf{fix}\ f[\vec{a}].m \Rightarrow d, \vec{s})$ is reducible of prop $P[\vec{a} \mapsto \vec{s}]$ for each $\vec{s}$ satisfying $s_0[\vec{a} \mapsto \vec{s}] < s_1$, where $m = (<, s_0)$.

PROPOSITION 9. *Assume that $\emptyset; \emptyset \vdash P \leq_{pr} P'$ is derivable and d is reducible of prop P. Then d is also reducible of prop P'.*

Proof By induction on the height of the derivation of $\emptyset; \emptyset \vdash P \leq_{pr} P'$. ∎

The following lemma is the key to establishing Theorem 11, the main theoretical result in this paper:

LEMMA 10 (Main). *Assume that $\Sigma; \overline{B}; \Pi \vdash d : P \ll \mu$ is derivable. Given Θ and θ such that $\Theta : \Sigma$ and $\theta : \Pi[\Theta]$, if we have*

1. *$\models \overline{B}[\Theta]$ holds, and*

2. *for each $x \in \mathbf{dom}(\Pi)$, $\theta(x)$ is reducible of prop $\Pi(x)$, and*

3. *for each $f \in \mathbf{dom}(\Pi)$, $\theta(f)$ is s-reducible of prop $\forall \vec{a} : \vec{\sigma}.\ P$ for $s = \mu(f)[\Theta]$, where $\Pi(f) = \forall \vec{a} : \vec{\sigma}.\ m \Rightarrow P$,*

then $d[\Theta][\theta]$ is reducible of prop $P[\Theta]$.

Proof By structural induction on the derivation of $\Sigma; \overline{B}; \Pi \vdash d : P \ll \mu$. Proposition 9 is needed for handling the rule **(pr-sub)**. Please see [34] for details in a closely related proof.[3] ∎

THEOREM 11 (Totality). *Assume that $\emptyset; \emptyset; \emptyset \vdash d : P$ is derivable. Then $d \rightarrow^* v$ holds for some value v.*

Proof By Lemma 10, we have that d is reducible of prop P. Then by the definition of reducibility, $d\downarrow$ holds, and by Theorem 6, we have $d \rightarrow^* v$ for some value v. ∎

4 Dataprops and Pattern Matching

We find that dataprops (similar to datatypes) and pattern matching are indispensable in practice. The following is some additional syntax we need for introducing these features:

$$
\begin{array}{llll}
\text{patterns} & p & ::= & x \mid dcc[\vec{a}](p_1, \ldots, p_n) \\
\text{dynamic terms} & d & ::= & \ldots \mid \mathbf{case}\ d_0\ \mathbf{of}\ p_1 \Rightarrow d_1 \mid \cdots \mid p_n \Rightarrow d_n \\
\text{eval. ctx.} & E & ::= & \ldots \mid \mathbf{case}\ E\ \mathbf{of}\ p_1 \Rightarrow d_1 \mid \cdots \mid p_n \Rightarrow d_n
\end{array}
$$

As usual, we require that any variable, static or dynamic, occur at most once in a pattern. Given a value v and a pattern p, we use a judgment of

[3] The proof of Lemma 3.9.

$$\frac{}{\Sigma \vdash x \Downarrow P \Rightarrow \emptyset; \emptyset; \emptyset, x : P} \textbf{(pat-var)}$$

$$\frac{\begin{array}{c}\vdash dcc : \forall \vec{a} : \vec{\sigma}.\ \overline{B}_0 \supset ((P_1, \ldots, P_n) \Rightarrow \delta(\vec{s}_0)) \\ \Sigma, \vec{a} : \vec{\sigma} \vdash p_i \Downarrow P_i \Rightarrow \Sigma_i; \overline{B}_i; \Pi_i \text{ for } 1 \leq i \leq n \\ \Sigma' = \Sigma_1, \ldots, \Sigma_n \ \ \overline{B}' = \overline{B}_1, \ldots, \overline{B}_n \ \ \Pi' = \Pi_1, \ldots, \Pi_n\end{array}}{\Sigma \vdash dcc[\vec{a}](p_1, \ldots, p_n) \Downarrow \delta(\vec{s}) \Rightarrow \vec{a} : \vec{\sigma}, \Sigma'; \overline{B}_0, \vec{s}_0 = \vec{s}, \overline{B}'; \Pi'} \textbf{(pat-dcc)}$$

$$\frac{\Sigma \vdash p \Downarrow P_1 \Rightarrow \Sigma'; \overline{B}'; \Pi' \quad \Sigma, \Sigma'; \overline{B}, \overline{B}'; \Pi, \Pi' \vdash d : P_2}{\Sigma; \overline{B}; \Pi \vdash p \Rightarrow d \Downarrow P_1 \Rightarrow P_2} \textbf{(pr-clause)}$$

$$\frac{\Sigma; \overline{B}; \Pi \vdash d_0 : P_1 \quad \Sigma; \overline{B}; \Pi \vdash p_i \Downarrow d_i : P_1 \Rightarrow P_2 \text{ for } 1 \leq i \leq n}{\Sigma; \overline{B}; \Pi \vdash (\textbf{case } d_0 \textbf{ of } p_1 \Rightarrow d_1 \mid \cdots \mid p_n \Rightarrow d_n) : P_2} \textbf{(pr-case)}$$

Figure 7. The proping rules for pattern matching

the form $v \Downarrow p \Rightarrow (\Theta; \theta)$ to indicate $v = p[\Theta][\theta]$. The rules for deriving such judgments are given as follows,

$$\frac{}{v \Downarrow x \Rightarrow ([]; [x \mapsto v])} \textbf{(vp-var)}$$

$$\frac{\begin{array}{c}v_i \Downarrow p_i \Rightarrow (\Theta_i; \theta_i) \text{ for } 1 \leq i \leq n \\ \Theta = [\vec{a} \mapsto \vec{s}] \cup \Theta_1 \cup \ldots \cup \Theta_n \qquad \theta = \theta_1 \cup \ldots \cup \theta_n\end{array}}{dcc[\vec{s}](v_1, \ldots, v_n) \Downarrow dcc[\vec{a}](p_1, \ldots, p_n) \Rightarrow (\Theta; \theta)} \textbf{(vp-dcc)}$$

and we say that v matches p if $v \Downarrow p \Rightarrow (\Theta; \theta)$ is derivable for some Θ and θ. Note that in the rule **(vp-dcc)**, the unions $\Theta_1 \cup \ldots \cup \Theta_n$ and $\theta_1 \cup \ldots \cup \theta_n$ are well-defined since any variable, either static or dynamic, can occur at most once in a pattern.

A dynamic term of the form **case** v **of** $p_1 \Rightarrow d_1 \mid \cdots \mid p_n \Rightarrow d_n$ is a redex if $v \Downarrow p_i \Rightarrow \theta$ holds for some $1 \leq i \leq n$, and its reduction is $d_i[\theta]$. Note that reducing such a redex may involve nondeterminism as v may match several patterns p_i.

The proping rules for pattern matching is given in Figure 7. The meaning of a judgment of the form $\Sigma \vdash p \Downarrow s \Rightarrow \Sigma'; \overline{B}'; \Pi'$ is formally captured by the following lemma.

LEMMA 12. *Assume* $\emptyset; \emptyset; \emptyset \vdash v : P$, $\emptyset \vdash p \Downarrow P \vdash \Sigma; \overline{B}; \Pi$ *and* $v \Downarrow p \Rightarrow (\Theta; \theta)$. *Then we have* $\Theta : \Sigma$, $\models \overline{B}[\Theta]$ *and* $\theta : \Pi[\Theta]$.

Proof By structural induction. ■

As an example, the judgment below is derivable,

$$m' : int, n' : int, p' : int \vdash MULind(x) \Downarrow \mathbf{MUL}(m', n', p') \Rightarrow \Sigma; \overline{B}; \Pi$$

where $MULind$ is assumed to have the following c-prop:

$$\forall m : int.\forall n : int.\forall p : int.\ (\mathbf{MUL}(m, n, p)) \Rightarrow \mathbf{MUL}(m + 1, n, p + n)$$

and $\Sigma = (m : int, n : int, p : int)$, $\overline{B} = (m + 1 = m', n = n', p + n = p')$ and $\Pi = (x : \mathbf{MUL}(m, n, p))$.

In order to guarantee that a closed well-typed dynamic term of the following form:

$$\textbf{case } v \textbf{ of } p_1 \Rightarrow d_1 \mid \cdots \mid p_n \Rightarrow d_n$$

is always a redex, we need to verify the exhaustiveness of pattern matching in the rule **(pr-case)**. We refer the reader to [33, 35] for more details on this issue. From now on, we assume that the exhaustiveness of pattern matching is properly verified when the rule **(pr-case)** is applied.

In the presence of pattern matching, nonterminating programs can be readily constructed if we impose no restrictions on dataprops. For instance, please see [34] for such an example. We thus need to propose some restrictions that can be imposed to disallow nonterminating programs.

DEFINITION 13. An occurrence δ in a prop P is ground if $P = \delta(\vec{s})$, or $P = B \wedge P_0$ and the occurrence is ground in P_0, or $P = \exists a : \sigma.\ P_0$ and the occurrence is ground in P_0. An occurrence δ in $\vec{P}$ is ground if the occurrence is ground in some P in $\vec{P}$.

Let us fix a recursive prop constructor δ_0 that takes static terms $\vec{s}$ of sorts $\vec{\sigma}_0$ to form a base prop $\delta_0(\vec{s})$. Also, we assume that dcc_k $(k = 1, \ldots, m)$ of c-props $\forall \vec{a}_k : \vec{\sigma}_k.\ (\vec{P}_k) \Rightarrow \delta_0(\vec{s}_k)$ are associated with δ_0. Then Theorem 11 still holds after δ_0 and dcc_k $(k = 1, \ldots, m)$ are added into ATS/LF if

1. all occurrences of δ_0 in $\vec{P}_k$ are ground, or
2. there is a metric $m = (<, s_0)$ such that $\vec{a} : \vec{\sigma}_0 \vdash s_0 : \sigma$ is derivable and for each occurrence of $\delta_0(\vec{s})$ in $\vec{P}_k$, $s_0[\vec{a} \mapsto \vec{s}] < s_0[\vec{a} \mapsto \vec{s}_k]$ holds.

The reason for this claim is that it is possible to define the notion of reducibility of prop $\delta_0(\vec{s})$ for all $\vec{s} : \vec{\sigma}_0$ if either of these two criteria is satisfied. The detailed justification for the first criterion can be found in [34]. As for the second criterion, its justification is rather similar to that of the first one. For instance, the first criterion is satisfied if δ_0 is **MUL**. In practice, it seems uncommon to encounter a need for the second criterion. A genuine case that does require the second criterion occurs in an encoding of the reducibility predicate [30] for the simply-typed lambda-calculus [13].

```
// [lemma1_commute] proves: m * 0 = p implies p = 0
prfun lemma1_commute {m:nat} {p:int} .<m>.
   (pf: MUL (m, 0, p)): [p == 0] void =
  case pf of
  | MULbas () => ()
  | MULind (pf) => let val _ = lemma1_commute (pf) in () end

// [lemma2_commute] proves: m * n = p implies m * (n-1) = p-m
prfun lemma2_commute {m,n:nat} {p:int} .<m>.
   (pf: MUL (m, n, p)): MUL (m, n-1, p-m) =
  case pf of
  | MULbas () => MULbas ()
  | MULind pf => MULind (lemma2_commute pf)

// [theorem_commute] proves:
//    m * n = p1 and n * m = p2 implies p1 = p2
prfun theorem_commute {m,n:nat} {p1,p2:int} .<m>.
  (pf1: MUL (m, n, p1), pf2: MUL (n, m, p2)): [p1==p2] void =
  case pf1 of
  | MULbas () => let
      prval () = lemma1_commute pf2
    in
      // empty
    end // end of [MULbas]
  | MULind pf1 => let
      prval pf2 = lemma2_commute pf2
    in
      theorem_commute (pf1, pf2)
    end // end of [MULind]
(*
  | MULneg _ =/=> () // this case is impossible!
*)
```

Figure 8. A proof of the commutativity of multiplication on natural numbers

5 *ATS/LF* as a (meta) logical framework

We use some short examples in this section to illustrate how deduction systems can be encoded in *ATS/LF*.

5.1 Arithmetic

In Figure 8, we present some code for proving that multiplication on natural numbers is commutative. We first establish two lemmas that prove the following:

1. $\forall m : nat.\forall p : int.\ m \times 0 = p \supset p = 0$, and

2. $\forall m : nat.\forall n : nat.\forall p : int.\ m \times n = p \supset m \times (n - 1) = p - m$, and

where the induction is on m. Then a theorem is proven that states:

$$\forall m : nat.\forall n : nat.\forall p_1 : int.\forall p_2 : int.\ m \times n = p_1 \wedge m \times n = p_2 \supset p_1 = p_2$$

In other words, the commutativity of multiplication on natural numbers is established. If we go a bit further, we can readily use the commutativity of multiplication to construct the following proof function and thus establish the irrationality of the square root of 2:

```
// for proving that the square root of 2 is irrational
prfun lemma_irrational {p,q:nat} {x:int}
  (pf1: MUL (p, p, x), pf2: MUL (q, q, 2*x)): [x == 0] void =
  ...
```

5.2 Sequent Calculus

We present some code in Figure 9 that illustrates an approach to encoding a fragment of intuitionistic sequent calculus in *ATS/LF*. This approach is largely adopted from [7]. The syntax for this fragment is given as follows:

formulas	α	$::=$	$\ldots \mid \alpha_1 \supset \alpha_2$
formula sequences	Γ	$::=$	$\emptyset \mid \Gamma, \alpha$

For brevity, we only support the implication logic connective here. We first declare a sort *form* for representing (propositional) formulas and another sort *forms* for representing sequences of formulas. Given the representations being first-order, we skip the issue of representation adequacy as it is evident. We then declare a dataprop **IN** such that given a formula A and a sequence G of formulas, $\mathbf{IN}(A, G)$ is a prop that indicates A occurring in G if a closed proof value of prop $\mathbf{IN}(A, G)$ can be constructed.

To represent derivations in sequent calculus, we declare a dataprop **DER**; given G, A and n, a derivation for $\Gamma \vdash \alpha$ of size n can be constructed if there

```
datasort form = imp of (form, form)
datasort forms = none | more of (forms, form)

datatype IN (form, forms) =
  | {G:forms; A:form} INone (A, more (G, A))
  | {G:forms; A1,A2:form} INshi (A1, more (G, A2)) of IN (A1, G)

datatype DER (forms, form, int) =
  | {G:forms; A:form} axiom (G, A, 0) of IN (A, G)
  | {G:forms; A1,A2,A3:form; n1,n2:nat}
    impl (G, A3, n1+n2+1) of
      (IN (imp (A1,A2),G), DER (G,A1,n1), DER (more (G,A2),A3,n2))
  | {G:forms; A1,A2:form; n:nat}
    impr (G, imp (A1, A2), n+1) of DER (more (G, A1), A2, n)

// [sup (G1, G2)] means that G1 contains G2
propdef SUP (G1:forms, G2:forms) = {A:form} IN (A, G2) -> IN (A, G1)

prfun shiSUP {G1,G2:forms; A:form} .<>.
   (f: SUP (G1, G2)): SUP (more (G1, A), more (G2, A)) =
  lam i => case i of INone () => INone () | INshi i => INshi (f i)

prfun supDER {G1,G2:forms; A:form; n:nat} .<n>.
   (f: SUP (G1, G2), d: DER (G2, A, n)): DER (G1, A, n) =
  case d of
  | axiom i => axiom (f i)
  | impl (i, d1, d2) =>
    impl (f i, supDER (f, d1), supDER (shiSUP f, d2))
  | impr (d) => impr (supDER (shiSUP f, d))

prfun cutElim {G:forms; A1,A2:form; n1,n2:nat} .<A1, n1, n2>.
  (d1: DER (G, A1), d2: DER (more (G, A1), A2)): DER (G, A2) =
  ...
```

Figure 9. An encoding of sequent calculus in ATS/LF

is a proof value of prop $\mathbf{DER}(G, A, n)$, where we assume G and A represent Γ and α, respectively. In particular, the three constructors *axiom*, *impl* and *impr* represent the following three rules **(axiom)**, **(impl)** and **(impr)**, respectively:

$$\frac{\Gamma \vdash \alpha}{\alpha \in \Gamma}\ \textbf{(axiom)}$$

$$\frac{\alpha_1 \supset \alpha_2 \in \Gamma \quad \Gamma \vdash \alpha_1 \quad \Gamma, \alpha_2 \vdash \alpha_3}{\Gamma \vdash \alpha_3}\ \textbf{(impl)}$$

$$\frac{\Gamma, \alpha_1 \vdash \alpha_2}{\Gamma \vdash \alpha_1 \supset \alpha_2}\ \textbf{(impr)}$$

As an example, we implement a function *supDER* of the following prop:

$$\forall G_1 : \mathit{forms}.\forall G_2 : \mathit{forms}.\forall A : \mathit{form}.\forall n : \mathit{nat}.$$
$$(\mathbf{SUP}(G_1, G_2), \mathbf{DER}(G_2, A, n)) \to \mathbf{DER}(G_1, A, n)$$

where $\mathbf{SUP}(G_1, G_2)$ stands for $\forall A : \mathit{form}.\ \mathbf{IN}(A, G_2) \to \mathbf{IN}(A, G_1)$. Therefore, *supDER* encodes a proof of the following statement: *If* Γ_1 *contains* Γ_2 *and there is a derivation for* $\Gamma_2 \vdash \alpha$ *of size n, then there is also a derivation for* $\Gamma_1 \vdash \alpha$ *of size n.* With this, the admissibility of following structural rules can be readily established:

$$\frac{\Gamma \vdash \alpha_2}{\Gamma, \alpha_1 \vdash \alpha_2}\ (\mathbf{weak.}) \quad \frac{\Gamma, \alpha_1, \alpha_1 \vdash \alpha_2}{\Gamma, \alpha_1 \vdash \alpha_2}\ (\mathbf{contr.}) \quad \frac{\Gamma, \alpha_1, \alpha_2 \vdash \alpha_3}{\Gamma, \alpha_2, \alpha_1 \vdash \alpha_3}\ (\mathbf{exch.})$$

Clearly, the availability of higher-order functions in *ATS/LF* can be easily appreciated in this example. It is straightforward to construct a function *cutElim* of the following prop:

$$\forall G : \mathit{forms}.\forall A_1 : \mathit{form}.\forall A_2 : \mathit{form}.\forall n_1 : \mathit{nat}.\forall n_2 : \mathit{nat}.$$
$$(\mathbf{DER}(G, A_1, n_1), \mathbf{DER}(G, A_2, n_2)) \to \exists n : \mathit{nat}.\ \mathbf{DER}(G, A_1, n)$$

That is, *cutElim* encodes a proof of the cut elimination theorem for intuitionistic sequent calculus. In particular, the metric for verifying the termination of *cutElim* is the triple $\langle A_1, n_1, n_2 \rangle$, lexicographically ordered (the ordering on formulas is the standard structural ordering). The interested reader may find further details in [7].

5.3 Lambda-Calculus

We now present an interesting example that involves the use of higher-order abstract syntax [8, 25, 26]. In Figure 10, we declare a sort *tm* for representing pure λ-terms in the statics of *ATS/LF*. For instance, the lambda-term $\lambda x.\lambda y.y(x)$ is represented as:

$$\mathit{TMlam}(\lambda a_1 : \mathit{tm}.\mathit{TMlam}(\lambda a_2 : \mathit{tm}.\mathit{TMapp}(a_2, a_1)))$$

The essence in this representation is that a variable at the object level (the lambda-calculus) is represented by a variable at the meta level (the statics of *ATS/LF*). A particularly appealing feature of this representation is that a substitution function at object-level can often be readily supported by a (built-in) substitution function at meta level. For instance, if *TMlam*(f) and t represent $\lambda x.M_1$ and M_2, then $f(t)$ represents $M_1[x \mapsto M_2]$. Of course, the adequacy of such a representation needs to be formally justified, which can be found in [24]. As an example, we declare a prop constructor

```
datasort tm = TMlam of (tm -> tm) | TMapp of (tm, tm)

dataprop EVAL (tm, tm) =
  | {f:tm -> tm} EVALlam (TMlam f, TMlam f)
  | {t1,t2,t3:tm; f: tm -> tm}
    EVALapp (TMapp (t1, t2), t3) of
      (EVAL (t1, TMlam f), EVAL (f t2, t3))

datasort tp = TPzero | TPfun of (tp, tp)

// This is a representation *not* to be used
dataprop TPDER0 (tm, tp) =
  | {f: tm -> tm; T1,T2:tp}
    TPDER0lam (TMlam f, TPfun (T1, T2)) of
      {x:tm} TPDER0 (x, T1) -> TPDER0 (f x, T2)
  | {t1,t2:tm; T1,T2:tp}
    TPDER0app (TMapp (t1, t2), T2) of
      (TPDER0 (t1, TPfun (T1, T2)), TPDER0 (t2, T1))

datasort ctx = CTXemp | CTXmore of (ctx, tm, tp)

dataprop IN (tm, tp, ctx) =
  | {G:ctx; t:tm; T:tp} INone (t, T, CTXmore (G, t, T))
  | {G:ctx; t,t':tm; T,T':tp}
    INshi (t, T, CTXmore (G, t', T')) of IN (t, T, G)

// This is the representation to be actually used
dataprop TPDER (ctx, tm, tp, int) =
  | {G:ctx; t:tm; T:tp} TPDERhyp (G, t, T, 0) of IN (t, T, G)
  | {G:ctx; f:tm -> tm; T1,T2:tp; n:nat}
    TPDERlam (G, TMlam f, TPfun (T1, T2), n+1) of
      {x:tm} TPDER (CTXmore (G, x, T1), f x, T2, n)
  | {G:ctx; t1,t2:tm; T1,T2:tp; n1,n2:nat}
    TPDERapp (G, TMapp (t1, t2), T2, n1+n2+1) of
      (TPDER (G, t1, TPfun (T1, T2), n1), TPDER (G, t2, T1, n2))

prfun SubstitutionLemma {G:ctx} {t1,t2:tm} {T1,T2:tp}
  (d1: TPDER (G, t1, T1), d2: TPDER (CTXmore (G, t1, T1), t2, T2))
  : TPDER (G, t2, T2) = ...

prfun TypePreservation {t1,t2:tm} {T:tp}
  (eval (t1, t2), d: TPDER (CTXemp, t1, T))
  : TPDER (CTXemp, t2, T) = ...
```

Figure 10. An encoding of lambda-calculus in *ATS/LF*

EVAL that takes two static terms s_1 and s_2 of the sort *tm* to form a prop $\mathbf{EVAL}(s_1, s_2)$; if a closed value of prop $\mathbf{EVAL}(s_1, s_2)$ can be constructed, then the weak head normal form (WHNF) of the lambda-term represented by s_1 is the lambda-term represented by s_2.

Suppose that we now want to assign simple types to lambda-terms. We declare a datasort *tp* in Figure 10 for representing simple types; *TPzero* represents a base type and $\mathit{TPfun}(\cdot,\cdot)$ represents a function type.

In order to represent typing derivations, we declare a prop constructor $\mathbf{TPDER}_0$ that takes two static terms t and T of sorts *tm* and *tp*, respectively, to form a prop $\mathbf{TPDER}_0(t, T)$; a proof value of prop $\mathbf{TPDER}_0(t, T)$ is *intended* to represent a typing derivation that assigns the type represented by T to the lambda-term represented by t. Note that this is a higher-order representation as *TPDER0lam* is assigned the following c-prop:

$$\begin{array}{l}\forall f : \mathit{tm} \rightarrow \mathit{tm}.\forall T_1 : \mathit{tp}.\forall T_2 : \mathit{tp}.\\ \quad (\forall t : \mathit{tm}.\ \mathbf{TPDER}_0(t, T_1) \rightarrow \mathbf{TPDER}_0(f(t), T_2)) \Rightarrow\\ \quad \mathbf{TPDER}_0(\mathit{TMlam}(t), \mathit{TPfun}(T_1, T_2))\end{array}$$

There are some rather serious problems with this representation. First and foremost, the adequacy of this representation is difficult to establish (even if it holds, which we strongly doubt). Second, neither of the two criteria in Section 4 is satisfied, and thus we need additional techniques for proving the totality of functions that involve the use of $\mathbf{TPDER}_0$.

To avoid these difficult issues, we introduce a first-order representation for typing derivations that assign simple types to lambda-terms. We first declare a sort *ctx* for representing contexts; *CTXemp* represents the empty context and $\mathit{CTXmore}(G, t, T)$ represents the context that extends G with a declaration for a typing derivation that assigns the type represented by T to the lambda-term represented by t. We then declare a prop constructor **TPDER** that takes four static terms G, t, T and n of sorts *ctx*, *tm*, *tp* and *int*, respectively, to form a prop $\mathbf{TPDER}(G, t, T, n)$; a value of prop $\mathbf{TPDER}(G, t, T, n)$ represents a typing derivation of size n that assigns the type represented by T to the lambda-term represented by t, where G indicates that the derivation may contain indeterminates (of the form $\mathit{TPDERhyp}(\cdots)$) standing for typing derivations declared in G. With this representation for typing derivations, the substitution lemma and the type preservation theorem for simply typed lambda-calculus can be readily encoded in *ATS/LF* as two functions of the following props, respectively:

$$\forall G : ctx.\forall t_1 : tm.\forall t_2 : tm.\forall T_1 : tp.\forall T_2 : tp.$$
$$(\mathbf{TPDER}^*(G, t_1, T_1), \mathbf{TPDER}^*(CTXmore(G, t_1, T_1), t_2, T_2)) \rightarrow$$
$$\mathbf{TPDER}^*(G, t_2, T_2)$$

$$\forall t_1 : tm.\forall t_2 : tm.\forall T : tp.$$
$$(\mathbf{EVAL}(t_1, t_2), \mathbf{TPDER}^*(CTXemp, t_1, T)) \rightarrow$$
$$\mathbf{TPDER}^*(CTXemp, t_2, T)$$

where $\mathbf{TPDER}^*(G, t, T) = \exists n : nat.\ \mathbf{TPDER}(G, t, T, n)$. The interested reader should have no difficulty in filling out the details.

6 Conclusion

We have presented a type system *ATS/LF* rooted in the framework *Applied Type System* [36, 31]. In *ATS/LF*, every well-typed program is guaranteed to be total. When constructing a recursive function in *ATS/LF*, the programmer is required to provide a metric for verifying that the recursive function is terminating. This is essentially the approach to program termination advocated in [34]. We, however, have given an account of this approach in a more general setting. While the primary motivation for developing *ATS/LF* is to support a programming paradigm that allows programs to be combined with proofs, we argue that *ATS/LF* can also be used as a logical framework for encoding deduction systems and their properties. This application is not attempted in [34]. As of now, there is no support for parametric polymorphism in *ATS/LF*. We expect to incorporate this feature into *ATS/LF*, and we consider such an incorporation to be largely straightforward.

Acknowledgment The author thanks Chiyan Chen for his time and effort in preparing, together with the author, an earlier draft on which the current paper is based.

BIBLIOGRAPHY

[1] Peter Aczel. An Introduction to Inductive Definition. In Jon Barwise, editor, *Handbook of Mathematical Logic*, pages 739–782. North Holland Publishing Company, 1977.

[2] Peter B Andrews. General Models, Descriptions and Choice in Type Theory. *Journal of Symbolic Logic*, 37:385–394, 1972.

[3] Peter B. Andrews. *An Introduction to Mathematical Logic: To Truth through Proof.* Academic Press, Inc., Orlando, Florida, 1986.

[4] Hendrik Pieter Barendregt. Lambda Calculi with Types. In S. Abramsky, Dov M. Gabbay, and T.S.E. Maibaum, editors, *Handbook of Logic in Computer Science*, volume II, pages 117–441. Clarendon Press, Oxford, 1992.

[5] Ana Bove. *General Recursion in Type Theory.* Ph. D. Dissertation, Chalmers University of Technology, November 2002.

[6] Chiyan Chen and Hongwei Xi. Combining Programming with Theorem Proving. In *Proceedings of the Tenth ACM SIGPLAN International Conference on Functional Programming*, pages 66–77, Tallinn, Estonia, September 2005.

[7] Chiyan Chen, Dengping Zhu, and Hongwei Xi. Implementing Cut Elimination: A Case Study of Simulating Dependent Types in Haskell. In *Proceedings of the 6th International Symposium on Practical Aspects of Declarative Languages*, Dallas, TX, June 2004. Springer-Verlag LNCS vol. 3057.

[8] Alonzo Church. A formulation of the simple type theory of types. *Journal of Symbolic Logic*, 5:56–68, 1940.

[9] Robert L. Constable et al. *Implementing Mathematics with the NuPrl Proof Development System.* Prentice-Hall, Englewood Cliffs, New Jersey, 1986.

[10] Robert L. Constable and Scott Fraser Smith. Partial Objects In Constructive Type Theory. In *Proceedings of Symposium on Logic in Computer Science*, pages 183–193, Ithaca, New York, June 1987.

[11] Thierry Coquand and Gérard Huet. The Calculus of Constructions. *Information and Computation*, 76(2–3):95–120, February–March 1988.

[12] G.B. Dantzig and B.C. Eaves. Fourier-Motzkin elimination and its dual. *Journal of Combinatorial Theory (A)*, 14:288–297, 1973.

[13] Kevin Donnelly and Hongwei Xi. A Formalization of Strong Normalization for Simply Typed Lambda-Calculus and System F. In *Proceedings of Workshop on Logical Frameworks and Meta-Languages: Theory and Practice*, pages 109–125. ENTCS 174(5), 2006.

[14] Gilles Dowek, Amy Felty, Hugo Herbelin, Gérard Huet, Chet Murthy, Catherine Parent, Christine Paulin-Mohring, and Benjamin Werner. The Coq Proof Assistant User's Guide. Rapport Technique 154, INRIA, Rocquencourt, France, 1993. Version 5.8.

[15] Susumu Hayashi and Hiroshi Nakano. *PX: A Computational Logic.* The MIT Press, 1988.

[16] Leon Henkin. Completeness in the theory of types. *Journal of Symbolic Logic*, 15:81–91, 1950.

[17] Furio Honsell, Ian A. Mason, Scott Smith, and Carolyn Talcott. A Variable Typed Logic of Effects. *Information and Computation*, 119(1):55–90, 15 May 1995.

[18] Per Martin-Löf. *Intuitionistic Type Theory.* Bibliopolis, Naples, Italy, 1984.

[19] N.P. Mendler. Recursive Types and Type Constraints in Second-Order Lambda Calculus. In *Proceedings of Symposium on Logic in Computer Science*, pages 30–36, Ithaca, New York, June 1987. The Computer Society of the IEEE.

[20] Robin Milner, Mads Tofte, Robert W. Harper, and D. MacQueen. *The Definition of Standard ML (Revised).* MIT Press, Cambridge, Massachusetts, 1997.

[21] Bengt Nordström. Terminating General Recursion. *BIT*, 28(3):605–619, October 1988.

[22] Bengt Nordström, Kent Petersson, and Jan M. Smith. *Programming in Martin-Löf's Type Theory*, volume 7 of *International Series of Monographs on Computer Science.* Clarendon Press, Oxford, 1990.

[23] Lawrence Paulson. *Isabelle: A Generic Theorem Prover.* Springer-Verlag LNCS 828, 1994.

[24] Frank Pfenning. *Computation and Deduction.* Lecture Notes, 2002.

[25] Frank Pfenning and Conal Elliott. Higher-order abstract syntax. In *Proceedings of the ACM SIGPLAN '88 Symposium on Language Design and Implementation*, pages 199–208, Atlanta, Georgia, June 1988.

[26] Frank Pfenning and Peter Lee. A Language with Eval and Polymorphism. In *International Joint Conference on Theory and Practice in Software Development*, pages 345–359, Barcelona, Spain, March 1989. Springer-Verlag LNCS 352.

[27] Frank Pfenning and Carsten Schürmann. System description: Twelf - a meta-logical framework for deductive systems. In H. Ganzinger, editor, *Proceedings of the 16th International Conference on Automated Deduction (CADE-16)*, pages 202–206, Trento, Italy, July 1999. Springer-Verlag LNAI 1632.

[28] Brigitte Pientka and Frank Pfenning. Termination and Reduction Checking in the Logical Framework. In *Proceedings of Workshop on Automation of Proofs by Mathematical Induction*, Pittsburgh, PA, June 2000.

[29] Carsten Schürmann et al. Delphin Project. Available at: `http://www.cs.yale.edu/~delphin`.

[30] W. W. Tait. Intensional Interpretations of Functionals of Finite Type I. *Journal of Symbolic Logic*, 32(2):198–212, June 1967.

[31] Hongwei Xi. Applied Type System. Available at: `http://www.cs.bu.edu/~hwxi/academic/drafts/ATS.ps`.

[32] Hongwei Xi. Applied Type System. Available at: `http://www.ats-lang.org/`.

[33] Hongwei Xi. Dead Code Elimination through Dependent Types. In *The First International Workshop on Practical Aspects of Declarative Languages*, San Antonio, January 1999. Springer-Verlag LNCS vol. 1551.

[34] Hongwei Xi. Dependent Types for Program Termination Verification. *Journal of Higher-Order and Symbolic Computation*, 15(1):91–132, March 2002.

[35] Hongwei Xi. Dependently Typed Pattern Matching. *Journal of Universal Computer Science*, 9(8):851–872, 2003.

[36] Hongwei Xi. Applied Type System (extended abstract). In *post-workshop Proceedings of TYPES 2003*, pages 394–408. Springer-Verlag LNCS 3085, 2004.

[37] Hongwei Xi. Dependent ML: an approach to practical programming with dependent types. *Journal of Functional Programming*, 17(2):215–286, 2007.

[38] Hongwei Xi and Frank Pfenning. Dependent Types in Practical Programming. In *Proceedings of 26th ACM SIGPLAN Symposium on Principles of Programming Languages*, pages 214–227, San Antonio, Texas, January 1999. ACM press.

[39] Dengping Zhu and Hongwei Xi. Safe Programming with Pointers through Stateful Views. In *Proceedings of the 7th International Symposium on Practical Aspects of Declarative Languages*, Long Beach, CA, January 2005. Springer-Verlag LNCS, 3350.

LaVergne, TN USA
31 March 2010
177735LV00001B/13/P